Classroom Manual for

Automotive Electricity and Electronics

Second Edition

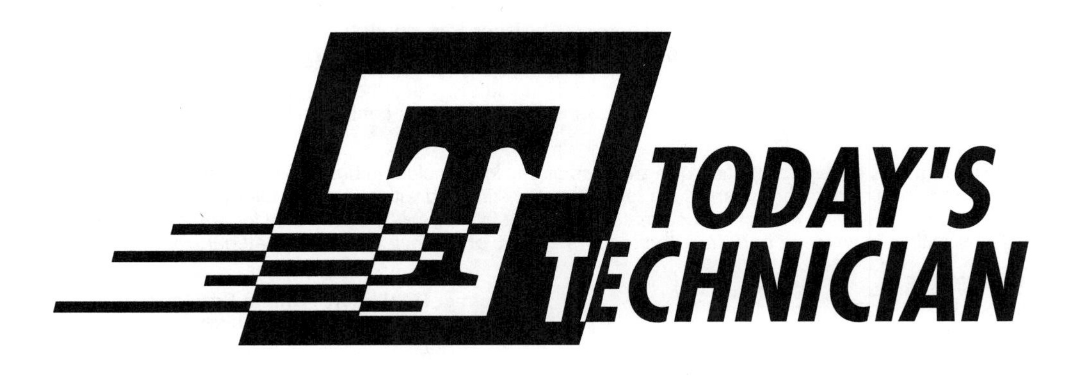

Classroom Manual for Automotive Electricity and Electronics

Second Edition

Barry Hollembeak
Technical Training Inc.
Rochester Hills, Michigan

Jack Erjavec
Series Advisor
Columbus State Community College
Columbus, Ohio

Delmar Publishers
I(T)P® An International Thomson Publishing Company

Albany • Bonn • Boston • Cincinnati • Detroit • London • Madrid • Melbourne
Mexico City • New York • Pacific Grove • Paris • San Francisco • Singapore • Tokyo
Toronto • Washington

NOTICE TO THE READER

DELMAR STAFF
Publisher: Robert D. Lynch
Acquisitions Editor: Vernon Anthony
Developmental Editor: Catherine Wein
Project Editors: Thomas Smith
and Christopher Chien
Production Coordinator: Karen Smith
Art/Design Coordinator: Michael Prinzo

Printed in the United States of America

For information, contact:

Delmar Publishers
3 Columbia Circle, Box 15015
Albany, New York 12212-5015

International Thomson Publishing Europe
Berkshire House 168-173
High Holborn
London, WC1V7AA
England

Thomas Nelson Australia
102 Dodds Street
South Melbourne, 3205
Victoria, Australia

Nelson Canada
1120 Birchmont Road
Scarborough, Ontario
Canada M1K 5G4

International Thomson Editores
Campos Eliseos 385, Piso 7
Col Polanco
11560 Mexico DF Mexico

International Thomson Publishing GmbH
Königswinterer Strasse 418
53227 Bonn
Germany

International Thomson Publishing Asia
221 Henderson Road
#05-10 Henderson Building
Singapore 0315

International Thomson Publishing-Japan
Hirakawacho Kyowa Building, 3F
2-2-1 Hirakawacho
Chiyoda-ku, Tokyo 102
Japan

7 8 9 10 XXX 02 01

Library of Congress Cataloging-in-Publication Data

Hollembeak, Barry.
Automotive electricity and electronics / Barry Hollembeak. — 2nd ed.
p. cm. — (Today's technician)
Vol. 1, rev. ed. of Santini, Al. Automotive electricity and electronics. 1988.
Vol. 2, rev. ed. of Hollembeak, Barry. Shop manual for automotive electricity and electronics. 1994.
Includes index.
Contents: [1] Shop manual — [2] Classroom manual.
ISBN 0-8273-7635-9
1. Automobiles — Electric equipment — Maintenance and repair. 2. Automobiles — Electronic equipment — Maintenance and repair. I. Hollembeak, Barry. Shop manual for automotive electricity and electronics. II. Santini, Al. Automotive electricity and electronics. III. Title. IV. Series.
TL272.H62 1996
629.25'4'0288 — dc20 96-42050
CIP

CONTENTS

PREFACE

Thanks to the support the *Today's Technician* series has received from those who teach automotive technology, Delmar Publishers is able to live up to its promise to provide new editions every three years. We have listened to our critics and our fans and present this new revised edition. By revising our series every three years, we can and will respond to changes in the industry,changes in the certification process, and to the ever-changing needs of those who teach automotive technology.

The *Today's Technician* series, by Delmar Publishers, features textbooks that cover all mechanical and electrical systems of automobiles and light trucks. Principal titles correspond with the eight major areas of ASE (National Institute for Automotive Service Excellence) certification. Additional titles include remedial skills and theories common to all of the certification areas and advanced or specialized subject areas that reflect the latest technological trends.

Each title is divided into two manuals: a Classroom Manual and a Shop Manual. Dividing the material into two manuals provides the reader with the information needed to begin a successful career as an automotive technician without interrupting the learning process by mixing cognitive and performance-based learning objectives.

Each Classroom Manual contains the principles of operation for each system and subsystem. It also discusses the design variations used by different manufacturers. The Classroom Manual is organized to build upon basic facts and theories. The primary objective of this manual is to allow the reader to gain an understanding of how each system and subsystem operates. This understanding is necessary to diagnose the complex automobile systems.

The understanding acquired by using the Classroom Manual is required for competence in the skill areas covered in the Shop Manual. All of the high priority skills, as identified by ASE, are explained in the Shop Manual. The Shop Manual also includes step-by-step instructions for diagnostic and repair procedures. Photo Sequences are used to illustrate many of the common service procedures. Other common procedures are listed and are accompanied with fine-line drawings and photographs that allow the reader to visualize and conceptualize the finest details of the procedure. The Shop Manual also contains the reasons for performing the procedures, as well as when that particular service is appropriate.

The two manuals are designed to be used together and are arranged in corresponding chapters. Not only are the chapters in the manuals linked together, the contents of the chapters are also linked. Both manuals contain clear and thoughtfully selected illustrations. Many of the illustrations are original drawings or photos prepared for inclusion in this series. This means that the art is a vital part of each manual.

The page layout is designed to include information that would otherwise break up the flow of information presented to the reader. The main body of the text includes all of the "need-to-know" information and illustrations. In the side margins are many of the special features of the series. Items such as definition of new terms, common trade jargon, tools list, and cross-referencing are placed in the margin, out of the normal flow of information so as not to interrupt the thought process of the reader.

Highlights of this Edition-Shop Manual

The text was updated throughout, to include the latest developments. Some of these new topics include dual-mass flywheels, differential designs, six-speed transmissions, and all-wheel-drive systems. We also added a new chapter that covers transmission-related electrical systems. This chapter includes basic electrical diagnosis and repairs, switches, speed sensors, solenoids, electromagnetic clutches, and electronic circuits.

Located at the end of each chapter are two new features: Job Sheets and ASE Challenge Questions. The Job Sheets provide a format for students to perform some of the tasks covered in the chapter. In addition to walking a student through a procedure, step-by-step, these Job Sheets challenge the student by asking why or how something should be done, thereby making the students think about what they are doing.

Speaking of challenging questions, each chapter ends with a group of questions that reflect the content of an ASE exam. These questions are not merely end-of-chapter questions, they represent the content of an ASE test. These questions, of course, are in addition to the ASE style end-of-chapter questions that were in the first edition.

Highlights of this Edition-Classroom Manual

The text was updated throughout, to include the latest developments. Some of these new topics include dual-mass flywheels, diffential designs, six-speed transmissions, and all-wheel-drive systems. We also added a new chapter that covers transmission-related electrical systems. This chapter includes basic electrical and electronic theory and the various applications for switches, speed sensors, solenoids, electromagnetic clutches, and electronic circuits.

Jack Erjavec

Classroom Manual

To stress the importance of safe work habits, the Classroom Manual dedicates one full chapter to safety. Included in this chapter are common safety practices, safety equipment, and safe handling of hazardous materials and wastes. This includes information on MSDS sheets and OSHA regulations. Other features of this manual include:

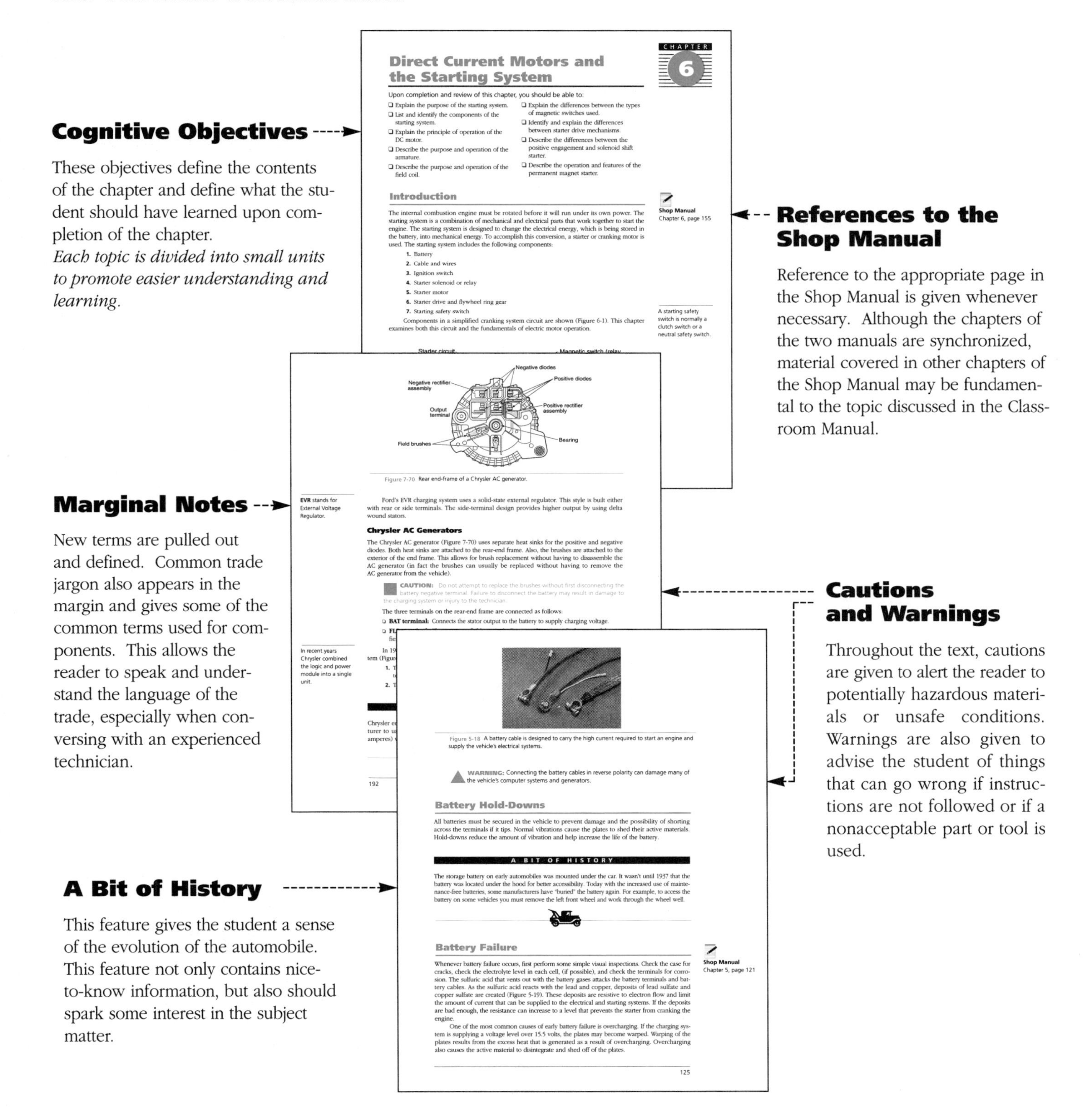

CHAPTER 6

Direct Current Motors and the Starting System

Upon completion and review of this chapter, you should be able to:

- ❑ Explain the purpose of the starting system.
- ❑ List and identify the components of the starting system.
- ❑ Explain the principle of operation of the DC motor.
- ❑ Describe the purpose and operation of the armature.
- ❑ Describe the purpose and operation of the field coil.
- ❑ Explain the differences between the types of magnetic switches used.
- ❑ Identify and explain the differences between starter drive mechanisms.
- ❑ Describe the differences between the positive engagement and solenoid shift starter.
- ❑ Describe the operation and features of the permanent magnet starter.

Introduction

Shop Manual
Chapter 6, page 155

The internal combustion engine must be rotated before it will run under its own power. The starting system is a combination of mechanical and electrical parts that work together to start the engine. The starting system is designed to change the electrical energy, which is being stored in the battery, into mechanical energy. To accomplish this conversion, a starter or cranking motor is used. The starting system includes the following components:

1. Battery
2. Cable and wires
3. Ignition switch
4. Starter solenoid or relay
5. Starter motor
6. Starter drive and flywheel ring gear
7. Starting safety switch

A starting safety switch is normally a clutch switch or a neutral safety switch.

Components in a simplified cranking system circuit are shown (Figure 6-1). This chapter examines both this circuit and the fundamentals of electric motor operation.

Figure 7-70 Rear end-frame of a Chrysler AC generator.

EVR stands for External Voltage Regulator.

Ford's EVR charging system uses a solid-state external regulator. This style is built either with rear or side terminals. The side-terminal design provides higher output by using delta wound stators.

Chrysler AC Generators

The Chrysler AC generator (Figure 7-70) uses separate heat sinks for the positive and negative diodes. Both heat sinks are attached to the rear-end frame. Also, the brushes are attached to the exterior of the end frame. This allows for brush replacement without having to disassemble the AC generator (in fact the brushes can usually be replaced without having to remove the AC generator from the vehicle).

CAUTION: Do not attempt to replace the brushes without first disconnecting the battery negative terminal. Failure to disconnect the battery may result in damage to the charging system or injury to the technician.

The three terminals on the rear-end frame are connected as follows:

- ❑ **BAT terminal:** Connects the stator output to the battery to supply charging voltage.

In recent years Chrysler combined the logic and power module into a single unit.

192

Figure 5-18 A battery cable is designed to carry the high current required to start an engine and supply the vehicle's electrical systems.

WARNING: Connecting the battery cables in reverse polarity can damage many of the vehicle's computer systems and generators.

Battery Hold-Downs

All batteries must be secured in the vehicle to prevent damage and the possibility of shorting across the terminals if it tips. Normal vibrations cause the plates to shed their active materials. Hold-downs reduce the amount of vibration and help increase the life of the battery.

A BIT OF HISTORY

The storage battery on early automobiles was mounted under the car. It wasn't until 1937 that the battery was located under the hood for better accessibility. Today with the increased use of maintenance-free batteries, some manufacturers have "buried" the battery again. For example, to access the battery on some vehicles you must remove the left front wheel and work through the wheel well.

Battery Failure

Shop Manual
Chapter 5, page 121

Whenever battery failure occurs, first perform some simple visual inspections. Check the case for cracks, check the electrolyte level in each cell, (if possible), and check the terminals for corrosion. The sulfuric acid that vents out with the battery gases attacks the battery terminals and battery cables. As the sulfuric acid reacts with the lead and copper, deposits of lead sulfate and copper sulfate are created (Figure 5-19). These deposits are resistive to electron flow and limit the amount of current that can be supplied to the electrical and starting systems. If the deposits are bad enough, the resistance can increase to a level that prevents the starter from cranking the engine.

One of the most common causes of early battery failure is overcharging. If the charging system is supplying a voltage level over 15.5 volts, the plates may become warped. Warping of the plates results from the excess heat that is generated as a result of overcharging. Overcharging also causes the active material to disintegrate and shed off of the plates.

125

Cognitive Objectives

These objectives define the contents of the chapter and define what the student should have learned upon completion of the chapter.
Each topic is divided into small units to promote easier understanding and learning.

References to the Shop Manual

Reference to the appropriate page in the Shop Manual is given whenever necessary. Although the chapters of the two manuals are synchronized, material covered in other chapters of the Shop Manual may be fundamental to the topic discussed in the Classroom Manual.

Marginal Notes

New terms are pulled out and defined. Common trade jargon also appears in the margin and gives some of the common terms used for components. This allows the reader to speak and understand the language of the trade, especially when conversing with an experienced technician.

Cautions and Warnings

Throughout the text, cautions are given to alert the reader to potentially hazardous materials or unsafe conditions. Warnings are also given to advise the student of things that can go wrong if instructions are not followed or if a nonacceptable part or tool is used.

A Bit of History

This feature gives the student a sense of the evolution of the automobile. This feature not only contains nice-to-know information, but also should spark some interest in the subject matter.

SYNC 7

Cylinders 1 and 4 6

1

5

2 Cylinders 2 and 5

3

4 Cylinders 3 and 6

Figure 8-57 The timing wheel for a 6-cylinder engine equipped with a magnetic pulse generator. The seventh slot is the synchronizing slot.

Terms to Know

A list of new terms appears next to the Summary. Definitions for these terms can be found in the Glossary at the end of the manual.

Terms to Know

ATDC
Ballast resistor
Barometric pressure switch
Basic timing
Breaker plate
Breaker point
BTDC
Centrifugal advance
Contact points
Control module
DIS
Distributor
Fast start
Firing order
Heat range
Ignition timing
Inductive reluctance
Look-up tables
Photoelectric sensor
Primary circuit
Pulse transformer
Rotor
Secondary circuit

Summary

- ❑ The ignition system supplies high voltage to the spark plugs to ignite the air/fuel mixture in the combustion chambers.
- ❑ The arrival of the spark is timed to coincide with the compression stroke of the piston. This basic timing can be advanced or retarded under certain conditions, such as high engine rpm or extremely light or heavy engine loads.
- ❑ The ignition system has two interconnected electrical circuits: a primary circuit and a secondary circuit.
- ❑ The primary circuit supplies low voltage to the primary winding of the ignition coil. This creates a magnetic field in the coil.
- ❑ A switching device interrupts primary current flow, collapsing the magnetic field, and creating a high-voltage surge in the ignition coil secondary winding.
- ❑ The switching device used in electronic systems is an NPN transistor. Old ignitions use mechanical breaker point switching.
- ❑ The secondary circuit carries high-voltage surges to the spark plugs. On some systems, the circuit runs from the ignition coil, through a distributor, to the spark plugs.
- ❑ The distributor may house the switching device plus centrifugal or vacuum timing advance mechanisms. Some systems locate the switching device outside the distributor housing.
- ❑ Ignition timing is directly related to the position of the crankshaft. Magnetic pulse generators and Hall-effect sensors are the most widely used engine position sensors. They generate an electrical signal at certain times during crankshaft rotation. This signal triggers the electronic switching device to control ignition timing.
- ❑ Direct ignition systems eliminate the distributor. Each spark plug, or in some cases, pair of spark plugs, has its own ignition coil. Primary circuit switching and timing control are done using a special ignition module tied into the vehicle control computer.
- ❑ Computer-controlled ignition eliminates centrifugal and vacuum timing mechanisms. The computer receives input from numerous sensors. Based on this data, the computer determines the optimum firing time and signals an ignition module to activate the secondary

Summaries

Each chapter concludes with summary statements that contain the important topics of the chapter. These are designed to help the reader review the contents.

Review Questions

Short answer essay, fill-in-the-blank, and multiple-choice type questions follow each chapter. These questions are designed to accurately assess the student's competence in the stated objectives at the beginning of the chapter.

Review Questions

Short Answer Essays

1. List the major components of the charging system.
2. List and explain the function of the major components of the AC generator.
3. How does the regulator control the charging system's output?
4. What is the relationship between field current and AC generator output?
5. Identify the differences between A, B, and isolated circuits.
6. Explain the operation of charge indicator lamps.
7. Describe the two styles of stators.
8. What is the difference between half-wave and full-wave rectification?
9. Describe how AC voltage is rectified to DC voltage in the AC generator.
10. What is the purpose of the charging system?

Fill-in-the-Blanks

1. The charging system converts the ________________ energy of the engine into ________________ energy to recharge the battery and run the electrical accessories.
2. All charging systems use the principle of ________________ ________________ to generate the electrical power.
3. The ________________ creates the rotating magnetic field of the AC generator.
4. ________________ are electrically conductive sliding contacts, usually made of copper and carbon.
5. In the ________________ connection stator, one lead from each winding is connected to one common junction.
6. The ________________ ________________ controls the output voltage of the AC generator, based on charging system demands, by controlling ________________ current.
7. In an electronic regulator, ________________ ________________ controls AC gen[...] the field coil is energized.

The J1930 List of Terminology

Located in the appendix, this list serves as a reference to the acceptable industry terms as defined by SAE.

SAE J1930 Revised SEP95

TABLE 1—CROSS REFERENCE AND LOOK UP

Existing Usage	Acceptable Usage	Acceptable Acronized Usage
A/C (Air Conditioning)	Air Conditioning	A/C
A/C Cycling Switch	Air Conditioning Cycling Switch	A/C Cycling Switch
A/T (Automatic Transaxle)	Automatic Transaxle[1]	A/T[1]
A/T (Automatic Transmission)	Automatic Transmission[1]	A/T[1]
AAT (Ambient Air Temperature)	**Ambient Air Temperature**	**AAT**
AC (Air Conditioning)	Air Conditioning	A/C
ACC (Air Conditioning Clutch)	Air Conditioning Clutch	A/C Clutch
Accelerator	Accelerator Pedal	AP
Accelerator Pedal Position	**Accelerator Pedal Position[1]**	**APP[1]**
ACCS (Air Conditioning Cyclic Switch)	Air Conditioning Cycling Switch	A/C Cycling Switch
ACH (Air Cleaner Housing)	Air Cleaner Housing[1]	ACL Housing1
ACL (Air Cleaner)	Air Cleaner[1]	ACL[1]
ACL (Air Cleaner) Element	Air Cleaner Element[1]	ACL Element[1]
ACL (Air Cleaner) Housing	Air Cleaner Housing[1]	ACL Housing[1]
ACL (Air Cleaner) Housing Cover	Air Cleaner Housing Cover[1]	ACL Housing Cover[1]
ACS (Air Conditioning System)	Air Conditioning System	A/C System
ACT (Air Charge Temperature)	Intake Air Temperature[1]	IAT[1]
Adaptive Fuel Strategy	Fuel Trim[1]	FT[1]
AFC (Air Flow Control)	Mass Air Flow	MAF
AFC (Air Flow Control(	Volume Air Flow	VAF
AFS (Air Flow Sensor)	Mass Air Flow Sensor	MAF Sensor
AFS (Air Flow Sensor)	Volume Air Flow Sensor	VAF Sensor
After Cooler	Charge Air Cooler[1]	CAC[1]
AI (Air Injection)	Secondary Air Injection[1]	AIR[1]
AIP (Air Injection Pump)	Secondary Air Injection Pump[1]	AIR Pump[1]
AIR (Air Injection Reactor)	Pulsed Secondary Air Injection[1]	PAIR[1]
AIR (Air Injection Reactor)	Secondary Air Injection[1]	AIR[1]
AIRB (Secondary Air Injection Bypass)	Secondary Air Injection Bypass[1]	AIR Bypass[1]
AIRD (Secondary Air Injection Diverter)	Secondary Air Injection Diverter[1]	AIR Diverter[1]
Air Cleaner	Air Cleaner[1]	ACL[1]
Air Cleaner Element	Air Cleaner Element[1]	ACL Element[1]
Air Cleaner Housing	Air Cleaner Housing[1]	ACL Housing[1]
Air Cleaner Housing Cover	Air Cleaner Housing Cover[1]	ACL Housing Cover[1]
Air Conditioning	Air Conditioning	A/C
Air Conditioning Sensor	Air Conditioning Sensor	A/C Sensor
Air Control Valve	Secondary Air Injection Control Valve[1]	AIR Control Valve[1]
Air Flow Meter	Mass Air Flow Sensor[1]	MAF Sensor[1]
Air Flow Meter	Volume Air Flow Sensor[1]	VAF Sensor[1]
Air Intake System	Intake Air System[1]	IA System[1]
Air Flow Sensor	Mass Air Flow Sensor[1]	MAF Sensor[1]
Air Management 1	Secondary Air Injection Bypass[1]	AIR Bypass[1]
Air Management 2	Secondary Air Injection Diverter[1]	AIR Diverter[1]
Air Temperature Sensor	Intake Air Temperature Sensor[1]	IAT Sensor[1]
Air Valve	Idle Air Control Valve[1]	IAC Valve[1]
AIV (Air Injection Valve)	Pulsed Secondary Air Injection[1]	PAIR[1]
ALCL (Assembly Line Communication Link)	Data Link Connector[1]	DLC[1]
Alcohol Concentration Sensor	Flexible Fuel Sensor[1]	FF Sensor[1]
ALDL (Assembly Line Diagnostic Link)	Data Link Connector[1]	DLC[1]

236

Shop Manual

To stress the importance of safe work habits, the Shop Manual also dedicates one full chapter to safety. Other important features of this manual include:

Performance Objectives

These objectives define the contents of the chapter and define what the student should have learned upon completion of the chapter. These objectives also correspond with the list of required tasks for ASE certification. *Each ASE task is addressed.*

Although this textbook is not designed to simply prepare someone for the certification exams, it is organized around the ASE task list. These tasks are defined generically when the procedure is commonly followed and specifically when the procedure is unique for specific vehicle models. Imported and domestic model automobiles and light trucks are included in the procedures.

CHAPTER 3

Basic Electrical Troubleshooting and Service

Upon completion and review of this chapter, you should be able to:

- ❑ Describe how different electrical problems cause changes in an electrical circuit.
- ❑ Diagnose and repair circuit protection devices.
- ❑ Test switches with a variety of test instruments.
- ❑ Test relays and relay circuits for proper operation.
- ❑ Identify and test fixed and variable resistors with a lab scope, voltmeter, or ohmmeter.
- ❑ Diagnose diodes for opens, shorts, and other defects.
- ❑ Locate and repair opens in a circuit.
- ❑ Locate and repair shorts in a circuit.
- ❑ Locate and repair the cause of unwanted high resistance in a circuit.

Basic Tools

Basic mechanic's tool set

Service manual

Introduction

Troubleshooting electrical problems involves the same tools and methods, regardless of which circuit has the problem. All electrical circuits must have voltage, current, and resistance. Testing for the presence of these, measuring them, and comparing your measurements to specifications is the key to effective diagnosis. To do this you must have a solid understanding of these basic electrical properties.

Voltage is the electrical pressure that causes electrons to move provided there is a complete path for them to do so. Current is the aggregate flow of electrons through a wire and can be defined as the rate of electron flow. Resistance is defined as opposition to current flow. An electrical circuit must have resistance in it in order to change electrical energy to light, heat, or movement.

An electrical circuit may develop an open, a short, or an excessive voltage drop that will cause it to operate improperly. An open circuit is a circuit in which there is a break in continuity. The open can be on either the insulated side or the ground side. A shorted circuit decreases the resistance of the circuit. This happens by shorting across to another circuit or by shorting to a ground. When there is a circuit to circuit short, one of the circuits is not controlled by its switch. Since the shorted circuit becomes a new parallel leg to the circuit, the entire parallel circuit will turn on and off with the switch controlling the other circuit. With this type of problem many strange things can happen. When a circuit is shorted to ground, a new parallel leg is present. This new leg has very low resistance and causes the current in the circuit to increase drastically.

The term **circuit** means a circle and is the path of electron flow consisting of the voltage source, conductors, load component, and return path to the voltage source.

High resistance problems can occur anywhere in the circuit. However, the effect of high resistance is the same regardless of where it is. Additional or unwanted resistance in series with a circuit will always reduce the current in the circuit and will reduce the amount of voltage drop by the component in the circuit.

Basic Electrical Troubleshooting

Troubleshooting electrical problems involves using meters, test lights, and jumper wires to determine if any part of the circuit is open or shorted, or if there is unwanted resistance.

To troubleshoot a problem, always begin by verifying the customer's complaint. Then operate the system and others, to get a complete understanding of the problem. Often there are other problems, which are not as evident or bothersome to the customer, that will provide helpful

51

Tools Lists

Each chapter begins with a list of the Basic Tools needed to perform the tasks included in the chapter. Whenever a Special Tool is required to complete a task, it is listed in the margin next to the procedure.

Marginal Notes

Page numbers for cross-referencing appear in the margin. Some of the common terms used for components, and other bits of information, also appear in the margin. This provides an understanding of the language of the trade and helps when conversing with an experienced technician.

Photo Sequences

Many procedures are illustrated in detailed Photo Sequences. These detailed photographs show the students what to expect when they perform particular procedures. They also can provide a student a familiarity with a system or type of equipment, which the school may not have.

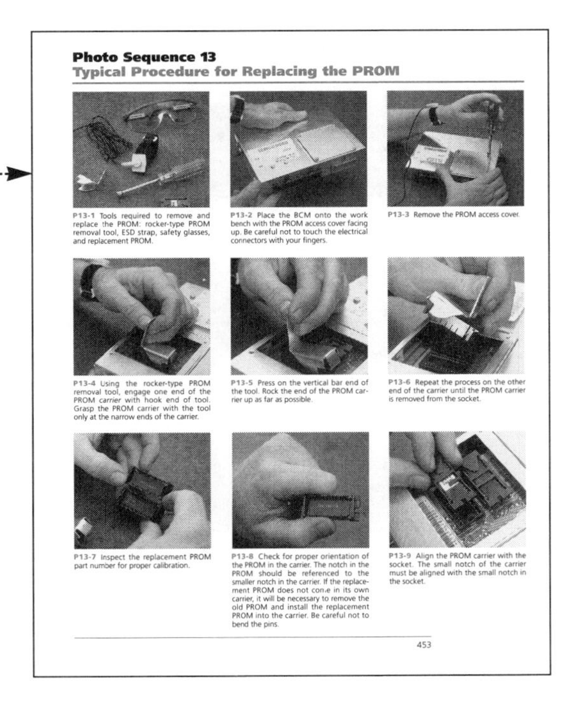

Photo Sequence 13
Typical Procedure for Replacing the PROM

P13-1 Tools required to remove and replace the PROM: rocker-type PROM removal tool, ESD strap, safety glasses, and replacement PROM.

P13-2 Place the BCM onto the work bench with the PROM access cover facing up. Be careful not to touch the electrical connectors with your fingers.

P13-3 Remove the PROM access cover.

P13-4 Using the rocker-type PROM removal tool, engage one end of the PROM *carrier* with hook end of tool. Grasp the PROM carrier with the tool only at the narrow ends of the carrier.

P13-5 Press on the vertical bar end of the tool. Rock the end of the PROM carrier up as far as possible.

P13-6 Repeat the process on the other end of the carrier until the PROM carrier is removed from the socket.

P13-7 Inspect the replacement PROM part number for proper calibration.

P13-8 Check for proper orientation of the PROM in the carrier. The notch in the PROM should be referenced to the smaller notch in the carrier. If the replacement PROM does not come in its own carrier, it will be necessary to remove the old PROM and install the replacement PROM into the carrier. Be careful not to bend the pins.

P13-9 Align the PROM carrier with the socket. The small notch of the carrier must be aligned with the small notch in the socket.

453

Service Tips

Whenever a short-cut or special procedure is appropriate, it is described in the text. These tips are generally those things commonly done by experienced technicians.

Cautions and Warnings

Throughout the text, cautions are given to alert the reader to potentially hazardous materials or unsafe conditions. Warnings are also given to advise the student of things that can go wrong if instructions are not followed or if a nonacceptable part or tool is used.

8. Excessive current drain caused by a light or other electrical component remaining on after the ignition switch is turned off.
9. Check for symptoms of undercharging. These include slow cranking, discharged battery, low instrument panel ammeter or voltmeter readings, and charge indicator lamp on.
10. Check for symptoms of overcharging. These include high ammeter and voltmeter readings, battery boiling, and charge indicator lamp on.

WARNING: Do not overtighten the drive belt. Early bearing failure can occur if the belt is tightened beyond manufacturer's specifications.

SERVICE TIP: To check the fusible link to the AC generator, use a voltmeter and test for voltage at the BAT terminal. If the battery is good, voltage should be present. If there is no voltage, the fusible link is probably burned out. A better test would be to measure the voltage drop across the link; this will identify any high resistance in the circuit.

The manufacturer of the vehicle you are working on may have several additional tests to perform. It is important to always follow the procedures outlined by the manufacturer for the vehicle being tested.

CAUTION: Many charging system tests require that the vehicle be operated in the shop area. Always place wheel blocks against the drive wheels. Be sure there is proper ventilation of the vehicle's exhaust. Also, be aware of the drive belts and cooling fan. Be sure of where your hands and tools are at all times.

Charging System Service Cautions

The following are some of the general rules when servicing the charging system:

1. Do not run the vehicle with the battery disconnected. The battery acts as a buffer and stabilizes any voltage spikes that may cause damage to the vehicle's electronics.
2. Do not allow output voltage to increase over 16 volts when performing charging system tests.
3. If the battery needs to be recharged, disconnect the cables while charging.
4. Do not attempt to remove electrical components from the vehicle with the battery connected.
5. Before connecting or disconnecting any electrical connections, the ignition switch must be in the OFF position.
6. Avoid contact with the BAT terminal of the AC generator while the battery is connected. Battery voltage is always present at this terminal.

AC Generator Noises

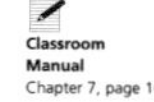

Classroom Manual
Chapter 7, page 163

Noises that come from the AC generator can be from three sources. The causes of the noises are identifiable by the type of noise they make. A loose belt will make a squealing noise. Check the belt condition and tension. Replace the belt if necessary.

SERVICE TIP: With the engine off, rub a piece of bar soap on the pulley surface

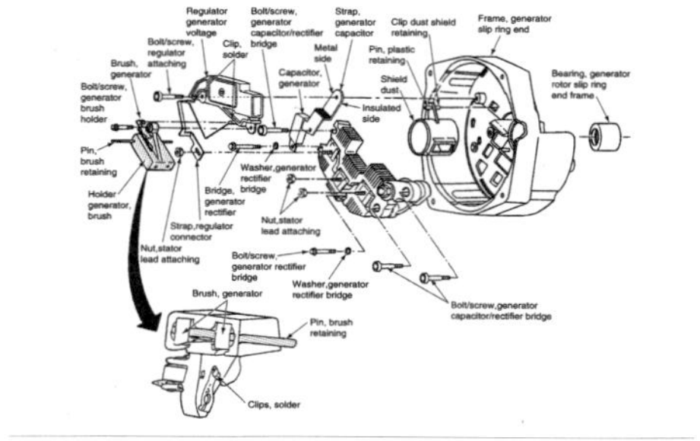

Figure 7-36 When assembling a generator, use a pin to hold the brushes back so that the rotor can be inserted into the brush holder. (Courtesy of General Motors Corporation)

Customer Care

This feature highlights those little things a technician can do or say to enhance customer relations.

References to the Classroom Manual

Reference to the appropriate page in the Classroom Manual is given whenever necessary. Although the chapters of the two manuals are synchronized, material covered in other chapters of the Classroom Manual may be fundamental to the topic discussed in the Shop Manual.

Diode Pattern Testing

Classroom Manual
Chapter 7, page `72

Special Tools
Oscilloscope
Carbon pile

CUSTOMER CARE: It is good practice to check the diode pattern of the AC generator anytime an electronic component fails. Because the electronics of the vehicle cannot accept AC current, the damage to the replaced component could have been the result of a bad diode. By performing this check it is possible to find the cause of the problem.

Set an oscilloscope on the lowest scale available. Connect the primary test leads on the AC generator output terminal and ground. Start the engine and place a moderate load on the charging system (15 to 20 amperes). Different patterns may appear. What is considered normal depends on the load placed on the system.

The diode pattern (Figure 7-37) illustrates a good pattern. However, the second pattern shown (Figure 7-38) is also a good pattern if the AC generator is under a full load. The third pattern shown (Figure 7-39) is a good pattern for some AC generators.

Patterns that have high resistance, open, and shorted diodes are illustrated (Figures 7-40 through 7-43). Remember to check the waveforms for noise. If the diodes don't rectify all of the AC, some will ride on the DC output.

Job Sheets

Located at the end of each chapter, the Job Sheets provide a format for students to perform procedures covered in the chapter. A reference to the ASE Task addressed by the procedure is referenced on the Job Sheet.

Job Sheet 5

5

Name ______________________ Date __________

Use of a Voltmeter

Upon completion of this job sheet, you will be able to measure available voltage and voltage drop.

ASE Correlation

This job sheet is related to the ASE Electrical/Electronic Systems Test's content area: General Electrical/Electronic System Diagnosis, task: Check applied voltages and voltage drops in electrical/electronic circuits and components with a voltmeter; determine needed repairs.

Tools and Materials

A vehicle — Wiring diagram for vehicle
A DMM — Basic hand tools

Procedure

1. Set the DMM to the appropriate scale to read 12 volts DC.
2. Connect the meter across the battery (positive to positive and negative to negative).
   What is your reading on the meter? __________ volts
3. With the meter still connected across the battery, turn on the headlights of the vehicle.
   What is your reading on the meter? __________ volts
4. Keep the headlights on. Connect the positive lead of the meter to the point on the vehicle where the battery's ground cable attaches to the frame. Keep the negative lead where it is.
   What is your reading on the meter? __________ volts.
   What is being measured? __________
5. Disconnect the meter from the battery and turn off the headlights.
6. Refer to the correct wiring diagram and determine what wire at the right headlight delivers current to the lamp when the headlights are on and low beams selected.
   Color of the wire __________
7. From the wiring diagram identify where the headlight is grounded.
   Place of ground __________
8. Connect the negative lead of the meter to the point where the headlight is grounded.
9. Connect the positive lead of the meter to the power input of the headlight.
10. Turn on the headlights.
   What is your reading on the meter? __________ volts
   What is being measured? __________

79

Case Studies

Case Studies concentrate on the ability to properly diagnose the systems. Each chapter ends with a case study in which a vehicle has a problem, and the logic used by a technician to solve the problem is explained.

CASE STUDY

A customer says the cornering lights on the vehicle are working only part of the time. The technician turns on the turn signals and the cornering lights illuminate. He then shakes the vehicle and notices that the cornering lights stop working. The lights come on when the vehicle is shaken again.

The technician checks all of the connections at the battery and at the light sockets. All are good. The technician then shakes the vehicle until the cornering lights go out. Next he uses a test light to check for voltage at the light sockets. The test light does not come on. Tracing the circuit backward toward the turn signal switch, the technician discovers a voltage into the switch, but not out of it. Once the turn signal switch is removed from the steering column, the cause of the problem is easily found. The spring that maintains pressure on the contacts of the cornering light switch has been dislodged. The contacts intermittently have continuity until the vehicle goes over a bump in the road.

Terms to Know

Bezel
Courtesy lights
Curb height
Dimmer switch
Double filament lamp
Feedback
Filament
Flasher
Multifunction switch
Prism

ASE Style Review Questions

1. The dimmer switch is being discussed:
Technician A says all dimmer switches are mounted on the floor board.
Technician B says the dimmer switch can be incorporated into the multifunction switch.
Who is correct?
A. A only
B. B only
C. Both A and B
D. Neither A nor B

2. *Technician A* says repairs to the lighting system must assure vehicle safety.
Technician B says the lighting system repairs must meet all applicable laws.
Who is correct?
A. A only
B. B only
C. Both A and B
D. Neither A nor B

3. A customer complains that the headlights are brighter than normal and that she has to replace the lamps regularly:
Technician A says this can be caused by too high generator output.
Technician B says this can be caused by excessive voltage drop in the circuit.
Who is correct?
A. A only
B. B only
C. Both A and B
D. Neither A nor B

4. A customer complains that none of the external parking and headlights turn on:
Technician A says the circuit from the battery to the switch may be faulty.
Technician B says the headlight switch can be at fault.
Who is correct?
A. A only

Terms to Know

Terms in this list can be found in the Glossary at the end of the manual.

ASE Style Review Questions

Each chapter contains ASE style review questions that reflect the performance objectives listed at the beginning of the chapter. These questions can be used to review the chapter as well as to prepare for the ASE certification exam.

Table 9-2 ASE TASK

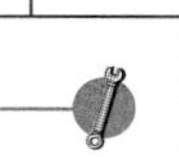

Test, replace, and aim headlights/bulbs.

Problem Area	Symptoms	Possible Causes	Classroom Manual	Shop Manual
Headlights burned out or out of adjustment	Improper road illumination in front of the vehicle.	1. Burned out headlight	138	92
		2. Improper headlight aiming	139	99
		3. Open circuit	80, 144	103
		4. Short in insulated circuit	144	103
		5. Excessive resistance	82	103
		6. Poor ground connection	22	103
		7. Improper bulb application	138	92

Table 9-3 ASE TASK

Inspect, test, and repair or replace headlight and dimmer switches, relays, sockets, connectors, and wires of headlight circuits.

Problem Area	Symptoms	Possible Causes	Classroom Manual	Shop Manual
Open or defective components	Improper or no headlight operation.	1. Defective headlight switch	142	107
		2. Defective relay	144	23
		3. Defective dimmer switch	144	116
		4. Open circuit	80, 144	103
		5. Poor connections	90	103

Table 9-4 ASE TASK

Diagnose the cause of intermittent, slow, or no retractable headlight assembly operation. Inspect, test, and repair or replace motors, switches, relays, connectors, and wires of retractable headlight assembly circuits.

Problem Area	Symptoms	Possible Causes	Classroom Manual	Shop Manual
Concealed headlight door motors	Headlight doors fail to operate properly.	1. Loose or broken vacuum connections	90	105
		2. Defective vacuum motors	148	105

Diagnostic Chart

Chapters include detailed diagnostic charts linked with the appropriate ASE task. These charts list common problems and most probable causes. They also list a page reference in the Classroom Manual for better understanding of the system's operation and a page reference in the Shop Manual for details on the procedure necessary for correcting the problem.

APPENDIX A

ASE PRACTICE EXAMINATION

1. A customer says he hears the sound of gears clashing when he attempts to shift into reverse gear immediately after disengaging the clutch; he does not hear any noises when he shifts into first gear.
Technician A says that the customer appears to be shifting into reverse gear too soon after disengaging the clutch.
Technician B says that there may be a problem with reverse gear.
Who is correct?
A. Technician A **C.** Both A and B
B. Technician B **D.** Neither A nor B

2. A customer says the clutch of her car appears to be slipping; after shifting into first gear the vehicle does not begin to move until the clutch is almost completely disengaged.
Technician A says there may be excessive clutch pedal free-play.
Technician B says the clutch master cylinder primary seal may be leaking.
Who is correct?
A. Technician A **C.** Both A and B
B. Technician B **D.** Neither A nor B

3. Clutch actuation systems are being discussed.
Technician A says normal clutch wear will result in a decrease in clutch pedal free-play on a vehicle equipped with linkage-type actuation.
Technician B says a worn release bearing fork pivot will result in excessive clutch pedal free-play on linkage-type clutch systems.
Who is correct?
A. Technician A **C.** Both A and B
B. Technician B **D.** Neither A nor B

4. *Technician A* says clutch chatter can be caused by a clutch disc that has been saturated with oil.
Technician B says a flywheel that has insufficient lateral runout can cause clutch chatter.
Who is correct?
A. Technician A **C.** Both A and B
B. Technician B **D.** Neither A nor B

5. A vehicle has been towed into the shop because the customer claims the transmission will not shift into any gear while the engine is running. Which of the following could cause this problem?
A. A clutch disc with worn friction material.
B. A pressure plate with a weak spring diaphgram.
C. A frozen clutch release bearing.
D. A frozen clutch pilot bearing.

6. A severe pulsation is felt at the clutch pedal the instant the clutch pedal is touched.
Technician A says a flywheel with excessive lateral runout could cause this problem.
Technician B says this problem could be caused by a worn transmission input shaft bearing.
Who is correct?
A. Technician A **C.** Both A and B
B. Technician B **D.** Neither A nor B

7. A clutch master cylinder is being replaced.
Technician A says this procedure does not require any adjustments.
Technician B says failure to bleed the hydraulic system properly can result in excessive clutch pedal free-play.
Who is correct?
A. Technician A **C.** Both A and B
B. Technician B **D.** Neither A nor B

8. During an engine replacement an oily fluid is found all around the clutch components.
Technician A says the engine rear main oil seal could have been leaking.
Technician B says if the parts are not worn out they can be washed in cleaning solvent and then reused.
Who is correct?
A. Technician A **C.** Both A and B
B. Technician B **D.** Neither A nor B

9. The alignment of the clutch bellhousing to the engine block is being discussed.
Techncian A says a typical misalignment limit is .500".
Technician B says misalignment can be corrected by installing shims between the bellhousing and the engine.
Who is correct?
A. Technician A **C.** Both A and B
B. Technician B **D.** Neither A nor B

407

ASE Practice Examination

A 50 question ASE practice exam, located in the appendix, is included to test students on the content of the complete Shop Manual.

Reviewers

I would like to extend a special thanks to those who saw things I overlooked and for their contributions:

Randy Briggs
Western Iowa Technical Community College
Elkpoint, SD

Thomas G. Broxholm
Skyline College
San Bruno, CA

Ronnie Bush
Tennessee Technology Center at Jackson
Jackson, TN

Jon Clausen
SE Technical Institute
Sioux Falls, SD

Steve Michener
Mt. Hood Community College
Gresham, OR

Jerry L. Mumm
Ricks College
Rexburg, ID

Al Playter
Centenial College
Scarborough, Ont. CANADA

Fred Raadsheer
British Columbia Institute of Technology
Burnaby, B.C. CANADA

Patrick J. TieKamp
University of New Mexico - Gallup
Gallup, NM

Gus Winchester
DeKalb Technical Institute
Clarkston, GA

Contributing Companies

I would also like to thank these companies who provided technical information and art for this edition:

Alfa Electronics
American Honda Motor Co., Inc.
American Isuzu Motors, Inc.
Arrow Safety
Battery Council International
Blackhawk Automotive, Inc.
Chilton Book Company
Chrysler Corporation
CRC Chemicals
Delco Remy Division-GMC
Ford Motor Company
General Motors Corporation, Service Technology Group
Gentex Corporation
Goodson Shop Supplies
Hyundai Motor America
Lincoln
Matco Tools
Mazda Ltd.
Mitsubishi Motor Sales of America, Inc.
Nissan North America Inc.
Optima Batteries, Inc.
Robert Bosch Corporation
Snap-on Tools Corporation
Sun Electric Company
Tif Instruments, Inc.
Toyota Motor Sales USA Inc.
Volkswagen of America

CHAPTER 1

Safety

Upon completion and review of this chapter, you should be able to:

- ❑ Explain how safety is a part of professionalism.
- ❑ Describe the basic safety rules of servicing electrical systems.
- ❑ Properly lift objects.
- ❑ Properly work around batteries.
- ❑ Explain the safety precautions associated with charging and starting system service.
- ❑ List and describe personal safety responsibilities.
- ❑ Classify fires and fire extinguishers.
- ❑ Describe the proper use of floor jacks and safety stands.
- ❑ Explain the safe use of hoists.
- ❑ Properly run the engine in the shop.
- ❑ List the safety precautions associated with servicing the air bag system.
- ❑ Explain the safety precautions that are necessary when servicing the antilock brake system.

Introduction

Being a professional technician is more than being knowledgeable about automotive systems; it is also an attitude. You should know how to deal with customers, fellow workers, fellow students, and your instructor or boss. You should also work and behave as a professional. One of the most obvious traits of a professional is the ability to work productively and safely. This is where knowledge becomes very important. You need it to be productive and you need it to ensure your own safety and the safety of others.

Personal Safety

Personal safety depends on what you wear and what you don't wear while working on an automobile. It also depends on your awareness of certain hazards. Personal safety is also the result of your actions and your attitude while working.

Dress and Appearance

Shop Manual
Chapter 1, page 1

Nothing displays professional pride and a positive attitude more than the way you dress. Customers are demanding a professional atmosphere in the service shop. Your appearance instills customer confidence, as well as expresses your attitude toward safety. The wearing of proper and clean clothes can prevent injuries.

Long sleeve shirts should have their cuffs buttoned or rolled up tightly. Shirt tails should be tucked in at all times. If a necktie is worn in the shop area, tuck it inside your shirt. Clip-on ties are recommended if you must wear a tie.

Long hair is a serious safety concern. Very serious injury can result if hair becomes caught in rotating machinery, fan belts, or fans. If your hair is long enough to touch the bottom of your shirt collar, it should be tied back and tucked under a hat.

A **conductor** is capable of supporting the flow of electricity through it.

Jewelry has no place in the automotive shop, especially when working on the electrical systems. Rings, watches, bracelets, necklaces, earrings, and so forth, can cause serious injury. The gold, silver, and other metals used in jewelry are excellent conductors of electricity. Your body is also a good conductor. When electrical current flows through a conductor, it generates heat. The heat is great enough to cause severe burns. Jewelry can also get caught in moving parts. This can cause serious cuts. Necklaces can cause serious injury or even death if they become caught in moving equipment.

You should wear shoes or boots that will protect your feet, just in case something falls or you stumble into something. It is a good idea to wear safety shoes or boots with steel toes and shanks. Most safety shoes also have slip-resistant soles.

Eye Protection

A **short** increases electrical current in a circuit.

An **insulator** is not capable of supporting the flow of electricity.

Occupational safety glasses are designed with special high impact lens and frames, and provide for side protection.

Safety goggles provide eye protection from all sides because they fit against the face and forehead to seal off the eyes from outside elements.

The importance of wearing proper eye protection cannot be overemphasized. Every working day there are over 1,000 eye injuries, many resulting in blindness. Almost all of these are preventable. The safest and surest way to protect your eyes is to wear the proper eye protection anytime you enter the shop. At a minimum, wear eye protection when grinding, using power tools, hammering, cutting, chiseling, or performing service under the car. In addition, wear eye protection when doing any work that can cause sparks, dirt, or rust to enter your eyes, and when you are working around chemicals. Remember, just because you are not doing the work yourself does not mean you cannot suffer an eye injury. Many eye injuries are caused by a co-worker. Wear eye protection anytime you are near an eye hazard.

The key to protecting your eyes is the use of *proper* eye protection. Prescription glasses do not provide adequate protection. Regular glasses are designed to impact standards that are far below that required in the workplace. A flying object may be stopped by the lens, but the frame may allow the lens to pop out and hit your face, causing injury. In addition, regular glasses do not provide side protection.

There are many types of eye protection (Figure 1-1). One of the best ways to protect your eyes is to wear occupational safety glasses. These glasses are light and comfortable. They are constructed of tempered glass or safety plastic lens, and have frames that prevent the lens from being pushed out upon impact. They have side shields to prevent the entry of objects from the side. Occupational safety glasses are available in prescription lens so they can be worn instead of regular corrective lens glasses.

Safety goggles fit snugly around the area of your eyes to prevent the entry of objects and to provide protection from liquid splashes. The force of impact on the lens is distributed throughout

Figure 1-1 Types of eye protection.

the entire area where the safety goggles are in contact with your face and forehead. Safety goggles are designed to fit over regular glasses.

Face shields are used when there is potential for sparks, flying objects, or splashed liquids, which can cause neck, facial, and eye injuries. The plastic is not as strong or impact resistant as occupational safety glasses or safety goggles. If there is a danger of high impact objects hitting the face shield, wear safety glasses under the face shield.

A **face shield** is a clear plastic shield that protects the entire face.

CAUTION: Before removing your eye protection, close your eyes. Pieces of metal, dirt, or other foreign material may have accumulated on the outside. These could fall into your eyes when you remove your glasses or shield.

Safety glasses provide little or no protection against chemicals. When working with chemicals, such as battery acid, refrigerants, cleaning solutions, etc., safety goggles should be worn. Full face shields are not intended to provide primary protection for your eyes. They are designed to provide primary protection for your face and neck and should be worn in addition to eye protection.

Rotating Belts and Pulleys

Many times the technician must work around rotating parts such as generators, power steering pumps, air pumps, water pumps, and air conditioner (A/C) compressors. Always think before acting. Be aware of where you are placing your hands and fingers at all times. Do not place rags, tools, or test equipment near moving parts. In addition, make sure you are not wearing any loose clothing or jewelry that can get caught.

Other rotating equipment or components of concern include tire changers, spin balancers, drills, bench grinders, and drive shafts.

Electric Cooling Fans

Be very cautious around electric cooling fans. Many of these fans will operate even if the ignition switch is turned off. They are controlled by a temperature sensing unit in the engine block or radiator and will turn on anytime the coolant temperature reaches a certain point. Before working on or around an electric cooling fan, disconnect the electrical connector to the fan motor or the negative battery cable.

WARNING: Reconnect the fan connector before returning the vehicle to service. Failure to do so can allow the engine to overheat.

Lifting

Back injuries are one of the most crippling injuries in the industry, yet most of them are preventable. Most occupational back injuries are caused by improper lifting practices. These injuries can be avoided by following a few simple lifting guidelines:

1. Do not lift a heavy object by yourself. Get help from someone else.
2. Do not lift more than you can handle. If the object is too heavy, use the proper equipment to lift it.
3. Do not attempt to lift an object if there is not a good way to hold onto it. Study the object to determine the best balance and grip points.
4. Do not lift with your back. Your legs have some of the strongest muscles in your body. Use them.
5. Place your body close to the object. Keep your back and elbows straight (Figure 1-2).
6. Make sure you have a good grip on the object. Do not attempt to readjust the load once you have lifted it. If you are not comfortable with your balance and grip, lower the object and reposition yourself.

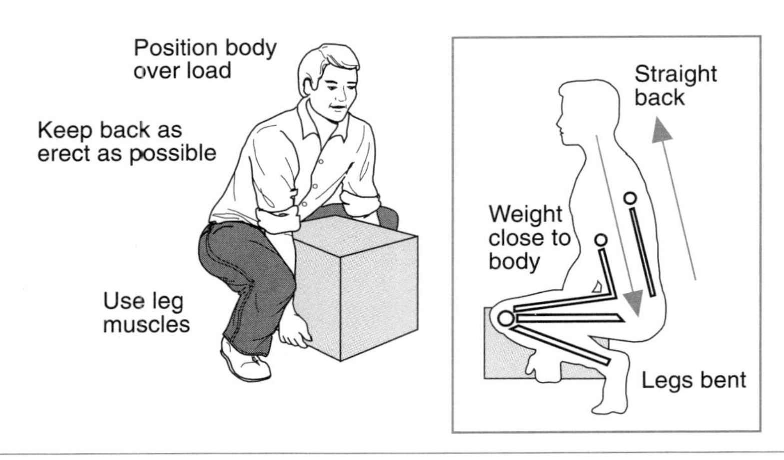

Figure 1-2 When lifting a heavy object, keep your back straight and lift with your legs.

7. When lifting, keep the object as close to your body as possible. Keep your back straight and lift with your legs.
8. While carrying the object, do not twist your body to change directions. Use your feet to turn your whole body into the new direction.
9. To set the load down, keep the object close to your body. Bend at the knees and keep your back straight. Do not bend forward or twist.
10. If you need to place the object onto a shelf or bench top, place an edge of the object on the surface and slide it into place. Do not lean forward.

Tool and Equipment Safety

Technicians would not be able to do their job without tools and equipment. Most injuries caused by tools and equipment are the result of improper use, improper maintenance, and/or carelessness.

Hand Tools

Hand tools use only the force generated from the body to operate. They multiply the force received through leverage to accomplish the work.

Here are some very simple steps that you can take to help assure safe hand tool use:

1. Do not use tools that are worn out or broken.
2. Do not use a tool to do something that it was not designed for. Use the proper tool for the job.
3. Keep your tools clean and in good condition.
4. Point sharp edges of tools away from yourself.
5. Do not hold small components in your hands while using tools such as screwdrivers. The tool may slip and cause injury to your hand.
6. Examine your work area for things that can cause injury if a tool slips or a fastener breaks loose quickly. Readjust yourself or the tool to avoid them.
7. Do not put sharp tools into your pockets.

Power Tool Safety

Shop Manual
Chapter 1, page 3

Many times a technician will be required to use power tools when performing electrical service. Drills and hole saws will be used to install new accessories onto the vehicle or to drill holes for wiring to pass through. Grinders, drill presses, and hydraulic presses may be used to help fabricate or modify components. Pneumatic tools are used to remove or fasten components. All of these tools can cause injury if not used properly. Use the following guidelines when working with power tools:

Power tools use forces other than those generated by the body. They can use compressed air, electricity, or hydraulic pressure to generate and multiply force.

Pneumatic tools are powered by compressed air.

1. Ask your instructor if you are not sure of the correct operation of a tool.
2. Always wear proper eye protection when using power tools.
3. Check that all safety guards and safety equipment are installed on the tool.
4. Before using an electrical tool, check the condition of the plug and cord. The plug should be a three-prong plug. Never cut off the grounding prong. Do not use the tool if the wires are frayed or broken. Plug the tool only into a grounded receptacle.
5. Before using an air tool, check the condition of the air hose. Do not use the tool if the hose shows signs of weakness such as bulges or fraying. Also the tool should be properly oiled.
6. Before using a hydraulic tool, check the condition of all hoses and gauges. Do not use the tool if any of these are defective.
7. Make sure other people are not in the area when you turn the tool on.
8. Do not leave the area with the tool still running. Stay with the tool until it stops. Then disconnect it.
9. Make all adjustments to the tool before turning it on.
10. If the tool is defective or does not pass your safety inspection, put a sign on it and report it to your supervisor.

Compressed Air Safety

Compressed air is used in an automotive shop to do many things; however, cleaning off your clothes is not one of them. Dirt and other objects blown off of your clothes can cause serious injury to yourself and others. There may be dirt in the nozzle or hose that will be ejected at a high rate of speed and can be forced into someone's skin or eyes. In addition, most shops have compressed air systems equipped with automatic oilers. The presssure can push the oil and air bubbles through your skin and into your blood.

Use only approved safety nozzles when using compressed air to dry parts that have been cleaned. Safety nozzles have a relief passage that prevents high pressures from being expelled directly out the front. However, it is best not to use compressed air to dry parts. There are instances where air must be used to dry small passages. In these cases, only use air pressure that has been regulated down to about 25 psi.

Check the air hoses for signs of wear. Do not use them if they are bulging, frayed, or if the couplers are damaged.

Lifting the Vehicle Safely

There will be times when it will be necessary to raise the vehicle off the floor to perform some work. The vehicle can be lifted by use of floor jacks or a hoist. Each method has its own set of safety precautions and rules that must be followed.

Shop Manual
Chapter 1, page 4

Jack and Jack Stand Safety

A **floor jack** is a portable hydraulic tool used to raise and lower a vehicle.

Floor jacks are used to lift the vehicle a short distance off the floor or when only a portion of the vehicle needs to be raised (Figure 1-3). Before using a floor jack, check it for signs of hydraulic fluid leaks and for damage that would compromise its safe use. Before lifting the vehicle, place wheel blocks in front of and behind one of the tires that will remain on the ground.

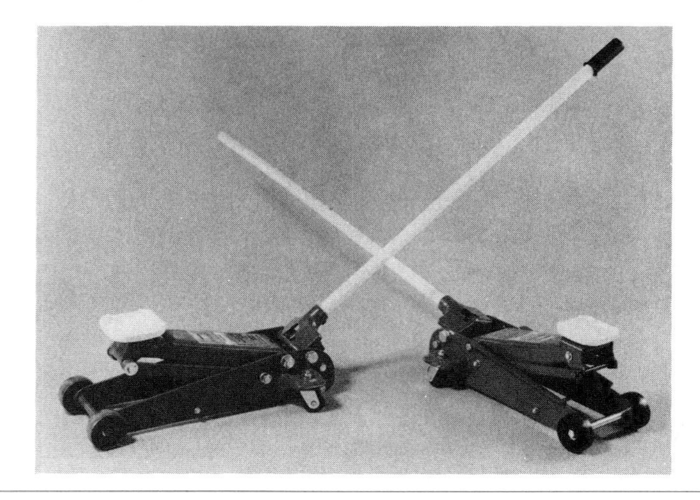

Figure 1-3 Floor jacks are used to raise a vehicle a short distance off the floor. (Courtesy of Blackhawk Automotive, Inc.)

CAUTION: Place the wheel blocks around a tire that is going to be lifted last. If the front of the vehicle is being lifted, place the wheel blocks around one of the rear tires. When the vehicle is being lowered, place blocks around one of the tires on the floor before lowering the other end of the vehicle.

Jack stands, or safety stands, are support devices used to hold the vehicle off of the floor after it has been raised by the floor jack.

Many jack manufacturers and service manuals provide illustrations for the proper lift points on a vehicle (Figure 1-4). If this information is not available, always place the floor jack on major strength parts. These areas include the frame, cross member, or differential. If you are in doubt about the proper lift point, ask your instructor. Never lift on sheet metal or plastic parts.

The floor jack is to be used only to lift the vehicle off of the floor. It is not intended to support the vehicle while someone is under it. Use jack stands to support the vehicle (Figure 1-5). Use one jack stand for each quarter of the vehicle that is lifted (Figure 1-6). Place the jack stand under the frame or a major support component of the vehicle. When the vehicle is lowered onto the stands, make sure that they do not tilt.

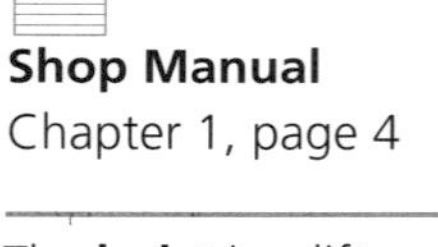

Shop Manual
Chapter 1, page 4

Hoist Safety

The **hoist** is a lift used to raise the entire vehicle.

Hoists are used when the entire vehicle needs to be raised, usually high enough for the technician to stand underneath the vehicle (Figure 1-7). When a vehicle is placed on the hoist, it must be centered. The balance of the vehicle must be taken into consideration, as well as the effects on balance if a heavy component is removed from the vehicle.

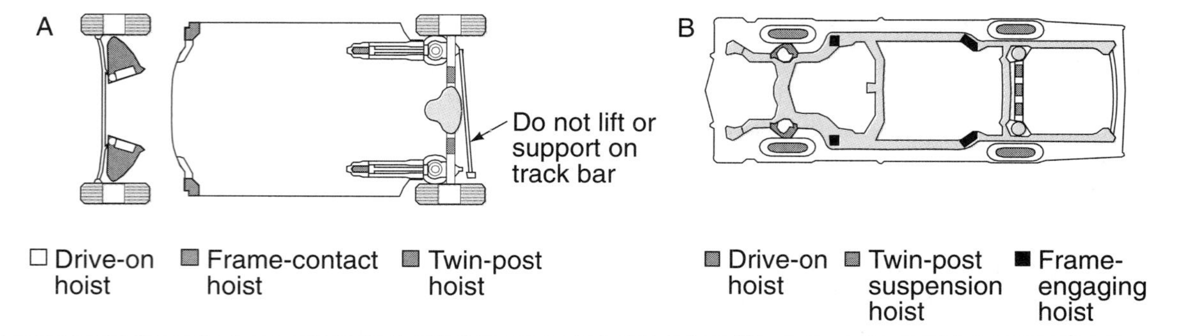

Figure 1-4 Examples of illustrations showing proper lift points on a vehicle.

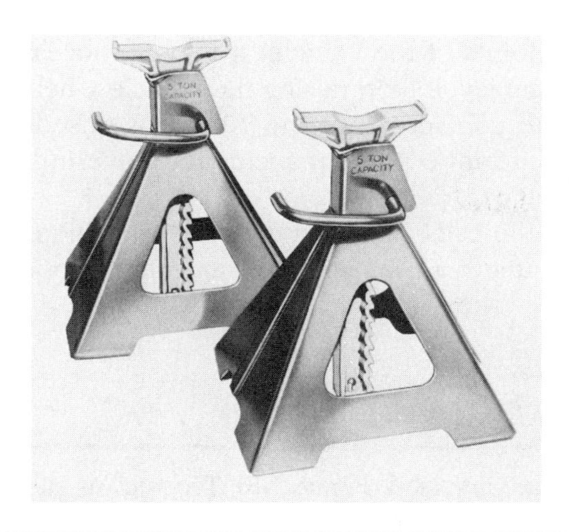

Figure 1-5 Jack stands are used to support the vehicle after it has been raised by a floor jack. (Courtesy of Lincoln, St. Louis)

Figure 1-6 Do not trust the floor jack to support a vehicle. Never go under a raised vehicle that is not supported by jack stands.

Figure 1-7 One type of hoist that can be found in an automotive shop. The hoist is used to lift the entire vehicle.

Place the hoist pads under proper lift points of the vehicle. It may be necessary to adjust the pad height in order to lift the vehicle level. Stop raising the vehicle when it is a few inches off the floor to confirm good pad contact. Once the vehicle is raised to the desired height, use the hoist's locking mechanism to prevent accidental lowering of the vehicle.

If the vehicle is not level on the pads, or the pads are not in the proper position, lower the vehicle and readjust as needed. Never get under a vehicle that is not sitting properly on the hoist.

Fires and Fire Extinguishers

Shop Manual
Chapter 1, page 8

Fires are classified by the types of materials that are involved (Figure 1-8). Technicians should be able to locate the correct fire extinguisher to control the fire types that they are likely to experience. Technicians must also be able to fight a fire in an emergency.

CAUTION: Do not risk your life in fighting a fire. If it is evident that the fire is out of control, get out. Always be aware of where you are and the location of the closest exit. Do not open the garage doors in the event of a fire.

Labels on the fire extinguisher will indicate the types of fires that it will put out. Become familiar with the use of a fire extinguisher.

Gasoline

Gasoline is so commonly found in automotive repair shops that its dangers are often forgotten. A slight spark or an increase in heat can cause a fire or explosion. Gasoline fumes are heavier than air. Therefore, when an open container of gasoline is sitting about, the fumes spill out over the sides of the container onto the floor. These fumes are more flammable than liquid gasoline and can easily explode.

Never smoke around gasoline or in a shop filled with gasoline fumes because even the droppings of hot ashes can ignite the gasoline. If an engine has a gasoline leak or you have caused a leak by disconnecting a fuel line, wipe it up immediately and stop the leak. While stopping the leak, be extra careful not to cause sparks. The rags used to wipe up the gasoline should be taken outside to dry. Immediately wipe up any gasoline spilled on the floor. If vapors are present in the shop, have the doors open and the ventilating system turned on to get rid of the fumes. Remember that it only takes a small amount of fuel to cause a fire.

Gasoline should always be stored in approved containers and never in glass containers. If the glass container is knocked over or dropped, a terrible explosion can occur. Never use gasoline to clean parts. Also never pour gasoline down a drain.

CAUTION: Never siphon gasoline or diesel fuel with your mouth. These liquids are poisonous and can make you sick or fatally ill.

Electrical System Safety

There are many safety requirements that must be followed when working on the vehicle's electrical system. In addition to personal safety, there is the concern of damaging the electrical system with improper service techniques. The following are a few of the safety rules. Throughout the text there will be cautions and warnings to alert you to potential hazards associated with working on a particular electrical system or component.

Class of Fire		Typical Fuel Involved	Type of Extinguisher
Class A Fires (green)	**For Ordinary Combustibles** Put out a class A fire by lowering its temperature or by coating the burning combustibles.	Wood Paper Cloth Rubber Plastics Rubbish Upholstery	Water*[1] Foam* Multipurpose dry chemical[4]
Class B Fires (red)	**For Flammable Liquids** Put out a class B fire by smothering it. Use an extinguisher that gives a blanketing, flame-interrupting effect; cover whole flaming liquid surface.	Gasoline Oil Grease Paint Lighter fluid	Foam* Carbon dixoide[5] Halogenated agent[6] Standard dry chemical[2] Purple K dry chemical[3] Multipurpose dry chemical[4]
Class C Fires (blue)	**For Electrical Equipment** Put out a class C fire by shutting off power as quickly as possible and by always using a nonconducting extinguishing agent to prevent electric shock.	Motors Appliances Wiring Fuse boxes Switchboards	Carbon dioxide[5] Halogenated agent[6] Standard dry chemical[2] Purple K dry chemical[3] Multipurpose dry chemical[4]
Class D Fires (yellow)	**For Combustible Metals** Put out a class D fire of metal chips, turnings, or shavings by smothering or coating with a specially designed extinguishing agent.	Aluminum Magnesium Potassium Sodium Titanium Zirconium	Dry powder extinguishers and agents only

*Cartridge-operated water, foam, and soda-acid types of extinguishers are no longer manufactured. These extinguishers should be removed from service when they become due for their next hydrostatic pressure test.

Notes:

(1) Freezes in low temperatures unless treated with antifreeze solution, usually weighs over 20 pounds, and is heavier than any other extinguisher mentioned.

(2) Also called ordinary or regular dry chemical (sodium bicarbonate).

(3) Has the greatest initial fire-stopping power of the extinguishers mentioned for class B fires. Be sure to clean residue immediately after using the extinguisher so sprayed surfaces will not be damaged (potassium bicarbonate).

(4) The only extinguishers that fight A, B, and C classes of fires. However, they should not be used on fires in liquefied fat or oil of appreciable depth. Be sure to clean residue immediately after using the extinguisher so sprayed surfaces will not be damaged (ammonium phosphates).

(5) Use with caution in unventilated, confined spaces.

(6) May cause injury to the operator if the extinguishing agent (a gas) or the gases produced when the agent is applied to a fire is inhaled.

Figure 1-8 A Guide to Fire Extinguisher Selection.

Battery Safety

Shop Manual
Chapter 1, page 9

Before attempting to do any type of work on or around the battery, you must be aware of certain precautions. To avoid personal injury or property damage, follow these precautions:

1. Battery acid is very corrosive. Do not allow it to come into contact with your skin, eyes,or clothing. If battery acid should get into your eyes, rinse them thoroughly with clean water and seek immediate medical attention. If battery acid comes into contact with your skin, wash with clean water. Baking soda added to the water will neutralize the acid. If the acid is swallowed, drink large quantities of water or milk followed by milk of magnesia and a beaten egg or vegetable oil.
2. When making connections to a battery, be careful to observe polarity, positive to positive and negative to negative.

3. When disconnecting battery cables, always disconnect the negative (ground) cable first.
4. When connecting battery cables, always connect the negative cable last.
5. Avoid any arcing or open flames near a battery. The vapors produced by the cycling of a battery are very explosive. Do not smoke around a battery.
6. Follow manufacturers' instructions when charging a battery. Charge the battery in a well-ventilated area. Do not connect or disconnect the charger leads while the charger is turned on.
7. Do not add electrolyte to the battery if it is low. Add only distilled water.
8. Do not wear any jewelry while servicing the battery. These items are excellent conductors of electricity. Severe burns may result if current flows through them by accidental contact with the battery positive terminal and a ground.
9. Never lay tools across the battery. They may come into contact with both terminals, shorting out the battery and causing it to explode, damaging both the tools and the battery.
10. Wear safety glasses and/or a face shield when servicing the battery.

WARNING: Always double-check the polarity of the battery charger's connections and leads before turning the charger on. Incorrect polarity can damage the battery or cause it to explode.

Starting System Service Safety

Before testing or servicing the starter system, become familiar with these precautions that should be observed:

1. Refer to the recommendations given in the service manual for correct procedures for disconnecting a battery. Some vehicles with on-board computers must be supplied with an auxiliary power source.
2. Disconnect the battery ground cable before disconnecting any of the starter circuit's wires or removing the starter motor.
3. Be sure the vehicle is properly positioned on the hoist or on safety jack stands.
4. Before performing any cranking test, be sure the vehicle transmission is in park or neutral and the parking brakes are applied. Put wheel blocks in front of and behind one tire.
5. Follow the manufacturer's directions for disabling the ignition system.
6. Be sure the test leads are clear of any moving engine components.
7. Never clean any electrical components in solvent or gasoline. Clean with denatured alcohol or wipe with clean rags only.

Charging System Service Safety

The following are some general rules for servicing the charging system:

1. Do not run the vehicle with the battery disconnected. The battery acts as a buffer and stabilizes any voltage spikes that may cause damage to the vehicle's electronics.
2. When performing charging system tests, do not allow output voltage to increase over 16 volts.
3. If the battery needs to be recharged, disconnect the battery cables while charging.

4. Do not attempt to remove electrical components from the vehicle with the battery connected.
5. Before connecting or disconnecting any electrical connections, the ignition switch must be in the "off" position.
6. Avoid contact with the BAT terminal of the generator while the battery is connected. Battery voltage is always present at this terminal.

BAT is the terminal identifier for the conductor from the generator to the battery positive terminal.

Air Bag Safety

An air bag system demands that the technician pay close attention to safety warnings and precautions when working on or around it. Most air bags are deployed by an explosive charge. Accidental deployment of the air bag can result in serious injury. When working on or around the steering wheel or air bag module, be aware of your hands and arms. Do not place your arm over the module. If the air bag deploys, injury can result.

CAUTION: Obey the warnings in the service manual when working on the air bag system. Failure to follow these warnings may result in air bag deployment and injury.

Shop Manual
Chapter 1, page 10

Air bag systems contain a means of deploying the bag even if the battery is disconnected. This system is needed in the event that the battery is damaged or disconnected during an accident. The reserve energy can be stored for over 30 minutes after the battery is disconnected. Follow the service manual procedures for disabling the system.

The air bag system is designed as a supplemental restraint that in the case of an accident will deploy a bag out of the steering wheel or passenger side dash panel to provide additional protection against head and face injuries.

When carrying the air bag module, carry it so that the bag and trim are facing away from your body (Figure 1-9). In the event of accidental deployment, the charge will be away from you. Do not face the module toward any other people.

When you place the module on the bench, face the bag and trim up (Figure 1-10). This provides a free space for the bag to expand if it deploys. If the module will be stored for any period of time, place it in the car's trunk and close the trunk.

While troubleshooting the air bag system, do not use electrical testers such as battery-powered or AC-powered voltmeters, ohmmeters, and so on, or any other equipment except those specified in the service manual. Do not use a test light to troubleshoot the system.

When it is necessary to make a repair or replace a component in the air bag system, always disconnect the negative battery cable before making the repair. It is a good practice to insulate

The **air bag module** is the air bag and inflator assembly together in a single package.

Figure 1-9 Carry an air bag module so that the bag and trim are away from your body. (Courtesy of General Motors Corporation)

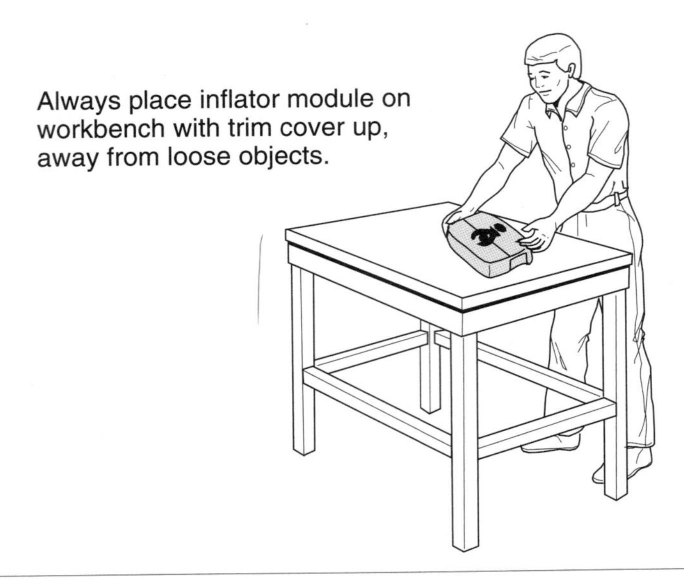

Figure 1-10 Place the air bag module on a bench with the bag and trim facing up. (Courtesy of General Motors Corporation)

the terminal with tape or a rubber hose to prevent it from coming into contact with the battery post. Some manufacturers recommend that the air bag inflator module(s) be disconnected, in addition to the negative battery cable.

Antilock Brake Service Safety

Shop Manual
Chapter 1, page 11

Antilock brake systems (ABS) automatically pulsate the brakes to prevent wheel lock-up under panic stop and poor traction conditions.

Antilock brakes (ABS) are available on most of today's vehicles. There are many different systems used and each has it own safety requirements regarding servicing the system. Become familiar with the warnings and cautions associated with the system you are working on by studying the service manual before performing any service.

Certain components of the ABS are not intended to be serviced individually. Do not attempt to remove or disconnect these components. Only those components with approved removal and installation procedures in the service manual should be serviced.

Some operations require that the tubes, hoses, and fittings be disconnected. Many anti-lock brake systems use high hydraulic pressures and an accumulator to store this pressurized fluid. Before disconnecting any lines or fittings, the accumulator must be fully depressurized. Follow the service manual procedures for depressurizing the system.

Many late-model ABS systems do not use an accumulator. Therefore these systems do not require depressurizing. However, always refer to the correct service manual before servicing a brake system.

Summary

Terms to Know

Air bag module

Antilock brake system (ABS)

Carbon monoxide

- Being a professional technician means more than having knowledge of vehicle systems. It also requires an understanding of all the hazards in the work area.
- As a professional technician, you should work responsibly to protect yourself and the people around you.
- Technicians must be aware that it is their responsibility to prevent injuries in the shop, and their actions and attitudes reflect how seriously they accept that responsibility.

- ❑ Long sleeve shirts should have their cuffs buttoned or rolled up tightly, and shirt tails should be tucked in at all times. Neckties should be tucked inside your shirt and only clip-on ties should be worn.
- ❑ Long hair should be tied back and tucked under a hat.
- ❑ Jewelry has no place in the automotive shop.
- ❑ The safest and surest method of protecting your eyes is to wear proper eye protection anytime you enter the shop.
- ❑ When working around rotating pulleys and belts, be aware of where you are placing your hands and fingers at all times.
- ❑ Most occupational back injuries are caused by improper lifting practices.
- ❑ Most injuries caused by tools and equipment are the result of improper use, improper maintenance, and carelessness.
- ❑ Never use compressed air for cleaning off your clothes.
- ❑ The floor jack is to be used only to lift the vehicle off the floor. Use jack stands to support the vehicle after it is lifted.
- ❑ Fires are classified by the types of materials involved. Fire extinguishers are classified by the type of fire they will extinguish.
- ❑ Batteries can cause serious injury if all safety rules are not followed when working on or around them.
- ❑ The air bag system demands that the technician pay close attention to safety warnings and precautions when working on or around them.
- ❑ Air bag systems contain a means of deploying the bag even if the battery is disconnected. The reserve energy can be stored for over 30 minutes after the battery is disconnected.
- ❑ Carry the air bag module so that the bag and trim are away from your body and place the module on the bench with the bag and trim facing up.
- ❑ Do not use a test light to troubleshoot an air bag system.
- ❑ Become familiar with the warnings and cautions associated with the antilock brake system that you are working on by studying the service manual before performing any service.

Terms to Know (continued)

Conductor
Face shield
Floor jack
Hand tools
Hoist
Insulator
Jack stands
Occupational safety glasses
Personal safety
Pneumatic tools
Power tools
Safety goggles

Review Questions

Short Answer Essays

1. Explain how safety is a part of professionalism.
2. List the basic safety rules of proper lifting.
3. List three safety rules that should be followed when working around the battery.
4. What type of materials are involved in a Class B fire?
5. Describe the proper use of floor jacks and safety stands.
6. How should gasoline be stored?
7. What types of tools are never to be used in troubleshooting the air bag system?

8. What must be done before tubes, hoses, and fittings are disconnected on an ABS?
9. What are the safety regulations that should be followed when lifting a vehicle on a hoist?
10. What must be done before disconnecting any of the wires to the starter or generator?

Fill-in-the-Blanks

1. When working with chemicals, you should always wear ____________________ ____________________.
2. As a professional technician, you should work in such a manner as to protect ____________________ and the people around you.
3. The safest and surest method of protecting your eyes is to wear proper eye protection ____________________ you enter the shop.
4. Most occupational back injuries are caused by ____________________ lifting practices.
5. Never use ____________________ ____________________ for cleaning off your clothes.
6. Carry the air bag module so that the bag and trim are facing ____________________ from your body.
7. Do not use a ____________________ ____________________ to troubleshoot the air bag system.
8. Become familiar with the warnings and cautions associated with the anti-lock brake system that you are working on by studying the ____________________ ____________________ before performing any service.
9. Long hair should be ____________________ and ____________________ under a hat.
10. Most injuries that are caused by tools and equipment are the result of improper ____________________, improper ____________________, and ____________________.

ASE Style Review Questions

1. The use of power equipment and tools is being discussed:
   Technician A says eye protection should be worn.
   Technician B says to check the condition of the tool before using it.
   Who is correct?
   A. A only
   B. B only
   C. Both A and B
   D. Neither A nor B

2. The battery is going to be removed from the vehicle:
   Technician A says to disconnect the positive cable first.
   Technician B says not to lay tools across the top of the battery.
   Who is correct?
   A. A only
   B. B only
   C. Both A and B
   D. Neither A nor B

3. A vehicle is in the shop with the engine running:
   Technician A says it is safe to work around the electric cooling fan when it is off and the engine is running.
   Technician B says you should never stand in front of a vehicle when its engine is running.
   Who is correct?
   A. A only
   B. B only
   C. Both A and B
   D. Neither A nor B

4. Fire extinguisher use is being discussed:
   Technician A says electrical fires are extinguished with Class A fire extinguishers.
   Technician B says gasoline is extinguished with Class B fire extinguishers.
   Who is correct?
   A. A only
   B. B only
   C. Both A and B
   D. Neither A nor B

5. Service of an air bag system is being discussed:
   Technician A says to use a test light to check for applied voltage.
   Technician B says to disconnect the battery negative cable and wait 30 minutes before repairing the system.
   Who is correct?
   A. A only
   B. B only
   C. Both A and B
   D. Neither A nor B

6. Servicing the antilock brake system is being discussed:
   Technician A says the system must be depressurized before disconnecting lines and fittings.
   Technician B says many systems have very high pressures.
   Who is correct?
   A. A only
   B. B only
   C. Both A and B
   D. Neither A nor B

7. While discussing proper eye protection:
   Technician A says safety glasses should be worn when working with battery acid and refrigerants.
   Technician B says full face shields are designed to provide protection for the face, neck, and eyes of the technician.
   Who is correct?
   A. A only
   B. B only
   C. Both A and B
   D. Neither A nor B

8. Proper dress in an auto repair shop is being discussed:
   Technician A says the way people dress reflects very little about their attitude toward safety.
   Technician B says jewelry can be a personal safety hazard and should be removed while working in an auto repair shop.
   Who is correct?
   A. A only
   B. B only
   C. Both A and B
   D. Neither A nor B

9. The vehicle is being raised by a floor jack:
Technician A says the floor jack will support the vehicle safely.
Technician B says to use jack stands at each corner of the vehicle that is lifted.
Who is correct?
A. A only
B. B only
C. Both A and B
D. Neither A nor B

10. Battery service is being discussed:
Technician A says to charge the battery in a well-ventilated area.
Technician B says do not add electrolyte to the battery if its level is low.
Who is correct?
A. A only
B. B only
C. Both A and B
D. Neither A nor B

Basic Theories

Upon completion and review of this chapter, you should be able to:

- ❑ Explain the theories and laws of electricity.
- ❑ Describe the difference between insulators, conductors, and semiconductors.
- ❑ Define voltage, current, and resistance.
- ❑ Define and use Ohm's law correctly.
- ❑ Explain the basic concepts of capacitance.
- ❑ Explain the difference between AC and DC currents.
- ❑ Define and illustrate series, parallel, and series-parallel circuits and the electrical laws that govern them.
- ❑ Explain the basic theory of semiconductors.
- ❑ Explain the theory of electromagnetism.
- ❑ Explain the principles of induction.

Introduction

The electrical systems used in today's vehicles can be very complicated. However, through an understanding of the principles and laws that govern electrical circuits, technicians can simplify their job of diagnosing electrical problems. In this chapter you will learn the laws that dictate electrical behavior, how circuits operate, the difference between types of circuits, and how to apply Ohm's law to each type of circuit. You will also learn the basic theories of semiconductor construction. Because magnetism and electricity are closely related, a study of electromagnetism and induction is included in this chapter.

Basics of Electron Flow

Because electricity is an energy form that cannot be seen, some technicians regard the vehicle's electrical system as being more complicated than it is. These technicians approach the vehicle's electrical system with some reluctance. It is important for today's technician to understand that electrical behavior is confined to definite laws that produce predictable results and effects. To facilitate the understanding of the laws of electricity, a short study of atoms is presented.

An **atom** is the smallest part of a chemical element that still has all the characteristics of that element.

Atomic Structure

An atom is constructed of a fixed arrangement of electrons in orbit around a nucleus—much like planets orbiting the sun (Figure 2-1). The two types of particles that make up the nucleus are

Electron
Proton
Neutron

Figure 2-1 Basic construction of an atom.

Electrons are negatively charged particles. The **nucleus** contains positively charged particles called **protons** and the particles that have no charge are called **neutrons.**

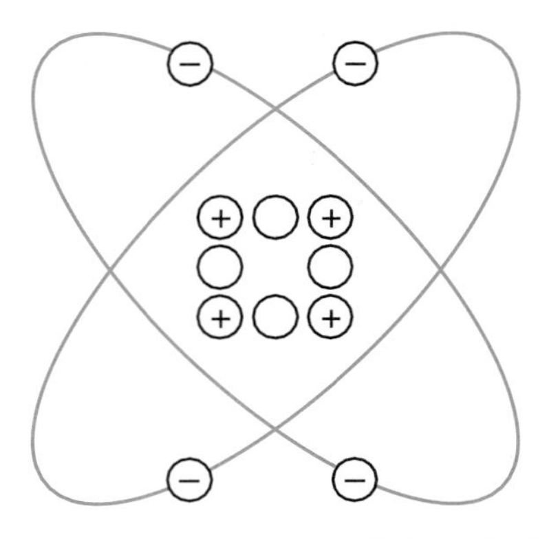
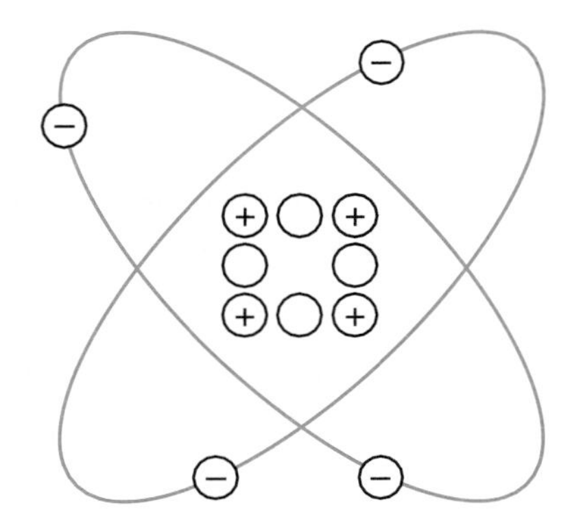

Figure 2-2 If the number of electrons and protons in an atom are the same, the atom is balanced.

Like charges repel each other; unlike charges attract each other.

An atom that has an equal amount of protons and electrons is called **balanced** or **neutral** (Figure 2-2).

The electron orbits around the nucleus are called **shells.**

The outermost orbit of the atom is called the **valence ring.**

A **conductor** is something that supports the flow of electricity through it.

An **insulator** is not capable of supporting the flow of electricity.

tightly bound together. The electrons are free to move within their orbits at fixed distances around the nucleus. The attraction between the negative electrons and the positive protons causes the electrons to orbit the nucleus. All of the electrons surrounding the nucleus are negatively charged, so they repel each other when they get close. The electrons attempt to stay as far away from each other as possible without leaving their orbits.

Simply put, atoms like to have the same number of electrons and protons. This makes them balanced. To remain balanced, an atom will give up an electron or pull an electron from another atom. The electrons orbit around the nucleus of the atom. A specific number of electrons are in each of the electron orbit paths. The orbit closest to the nucleus has room for two electrons; the second orbit holds up to 8 electrons; the third holds up to 18; and the fourth and fifth hold up to 32 each. The number of orbits depends on the number of electrons the atom has.

Conductors and Insulators

To help explain why you need to know about these electrons and their orbits, let's look at the atomic structure of copper. Copper is a metal and is the most commonly used conductor of electricity. A copper atom contains 29 electrons, 2 in the first orbit, 8 in the second, 18 in the third orbit, and 1 in the fourth (Figure 2-3). The outer orbit, or shell as it is sometimes called, is referred to as the **valence ring**. This is the orbit we care about in our study of electricity.

There is only one electron in the valence ring of a copper atom. This is why copper is used as a conductor of electricity. Copper, silver, gold, and other good conductors of electricity have only one or two electrons in their valence ring. These atoms can be made to give up the electrons in their valence ring with little effort.

Since electricity is the movement of electrons from one atom to another, atoms that have one electron in their valence ring support electricity. They allow the electron to easily move from the valence ring of one atom to the valence ring of another atom. Therefore, if we have a wire made

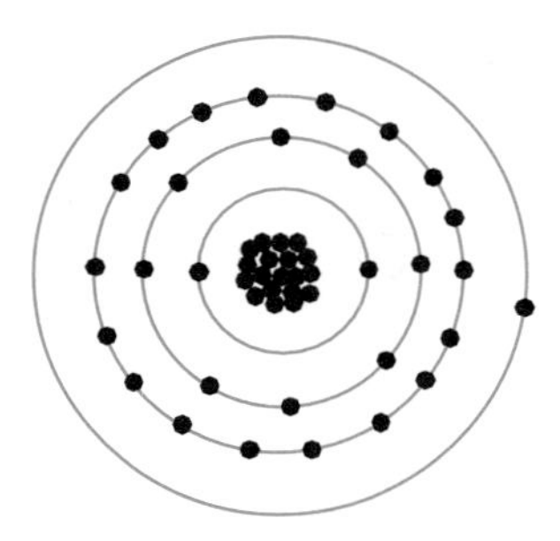

Figure 2-3 Basic structure of a copper atom.

of millions of copper atoms, we have a good conductor of electricity. To have electricity, we simply need to add one electron to one of the copper atoms. That atom will shed the electron it had to another atom, which will shed its original electron to another, and so on. As the electrons move from atom to atom, a force is released. This force is what we use to light lamps, run motors, and so on. As long as we keep the electrons moving in the conductor, we have electricity.

A BIT OF HISTORY

Electricity was discovered by the Greeks over 2,500 years ago. They noticed that when amber was rubbed with other materials it was charged with an unknown force that had the power to attract objects, such as dried leaves and feathers. The Greeks called amber "elektron." The word *electric* is derived from this word and means "to be like amber."

Semiconductors are neither conductors nor insulators.

Insulators are materials that don't allow electrons to flow through them easily. Insulators are atoms that have seven or eight electrons in their valence ring. The electrons are held tightly around the atom's nucleus and they can't be moved easily. Insulators are used to prevent electron flow or to contain it within a conductor. Insulating material covers the outside of most conductors to keep the moving electrons within the conductor.

The term **circuit** means a circle and is the path of electron flow consisting of the voltage source, conductors, load component, and return path to the voltage source.

In summary, the number of electrons in the valence ring determines whether an atom is a good conductor or insulator. Some atoms are not good insulators or conductors; these are called semiconductors.

1. three or less electrons—conductor
2. five or more electrons—insulator
3. four electrons—semiconductor

The electrons in the atoms of a conductor can be freed from their outer orbits by forces such as heat, friction, light, pressure, chemical reaction, and magnetism. When electrons are moved out of their orbit, they form an electrical current under proper conditions. Insulators are required to control the routing of the flow of electricity.

The electrical term **open** is used to mean that current flow is stopped. The path for electron flow is broken by opening the circuit. **Closed circuit** means there are no breaks in the path and current will flow.

WARNING: Any broken, frayed or damaged insulation material requires replacement or repair to the conductor. Exposed conductors can result in a safety hazard and circuit component damage.

CAUTION: The human body is a conductor of electricity. When performing service on electrical systems, be aware that electrical shock is possible. Although the shock is usually harmless, the reaction to the shock can cause injury. Observe all safety rules associated with electricity. Never wear jewelry when servicing or testing the electrical system.

An **ion** is an atom or group of atoms that has an electrical charge, in contrast to a balanced atom, which has no charge.

Electricity Defined

Electricity is the movement of electrons through a conductor (Figure 2-4). Electrons are attracted to protons. Since we have excess electrons on the other end of the conductor, we have many electrons being attracted to the protons. This attraction sort of pushes the electrons toward the protons. This push is normally called electrical pressure. The amount of electrical pressure is determined by the number of electrons that are attracted to protons. The electrical pressure or electromotive force (EMF) attempts to push an electron out of its orbit and toward the excess

Electromotive force (EMF) is referred to as voltage or electrical pressure.

Random movement of electrons is not electric current; the electrons must move in the same direction.

The speed of light is 186,000 miles per second (299,000 kilometers per second).

The **electron theory** defines the movement of electrical current as from negative to positive. This theory is based on the flow of electrons.

The **conventional theory** of current flow states that current moves from a positive point to a less positive point.

The term **voltage** is the difference or electrical potential that exists between the negative and positive sides of a circuit.

The **ground circuit** used in most automotive electrical systems is through the vehicle chassis and/or engine block. The term *ground* means the common negative connection of the electrical system. The symbol for ground is ⏚.

Voltage is the electrical pressure that causes electrons to move through a circuit.

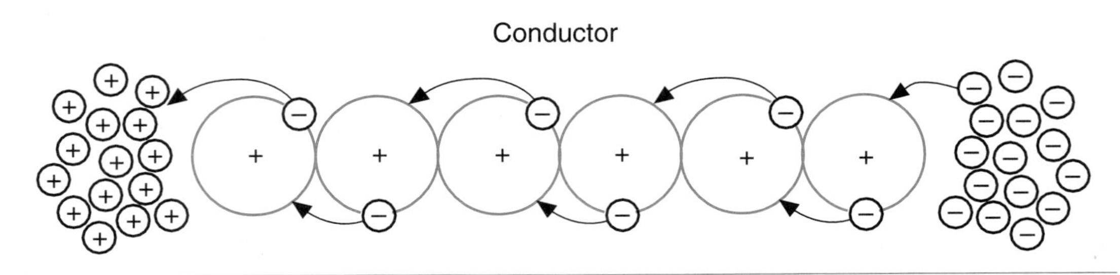

Figure 2-4 As electrons flow in one direction from one atom to another, an electrical current is developed.

protons. If an electron is freed from its orbit, the atom acquires a positive charge because it now has one more proton than it has electrons. The unbalanced atom or ion attempts to return to its balanced state so it will attract electrons from the orbit of other balanced atoms. This starts a chain reaction as one atom captures an electron and another releases an electron. As this action continues to occur, electrons will flow through the conductor. A stream of free electrons forms and an electrical current is started. This does not mean a single electron travels the length of the insulator; it means the overall effect is electrons moving in one direction. All this happens at the speed of light. The strength of the electron flow is dependant on the potential difference or voltage.

The three elements of electricity are voltage, current, and resistance. How these three elements interrelate governs the behavior of electricity. Once the technician comprehends the laws that govern electricity, understanding the function and operation of the various automotive electrical systems is an easier task. This knowledge will assist the technician in diagnosis and repair of automotive electrical systems.

Voltage

Voltage can be defined as an electrical pressure (Figure 2-5) and is the electromotive force (EMF) that causes the movement of the electrons in a conductor. In Figure 2-4, voltage is the force of attraction between the positive and negative charges. An electrical pressure difference is created when there is a mass of electrons at one point in the circuit, and a lack of electrons at another point in the circuit. In the automobile, the battery or generator is used to apply the electrical pressure.

The amount of pressure applied to a circuit is stated in the number of volts. If a voltmeter is connected across the terminals of an automobile battery it may indicate 12.6 volts. This is actually indicating that there is a difference in potential of 12.6 volts. There is 12.6 volts of electrical pressure between the two battery terminals.

In a circuit that has current flowing, voltage will exist between any two points in that circuit (Figure 2-6). The only time voltage does not exist is when the potential drops to zero. In Figure 2-6 the voltage potential between points A and C and B and C is 12.6 volts. However, between points A and B the pressure difference is zero and the voltmeter will indicate 0 volts.

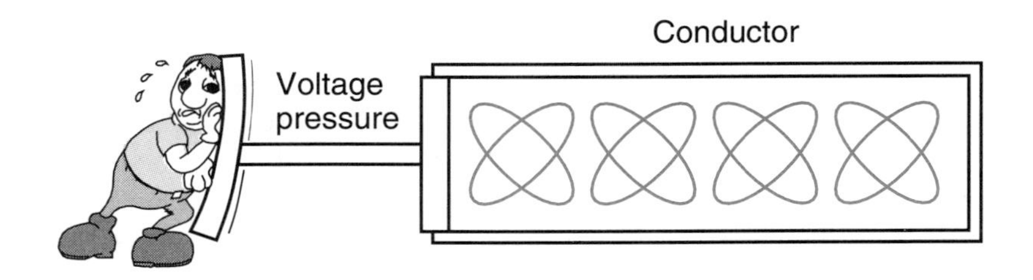

Figure 2-5 Voltage is the pressure that causes the electrons to move.

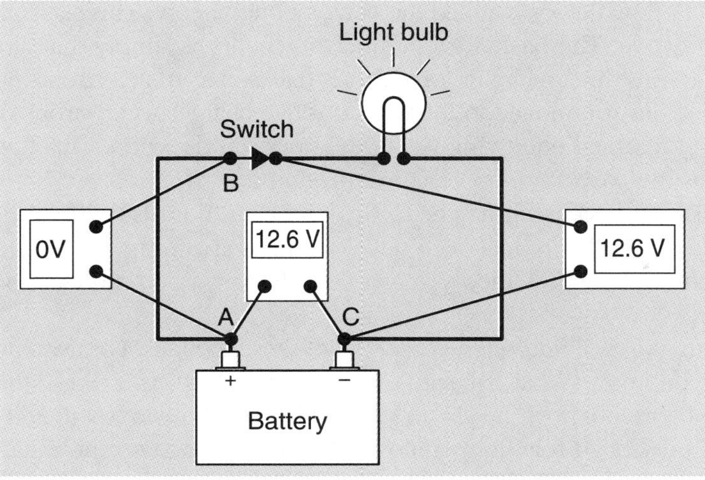

Figure 2-6 A simplified light circuit illustrating voltage potential.

An *E* can be used for the symbol to designate voltage (electromotive force). A *V* is also used as a symbol for voltage.

One volt (V) is the amount of pressure required to move one ampere of current through one ohm of resistance.

Current

Current can be defined as the rate of electron flow (Figure 2-7) and is measured in amperes. Current is a measurement of the electrons passing any given point in the circuit in one second. Because the flow of electrons is at nearly the speed of light, it would be impossible to physically see electron flow. However, the rate of electron flow can be measured. Current will increase as pressure or voltage is increased—provided circuit resistance remains constant.

Current is the aggregate flow of electrons through a wire.

One ampere (A) represents the movement of 6.25 billion billion electrons (or one **coulomb**) past one point in a conductor in one second.

A BIT OF HISTORY

The ampere is named after André Ampère, who in the late 1700s worked with magnetism and current flow to develop some foundations for understanding the behavior of electricity.

So far we have described current as the movement of electrons through a conductor. Electrons are negatively charged particles that move toward something that is positively charged. Electrons move because of this potential difference. This describes one of the common theories about current flow. The electron theory states that since electrons are negatively charged, current flows from the most negative to the most positive point within an electrical circuit. In other words, current flows from negative to positive. This theory is widely accepted by the electronic industry.

The symbol for current is *I*.

The ohm (Ω) is the resistance of a conductor such that a constant current of 1 ampere in it produces a voltage of 1 volt between its ends.

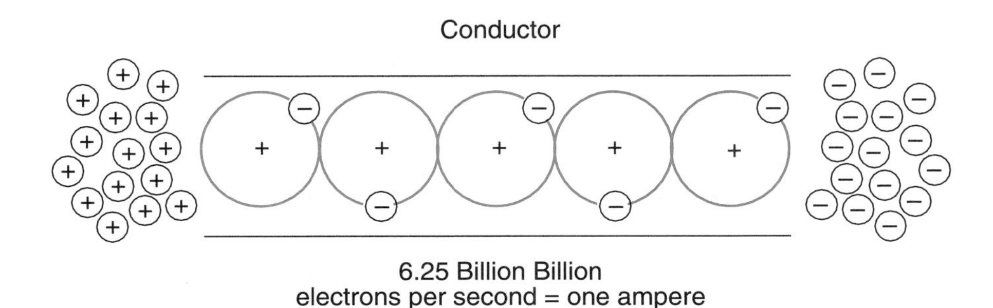

Figure 2-7 The rate of electron flow is called current and is measured in amperes.

Another current flow theory is called the conventional theory. This states that current flows from positive to negative. The basic idea behind this theory is simply that although electrons move toward the protons, the energy or force that is released as the electrons move begins at the point where the first electron moved to the most positive charge. As electrons continue to move in one direction, the released energy moves in the opposite direction. This theory is the oldest theory and serves as the basis for most electrical diagrams.

Trying to make sense of it all may be difficult for you. It is also difficult for scientists and engineers. In fact, another theory has been developed to explain the mysteries of current flow. This theory is called the hole-flow theory and is actually based on both electron theory and the conventional theory.

As a technician, you will find references to all of these theories. Fortunately, it really doesn't mattter as long as you know what current flow is and what affects it. From this understanding you will be able to figure out how the circuit basically works, how to test it, and how to repair it. In this text, we will present current flow as moving from positive to negative and electron flow as moving from negative to positive. Remember that current flow is the result of the movement of electrons, regardless of the theory.

Resistance

The third component in electricity is resistance. **Resistance** is the opposition to current flow and is measured in ohms. The size, type, length, and temperature of the material used as a conductor will determine the resistance of the conductor. Devices that use electricity to operate (motors and lights) have a greater amount of resistance than the conductor.

The symbol for resistance is *R*.

A complete electrical circuit consists of the following: (1) a power source, (2) a load or resistance unit, and (3) conductors. Resistance (load) is required to change electrical energy to light, heat, or movement. There is resistance in any working device of a circuit, such as a lamp, motor, relay, or other load component.

There are five basic characteristics that determine the amount of resistance in any part of a circuit:

1. The atomic structure of the material: The fewer the number of electrons in the outer valence ring, the higher the resistance of the conductor.
2. The length of the conductor: The longer the conductor, the higher the resistance.
3. The diameter of the conductor: The smaller the cross-sectional area of the conductor, the higher the resistance.
4. Temperature: A change in the temperature of the conductor causes a change in the resistance.
5. Physical condition of the conductor: If the conductor is damaged by nicks or cuts, the resistance will increase because the conductor's diameter is decreased by these.

There may be unwanted resistance in a circuit. This could be in the form of a corroded connection or a broken conductor. In these instances the resistance may cause the load component to operate at reduced efficiency or to not operate at all.

It does not matter if the resistance is from the load component or from unwanted resistance. There are certain principles that dictate its impact in the circuit:

1. Voltage always drops as current flows through the resistance.
2. An increase in resistance causes a decrease in current.
3. All resistances change the electrical energy into heat energy to some extent.

Voltage Drop

Voltage drop occurs when current flows through a load component or resistance. Voltage drop is the amount of electrical pressure lost or consumed as it pushes current flow through a resistance. After a resistance, the voltage is lower than it was before the resistance.

There must be a voltage present for current to flow through a resistor. Kirchhoff's law basically states that the sum of the voltage drops in an electrical circuit will always equal source voltage. In other words, all of the source's voltage is used by the circuit.

A BIT OF HISTORY

Gustav Kirchhoff was a German scientist who in the 1800s discovered two facts about the characteristics of electricity. One is called his voltage law, which states "The sum of the voltage drops across all resistances in a circuit must equal the voltage of the source." His law on current states "The sum of the currents flowing into any point in a circuit equals the sum of the currents flowing out of the same point." These laws describe what happens when electricity is applied to a load. Voltage drops while current remains constant; current does not drop.

Ohm's Law

Ohm's law defines the relationship between current, voltage, and resistance.

Understanding Ohm's law is the key to understanding how electrical circuits work. The law states that it takes one volt of electrical pressure to push one ampere of electrical current through one ohm of electrical resistance. This law can be expressed mathematically as:

1 Volt = 1 Ampere times 1 Ohm

This formula is most often expressed as: $E = I \times R$. Whereas E stands for EMF or electrical pressure (voltage), I stands for intensity of electron flow (current), and R represents resistance. This formula is often used to find the amount of one electrical characteristic when the other two are known. As an example: if we have 2 amps of current and 6 ohms of resistance in a circuit, we must have 12 volts of electrical pressure.

$E = 2 \text{ Amps} \times 6 \text{ Ohms}$ $\quad E = 2 \times 6$ $\quad E = 12 \text{ volts}$

If we know the voltage and resistance but not the current of a circuit, we can quickly calculate it by using Ohm's law. Since $E = I \times R$, I would equal E divided by R. Let's supply some numbers to this. If we have a 12 volt circuit with 6 ohms of resistance, we can determine the amount of current in this way:

$$I = \frac{E}{R} \text{ or } \frac{12 \text{ Volts}}{6 \text{ Ohms}} \text{ or } I = 2 \text{ Amps}$$

We can use the same logic to calculate resistance when we know voltage and current. $R = E/I$. One easy way to remember the formulas of Ohm's law is to draw a circle and divide it into three parts as shown in Figure 2-8. Simply cover the value you want to calculate. The formula you need to use is all that shows.

The resistance of an actual lamp in an automotive application will change when current passes through it because its temperature changes.

To show how easy this works, consider the 12-volt circuit in Figure 2-9. This circuit contains a 3-ohm light bulb. We want to find the current in the circuit. By covering the I in the circle we see the formula we need, $I = E/R$. Then we plug in the numbers, $I = 12/3$. Therefore our circuit current is 4 amps.

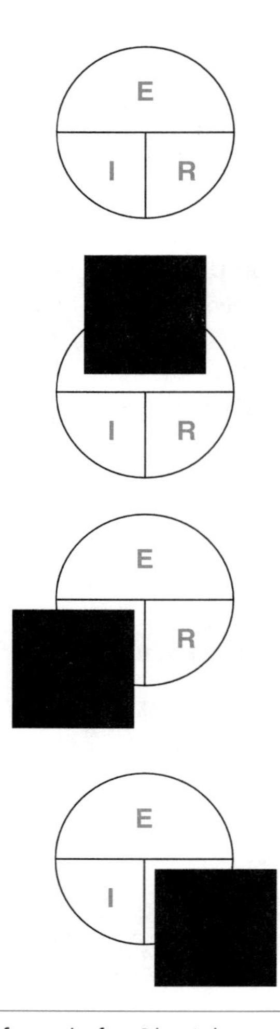

Figure 2-8 The mathematical formula for Ohm's law using a circle to help understand the different formulas that can be derived from it. To identify which formula to use, cover the unknown value. The exposed formula is the one to use to calculate the unknown.

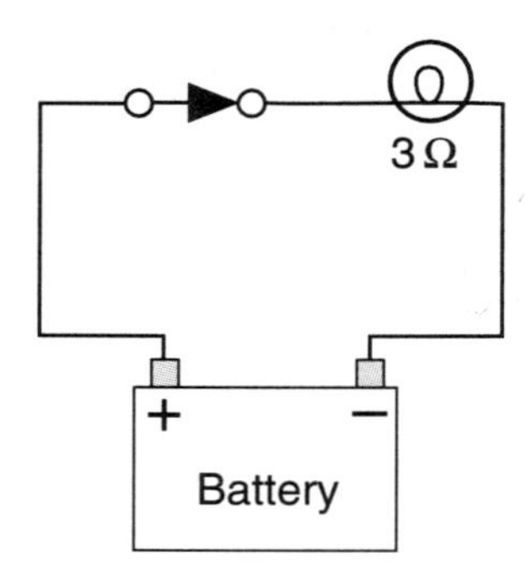

Figure 2-9 Simplified light circuit with 3 ohms of resistance in the lamp.

A BIT OF HISTORY

Georg S. Ohm was a German scientist in the 1800s who discovered that all electrical quantities are proportional to each other and therefore have a mathematical relationship.

Ohm's law is important to your understanding of electricity in more ways than your ability to calculate values. Based on Ohm's law we can see that current will increase if we decrease the resistance and not change the voltage. Likewise, if we increase the resistance, current will decrease. Ohm's law explains what happens when we do things to a circuit and when something goes wrong in a circuit. As an example: Look at Figure 2-10; on the left side is a 12-volt circuit with a 3-ohm light bulb. This circuit will have 4 amps of current flowing through it. If we add a 1-ohm resistor to the same circuit (as shown to the right in Figure 2-10), we now have a total resistance of 4 ohms. Because of the increased resistance, our current dropped to 3 amps. Our light bulb will be powered by less current and will be less bright than it was before we added the additional resistance.

Another point to consider is voltage drop. Before we added the 1-ohm resistor, the source voltage (12 volts) was dropped by the light bulb. With the additional resistance, the voltage drop of the light bulb decreased to 9 volts. The remaining 3 volts were dropped by the 1-ohm resistor. We know this from using Ohm's law. When the circuit current was 4 amps, the light bulb had 3 ohms of resistance. To find the voltage drop we multiply the current by the resistance.

$$E = I \times R \quad \text{or} \quad E = 4 \times 3 \quad \text{or} \quad E = 12$$

When we added the resistor to the circuit, the light bulb still had 3 ohms of resistance, but the current in the circuit decreased to 3 amps. Again we can determine the voltage drop by multiplying the current by the resistance.

$$E = I \times R \quad \text{or} \quad E = 3 \times 3 \quad \text{or} \quad E = 9$$

The voltage drop of the additional resistor is calculated in the same way: $E = I \times R$ or $E = 3$ volts. The total voltage drop of the circuit is the same for both circuits; however, the voltage drop at the light bulb changed. This also would cause the light bulb to be dimmer.

Electrical Power

Electrical power is another term used to describe electrical activity. Power is expressed in **watts**. A watt is equal to one volt multiplied by one ampere. There is another mathematical formula that expresses the relationship between voltage, current, and power. It is simply: $P = E \times I$. Power measurements are measurements of the rate at which electricity is doing work.

Power (*P*) is the rate of doing electrical work.

The best examples of power are light bulbs. Household light bulbs are sold by wattage. A 100-watt bulb is brighter and uses more electricity than a 60-watt bulb. Seldom do technicians worry about wattage when working on cars. However, an understanding of electrical power will help in the understanding of electrical circuits.

It is possible to convert horsepower ratings to electrical power rating using the conversion factor: 1 horsepower equals 746 watts.

Let's go back to Figure 2-10. The light bulb in the circuit on the left had a 12-volt drop at 4 amps of current. We can calculate the power used by the bulb by multiplying the voltage and the current.

$$P = E \times I \quad \text{or} \quad P = 12 \times 4 \quad \text{or} \quad P = 48$$

The bulb produced 48 watts of power. Now let's look at the other circuit. This time the bulb dropped 9 volts at 3 amps of current. The power of the bulb is calculated in the same way as before.

$$P = E \times 1 \quad \text{or} \quad P = 9 \times 3 \quad \text{or} \quad P = 27$$

This bulb produced 27 watts of power, a little more than half of the original. It would be almost half as bright. The key to understanding what happened is to remember the light bulb didn't change; the circuit changed.

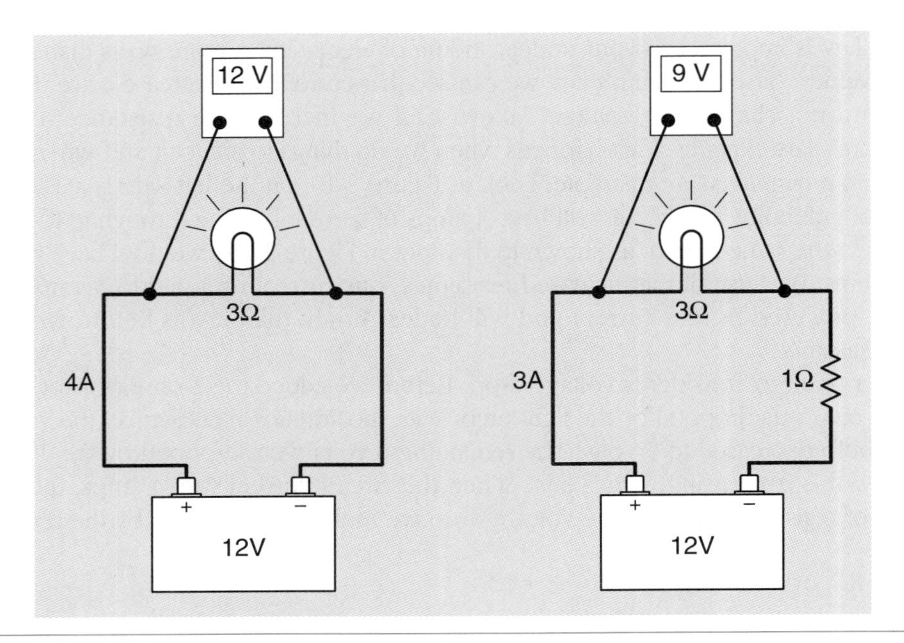

Figure 2-10 The light circuit in Figure 2-9 shown with normal circuit values and with added resistance in series.

Capacitance

Capacitance (*C*) is the ability of two conducting surfaces to store voltage. The two surfaces must be separated by an insulator.

Capacitors are rated in units called **farads.** A farad is a large unit, and most commonly used capacitors are rated in microfarads.

Some automotive electrical systems will use a capacitor or condenser to store electrical charges (Figure 2-11). A capacitor does not consume any power. All of the voltage stored in the capacitor is returned to the circuit when the capacitor discharges. Because the capacitor stores voltage, it will also absorb voltage changes in the circuit. By providing for this storage of voltage, damaging voltage spikes can be controlled. They are also used to reduce radio noise.

Most capacitors are connected in parallel across the circuit (Figure 2-12). Capacitors operate on the principle that opposite charges attract each other, and that there is a potential voltage between any two oppositely charged points. When the switch is closed, the protons at the positive battery terminal will attract some of the electrons on one plate of the capacitor away from the area near the dielectric material. As a result, the atoms of the positive plate are unbalanced because there are more protons than electrons in the atom. This plate now has a positive charge because of the shortage of electrons (Figure 2-13). The positive charge of this plate will attract electrons on the other plate. The dielectric keeps the electrons on the negative plate from crossing over to the positive plate, resulting in a storage of electrons on the negative plate (Figure 2-14). The movement of electrons to the negative plate and away from the positive plate is an electrical current.

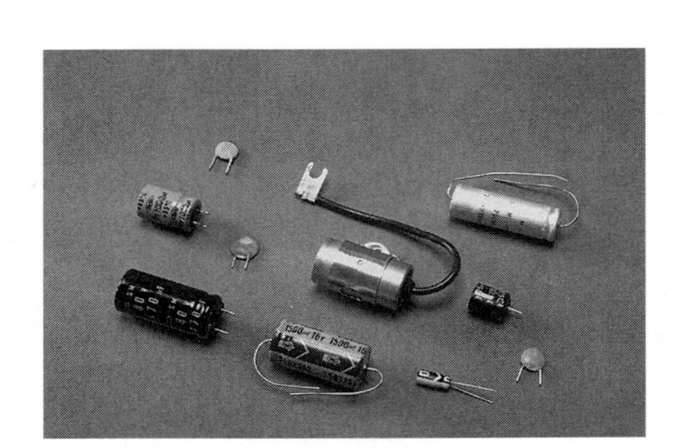

Figure 2-11 Capacitors that can be found in automotive electrical circuits.

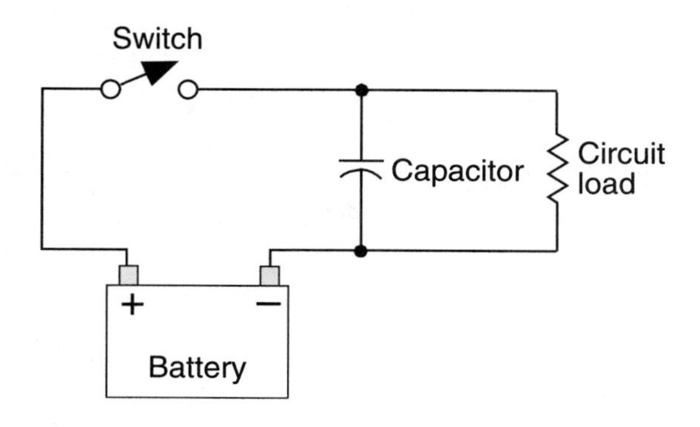

Figure 2-12 A capacitor connected to a circuit.

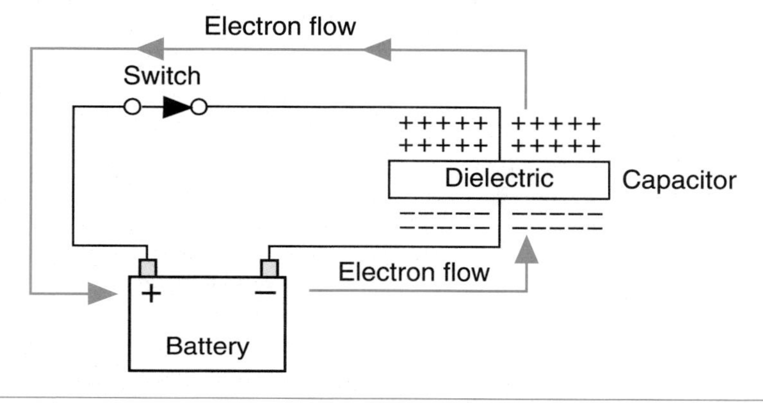

Figure 2-13 The positive plate sheds its electrons.

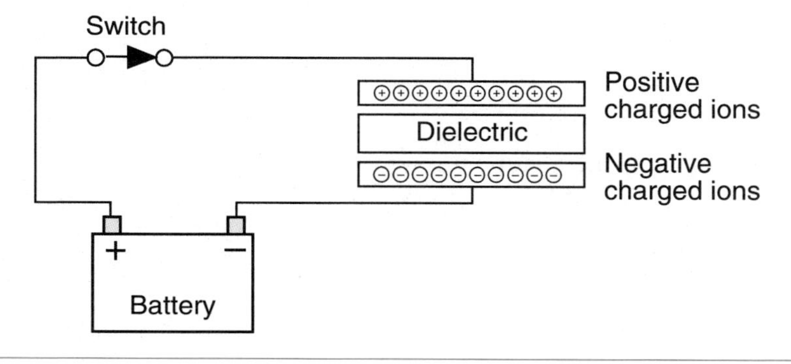

Figure 2-14 The electrons will be stored on the negative plate.

The plate connected to the positive battery terminal is the **positive plate.**

An atom that has less electrons than protons is said to be a **positive ion.**

The insulator in a capacitor is called a **dielectric.** The dielectric can be made of some insulator material such as ceramic, glass, paper, plastic, or even the air between the two plates.

The field that is between the two oppositely charged plates is called the **electrostatic field.**

Static electricity is electricity that is not in motion.

Current will flow "through" the capacitor until the voltage charges across the capacitor and across the battery are equalized. Current flow through a capacitor is only the effect of the electron movement onto the negative plate and away from the positive plate. Electrons do not actually pass through the capacitor from one plate to another. The charges on the plates do not move through the electrostatic field. They are stored on the plates as static electricity. When the charges across the capacitor and battery are equalized, there is no potential difference and no more current will flow "through" the capacitor (Figure 2-15). Current will now flow through the load components in the circuit (Figure 2-16).

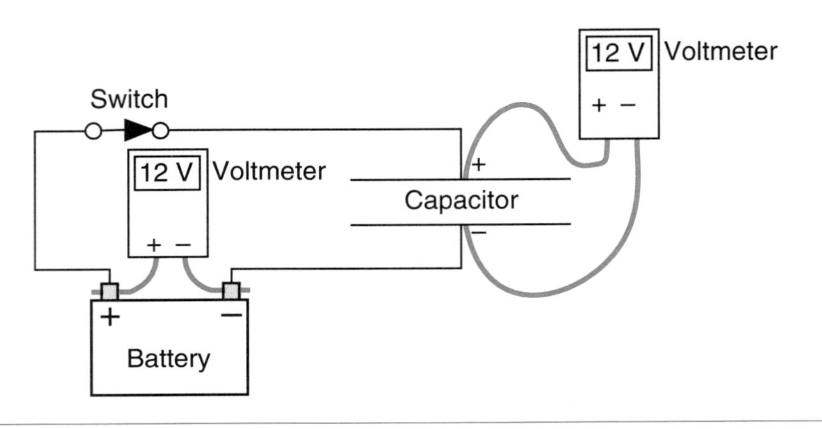

Figure 2-15 A capacitor that is fully charged.

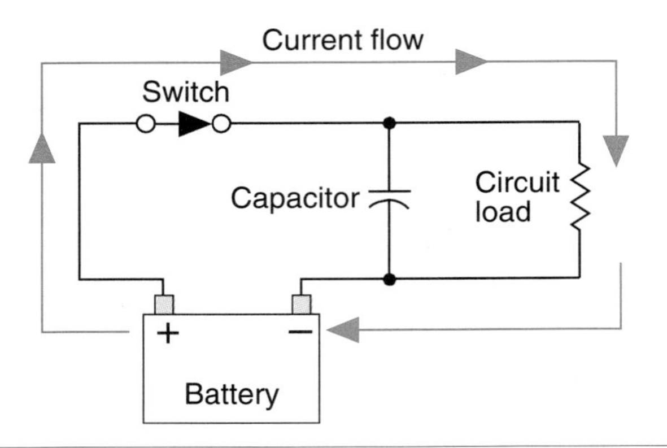

Figure 2-16 Current flow with a fully charged capacitor.

When the switch is opened, current flow from the battery through the resistor is stopped. However, the capacitor has a storage of electrons on its negative plate. Because the negative plate of the capacitor is connected to the positive plate through the resistor, the capacitor acts as a battery. The capacitor will discharge the electrons through the resistor until the atoms of the positive plate and negative plate return to a balanced state (Figure 2-17).

In the event that a high voltage spike occurs in the circuit, the capacitor will absorb the additional voltage before it is able to damage the circuit components. A capacitor can also be used to stop current flow quickly when a circuit is opened (such as in the ignition system). It can also store a high voltage charge and then discharge it when a circuit needs the voltage (such as in some air bag systems).

Types of Current

There are two classifications of electrical current flow; direct current (DC) and alternating current (AC). The type of current flow is determined by the direction it flows and by the type of voltage that drives them.

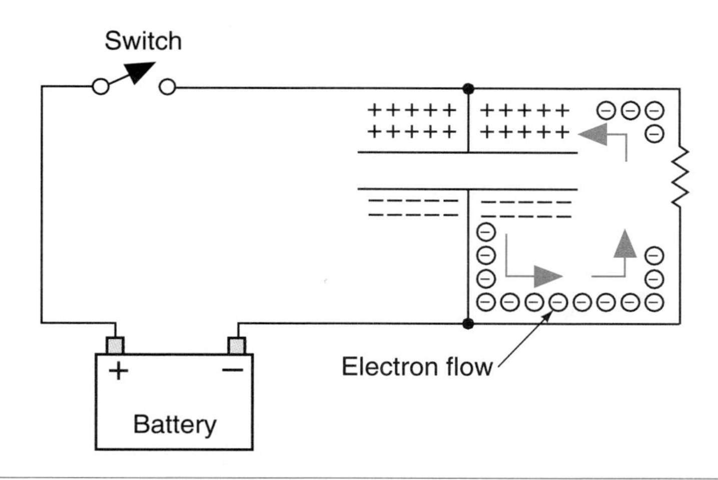

Figure 2-17 Current flow with the switch open and the capacitor discharging.

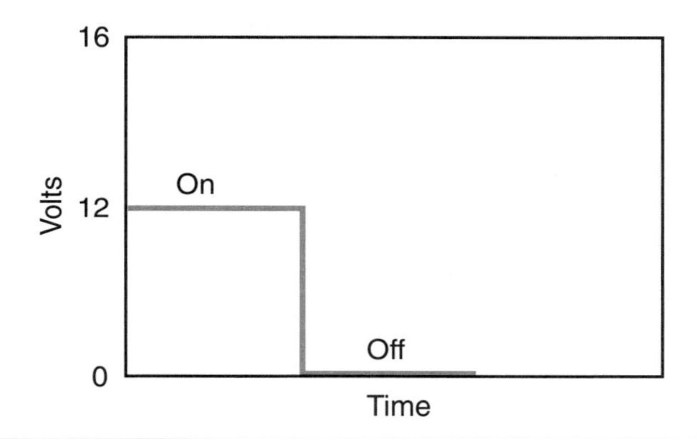

Figure 2-18 Direct current flows in the same direction and is the same throughout the circuit.

DC voltage is created and stored in the automotive battery.

The unrectified current produced within a generator is the most common example of alternating current found in the automobile. Generator circuits convert AC current to DC current.

The AC voltage shown in Figure 2-19 is called a **sine wave.**

One **cycle** is completed when the voltage has gone positive, returned to zero, gone negative, and returned to zero.

Direct Current

Direct current (DC current) can only be produced by a battery and has a current that is the same throughout the circuit and flows in the same direction (Figure 2-18). Voltage and current are constant if the switch is turned on or off. Most of the electrically controlled units in the automobile require direct current.

Alternating Current

Alternating current (AC) is produced anytime a conductor moves through a magnetic field. In an alternating current circuit, voltage and current do not remain constant. Alternating current changes directions from positive to negative. The voltage in an AC circuit starts at zero and rises to a positive value. Then it falls back to zero and goes to a negative value. Finally it returns to zero (Figure 2-19).

The portion of the circuit from the positive side of the source to the load component is called the **insulated side** or "hot" side of the circuit. The portion of the circuit that is from the load component to the negative side of the source is called the **ground side** of the circuit.

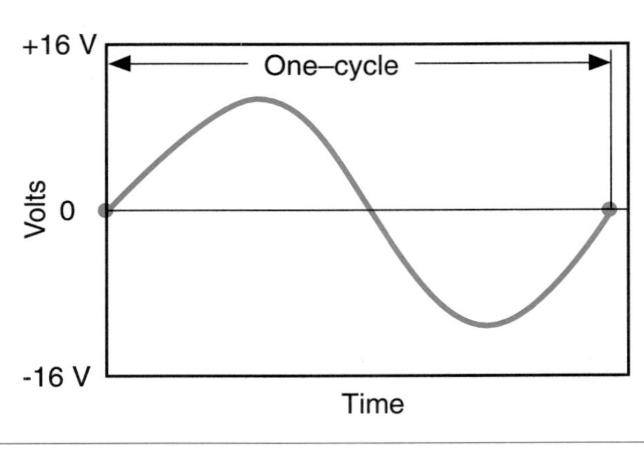

Figure 2-19 Alternating current reverses direction.

Electrical Circuits

The electrical term *continuity* refers to the circuit being continuous. For current to flow, the electrons must have a continuous path from the source voltage to the load component and back to the source. A simple automotive circuit is made up of three parts:

1. Battery (power source).
2. Wires (conductors).
3. Load (light, motor, etc.).

The basic circuit shown (Figure 2-20) includes a switch to turn the circuit on and off, a protection device (fuse), and a load. In this instance, with the switch closed, current flows from the positive terminal of the battery through the light and returns to the negative terminal of the battery. To have a complete circuit the switch must be closed or turned on. The effect of opening and closing the switch to control electrical flow would be the same if the switch was installed on the ground side of the light.

There are basically three different types of electrical circuits: (1) the series circuit, (2) the parallel circuit, and (3) the series-parallel circuit.

Series Circuit

The **series circuit** provides a single path for current flow from the electrical source through all the circuit's components, and back to the source.

A series circuit consists of one or more resistors (or loads) with only one path for current to flow. If any of the components in the circuit fails, the entire circuit will not function. All of the current that comes from the positive side of the battery must pass through each resistor, then back to the negative side of the battery.

C
B
A
+ −
Battery

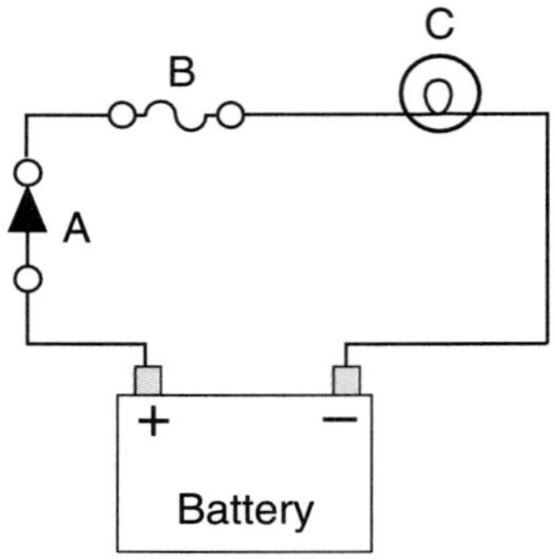

Figure 2-20 A basic electrical circuit including (A) a switch, (B) a fuse, and (C) a lamp.

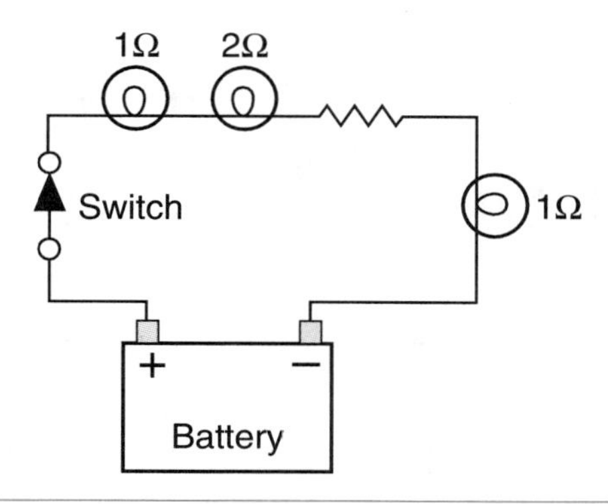

Figure 2-21 The total resistance in a series circuit is the sum of all resistances in the circuit.

The total resistance of a series circuit is calculated by simply adding the resistances together. As an example, refer to Figure 2-21. Here is a series circuit with three light bulbs; one bulb has 2 ohms of resistance, and the other two have 1 ohm each. The total resistance of this circuit is 2 + 1 + 1 or 4 ohms.

The characteristics of a series circuit are:

1. The total resistance is the sum of all resistances.
2. The current through each resistor is the same.
3. The current is the same throughout the circuit.
4. The voltage drop across each resistor will be different if the resistor values are different (Figure 2-22).
5. The sum of the voltage drop of each resistor equals the source voltage.

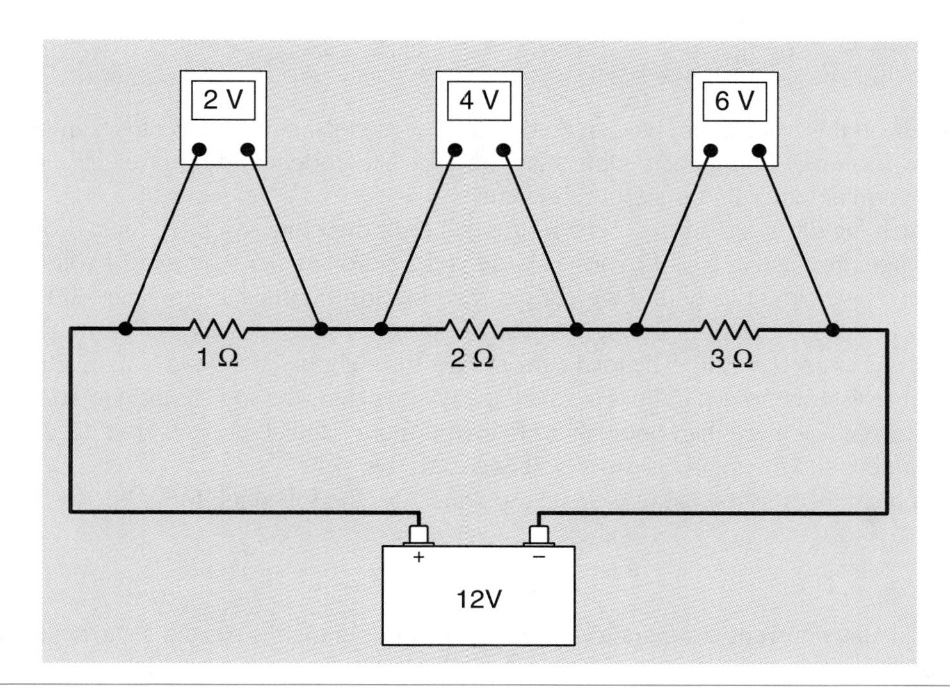

Figure 2-22 The voltage drop across each resistor in series will be different if their resistance values are different.

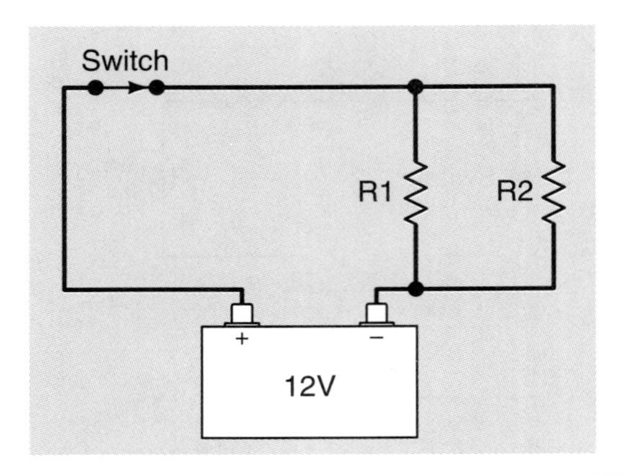

Figure 2-23 In a parallel circuit, current can flow through more than one parallel leg at a time.

Parallel Circuit

In a parallel circuit each path of current flow has separate resistances that operate either independently or in conjunction with each other (depending on circuit design). In a parallel circuit, current can flow through more than one parallel leg at a time (Figure 2-23). In this type of circuit, failure of a component in one parallel leg does not affect the components in other legs of the circuit.

A **parallel circuit** provides two or more paths for electricity to flow.

The total resistance of a parallel circuit with two legs or two paths for current flow is calculated by using this formula:

$$R_T = \frac{R_1 \times R_2}{R_1 + R_2}$$

The legs of a parallel circuit are also called parallel branches or **shunt circuits.**

If the value of R_1 in Figure 2-23 was 3 ohms and R_2 had a value of 6 ohms, the total resistance can be found.

$$R_T = \frac{R_1 \times R_2}{R_1 + R_2} \text{ or } R_T = \frac{3 \times 6}{3 + 6} \text{ or } R_T = \frac{18}{9} \text{ or } R_T = 2$$

Based on this calculation, we can determine that the total circuit current is 6 amps (12 volts divided by 2 ohms). Using basic Ohm's law and a basic understanding of electricity, we can quickly determine other things about this circuit.

Each leg of the circuit has 12 volts applied to it; therefore, each leg must drop 12 volts. So the voltage drop across R_1 is 12 volts, and the voltage drop across R_2 is also 12 volts. Using the voltage drops we can quickly find the current that flows through each leg. Since R_1 has 3 ohms and drops 12 volts, the current through it must be 4 amps. R_2 has 6 ohms and drops 12 volts and its current is 2 amps ($I = E/R$). The total current flow through the circuit is 4 + 2 or 6 amps.

Total resistance in a parallel circuit is always less than the lowest individual resistance because current has more than one path to follow. If more parallel resistors are added, more circuits are added, and the total resistance will decrease.

If all resistances in the parallel circuit are equal, use the following formula:

$$R_T = \frac{\text{Value of one resistor}}{\text{Total number of resistors}}$$

To calculate current in a parallel circuit, each shunt branch is treated as an individual circuit. Applied voltage is the same to all branches. To determine the branch current simply divide the source voltage by the shunt branch resistance:

$$I = E \div R$$

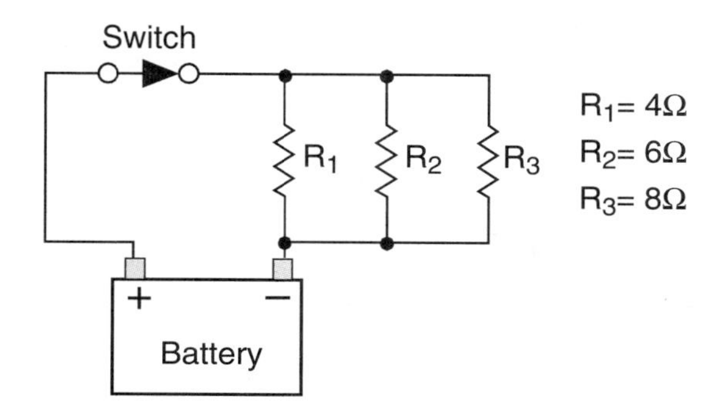

Figure 2-24 A parallel circuit with different resistances in each branch.

Referring to Figure 2-24, the total resistance of a circuit with more than two legs can be calculated with the following formula:

$$R_T = \frac{1}{\frac{1}{R_1} + \frac{1}{R_2} + \frac{1}{R_3} \cdots \frac{1}{R_n}}$$

Often it is much easier to calculate total resistance of a parallel circuit by using total current. Begin by finding the current through each leg of the parallel circuit; then add them together to find total current. Use basic Ohm's law to calculate the total resistance.

$R = E/I$ or $R = 12/6.5$ amps or $R = 1.85$ ohms

The characteristics of a parallel circuit are:

1. The voltage applied to each parallel leg is the same.
2. The voltage dropped across each parallel leg will be the same; however, if the leg contains more than one resistor, the voltage drop across each of them will depend on the resistance of each resistor in that leg.
3. The total resistance of a parallel circuit will always be less than the resistance of any of its legs.
4. The current flow through the legs will be different if the resistance is different.
5. The sum of the current in each leg equals the total current of the parallel circuit.

Series-Parallel Circuits

The series-parallel circuit has some loads that are in series with each other and some that are in parallel (Figure 2-25). To calculate the total resistance in this type of circuit, calculate the equivalent series loads of the parallel branches first. Next, calculate the series resistance and add it to the equivalent series load. For example, if the parallel portion of the circuit has two branches with 4 Ω resistance each, and the series portion has a single load of 10 Ω, use the following method to calculate the equivalent resistance of the parallel circuit:

$$R_T = \frac{R_1 \times R_2}{R_1 + R_2} \text{ or } \frac{4 \times 4}{4 + 4} \text{ or } \frac{16}{8} \text{ or } 2 \text{ ohms}$$

A **series-parallel circuit** is a combination of the series and parallel circuits.

Then add this equivalent resistance to the actual series resistance to find the total resistance of the circuit.

2 ohms + 10 ohms = 12 ohms

The **equivalent series load,** or **equivalent resistance,** is the equivalent resistance of a parallel circuit plus the resistance in series and is equal to the resistance of a single load in series with the voltage source.

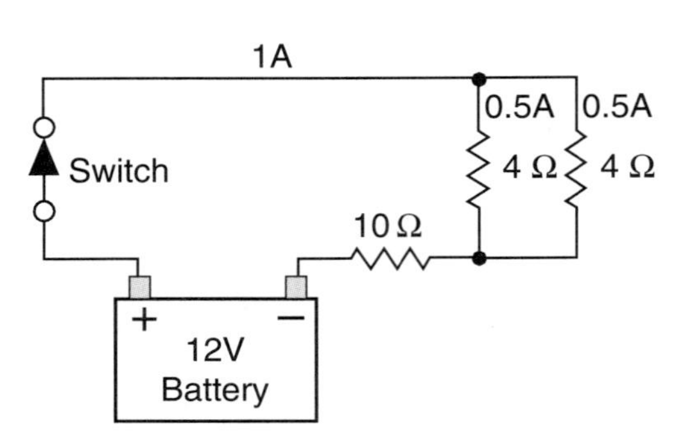

Figure 2-25 A series-parallel circuit with known resistance values.

With the total resistance now known, total circuit current can be calculated. Because the source voltage is 12 volts, 12 is divided by 12 ohms.

$I = E/R$ or $I = 12/12$ or $I = 1$ amp

The current flow through each parallel leg is calculated by using the resistance of each leg and voltage drop across that leg. To do this you must first find the voltage drops. Since all 12 volts are dropped by the circuit, we know that some are dropped by the parallel circuit and the rest by the resistor in series. We also know that the circuit current is 1 amp, that the equivalent resistance value of the parallel circuit is 2 ohms, and the resistance of the series resistor is 10. Using Ohm's law we can calculate the voltage drop of the parallel circuit:

$E = I \times R$ or $E = 1 \times 2$ or $E = 2$

Two volts are dropped by the parallel circuit. This means 2 volts are dropped by each of the 4 ohm resistors. Using our voltage drop, we can calculate our current flow through each parallel leg.

$I = E/R$ or $I = 2/4$ or $I = 0.5$ amps

Since the resistance on each leg is the same, each leg has 0.5 amps through it. If we did this right, the sum of the amperages will equal the current of the circuit. It does: 0.5 + 0.5 = 1.

It is important to realize that the actual or measured values of current, voltage, and resistance may be somewhat different than the calculated values. The change is caused by the effects of heat on the resistances. As the voltage pushes current through a resistor, the resistor heats up. The resistor changes the electrical energy into heat energy. This heat may cause the resistance to increase or decrease depending on the material it is made of. The best example of a resistance changing electrical energy into heat energy is a light bulb. A light bulb gives off light because the conductor inside the bulb heats up and glows when current flows through it.

Technicians will seldom have the need to calculate the values in an electrical circuit. The primary importance of being able to use Ohm's law is to predict what will happen if something else happens. Technicians use electrical meters to measure current, voltage, and resistance. When a measured value is not within specifications, you should be able to explain why. Ohm's law is used to do that.

Using Ohm's Law

For the most part, automotive electrical systems are connected in parallel. Actually the system is made up of a number of series circuits wired in parallel. This allows each electrical component to work independently of the others. When one component is turned on or off, the operation of the other components should not be affected. Let's look at Figure 2-26.

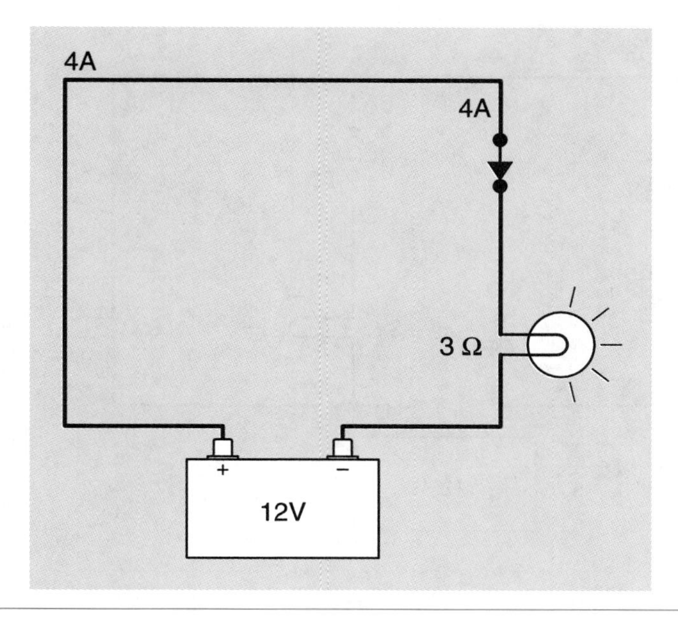

Figure 2-26 A simple light circuit.

Here we have a 12-volt circuit with one 3-ohm light bulb. The switch controls the operation of the light bulb. When the switch is closed, current flows and the bulb is lit. Four amps will flow through the circuit and the bulb.

I = E/R or I = 12/3 or I = 4 amps

Now let's add a 6-ohm light bulb in parallel to the 3-ohm light bulb (Figure 2-27). With the switch for the new bulb closed, 2 amps will flow through that bulb. The 3-ohm bulb is still receiving 12 volts and has 4 amps flowing through it; it will operate in the same way and with the same brightness as it did before we added the 6-ohm light bulb. The only thing that changed was circuit current; it is now 6 (4 + 2) amps.

Leg #1 I = E/R or I = 12/3 or I = 4 amps
Leg #2 I = E/R or I = 12/6 or I = 2 amps

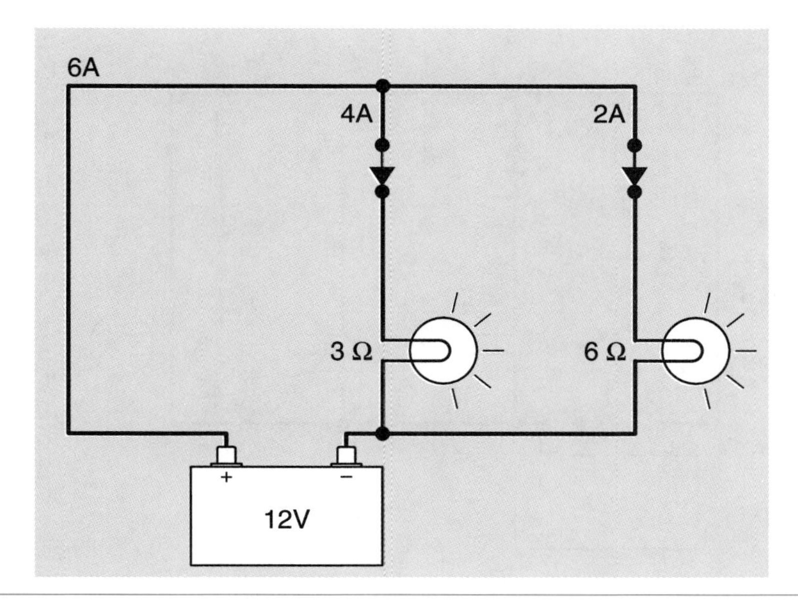

Figure 2-27 Two light bulbs connected in parallel.

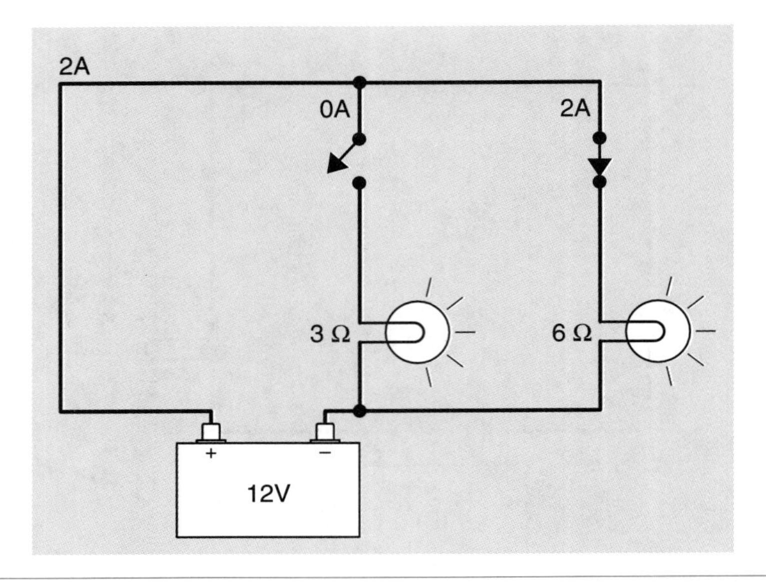

Figure 2-28 Two light bulbs connected in parallel; one switched on, the other switched off.

If we open the switch to the 3-ohm bulb (Figure 2-28), the 6-ohm bulb works in the same way and with the same brightness as it did before we opened the switch. In this case two things happened, the 3-ohm bulb no longer is lit, and the circuit current dropped 2 amps.

Let's add another bulb in parallel. Figure 2-29 is the same circuit as Figure 2-28 except a 1-ohm light bulb and switch was added in parallel to the circuit. With the switch for the new bulb closed, 12 amps will flow through that circuit. The other bulbs are working in the same way and with the same brightness as before. Again the only thing that changed is the total circuit current, which is now 18 amps.

Leg #1	I = E/R	or	I = 12/3	or	I = 4 amps
Leg #2	I = E/R	or	I = 12/6	or	I = 2 amps
Leg #3	I = E/R	or	I = 12/1	or	I = 12 amps

Total current = 4 + 2 + 12 or 18 amps

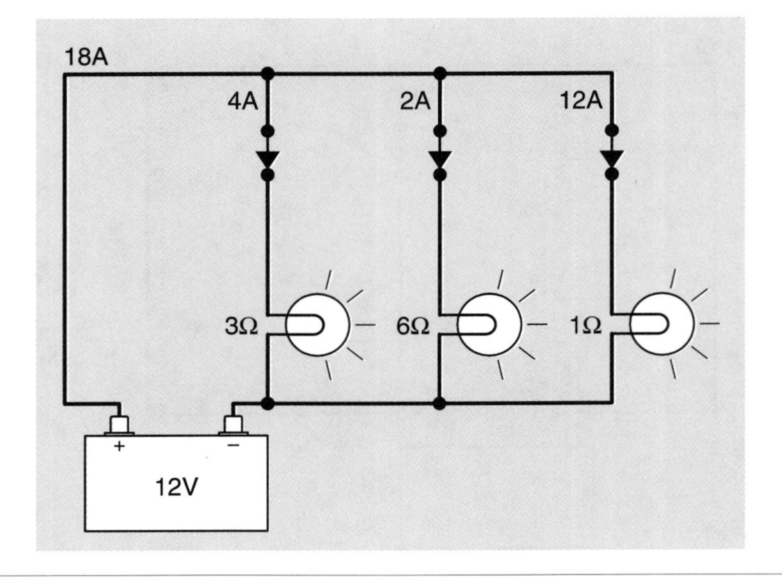

Figure 2-29 Three light bulbs connected in parallel.

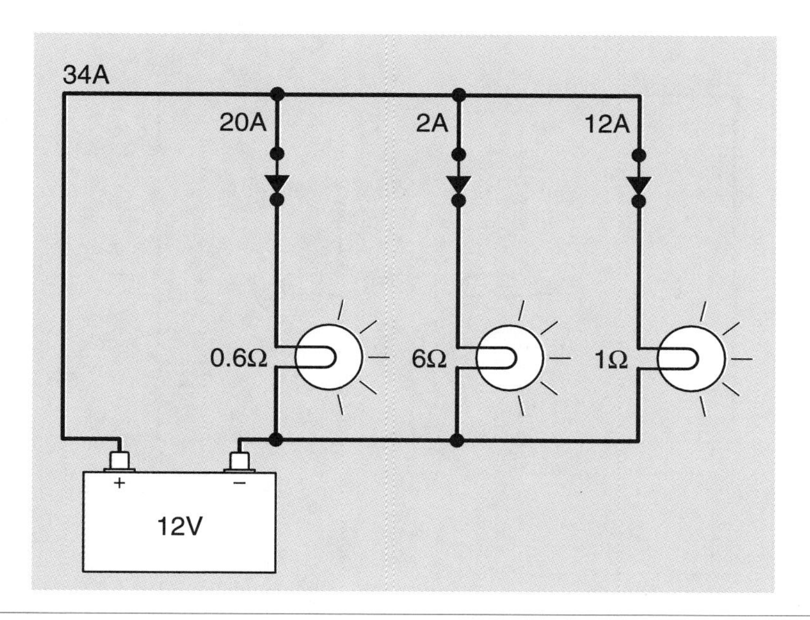

Figure 2-30 Three light bulbs connected in parallel.

When the switch for any of these bulbs is opened or closed, the only things that happen are the bulbs either turn off or on, and the total current through the circuit changes. Notice as we add more parallel legs, total circuit current goes up. There is a commonly used statement, "Current always takes the path of least resistance," that is misused. As you can see in these circuits, current flows to all of the bulbs regardless of the bulbs' resistance. The resistances with lower values will draw higher currents, but all of the resistances will receive the current they allow. The statement should be, "Larger amounts of current will flow through lower resistances." This is very important to remember when diagnosing electrical problems.

From Ohm's law, we know that when resistance decreases, current increases. If we put a 0.6-ohm light bulb in place of the 3-ohm bulb (Figure 2-30), the other bulbs will work in the same way and with the same intensity as they did before. However, 20 amps of current will flow through the 0.6-ohm bulb. This will raise our circuit current to 34 amps. Lowering the resistance on the one leg of the parallel circuit does only one thing; it greatly increases the current through the circuit. This high current may do something to the circuit; it may burn the wires. But which wires? Only the wires that would carry the 34 amps or the 20 amps to the bulb, not the wires to the other bulbs.

Leg #1	I = E/R	or	I = 12/0.6	or	I = 20 amps
Leg #2	I = E/R	or	I = 12/6	or	I = 2 amps
Leg #3	I = E/R	or	I = 12/1	or	I = 12 amps

Total current = 20 + 2 + 12 or 34 amps

Let's see what happens when we add resistance to one of the parallel legs. An increase in resistance should cause a decrease in current. In Figure 2-31, a 1-ohm resistor was added after the 1-ohm light bulb. This resistor is in series with the light bulb, and the total resistance of that leg is now 2 ohms. The current through that leg is now 6 amps. Again, the other bulbs were not affected by the change. The only change to the whole circuit was in total circuit current, which now drops to 12 amps. The added resistance lowered total circuit current and changed the way the 1-ohm bulb works. This bulb will now drop only 6 volts. The remaining 6 volts will be dropped by the added resistor. The 1-ohm bulb will be much dimmer than before; its power rating dropped from 144 watts to 36 watts. Additional resistance causes the bulb to be dimmer. The

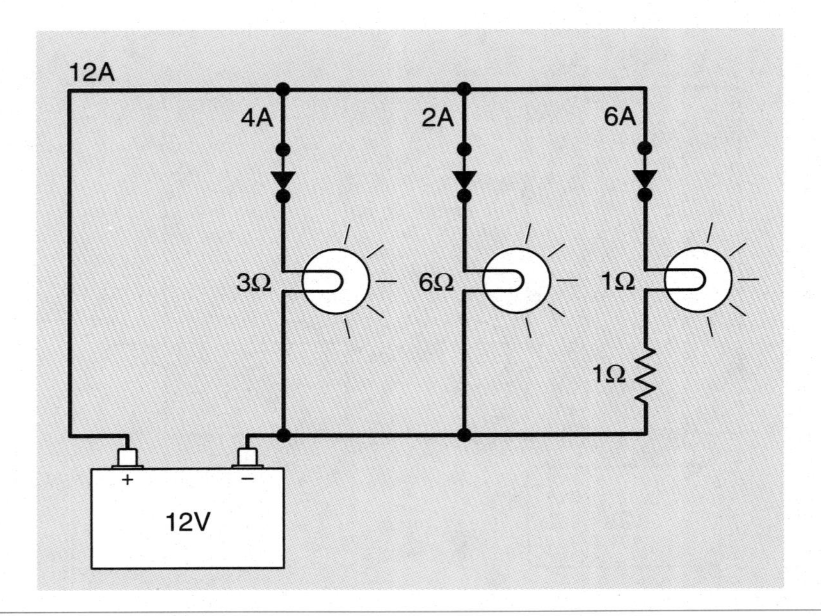

Figure 2-31 A series circuit contained in a leg of a parallel circuit.

bulb itself wasn't changed, only the resistance of that leg changed. The dimness is caused by the circuit, not the bulb.

Leg #1	I = E/R	or	I = 12/3	or	I = 4 amps			
Leg #2	I = E/R	or	I = 12/6	or	I = 2 amps			
Leg #3	I = E/R	or	I = 12/1+1	or	I = 12/2	or	I = 6 amps	

Total current = 4 + 2 + 6 or 12 amps

Now let's see what happens when we add a resistance that is common to all of the parallel legs. In Figure 2-32 we added a 0.333 ohm resistor (0.333 was chosen to keep the math simple!) to the negative connection at the battery. This will cause the circuit's current to decrease; it will also change the operation of the bulbs in the circuit. The total resistance of the bulbs in parallel is 0.667 ohms.

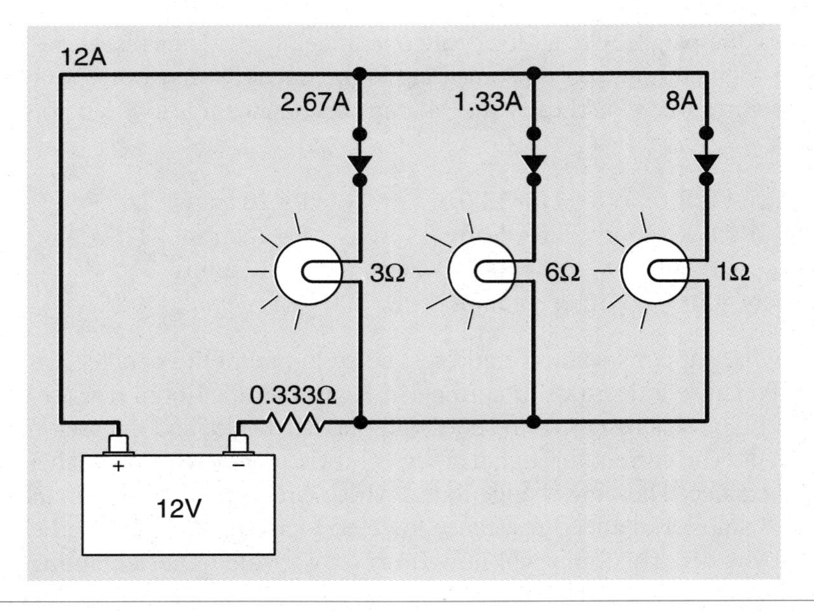

Figure 2-32 A resistor in series with a parallel circuit.

$$R_T = \frac{1}{\frac{1}{3} + \frac{1}{6} + \frac{1}{1}} \text{ or } R_T = \frac{1}{0.333 + 0.167 + 1} \text{ or } R_T = \frac{1}{1.5} \text{ or } R_T = 0.667$$

The total resistance of the circuit is 1 ohm (0.667 + 0.333), which means the circuit current is now 12 amps. Because there will be a voltage drop across the 0.333 ohm resistor, each of the parallel legs will have less than source voltage. To find the amount of voltage dropped by the parallel circuit we multiply the amperage by the resistance. Twelve amps multiplied by 0.667 equal 8. So 8 volts will be dropped by the parallel circuit; the remaining 4 volts will be dropped by the 0.333 resistor. The amount of current through each leg can be calculated by taking the voltage drop and dividing it by the resistance of the leg.

Leg #1 $I = E/R$ or $I = 8/3$ or $I = 2.667$ amps
Leg #2 $I = E/R$ or $I = 8/6$ or $I = 1.333$ amps
Leg #3 $I = E/R$ or $I = 8/1$ or $I = 8$ amps
Total circuit current = 2.667 + 1.333 + 8 or 12 amps

The added resistance affected the operation of all the bulbs, because it was added to a point that was common to all of the bulbs. All of the bulbs would be dimmer, and circuit current would be lower.

Semiconductors

As discussed earlier, electrical materials are classified as conductors, insulators, or semiconductors. Semiconductors include diodes, transistors, and silicon-controlled rectifiers. These semiconductors are often called solid-state devices because they are constructed of a solid material. The most common materials used in the construction of semiconductors are silicon or germanium. Both of these materials are classified as crystals.

Silicon and germanium have four electrons in their outer orbits. Because of their crystal-type structure, each atom shares an electron with four other atoms (Figure 2-33). As a result of this covalent bonding, each atom will have eight electrons in its outer orbit. All the orbits are filled and there are no free electrons, thus the material (as a category of matter) falls somewhere between conductor and insulator.

Semiconductors are materials that don't conduct electricity well, nor do they insulate well.

A **crystal** is the term used to describe a material that has a definite atom structure.

When atoms share electrons with other atoms it is called **covalent bonding.**

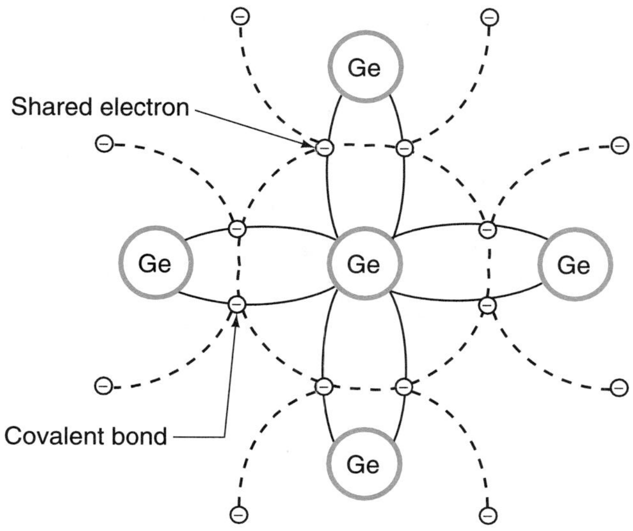

Figure 2-33 Crystal structure of germanium.

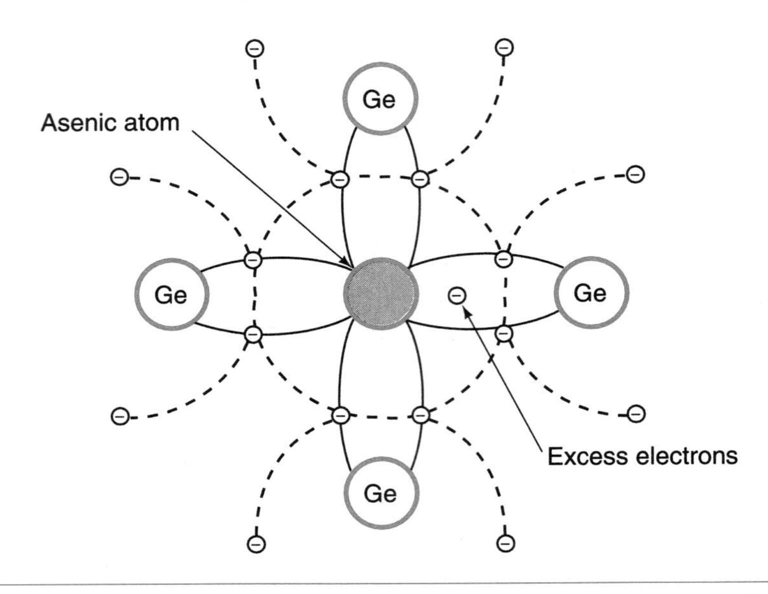

Figure 2-34 Germanium crystal doped with an arsenic atom to produce an N-type material.

When there are free electrons the material is called an **N-type material.** This is the semiconductor material that has been doped with an impurity (which is a small amount of another element) that has five electrons in its outer shell. The free electrons in the impurity give the material a negative charge.

The absence of an electron is called **a hole.** These holes are said to be positively charged because they have a tendency to attract free electrons into the hole.

Perfect crystals are not used for manufacturing semiconductors. They are doped with impurity atoms. This doping adds a small percentage of another element to the crystal. The doping element can be arsenic, antimony, phosphorous, boron, aluminum, or gallium.

If the crystal is doped by using arsenic, antimony, or phosphorous the result is a material with free electrons (Figure 2-34). Materials such as arsenic have five electrons, which leaves one electron left over. This doped material becomes negatively charged. Under the influence of an EMF, it will support current flow.

If boron, aluminum, or gallium are added to the crystal, a P-type material is produced. Materials like boron have three electrons in their outermost orbit. Because there is one fewer electron, there is an absence of an electron that produces a hole (Figure 2-35) and becomes positively charged.

By putting N-type and P-type materials together in a certain order, solid-state components are built that can be used for switching devices, voltage regulators, electrical control, and so on.

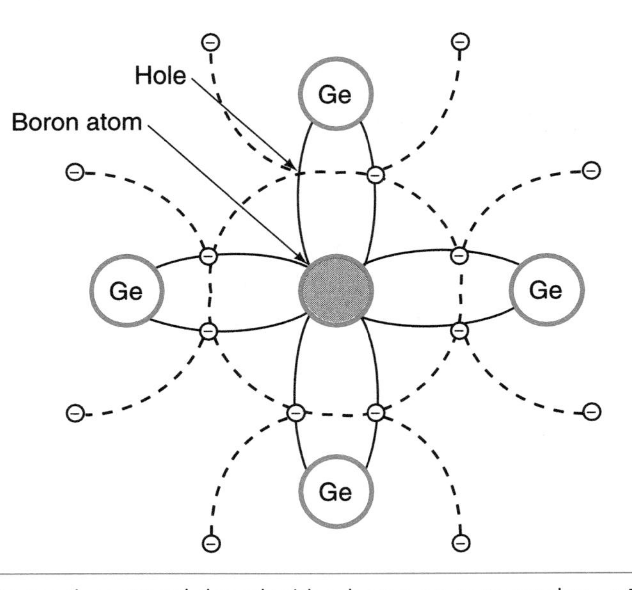

Figure 2-35 Germanium crystal doped with a boron atom to produce a P-type material.

Magnetism Principles

Magnetism is a force that is used to produce most of the electrical power in the world. It is also the force used to create the electricity to recharge a vehicle's battery, make a starter work, and produce signals for various operating systems. A magnet is a material that attracts iron, steel, and a few other materials.

The concentration of the magnetic lines of force is called **magnetic flux density** (Figure 2-37).

A BIT OF HISTORY

The force of a magnet was first discovered over 2,000 years ago by the Greeks. They noticed that a type of stone, now called magnetite, was attracted to iron. During the Dark Ages, the strange powers of magnetite were believed to be caused by evil spirits.

Permeability is the term used to indicate the magnetic conductivity of a substance compared with the conductivity of air. The greater the permeability, the greater the magnetic conductivity and the easier a substance can be magnetized or the more attracted it is to a magnet.

There are two types of magnets used on automobiles, permanent magnets and electromagnets. Permanent magnets are magnets that do not require any force or power to keep their magnetic field. Electromagnets depend on electrical current flow to produce and, in most cases, keep their magnetic field.

Magnets

All magnets have polarity. A magnet that is allowed to hang free will align itself north and south. The end facing north is called the north seeking pole, and the end facing south is called the south seeking pole. Like poles will repel each other, and unlike poles will attract each other. These principles are shown (Figure 2-36). The magnetic attraction is the strongest at the poles.

Reluctance is the term used to indicate a material's resistance to the passage of flux lines. High reluctant materials are not attracted to magnets.

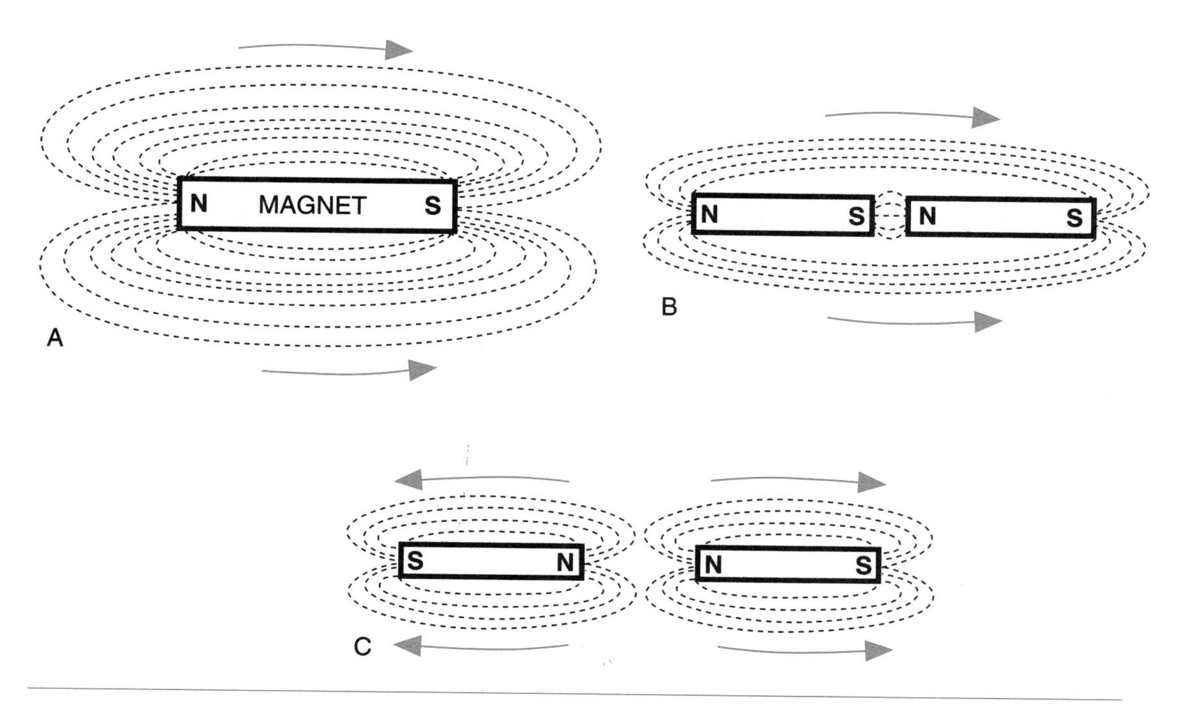

Figure 2-36 Magnetic principles: (A) all magnets have poles, (B) unlike poles attract each other, and (C) like poles repel.

Figure 2-37 Iron filings indicate the lines of magnetic flux.

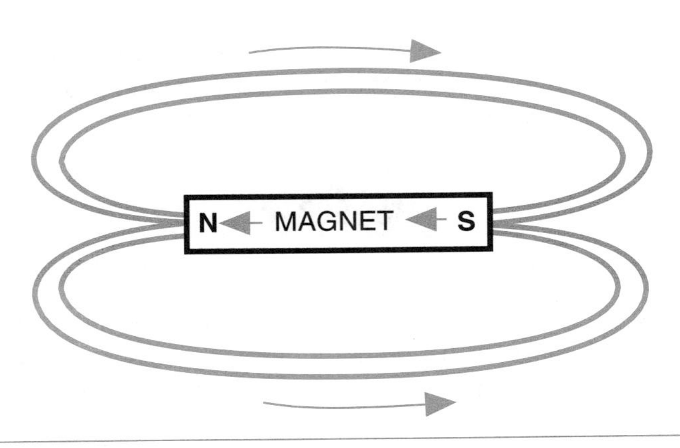

Figure 2-38 Lines of force through a magnet.

A strong magnet produces many lines of force, and a weak magnet produces fewer lines of force. Invisible lines of force leave the magnet at the north pole and enter again at the south pole. While inside the magnet, the lines of force travel from the south pole to the north pole (Figure 2-38).

The field of force (or magnetic field) is all the space, outside the magnet, that contains lines of magnetic force. Magnetic lines of force penetrate all substances; there is no known insulation against magnetic lines of force. The lines of force may be deflected only by other magnetic materials or by another magnetic field.

Electromagnetism

Electromagnetism is a form of magnetism that occurs when current flows through a conductor.

Whenever an electrical current flows through a conductor, a magnetic field is formed around the conductor (Figure 2-39). The number of lines of force, and the strength of the magnetic field produced, will be in direct proportion to the amount of current flow.

The direction of the lines of force is determined by the right-hand rule. Using the conventional theory of current flow being from positive to negative, the right hand is used to grasp the wire with the thumb pointing in the direction of current flow. The fingers will point in the direction of the magnetic lines of force (Figure 2-40).

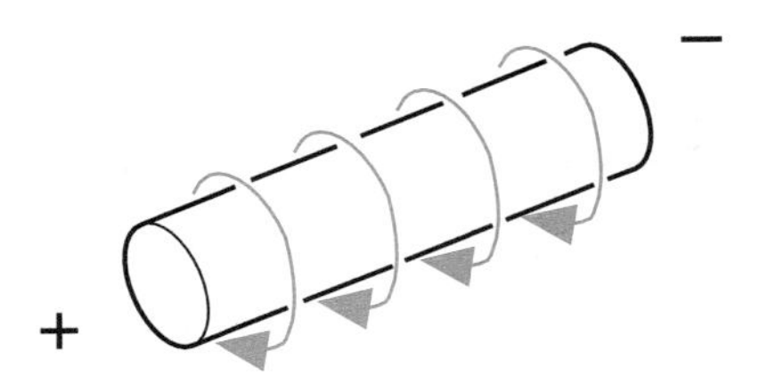

Figure 2-39 A magnetic field surrounds a conductor that has current flowing through it.

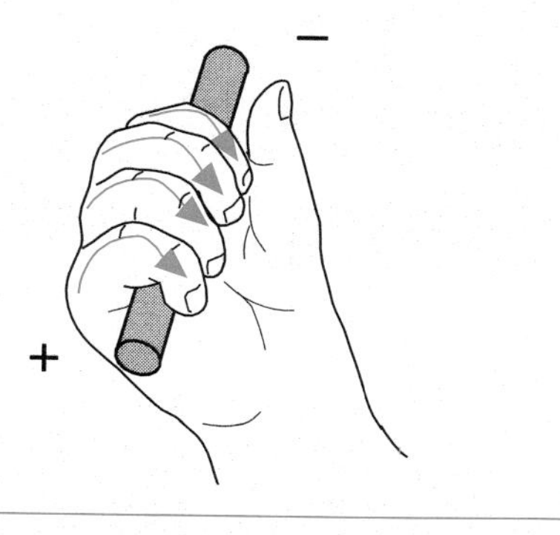

Figure 2-40 Right-hand rule to determine direction of magnetic lines.

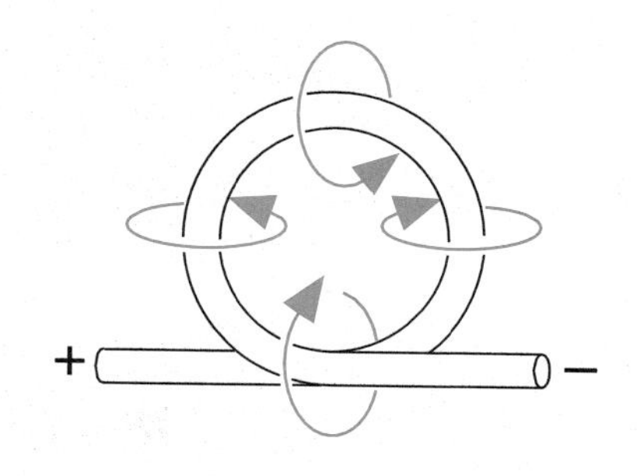

Figure 2-41 Looping the conductor increases the magnetic field.

André Marie Ampère noted that current flowing in the same direction through two nearby wires will cause the wires to be attracted to one another. Also, he observed that if current flow in one of the wires is reversed, the wires will repel one another. In addition, he found that if a wire is coiled with current flowing through the wire, the same magnetic field that surrounds a straight wire combines to form one larger magnetic field. This magnetic field has true north and south poles (Figure 2-41). Looping the wire doubles the flux density where the wire is running parallel to itself. The illustration (Figure 2-42) shows how these lines of force will join and add to each other.

The north pole can be determined in the coil by use of the right-hand rule. Grasp the coil with the fingers pointing in the direction of current flow (+ to -) and the thumb will point toward the north pole (Figure 2-43).

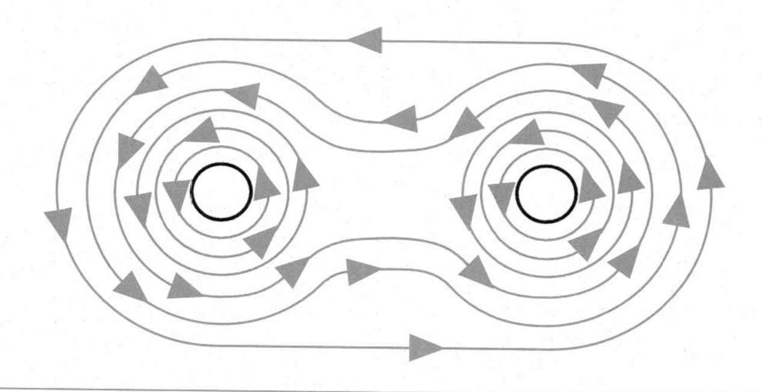

Figure 2-42 Lines of force join together and attract each other.

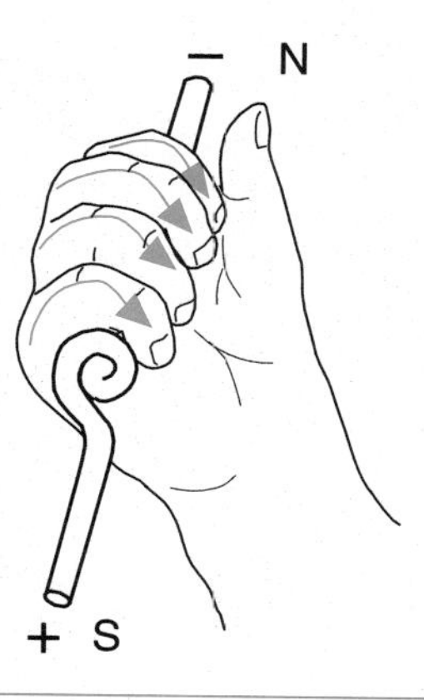

Figure 2-43 Right-hand rule to determine magnetic poles.

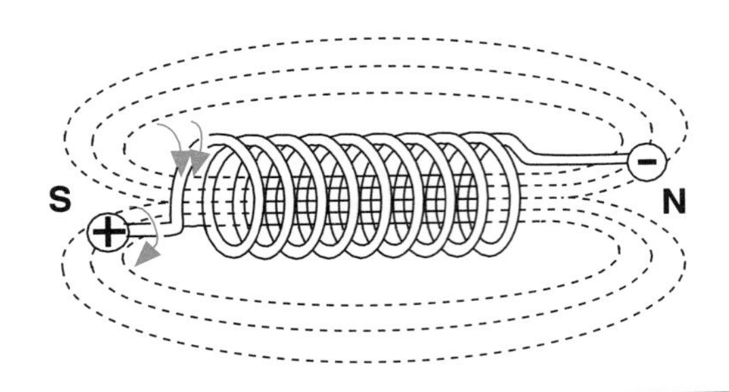

Figure 2-44 Adding more loops of wire increases the magnetic flux density.

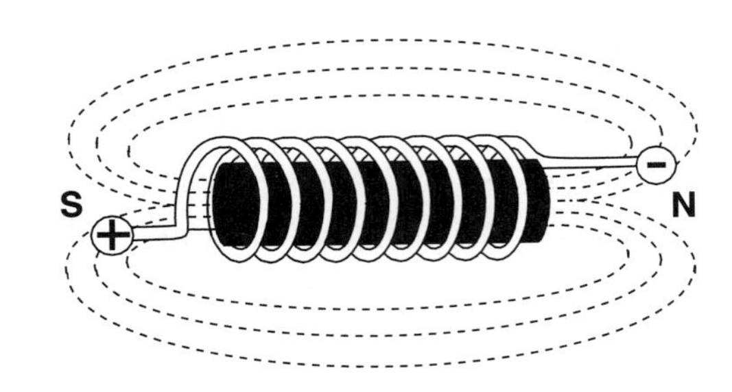

Figure 2-45 The addition of an iron core concentrates the flux density.

As more loops are added, the fields from each loop will join and increase the flux density (Figure 2-44). To make the magnetic field even stronger, an iron core can be placed in the center of the coil (Figure 2-45). The soft iron core is a material that has high permeability and provides an excellent conductor for the magnetic field that travels through the center of the wire coil.

The strength of an electromagnetic coil is affected by the following factors:

1. The amount of current flowing through the wire.
2. The number of windings or turns.
3. The size, length, and type of core material.
4. The direction and angle at which the lines of force are cut.

The strength of the magnetic field is measured in ampere-turns:

$$\text{ampere-turns} = \text{amperes} \times \text{number of turns}$$

The magnetic field strength is measured by multiplying the current flow in amperes through a coil by the number of complete turns of wire in the coil. For example, in the illustration (Figure 2-46), a 1,000 turn coil with 1 ampere of current would have a field strength of 1,000 ampere-turns. This coil would have the same field strength as a coil with 100 turns and 10 amperes of current.

Induction is the magnetic process of producing a current flow in a wire without any actual contact to the wire.

Theory of Induction

Electricity can be produced by magnetic induction. Magnetic induction occurs when a conductor is moved through the magnetic lines of force (Figure 2-47) or when a magnetic field is moved across a conductor. A difference of potential is set up between the ends of the conductor and a voltage is induced. This voltage exists only when the magnetic field or the conductor is in motion.

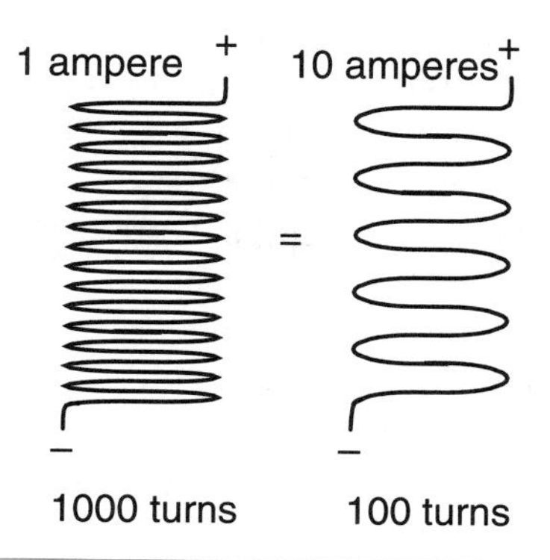

Figure 2-46 Magnetic field strength is determined by the amount of amperage and the number of coils.

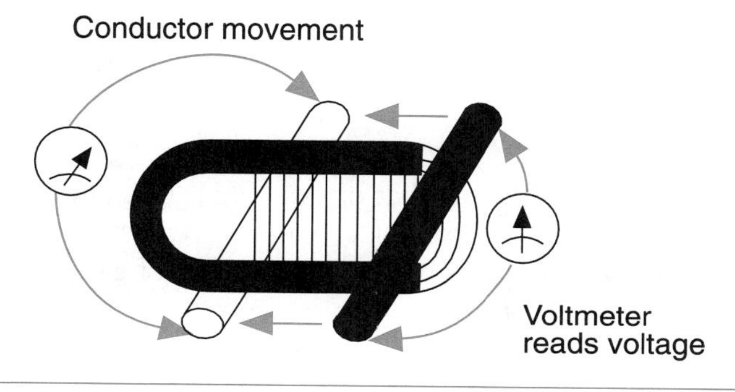

Figure 2-47 Moving a conductor through a magnetic field induces an electrical potential difference.

The induced voltage can be increased by either increasing the speed in which the magnetic lines of force cut the conductor, or by increasing the number of conductors that are cut. It is this principle that is behind the operation of all ignition systems, starter motors, charging systems, and relays.

Saturation is the point at which the magnetic strength eventually levels off, and where current will no longer increase as it passes through the coil.

A common induction device is the ignition coil. As the current increases, the coil will reach a point of saturation. The magnetic lines of force, which represent stored energy, will collapse when the applied voltage is removed. When the lines of force collapse, the magnetic energy is returned back to the wire as electrical energy.

If voltage is induced in the wires of a coil when current is first connected or disconnected, it is called self-induction. The resulting current is in the opposite direction of the applied current and tends to reduce the magnetic force. Self-induction is governed by Lenz's law, which states:

A desirable induction is called a **mutual induction.**

An induced current flows in a direction to oppose the magnetic field that produced it.

Self induction is generally not wanted in automotive circuits. For example, when a switch is opened, self-induction tends to continue to supply current in the same direction as the original current because as the magnetic field collapses, it induces voltage in the wire. According to Lenz's law, voltage induced in a conductor tends to oppose a change in current flow. Self induction can cause an electrical arc to occur across an opened switch. The arcing may momentarily bypass the switch and allow the circuit that was turned off to operate for a short period of time. The arcing will also burn the contacts of the switch.

Self-induction is also referred to as counter EMF (**CEMF**) or as a voltage spike.

Self-induction is commonly found in electrical components that contain a coil or an electric motor. To help reduce the arc across contacts, a capacitor or clamping diode may be connected to the circuit. The capacitor will absorb the high voltage arcs and prevent arcing across the contacts. Diodes are semiconductors that allow current flow in only one direction. A clamping diode can be connected in parallel to the coil and will prevent current flow from the self-induction coil to the switch.

Magnetic induction is also the basis for a generator and many of the sensors on today's vehicles. In a generator, a magnetic field rotates inside a set of conductors. As the magnetic field crosses the wires, a voltage is induced. The amount of voltage induced by this action depends on the speed of the rotating field, the strength of the field, and the number of conductors the field cuts through.

Magnetic sensors are used to measure speeds, such as engine, vehicle, and shaft speeds. These sensors typically use a permanent magnet. Rotational speed is determined by the passing of blades or teeth in and out of the magnetic field. As a tooth moves in and out of the magnetic field, the strength of the magnetic field is changed and a voltage signal is induced. This signal is sent to a control device, where it is interpreted.

EMI Suppression

Electromagnetic Interference (EMI) is an undesirable creation of electromagnetism whenever current is switched on and off.

A choke is an inductor in series with a circuit.

As manufacturers began to increase the number of electronic components and systems in their vehicles, the problem of EMI had to be controlled. The low power integrated circuits used on modern vehicles are sensitive to the signals produced as a result of EMI. EMI is produced as current in a conductor is turned on and off. EMI is also caused by static electricity that is created by friction. The friction is a result of the tires and their contact with the road, or from fan belts contacting the pulleys.

EMI can disrupt the vehicle's computer systems by inducing false messages to the computer. The computer requires messages to be sent over circuits in order to communicate with other computers, sensors, and actuators. If any of these signals are disrupted, the engine and/or accessories may turn off.

EMI can be suppressed by any one of the following methods:

1. Adding a resistance to the conductors. This is usually done to high-voltage systems such as the secondary circuit of the ignition system.
2. Connecting a capacitor in parallel and a choke coil in series with the circuit.
3. Shielding the conductor or load components with a metal or metal impregnated plastic.
4. Increasing the number of paths to ground by using designated ground circuits. This provides a clear path to ground that is very low in resistance.
5. Adding a clamping diode in parallel to the component.
6. Adding an isolation diode in series to the component.

Summary

- ❑ An atom is constructed of a complex arrangement of electrons in orbit around a nucleus. If the number of electrons and protons are equal, the atom is balanced or neutral.
- ❑ A conductor allows electricity to easily flow through it.
- ❑ An insulator does not allow electricity to easily flow through it.
- ❑ Electricity is the movement of electrons from atom to atom. In order for the electrons to move in the same direction, an electromotive force (EMF) must be applied to the circuit.
- ❑ The electron theory defines electron flow as motion from negative to positive.
- ❑ The conventional theory of current flow states that current flows from a positive point to a less positive point.
- ❑ Voltage is defined as an electrical pressure and is the difference between the positive and negative charges.
- ❑ Current is defined as the rate of electron flow and is measured in amperes. Amperage is the amount of electrons passing any given point in the circuit in one second.
- ❑ Resistance is defined as opposition to current flow and is measured in ohms (Ω).
- ❑ Ohm's law defines the relationship between current, voltage, and resistance. It is the basic law of electricity and states that the amount of current in an electric circuit is inversely proportional to the resistance of the circuit, and is directly proportional to the voltage in the circuit.
- ❑ Wattage represents the measure of power (P) used in a circuit. Wattage is measured by using the power formula, which defines the relationship between amperage, voltage, and wattage.
- ❑ Capacitance is the ability of two conducting surfaces to store voltage.
- ❑ Direct current results from a constant voltage and a current that flows in one direction.
- ❑ In an alternating current circuit, voltage and current do not remain constant. AC current changes direction from positive to negative and negative to positive.
- ❑ For current to flow, the electrons must have a complete path from the source voltage to the load component and back to the source.
- ❑ The series circuit provides a single path for current flow from the electrical source through all the circuit's components, and back to the source.
- ❑ A parallel circuit provides two or more paths for current to flow.
- ❑ A series-parallel circuit is a combination of the series and parallel circuits.
- ❑ The equivalent series load is the total resistance of a parallel circuit plus the resistance of the load in series with the voltage source.
- ❑ Voltage drop is caused by a resistance in the circuit that reduces the electrical pressure available after the resistance.
- ❑ Kirchhoff's voltage law states that the total voltage drop in an electrical circuit will always equal the available voltage at the source.

Terms to Know

Capacitance
Circuit
Conductor
Electromagnetism
Electromotive force (emf)
Equivalent series load
Induction
Insulator
Parallel circuit
Potential
Power formula
Semiconductors
Series circuit
Series-parallel circuit
Valence ring

Review Questions

Short Answer Essays

1. List and define the three elements of electricity.
2. Describe the use of Ohm's law.
3. List and describe the three types of circuits.
4. Explain the principle of electromagnetism.
5. Describe the principle of induction.
6. Describe the basics of electron flow.
7. Define the two types of electrical current.
8. Describe the difference between insulators, conductors, and semiconductors.
9. Explain the basic concepts of capacitance.
10. What does the measurement of "Watt" represent?

Fill-in-the-Blanks

1. ____________________ are negatively charged particles. The nucleus contains positively charged particles called ____________________ and particles that have no charge called ____________________.

2. A ____________________ allows electricity to easily flow through it. An ____________________ does not allow electricity to easily flow through it.

3. For the electrons to move in the same direction, there must be an ____________________ applied.

4. The ____________________ ____________________ of current flow states that current flows from a positive point to a less positive point.

5. Resistance is defined as ____________________ to current flow and is measured in ____________________.

6. ____________________ is the ability of two conducting surfaces to store voltage.

7. Kirchhoff's voltage law states that the ____________________ ____________________ ____________________ in an electrical circuit will always ____________________ available voltage at the source.

8. The ____________________ of all the resistors in series is the total resistance of that series circuit.

9. ____________________ is defined as an electrical pressure.

10. ____________________ is defined as the rate of electron flow.

ASE Style Review Questions

1. The methods that can be used to form an electrical current are being discussed:
Technician A says electricity can be generated by magnetic induction.
Technician B says electricity can be produced by a battery.
Who is correct?
A. A only
B. B only
C. Both A and B
D. Neither A nor B

2. Circuit resistance is being discussed:
Technician A says in a series circuit total resistance is figured by adding together all of the resistances in the circuit.
Technician B says in a parallel circuit the total resistance is less than the lowest resistor.
Who is correct?
A. A only
B. B only
C. Both A and B
D. Neither A nor B

3. While discussing voltage drop:
Technician A says all of the voltage from the source is dropped in a complete circuit.
Technician B says corrosion is not a contributor to voltage drop.
Who is correct?
A. A only
B. B only
C. Both A and B
D. Neither A nor B

4. *Technician A* says voltage is the electrical pressure that causes electrons to move.
Technician B says voltage will exist between any two points in that circuit unless the potential drops to zero.
Who is correct?
A. A only
B. B only
C. Both A and B
D. Neither A nor B

5. *Technician A* says wattage is a measure of the total electrical work being performed.
Technician B says the power formula is expressed as $P = R \times I$.
Who is correct?
A. A only
B. B only
C. Both A and B
D. Neither A nor B

6. *Technician A* says a capacitor consumes electrical power.
Technician B says a capacitor induces voltage.
Who is correct?
A. A only
B. B only
C. Both A and B
D. Neither A nor B

7. Types of electrical currents are being discussed:
Technician A says alternating current is produced from a voltage and current that remain constant and flow in the same direction.
Technician B says direct current changes directions from positive to negative.
Who is correct?
A. A only
B. B only
C. Both A and B
D. Neither A nor B

8. The principles of induction are being discussed:
Technician A says induction is the magnetic process of producing a current flow in a wire without any actual contact to the wire.
Technician B says induced voltage only exists when the magnetic field or the conductor is in motion.
Who is correct?
A. A only
B. B only
C. Both A and B
D. Neither A nor B

9. *Technician A* says if the resistance increases and the voltage remains constant, the amperage will increase.
Technician B says Ohm's law can be stated as $I = E \div R$.
Who is correct?
A. A only
B. B only
C. Both A and B
D. Neither A nor B

10. *Technician A* says two 4-ohm resistors in parallel have an equivalent resistance of 2 ohms.
Technician B says two 4-ohm resistors in series have an equivalent resistance of 8 ohms.
Who is correct?
A. A only
B. B only
C. Both A and B
D. Neither A nor B

Electrical Components

Upon completion and review of this chapter, you should be able to:

- ❑ Explain the purpose of a circuit protection device. Describe the most common types in use.
- ❑ Describe the common types of electrical system components used and how they affect the electrical system.
- ❑ Explain the operation of the electrical controls, including switches, relays, and variable resistors.
- ❑ Describe the basic operating principles of electronic components.
- ❑ Explain the use of electronic components in the circuit.
- ❑ Define circuit defects including opens, shorts, grounds, and excessive voltage drops.
- ❑ Explain the effects that each type of circuit defect has on the operation of the electrical system.

Introduction

In this chapter you will be introduced to electrical and electronic components. These components include circuit protection devices, switches, relays, variable resistors, diodes, and different forms of transistors. Today's technician must comprehend the operation of these components and the ways they affect electrical system operation. With this knowledge the technician will be able to accurately and quickly diagnose many electrical failures.

To be able to properly diagnose the components and circuits, the technician must be able to use the test equipment that is designed for electrical system diagnosis. In this chapter you will learn about the various types of test equipment used for diagnosing electrical systems. You will learn the appropriate equipment to use to locate the fault based on the symptoms. In addition, the various types of defects that cause the system to operate improperly are discussed.

Circuit Protection Devices

Shop Manual
Chapter 3, page 53

A **protection device** is designed to "turn off" the system it protects. This is done by creating an open (like turning off a switch) to prevent a complete circuit.

Most automotive electrical circuits are protected from high current flow that would exceed the capacity of the circuit's conductors and/or loads. Excessive current results from a decrease in the circuit's resistance. Circuit resistance will decrease when too many components are connected in parallel or when a component or wire becomes shorted. A short is an undesirable, low resistance path for current flow. When the circuit's current reaches a predetermined level, most circuit protection devices open and stop current flow in the circuit. This action prevents damage to the wires and the circuit's components.

A few late-model vehicles use a thermistor as a protection device in some of their circuits. This type of component changes its resistance according to changes in heat. Thermistors used to protect circuits increase their resistance with an increase in temperature. Therefore, as current flow increases the resistance in the circuit increases. This lowers the current flow to protect the circuit.

Fuses

Excess current flow in a circuit is called an **overload.**

The most commonly used circuit protection device is the fuse (Figure 3-1). A fuse contains a metal strip that will melt when the current flowing through it exceeds its rating. The thickness of the metal strip determines the rating of the fuse. When the metal strip

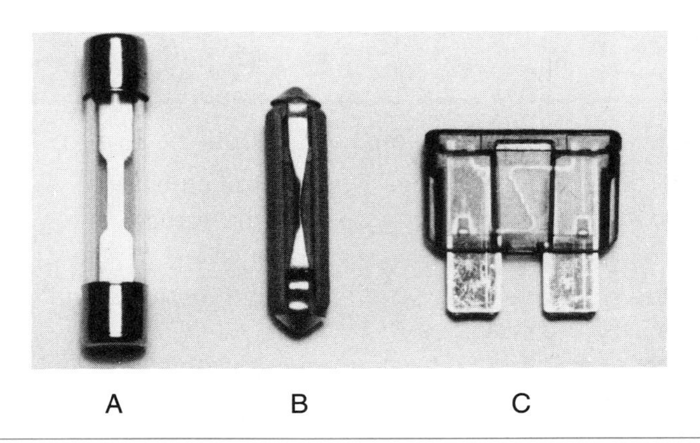

Figure 3-1 Three basic types of automotive fuses: (A) glass cartridge, (B) ceramic, and (C) blade.

A **fuse** is a replaceable element that will melt if the current passing through it exceeds the fuse rating.

Fuse box is the term used to indicate the central location of the fuses contained in a single holding fixture.

melts, excessive current is indicated. The cause of the excessive current must be found and repaired; then a new fuse of the same rating should be installed. The most commonly used automotive fuses are rated from 4 to 30 amps.

There are three basic types of fuses: glass or ceramic fuses, blade-type fuses, and bullet or cartridge fuses. Glass and ceramic fuses are found mostly on older vehicles. Sometimes, however, you can find them in a special holder connected in series with a circuit. **Glass fuses** are small glass cylinders with metal caps. The metal strip connects the two caps. The rating of the fuse is normally marked on one of the caps.

Blade-type fuses are flat plastic units and are available in three different physical sizes: mini, standard, and maxi (Figure 3-2). The plastic housing is formed around two male blade-type connectors. The metal strip connects these connectors inside the plastic

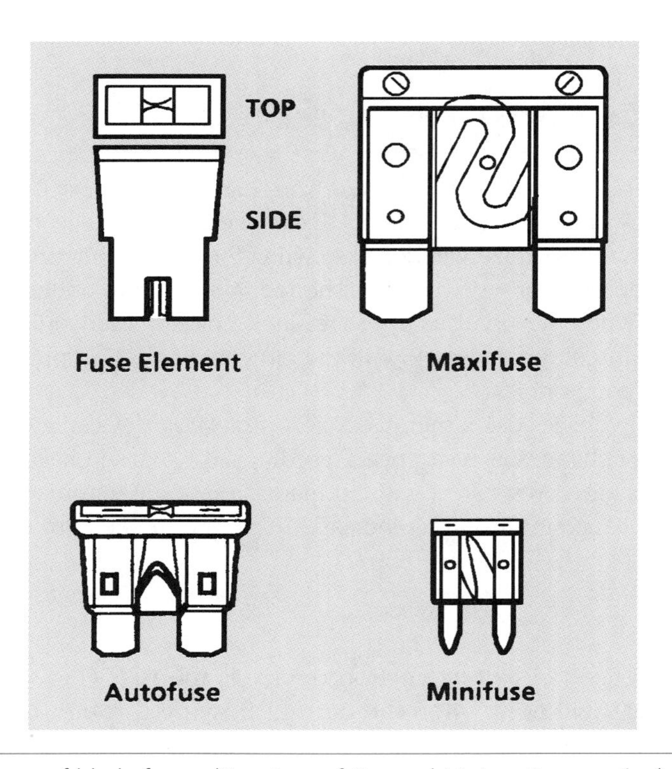

Figure 3-2 Types of blade fuses. (Courtesy of General Motors Corporation)

housing. The rating of these fuses is on top of the plastic housing and the plastic is color coded (Figure 3-3).

Cartridge-type fuses are used in many European vehicles. These fuses are made of plastic or a ceramic material. They have pointed ends and the metal strip rounds from end to end. This type fuse is much like a glass fuse except the metal strip is not enclosed.

Fuses are typically located in a central fuse block or power distribution box. However, fuses may also be found in relay boxes and electrical junction boxes. Power distribution boxes are normally located in the engine compartment and house fuses and relays. A common location for a fuse box is under the instrumental panel (Figure 3-4). The fuse box may also be located behind kick panels, in the glove box, in the engine compartment, or in a variety of other places on the vehicle. Fuse ratings and the circuits they protect are normally marked on the cover of the fuse or power distribution box. Of course, this information can also be found in the vehicle's owner's manual and the service manual.

Bus bar is a common electrical connection to which all of the fuses are attached. The bus bar is connected to battery voltage.

AUTOFUSE

CURRENT RATING	COLOR
3	VIOLET
5	TAN
7.5	BROWN
10	RED
15	BLUE
20	YELLOW
25	NATURAL
30	GREEN

MAXIFUSE

CURRENT RATING	COLOR
20	YELLOW
30	GREEN
40	AMBER
50	RED
60	BLUE
70	BROWN
80	NATURAL

MINIFUSE

CURRENT RATING	COLOR
5	TAN
7.5	BROWN
10	RED
15	BLUE
20	YELLOW
25	WHITE
30	GREEN

Figure 3-3 Color coding for blade-type fuses. An autofuse is a standard blade-type fuse. (Courtesy of General Motors Corporation)

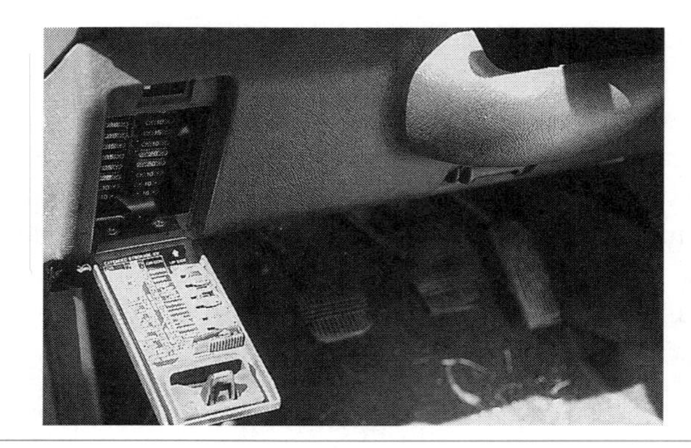

Figure 3-4 Fuse boxes are normally located under the dash or in the engine compartment.

A "blown" fuse is identified by a burned-through metal wire in the capsule (Figure 3-5).

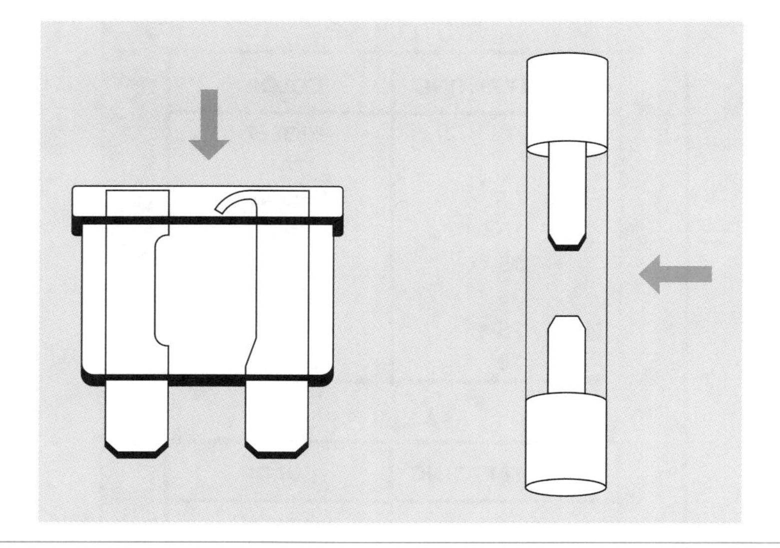

Figure 3-5 Blown fuses.

A fuse is connected in series with the circuit. Normally the fuse is located before all of the loads of the circuit (Figure 3-6). However, it may be placed before an individual load (Figure 3-7).

When adding accessories to the vehicle, the correct fuse rating must be selected. Use the power formula to determine the correct fuse rating (watts ÷ volts = amperes). The fuse selected should be rated slightly higher than the actual current draw to allow for current surges (5% to 10%).

WARNING: Fuses are rated by amperage and voltage. Never install a larger rated fuse into a circuit than the one that was designated by the manufacturer. Doing so may damage or destroy the circuit.

A **fusible link** is made of meltable conductor material with a special heat resistant insulation. When there is an overload in the circuit, the conductor link melts and opens the circuit. To properly test a fusible link, use an ohmmeter or continuity tester.

Fusible Links

A vehicle may have one or several fusible links to provide protection for the main power wires before they are divided into smaller circuits at the fuse box. The fusible links are usually

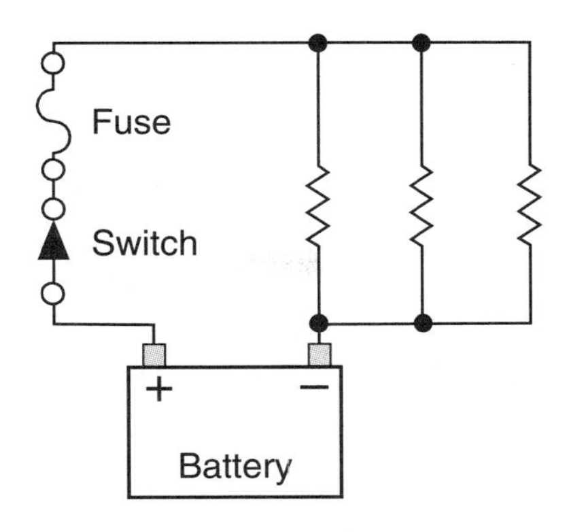

Figure 3-6 One fuse to protect an entire parallel circuit.

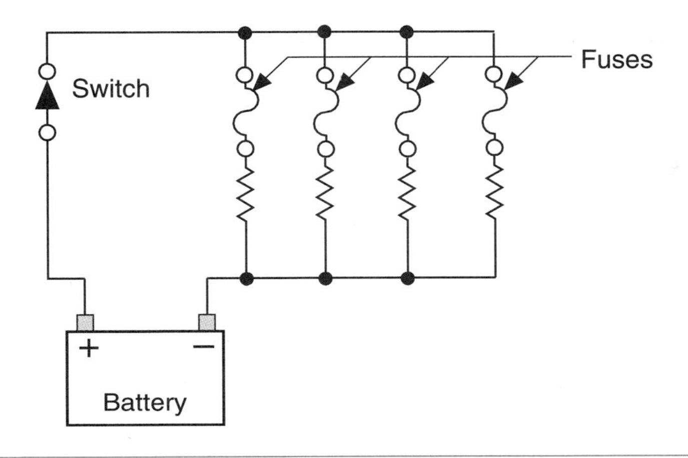

Figure 3-7 Fuses used to protect each branch of a parallel circuit.

located at a main connection near the battery or starter solenoid (Figure 3-8). The current capacity of a fusible link is determined by its size. A fusible link is usually four wire sizes smaller (four numbers larger) than the circuit it protects. The smaller the wire, the larger its number. A circuit that uses 14-gauge wire would require an 18-gauge fusible link for protection.

Some GM vehicles have the fusible link located at the main connection near the starter motor.

WARNING: Do not replace a fusible link with a resistor wire or vice versa.

A "blown" fusible link is usually identified by bubbling of the insulator material around the link.

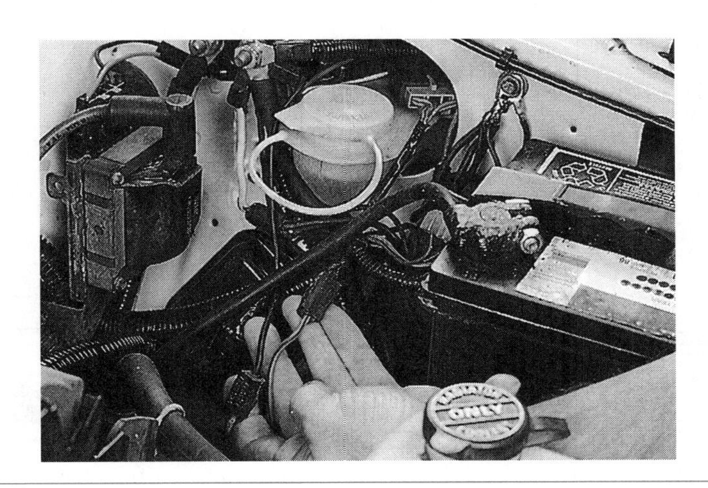

Figure 3-8 Fusible links located near the battery.

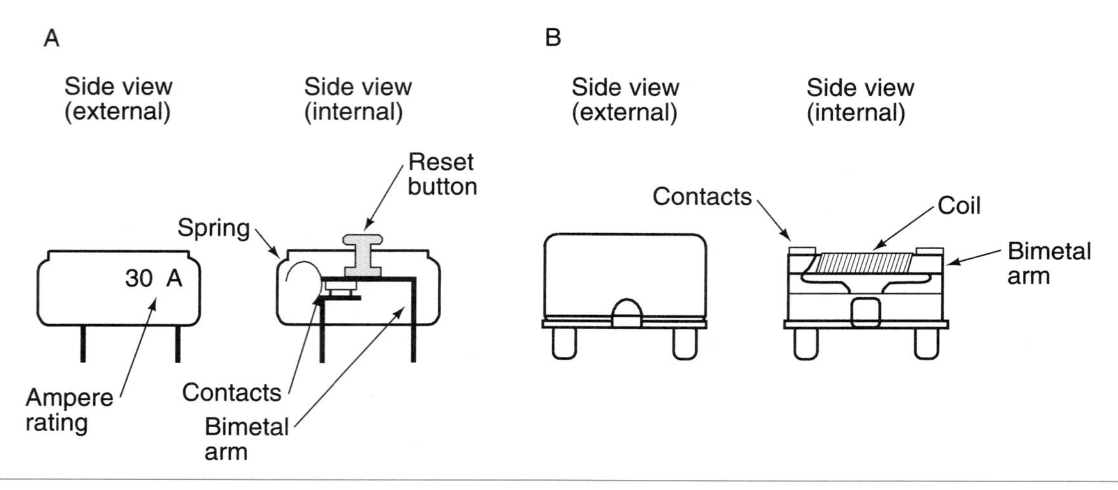

Figure 3-9 Nonself-resetting circuit breakers: (A) manual reset-type; and (B) a circuit breaker that requires removal from power to be reset.

Maxi-fuses

A **maxi-fuse** looks similar to a blade-type fuse except it is larger and has a higher current capacity. It is also referred to as a **cartridge fuse.**

In place of fusible links, many manufacturers use a maxi-fuse. By using maxi-fuses, manufacturers are able to break down the electrical system into smaller circuits. If a fusible link burns out, many of the vehicle's electrical systems may be affected. By breaking down the electrical system into smaller circuits and installing maxi-fuses, the consequence of a circuit defect will not be as severe as it would have been with a fusible link. In place of a single fusible link, there may be many maxi-fuses, depending on how the circuits are divided. This makes the technician's job of diagnosing a faulty circuit much easier.

Maxi-fuses are used because they are less likely to cause an underhood fire when there is an overload in the circuit. If the fusible link is burned in two, it is possible that the "hot" side of the fuse can come into contact with the vehicle frame and the wire can catch on fire.

Circuit Breakers

A **bimetallic strip** consists of two different types of metals. One strip will react more quickly to heat than the other, which causes the strip to flex in proportion to the amount of current flow.

A circuit that is susceptible to an overload on a routine basis is usually protected by a circuit breaker. Some circuit breakers require manual resetting by pressing a button. Others must be removed from the power to be reset (Figure 3-9). Some circuit breakers are self-resetting. This type of circuit breaker (Figure 3-10) uses a bimetallic strip that reacts to excessive current.

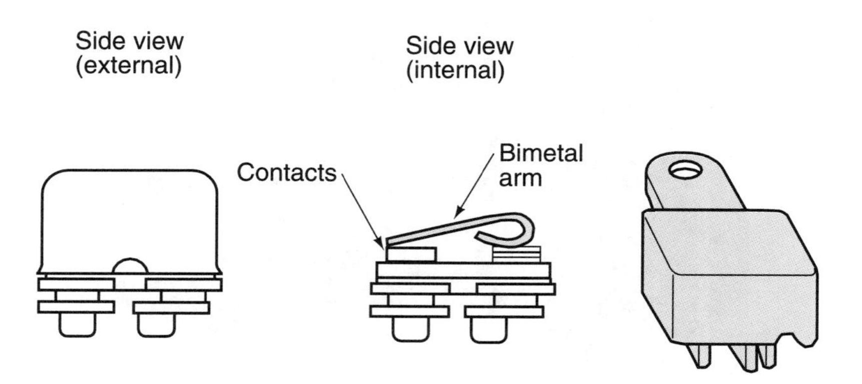

Figure 3-10 A self-resetting circuit breaker uses a bimetallic strip that opens if current draw is excessive.

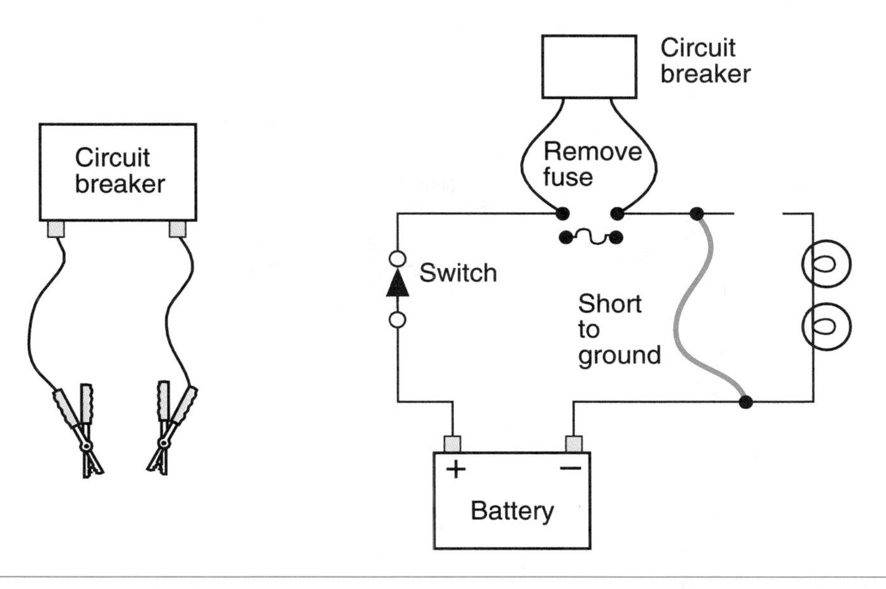

Figure 3-11 A circuit breaker fitted with alligator clips can be used to diagnose electrical problems.

When an overload or circuit defect occurs that causes an excessive amount of current draw, the current flowing through the bimetallic strip causes it to heat. As the strip heats, it bends and opens the contacts. Once the contacts are opened, current can no longer flow. With no current flowing, the strip cools and closes again. If the excessive current cause is still in the circuit, the breaker will open again. The circuit breaker will continue to open and close as long as the overload is in the circuit.

SERVICE TIP: A circuit breaker, fitted with alligator clips, is useful to bypass a fuse that keeps blowing. The circuit breaker will keep the current flowing through the circuit so that it may be checked for the cause of high current draw while protecting the circuit by cycling on and off (Figure 3-11). This diagnostic tool should not be used on electronic circuits.

An example of the use of a circuit breaker is in the power window circuit. Because the window is susceptible to jams due to ice buildup on the window, a current overload is possible. If this should occur, the circuit breaker will heat up and open the circuit before the window motor is damaged. If the operator continues to attempt to operate the power window, the circuit breaker will open and close until the cause of the jam is removed.

Some accessories, such as power window motors, are protected by a solid state-type circuit breaker typically called an electronic circuit breaker (**ECB**). The solid state material that is the basis for this type circuit breaker has a positive temperature coefficient and greatly increases its resistance when excessive current passes through it. The excessive current heats the ECB. As the ECB heats up, its resistance increases, which causes a decrease in current. When there is a great amount of current passing through the ECB, the resistance is so great that only a small amount of current flows and the protected device does not operate. The ECB will not reset until the circuit is opened. After one to two seconds of no current flow, the circuit breaker resets.

Electrical Components

Electrical circuits require different components depending on the type of work they do and how they are to perform it. A light may be wired directly to the battery, but it will remain on until the

Figure 3-12 Common types of switches used in automotive electrical systems.

battery drains. A switch will provide for control of the light circuit. However, if variable dimming of the light is also required, a rheostat is also needed.

There are several electrical components that may be incorporated into a circuit to achieve the desired results from the system. These components include switches, relays, buzzers, and various types of resistors.

Shop Manual
Chapter 3, page 57

Switch controls are designated as to whether they are **normally open (NO)** or **normally closed (NC).**

The term **throw** refers to the number of output circuits. The term **pole** refers to the number of input circuits.

Switches

A switch is the most common means of providing control of electrical current flow to an accessory (Figure 3-12). A switch can control the on/off operation of a circuit or direct the flow of current through various circuits. The contacts inside the switch assembly carry the current when they are closed. When they are open, current flow is stopped.

A normally open switch will not allow current flow when it is in its rest position. The contacts are open until they are acted on by an outside force that closes them to complete the circuit. A normally closed switch will allow current flow when it is in its rest position. The contacts are closed until they are acted on by an outside force that opens them to stop current flow.

The simplest type of switch is the single-pole, single-throw (**SPST**) switch (Figure 3-13). This switch controls the on/off operation of a single circuit. The most common type of SPST switch design is the hinged pawl. The pawl acts as the contact and changes position as directed to open or close the circuit.

Some SPST switches are designed to be a momentary contact switch. This switch usually has a spring that holds the contacts open until an outside force is applied and closes them. The horn button on most vehicles is of this design.

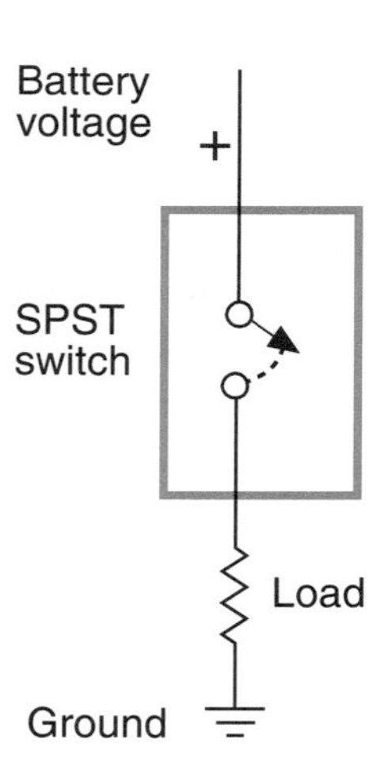

Figure 3-13 A simplified illustration of a SPST switch.

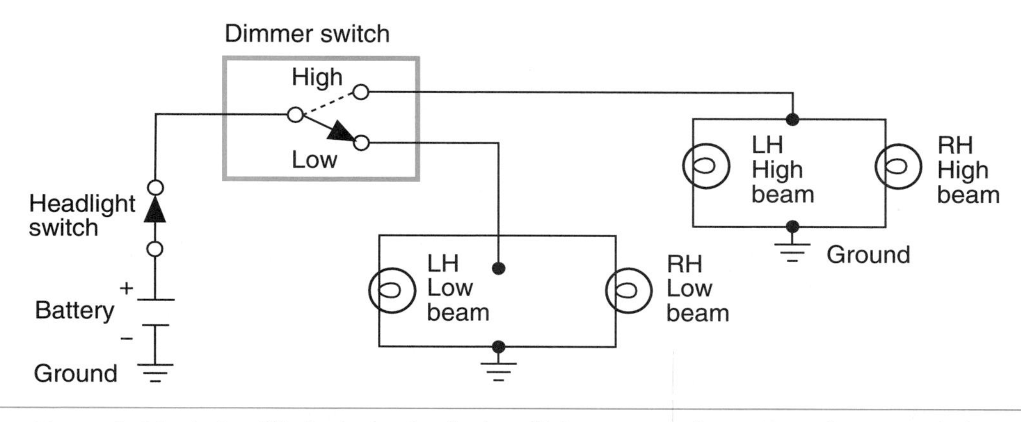

Figure 3-14 A simplified schematic of a headlight system using a SPDT dimmer switch.

Some electrical systems may require the use of a single-pole, double-throw switch (**SPDT**). The dimmer switch used in the headlight system is usually a SPDT switch. This switch has one input circuit with two output circuits. Depending on the position of the contacts, voltage is applied to the high beam circuit or to the low beam circuit (Figure 3-14).

One of the most complex switches is the **ganged switch**. This type of switch is commonly used as an ignition switch. In Figure 3-15, the five wipers are all ganged together and will move together. Battery voltage is applied to the switch from the starter relay terminal. When the ignition key is turned to the start position, all wipers move to the "S" position. Wipers D and E will complete the circuit to ground to test the instrument panel warning lamps. Wiper B provides battery voltage to the ignition coil. Wiper C supplies battery voltage to the starter relay and the ignition module. Wiper A has no output.

The dotted lines used in the symbol indicate the movement of the switch pawl from one position to the other.

Ganged means the wipers all move together.

The **make-before-break wiper** prevents any break in voltage to the ignition coil when the switch is moved from the S position to the R position.

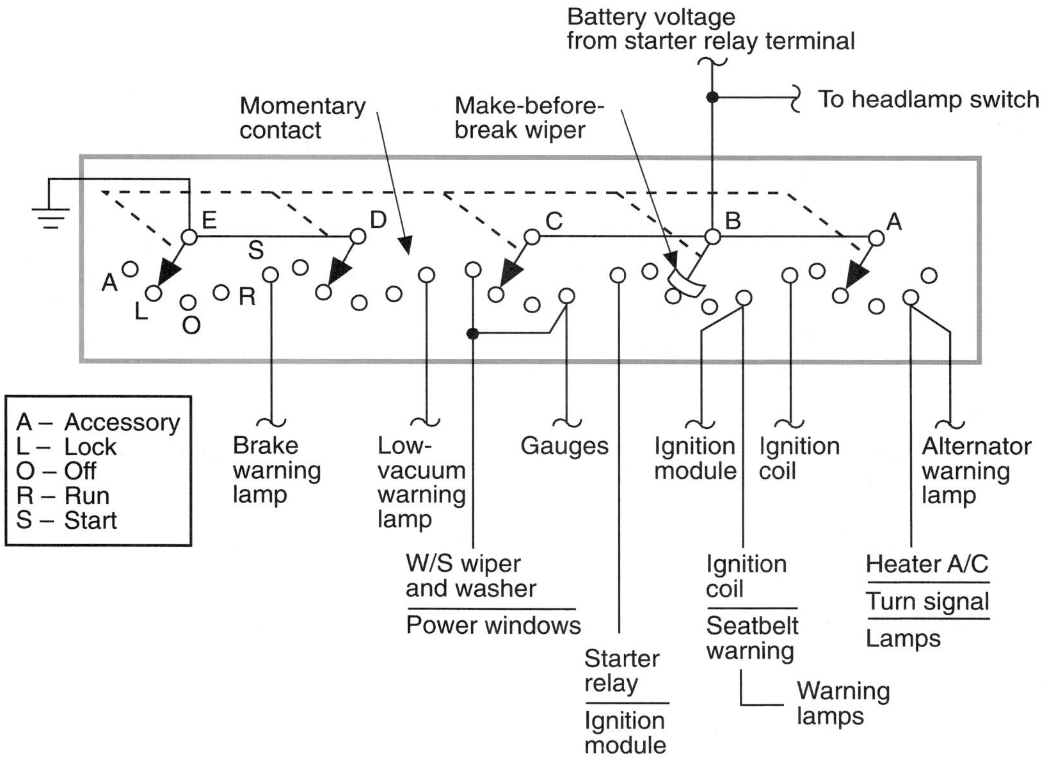

Battery voltage from starter relay terminal
To headlamp switch
Momentary contact
Make-before-break wiper
E D C B A
S
A L O R
A – Accessory
L – Lock
O – Off
R – Run
S – Start
Brake warning lamp
Low-vacuum warning lamp
Gauges
Ignition module
Ignition coil
Alternator warning lamp
W/S wiper and washer
Power windows
Starter relay
Ignition module
Ignition coil
Seatbelt warning
Warning lamps
Heater A/C
Turn signal
Lamps

Figure 3-15 Illustration of an ignition switch.

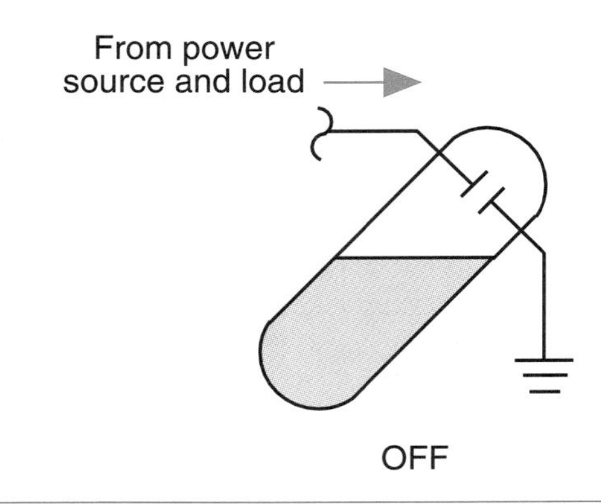

Figure 3-16 A mercury switch in the open position. The mercury is not covering the points.

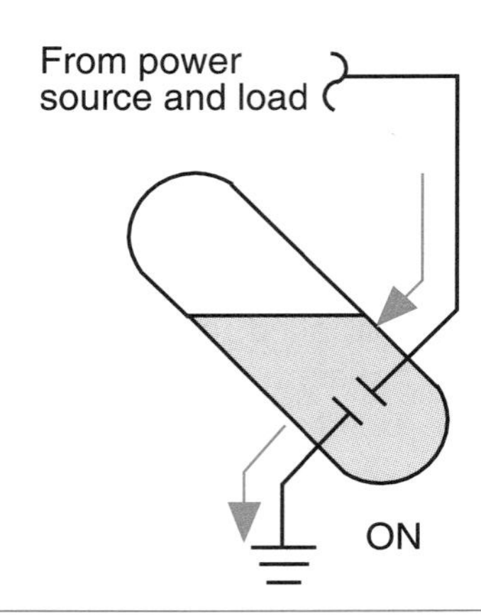

Figure 3-17 When the mercury switch is tilted, the mercury covers the points and completes the circuit.

Once the engine starts, the wipers are moved to the RUN position. Wipers D and E are moved out of contact with any output terminals. Wiper A supplies battery voltage to the comfort controls and turn signals, wiper B supplies battery voltage to the ignition coil and other accessories, and wiper C supplies battery voltage to other accessories. The jumper wire between terminals A and R of wiper C indicate that those accessories listed can be operated with the ignition switch in the RUN or ACC position.

Mercury switches are used by many vehicle manufacturers to detect motion. This switch uses a capsule that is partially filled with mercury and has two electrical contacts located at one end. If the switch is constructed as a normally open switch, the contacts are located above the mercury level (Figure 3-16). Mercury is an excellent conductor of electricity. So if the capsule is moved so the mercury touches both of the electrical contacts, the circuit is completed (Figure 3-17). This type of switch is used to illuminate the engine compartment when the hood is opened. While the hood is shut, the capsule is tilted in a position such that the mercury is not able to complete the circuit. Once the hood is opened, the capsule tilts with the hood and the mercury completes the circuit and the light turns on.

Shop Manual
Chapter 3, page 59

Relays

A **relay** is a device that uses low current to control a high current circuit.

Some circuits utilize electromagnetic switches called relays (Figure 3-18). The coil in the relay has a very high resistance, thus it will draw very low current. This low current is used to produce a magnetic field that will close the contacts. Normally open relays have their points closed by the

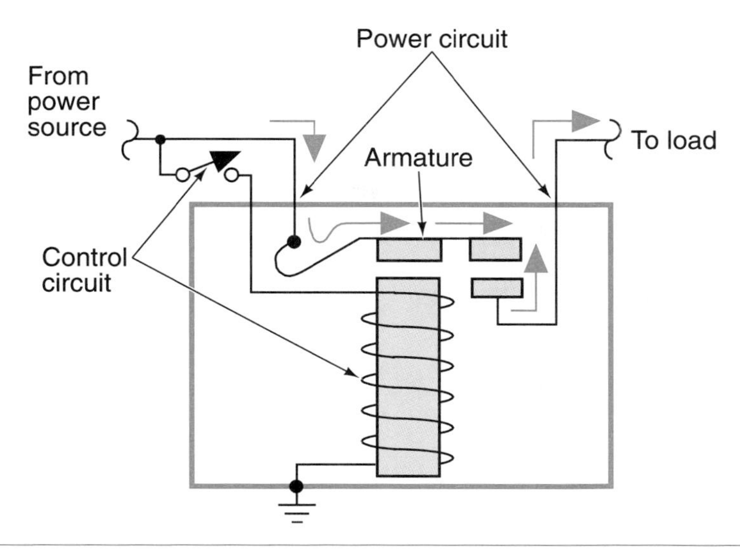

Figure 3-18 A relay uses electrical current to create a magnetic field to draw the contact points together to close the circuit.

electromagnetic field. Normally closed relays have their points opened by the magnetic field. The contacts are designed to carry the high current required to operate the load component. When current is applied to the coil, the contacts close and heavy battery current flows to the load component that is being controlled.

The illustration (Figure 3-19) shows a relay application in a horn circuit. Battery voltage is applied to the coil. Because the horn button is a normally open-type switch, the current flow to ground is open. Pushing the horn button will complete the circuit, allowing current flow through the coil. The coil develops a magnetic field, which closes the contacts. With the contacts closed, battery voltage is applied to the horn (which is grounded). Used in this manner, the horn relay becomes a control of the high current necessary to blow the horn. The control circuit may be wired with very thin wire because it will have low current flowing through it. The control unit may have only .25 ampere flowing through it, and the horn may require 24 or more amperes.

Solenoids

A **solenoid** is an electromagnetic device and operates in the same way as a relay; however, a solenoid uses a movable iron core. Solenoids can do mechanical work, such as switch electrical, vacuum, and liquid circuits. The iron core inside the coil of the solenoid is spring loaded. When

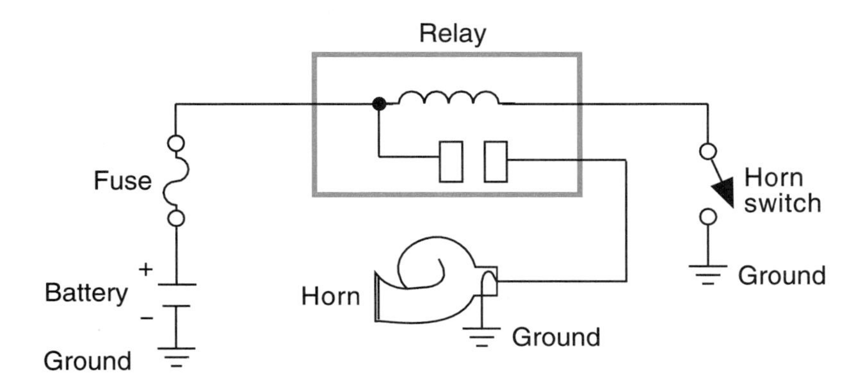

Figure 3-19 A relay can be used in a horn circuit to reduce the required size of the conductors installed in the steering column.

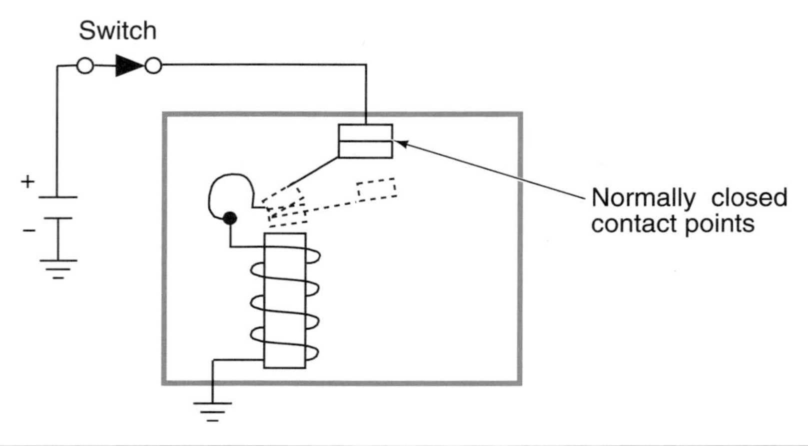

Figure 3-20 A buzzer reacts to the current flow to open and close rapidly, which creates a noise.

A **buzzer**, or **sound generator,** is sometimes used to warn the driver of possible safety hazards by emitting an audio signal (such as when the seat belt is not buckled).

Shop Manual
Chapter 3, page 61

A **stepped resistor** has two or more fixed resistor values. The stepped resistor can have an integral switch or have a switch wired in series.

A stepped resistor is also known as a tapped resistor.

Shop Manual
Chapter 3, page 61

A **variable resistor** provides for an infinite number of resistance values within a range.

A **rheostat** is a two-terminal variable resistor used to regulate the strength of an electrical current.

current flows through the coil, the magnetic field created around the coil attracts the core and moves it into the coil. To do work, the core is attached to a mechanical linkage, which causes something to move. When current flow through the coil stops, the spring pushes the core back to its original position. Some power door locks use solenoids to work the locking devices. Solenoids may also switch a circuit on or off, in addition to causing a mechanical action. Such is the case with some starter solenoids. These devices move the starter gear in and out of mesh with the flywheel. At the same time they complete the circuit from the battery to the ignition circuit. Both of these actions are necessary to start an engine.

Buzzers

A buzzer is similar in construction to a relay except for the internal wiring (Figure 3-20). The coil is supplied current through the normally closed contact points. When voltage is applied to the buzzer, current flows through the contact points to the coil. When the coil is energized, the contact arm is attracted to the magnetic field. As soon as the contact arm is pulled down, the current flow to the coil is opened, and the magnetic field is dissipated. The contact arm then closes again, and the circuit to the coil is closed. This opening and closing action occurs very rapidly. It is this movement that generates the vibrating signal.

Stepped Resistors

A stepped resistor is commonly used to control electrical motor speeds (Figure 3-21). By changing the position of the switch, resistance is increased or decreased within the circuit. If the current flows through a low resistance, then higher current flows to the motor and its speed is increased. If the switch is placed in the low speed position, additional resistance is added to the circuit. Less current flows to the motor, which causes it to operate at a reduced speed.

A stepped resistor is also used to convert digital to analog signals in a computer circuit. This is accomplished by converting the on/off digital signals into a continuously variable analog signal.

Variable Resistors

The most common types of variable resistors are rheostats and potentiometers. A rheostat has one terminal connected to the fixed end of a resistor and a second terminal connected to a moveable contact called a wiper (Figure 3-22). By changing the position of the wiper on the resistor, the amount of resistance can be increased or decreased. The most common use of the rheostat is in the instrument panel lighting switch. As the switch knob is turned, the instrument lights dim or brighten depending on the resistance value.

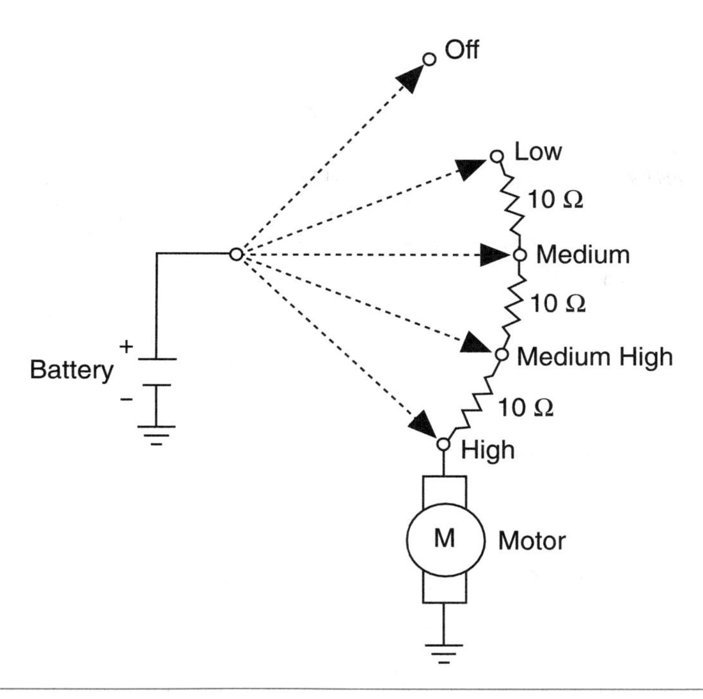

Figure 3-21 A stepped resistor is commonly used to control motor speeds. The total resistance is 30 ohms in the low position, 20 ohms in the medium position, 10 ohms in the medium-high postion, and 0 ohms in the high position.

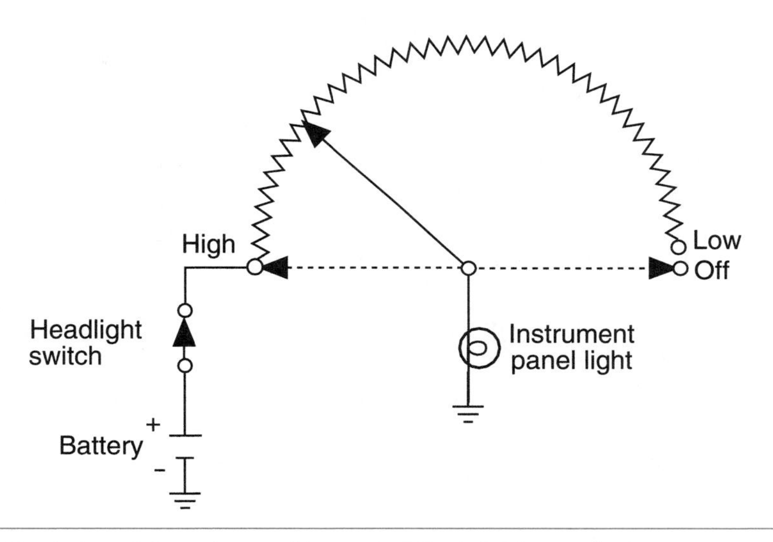

Figure 3-22 A rheostat can be used to control the brightness of a lamp.

A **potentiometer** is a three-wire variable resistor that acts as a voltage divider to produce a continuously variable output signal proportional to a mechanical position.

When a potentiometer is installed into a circuit, one terminal is connected to a power source at one end of the resistor. The second wire is connected to the opposite end of the resistor and is the ground return path. The third wire is connected to the wiper contact (Figure 3-23). The wiper senses a variable voltage drop as it is moved over the resistor. Because the current always flows through the same amount of resistance, the total voltage drop measured by the potentiometer is very stable. For this reason, the potentiometer is a common type of input sensor for the vehicle's onboard computers.

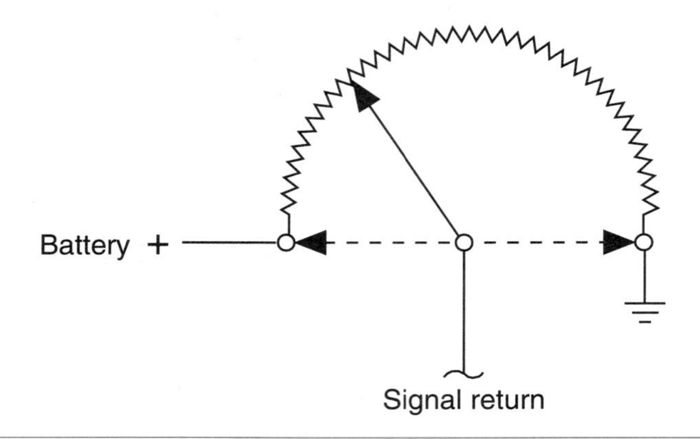

Figure 3-23 A potentiometer is used to send a voltage signal from the switch's wiper.

Shop Manual
Chapter 3, page 63

Electronic Components

Because a semiconductor material can operate as both a conductor and an insulator, it is very useful as a switching device. How a semiconductor material works depends on the way current flows, or tries to flow, through it.

Diodes

A **diode** is an electrical one-way check valve that will allow current to flow in one direction only.

A diode is the simplest semiconductor device. It is formed by joining P-type semiconductor material with N-type material. The N (negative) side of a diode is called the cathode and the P (positive) side, the anode (Figure 3-24). The point where the cathode and anode join together is called the PN junction. A diode allows current flow in one direction only. It can function as a switch, acting as an insulator or a conductor, depending on the direction of voltage bias.

The action of a diode depends on which side receives a positive voltage. This can get confusing but recall the difference between AC and DC. In AC, current flows in both directions and the voltage varies between positive and negative. Diodes are used to separate the positive and negative voltages in AC. An example of this, and something that is covered in greater detail in a later chapter, is the use of a diode in a generator. A generator produces AC current. Since the battery is a DC power source, DC is required to recharge it. The diodes of a generator change the AC current that is produced into DC so that the battery can be recharged.

In DC circuits, positive pressure or voltage always comes from the positive side of the battery. Therefore, a diode's action depends on whether the anode or the cathode is connected to the positive side of the battery. When positive voltage is present on the P-side or the anode, the

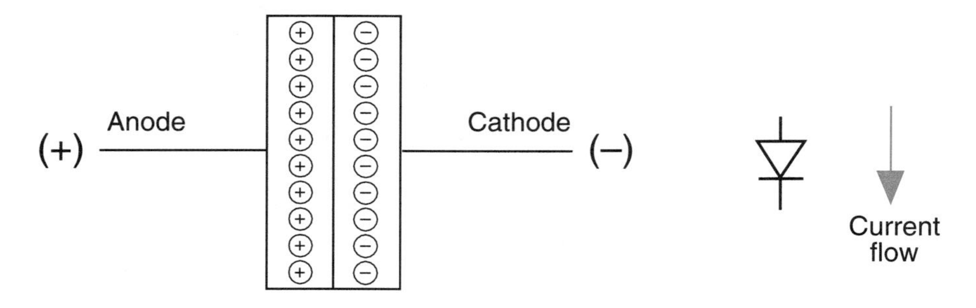

Figure 3-24 A diode and its symbol.

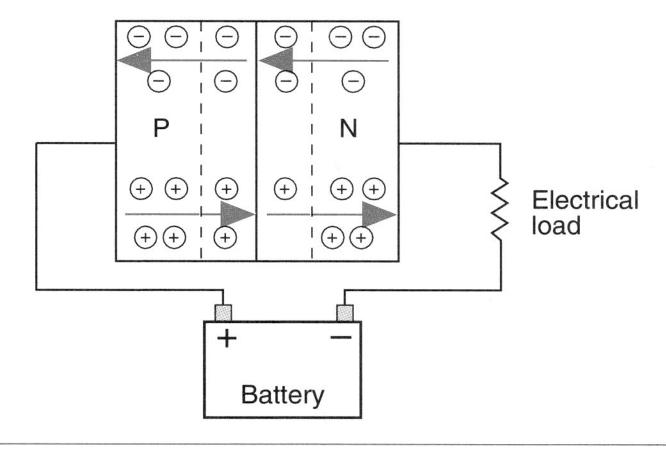

Figure 3-25 Forward-biased voltage causes current flow.

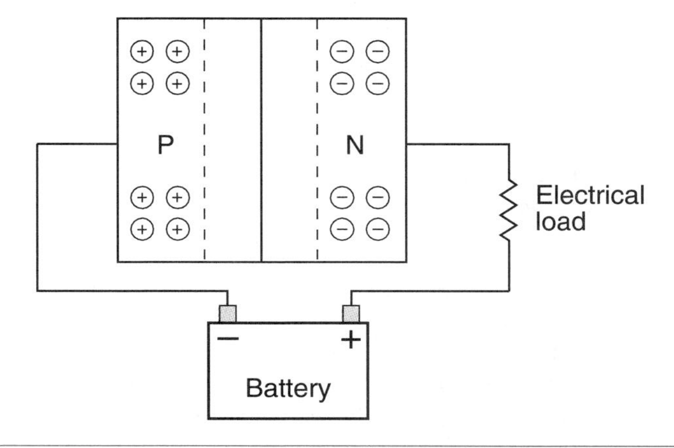

Figure 3-26 Reverse-biased voltage prevents current flow.

diode is forward biased and current will flow through it (Figure 3-25). When positive voltage is present at the cathode or N-side, the diode is reverse biased and current flow is prevented (Figure 3-26). When a diode is forward biased, it is a conductor. It is an insulator when it is reverse biased. Because a diode is a semiconductor there will always be a voltage drop across a diode.

Forward bias means that a positive voltage is applied to the P-type material and negative voltage to the N-type material.

Reversed bias means that positive voltage is applied to the N-type material and negative voltage is applied to the P-type material.

Zener Diodes

As stated, if a diode is reverse biased it will not conduct current. However, if the reverse voltage is increased, a voltage level will be reached at which the diode will conduct in the reverse direction. Reverse current can destroy a simple PN-type diode. But the diode can be doped with materials that will withstand reverse current.

A zener diode is designed to operate in reverse bias at the breakdown region (Figure 3-27). At the point that breakdown voltage is reached, a large current flows in reverse bias. This prevents the voltage from climbing any higher. This makes the zener diode an excellent component for regulating voltage. If the zener diode is rated at 15 volts, it will not conduct in reverse bias when the voltage is below 15 volts. At 15 volts it will conduct and the voltage will not increase over 15 volts.

The voltage reached when the diode conducts in reverse direction is called **zener voltage.**

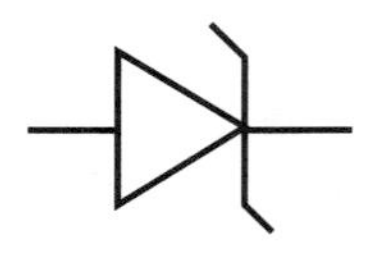

Figure 3-27 Symbol for a zener diode.

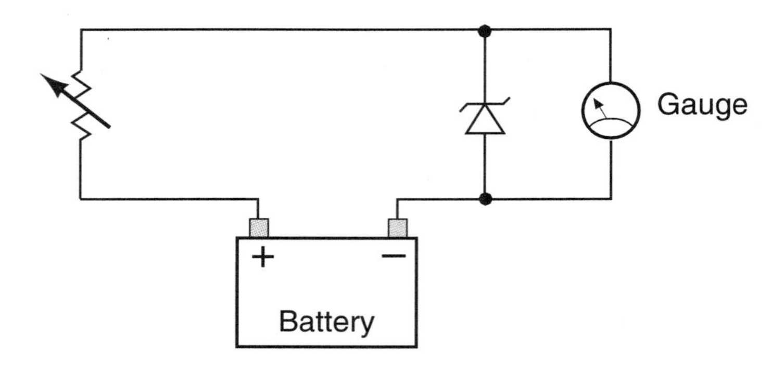

Figure 3-28 Simplified instrument gauge circuit that uses a zener diode to maintain a constant voltage to the gauge.

The illustration (Figure 3-28) shows a simplified circuit that has a zener diode in it to provide a constant voltage level to the instrument gauge. In this example, the zener diode is connected in series with the resistor and in parallel to the gauge. If the voltage to the gauge must be limited to 7 volts, the zener diode used would be rated at 7 volts. The zener diode maintains a constant voltage drop, and the total voltage drop in a series circuit must equal the amount of source voltage, thus voltage that is greater than the zener voltage must be dropped over the resistor. Even though source voltage may vary (as a normal result of the charging system), causing different currents to flow through the resistor and zener diode, the voltage dropped by the zener diode remains the same.

The zener breaks down when system voltage reaches 7 volts. At this point the zener diode conducts reverse current, causing an additional voltage drop across the resistor. The amount of voltage to the instrument gauge will remain at 7 volts because the zener diode "makes" the resistor drop the additional voltage to maintain this limit.

Light-Emitting Diodes

A **light-emitting diode** (**LED**) is similar in operation to the diode, except the LED emits light when it is forward biased.

A light-emitting diode (**LED**) has a small lens built into it so that light can be seen when current flows through it (Figure 3-29). When the LED is forward biased, there is current flow and light radiates from the junction of the diode. Normally an LED requires 1.5 to 2.2 volts to light. The light from an LED is not heat energy as is the case with other lights; it is electrical energy. Because of this, LEDs do not burn out after being used for a while.

Clamping Diodes

Whenever the current flow through a coil (such as used in a relay or solenoid) is discontinued, a voltage surge or spike is produced. This surge results from the collapsing of the magnetic field

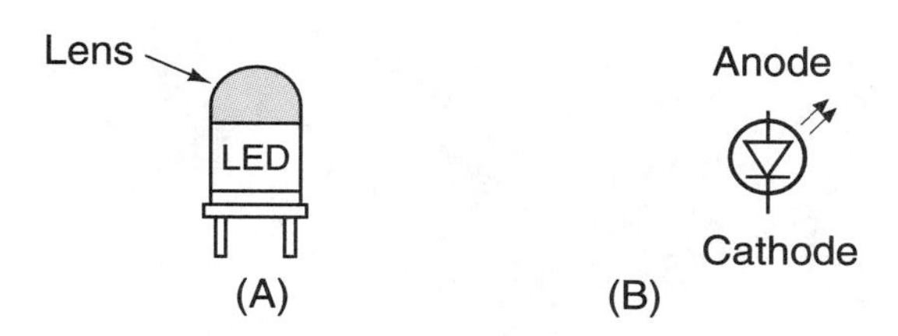

Figure 3-29 (A) A light-emitting diode uses a lens to emit the generated light; (B) the symbol for an LED.

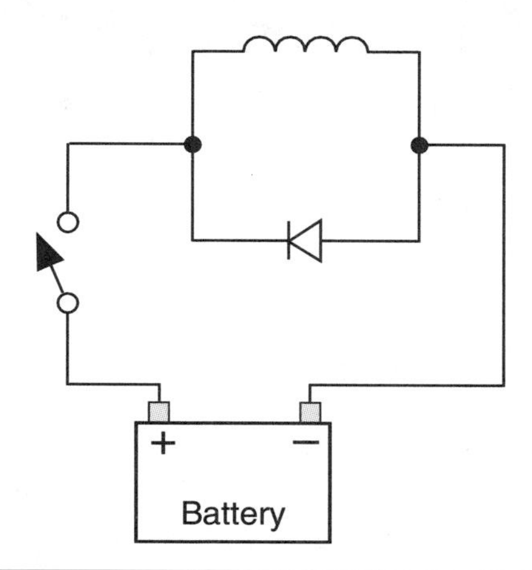

Figure 3-30 A clamping diode in parallel to a coil prevents voltage spikes when the switch is opened.

around the coil. The movement of the field across the windings induces a very high voltage spike, which can damage electronic components as it flows through the system. In some circuits a capacitor can be used as a shock absorber to prevent component damage from this surge. In today's complex electronic systems, a **clamping diode** is commonly used to prevent the voltage spike. By installing a clamping diode in parallel with the coil, a bypass is provided for the electrons during the time that the circuit is open (Figure 3-30).

An example of the use of clamping diodes is on some air conditioning compressor clutches (Figure 3-31). Because the clutch operates by electromagnetism, opening the clutch coil circuit produces a voltage spike. If this voltage spike was left unchecked, it could damage the vehicle's onboard computers. The installation of the clamping diode prevents the voltage spike from reaching the computers. The clamping diode must be connected to the circuit in reverse bias.

Figure 3-31 A clamping diode connected across the terminals of an air-conditioning compressor clutch.

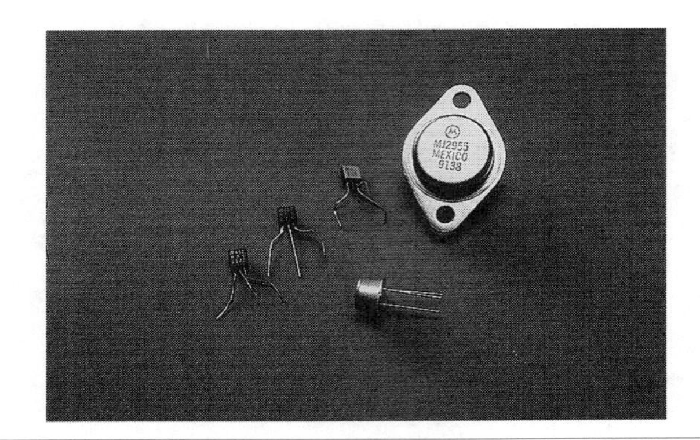

Figure 3-32 Transistors are used in automotive applications.

Transistors

A **transistor** is a three-layer semiconductor (Figure 3-32). It is used as a very fast switching device.

The word *transistor* is a combination of two words, *transfer* and *resist*. The transistor is used to control current flow in the circuit. It can be used to allow a predetermined amount of current flow or to resist this flow.

The most harmful things to a transistor are excess voltage and heat. If you have a problem with a module that stops operating after the engine is warm, test it by heating it slowly with a hair drier. Two principle uses for transistors are as switches and amplifiers. The average transistor should last 100 years unless it receives some abuse.

Transistors are made by combining P-type and N-type materials in groups of three. The two possible combinations are NPN (Figure 3-33) and PNP (Figure 3-34). In effect, a transistor is two diodes that share a common center layer. When the transistor is properly connected, the emitter-base junction will be forward biased and the collector-base junction will be reverse biased.

WARNING: The NPN-type transistor is the most commonly used transistor in automotive electronics. A PNP transistor cannot be replaced with an NPN transistor.

The three layers of the transistor are designated as emitter, collector, and base. The emitter is the outside layer of the forward-biased diode that has the same polarity as the circuit side to which it is applied. The arrow on the transistor symbol refers to the emitter lead and points in the direction of positive current flow and to the N material. The collector is the outside layer of the

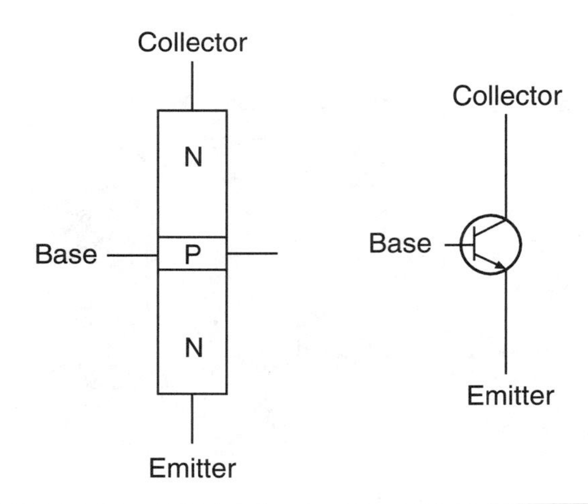

Figure 3-33 A NPN transistor and its symbol.

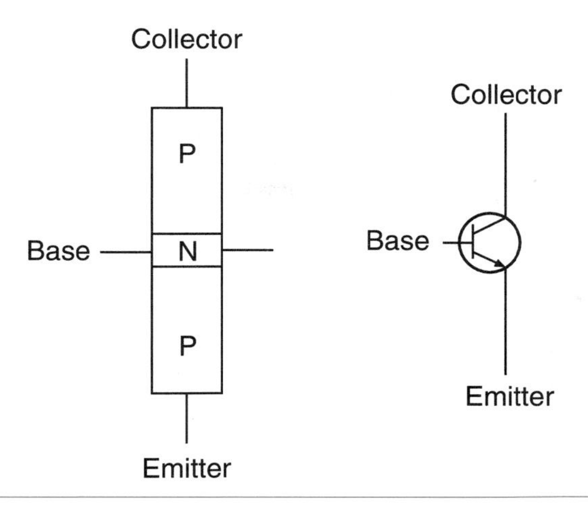

Figure 3-34 A PNP transistor and its symbol.

reverse-biased diode. The base is the shared middle layer. Each of these different layers has its own lead for connecting to different parts of the circuit.

In the NPN transistor, the emitter conducts current flow to the collector when the base is forward biased. The transistor cannot conduct unless the voltage applied to the base leg exceeds the emitter voltage by approximately 0.7 volt. This means both the base and collector must be positive with respect to the emitter. With less than 0.7 volt applied to the base leg (compared to the voltage at the emitter), the transistor acts as an opened switch. When the voltage difference is greater than 0.7 volt at the base, compared to the emitter voltage, the transistor acts as a closed switch (Figure 3-35).

When an NPN transistor is used in a circuit, it normally has a reverse bias applied to the base-collector junction. If the emitter-base junction is also reverse biased, no current will flow through the transistor (Figure 3-36). If the emitter-base junction is forward biased (Figure 3-37), current flows from the emitter to the base. Because the base is a thin layer and a positive voltage is applied to the collector, electrons flow from the emitter to the collector.

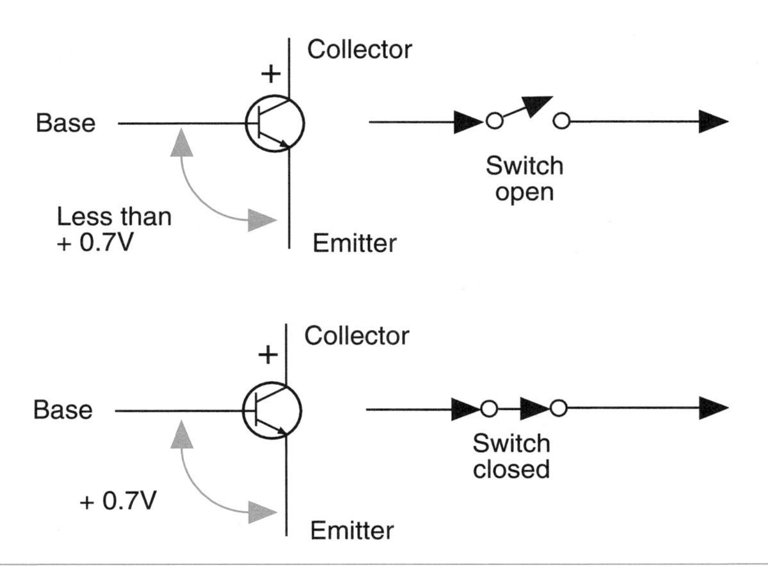

Figure 3-35 NPN transistor action.

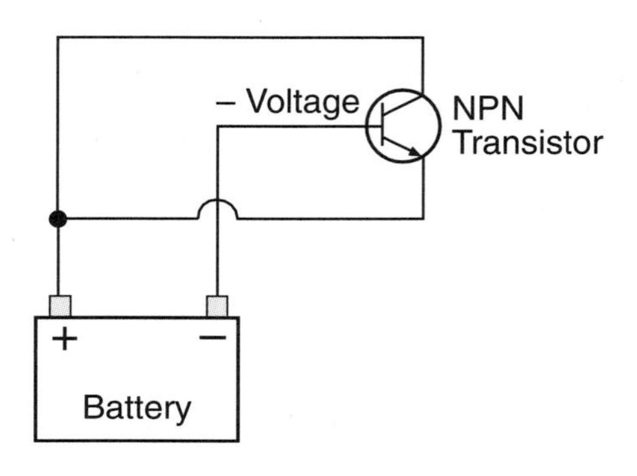

Figure 3-36 NPN transistor with reverse-biased voltage applied to its base. No current flow.

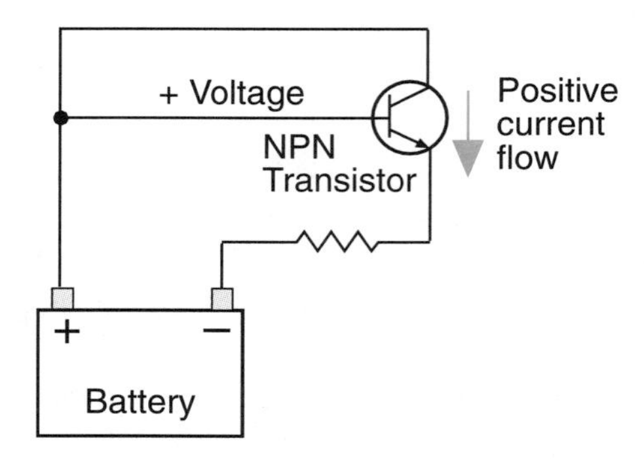

Figure 3-37 NPN transistor with forward-biased voltage applied to its base. Current flows.

Current flow through transistors is always based on hole and/or electron flow.

In the PNP transistor, current will flow from the emitter to the collector when the base leg is forward biased with a voltage that is more negative than that at the emitter (Figure 3-38). For current to flow through the emitter to the collector, both the base and the collector must be negative in respect to the emitter. Current flow through an NPN is in the opposite direction as a PNP.

Current can be controlled through a transistor. Thus transistors can be used as a very fast electrical switch. It is also possible to control the amount of current flow through the collector. This is because the output current is proportional to the amount of current through the base leg.

A transistor has three operating conditions:

1. **Cutoff:** When reverse-biased voltage is applied to the base leg of the transistor. In this condition the transistor is not conducting and no current will flow.
2. **Conduction:** Bias voltage difference between the base and the emitter has increased to the point that the transistor is switched on. In this condition the transistor is conducting. Output current is proportional to that of the current through the base.
3. **Saturation:** This occurs when the collector to emitter voltage is reduced to near zero by a voltage drop across the collector's resistor.

FET-type transistors are common in the onboard computer systems of today's vehicles.

These types of transistors are called bipolar because they have three layers of silicon; two of these layers are the same. Another type of transistor is the field-effect transistor (FET). The FET's leads are listed as source, drain, and gate. The source supplies the electrons and is similar

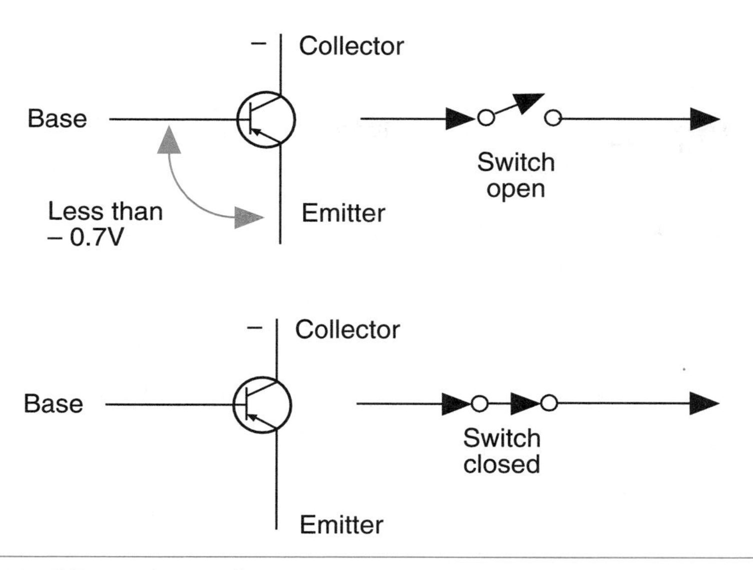

Figure 3-38 PNP transistor action.

to the emitter in the bipolar transistor. The drain collects the current and is similar to the collector. The gate creates the electro-static field that allows electron flow from the source to the drain. It is similar to the base.

The FET transistor needs a voltage applied to the gate terminal to get electron flow from the source to the drain. The source and drain are constructed of the same type of doped material. They can be either N-type or P-type materials. The source and drain are separated by a thin layer of either N-type or P-type material the opposite of the gate and drain.

Using the illustration (Figure 3-39), if the source voltage is held at 0 volts and 6 volts are applied to the drain, no current will flow between the two. However, if a lower positive voltage is applied to the gate, the gate forms a capacitive field between the channel and itself. The voltage of the capacitive field attracts electrons from the source and current will flow through the channel to the higher positive voltage of the drain.

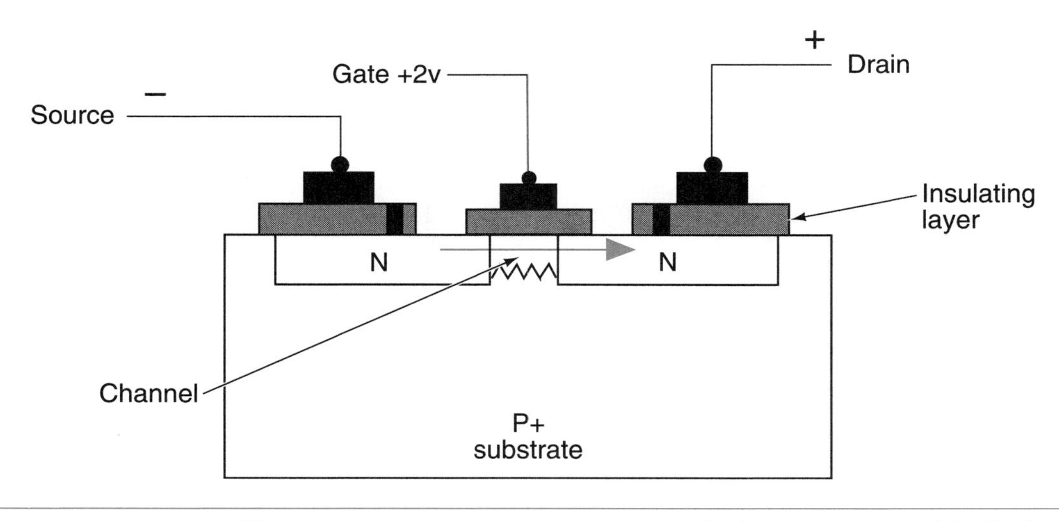

Figure 3-39 An FET uses a positive voltage to the gate terminal to create a capacitive field to allow electron flow.

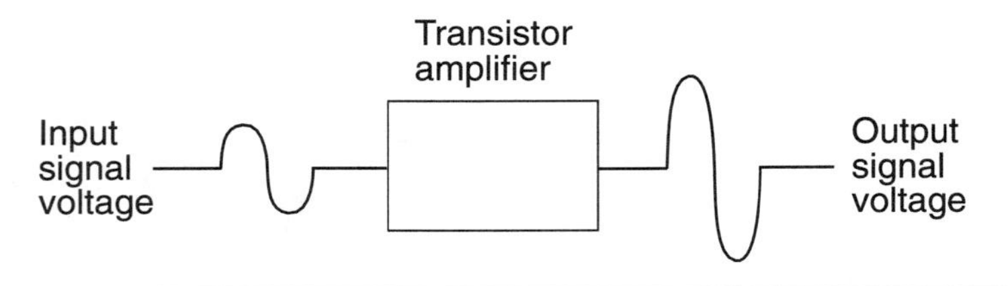

Figure 3-40 A simplified amplifier circuit.

While electrons are flowing from the source to the drain (electron theory), positive charges are flowing from the drain to the source (conventional theory).

This type of FET is called an enhancement-type FET because the field effect improves current flow from the source to the drain. This operation is similar to that of a normally open switch. A depletion-type FET is like a normally closed switch, whereas the field effect cuts off current flow from the source to the drain.

WARNING: Some forms of FETs have a very thin insulation layer between the gate and channel that static electricity from your hands is able to burn through. Be careful not to touch the connector pins of the computer or the integrated circuits inside of the computer. Always use a static strap.

A static strap is a wrist strap and wire with an alligator clip to allow technicians to ground themselves.

Transistor Amplifiers

A transistor can be used in an amplifier circuit to amplify the voltage. This is useful when using a very small voltage for sensing computer inputs, but needing to boost that voltage to operate an accessory (Figure 3-40). The waveform showing the small signal voltage that is applied to the base leg of a transistor may look like that shown (Figure 3-41A). The waveform showing the corresponding signal through the collector will be inverted (Figure 3-41B). Three things happen in an amplified circuit:

1. The amplified voltage at the collector is greater than that of the base voltage.
2. The input current increases.
3. The pattern has been inverted.

The first transistor in a Darlington pair is used as a preamplifier to produce a large current to operate the second transistor. The second transistor is isolated from the control circuit and is the final amplifier. The second transistor boosts the current to the amount required to operate the load component. The Darlington pair is utilized by most control modules used in electronic ignition systems.

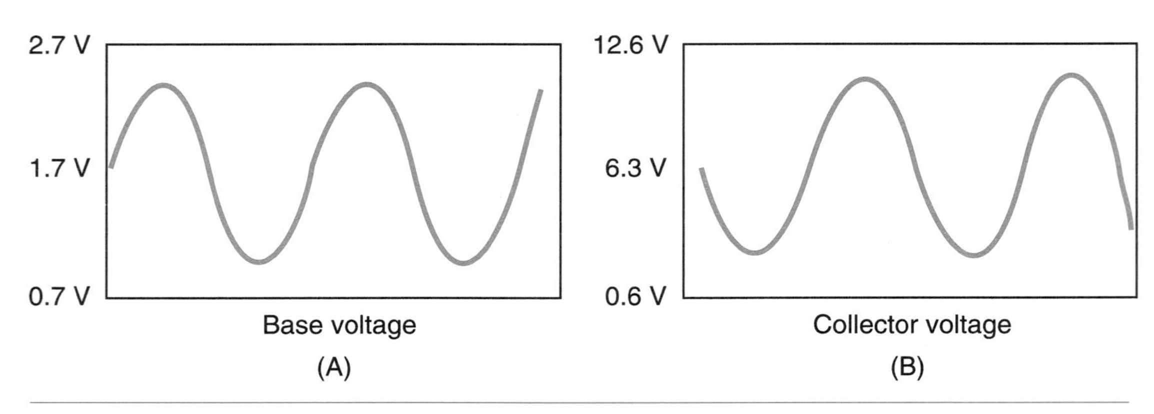

Figure 3-41 (A) Base voltage, (B) Collector voltage.

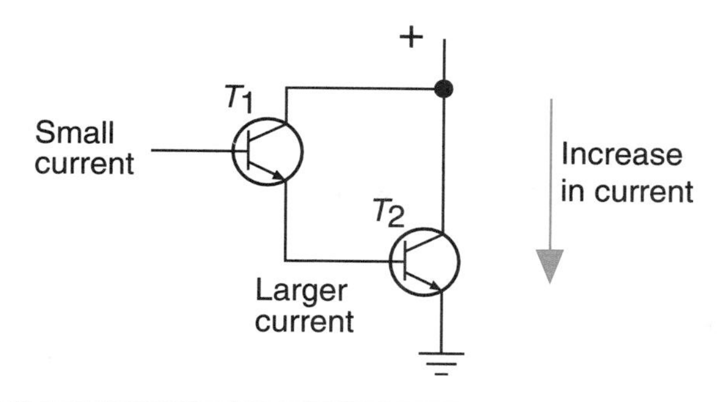

Figure 3-42 A Darlington pair used to amplify current. T_1 acts as a preamplifier that creates a larger base current for T_2, which is the final amplifier that allows a larger current.

Some amplifier circuits use a **Darlington pair,** which is two transistors that are connected together (Figure 3-42).

Figure 3-43 Phototransistor.

Phototransistors

In a phototransistor, a small lens is used to focus incoming light onto the sensitive portion of the transistor. When light strikes the transistor, holes and free electrons are formed. These increase current flow through the transistor according to the amount of light. The stronger the light intensity, the more current that will flow. This type of phototransistor is often used in automatic headlight dimming circuits.

A **phototransistor** is a transistor that is sensitive to light (Figure 3-43).

Thyristors

The most common type of thyristor used in automotive applications is the silicon-controlled rectifier (SCR). Like the transistor, the SCR has three legs. However, it consists of four regions arranged PNPN (Figure 3-44). The three legs of the SCR are called the anode (or P-terminal), the cathode (or N-terminal), and the gate (one of the center regions).

The SCR requires only a trigger pulse (not a continuous current) applied to the gate to become conductive. Current will continue to flow through the anode and cathode as long as the voltage remains high enough.

The SCR can be connected into a circuit either in the forward or reverse direction. Using Figure 3-44 of a forward direction connection, the P-type anode is connected to the positive side of the circuit and the N-type cathode is connected to the negative side. The center PN junction blocks current flow through the anode and cathode.

A **thyristor** is a semiconductor switching device composed of alternating N and P layers. It can also be used to rectify current from AC to DC.

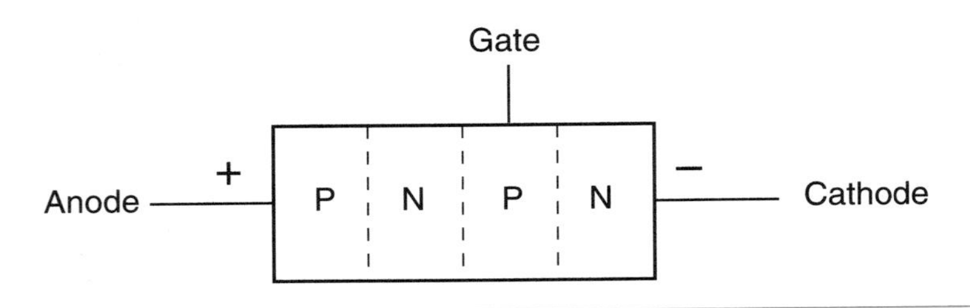

Figure 3-44 A forward direction SCR.

Once a positive voltage pulse is applied to the gate, the SCR turns on. Even if the positive voltage pulse is removed, the SCR will continue to conduct.

The SCR will also block any reverse current from flowing from the cathode to the anode. Because current can flow only in one direction through the SCR, it can rectify AC current to DC current.

Integrated Circuits

Electrical components that are made separately and have wire leads for connections to the circuit are called **discrete devices.**

An integrated circuit is a complex circuit of many transistors, diodes, resistors, capacitors, and other electronic devices that are formed onto a tiny silicon chip (Figure 3-45). As many as 30,000 transistors can be placed on a chip that is 1/4 inch (6.35 mm) square.

Integrated circuits are constructed by photographically reproducing circuit patterns onto a silicon wafer. The process begins with a large-scale drawing of the circuit. This drawing can be room size. Photographs of the circuit drawing are reduced until they are the actual size of the circuit. The reduced photographs are used as a mask. Conductive P-type and N-type materials,

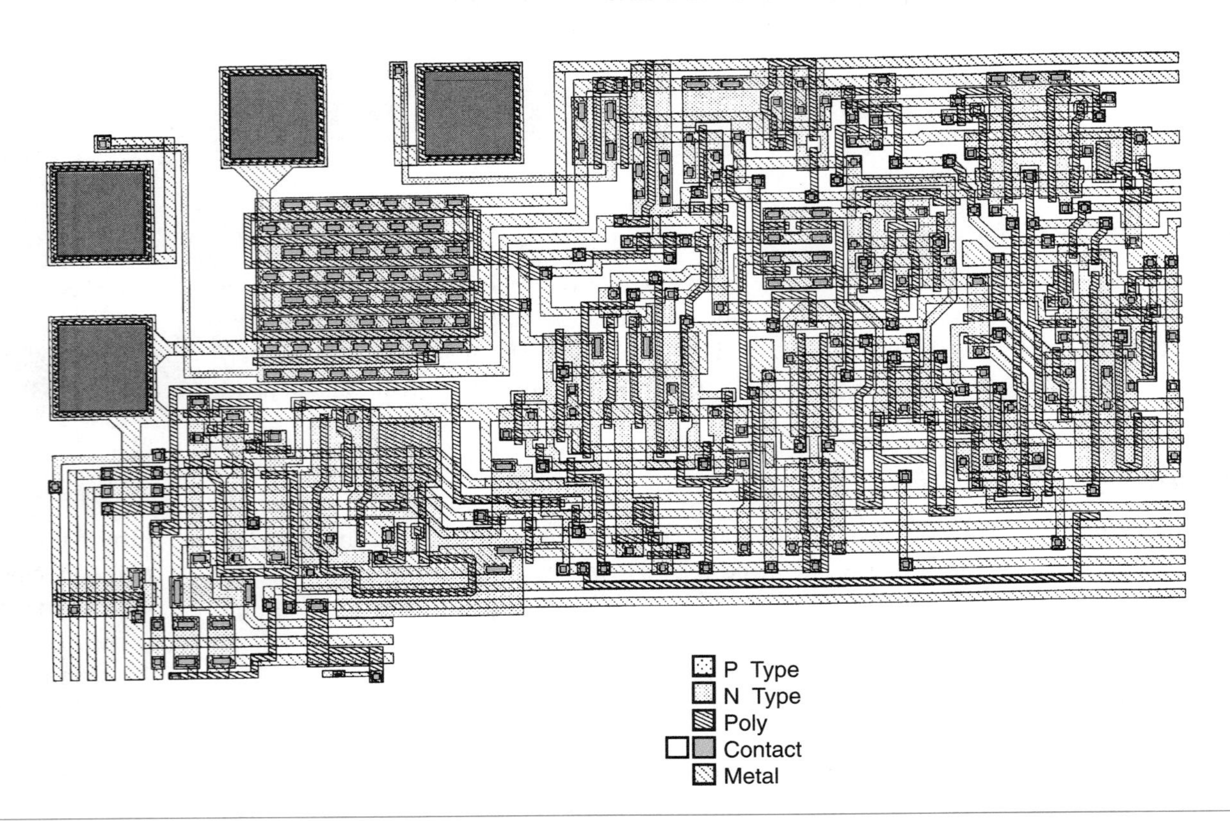

Figure 3-45 An enlarged illustration of an integrated circuit with thousands of transistors, diodes, resistors, and capacitors. The actual size can be less than 1/4 inch square.

along with insulating materials, are deposited onto the silicon wafer. The mask is placed over the wafer and selectively exposes the portion of material to be etched away or the portions requiring selective deposition. The entire process of creating an integrated circuit chip takes over 100 separate steps. Out of a single wafer 4 inches in diameter, thousands of integrated circuits can be produced.

The small size of the integrated chip has made it possible for the vehicle manufacturers to add several computer-controlled systems to the vehicle without taking up much space. Also a single computer is capable of performing several functions.

The integrated circuit is also called an **IC chip**.

WARNING: Integrated circuits can be damaged by static electricity. Use caution when working with these circuits. There are antistatic straps available for the technician to wear to reduce the possibility of destroying the integrated circuit.

WARNING: Do not connect or disconnect an IC to the circuit with the power on. The arc produced may damage the chip.

WARNING: Do not test with an ohmmeter.

Circuit Defects

All electrical problems can be classified as being one of three types of problems: an open, short, or high resistance. Each one of these will cause a component to operate incorrectly or not at all. Understanding what each of these problems will do to a circuit is the key to proper diagnosis of any electrical problem.

Open

An open is simply a break in the circuit (Figure 3-46). An open is caused by turning a switch off, a break in a wire, a burned out light bulb, a disconnected wire or connector, or anything that opens the circuit. When a circuit is open, current does not flow and the component doesn't work. Because there is no current flow, there are no voltage drops in the circuit. Source voltage is available everywhere in the circuit up to the point at which it is open. Source voltage is even available after a load, if the open is after that point.

Shop Manual
Chapter 3, page 64

An **open circuit** is a circuit in which there is a break in continuity.

Opens caused by a blown fuse will still cause the circuit not to operate, but the cause of the problem is the excessive current that blew the fuse. Nearly all other opens are caused by a break in the continuity of the circuit. These breaks can occur anywhere in the circuit.

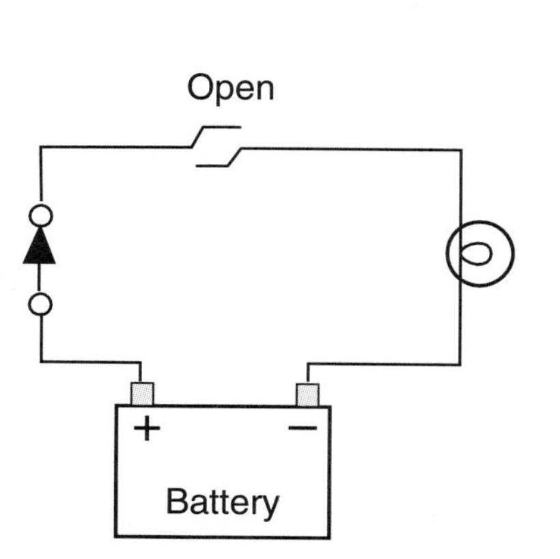

Figure 3-46 An open circuit stops all current flow.

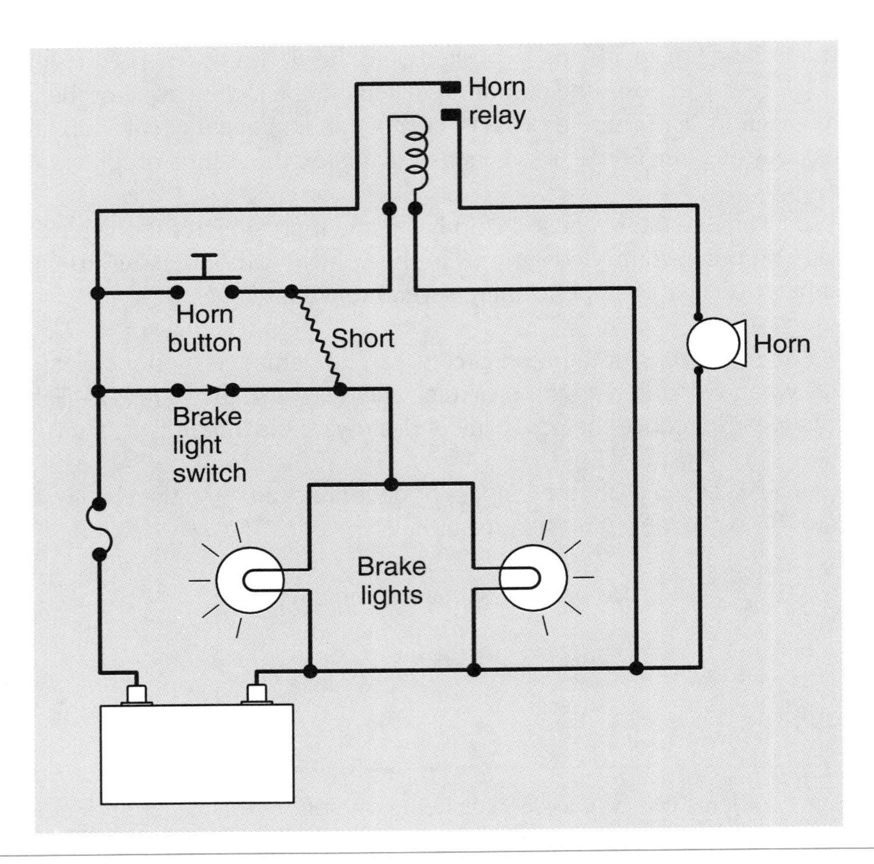

Figure 3-47 A wire-to-wire short.

Shop Manual
Chapter 3, page 65

A **shorted circuit** is a circuit that allows current to bypass part of the normal path.

Shorts

A short results from an unwanted path for current. Shorts cause an increase in current flow. This increased current flow can burn wires or components. Sometimes two circuits become shorted together. When this happens, one circuit powers another. This may result in strange happenings, such as the horn sounding every time the brake pedal is depressed. In this case, the brake light circuit is shorted to the horn circuit (Figure 3-47). Improper wiring or damaged insulation are the two major causes of short circuits.

A short can also be an unwanted short to ground (Figure 3-48). This problem provides a low resistance path for current to travel. Let's look at Figure 3-49 to see what happens with a short. Using the previous circuit, we'll give some resistance values to the two light bulbs. One has 3 ohms of resistance and the other has 6. Since these two are in parallel, the total resistance is 2 ohms. Our short made a path from the power side of one bulb to the return path to our battery. The short was caused by a screw that went through the dash into the wire for the bulb. This created a low resistance path. So, if we give it a low resistance value of 0.001 ohms, we can calculate what would happen to the current in this circuit. The short becomes another leg in the parallel circuit. Since the total resistance of a parallel circuit is always lower than the lowest resistance, we know the total resistance of the circuit is now less than 0.001 ohms. Using Ohm's law we can calculate the current flow through the circuit.

$$I = E/R \quad \text{or} \quad I = 12/.001 \quad \text{or} \quad I = 12{,}000 \text{ Amps}$$

Needless to say, it would take a large wire to carry that kind of amperage. Our 10-amp fuse would melt quickly when the short occurred. This would protect our wires and light bulbs. Some call this type of problem a grounded circuit.

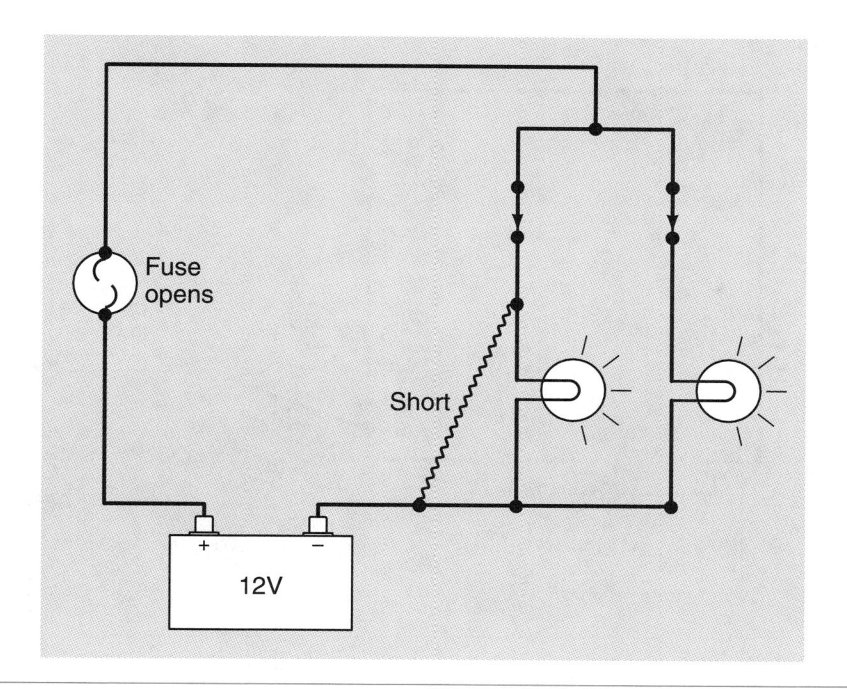

Figure 3-48 A short to ground.

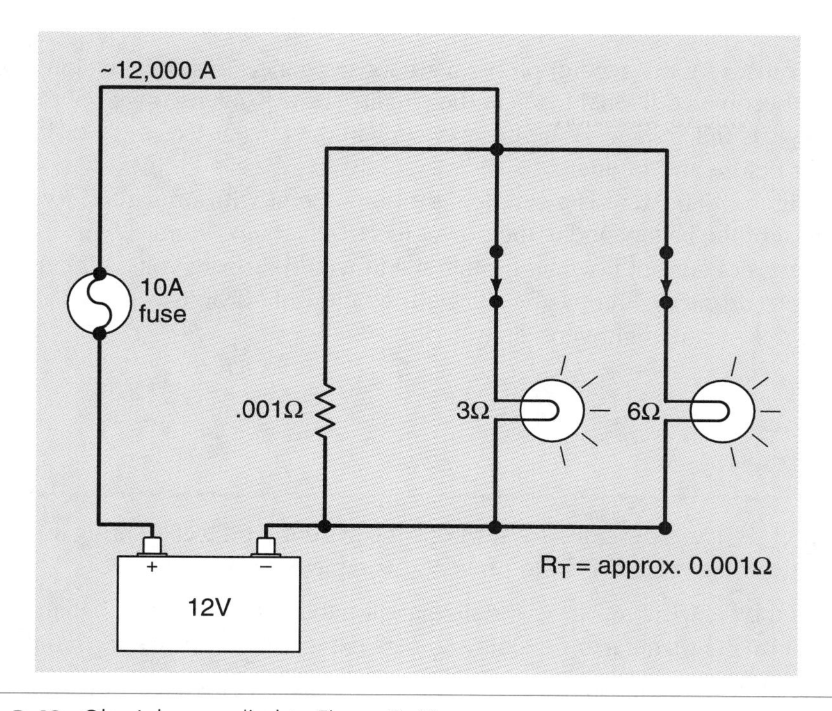

Figure 3-49 Ohm's law applied to Figure 3-48.

High Resistance

High resistance problems occur when there is unwanted resistance in the circuit. The higher than normal resistance causes the current flow to be lower than normal, and the components in the circuit are unable to operate properly. Components in high resistant circuits work weaker than they should. Dim light bulbs are an excellent example of this type of problem.

Shop Manual
Chapter 3, page 68

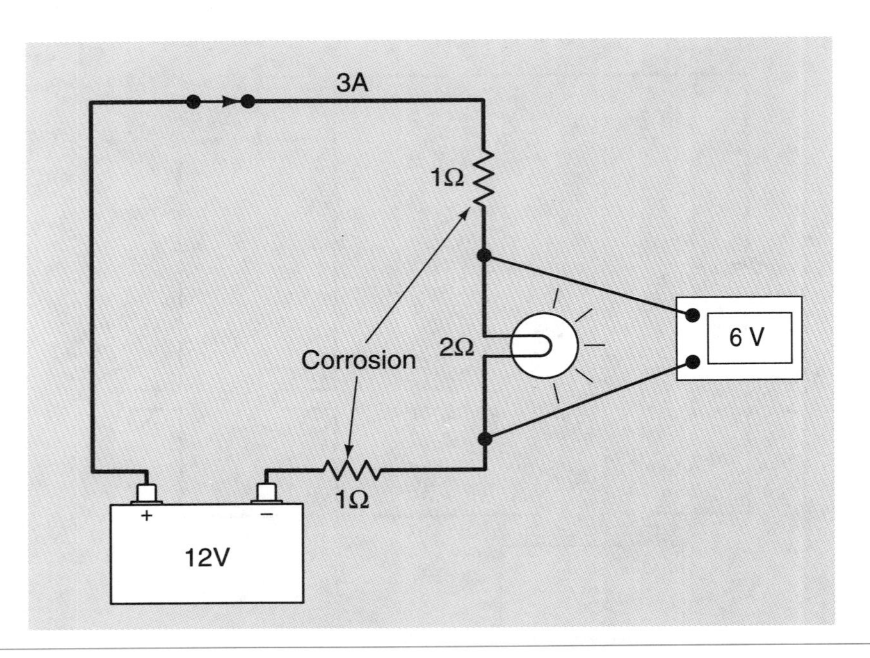

Figure 3-50 A simple light circuit with unwanted resistances.

Common causes for this type of problem are loose connections or corrosion at a connector. The resistances become additional loads in the circuit. These loads use some of the circuit's voltage, which prevents full voltage to the normal loads in the circuit. Excessive or unwanted resistance can occur before and/or after a load.

Look at Figure 3-50. Here is a simple light bulb circuit with unwanted resistances at the negative terminal of the battery and at the power feed to the bulb. Normally this 2-ohm light bulb would have 6 amps of current flowing through it and would drop 12 volts. With the added resistance, the current drops to 3 amps and the bulb would only drop 6 volts. As a result of these resistances, the bulb would light very dimly.

Summary

Terms to Know

Bimetallic
Bus bar
Clamping diode
Diode
Forward bias
Fuse
Fuse box
Grounded circuit
Light-emitting diode (LED)
Normally closed (NC)

- ❑ The protection device is designed to "turn off" the system it protects. This is done by creating an open (like turning off a switch) to prevent a complete circuit.
- ❑ Fuses are rated by amperage. Never install a larger rated fuse into a circuit than the one that was designed by the manufacturer. Doing so may damage or destroy the circuit.
- ❑ A switch can control the on/off operation of a circuit or direct the flow of current through various circuits.
- ❑ A normally open switch will not allow current flow when it is in its rest position. A normally closed switch will allow current flow when it is in its rest position.
- ❑ A relay is a device that uses low current to control a high current circuit.
- ❑ A buzzer is sometimes used to warn the driver of possible safety hazards by emitting an audio signal (such as when the seat belt is not buckled).
- ❑ A stepped resistor has two or more fixed resistor values. It is commonly used to control electrical motor speeds.

- ❑ A variable resistor provides for an infinite number of resistance values within a range. A rheostat is a two-terminal variable resistor used to regulate the strength of an electrical current. A potentiometer is a three-wire variable resistor that acts as a voltage divider to produce a continuously variable output signal proportional to a mechanical position.
- ❑ A diode is an electrical one-way check valve that will allow current to flow in one direction only.
- ❑ Forward bias means that a positive voltage is applied to the P-type material and negative voltage to the N-type material. Reversed bias means that positive voltage is applied to the N-type material and negative voltage is applied to the P-type material.
- ❑ A transistor is a three-layer semiconductor that is commonly used as a very fast switching device.
- ❑ An integrated circuit is a complex circuit of thousands of transistors, diodes, resistors, capacitors, and other electronic devices that are formed onto a tiny silicon chip.
- ❑ An open circuit is a circuit in which there is a break in continuity.
- ❑ A shorted circuit is a circuit that allows current to bypass part of the normal path.
- ❑ A short to ground is a condition that allows current to return to ground before it has reached the intended load component.

Terms to Know (continued)

Normally open (NO)
Open circuit
Overload
Phototransistor
Potentiometer
Relay
Reverse bias
Rheostat
Saturation
Shorted circuit
Stepped resistor
Thyristor
Transistor
Variable resistor
Zener diode

Review Questions

Short Answer Essays

1. Describe the use of three types of semiconductors.
2. What types of variable resistors are used on automobiles?
3. Define what is meant by opens, shorts, grounds, and excessive resistance.
4. Explain the effects that each type of circuit defect will have on the operation of the electrical system.
5. Explain the purpose of a circuit protection device.
6. Describe the most common types of circuit protection devices.
7. Describe the common types of electrical system components used and how they affect the electrical system.
8. Describe the difference between a rheostat and a potentiometer.
9. Explain the difference between normally open (NO) and normally closed (NC) switches.
10. Explain the differences between forward biasing and reverse biasing a diode.

Fill-in-the-Blanks

1. Never install a larger rated ____________________ into a circuit than the one that was designed by the manufacturer.

2. A ____________________ can control the on/off operation of a circuit or direct the flow of current through various circuits.

3. A normally ____________________ switch will not allow current flow when it is in its rest position. A normally ____________________ switch will allow current flow when it is in its rest position.

4. An ____________________ ____________________ is a complex circuit of many transistors, diodes, resistors, capacitors, and other electronic devices that are formed onto a tiny silicon chip.

5. When a ____________________ voltage is applied to the P-material of a diode and ____________________ voltage is applied to the N-material, the diode is reverse biased. When a ____________________ voltage is applied to the N-material of a diode and ____________________ voltage is applied to the P-material, the diode is forward biased.

6. A ____________________ is used in electronic circuits as a very fast switching device.

7. A ____________________ is an electrical one-way check valve that will allow current to flow in one direction only.

8. A ____________________ is a device that uses low current to control a high current circuit.

9. A ____________________ ____________________ provides for an infinite number of resistance values within a range. A ____________________ is a two-terminal variable resistor used to regulate the strength of an electrical current.

10. The ____________________ requires only a trigger pulse applied to the gate to become conductive.

ASE Style Review Questions

1. Electrical shorts are being discussed:
   Technician A says a short is an electrical problem that adds a parallel leg to the circuit, which lowers the entire circuit's resistance.
   Technician B says a short causes unwanted voltage drops.
   Who is correct?
   A. A only
   B. B only
   C. Both A and B
   D. Neither A nor B

2. While discussing protection devices:
   Technician A says a fuse automatically resets after the cause of the overload is repaired.
   Technician B says the protection device creates an open when an overload occurs.
   Who is correct?
   A. A only
   B. B only
   C. Both A and B
   D. Neither A nor B

3. While discussing circuit components:
   Technician A says a switch can control the on/off operation of a circuit or direct the flow of current through various circuits.
   Technician B says a relay can be a SPDT-type switch.
   Who is correct?
   A. A only
   B. B only
   C. Both A and B
   D. Neither A nor B

4. While discussing diodes:
   Technician A says the zener diode is an excellent component for regulating voltage.
   Technician B says installing a clamping diode provides a bypass for the electrons when the circuit is opened suddenly.
   Who is correct?
   A. A only
   B. B only
   C. Both A and B
   D. Neither A nor B

5. While discussing circuit protection devices:
   Technician A says blade-type fuses are the most commonly used fuse on today's vehicles.
   Technician B says some manufacturers use a type of diode as a circuit protection device.
   Who is correct?
   A. A only
   B. B only
   C. Both A and B
   D. Neither A nor B

6. While discussing circuit defects:
   Technician A says an open means there is continuity in the circuit.
   Technician B says a short may increase circuit current.
   Who is correct?
   A. A only
   B. B only
   C. Both A and B
   D. Neither A nor B

7. While discussing the use of transistors:
   Technician A says a transistor can be used to control the switching on/off of a circuit.
   Technician B says a transistor can be used to amplify voltage.
   Who is correct?
   A. A only
   B. B only
   C. Both A and B
   D. Neither A nor B

8. Circuit defects are being discussed:
   Technician A says extra resistance can cause a lamp in a parallel circuit to burn brighter than normal.
   Technician B says unwanted voltage drops may appear on either the insulated or grounded return side of a circuit.
   Who is correct?
   A. A only
   B. B only
   C. Both A and B
   D. Neither A nor B

9. Diodes are being discussed:
   Technician A says diodes are aligned to allow current flow in one direction only.
   Technician B says diodes are used in generators to change DC voltages into AC voltages.
   Who is correct?
   A. A only
   B. B only
   C. Both A and B
   D. Neither A nor B

10. Circuit protection devices are being discussed:
    Technician A says a "blown" fuse is identified by bubbling of the insulator material around the link.
    Technician B says a "blown" fusible link is identified by a burned-through metal wire in the capsule.
    Who is correct?
    A. A only
    B. B only
    C. Both A and B
    D. Neither A nor B

Wiring and Circuit Diagrams

Upon completion and review of this chapter, you should be able to:

- ❑ Explain when single-stranded or multistranded wire should be used.
- ❑ Explain the use of resistive wires in a circuit.
- ❑ Describe the construction of spark plug wires.
- ❑ Explain how wire size is determined by the American Wire Gauge (AWG) and metric methods.
- ❑ Describe how to determine the correct wire gauge to be used in a circuit.
- ❑ Explain how temperature affects resistance and wire size selection.
- ❑ Explain the purpose and use of printed circuits.
- ❑ Explain why wiring harnesses are used and how they are constructed.
- ❑ Explain the purpose of wiring diagrams.
- ❑ Identify the common electrical symbols that are used.
- ❑ Explain the purpose of the component locator.

Introduction

Today's vehicles are equipped with a maze of wires. These wires come in different gauge sizes and lengths. Most wires have color traces for easy identification. If all of the wires contained within one modern vehicle were connected in one length, you would have about half a mile or more of wire. Trying to locate the cause of an electrical problem can be quite difficult if you do not have a good understanding of wiring systems and diagrams.

In this chapter you will learn how wiring harnesses are made, how to read the wiring diagram, how to interpret the symbols used, and how terminals are used. This will reduce the amount of confusion you may experience when repairing an electrical circuit. It is also important to understand how to determine the correct type and size of wire to carry the anticipated amount of current. It is possible to cause an electrical problem by simply using the wrong gauge size of wire. A technician must understand the three factors that cause resistance in a wire—length, diameter, and temperature—to perform repairs correctly.

Automotive Wiring

Primary wiring is the term used for conductors that carry low voltage. The insulation of primary wires is usually thin. Secondary wiring refers to wires used to carry high voltage, such as ignition spark plug wires. Secondary wires have extra thick insulation.

Most of the primary wiring conductors used in the automobile are made of several strands of copper wire wound together and covered with a polyvinyl chloride (PVC) insulation (Figure 4-1). Copper has low resistance and can be connected to easily by using crimping connectors or soldered connections. Other types of conductor materials used in automobiles include silver, gold, aluminum, and tin-plated brass.

Copper is used mainly because of its low cost and availability.

Aluminum was used in some GM rear wiring harnesses.

▲ **WARNING:** A solid copper wire may be used in low-voltage, low-current circuits where flexibility is not required. Do not use solid wire where high voltage, high current, or flexibility is required, unless solid wire was used by the manufacturer.

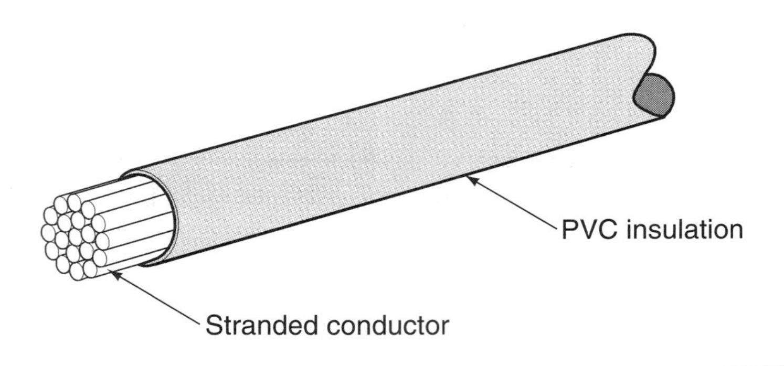

Figure 4-1 Stranded primary wiring.

General Motors has used single-stranded aluminum wire in limited applications where no flexing of the wire is expected. For example, it is used in the taillight circuits.

Resistance wire is designed with a certain amount of resistance per foot.

The resistance value of the ballast resistor is usually between 0.8 and 1.2 ohms.

Spark plug wires are often referred to as **television-radio-suppression (TVRS) cables.**

Stranded wire is used because it is very flexible, and current flows on the surface of the conductors. Because there is more surface area exposed in a stranded wire (each strand has its own surface), there is less resistance in the stranded wire than in the solid wire (Figure 4-2). The PVC insulation is used because it can withstand temperature extremes and corrosion. PVC insulation is also capable of withstanding battery acid, antifreeze, and gasoline. The insulation protects the wire from shorting to ground and from corrosion.

A ballast resistor is used to protect the ignition primary circuit from excessive voltage. It reduces the current flow through the coil's primary windings and provides a stable voltage to the coil.

Some automobiles use a resistance wire in the ignition system instead of a ballast resistor. It is most commonly used to limit the amount of current flow through the primary windings of the ignition coil. This wire is called the ballast resistor wire and is located between the ignition switch and the ignition coil (Figure 4-3) in the ignition "RUN" circuit.

Spark plug wires are also resistance wires. The resistance lowers the current flow through the wires. By keeping current flow low, the magnetic field created around the wires is kept to a minimum. The magnetic field needs to be controlled because it causes radio interference. The result of this interference is noise on the vehicle's radio and all nearby radios and televisions. The noise can interfere with emergency broadcasts and the radios of emergency vehicles. Because of this concern, all ignition systems are designed to minimize radio interference; most do so with resistance-type spark plug wires. Spark plug wires are targeted because they carry high voltage pulses. The lower current flow has no adverse effect on the firing of the spark plug.

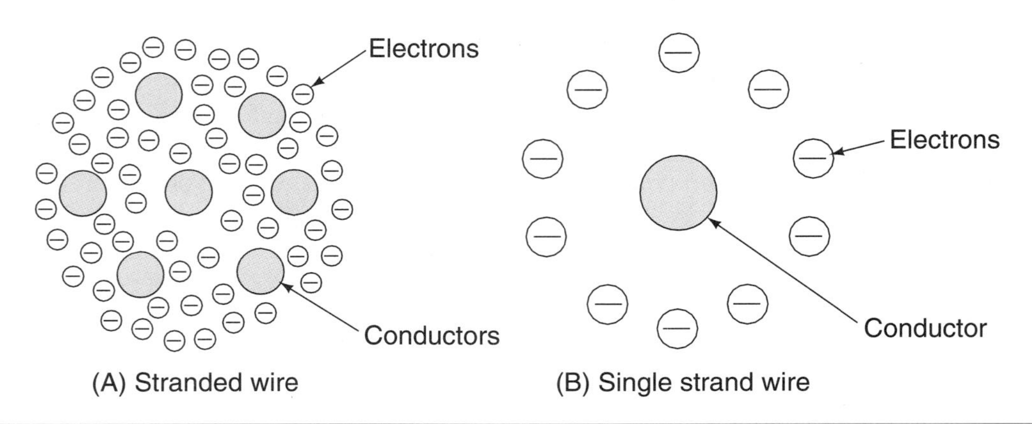

Figure 4-2 Stranded wire provides flexibility and more surface area for electron flow than a single-strand solid wire.

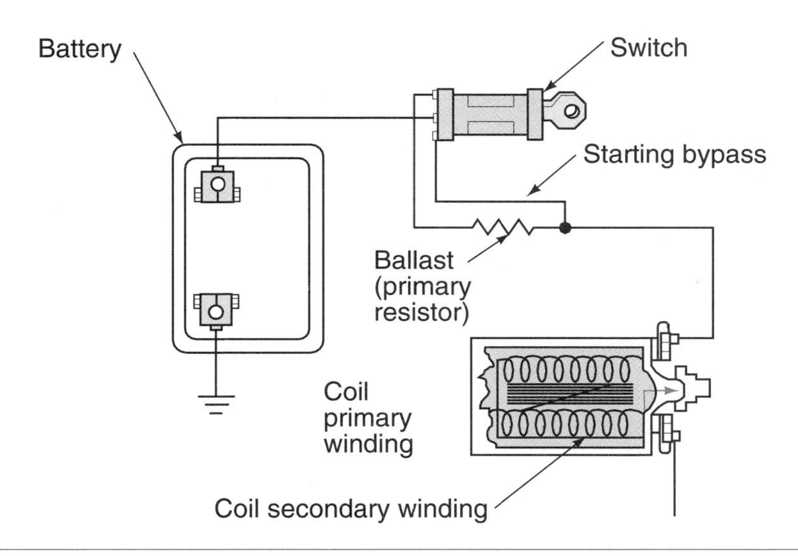

Figure 4-3 Ballast resistor used in some ignition primary circuits.

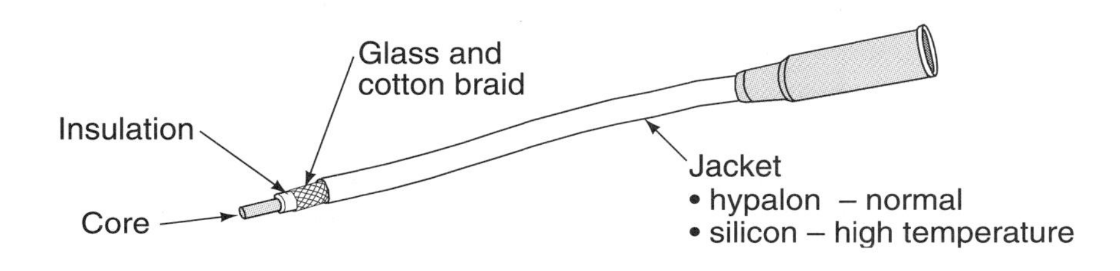

Figure 4-4 Typical spark plug wire.

Most spark plug wire conductors are made of nylon, rayon, fiberglass, or aramid thread impregnated with carbon. This core is surrounded by rubber (Figure 4-4). The carbon-impregnated core provides sufficient resistance to reduce RFI, yet does not affect engine operation. As the spark plug wires wear because of age and temperature changes, the resistance in the wire will change. Most plug wires have a resistance value of 3,000 Ω to 6,000 Ω per foot. However, some have between 6,000 Ω and 12,000 Ω. The accepted value when testing is 10,000 Ω per foot as a general specification.

The interference caused by electromagnetic energy is called **radiofrequency interference (RFI).**

Because the high voltage within the plug wires can create electromagnetic induction, proper wire routing is important to eliminate the possibility of cross-fire. To prevent cross-fire, the plug wires must be installed in the proper separator. Any two parallel wires next to each other in the firing order should be positioned as far away from each other as possible (Figure 4-5). When induction cross-fire occurs, no spark is jumped from one wire to the other. The spark is the result of induction from another field. Cross-fire induction is most common in two parallel wires that fire one after the other in the firing order.

It is important to check the procedure for testing the plug wires because some manufacturers require that the distributor cap be connected to the wires.

Cross-fire is the electromagnetic induction spark that can be transmitted in another wire close to the wire carrying the current.

Wire Sizes

An additional amount of consideration must be given for some margin of safety when selecting wire size. There are three major factors that determine the proper size of wire to be used:

1. The wire must have a large enough diameter—for the length required—to carry the necessary current for the load components in the circuit to operate properly.

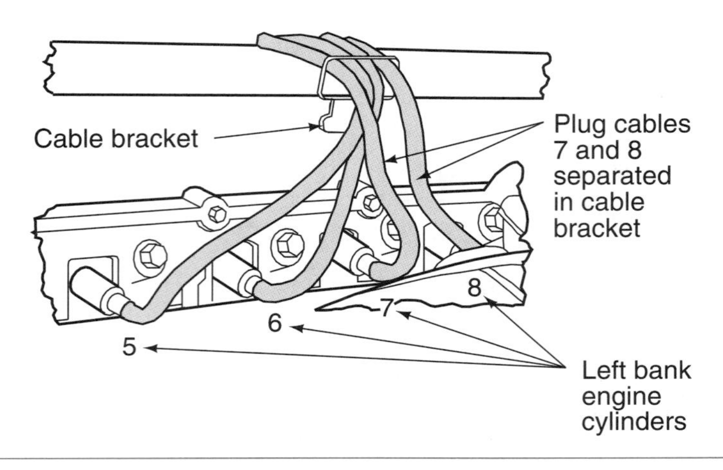

Figure 4-5 Proper spark plug wire routing to prevent cross-fire. (Reprinted with the permission of Ford Motor Company)

2. The wire must be able to withstand the anticipated vibration.
3. The wire must be able to withstand the anticipated amount of heat exposure.

The number assigned to a wire to indicate its size is referred to as **gauge.**

Wire size is based on the diameter of the conductor. The larger the diameter, the less the resistance. There are two common size standards used to designate wire size: American Wire Gauge (AWG) and metric.

The AWG standard assigns a number to the wire based on its diameter. The higher the number, the smaller the wire diameter. For example, 20-gauge wire is smaller in diameter than 10-gauge wire. Most electrical systems in the automobile use 14-, 16-, or 18-gauge wire. Most battery cables are 2-, 4-, or 6-gauge cable.

Some high current circuits will also use 10- or 12-gauge wire.

Both wire diameter and wire length affect resistance. Sixteen-gauge wire is capable of conducting 20 amperes for 10 feet with minimal voltage drop. However, if the current is to be carried for 15 feet, 14-gauge wire would be required. If 20 amperes were required to be carried for 20 feet, then 12-gauge wire would be required. The additional wire size is needed to prevent voltage drops in the wire. The illustration (Figure 4-6) lists the wire size required to carry a given amount of current for different lengths.

Another factor to wire resistance is temperature. An increase in temperature creates a similar increase in resistance. A wire may have a known resistance of 0.03 ohms per 10 feet at 70° F. When exposed to temperatures of 170° F, the resistance may increase to 0.04 ohms per 10 feet. Wires that are to be installed in areas that experience high temperatures, as in the engine compartment, must be of a size such that the increased resistance will not affect the operation of the load component. Also, the insulation of the wire must be capable of withstanding the high temperatures.

In the metric system, wire size is determined by the cross-sectional area of the wire. Metric wire size is expressed in square millimeters (mm^2). In this system the smaller the number, the smaller the wire conductor. The approximate equivalent wire size of metric to AWG is shown (Figure 4-7).

Terminals and Connectors

Shop Manual
Chapter 4, page 86

To perform the function of connecting the wires from the voltage source to the load component reliably, terminal connections are used. Today's vehicles can have as many as 500 separate circuit connections. The terminals used to make these connections must be able to perform with very low voltage drop. A loose or corroded connection can cause an unwanted voltage drop that results in poor operation of the load component.

Total Approximate Circuit Amperes	Wire Gauge (for Length in Feet)								
12 V	3	5	7	10	15	20	25	30	40
1.0	18	18	18	18	18	18	18	18	18
1.5	18	18	18	18	18	18	18	18	18
2	18	18	18	18	18	18	18	18	18
3	18	18	18	18	18	18	18	18	18
4	18	18	18	18	18	18	18	16	16
5	18	18	18	18	18	18	18	16	16
6	18	18	18	18	18	18	16	16	16
7	18	18	18	18	18	18	16	16	14
8	18	18	18	18	18	16	16	16	14
10	18	18	18	18	16	16	16	14	12
11	18	18	18	18	16	16	14	14	12
12	18	18	18	18	16	16	14	14	12
15	18	18	18	18	14	14	12	12	12
18	18	18	16	16	14	14	12	12	10
20	18	18	16	16	14	12	10	10	10
22	18	18	16	16	12	12	10	10	10
24	18	18	16	16	12	12	10	10	10
30	18	16	16	14	10	10	10	10	10
40	18	16	14	12	10	10	8	8	6
50	16	14	12	12	10	10	8	8	6
100	12	12	10	10	6	6	4	4	4
150	10	10	8	8	4	4	2	2	2
200	10	8	8	6	4	4	2	2	1

Note: 18 AWG as indicated above this line could be 20 AWG electrically.
18 AWG is recommended for mechanical strength.

Terminals are constructed of either brass or steel with a coating of tin or lead.

If as little as 10% voltage drop (1.2 V) occurs in a bad connection of a light circuit, it may result in a 30% loss of lighting efficiency.

Figure 4-6 The distance the current must be carried is a factor in determining the correct wire gauge to use.

Metric Size (mm²)	AWG (Gauge) Size	Ampere Capacity
0.5	20	4
0.8	18	6
1.0	16	8
2.0	14	15
3.0	12	20
5.0	10	30
8.0	8	40
13.0	6	50
19.0	4	60

Figure 4-7 Approximate AWG to metric equivalents.

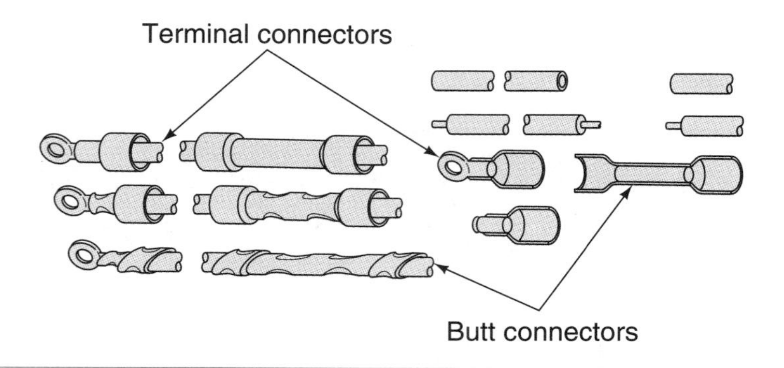

Figure 4-8 Primary wire terminals used in automotive applications.

Shop Manual
Chapter 4, page 94

Terminals can be either crimped or soldered to the conductor. The terminal makes the electrical connection and, it must be capable of withstanding the stress of normal vibration. The illustration (Figure 4-8) shows several different types of terminals used in the automotive electrical system. In addition, the following connectors are used on the automobile:

Shop Manual
Chapter 4, page 95

Shop Manual
Chapter 4, page 96

1. **Molded connector:** These connectors usually have one to four wires that are molded into a one-piece component (Figure 4-9).
2. **Multiple-wire hard-shell connector:** These connectors usually have a hard plastic shell that holds the connecting terminals of separate wires (Figure 4-10). The wire terminals can be removed from the shell to be repaired.
3. **Bulkhead connectors:** These connectors are used when several wires must pass through the bulkhead (Figure 4-11).
4. **Weather-Pack Connectors:** These connectors have rubber seals on the terminal ends and on the covers of the connector half (Figure 4-12). These connectors are used on computer circuits to protect the circuit from corrosion, which may result in a voltage drop.
5. **Metri-Pack Connectors:** These are like the weather-pack connectors, but do not have the seal on the cover half (Figure 4-13).
6. Heat shrink covered butt connectors are recommended for air bag applications by some manufacturers. Other manufacturers allow NO repairs to the circuitry, while still others require silver-soldered connections.

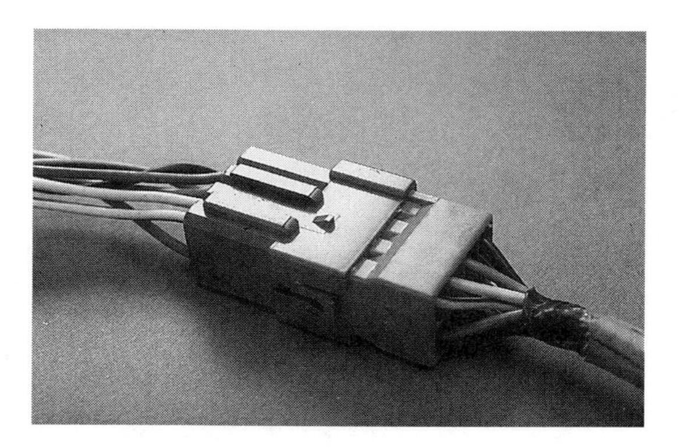

Figure 4-9 Multiple-circuit hard shell connector.

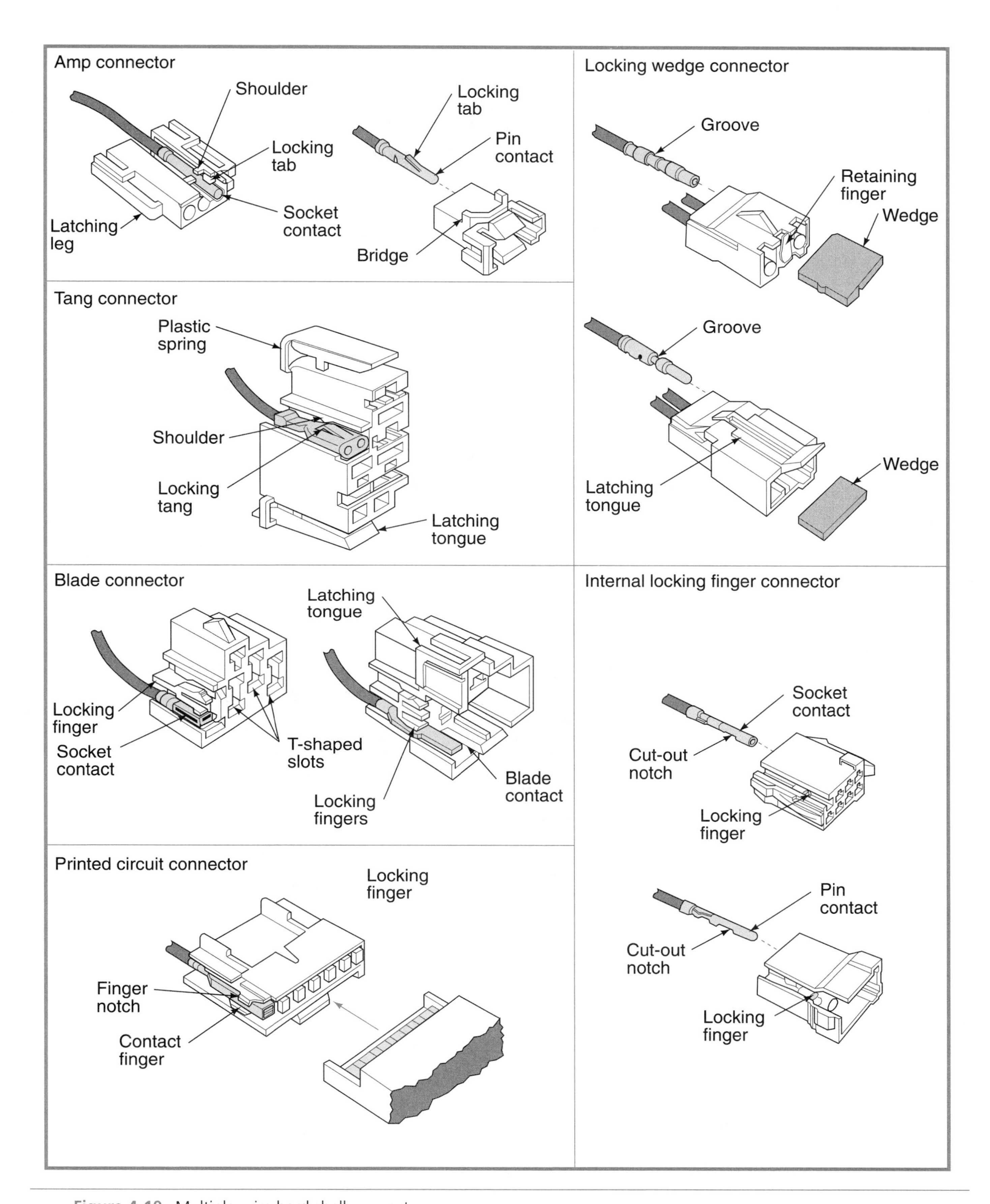

Figure 4-10 Multiple-wire hard shell connector.

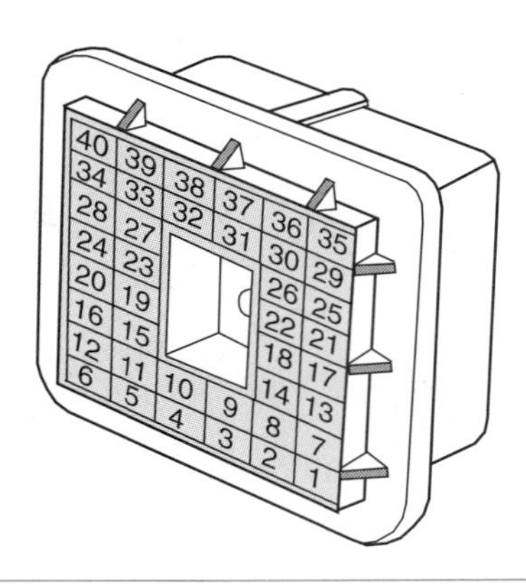

Figure 4-11 Bulkhead connector. (Courtesy of Chrysler Corporation)

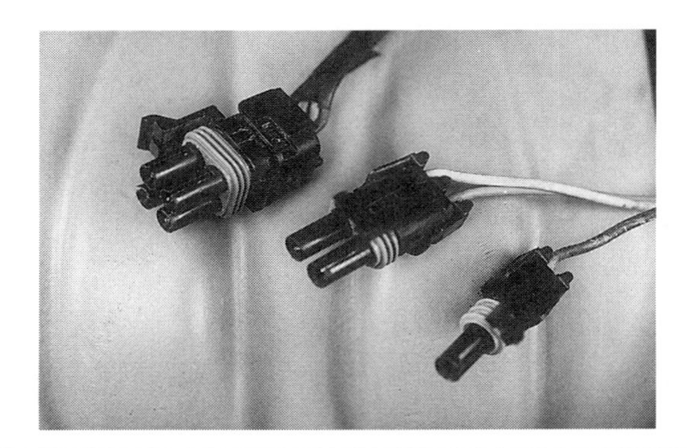

Figure 4-12 Weather-pack connector is used to prevent connector corrosion.

Figure 4-13 Metri-pack connector.

Common connections are used to share source of power or a common ground and are often called a splice.

Printed circuit boards are used to simplify the wiring of the circuits they operate. Other uses of printed circuit boards include the inside of radios, computers, and some voltage regulators.

To reduce the number of connectors in the electrical system, a common connection can be used (Figure 4-14). If there are several electrical components that are physically close to each other, a single common connection or splice eliminates using a separate connector for each wire.

Printed Circuits

Most instrument panels use printed circuit boards as circuit conductors. A printed circuit is made of a thin phenolic or fiberglass board that copper (or some other conductive material) has been deposited on. Portions of the conductive metal are then etched or eaten away by acid. The

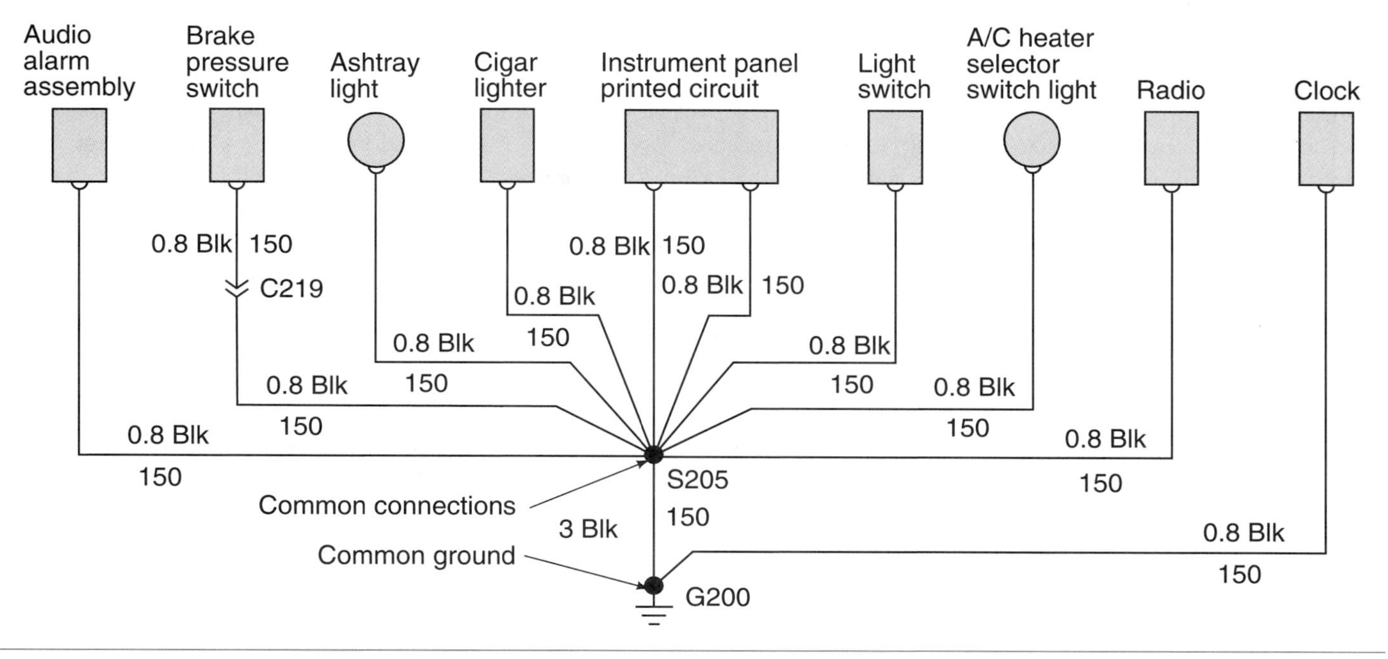

Figure 4-14 Common connections are used to reduce the number of connectors. (Courtesy of General Motors Corporation)

remaining strips of conductors provide the circuit path for the instrument panel lights, warning lights, indicator lights, and gauges of the instrument panel (Figure 4-15). The printed circuit board is attached to the back of the instrument panel housing. An edge connector joins the printed circuit board to the vehicle wiring harness.

Whenever it is necessary to perform repairs on or around the printed circuit board, it is important to follow these precautions:

1. When replacing light bulbs, be careful not to cut or tear the surface of the printed circuit board.
2. Do not touch the surface of the printed circuit with your fingers. The acid present in normal body oils can damage the surface.
3. If the printed circuit board needs to be cleaned, use a commercial cleaning solution designed for electrical use. If this solution is not available, it is possible to clean the board by **lightly** rubbing the surface with an eraser.

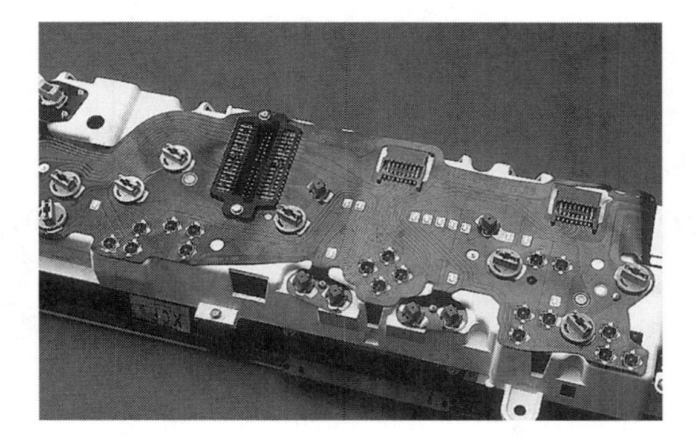

Figure 4-15 Printed circuits eliminate bulky wires behind the instrument panel.

Wiring Harness

A **wire harness** is an assembled group of wires that branch out to the various electrical components.

Most manufacturers use wiring harnesses to reduce the amount of loose wires hanging under the hood or dash of an automobile. The wiring harness provides for a safe path for the wires of the vehicle's lighting, engine, and accessory components. The wiring harness is made by grouping insulated wires and wrapping them together. The wires are bundled into separate harness assemblies that are joined together by connector plugs. The mutiple-pin connector plug may have more than 60 individual wire terminals.

There are several complex wiring harnesses in a vehicle, in addition to the simple harnesses. The engine compartment harness and the under dash harness are examples of a complex harness (Figure 4-16). Lighting circuits usually use a more simple harness (Figure 4-17). A complex harness serves many circuits. The simple harness services only a few circuits. Some individual circuit wires may branch out of a complex harness to other areas of the vehicle.

Most wiring harnesses now use a flexible tubing to provide for quick wire installation (Figure 4-18). The tubing has a seam that can be opened to accommodate the installation or removal of wires from the harness. The seam will close once the wires are installed, and will remain closed even if the tubing is bent.

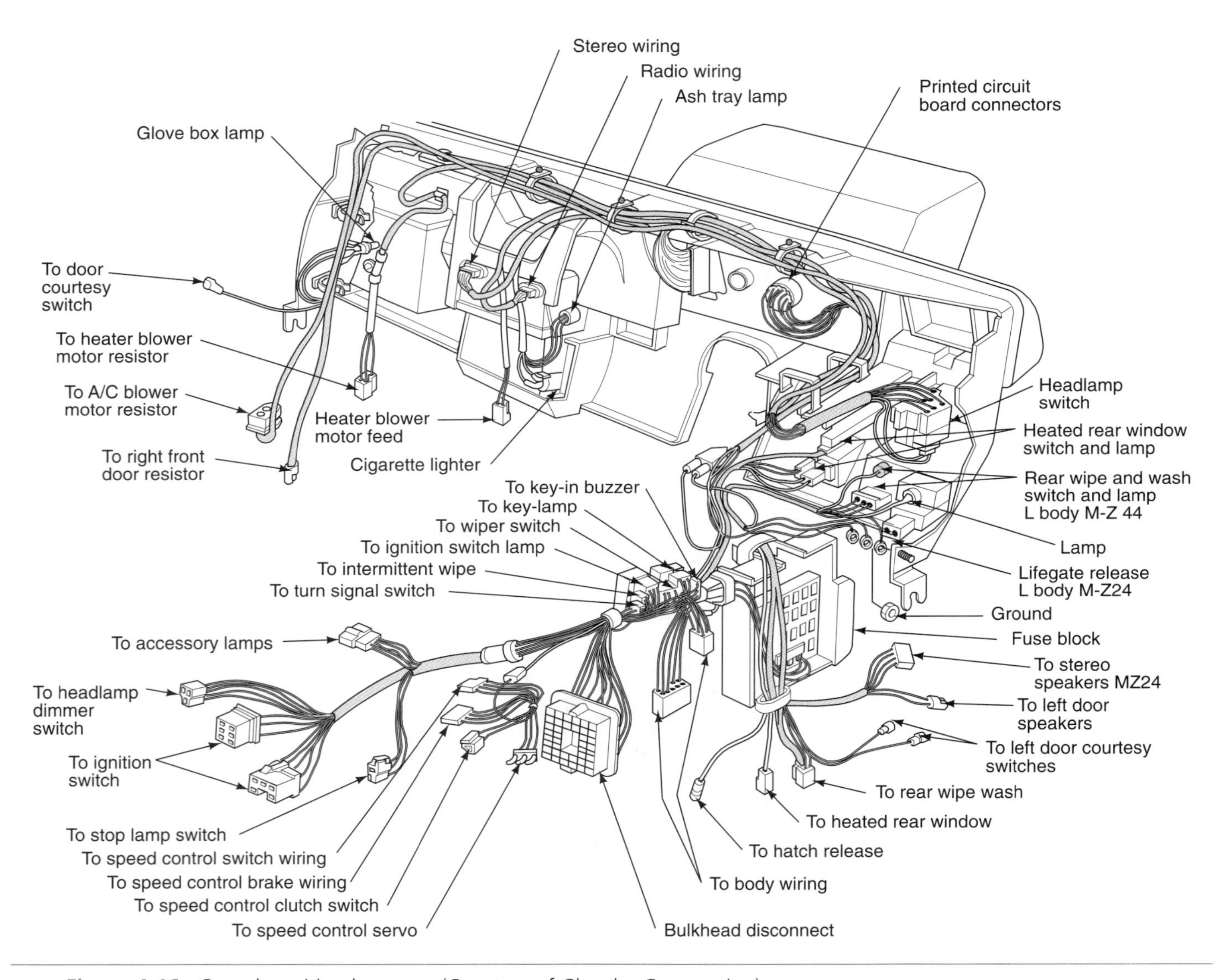

Figure 4-16 Complex wiring harness. (Courtesy of Chrysler Corporation)

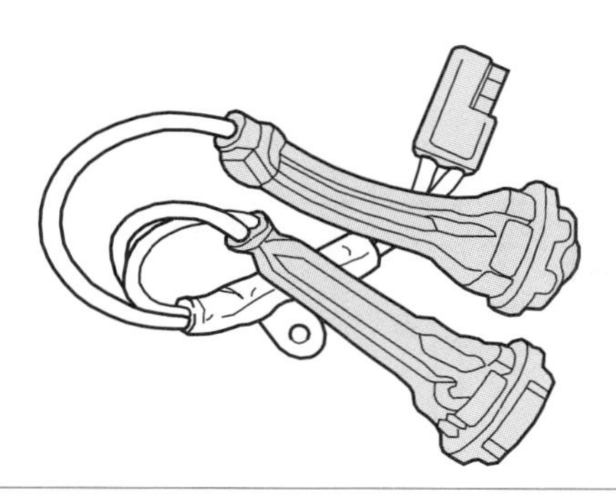

Figure 4-17 Simple wiring harness. (Courtesy of Chrysler Corporation)

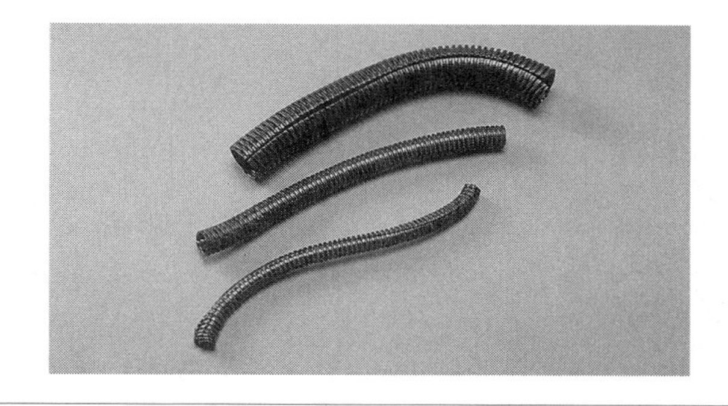

Figure 4-18 Flexible tubing used to make wiring harnesses.

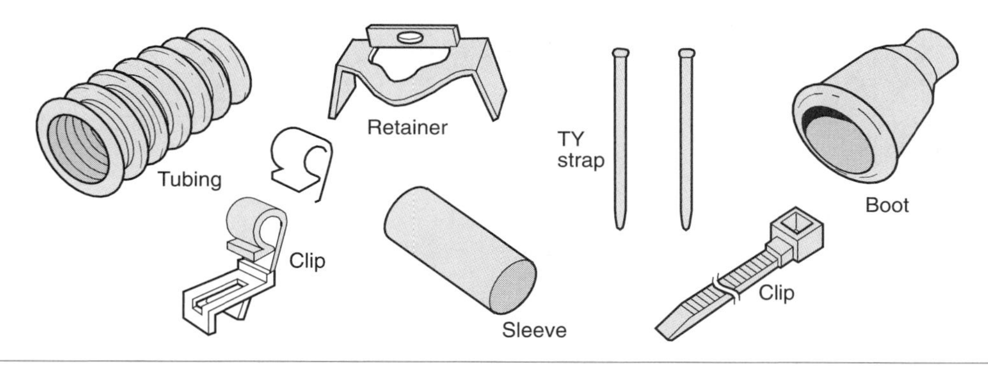

Figure 4-19 Typical wire protection devices. (Courtesy of Chrysler Corporation)

Wiring Protective Devices

Often overlooked, but very important to the electrical system, are proper wire protection devices (Figure 4-19). These devices prevent damage to the wiring by maintaining proper wire routing and retention. Special clips, retainers, straps, and supplementary insulators provide additional protection to the conductor over what the insulation itself is capable of providing. Whenever the technician must remove one of these devices to perform a repair, it is important that the device be reinstalled to prevent additional electrical problems.

Whenever it is necessary to install additional electrical accessories, try to support the primary wire in at least 1-foot intervals. If the wire must be routed through the frame or body, use rubber grommets to protect the wire.

WARNING: Do not use metal clamps to secure wires to the frame or body of the vehicle. The metal clamp may cut through the insulation and cause a short to ground. Use plastic clips in place of metal.

Wiring Diagrams

Shop Manual
Chapter 4, page 98

One of the most important tools for diagnosing and repairing electrical problems is a wiring diagram. These diagrams identify the wires and connectors from each circuit on a vehicle. They also show where different circuits are interconnected, where they receive their power, where the ground is located, and the colors of the different wires. All of this information is critical to proper diagnosis of electrical problems. Some wiring diagrams also give additional information that helps you understand how a circuit operates and how to identify certain components (Figure 4-20). Wiring diagrams do not explain how the circuit works; this is where your knowledge of electricity comes in handy.

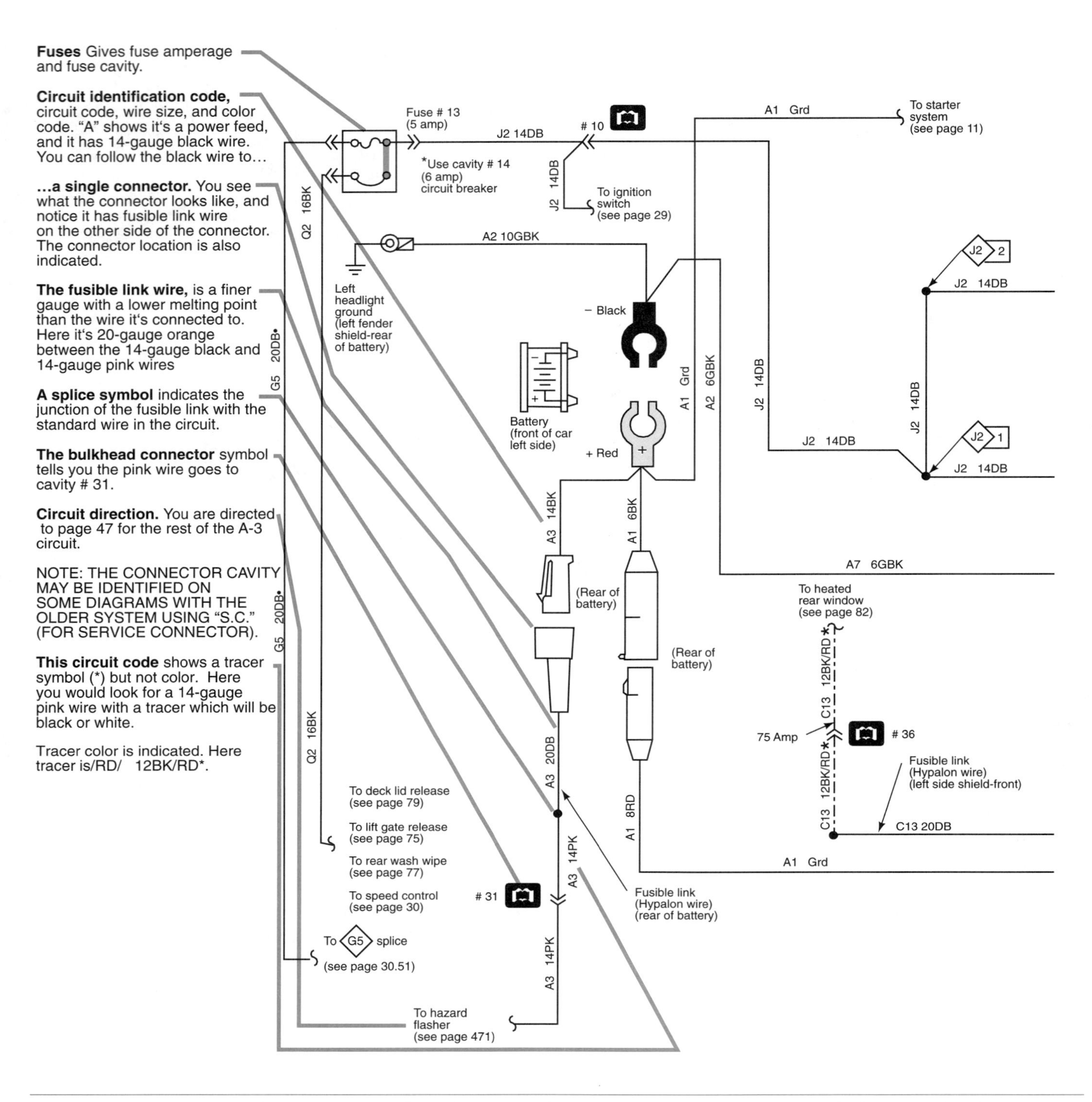

Figure 4-20 Wiring diagrams provide the technician with necessary information to accurately diagnose the electrical systems. (Courtesy of Chrysler Corporation)

A **wiring diagram** can show the wiring of the entire vehicle (Figure 4 -21) or a single circuit (Figure 4-22). These single circuit diagrams are also called block diagrams. In both types, the wire colors and the connectors are shown. Wiring diagrams of the entire vehicle tend to look

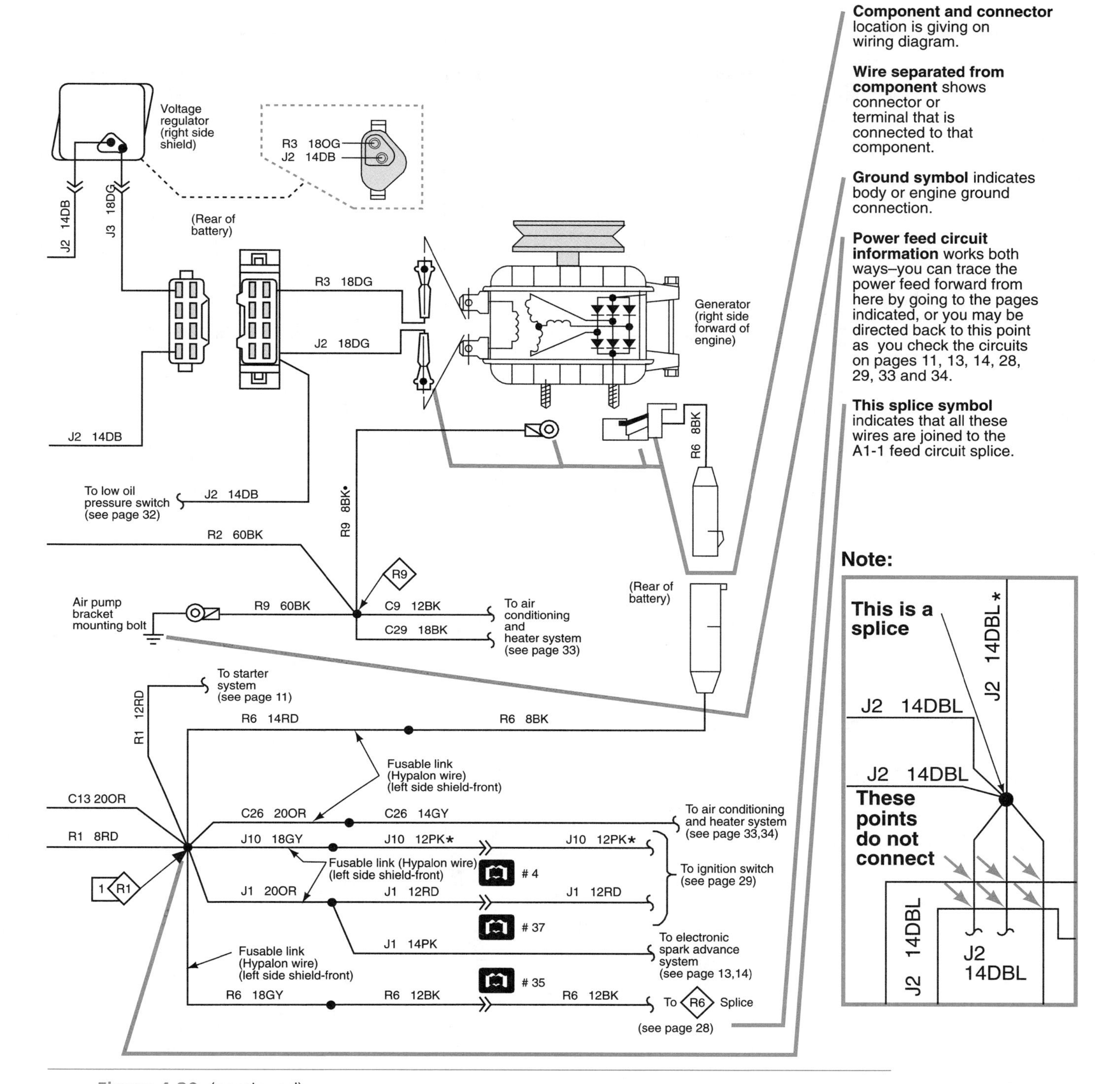

Figure 4-20 (continued)

more complex and threatening than block diagrams. However, once you simplify the diagram to only those wires, connectors, and components that belong to an individual circuit, they become less complex and more valuable.

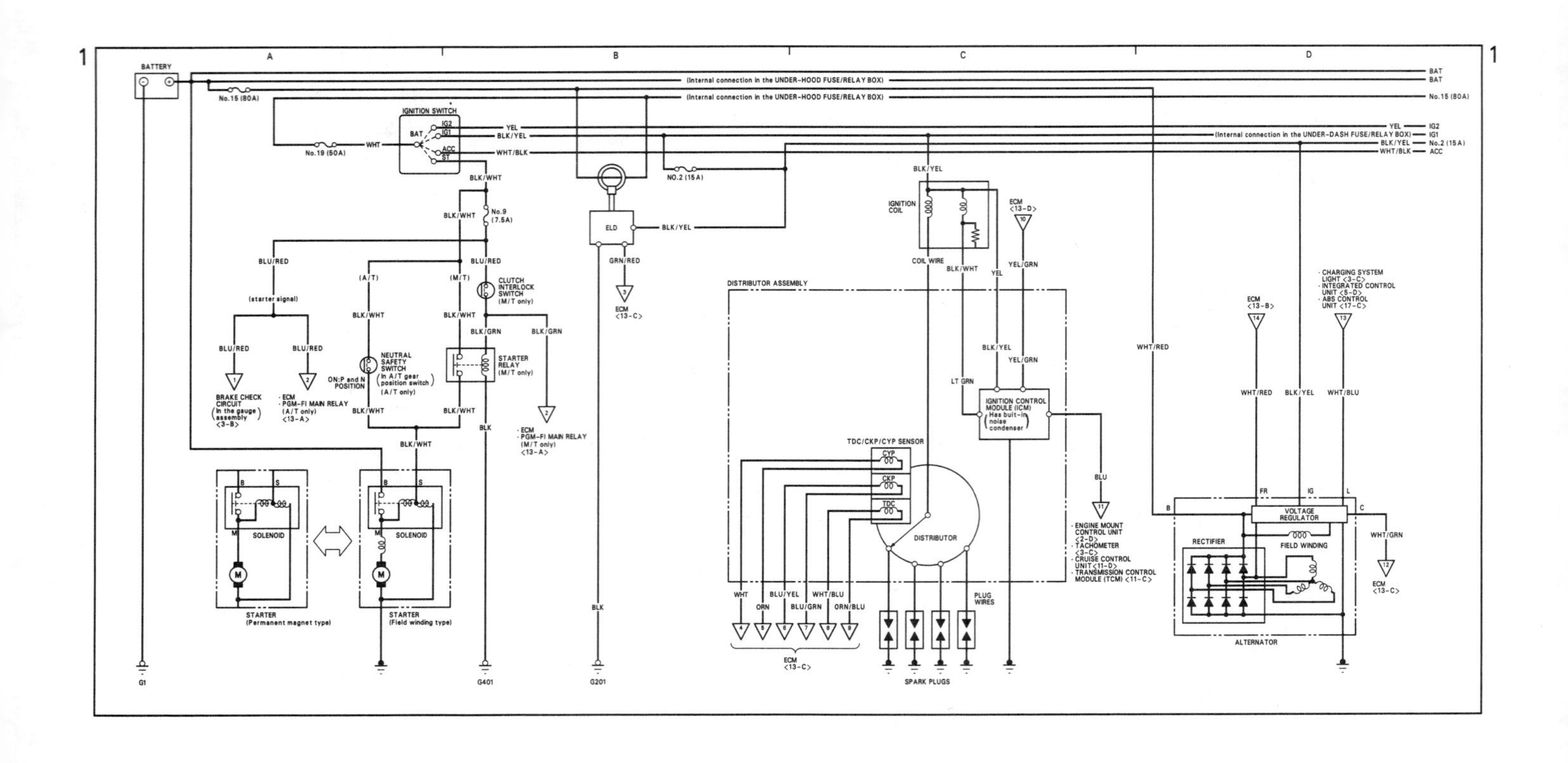

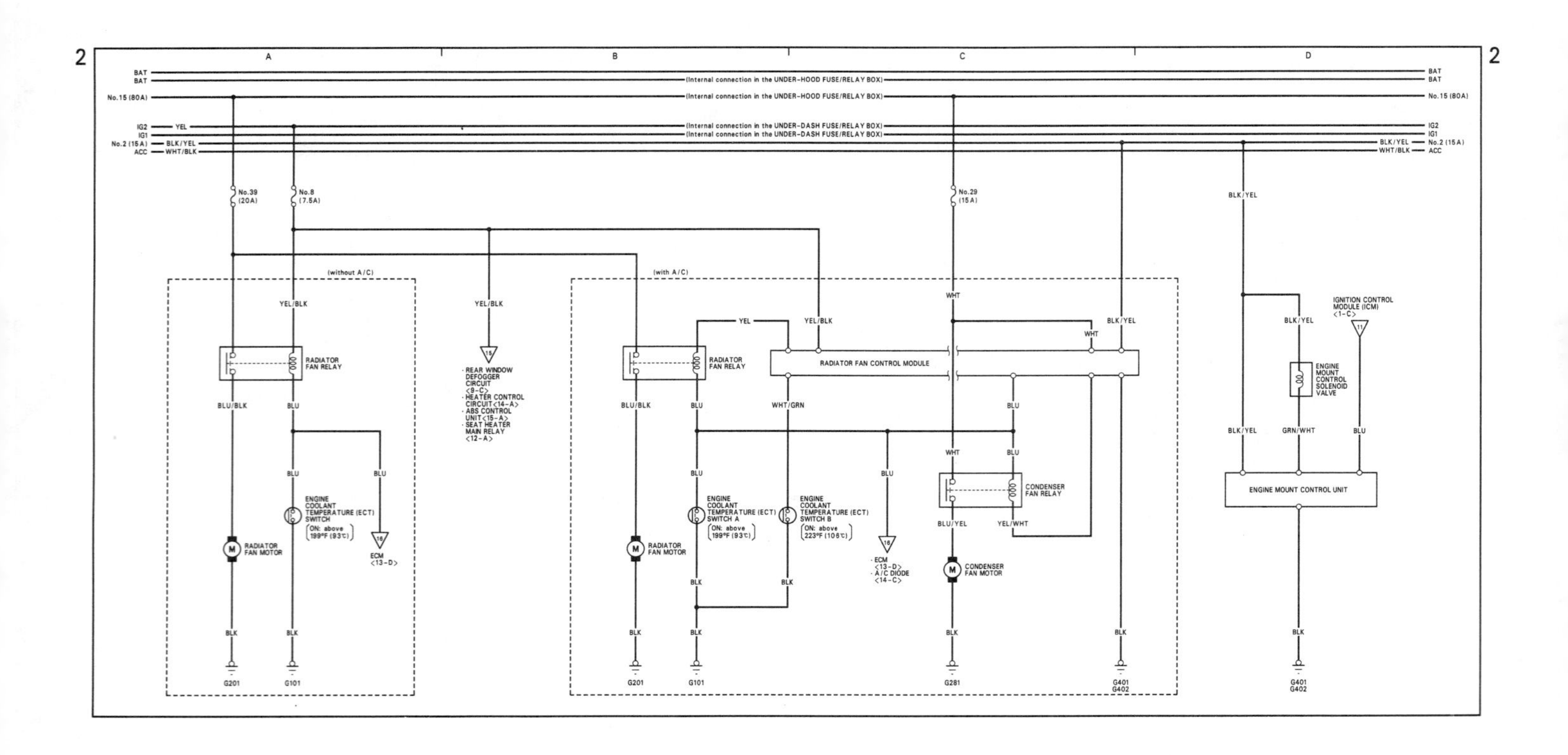

Figure 4-21 A complete system wiring diagram. (Courtesy of Honda Motor Company)

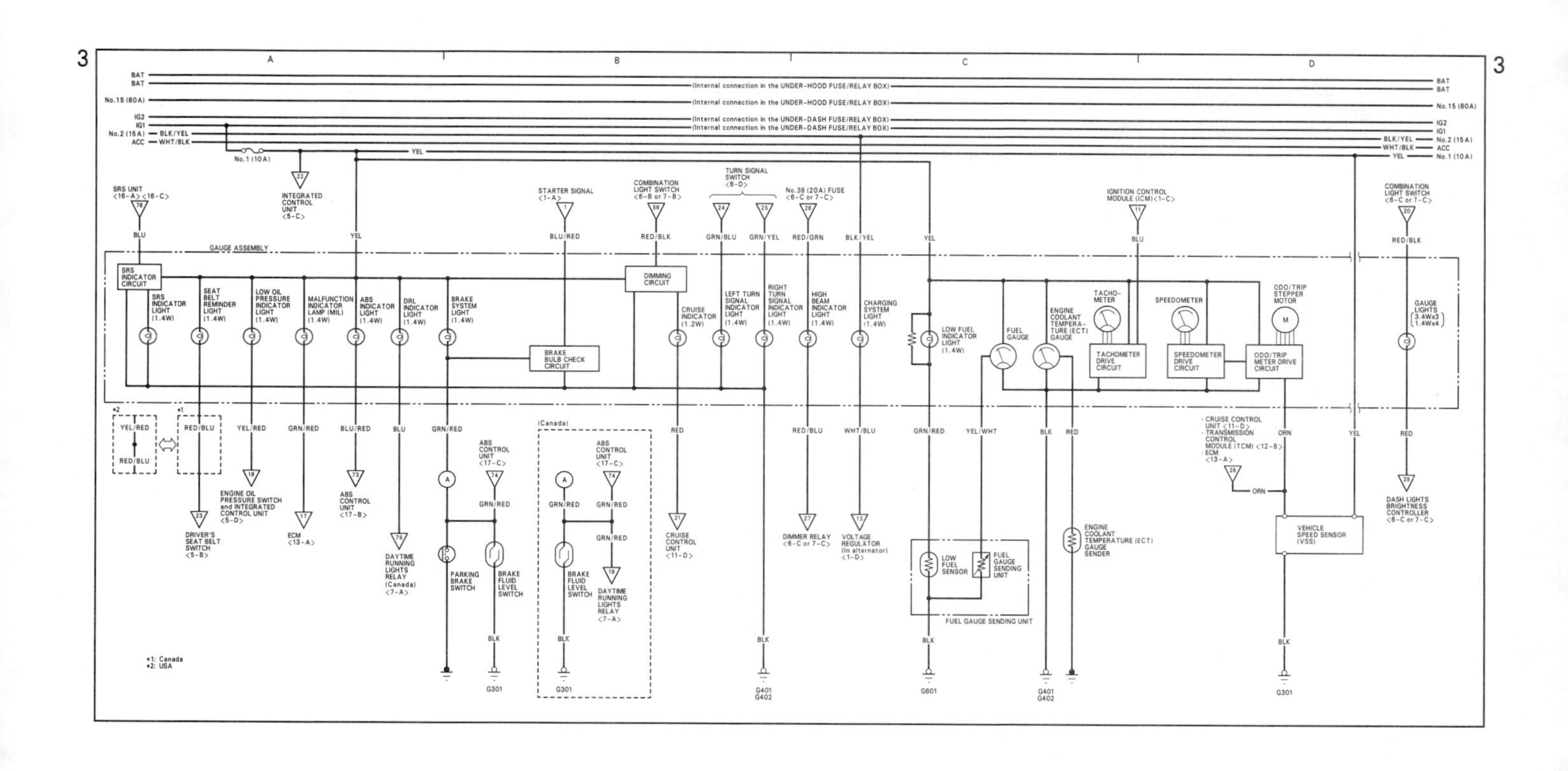

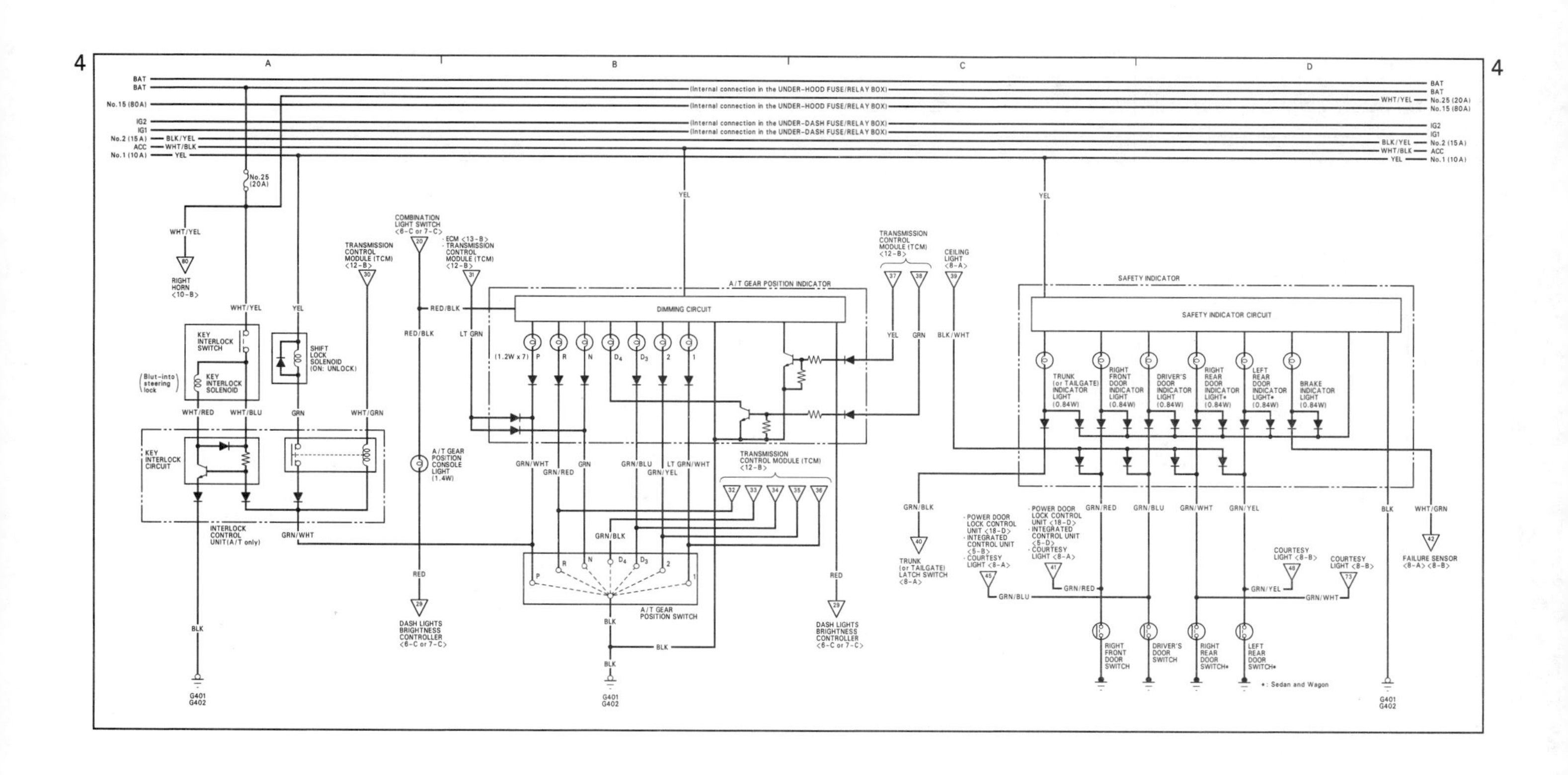

Figure 4-21 (continued)

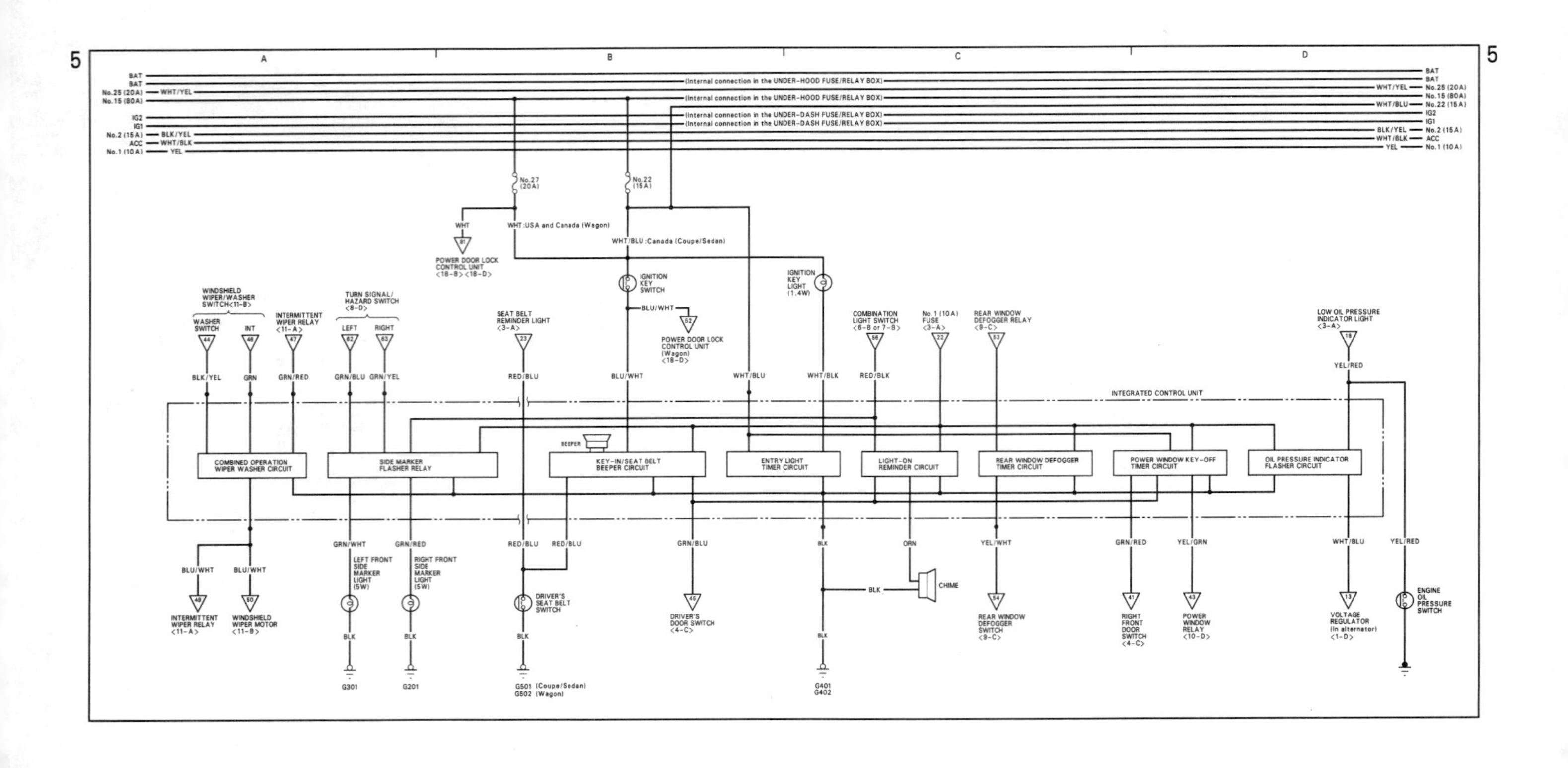

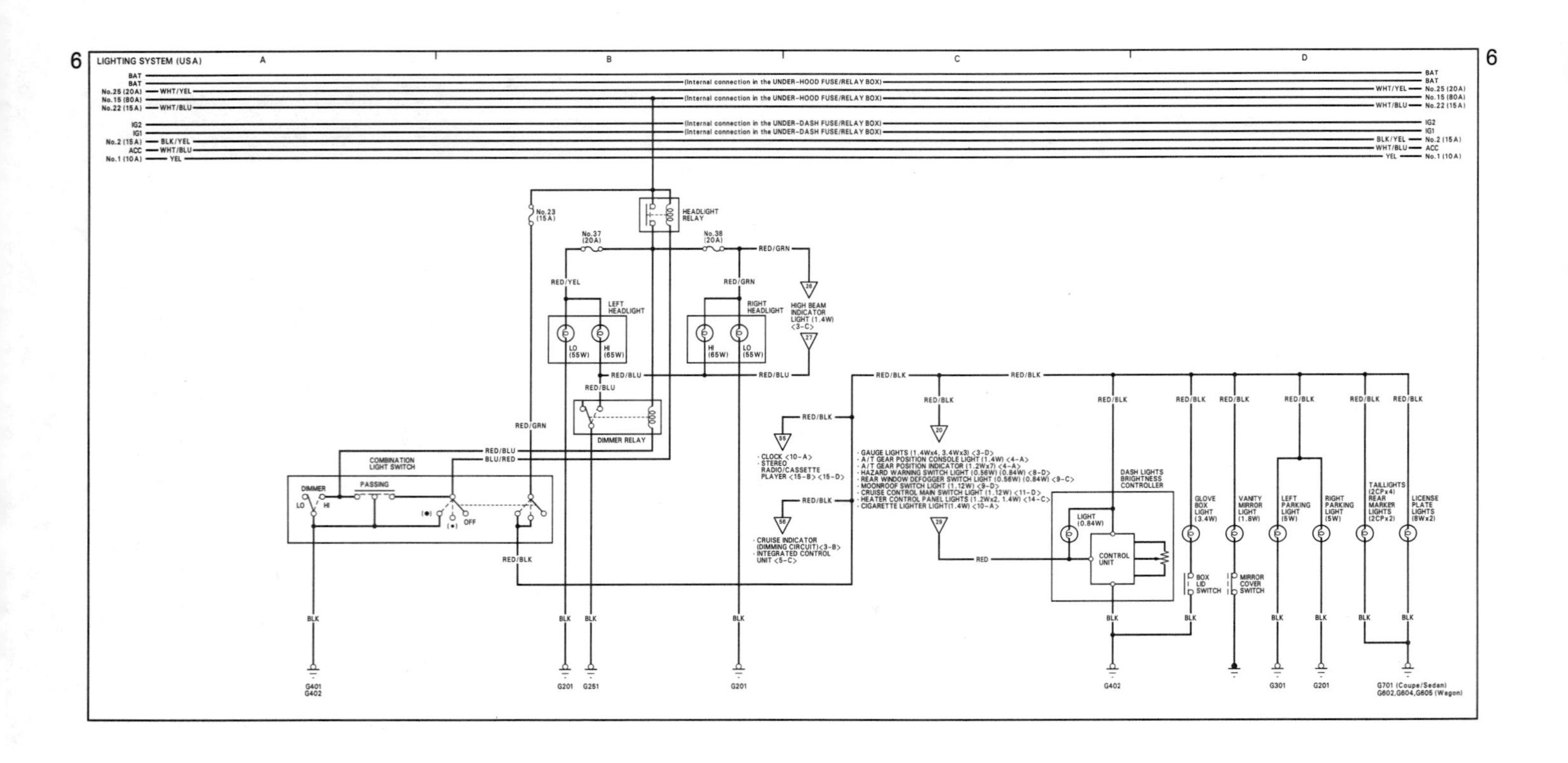

Figure 4-21 (continued)

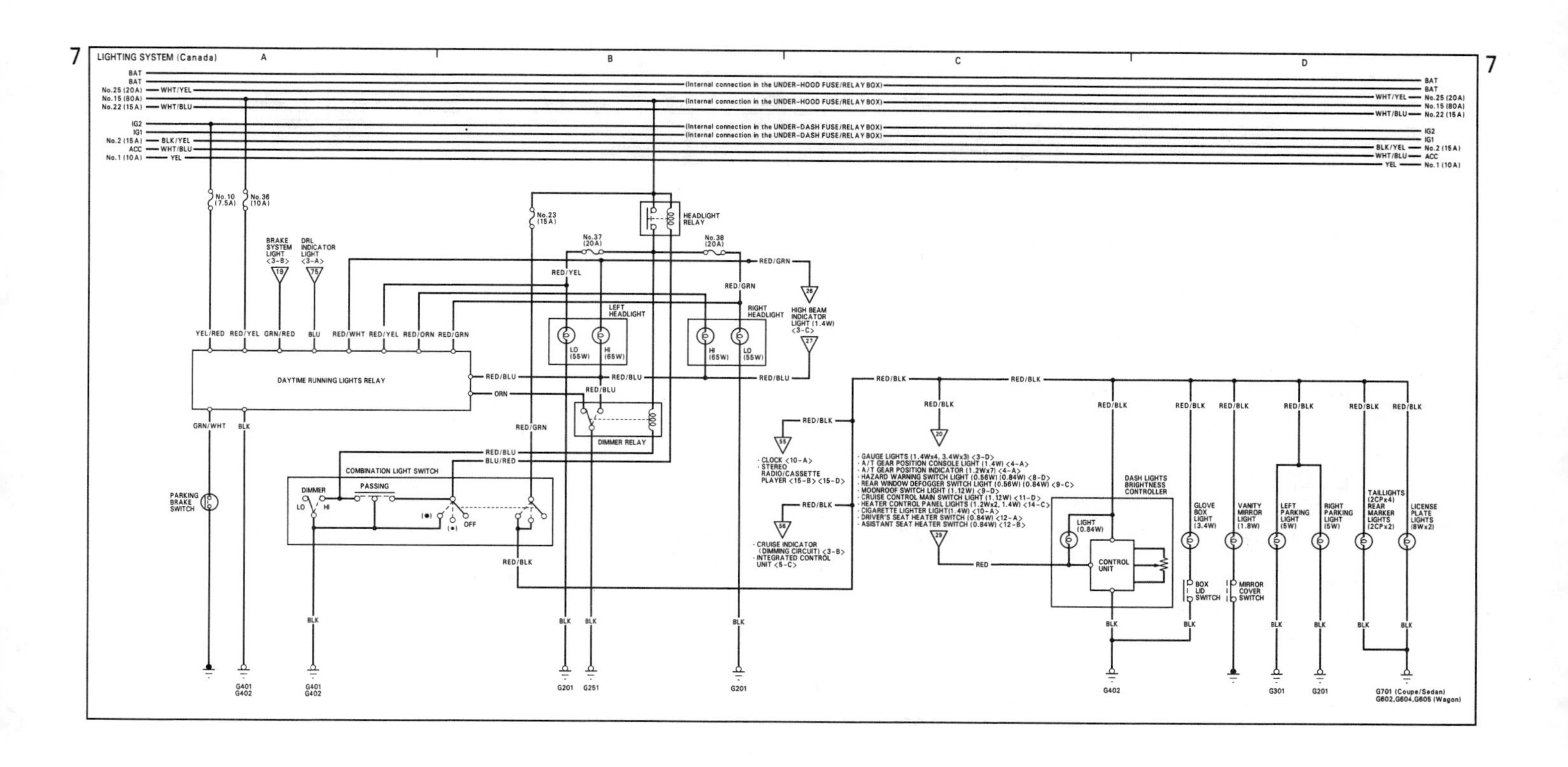

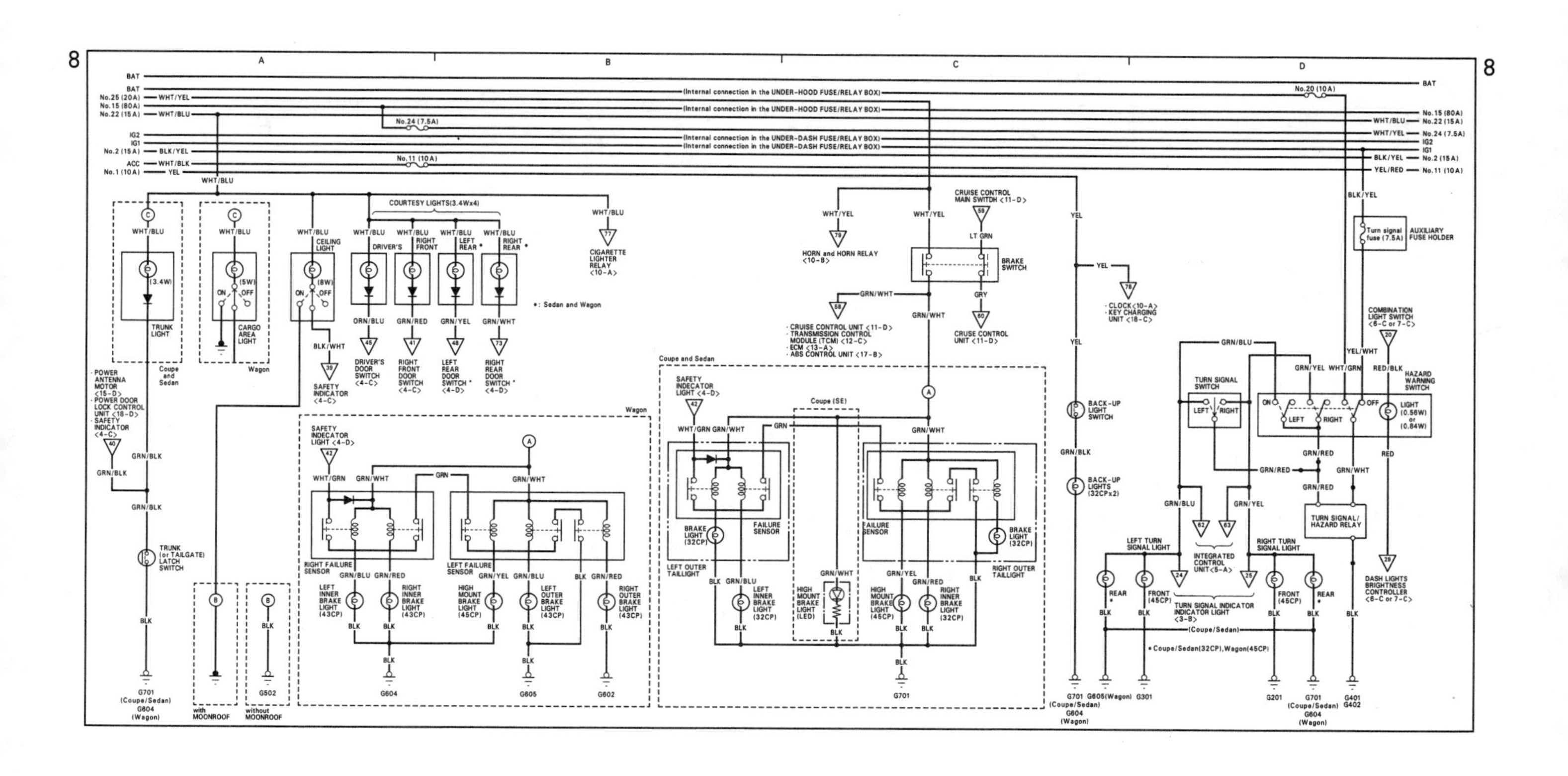

Figure 4-21 (continued)

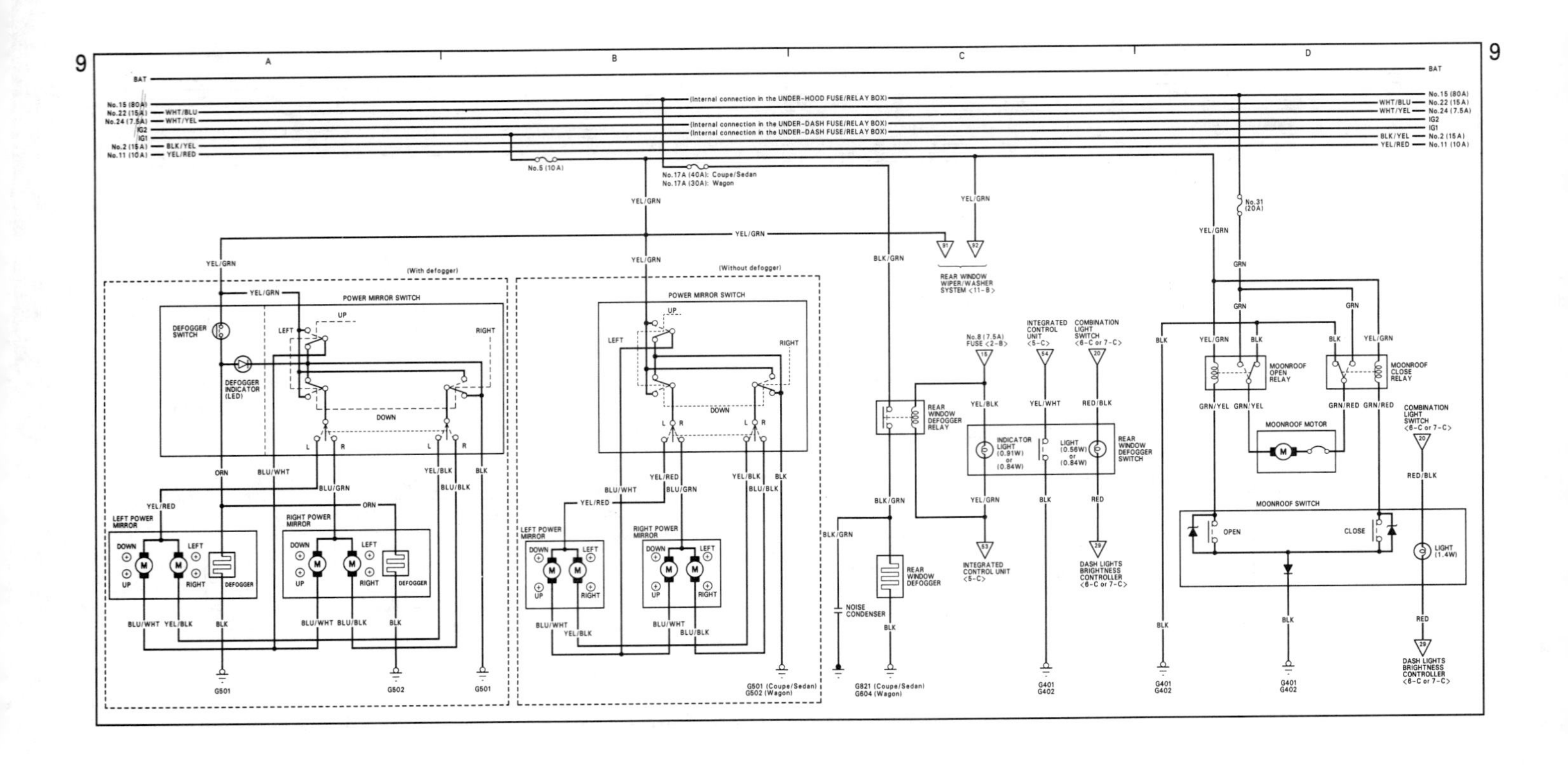

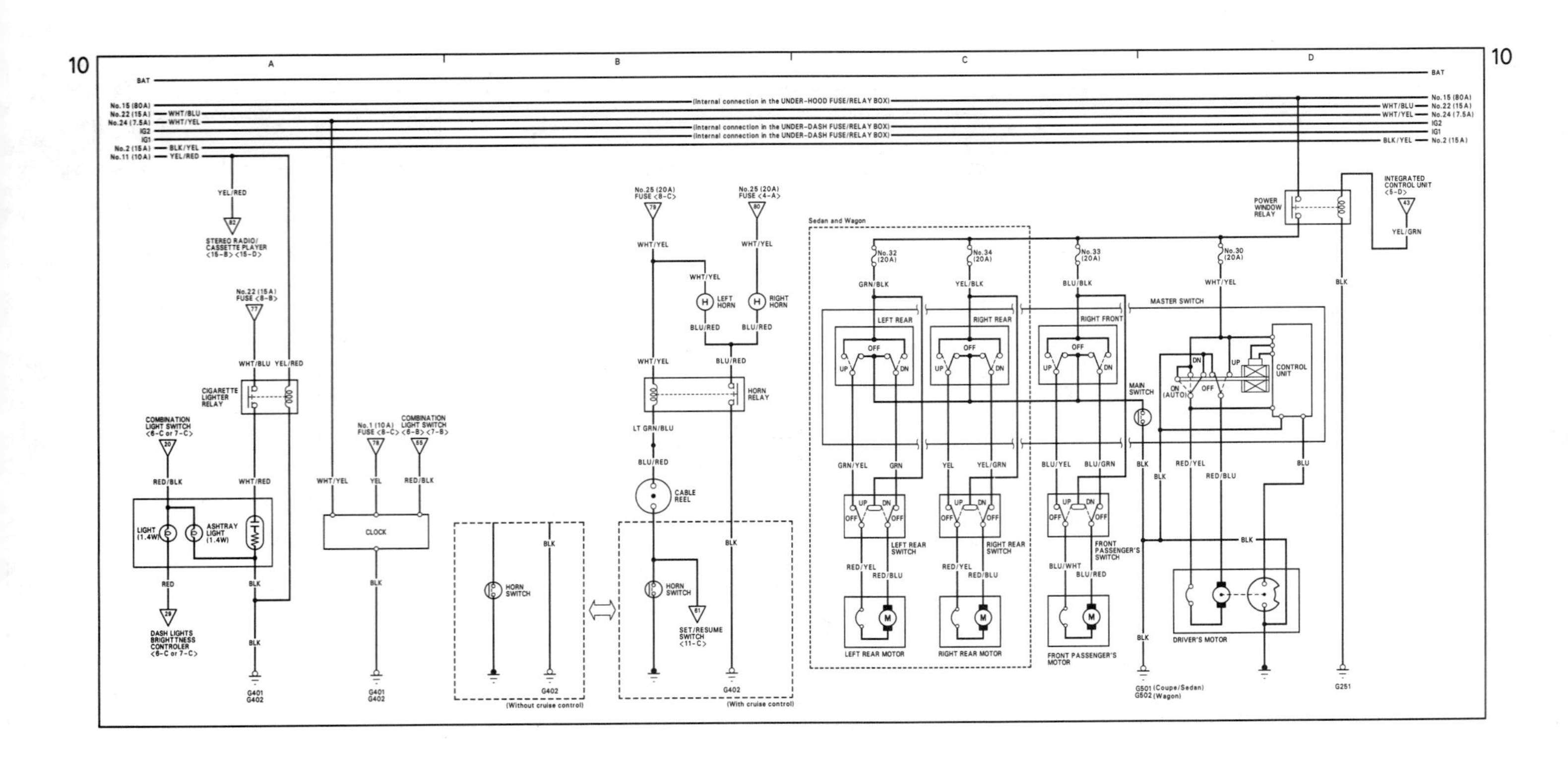

Figure 4-21 (continued)

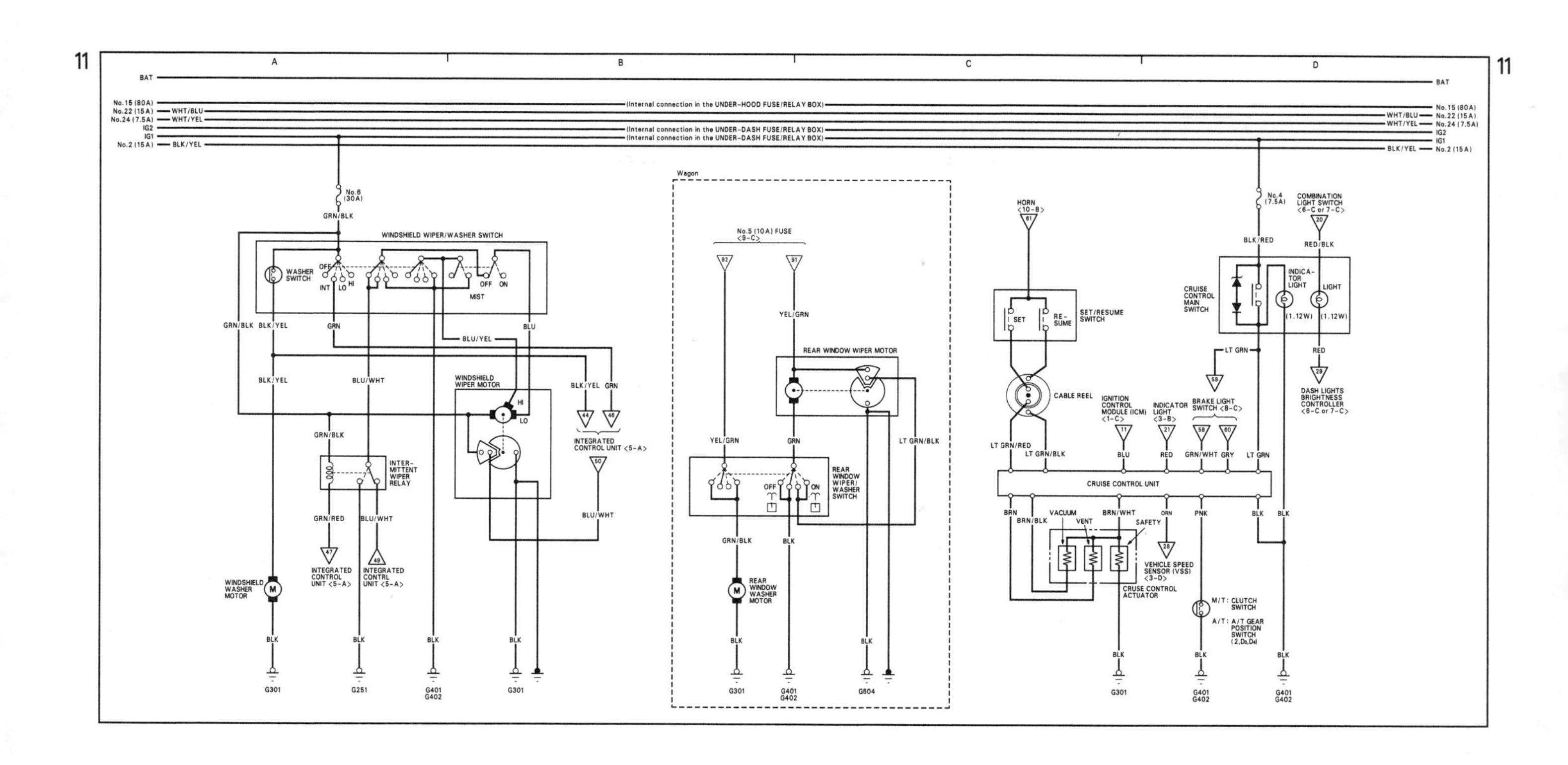

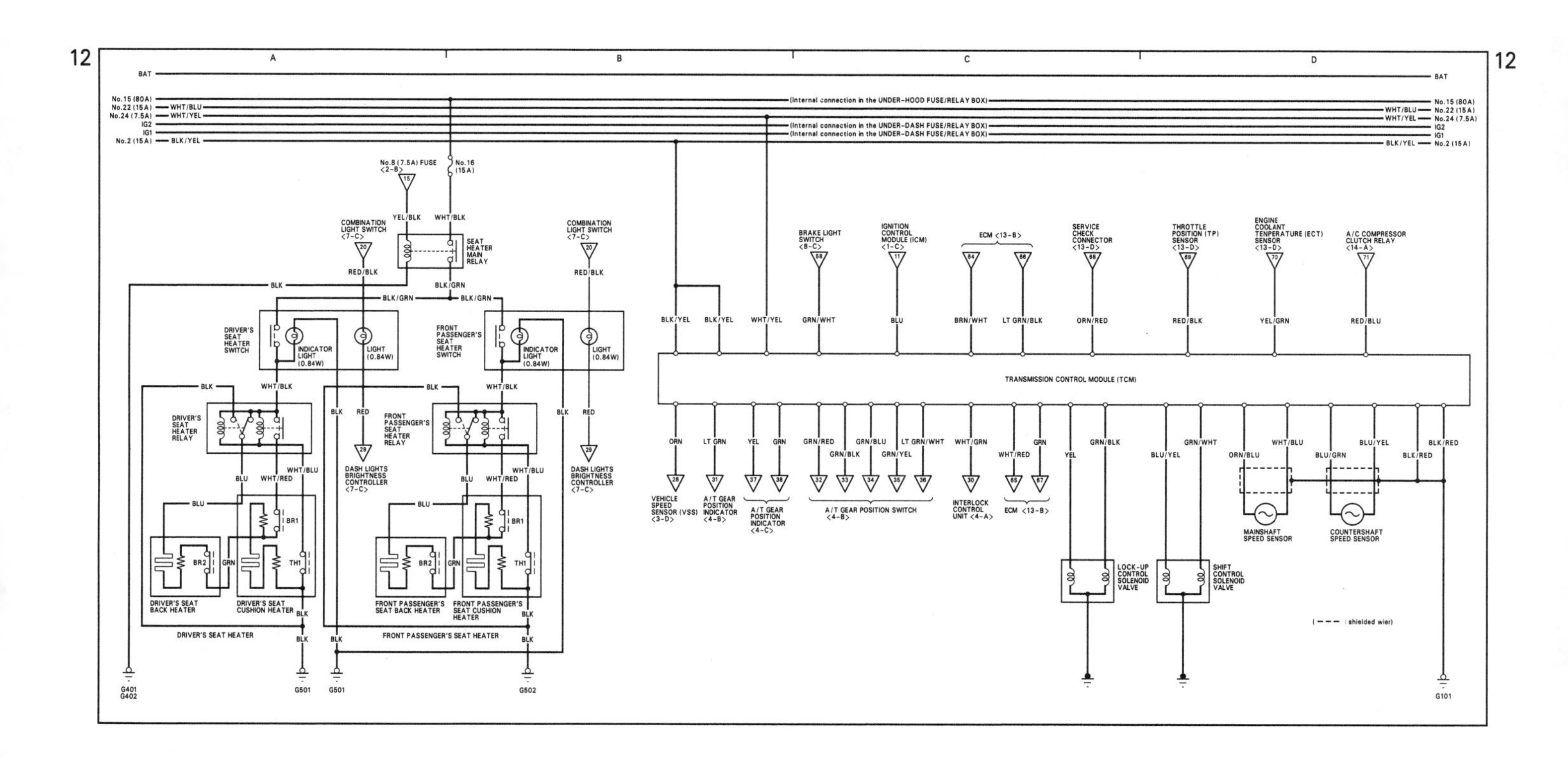

Figure 4-21 (continued)

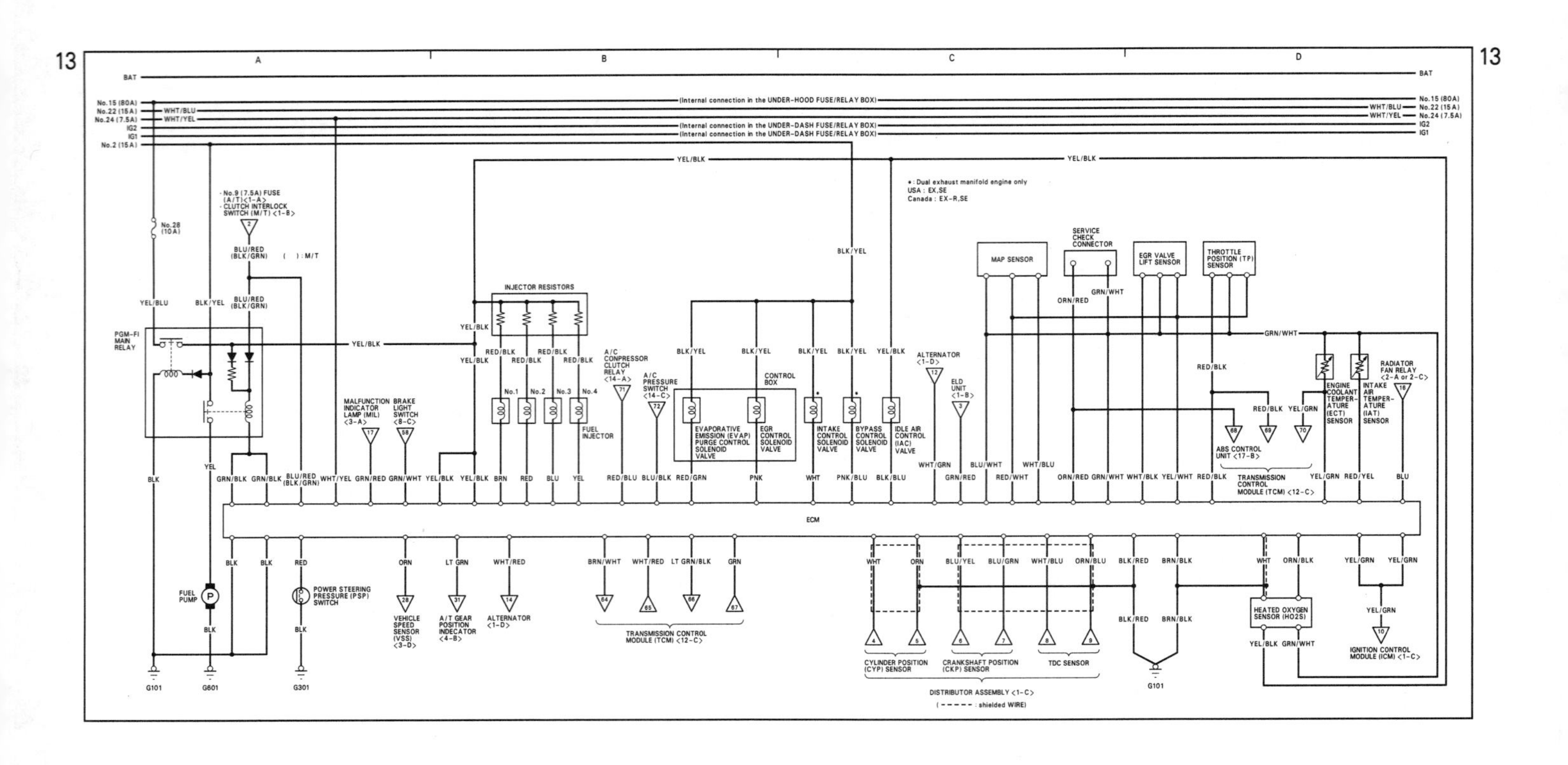

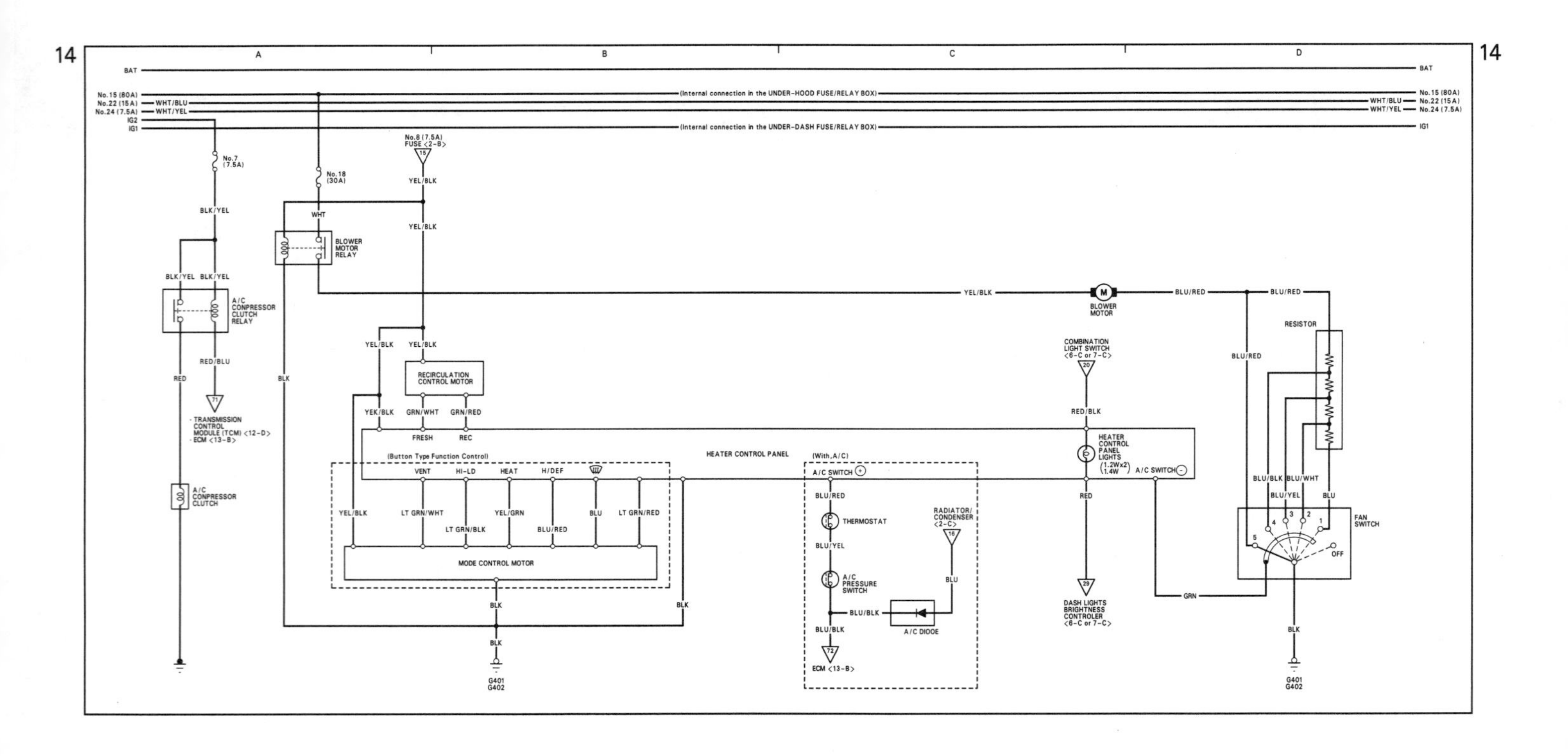

Figure 4-21 (continued)

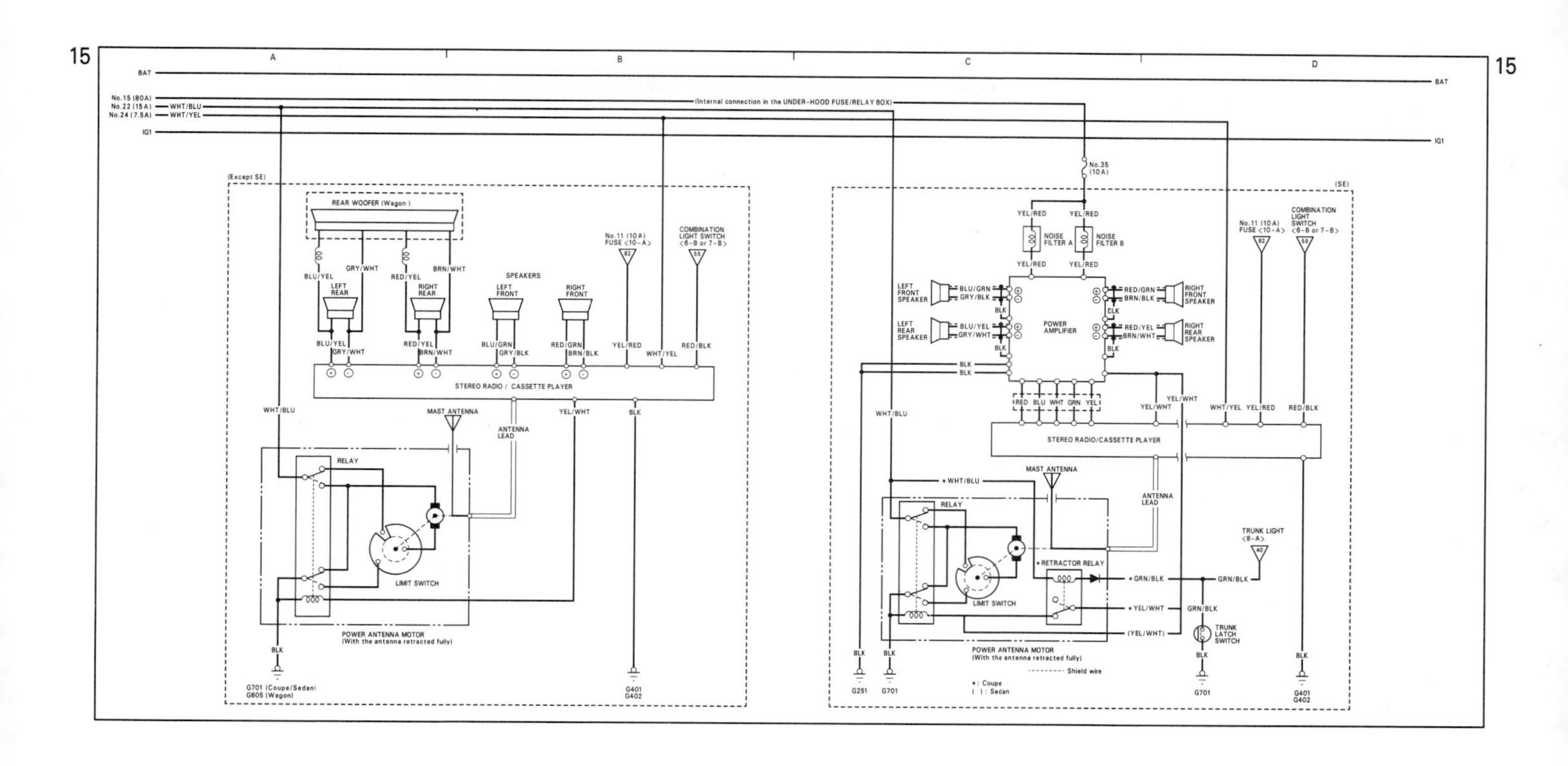

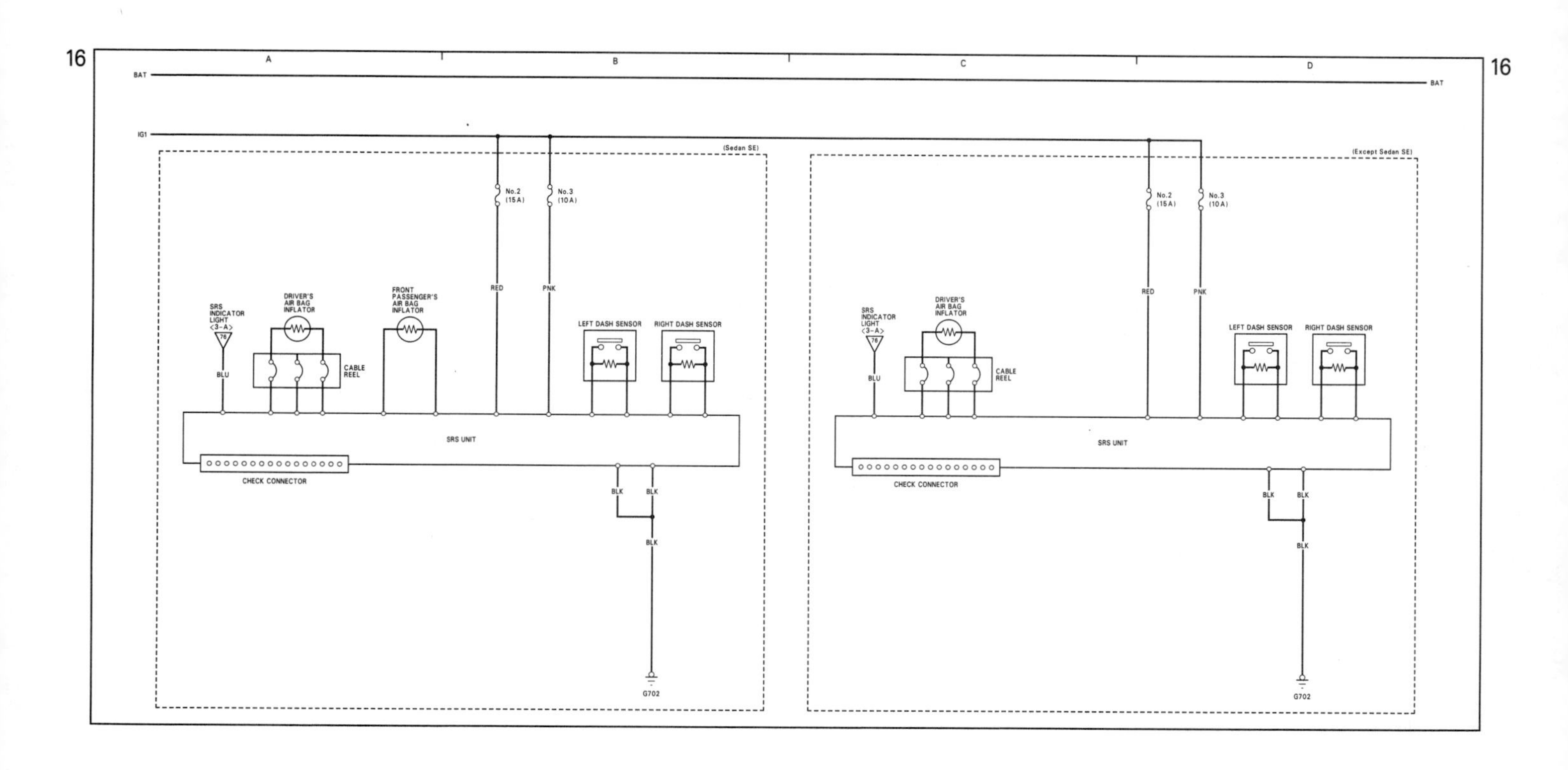

Figure 4-21 (continued)

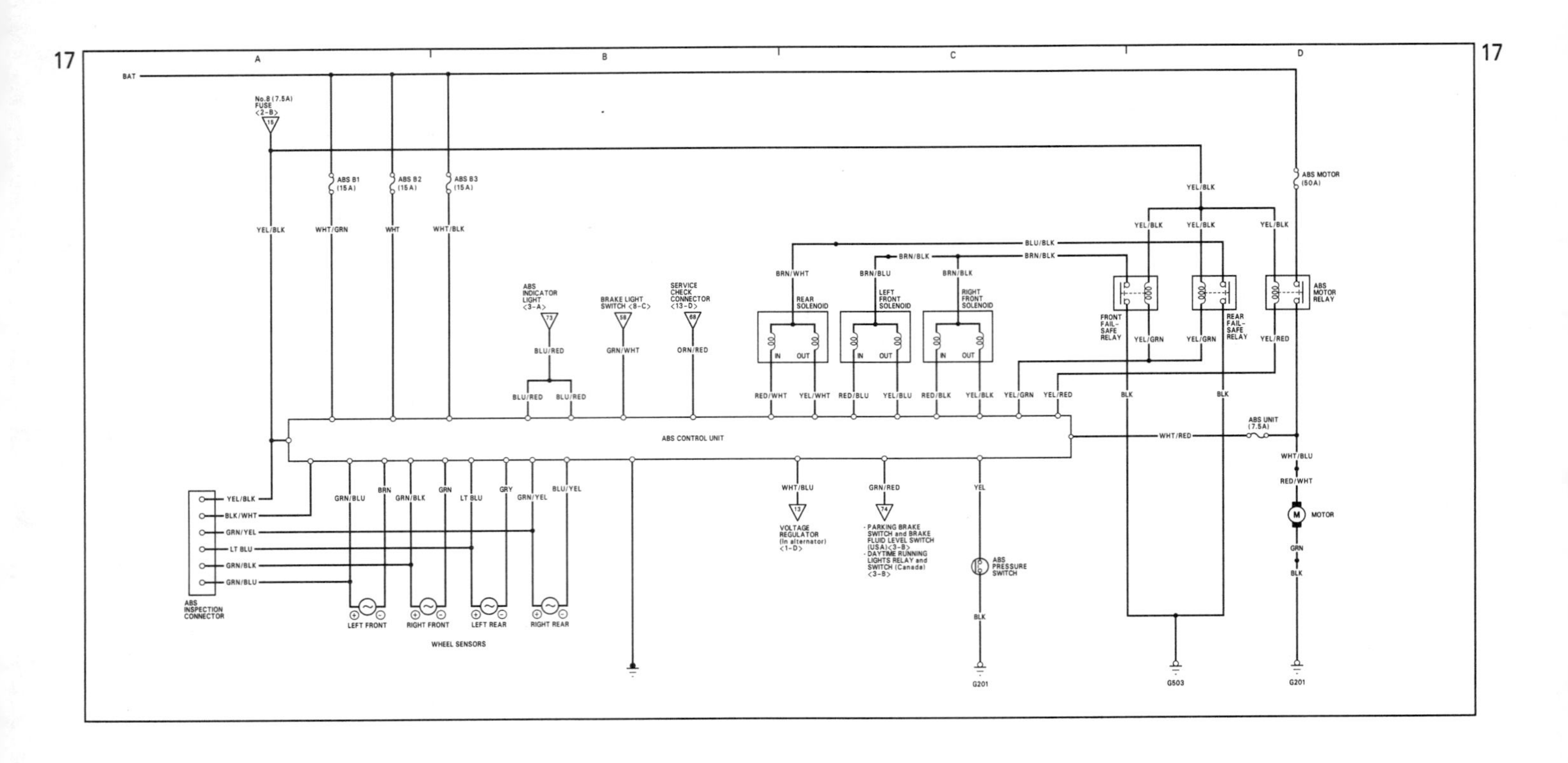

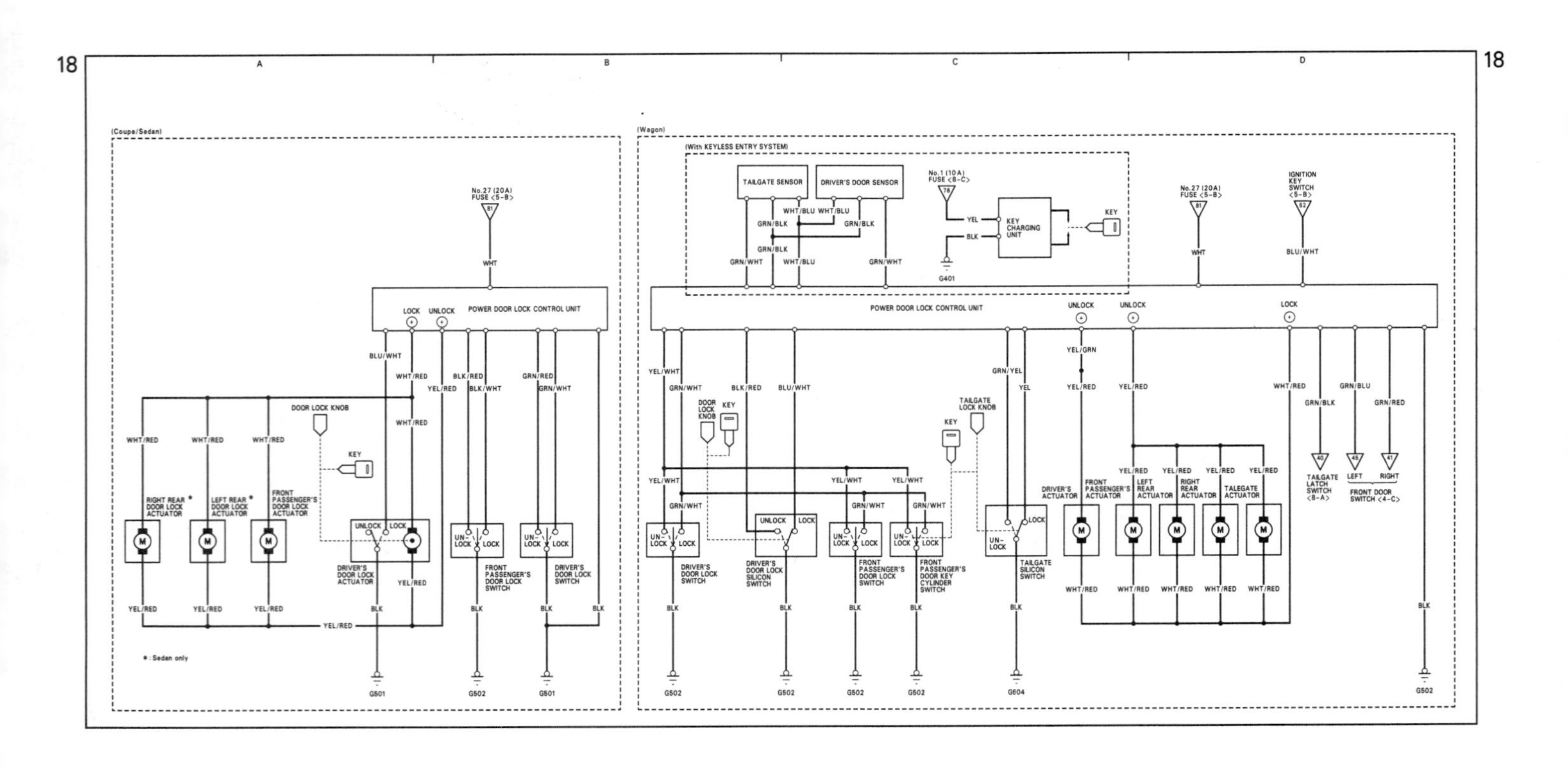

Figure 4-21 (continued)

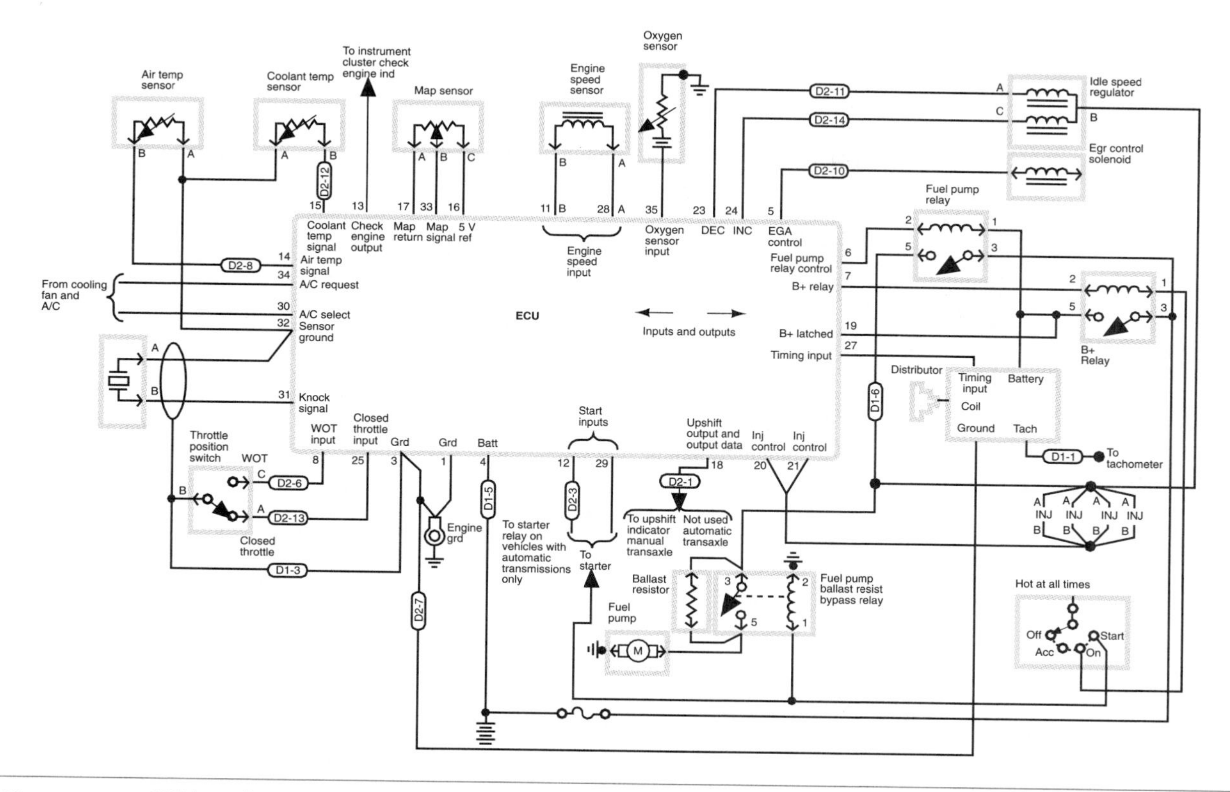

Figure 4-22 Wiring diagram illustrating only one specific circuit for easier reference. These are also known as block diagrams. (Courtesy of Chrysler Corporation)

Wiring diagrams show the wires, connections to switches and other components, and the type of connector used throughout the circuit. Total vehicle wiring diagrams are normally spread out over many pages of a service manual. Some are displayed on a single large sheet of paper that folds out of the manual. A system wiring diagram is actually a portion of the total vehicle diagram. The system and all related circuitry are shown on a single page. System diagrams are often easier to use than vehicle diagrams simply because there is less information to sort through.

A **wiring diagram** is an electrical schematic that shows a representation of actual electrical or electronic components (by use of symbols) and the wiring of the vehicle's electrical systems.

Remember that electrical circuits need a complete path in order to work. A wiring diagram shows the insulated side of the circuit and the point of ground. Also, when lines (or wires) cross on a wiring diagram, this does not mean they connect. If wires are connected, there will be a connector or a dot at the point where they cross. Most wiring diagrams do not show the location of the wires, connectors, or components in the vehicle. Some have location reference numbers displayed by the wires. After studying the wiring diagram you will know what you are looking for. Then you move to the car to find it.

In addition to entire vehicle and system specific wiring diagrams, there are other diagrams that may be used to diagnose electricity problems. An electrical **schematic** shows how the circuit is connected. It does not show the colors of the wires or their routing. Schematics are what have been used so far in this book. They display a working model of the circuit. These are especially handy when trying to understand how a circuit works. Schematics are typically used to show the internal circuitry of a component or to simplify a wiring diagram. One of the troubleshooting techniques used by good electrical technicians is to simplify a wiring diagram into a schematic.

Installation diagrams show where and how electrical components and wiring harnesses are installed in the vehicle. These are helpful when trying to locate where a particular wire or

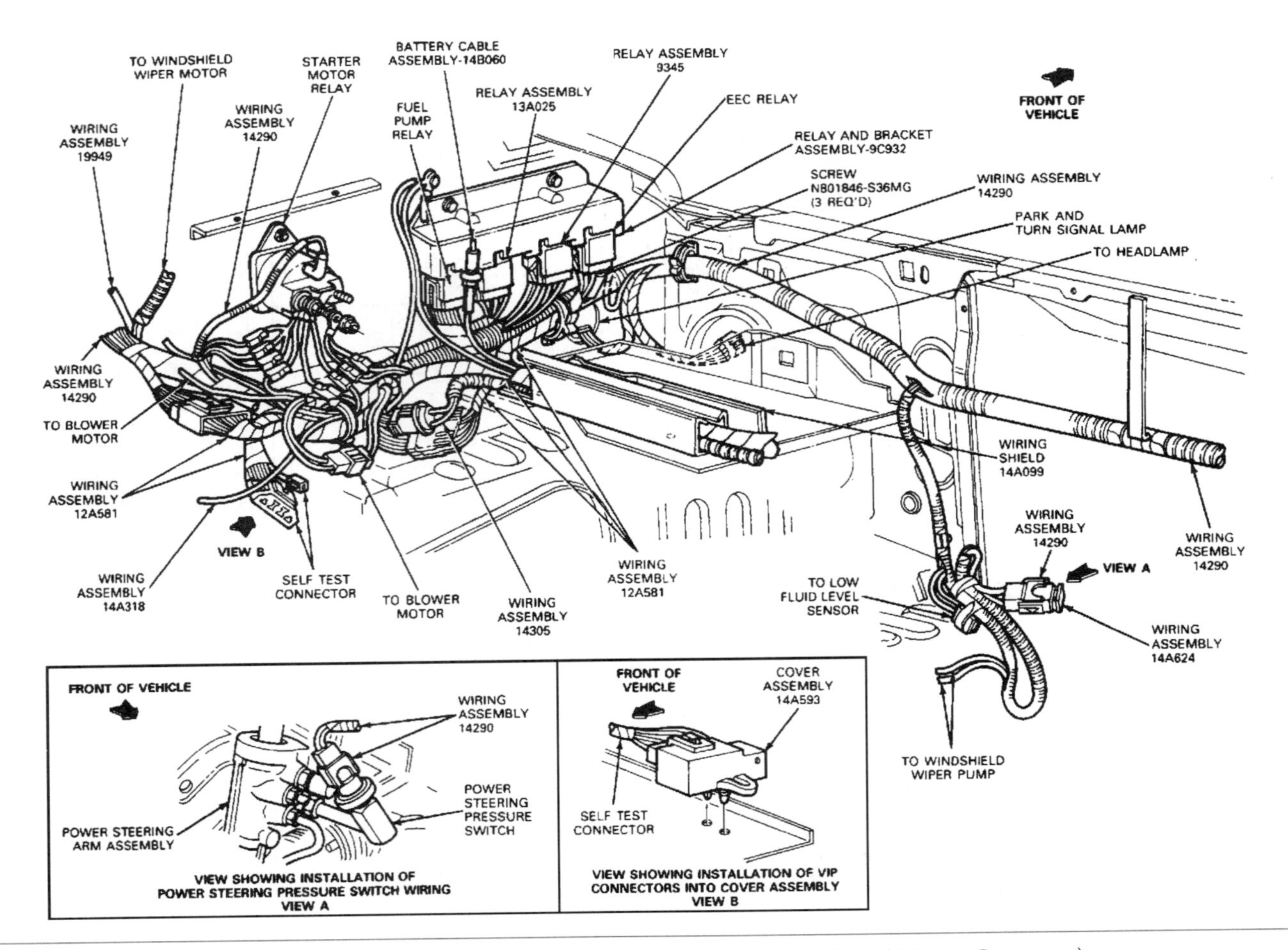

Figure 4-23 A typical installation diagram. (Reprinted with the permission of Ford Motor Company)

component may be in the car. These diagrams also may show how the component or wiring harness is attached to the vehicle (Figure 4-23).

Electrical Symbols

In place of actual pictures, a variety of **electrical symbols** are used to represent the components in the wiring diagram.

Most wiring diagrams do not show an actual drawing of the components. Rather they use symbols to represent the components. Often the symbol displays the basic operation of the component. Many different symbols have been used in wiring diagrams through the years. Figure 4-24 shows some of the commonly used symbols. The symbols may vary with each manufacturer, but most are close to those shown (Figure 4-25). You need to be familiar with all of the symbols; however, you don't need to memorize all of the variations. Wiring diagram manuals include a "legend" that helps you interpret the symbols.

A BIT OF HISTORY

The service manuals for early automobiles were hand drawn and labeled. They also had drawings of the actual components. As more and more electrical components were added to cars, this became impractical. Soon schematic symbols replaced the component drawings.

+	Positive
−	Negative
	Ground
	Fuse
	Circuit breaker
	Capacitor
Ω	Ohms
	Resistor
	Variable resistor
	Series resistor
	Coil
	Step up coil

	Connector
	Male connector
	Female connector
	Multiple connector
	Denotes wire continues elsewhere
	Splice
−2 2	Splice identification
	Optional Wiring with Wiring without
	Thermal element bimetal strip
	"Y" Windings
08:85	Digital readout
	Single filament lamp

	Open contact
	Closed contact
	Open switch
	Closed switch
	Closed ganged switch
	Open ganged switch
	Two pole single throw switch
	Pressure switch
	Solenoid switch
	Mercury switch
	Diode or rectifier
	Bi directional zener diode

	Dual filament lamp
	LED light emitting diode
	Thermistor
	Gauge
Timer	Timer
M	Motor
	Armature and brushes
	Denotes wire goes through grommet
	Denotes wire goes through 40 way disconnect
Steering column	Denotes wire goes through 25 way steering column connector
Instrument panel	Denotes wire goes through 25 way instrument panel connector

Figure 4-24 Common electrical symbols used in wiring diagrams. (Courtesy of Chrysler Corporation)

Color Codes and Circuit Numbering

Nearly all of the wires in an automobile are covered with colored insulation. These colors are used to identify wires and electrical circuits. The color of the wires is marked on a wiring diagram. Some wiring diagrams also include circuit numbers. These numbers, or letters and numbers, help identify a specific circuit. Both types of coding makes it easier to diagnose electrical problems. Unfortunately, not all manufacturers use the same method of wire identification.

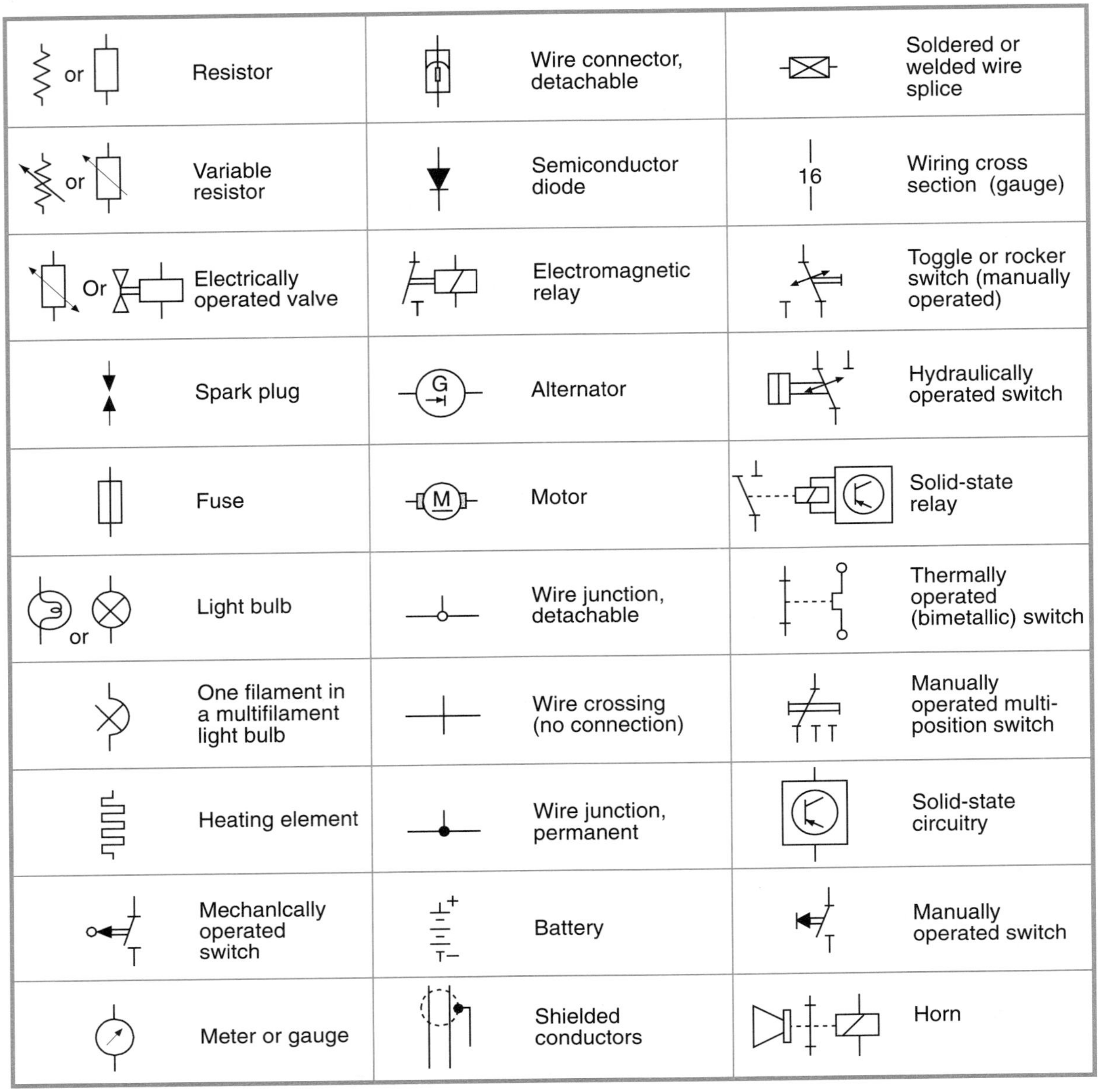

Resistor	Wire connector, detachable	Soldered or welded wire splice
Variable resistor	Semiconductor diode	Wiring cross section (gauge)
Electrically operated valve	Electromagnetic relay	Toggle or rocker switch (manually operated)
Spark plug	Alternator	Hydraulically operated switch
Fuse	Motor	Solid-state relay
Light bulb	Wire junction, detachable	Thermally operated (bimetallic) switch
One filament in a multifilament light bulb	Wire crossing (no connection)	Manually operated multi-position switch
Heating element	Wire junction, permanent	Solid-state circuitry
Mechanlcally operated switch	Battery	Manually operated switch
Meter or gauge	Shielded conductors	Horn

Figure 4-25 Electrical symbols used in some import manuals.

Figure 4-26 shows common color codes and their abbreviations. Most wiring diagrams list the appropriate color coding used by the manufacturer. Make sure you understand what color the code is referring to before looking for a wire.

In most color codes, the first group of letters designates the base color of the insulation. If a second group of letters is used, it indicates the color of the tracer. For example, a wire designated as WH/BLK would have a white base color with a black tracer.

Ford uses four methods of color coding its wires (Figure 4-27):

1. Solid color.
2. Base color with a stripe (tracer).
3. Base color with hash marks.
4. Base color with dots.

Color	Abbreviations		
Aluminum	AL		
Black	BLK	BK	B
Blue (Dark)	BLU DK	DB	DK BLU
Blue	BLU	B	L
Blue (Light)	BLU LT	LB	LT BLU
Brown	BRN	BR	BN
Glazed	GLZ	GL	
Gray	GRA	GR	G
Green (Green)	GRN DK	DG	DK GRN
Green (Light)	GRN LT	LG	LT GRN
Maroon	MAR	M	
Natural	NAT	N	
Orange	ORN	O	ORG
Pink	PNK	PK	P
Purple	PPL	PR	
Red	RED	R	RD
Tan	TAN	T	TN
Violet	VLT	V	
White	WHT	W	WH
Yellow	YEL	Y	YL

Figure 4-26 Common color codes used in automotive applications.

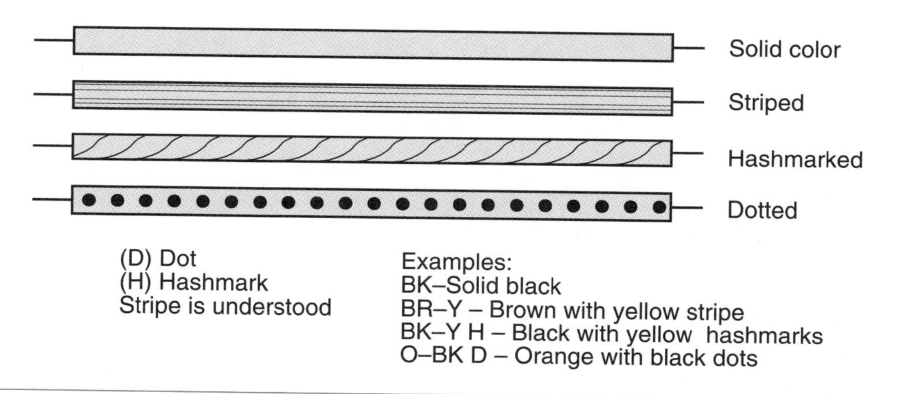

Figure 4-27 Four methods that Ford uses to color code its wires. (Reprinted with the permission of Ford Motor Company)

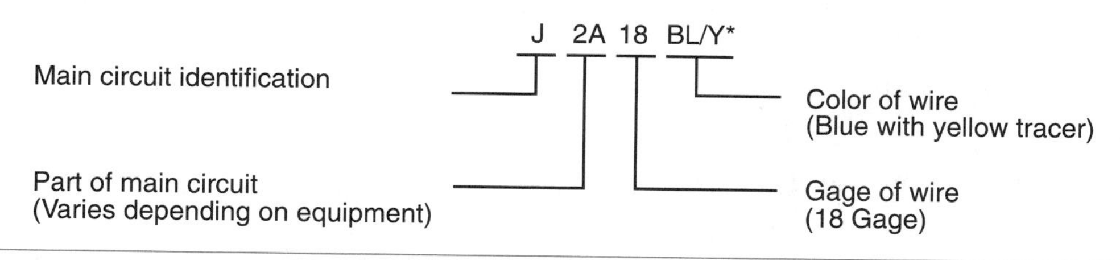

Figure 4-28 Chrysler wiring code identification. (Courtesy of Chrysler Corporation)

Chrysler uses a numbering method to designate the circuits on the wiring diagram (Figure 4-28). The circuit identification, wire gauge, and color of the wire are included in the wire number. Chrysler identifies the main circuits by using a main circuit identification code that corresponds to the first letter in the wire number (Figure 4-29).

General Motors uses numbers that include the wire gauge in metric millimeters, the wire color, the circuit number, splice number, and ground identification (Figure 4-30). In this example,

Main Circuit Identification Codes

Code	Circuit
A	BATTERY FEED (i.e. Fuselink Feeds, Starter Feeds, Starter Relay)
B	Brakes
C	Climate Control (A/C, Heater, E.B.L. and Heated Mirror Related Circuits)
D	Diagnostic Circuits
E	Dimming Illumination Circuits
F	Fused Circuits (Non-Dedicated Multi-System Feeds)
G	Monitoring Circuits (Gages, Clocks, Warning Devices)
H	**OPEN**
I	Not used as a circuit designator
J	**OPEN**
K	Engine Logic Module Control Circuits
L	Exterior Lighting Circuits
M	Interior Lighting Circuits (Dome, Courtesy Lamps, Cargo Lamps)
N	ESA Module Electronic Circuits
O	Not used as a circuit designator
P	Power Options (Battery Feed) (i.e. Seats, Door Locks, Mirrors, Deck Lid Release, etc.)
Q	Power Options (Ignition Feed) (i.e. Windows, Power Top, Power Sun Roof, etc.)
R	Passive Restraint
S	Suspension and Steering Circuits
T	Transmission/Transaxle, Differential, Transfer Case and Starter System Circuits
U	**OPEN**
V	Speed Control and Wash Wipe Circuit
W	**OPEN**
X	Sound Systems (i.e. Radio and Horn)
Y	**OPEN**
Z	Grounds (B−)

Figure 4-29 Chrysler circuit identification codes.

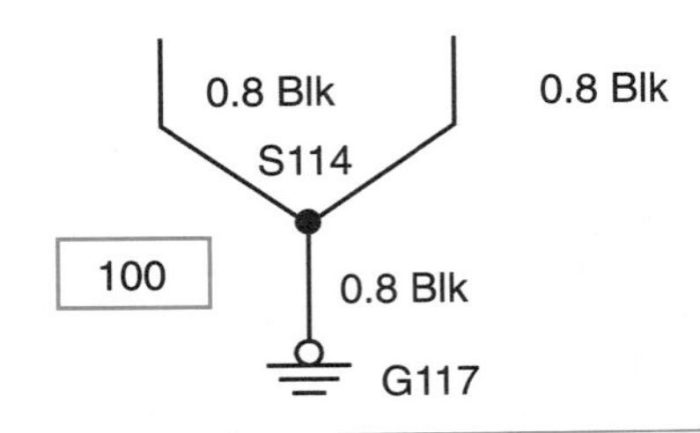

Figure 4-30 GM's method of circuit and wire identification.

A **tracer** is a thin or dashed line of a different color than the base color of the insulation.

the circuit is designated as 120, the wire size is 0.8 mm2, the insulation color is black, the splice is numbered S114, and the ground is designated as G117.

Most manufacturers also number connectors and terminals for identification.

Shop Manual
Chapter 4, page 99

Standardized Wiring Designations. The Society of Automotive Engineers (SAE) is attempting to standardize the circuit diagrams used by the various manufacturers. The system that is developed may be similar to the DIN used by import manufacturers. DIN assigns certain color codes to a particular circuit as follows:

- Red wires are used for direct battery-powered circuits and also ignition-powered circuits.
- Black wires are also powered circuits controlled by switches or relays.
- Brown wires are usually the grounds.
- Green wires are used for ignition primary circuits.

A combination of wire colors is used to identify subcircuits. The base color still identifies the circuit's basic purpose. In addition to standardized color coding, DIN attempts to standardize terminal identification and circuit numbering.

DIN is the abbreviation for Deutsche Institut füer Normung (German Institute for Standardization) and is the recommended standard for European manufacturers to follow.

Component Locators

The wiring diagrams in most service manuals may not indicate the exact physical location of the components of the circuit. Another shop manual called a component locator is used to find where a component is installed in the vehicle. The component locator may use both drawings and text to lead the technician to the desired component (Figure 4-31).

A **component locator** is used to determine the exact location of several of the electrical components.

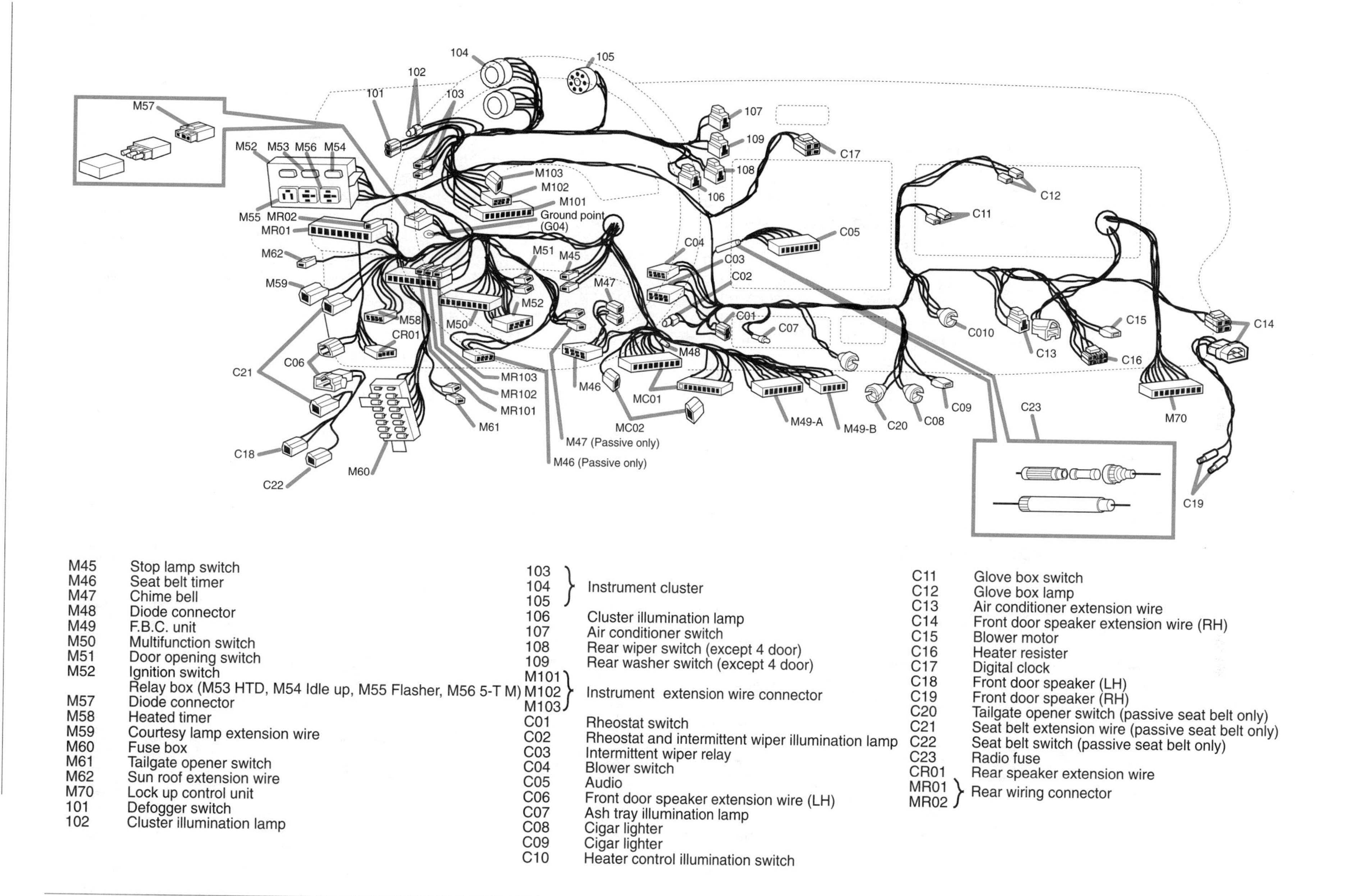

Figure 4-31 The component locator is used to find the actual location of a component. (Courtesy of Hyundai Motor America)

I-MARK

Component & Location	Page NO. 11–	Fig. No.
A/C Relay (1987 Turbo): Behind LH Side Of I/P	53	25
A/C Relay (1987 Turbo): RH Side Of Cowl	36	6
A/C Relay (1987): Behind LH Side Of I/P	36	7
A/C Relay (1988 Turbo): Behind LH Side Of I/P	56	33
A/C Relay (1988 Turbo): RH Side Of Cowl	55	31
A/C Relay (1988): Behind LH Side Of I/P	55	32
AIR Vacuum Switching Valve (1987 Turbo):RH Side Of Cowl	38	12
AIR Vacuum Switching Valve (1988 Turbo): RH Side Of Cowl	62	42
Alternator (1987 Exc. Turbo): RH Side Of Engine	39	14
Alternator (1987 Turbo): RH Side Of Engine	41	16
Alternator (1988 Exc. Turbo): RH Side Of Engine	57	37
Alternator (1988 Turbo): RH Side Of Engine	59	39
Auto Choke Relay (1987 Exc. Turbo): RH Side Of Cowl	36	5
Auto Choke Relay (1988 Exc. Turbo): RH Side Of Cowl	55	30
Canister Vacuum Switching Valve (1988 Turbo): RH Side Of Cowl	38	12
Canister Vacuum Switching Valve (1988 Turbo): RH Side Of Cowl	62	42
Car Speed Sensor (1987 Turbo): Inside Speedometer	38	11
Car Speed Sensor (1987 Turbo): Inside Speedometer	38	12
Car Speed Sensor (1988 Turbo): Inside Speedometer	61	41
Car Speed Sensor (1988 Turbo): Inside Speedometer	62	42
Carburetor (1987 Exc. Turbo): On Intake Manifold	37	10
Carburetor (1988 Exc. Turbo): On Intake Manifold	60	40
Condenser Fan Relay (1987 Turbo): RH Render Apron	36	6
Condenser Fan Relay (1988 Turbo): RH Fender Apron	55	31
Connector A1 (1987 Exc. Turbo): Behind Center Of I/P	49	23
Connector A1 (1987 Turbo): Behind Center Of I/P	51	24
Connector A1 (1988 Exc. Turbo): Behind Center Of I/P	70	49
Connector A1 (1988 Turbo): Behind Center Of I/P	72	50
Connector A2 (1987 Exc. Turbo): Behind Center Of I/P	49	23
Connector A2 (1987 Turbo): Behind Center Of I/P	51	24
Connector A2 (1988 Exc. Turbo): Behind Center Of I/P	70	49
Connector A2 (1988 Turbo): Behind Center Of I/P	72	50
Connector A3 (1987 Exc. Turbo): Behind Center Of I/P	49	23
Connector A3 (1987 Turbo): Behind Center Of I/P	51	24

I-MARK—Continued

Component & Location	Page NO. 11–	Fig. No.
Connector A3 (1988 Exc. Turbo): Behind Center Of I/P	70	49
Connector A3 (1988 Turbo): Behind Center Of I/P	72	50
Connector A4 (1987 Exc. Turbo): Behind Center Of I/P	49	23
Connector A4 (1987 Turbo): Behind Center Of I/P	51	24
Connector A4 (1988 Exc. Turbo): Behind Center Of I/P	70	49
Connector A4 (1988 Turbo): Behind Center Of I/P	72	50
Connector A5 (1987 Exc. Turbo): Behind Center Of I/P	49	23
Connector A5 (1988 Exc. Turbo): Behind Center Of I/P	70	49
Connector C1 (1987 Exc. Turbo): LH Side Of Cowl	46	21
Connector C1 (1987 Turbo): LH Side Of Cowl	47	22
Connector C1 (1988 Exc. Turbo): LH Side Of Cowl	67	47
Connector C1 (1988 Turbo): LH Side Of Cowl	68	48
Connector C2 (1987 Exc. Turbo): LH Side Of Cowl	46	21
Connector C2 (1987 Turbo): LH Side Of Cowl	47	22
Connector C2 (1988 Exc. Turbo): LH Side Of Cowl	67	47
Connector C2 (1988 Turbo): LH Side Of Cowl	68	48
Connector C3 (1987 Exc. Turbo): LH Side Of Cowl	46	21
Connector C3 (1987 Turbo): LH Fender Apron	47	22
Connector C3 (1988 Exc. Turbo): LH Side Of Cowl	67	47
Connector C3 (1988 Turbo): LH Fender Apron	68	48
Connector C4 (1987 Exc. Turbo): LH Side Of Cowl	46	21
Connector C4 (1987 Turbo): LH Side Of Cowl	47	22
Connector C4 (1988 Exc. Turbo): LH Side Of Cowl	67	47
Connector C4 (1988 Turbo): LH Side Of Cowl	68	48
Connector C5 (1987 Exc. Turbo): LH Side Of Cowl	46	21
Connector C5 (1987 Turbo): LH Side Of Cowl	47	22
Connector C5 (1988 Exc. Turbo): LH Side Of Cowl	67	47
Connector C5 (1988 Turbo): LH Side Of Cowl	68	48
Connector C6 (1987 Exc. Turbo): LH Side Of Cowl	46	21
Connector C6 (1987 Turbo): LH Side Of Cowl	47	22
Connector C6 (1988 Exc. Turbo): LH Side Of Cowl	67	47
Connector C6 (1988 Turbo): LH Side Of Cowl	68	48
Connector C7 (1987 Exc. Turbo): LH Fender Apron	46	21
Connector C7 (1987 Turbo): LH Fender Apron	47	22
Connector C7 (1988 Exc. Turbo): LH Fender Apron	67	47
Connector C7 (1988 Turbo): LH Fender Apron	68	48

Figure 4-31 (continued)

Many electrical components may be hidden behind kick panels, dash boards, fender wells, and under seats. The use of a component locator will save the technician time in finding the suspected defective unit.

Summary

- ❑ Most of the primary wiring conductors used in the automobile are made of several strands of copper wire wound together and covered with a polyvinyl chloride (PVC) insulation.
- ❑ Stranded wire is used because of its flexibility and current flows on the surface of the conductors. Because there is more surface area exposed in a stranded wire, there is less resistance in the stranded wire than in the solid wire.
- ❑ There are three major factors that determine the proper size of wire to be used: (1) The wire must be large enough diameter—for the length required—to carry the necessary current for the load components in the circuit to operate properly, (2) The wire must be able to withstand the anticipated vibration, and (3) The wire must be able to withstand the anticipated amount of heat exposure.
- ❑ Wire size is based on the diameter of the conductor.
- ❑ Factors that affect the resistance of the wire include the conductor material, wire diameter, wire length, and temperature.
- ❑ Terminals can be either crimped or soldered to the conductor. The terminal makes the electrical connection and it must be capable of withstanding the stress of normal vibration.
- ❑ Printed circuit boards are used to simplify the wiring of the circuits they operate. A printed circuit is made of a thin phenolic or fiberglass board that copper (or some other conductive material) has been deposited on.
- ❑ A wire harness is an assembled group of wires that branch out to the various electrical components. It is used to reduce the amount of loose wires hanging under the hood or dash. It provides for a safe path for the wires of the vehicle's lighting, engine, and accessory components.
- ❑ The wiring harness is made by grouping insulated wires and wrapping them together. The wires are bundled into separate harness assemblies that are joined together by connector plugs.
- ❑ A wiring diagram shows a representation of actual electrical or electronic components and the wiring of the vehicle's electrical systems.
- ❑ The technician's greatest helpmate in locating electrical problems is the wiring diagram. Correct use of the wiring diagram will reduce the amount of time a technician needs to spend tracing the wires in the vehicle.
- ❑ In place of actual pictures, variety of electrical symbols are used to represent the components in the wiring diagram.
- ❑ Color codes and circuit numbers are used to make tracing wires easier.
- ❑ In most color codes, the first group of letters designates the base color of the insulation. If a second group of letters is used, it indicates the color of the tracer.
- ❑ A component locator is used to determine the exact location of several of the electrical components.

Terms to Know

Ballast resistor
Bulkhead connectors
Common connections
Component locator
Cross-fire
Electrical symbols
Gauge
Metri-pack connectors
Molded connector
Multiple-wire connector
Primary wiring
Printed circuits
Radiofrequency interference (RFI)
Resistance wire
Secondary wiring
Stranded wire
Television-radio-suppression (TVRS) cables
Terminal connections
Weather-pack connectors
Wire harness
Wiring diagram

Review Questions

Short Answer Essays

1. Explain the purpose of wiring diagrams.
2. Explain how wire size is determined by the American Wire Gauge (AWG) and metric methods.
3. Explain the purpose and use of printed circuits.
4. Explain the purpose of the component locator.
5. Explain when single-stranded or multistranded wire should be used.
6. Explain how temperature affects resistance and wire size selection.
7. List the three major factors that determine the proper size of wire to be used.
8. List and describe the different types of terminal connectors used in the automotive electrical system.
9. What is the difference between a complex and a simple wiring harness?
10. Describe the methods the three domestic automobile manufacturers use for wiring code identification.

Fill-in-the-Blanks

1. There is ____________________ resistance in the stranded wire than in the solid wire.
2. ____________________ ____________________ is the electromagnetic induction spark that can be transmitted in another wire that is close to the wire carrying the current.
3. Wire size is based on the ____________________ of the conductor.
4. In the AWG standard, the ____________________ the number, the smaller the wire ____________________.
5. An increase in temperature creates a ____________________ ____________________ in resistance.
6. ____________________ connectors are used when several wires must pass through the bulkhead.
7. ____________________ ____________________ ____________________ are used to prevent damage to the wiring by maintaining proper wire routing and retention.
8. A wiring diagram is an electrical schematic that shows a ____________________ of actual electrical or electronic components (by use of symbols) and the ____________________ of the vehicle's electrical systems.
9. In most color codes, the first group of letters designates the ____________________ ____________________ of the insulation. The second group of letters indicates the color of the ____________________.
10. A ____________________ ____________________ is used to determine the exact location of several of the electrical components.

ASE Style Review Questions

1. Automotive wiring is being discussed:
   Technician A says most primary wiring is made of several strands of copper wire wound together and covered with an insulation.
   Technician B says the types of conductor materials used in automobiles include copper, silver, gold, aluminum, and tin-plated brass.
   Who is correct?
   A. A only
   B. B only
   C. Both A and B
   D. Neither A nor B

2. Stranded wire use is being discussed:
   Technician A says there is less exposed surface area for electron flow in a stranded wire.
   Technician B says there is more resistance in the stranded wire than in the same gauge solid wire.
   Who is correct?
   A. A only
   B. B only
   C. Both A and B
   D. Neither A nor B

3. Spark plug wires are being discussed:
   Technician A says RFI is controlled by using resistances in the conductor of the spark plug wire.
   Technician B says all late model ignition systems use resistance wires to control RFI.
   Who is correct?
   A. A only
   B. B only
   C. Both A and B
   D. Neither A nor B

4. Spark plug wire installation is being discussed:
   Technician A says there is little that can be done to prevent cross-fire.
   Technician B says the spark plug wires must be installed in the proper separator and any two parallel wires next to each other in the firing order should be positioned as far away from each other as possible.
   Who is correct?
   A. A only
   B. B only
   C. Both A and B
   D. Neither A nor B

5. The selection of the proper size of wire to be used is being discussed:
   Technician A says the wire must be large enough—for the length required—to carry the amount of current necessary for the load components in the circuit to operate properly.
   Technician B says temperature has little effect on resistance and it is not a factor in wire size selection.
   Who is correct?
   A. A only
   B. B only
   C. Both A and B
   D. Neither A nor B

6. Terminal connectors are being discussed:
   Technician A says good terminal connections will resist corrosion.
   Technician B says the terminals can be either crimped or soldered to the conductor.
   Who is correct?
   A. A only
   B. B only
   C. Both A and B
   D. Neither A nor B

7. Wire routing is being discussed:
   Technician A says to install additional electrical accessories it is necessary to support the primary wire in at least 10-foot intervals.
   Technician B says if the wire must be routed through the frame or body, use metal clips to protect the wire.
   Who is correct?
   A. A only
   B. B only
   C. Both A and B
   D. Neither A nor B

8. Printed circuit boards are being discussed:
   Technician A says printed circuit boards are used to simplify the wiring of the circuits they operate.
   Technician B says care must be taken not to touch the board with bare hands.
   Who is correct?
   A. A only
   B. B only
   C. Both A and B
   D. Neither A nor B

9. Wiring harnesses are being discussed:
Technician A says a wire harness is an assembled group of wires that branches out to the various electrical components.
Technician B says most underhood harnesses are simple harnesses.
Who is correct?
A. A only
B. B only
C. Both A and B
D. Neither A nor B

10. Wiring diagrams are being discussed:
Technician A says wiring diagrams give the exact location of the electrical components.
Technician B says a wiring diagram will indicate what circuits are interconnected, where circuits receive their voltage source, and what color of wires are used in the circuit.
Who is correct?
A. A only
B. B only
C. Both A and B
D. Neither A nor B

Automotive Batteries

Upon completion and review of this chapter, you should be able to:

- ❑ Explain the purposes of the battery.
- ❑ Describe the construction of conventional, maintenance-free, hybrid, and recombination batteries.
- ❑ Define the main elements of the battery.
- ❑ Explain the chemical action that occurs to produce current in a battery.
- ❑ Explain the chemical reaction that occurs in the battery during cycling.
- ❑ Describe the differences, advantages, and disadvantages between different types of batteries.
- ❑ Describe the different types of battery terminals used.
- ❑ Describe the methods used to rate batteries.
- ❑ Determine the correct battery to be installed into a vehicle.
- ❑ Explain the effects of temperature on battery performance.
- ❑ Describe the different loads or demands placed upon a battery during different operating conditions.
- ❑ Explain the major reasons for battery failure.
- ❑ Define battery-related terms such as deep cycle, electrolyte solution, and gassing.

Introduction

Shop Manual
Chapter 5, page 119

The battery does not store electricity as electrons. The battery stores energy in chemical form.

An automotive battery is an electrochemical device capable of producing electrical energy. When the battery is connected to an external load, such as a starter motor, an energy conversion occurs that results in an electrical current flowing through the circuit. Electrical energy is produced in the battery by the chemical reaction that occurs between two dissimilar plates that are immersed in an electrolyte solution. The automotive battery produces direct current (DC) electricity that flows in only one direction.

When discharging the battery (current flowing from the battery), the battery changes chemical energy into electrical energy. It is through this change that the battery releases stored energy. During charging (current flowing through the battery from the charging system), electrical energy is converted into chemical energy. As a result, the battery can store energy until it is needed.

The automotive battery has several important functions, including:

1. It operates the starting motor, ignition system, electronic fuel injection, and other electrical devices for the engine during cranking and starting.
2. It supplies all the electrical power for the vehicle accessories whenever the engine is not running or when the vehicle's charging system is not working.
3. It furnishes current for a limited time whenever electrical demands exceed charging system output.
4. It acts as a stabilizer of voltage for the entire automotive electrical system.
5. It stores energy for extended periods of time.

The largest demand placed on the battery occurs when it must supply current to operate the starter motor. The amperage requirements of a starter motor may be over several hundred amperes. This requirement is also affected by temperatures, engine size, and engine condition.

After the engine is started, the vehicle's charging system works to recharge the battery and to provide the current to run the electrical systems. Most AC generators have a maximum output

Many luxury vehicles, and vehicles equipped with many electrical accessories, may be equipped with a generator that is rated above 110 amperes.

Electrical loads that are still present when the ignition switch is in the OFF position are called **key-off** or **parasitic loads.**

The voltage regulator is also a key-off load on some General Motors vehicles because the sensing circuit is connected directly to the battery.

Vehicles that are not started for extended periods of time may discharge the battery as a result of parasitic loads.

The amount of time a battery can be discharged at a certain current rate until the voltage drops below a specified value is referred to as **reserve capacity.**

of 60 to 90 amperes. This is usually enough to operate all of the vehicle's electrical systems and meet the demands of these systems. However, under some conditions (such as engine running at idle) generator output is below its maximum rating. If there are enough electrical accessories turned on during this time (heater, wipers, headlights, and radio) the demand may exceed the AC generator output. The total demand may be 20 to 30 amperes. During this time the battery must supply the additional current.

Even with the ignition switch turned off, there are electrical demands placed on the battery. Clocks, memory seats, engine computer memory, body computer memory, and electronic sound system memory are all examples of key-off loads. The total current draw of key-off loads is usually less than 30 milliamperes.

In the event that the vehicle's charging system fails, the battery must supply all of the current necessary to run the vehicle. Most batteries will supply 25 amperes for approximately 120 minutes before discharging low enough to cause the engine to stop running.

The amount of electrical energy that a battery is capable of producing depends on the size, weight, active area of the plates, and the amount of sulfuric acid in the electrolyte solution.

In this chapter you will study the design and operation of different types of batteries currently used in automobiles. These include conventional batteries, maintenance-free batteries, hybrid batteries, and recombination batteries.

Conventional Batteries

The conventional battery is constructed of seven basic components:

1. Positive plates
2. Negative plates
3. Separators
4. Case
5. Plate straps
6. Electrolyte
7. Terminals

The difference between "3-year" and "5-year" batteries is the amount of material expanders used in the construction of the plates and the number of plates used to build a cell.

A plate, either positive or negative, starts with a grid. The grid has horizontal and vertical grid bars that intersect at right angles (Figure 5-1). An active material made from ground lead oxide, acid, and material expanders is pressed into the grid in paste form. The positive plate is

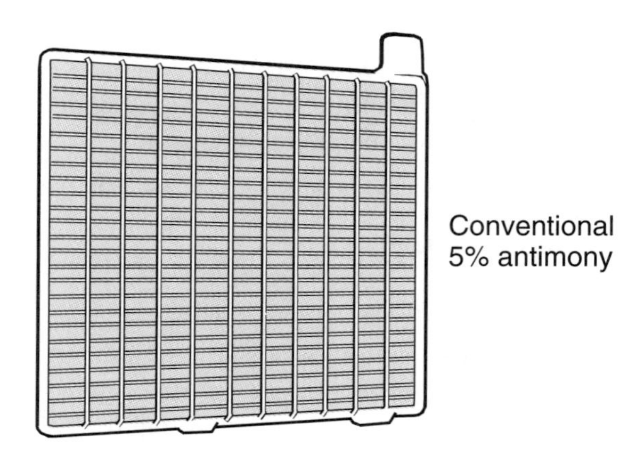

Figure 5-1 Conventional battery grid. (Reprinted with the permission of Ford Motor Company)

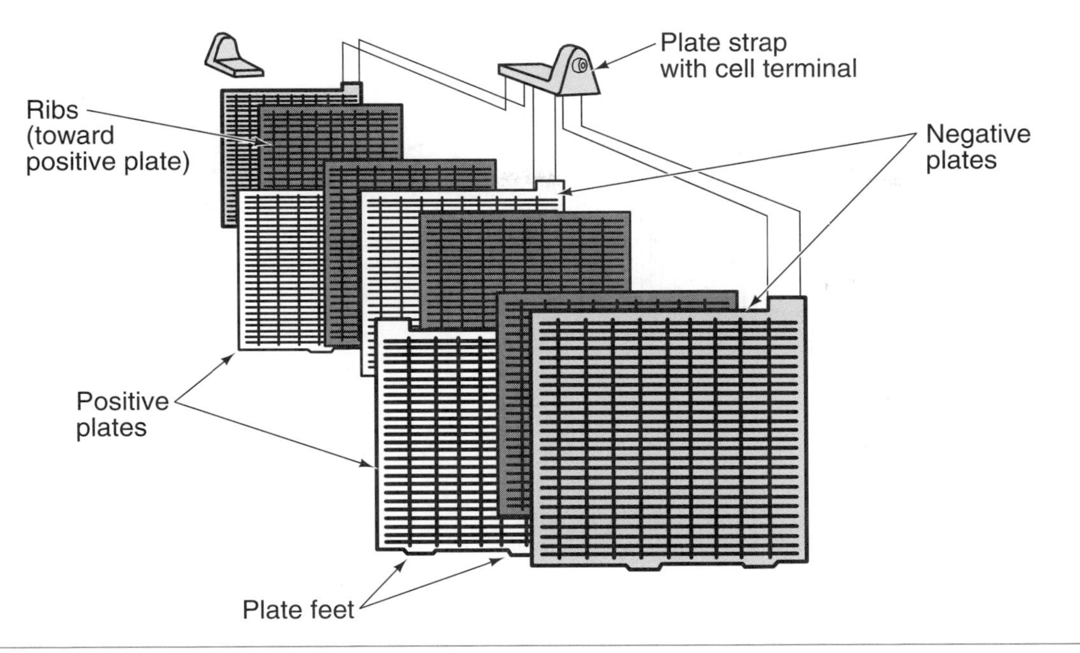

Figure 5-2 A battery cell consists of alternate positive and negative plates. (Reprinted with the permission of Ford Motor Company)

given a "forming charge" that converts the lead oxide paste into lead peroxide. The negative plate is given a "forming charge" that converts the paste into sponge lead.

The negative and positive plates are arranged alternately in each cell element (Figure 5-2). Each cell element can consist of 9 to 13 plates. The positive and negative plates are insulated from each other by separators made of microporous materials. The construction of the element is completed when all of the positive plates are connected to each other and all of the negative plates are connected to each other. The connection of the plates is by plate straps (Figure 5-3).

A typical 12-volt automotive battery is made up of six cells connected in series (Figure 5-4). This means the positive side of a cell element is connected to the negative side of the next cell element. This is repeated throughout all six cells. By connecting the cells in series the current

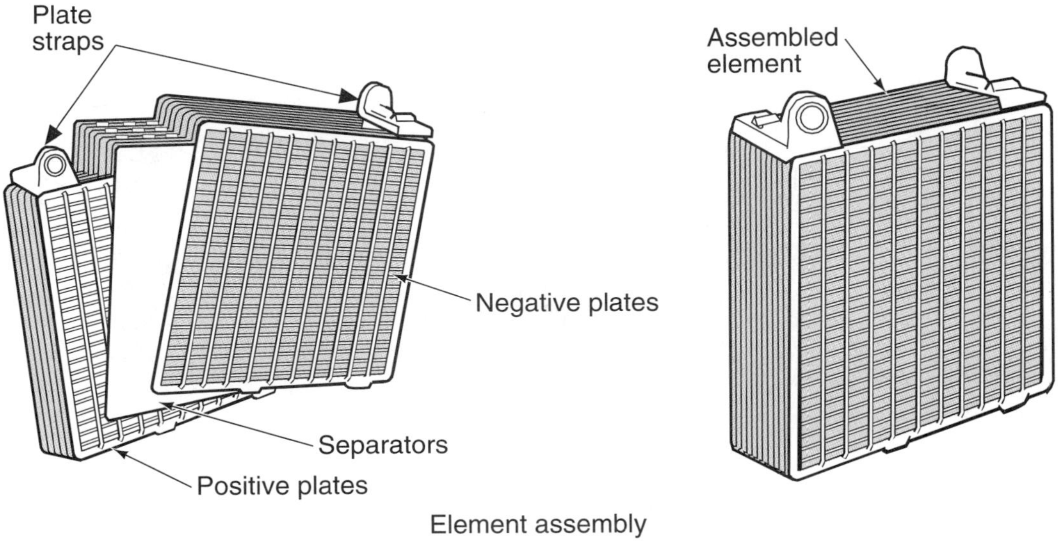

Plate
straps
Assembled
element
Negative plates
Separators
Positive plates
Element assembly

Figure 5-3 Construction of a battery element. (Reprinted with the permission of Ford Motor Company)

The condition of the battery should be the first test performed on a vehicle with an electrical problem. Without proper battery performance, the entire electrical system is affected.

Material expanders are fillers that can be used in place of the active materials. They are used to keep the manufacturing costs low.

Grids are generally made of lead alloys, usually antimony. About 5% to 6% antimony is added to increase the strength of the grid. The grid is the frame structure with connector tabs at the top.

The lead peroxide is composed of small grains of particles. This gives the plate a high degree of porosity, allowing the electrolyte to penetrate the plate.

The assembly of the positive plates, negative plates, and separators is called the **cell** or **element.**

Usually, negative plate groups contain one more plate than positive plate groups to help equalize the chemical activity.

Many batteries have envelope-type separators that retain active materials near the plates.

The most common connection used to connect cell elements is through the partition. It provides the shortest path and the least resistance.

Electrolyte consists of sulfuric acid diluted with water. The **electrolyte solution** used in automotive batteries consists of 64% water and 36% sulfuric acid, by weight. Electrolyte is both conductive and reactive.

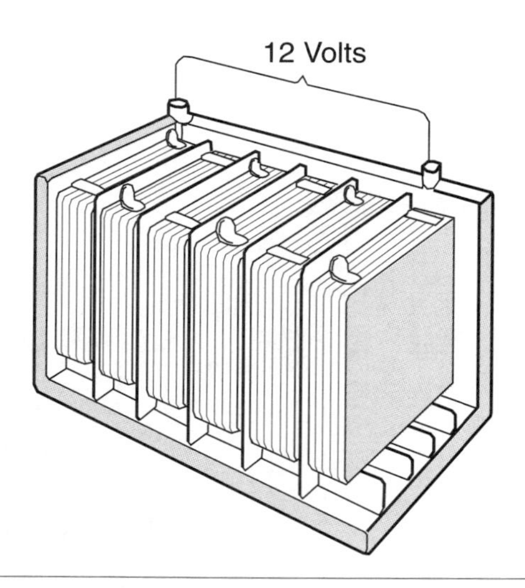

Figure 5-4 A 12-volt battery consists of six 2-volt cells that are connected in series. (Reprinted with the permission of Ford Motor Company)

capacity of the cell and cell voltage remain the same. The six cells produce 2.1 volts each. Wiring the cells in series produces the 12.6 volts required by the automotive electrical system. The plate straps provide a positive cell connection and a negative cell connection. The cell connection may be one of three types: through the partition, over the partition, or external (Figure 5-5). The cell elements are submerged in a cell case filled with electrolyte solution.

CAUTION: Electrolyte is very corrosive. It can cause severe injury if it comes in contact with skin and/or eyes. If you come into contact with battery acid, wash immediately with a water and baking soda solution (baking soda will neutralize the acid). If you get the acid in your eyes, flush them immediately with cool water or a commercial eye wash. Then seek medical treatment immediately. If you swallow any electrolyte, do not induce vomiting. Seek medical treatment immediately.

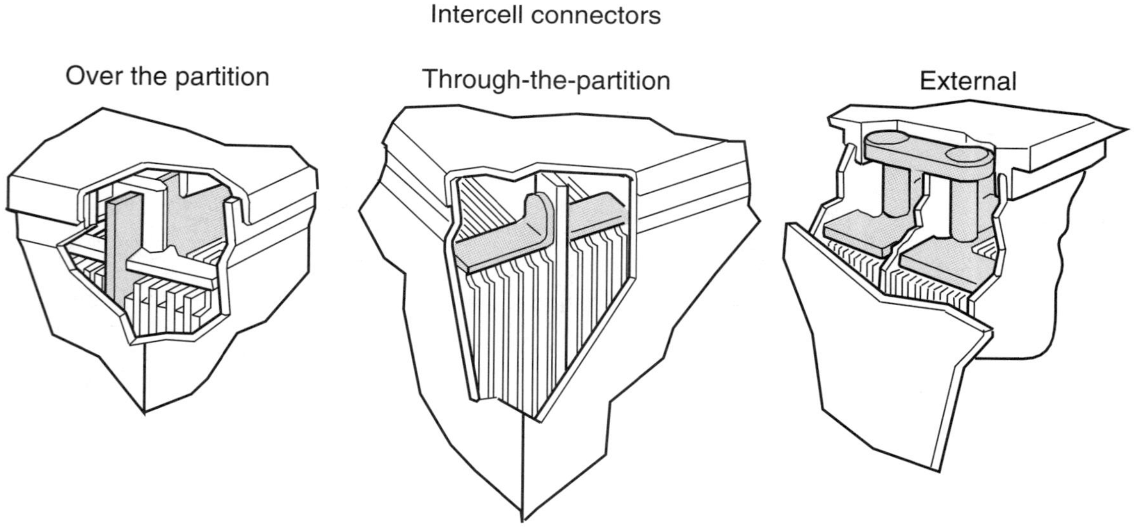

Intercell connectors

Over the partition

Through-the-partition

External

Figure 5-5 The cell elements can be connected using one of three different intercell connection methods. (Reprinted with the permission of Ford Motor Company)

WARNING: Do not add electrolyte to the battery (after the initial fill) if the cells are low. Add only distilled water. Distilled water has the minerals removed. The minerals in regular tap water can cause the battery cells to short out because they are conductive.

The battery case is made of polypropylene, hard rubber, and plastic base materials. The battery case must be capable of withstanding temperature extremes, vibration, and acid absorption. The cell elements sit on raised supports in the bottom of the case. By raising the cells, chambers are formed at the bottom of the case that trap the sediment that flakes off the plates. If the sediment was not contained in these chambers, it could cause a conductive connection across the plates and short the cell. The case is fitted with a one-piece cover.

Because the conventional battery releases hydrogen gas when it is being charged, the case cover will have vents. The vents are located in the cell caps of a conventional battery (Figure 5-6).

CAUTION: The hydrogen gases produced from a charging battery are very explosive. Exploding batteries are responsible for over 15,000 injuries per year that are severe enough to require hospital treatment. Do not smoke, have any open flames, or cause any sparks near the battery. Also, do not lay tools on the battery. They may short across the terminals causing the battery to explode. Always wear eye protection and proper clothing when working near the battery. Also, because most jewelry is an excellent conductor of electricity, do not wear it when performing work on or near the battery.

WARNING: A battery that has been rapidly discharged will create hydrogen gas. Do not attach jumper cables to a weak battery if starting the vehicle has been attempted. Wait for at least 10 minutes before connecting the jumper cable and attempting to start the vehicle.

Chemical Action

Shop Manual
Chapter 5, page 128

Activation of the battery is through the addition of electrolyte. This solution causes the chemical actions to take place between the lead peroxide of the positive plates and the sponge lead of the negative plates. The electrolyte is also the carrier that moves electric current between the positive and negative plates through the separators.

Figure 5-6 The vents of a conventional battery allow for the release of gases.

Specific gravity is the weight of a given volume of a liquid divided by the weight of an equal volume of water. Water has a specific gravity of 1.000.

The automotive battery has a fully charged specific gravity of 1.265 corrected to 80°F. Therefore, a specific gravity of 1.265 for electrolyte means it is 1.265 times heavier than an equal volume of water. As the battery discharges, the specific gravity of the electrolyte decreases.

Specific gravity is the weight of a given volume of a liquid divided by the weight of the same volume of water. Water, therefore, has a specific gravity of 1.000. The electrolyte, being heavier than water, has a specific gravity of more than 1.000. The electrolyte of a fully charged battery has a specific gravity of about 1.265. This decreases as the battery discharges because the electrolyte becomes more like water. The specific gravity of a battery can give you an indication of how charged a battery is.

Fully charged:	1.265 specific gravity
75% charged:	1.225 specific gravity
50% charged:	1.190 specific gravity
25% charged:	1.155 specific gravity
Discharged:	1.120 or lower specific gravity

These specific gravity values may vary slightly according to the design of the battery. However, regardless of the design, the specific gravity of the electrolyte in all batteries will decrease as the battery discharges. Temperature of the electrolyte will also affect its specific gravity. All specific gravity specifications are based on a standard temperature of 80°. When the temperature is above that temperature, the specific gravity is lower. When the temperature is below that standard, the specific gravity increases. Therefore all specific gravity measurements must be corrected for temperature. A general rule to follow is to add 0.004 for every 10 degrees above 80 degrees and subtract 0.004 for every 10 degrees below 80 degrees.

In operation, the battery is being partially discharged and then recharged. This represents an actual reversing of the chemical action that takes place within the battery. The constant cycling of the charge and discharge modes slowly wears away the active materials on the cell plates. This action eventually causes the battery plates to sulfate. The battery must be replaced once the sulfation of the plates has reached the point that there is insufficient active plate area.

In the charged state, the positive plate material is essentially pure lead peroxide, PbO_2. The active material of the negative plates is spongy lead, Pb. The electrolyte is a solution of sulfuric acid, H_2SO_4, and water. The voltage of the cell depends on the chemical difference between the active materials.

The illustration (Figure 5-7) shows what happens to the plates and electrolyte during discharge. The lead (Pb) from the positive plate combines with sulfate (SO_4) from the acid, forming

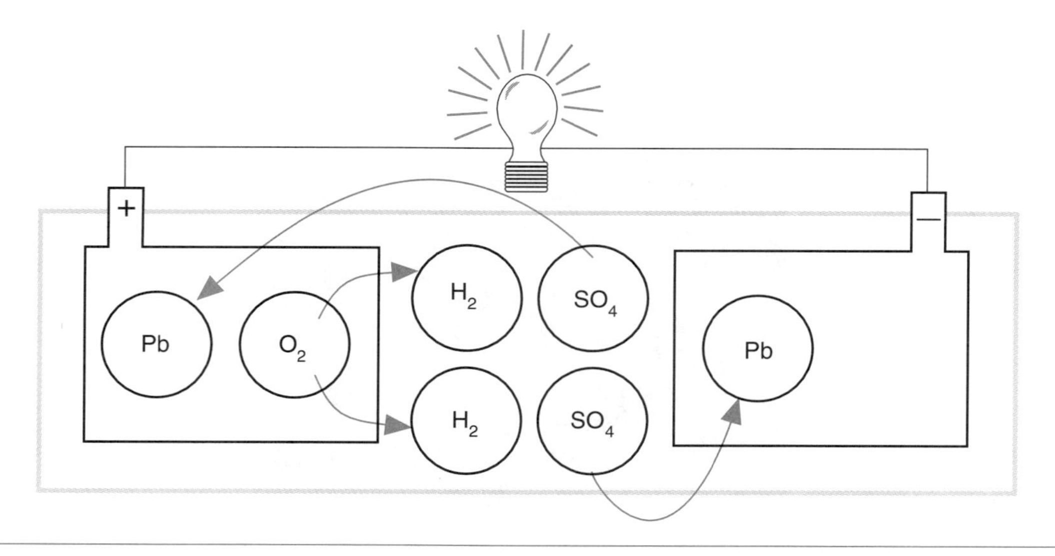

Figure 5-7 Chemical action that occurs inside the battery during the discharge cycle.

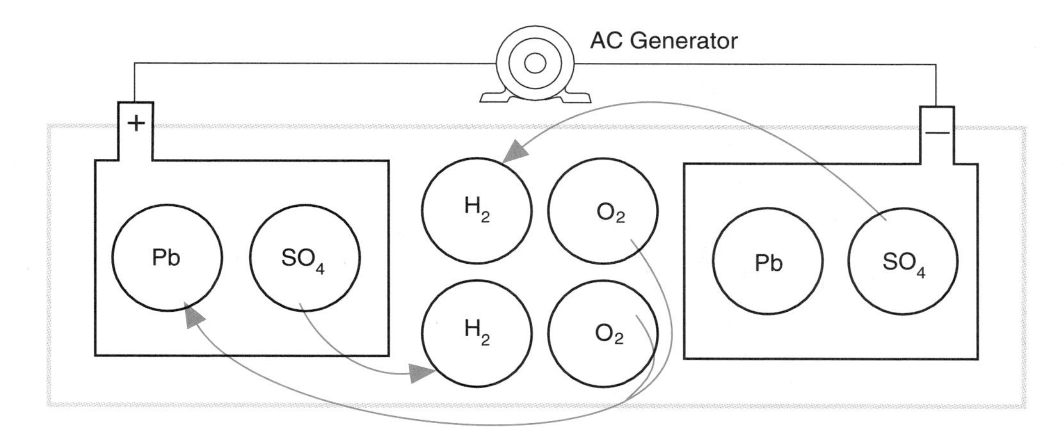

Figure 5-8 Chemical action inside the battery during the charge cycle.

lead sulfate ($PbSO_4$). While this is occurring, oxygen (O_2) in the active material of the positive plate joins with the hydrogen from the electrolyte forming water (H_2O). This water dilutes the acid concentration.

A similar reaction is occurring in the negative plate. Lead (Pb) is combining with sulfate (SO_4) forming lead sulfate ($PbSO_4$). The result of discharging is changing the positive plate from lead dioxide into lead sulfate and changing the negative plate into lead sulfate. Discharging a cell makes the positive and negative plates the same. Once they are the same, the cell is discharged.

The charge cycle is exactly the opposite (Figure 5-8). The lead sulfate ($PbSO_4$) in both plates is split into its original forms of lead (Pb) and sulfate (SO_4). The water in the electrolyte splits into hydrogen and oxygen. The hydrogen (H_2) combines with the sulfate to become sulfuric acid again (H_2SO_4). The oxygen combines with the positive plate to form the lead peroxide. This now puts the plates and the electrolyte back in their original form and the cell is charged.

Shop Manual
Chapter 5, page 121

Maintenance-free means there is no provision for the addition of water to the cells. The battery is sealed.

Gassing is the conversion of the battery water into hydrogen and oxygen gas. This process is also called **electrolysis.**

If electrolyte and dirt are allowed to accumulate on the top of the battery case it may create a conductive connection between the positive and negative terminals, resulting in a constant discharge on the battery.

Maintenance-Free Batteries

The maintenance-free battery contains cell plates made of a slightly different compound. The plate grids contain calcium, cadmium, or strontium to reduce gassing and self-discharge. The antimony used in conventional batteries is not used in maintenance-free batteries because it increases the breakdown of water into hydrogen and oxygen and because of its low resistance to overcharging. The use of calcium, cadmium, or strontium reduces the amount of vaporization that takes place during normal operation. The grid may be constructed with additional supports to increase its strength and to provide a shorter path, with less resistance, for the current to flow to the top tab (Figure 5-9).

Each plate is wrapped and sealed on three sides by an envelope design separator. The envelope is made from microporous plastic. By enclosing the plate in an envelope, the plate is insulated and reduces the shedding of the active material from the plate.

The battery is sealed except for a small vent so the electrolyte and vapors cannot escape. An expansion or condensation chamber allows the water to condense and drain back into the cells. Because the water cannot escape from the battery, it is not necessary to add water to the battery on a periodic basis. Containing the vapors also reduces the possibility of corrosion and discharge through the surface because of electrolyte on the surface of the battery. Vapors only leave the case when the pressure inside the battery is greater than atmospheric pressure.

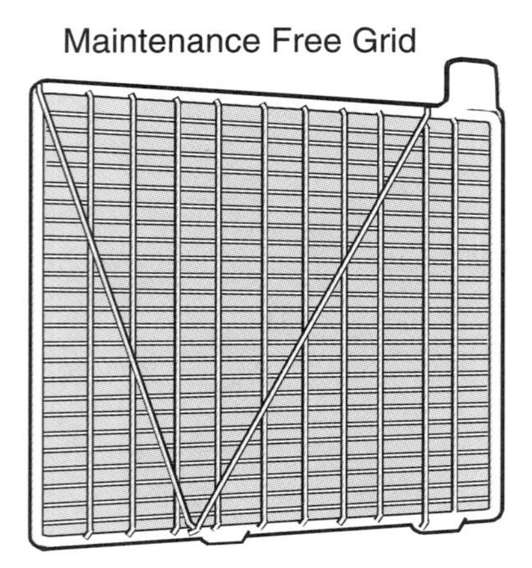

Figure 5-9 Maintenance-free battery grids with support bars give increased strength and faster electrical delivery. (Reprinted with the permission of Ford Motor Company)

A **hydrometer** is a test instrument that is used to check the specific gravity of the electrolyte to determine the battery's state of charge.

Some maintenance-free batteries have a built-in hydrometer that shows the state of charge (Figure 5-10). If the dot that is at the bottom of the hydrometer is green, then the battery is fully charged (more than 65% charged). If the dot is black, the battery state of charge is low. If the battery does not have a built-in hydrometer it cannot be tested with a hydrometer because the battery is sealed.

WARNING: If the dot is yellow or clear, do not atttempt to recharge the battery. A yellow or clear eye means the electrolyte is low; the battery must be replaced.

Many manufacturers have revised the maintenance-free battery to a "low maintenance battery," in that the caps are removable for testing and electrolyte level checks. Also the grid construction contains about 3.4% antimony. To decrease the distance and resistance of the path that current flows in the grid and to increase its strength, the horizontal and vertical grid bars do not intersect at right angles (Figure 5-11).

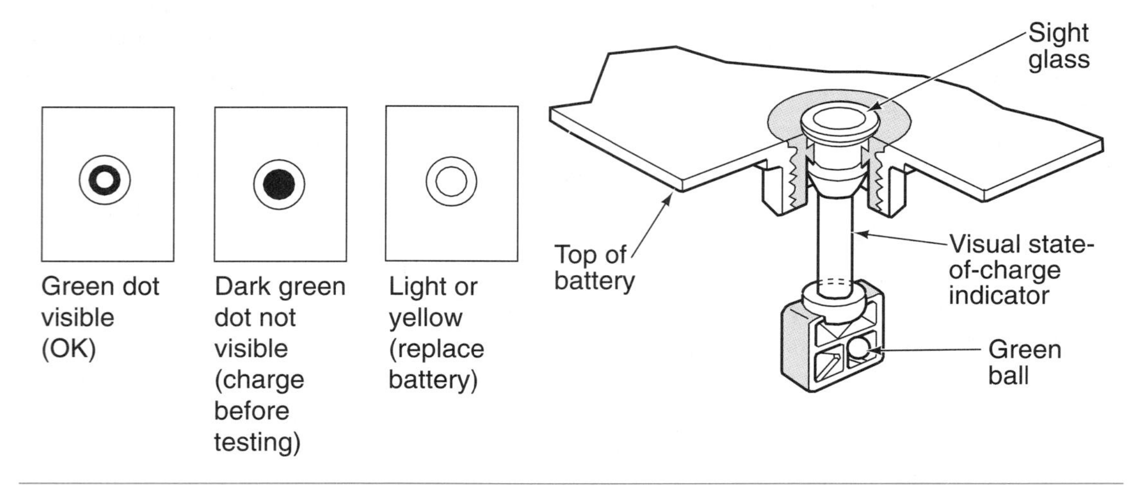

Figure 5-10 One cell of a maintenance-free battery has a built-in hydrometer, which gives an indication of overall battery condition. (Courtesy of Chrysler Corporation)

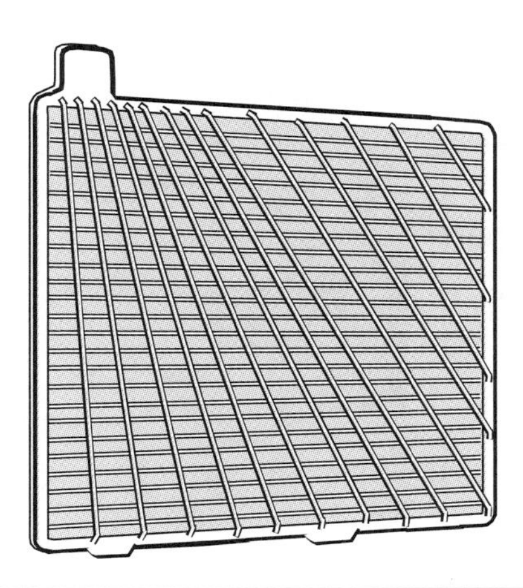

Figure 5-11 Low maintenance battery grid with vertical grid bars intersecting at an angle. (Reprinted with the permission Ford Motor Company)

Grid growth is a condition where the grid grows little metallic fingers that extend through the separators and short out the plates.

Deep cycling is to discharge the battery to a very low state of charge before recharging it.

The **hybrid battery** combines the advantages of the low maintenance and maintenance-free batteries.

Radial means branching out from a common center

Some manufacturers are using fiberglass separators.

The advantages of maintenance-free batteries over conventional batteries include:

1. A larger reserve of electrolyte above the plates.
2. Increased resistance to overcharging.
3. Longer shelf life (approximately 18 months).
4. Ability to be shipped with electrolyte installed, reducing the possibility of accidents and injury to the technician.
5. Higher cold cranking amps rating.

The major disadvantages of the maintenance-free battery include:

1. Grid growth when the battery is exposed to high temperatures.
2. Inability to withstand deep cycling.
3. Low reserve capacity.
4. Faster discharge by parasitic loads.
5. Shorter life expectancy.

A BIT OF HISTORY

Buick first introduced the storage battery as standard equipment in 1906.

Hybrid Batteries

The hybrid battery can withstand six deep cycles and still retain 100% of its original reserve capacity. The grid construction of the hybrid battery consists of approximately 2.75% antimony alloy on the positive plates and a calcium alloy on the negative plates. This allows the battery to withstand deep cycling while retaining reserve capacity for improved cranking performance. Also, the use of antimony alloys reduces grid growth and corrosion. The lead calcium has less gassing than conventional batteries.

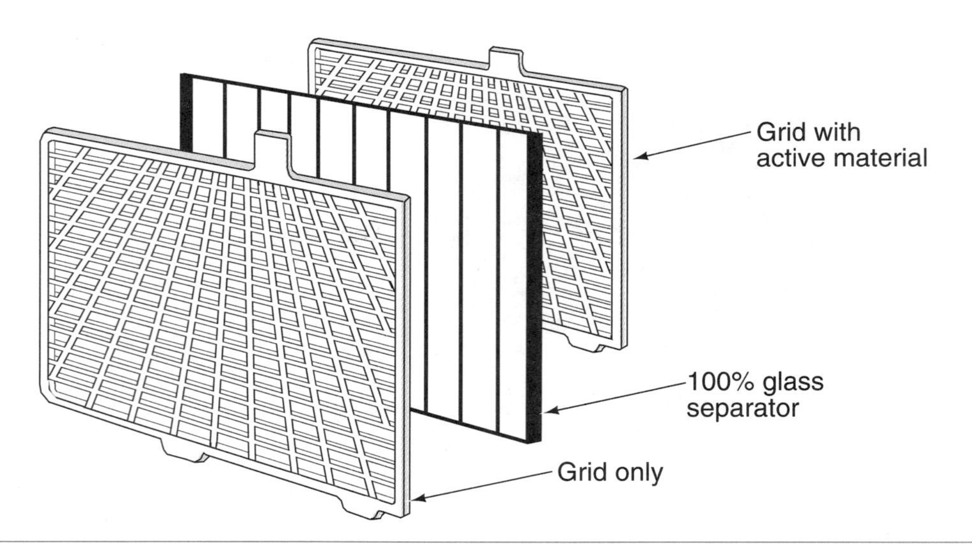

Figure 5-12 Hybrid and separator construction. (Reprinted with the permission of Ford Motor Company)

Grid construction differs from other batteries in that the plates have a lug located near the center of the grid. In addition, the vertical and horizontal grid bars are arranged in a radial pattern (Figure 5-12). By locating the lug near the center of the grid and using the radial design, the current has less resistance and a shorter path to follow to the lug (Figure 5-13). This means the battery is capable of providing more current at a faster rate.

The separators used are constructed of glass with a resin coating. The glass separators offer low electrical resistance with high resistance to chemical contamination. This type of construction provides for increased cranking performance and battery life.

Recombination Batteries

Recombination batteries are sometimes called dry-cell batteries because they do not use a liquid electrolyte solution.

One of the most recent variations of the automobile battery is the recombination battery (Figure 5-14). The recombination battery does not use a liquid electrolyte. Instead, it uses separators that hold a gel-type material. The separators are placed between the grids and have very low electrical resistance. Because of this design, output voltage and current are higher than in conventional batteries. The extra amount of available voltage (approximately 0.6 V) assists in cold weather starting. Also, gassing is virtually eliminated.

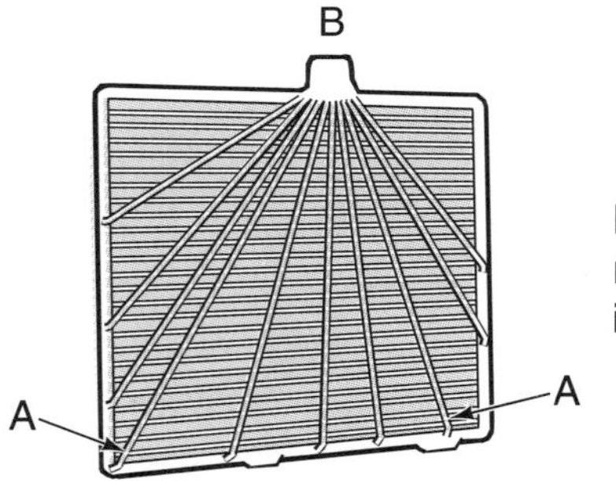

Figure 5-13 The hybrid battery grid construction allows for faster current delivery. Electrical energy at point "A" has a shorter distance to travel to get to the tab at point "B."

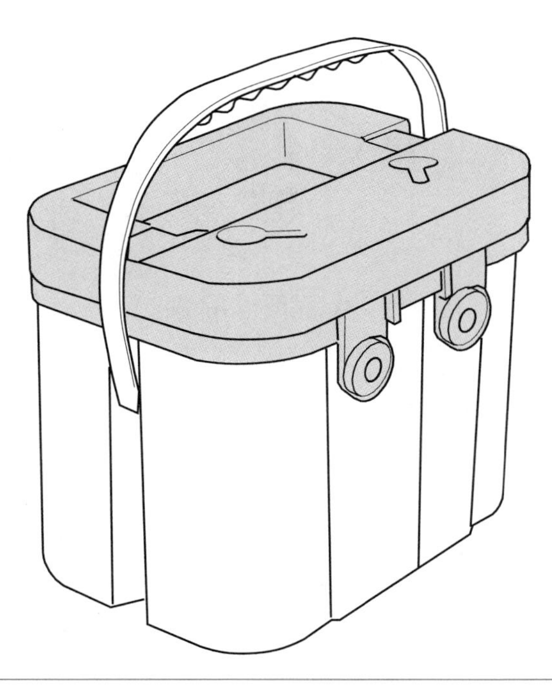

Figure 5-14 The recombination battery is one of the most recent advances in the automotive battery. (Courtesy of Optima Batteries, Inc.)

The following are some other safety features and advantages of the recombination battery:

1. Contains no liquid electrolyte. If the case is cracked, no electrolyte will spill.
2. Can be installed in any position, including upside down.
3. Is corrosion free.
4. Has very low maintenance because there is no electrolyte loss.
5. Can last as much as four times longer than conventional batteries.
6. Can withstand deep cycling without damage.
7. Can be rated over 800 cold cranking amperes.

WARNING: The recombination battery requires slightly different testing procedures. Refer to the manufacturer's manual before attempting to test the battery.

Battery Terminals

Shop Manual
Chapter 5, page 125

All automotive batteries have two terminals. One terminal is a positive connection, the other is a negative connection. The battery terminals extend through the cover or the side of the battery case. The following are the most common types of battery terminals (Figure 5-15):

1. **Post or top terminals:** Used on most automotive batteries. The positive post will be larger than the negative post to prevent connecting the battery in reverse polarity.
2. **Side terminals:** Positioned in the side of the container near the top. These terminals are threaded and require a special bolt to connect the cables. Polarity identification is by positive and negative symbols.
3. **L terminals:** Used on specialty batteries and some imports.

Battery terminals provide a means of connecting the battery plates to the vehicle's electrical system.

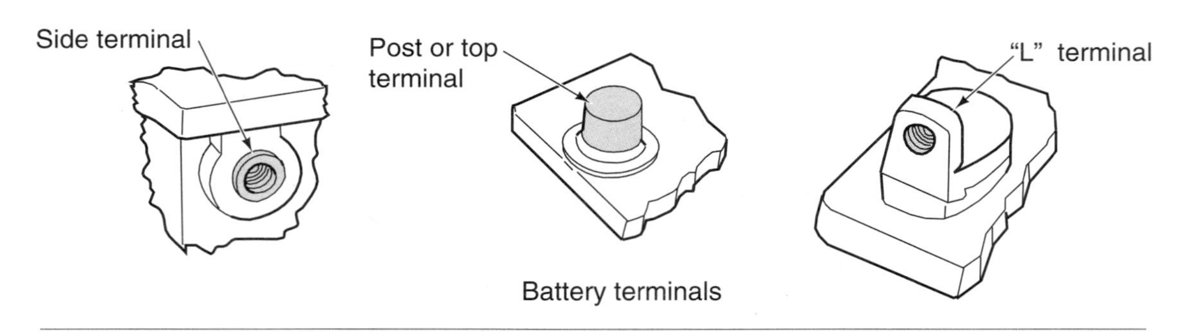

Figure 5-15 The most common types of automotive battery terminals. (Reprinted with the permission of Ford Motor Company)

Battery Ratings

Battery capacity ratings are established by the Battery Council International (BCI) in conjunction with the Society of Automotive Engineers (SAE). Battery cell voltage depends on the types of materials used in the construction of the battery. Current capacity depends on several factors:

1. The size of the cell plates. The larger the surface area of the plates, the more chemical action that can occur. This means a greater current is produced.
2. The weight of the positive and negative plate active materials.
3. The weight of the sulfuric acid in the electrolyte solution.

The battery's current capacity rating is an indication of its ability to deliver cranking power to the starter motor and of its ability to provide reserve power to the electrical system. The commonly used current capacity ratings are explained in the following sections.

The ampere-hour rating is also known as the 20-hour discharge rating.

Ampere-Hour Rating

The ampere-hour rating is the amount of steady current that a fully charged battery can supply for 20 hours at 80°F (26.7°C) without the cell voltage falling below 1.75 volts. For example, if a battery can be discharged for 20 hours at a rate of 4.0 amperes before its terminal voltage reads 10.5 volts, it would be rated at 80 ampere-hours.

The **cold cranking rating** indicates the battery's ability to deliver a specified amount of current to start an engine at low ambient temperatures.

Cold Cranking Rating

Cold cranking rating is the most common method of rating automotive batteries. It is determined by the load, in amperes, a battery is able to deliver for 30 seconds at 0°F (−17.7°C) without terminal voltage falling below 7.2 volts for a 12-volt battery. The cold cranking rating is given in total amperage and is identified as 300 CCA, 400 CCA, 500 CCA, and so on. Some batteries are rated as high as 1,100 CCA.

Cold cranking rating is also called **cold cranking amps (CCA)**

Reserve capacity rating is indicated in the amount of minutes, such as 60, 110, 120, and so on.

Reserve Capacity Rating

The reserve capacity rating is determined by the length of time, in minutes, that a fully charged battery can be discharged at 25 amperes before battery voltage drops below 10.5 volts. This rating gives an indication of how long the vehicle can be driven, with the headlights on, if the charging system should fail.

Temperature	% of Cranking Power
80°F (26.7°C)	100
32°F (0°C)	65
0°F (−17.8°C)	40

Figure 5-16 The effect temperature has on the cranking power of a battery.

Watt-Hour Rating

The starter motor converts the electrical power supplied by the battery into mechanical power, so some battery manufacturers rate their batteries using watt-hour rating. The watt-hour rating of the battery is determined at 0°F (−17.7°C) because the battery's capability to deliver wattage varies with temperature. Watt-hour rating is determined by calculating the ampere-hour rating of the battery times the battery voltage.

The term **watts** is the unit of measure of electrical power and is the equivalent of horsepower. One horsepower is equal to 746 watts.

Battery Size Selection

Some of the aspects that determine the battery rating required for a vehicle include engine size, engine type, climatic conditions, vehicle options, and so on. The requirement for electrical energy to crank the engine increases as the temperature decreases. Battery power drops drastically as temperatures drop below freezing (Figure 5-16). The engine also becomes harder to crank due to the tendancy of oils to thicken when cold, which results in increased friction. As a general rule, it takes 1 ampere of cold cranking power per cubic inch of engine displacement. Therefore, a 200 cubic inch displacement (CID) engine should be fitted with a battery of at least 200 CCA. To convert this into metric, it takes 1 amp of cold cranking power for every 16 cm^3 of engine displacement. A 1.6 liter engine should require at least a battery rated at 100 CCA. This rule may not apply to vehicles that have several electrical accessories. The best method of determining the correct battery is to refer to the manufacturer's specifications.

The battery that is selected should fit the battery holding fixture and the hold-down must be able to be installed. It is also important that the height of the battery not allow the terminals to short across the vehicle hood when it is shut. BCI group numbers are used to indicate the physical size and other features of the battery (Figure 5-17). This group number does not indicate the current capacity of the battery.

The largest current capacity rating that can be achieved in a given battery group size may be in the customer's best interests. However, be aware of overkill.

Battery Cables

Battery cables must be of a sufficient capacity to carry the current required to meet all demands (Figure 5-18). Normal 12-volt cable size is usually 4 or 6 gauge. Various forms of clamps and terminals are used to assure a good electrical connection at each end of the cable. Connections must be clean and tight to prevent arcing, corrosion, and high voltage resistance.

The positive cable is usually red (but not always) and the negative cable is usually black. The positive cable will fasten to the starter solenoid or relay. The negative cable fastens to ground on the engine block. Some manufacturers use a negative cable with no insulation. Sometimes the negative battery cable may have a body grounding wire to help assure that the vehicle body is properly grounded.

Grp. Size	Vlt.	Cold cranking power—amps for 30 secs. at 0°F*	No. of mo. warranted	Size of battery container in inches (incl. terminals)		
				Lgth.	Wd.	Ht.
17HF	6	400	24	7 1/4	6 3/4	9
21	12	450	60	8	6 3/4	8 1/2
22F	12	430	60	9	6 7/8	8 1/8
	12	380	55	9	6 7/8	8 1/8
	12	330	40	9	6 7/8	8 1/8
22NF	12	330	24	9 1/2	5 1/2	8 7/8
24	12	525	60	10 1/4	6 7/8	8 5/8
	12	450	55	10 1/4	6 7/8	8 5/8
	12	410	48	10 1/4	6 7/8	8 5/8
	12	380	40	10 1/4	6 7/8	8 5/8
	12	325	36	10 1/4	6 7/8	8 5/8
	12	290	30	10 1/4	6 7/8	8 5/8
24F	12	525	60	10 1/4	6 7/8	8 5/8
	12	450	55	10 1/4	6 7/8	8 5/8
	12	410	48	10 1/4	6 7/8	8 5/8
	12	380	40	10 1/4	6 7/8	8 5/8
	12	325	36	10 1/4	6 7/8	8 5/8
	12	290	30	10 1/4	6 7/8	8 5/8
27	12	560	60	12	6 7/8	8 5/8
27F	12	560	60	12	6 7/8	9
41	12	525	60	11 9/16	6 13/16	6 15/16
42	12	450	60	9 5/8	6 7/8	6 3/4
	12	340	40	9 5/8	6 7/8	6 3/4
45	12	420	60	9 1/2	5 1/2	8 7/8
46	12	460	60	10 1/4	6 7/8	8 5/8
48	12	440	60	12	6 7/8	7 1/2
49	12	600	60	14 1/2	6 7/8	7 1/2
56	12	450	60	10	6	8 3/8
	12	380	48	10	6	8 3/8
58	12	425	60	9 1/4	7 1/4	6 7/8
71	12	450	60	8	7 1/4	8 1/2
	12	395	55	8	7 1/4	8 1/2
	12	330	36	8	7 1/4	8 1/2
72	12	490	60	9	7 1/4	8 1/4
	12	380	48	9	7 1/4	8 1/4
74	12	585	60	10 1/4	7 1/4	8 3/4
	12	525	60	10 1/4	7 1/4	8 3/4
	12	505	60	10 1/4	7 1/4	8 3/4
	12	450	55	10 1/4	7 1/4	8 3/4
	12	410	48	10 1/4	7 1/4	8 3/4
	12	380	40	10 1/4	7 1/4	8 3/4
	12	325	36	10 1/4	7 1/4	8 3/4

*Meets or exceeds Battery Council International rating standards.

Figure 5-17 BCI battery group numbers indicate the size and features of a battery.

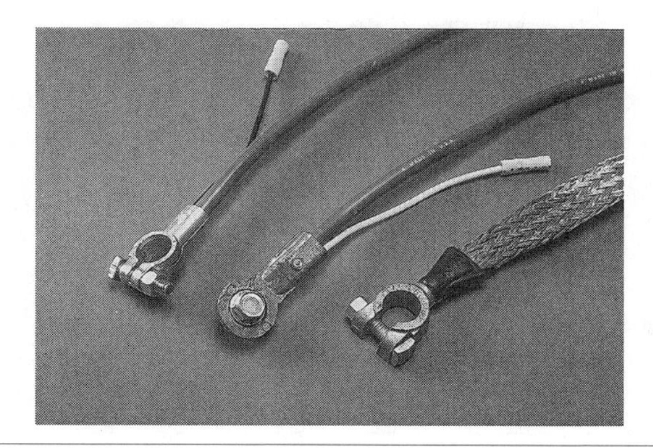

Figure 5-18 A battery cable is designed to carry the high current required to start an engine and supply the vehicle's electrical systems.

WARNING: Connecting the battery cables in reverse polarity can damage many of the vehicle's computer systems and generators.

Battery Hold-Downs

All batteries must be secured in the vehicle to prevent damage and the possibility of shorting across the terminals if it tips. Normal vibrations cause the plates to shed their active materials. Hold-downs reduce the amount of vibration and help increase the life of the battery.

A BIT OF HISTORY

The storage battery on early automobiles was mounted under the car. It wasn't until 1937 that the battery was located under the hood for better accessibility. Today with the increased use of maintenance-free batteries, some manufacturers have "buried" the battery again. For example, to access the battery on some vehicles you must remove the left front wheel and work through the wheel well.

Battery Failure

Shop Manual
Chapter 5, page 121

Whenever battery failure occurs, first perform some simple visual inspections. Check the case for cracks, check the electrolyte level in each cell, (if possible), and check the terminals for corrosion. The sulfuric acid that vents out with the battery gases attacks the battery terminals and battery cables. As the sulfuric acid reacts with the lead and copper, deposits of lead sulfate and copper sulfate are created (Figure 5-19). These deposits are resistive to electron flow and limit the amount of current that can be supplied to the electrical and starting systems. If the deposits are bad enough, the resistance can increase to a level that prevents the starter from cranking the engine.

One of the most common causes of early battery failure is overcharging. If the charging system is supplying a voltage level over 15.5 volts, the plates may become warped. Warping of the plates results from the excess heat that is generated as a result of overcharging. Overcharging also causes the active material to disintegrate and shed off of the plates.

Figure 5-19 Corroded battery terminals reduce the efficiency of a battery.

The sulfate that is not converted back to H_2SO_4 hardens on the plates. This results in **battery sulfation,** which permanently damages the battery.

If the charging system does not produce enough current to keep the battery charged, the lead sulfate can become crystalized on the plates. If this happens, the sulfate is difficult to remove and the battery will resist recharging. The recharging process converts the sulfate on the plates. If there is an undercharging condition, the sulfate is not converted and it will harden on the plates.

Vibration is another common reason for battery failure. If the battery is not secure, the plates will shed the active material as a result of excessive vibration. If enough material is shed, the sediment at the bottom of the battery can create an electrical connection between the plates. The shorted cell will not produce voltage, resulting in a battery that will have only 10.5 volts across the terminals. With this reduced amount of voltage, the starter usually will not be capable of starting the engine. To prevent this problem, make sure that proper hold-down fixtures are used.

During normal battery operation, the active materials on the plates will shed. The negative plate also becomes soft. Both of these events will reduce the effectiveness of the battery.

Summary

Terms to Know

Battery hold-downs
Battery terminals
Cell element
Cold cranking amps (CCA)
Cold cranking rating
Conventional batteries
Deep cycling
Electrochemical
Electrolysis
Electrolyte

- ❑ An automotive battery is an electrochemical device that provides for and stores electrical energy.
- ❑ Electrical energy is produced in the battery by the chemical reaction that occurs between two dissimilar plates that are immersed in an electrolyte solution.
- ❑ An automotive battery has the following important functions:
  1. It operates the starting motor, ignition system, electronic fuel injection, and other electrical devices for the engine during cranking and starting.
  2. It supplies all the electrical power for the vehicle accessories whenever the engine is not running or at low idle.
  3. It furnishes current for a limited time whenever electrical demands exceed charging system output.
  4. It acts as a stabilizer of voltage for the entire automotive electrical system.
  5. It stores energy for extended periods of time.

- ❑ Electrical loads that are still placed on the battery when the ignition switch is in the OFF position are called key-off or parasitic loads.
- ❑ The amount of electrical energy that a battery is capable of producing depends on the size, weight, and active area of the plates and the specific gravity of the electrolyte solution.
- ❑ The conventional battery is constructed of seven basic components:
  1. Positive plates
  2. Negative plates
  3. Separators
  4. Case
  5. Plate straps
  6. Electrolyte
  7. Terminals
- ❑ Electrolyte solution used in automotive batteries consists of 64% water and 36% sulfuric acid by weight.
- ❑ The electrolyte solution causes the chemical actions to take place between the lead dioxide of the positive plates and the sponge lead of the negative plates. The electrolyte is also the carrier that moves electric current between the positive and negative plates through the separators.
- ❑ The automotive battery has a fully charged specific gravity of 1.265 corrected to 80°F.
- ❑ Grid growth is a condition where the grid grows little metallic fingers that extend through the separators and short out the plates.
- ❑ Deep cycling is discharging the battery almost completely before recharging it.
- ❑ In a conventional battery, the positive plate is covered with lead peroxide and the negative plate is covered with sponge lead.
- ❑ In maintenance-free batteries, the cell plates contain calcium, cadmium, or strontium to reduce gassing and self-discharge.
- ❑ The grid construction of the hybrid battery consists of approximately 2.75% antimony alloy on the positive plates and a calcium alloy on the negative plates.
- ❑ The recombination battery uses separators that hold a gel-type material in place of liquid electrolyte.
- ❑ The three most common types of battery terminals are:
  1. Post or top terminals: Used on most automotive batteries. The positive post will be larger than the negative post to prevent connecting the battery in reverse polarity.
  2. Side terminals: Positioned in the side of the container near the top. These terminals are threaded and require a special bolt to connect the cables. Polarity identification is by positive and negative symbols.
  3. L terminals: Used on specialty batteries and some imports.
- ❑ The most common methods of battery rating are cold cranking, reserve capacity, ampere-hour, and watt-hour.

Terms to Know (Continued)

Gassing
Grids
Grid growth
Hybrid batteries
Hydrometer
Key-off loads
Low maintenance battery
Maintenance-free
Material expanders
Plates
Plate straps
Radial grid
Recombination batteries
Reserve capacity
Reserve capacity rating
Separators
Specific gravity
Sulfation
Watt

Review Questions

Short Answer Essays

1. Explain the purposes of the battery.
2. Describe how a technician can determine the correct battery to be installed into a vehicle.
3. Describe the methods used to rate batteries.
4. Describe the different types of battery terminals used.
5. Explain the effects that temperature has on battery performance.
6. Describe the different loads or demands that are placed upon a battery during different operating conditions.
7. List and describe the seven main elements of the conventional battery.
8. What are the major reasons that a battery fails?
9. List at least three safety concerns associated with working on or near the battery.
10. Describe the difference in construction of the hybrid battery as compared to the conventional battery.

Fill-in-the-Blanks

1. An automotive battery is an ____________________ device that provides for and stores ____________________ energy.
2. When discharging the battery, it changes ____________________ energy into ____________________ energy.
3. The assembly of the positive plates, negative plates, and separators is called the ____________________ or ____________________ .
4. The electrolyte solution used in automotive batteries consists of ____________________% water and ____________________% sulfuric acid.
5. A fully charged automotive battery has a specific gravity of ____________________ corrected to 80°F.
6. ____________________ ____________________ is a condition where the grid grows little metallic fingers that extend through the separators and shorts out the plates.

7. The ____________ ____________ rating indicates the battery's ability to deliver a specified amount of current to start an engine at low ambient temperatures.

8. The electrolyte solution causes the chemical actions to take place between the lead peroxide of the ____________ plates and the ____________ ____________ of the ____________ plates.

9. Some of the aspects that determine the battery rating required for a vehicle include engine ____________ , engine ____________ , ____________ conditions, and vehicle ____________ .

10. Electrical loads that are still present when the ignition switch is in the OFF position are called ____________ loads.

ASE Style Review Questions

1. *Technician A* says the battery provides electricity by releasing free electrons.
   Technician B says the battery stores energy in chemical form.
   Who is correct?
   A. A only
   B. B only
   C. Both A and B
   D. Neither A nor B

2. *Technician A* says the largest demand on the battery is when it must supply current to operate the starter motor.
   Technician B says the current requirements of a starter motor may be over one hundred amperes.
   Who is correct?
   A. A only
   B. B only
   C. Both A and B
   D. Neither A nor B

3. *Technician A* says even with the ignition switch turned off, there are electrical demands placed on the battery.
   Technician B says after the engine is started, the vehicle's charging system works to recharge the battery and to provide the current to run the electrical systems.
   Who is correct?
   A. A only
   B. B only
   C. Both A and B
   D. Neither A nor B

4. The current capacity rating of the battery is being discussed:
   Technician A says the amount of electrical energy that a battery is capable of producing depends on the size, weight, and active area of the plates.
   Technician B says the current capacity rating of the battery depends on the types of materials used in the construction of the battery.
   Who is correct?
   A. A only
   B. B only
   C. Both A and B
   D. Neither A nor B

5. The construction of the battery is being discussed:
Technician A says the 12-volt battery consists of positive and negative plates connected in parallel.
Technician B says the 12-volt battery consists of six 2-volt cells wired in series.
Who is correct?
   - **A.** A only
   - **B.** B only
   - **C.** Both A and B
   - **D.** Neither A nor B

6. Maintenance of a conventional battery is being discussed:
Technician A says electrolyte should be added to the battery if the cells are low.
Technician B says only distilled water should be used in batteries.
Who is correct?
   - **A.** A only
   - **B.** B only
   - **C.** Both A and B
   - **D.** Neither A nor B

7. Battery terminology is being discussed:
Technician A says grid growth is a condition where the grid grows little metallic fingers that extend through the separators and short out the plates.
Technician B says deep cycling is discharging the battery almost completely before recharging it.
Who is correct?
   - **A.** A only
   - **B.** B only
   - **C.** Both A and B
   - **D.** Neither A nor B

8. Battery rating methods are being discussed:
Technician A says the ampere-hour is determined by the load in amperes a battery is able to deliver for 30 seconds at 0°F (−17.7°C) without terminal voltage falling below 7.2 volts for a 12-volt battery.
Technician B says the cold cranking rating is the amount of steady current that a fully charged battery can supply for 20 hours at 80°F (26.7°C) without battery voltage falling below 10.5 volts.
Who is correct?
   - **A.** A only
   - **B.** B only
   - **C.** Both A and B
   - **D.** Neither A nor B

9. The hybrid battery is being discussed:
Technician A says the hybrid battery can withstand six deep cycles and still retain 100% of its original reserve capacity.
Technician B says the grid construction of the hybrid battery consists of approximately 2.75% antimony alloy on the positive plates and a calcium alloy on the negative plates.
Who is correct?
   - **A.** A only
   - **B.** B only
   - **C.** Both A and B
   - **D.** Neither A nor B

10. *Technician A* says connecting the battery cables in reverse polarity can damage many of the vehicle's computer systems.
Technician B says the battery must be secured in the vehicle to prevent internal damage and the possibility of shorting across the terminals if it tips.
Who is correct?
   - **A.** A only
   - **B.** B only
   - **C.** Both A and B
   - **D.** Neither A nor B

Direct Current Motors and the Starting System

Upon completion and review of this chapter, you should be able to:

- ❑ Explain the purpose of the starting system.
- ❑ List and identify the components of the starting system.
- ❑ Explain the principle of operation of the DC motor.
- ❑ Describe the purpose and operation of the armature.
- ❑ Describe the purpose and operation of the field coil.
- ❑ Explain the differences between the types of magnetic switches used.
- ❑ Identify and explain the differences between starter drive mechanisms.
- ❑ Describe the differences between the positive engagement and solenoid shift starter.
- ❑ Describe the operation and features of the permanent magnet starter.

Introduction

Shop Manual
Chapter 6, page 155

The internal combustion engine must be rotated before it will run under its own power. The starting system is a combination of mechanical and electrical parts that work together to start the engine. The starting system is designed to change the electrical energy, which is being stored in the battery, into mechanical energy. To accomplish this conversion, a starter or cranking motor is used. The starting system includes the following components:

1. Battery
2. Cable and wires
3. Ignition switch
4. Starter solenoid or relay
5. Starter motor
6. Starter drive and flywheel ring gear
7. Starting safety switch

A starting safety switch is normally a clutch switch or a neutral safety switch.

Components in a simplified cranking system circuit are shown (Figure 6-1). This chapter examines both this circuit and the fundamentals of electric motor operation.

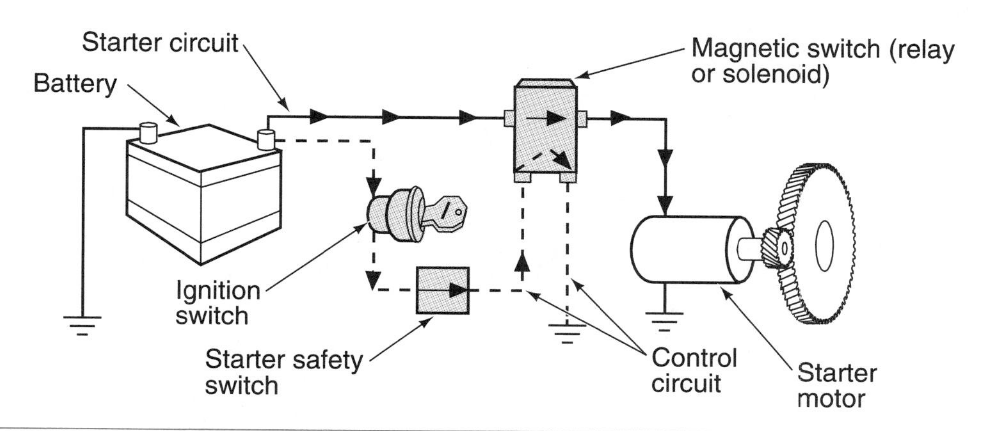

Figure 6-1 Major components of the starting system. The solid line represents the starting (cranking) circuit and the dashed line represents the starter control circuit.

A BIT OF HISTORY

In the early days of the automobile, the vehicle did not have a starter motor. The operator had to use a starting crank to turn the engine by hand. Charles F. Kettering invented the first electric "self-starter," which was developed and built by the Delco Electrical Plant. The self-starter first appeared on the 1912 Cadillac and was actually a combination starter and generator.

Motor Principles

Shop Manual
Chapter 6, page 175

The **armature** is the moveable component of the motor that consists of a conductor wound around a laminated iron core. It is used to create a magnetic field.

Pole shoes are made of high-magnetic permeability material to help concentrate and direct the lines of force in the field assembly.

DC motors use the interaction of magnetic fields to convert the electrical energy into mechanical energy. Magnetic lines of force flow from the north pole to the south pole of a magnet (Figure 6-2). If a current carrying conductor is placed within the magnetic field, two fields will be present. On the left side of the conductor the lines of force are in the same direction. This will concentrate the flux density of the lines of force on the left side. This will produce a strong magnetic field, because the two fields will reinforce each other. The lines of force oppose each other on the right side of the conductor. This results in a weaker magnetic field. The conductor will tend to move from the strong field to the weak field. This principle is used to convert electrical energy into mechanical energy in a starter motor by electromagnetism.

A simple electromagnet-style starter motor is shown (Figure 6-3). The inside windings are called the armature. The armature rotates within the stationary outside windings, called the field, which has windings coiled around pole shoes.

When current is applied to the field and the armature, both produce magnetic flux lines. The direction of the windings will place the left pole at a south polarity and the right side at a north polarity. The lines of force move from north to south in the field. In the armature, the flux lines circle in one direction on one side of the loop and in the opposite direction on the other side. Current will now set up a magnetic field around the loop of wire, which will interact with the north and south fields and put a turning force on the loop. This force will cause the loop to turn in the direction of the weaker field (Figure 6-4). However, the armature is limited in how far it is able to turn. When the armature is halfway between the shoe poles, the fields balance one another.

For the armature to continue rotating, the current flow in the loop must be reversed. To accomplish this, a split-ring commutator is in contact with the ends of the armature loops. Current enters and exits the armature through a set of brushes that slide over the commutator's sections. As the brushes pass over one section of the commutator to another, the current flow in the armature

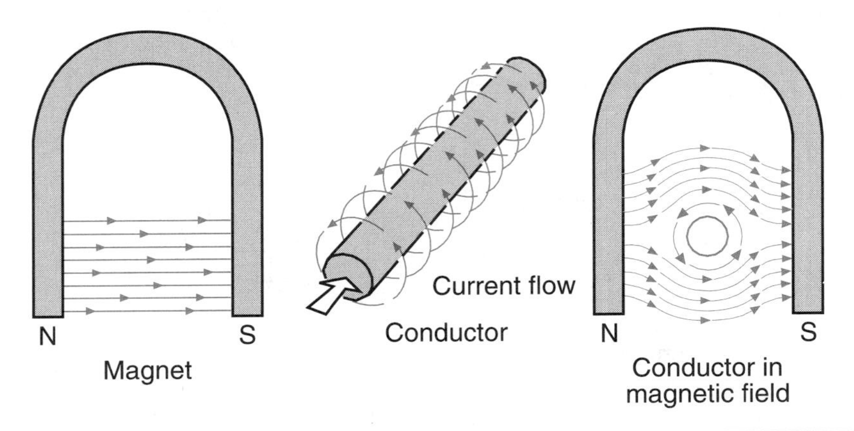

Figure 6-2 Magnetic field interaction.

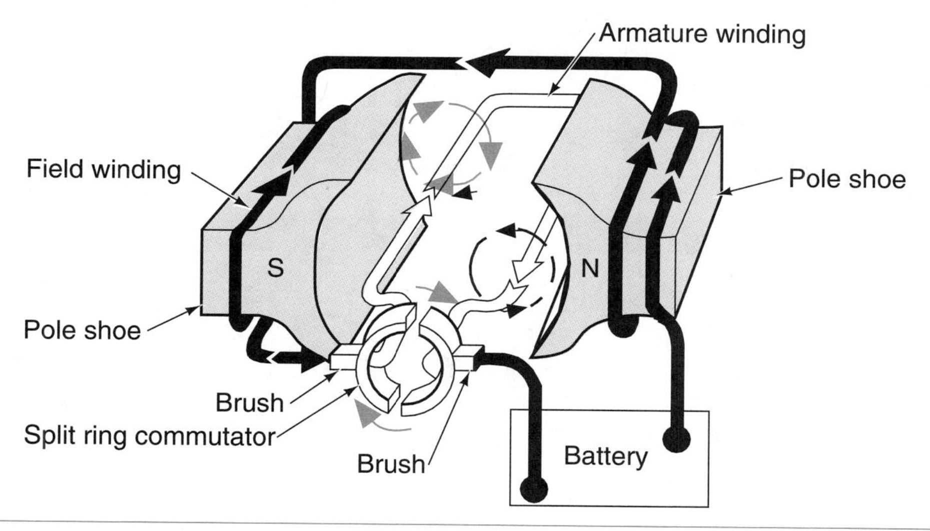

Figure 6-3 Simple electromagnetic motor.

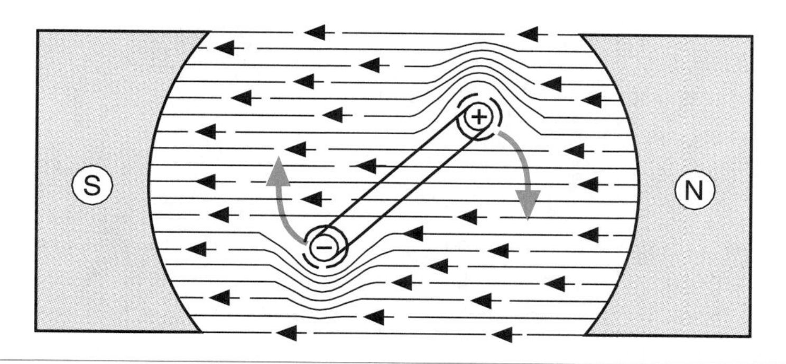

Figure 6-4 Rotation of the conductor is in the direction of the weaker field.

is reversed. The position of the magnetic fields are the same. However, the direction of current flow through the loop has been reversed. This will continue until the current flow is turned off.

A single loop motor would not produce enough torque to rotate an engine. Power can be increased by the addition of more loops or pole shoes. An armature with its many windings, with each loop attached to corresponding commutator sections is shown (Figure 6-5). In a typical starter motor (Figure 6-6) there are four brushes that make the electrical connections to the commutator. Two of the brushes are grounded to the starter motor frame, and two are insulated from the frame. Also, the armature is supported by bushings at both ends.

The point at which the fields balance is called the **static neutral point.**

The **commutator** is a series of conducting segments located around one end of the armature.

Brushes are electrically conductive sliding contacts, usually made of copper and carbon.

The number of field coils and brushes varies between manufacturers.

WARNING: The high amount of current required to operate the starter motor generates heat very quickly. Continuous operation of the starter motor for longer than 30 seconds causes serious heat damage. The starter motor should not be operated for more than 30 seconds at a time and should have a 2-minute wait between cranking attempts.

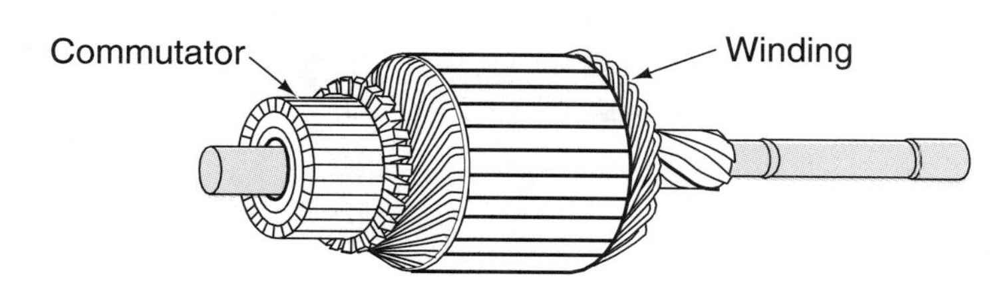

Figure 6-5 Starter armature.

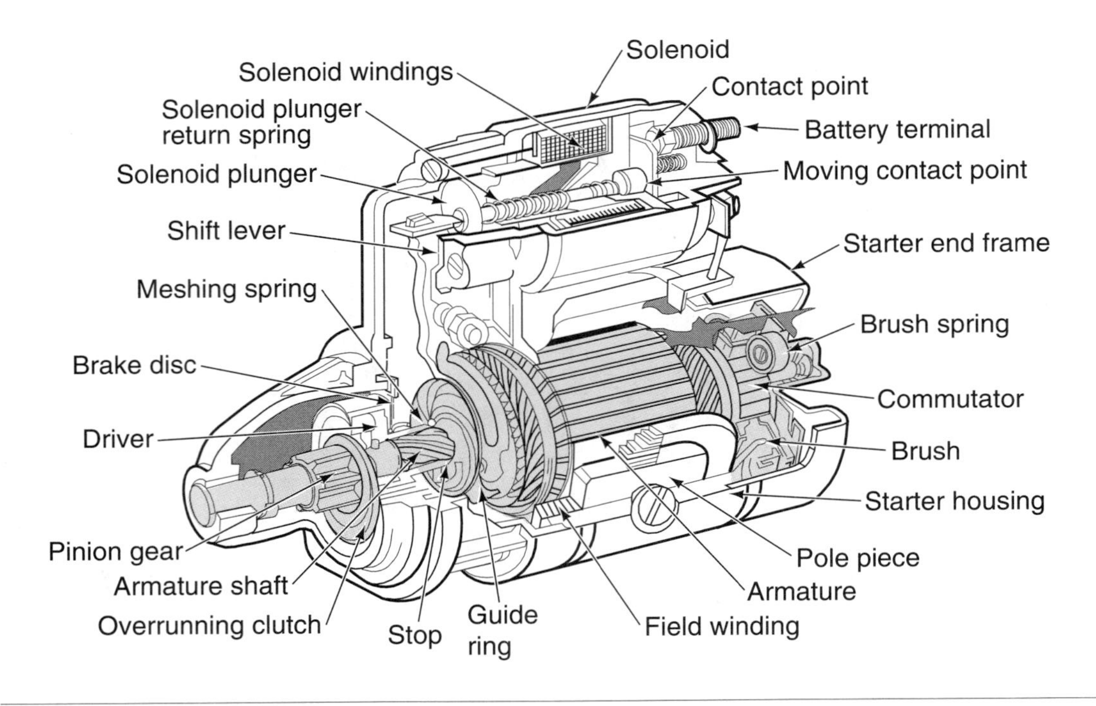

Figure 6-6 Starter and solenoid components. (Courtesy of Robert Bosch Corporation)

The stampings of the armature core are called **laminations.**

The construction of the armature from individual stampings is called **laminated construction.**

Shop Manual
Chapter 6, page 176

Counter voltages induced in a core are called **eddy currents**. These cause heat to build up in the core and waste energy.

The wave-winding pattern is the most commonly used due to its lower resistance.

Armature

The armature is constructed with a laminated core made of several thin iron stampings that are placed next to each other (Figure 6-7). Laminated construction is used because in a solid iron core the magnetic fields would generate eddy currents. By using laminated construction, eddy currents in the core are minimized.

The slots on the outside diameter of the laminations hold the armature windings. The windings loop around the core and are connected to the commutator. Each commutator segment is insulated from the adjacent segments. A typical armature can have more than 30 commutator segments.

A steel shaft is fitted into the center hole of the core laminations. The commutator is insulated from the shaft.

Two basic winding patterns are used in the armature: lap winding and wave winding. In the lap winding, the two ends of the winding are connected to adjacent commutator segments (Figure 6-8). In this pattern, the wires passing under a pole field have their current flowing in the same direction.

In the wave winding pattern, each end of the winding connects to commutator segments that are 90 or 180 degrees apart (Figure 6-9). In this pattern design some windings will have no

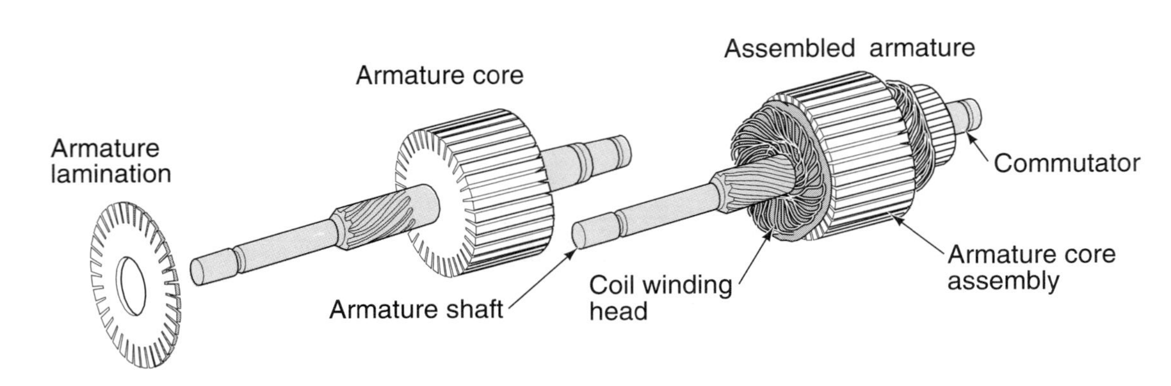

Figure 6-7 Lamination construction of a typical motor armature.

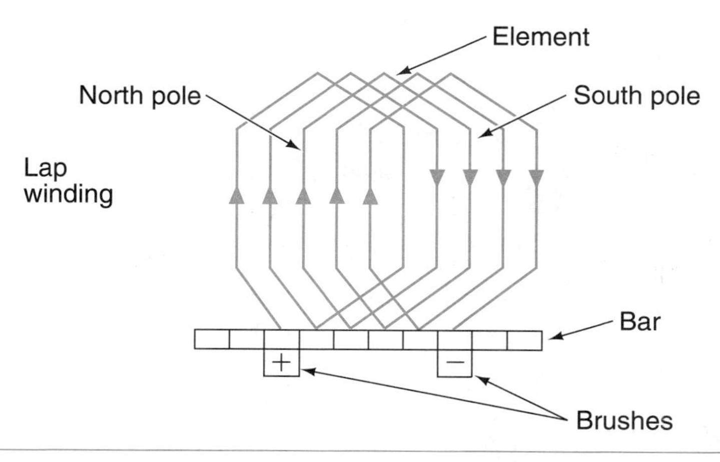

Figure 6-8 Lap winding diagram. (Courtesy of Delco-Remy Division—GMC)

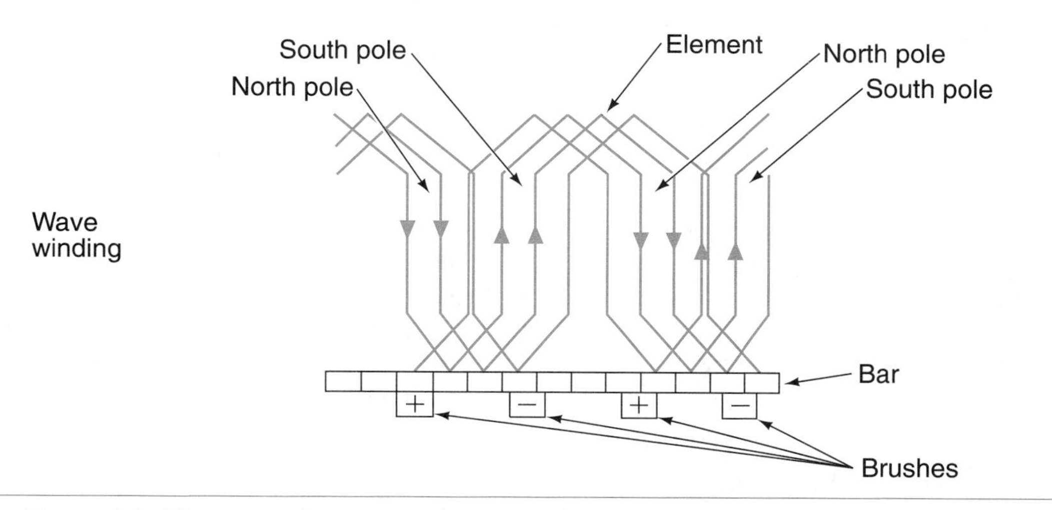

Figure 6-9 Wave-wound armature. (Courtesy of Delco-Remy Division—GMC)

current flow at certain positions of armature rotation. This occurs because the segment ends of the winding loop are in contact with brushes that have the same polarity.

Field Coils

The field coils are electromagnets constructed of wire ribbons or coils wound around a pole shoe (Figure 6-10). The field coils are attached to the inside of the starter housing (Figure 6-11). The

Shop Manual
Chapter 6, page 175

The pole shoes are constructed of heavy iron.

Most starter motors use four field coils.

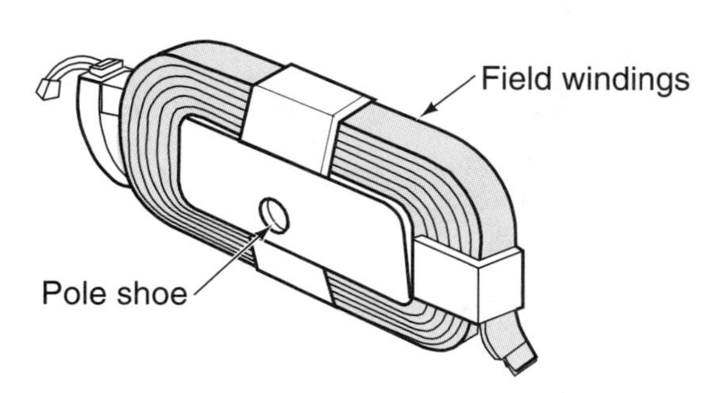

Figure 6-10 Field coil wound around a pole shoe.

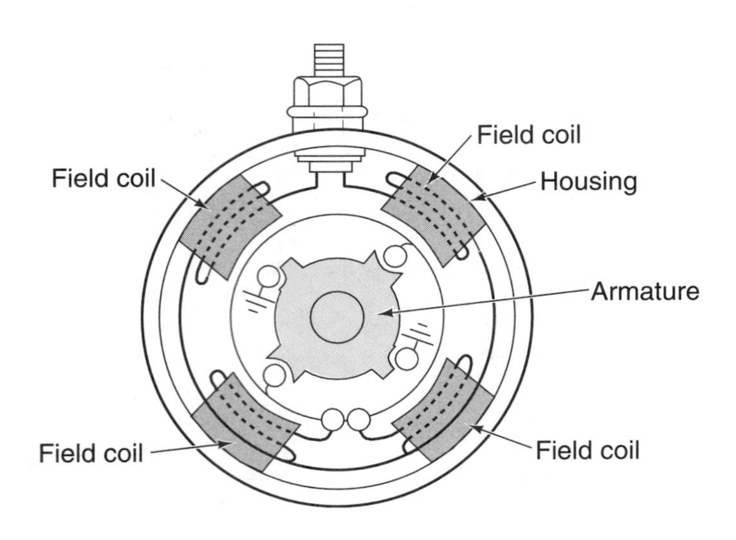

Figure 6-11 Field coils mounted to the inside of a starter housing.

iron pole shoes and the iron starter housing work together to increase and concentrate the field strength of the field coils (Figure 6-12).

When current flows through the field coils, strong stationary electromagnetic fields are created. The fields have a north and south magnetic polarity based on the direction the windings are wound around the pole shoes. The polarity of the field coils alternate to produce opposing magnetic fields.

In any DC motor, there are three methods of connecting the field coils to the armature: in series, in parallel (shunt), and a compound connection that uses both series and shunt coils.

Series-Wound Motors

Most starter motors are series-wound with current flowing first to the field windings, then to the brushes, through the commutator and the armature winding contacting the brushes at that time,

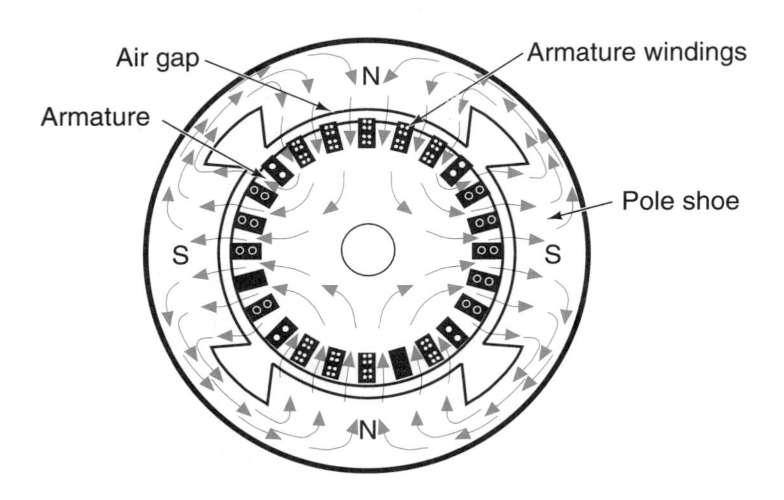

Figure 6-12 Magnetic fields in a four-pole starter motor. (Courtesy of Robert Bosch Corporation)

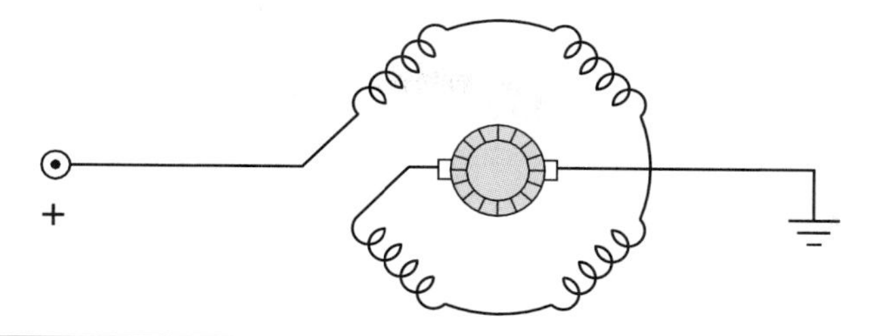

Figure 6-13 A schematic of a series-wound motor.

then through the grounded brushes back to the battery source (Figure 6-13). This design permits all of the current that passes through the field coils to also pass through the armature.

A series-wound motor will develop its maximum torque output at the time of initial start. As the motor speed increases, the torque output of the motor will decrease. This decrease of torque output is the result of counter electromotive force caused by self induction.

Counter electromotive force (CEMF) is voltage produced in the starter motor itself. This voltage acts against the supply voltage from the battery. CEMF is produced by electromagnetic induction.

Shunt-Wound Motors

Electric motors, or shunt motors, have the field windings wired in parallel across the armature (Figure 6-14). A shunt motor does not decrease in its torque output as speeds increase. This is because the CEMF produced in the armature does not decrease the field coil strength. Due to a shunt motor's inability to produce high torque, it is not typically used as a starter motor. However, shunt motors may be found as wiper motors, power window motors, power seat motors, and so on.

Shunt means there is more than one path for current to flow.

Compound Motors

In a compound motor some of the field coils are connected to the armature in series, and some field coils are connected in parallel with the battery and the armature (Figure 6-15). This configuration allows the compound motor to develop good starting torque and constant operating speeds. The field coil that is shunt wound is used to limit the speed of the starter motor. Also, on Ford's positive engagement starters, the shunt coil is used to engage the starter drive. This is possible because the shunt coil is energized as soon as battery voltage is sent to the starter.

A shunt-wound field is used to limit the speed that the motor can turn.

A starter motor that uses the characteristics of a series motor and a shunt motor is called a **compound motor.**

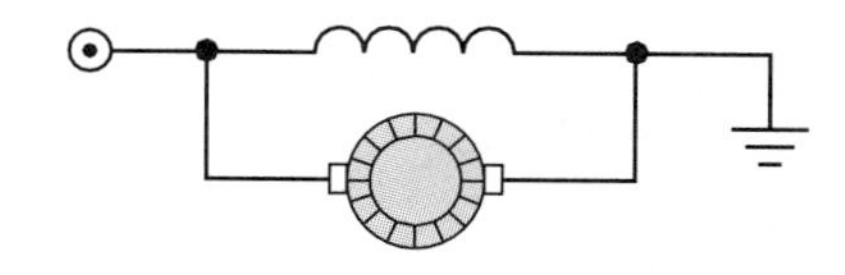

Figure 6-14 A schematic of a shunt-wound motor.

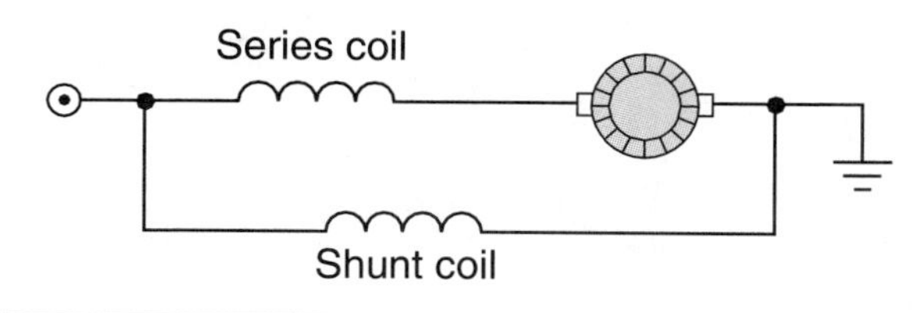

Figure 6-15 A compound motor uses both series and shunt coils.

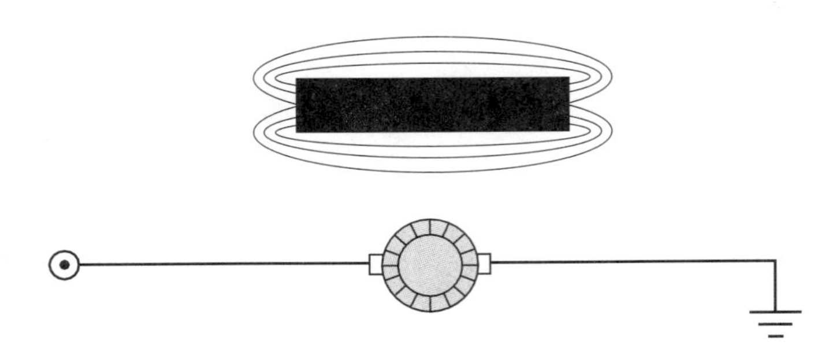

Figure 6-16 A permanent magnet motor has only an armature circuit, as the field is created by strong permanent magnets.

Permanent Magnet Motors

Some motors use permanent magnets in place of the field coils (Figure 6-16). These motors are used in many different applications, including starter motors. When a permanent magnet is used instead of coils, there is no field circuit in the motor. By eliminating this circuit, potential electrical problems have also been eliminated, such as field to housing shorts. Another advantage to using permanent magnets is weight savings; the weight of a typical starter motor is reduced by 50%. Most permanent magnet starters are gear reduction type starters.

Multiple permanent magnets are positioned in the housing, around the armature. These permanent magnets are an alloy of boron, neodymium, and iron. The field strength of these magnets is much greater than typical permanent magnets. The operation of these motors is the same as other electric motors, except there is no field circuit or windings.

Starter Drives

Shop Manual
Chapter 6, page 160

The **starter drive** is the part of the starter motor that engages the armature to the engine flywheel ring gear.

The **ratio** of the starter drive is determined by dividing the number of teeth on drive gear (pinion gear) into the number of teeth on the driven gear (flywheel).

A starter drive includes a pinion gear set that meshes with the flywheel ring gear on the engine's crankshaft (Figure 6-17). To prevent damage to the pinion gear or the ring gear, the pinion gear must mesh with the ring gear before the starter motor rotates. To help assure smooth engagement, the ends of the pinion gear teeth are tapered (Figure 6-18). Also, the action of the armature must always be from the motor to the engine. The engine must not be allowed to spin the armature. The ratio of the number of teeth on the ring gear and the starter drive pinion gear is usually

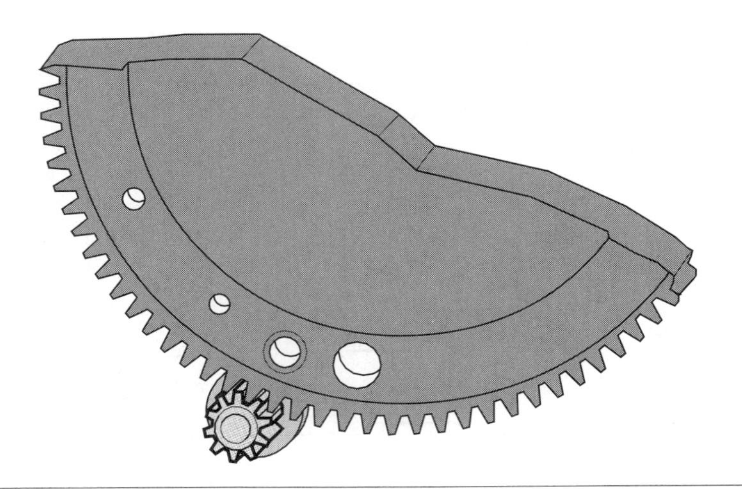

Figure 6-17 Starter drive pinion gear is used to turn the engine's flywheel.

Figure 6-18 The pinion teeth are tapered to allow for smooth engagement.

between 15:1 and 20:1. This means the starter motor is rotating 15 to 20 times faster than the engine. Normal cranking speed for the engine is about 200 rpm. If the starter drive had a ratio of 18:1, the starter would be rotating at a speed of 3,600 rpm. If the engine started and was accelerated to 2,000 rpm, the starter speed would increase to 36,000 rpm. This would destroy the starter motor if it was not disengaged from the engine.

Bendix Inertia Drive

The **bendix drive** depends on inertia to provide meshing of the drive pinion with the ring gear (Figure 6-19). The screwshaft threads are a part of the armature, and will turn at armature speed. At the end of the pinion and barrel is the pinion gear that will mesh with the ring gear. The pinion and barrel have internal threads that match those of the screwshaft. When current flows through the starter motor, the armature will begin to spin. Torque from the armature is transmitted via a shock-absorbing spring and drive head to the screwshaft. This causes the screwshaft to rotate. However, the barrel does not rotate. The barrel has a weight on one side to increase its inertial effect. The barrel tends to stay at rest, and the screwshaft rotates inside the barrel. As a result, the barrel is threaded down the length of the screwshaft to the end. At the end of the screwshaft, the pinion gear engages the ring gear. Here the pinion gear locks to the screwshaft and transfers torque from the armature to the ring gear and engine.

Once the engine starts and is running under its own power, it will rotate faster than the armature. This causes the barrel to screw back down the screwshaft and bring the pinion gear out of engagement with the ring gear

Inertia is the tendency of an object that is at rest to stay at rest, and an object that is in motion to stay in motion.

The purpose of the **drive spring** is to absorb the initial shock of engagement.

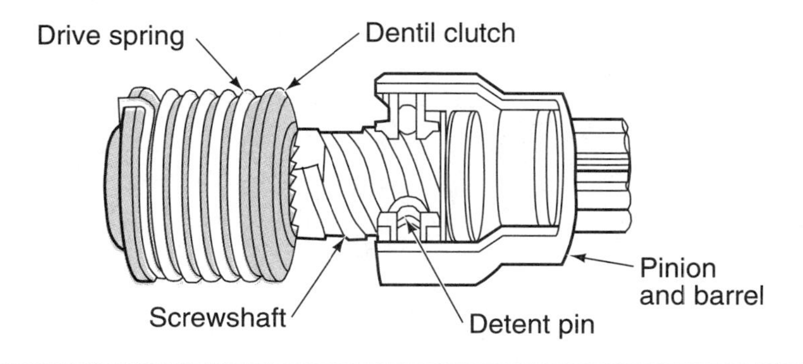

Figure 6-19 Bendix inertia drive. (Courtesy of Delco-Remy Division—GMC)

There are a couple different variations of the inertia drive. One of these variations is the barrel-type drive. The barrel-type starter drive is similar to the inertia drive except:

1. The pinion is mounted on the end of the barrel.
2. It has a higher gear ratio.
3. It works directly off screw threads at the end of the armature, instead of a screwshaft.

Another variation is the folo-thru drive. This drive is much like the barrel drive with the addition of a detent pin and a detent clutch.

When the pinion barrel has moved to the end of the screwshaft, the detent pin locks the barrel to the screwshaft. The detent pin operates on centrifugal force. As the shaft is turning rapidly, centrifugal force throws the pin into engagement between the barrel and the screwshaft.

The detent clutch disengages the sections of the screwshaft when the engine is running faster than the armature, and the drive is still engaged. Disengaging the clutch protects the motor from being damaged.

Overrunning Clutch Drive

The most common type of starter drive is the overrunning clutch. The overrunning clutch is a roller-type clutch that transmits torque in one direction only and freewheels in the other direction. This allows the starter motor to transmit torque to the ring gear, but prevents the ring gear from transferring torque to the starter motor.

In a typical overrunning-type clutch (Figure 6-20), the clutch housing is internally splined to the starter armature shaft. The drive pinion turns freely on the armature shaft within the clutch housing. When torque is transmitted through the armature to the clutch housing, the spring-loaded rollers are forced into the small ends of their tapered slots (Figure 6-21). They are then wedged tightly against the pinion barrel. The pinion barrel and clutch housing are now locked together; torque is transferred through the starter motor to the ring gear and engine.

When the engine starts and is running under its own power, the ring gear attempts to drive the pinion gear faster than the starter motor. This unloads the clutch rollers and releases the pinion gear to rotate freely around the armature shaft.

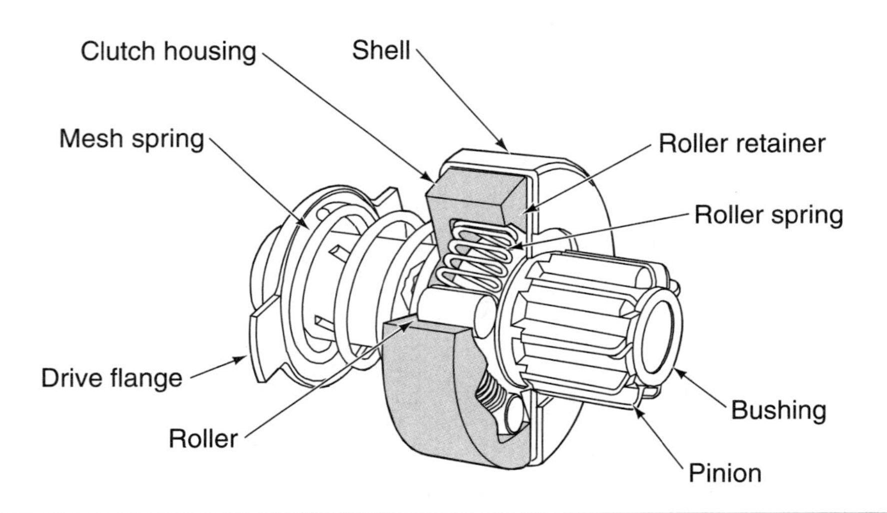

Figure 6-20 Overrunning clutch starter drive. (Reprinted with the permission of Ford Motor Company)

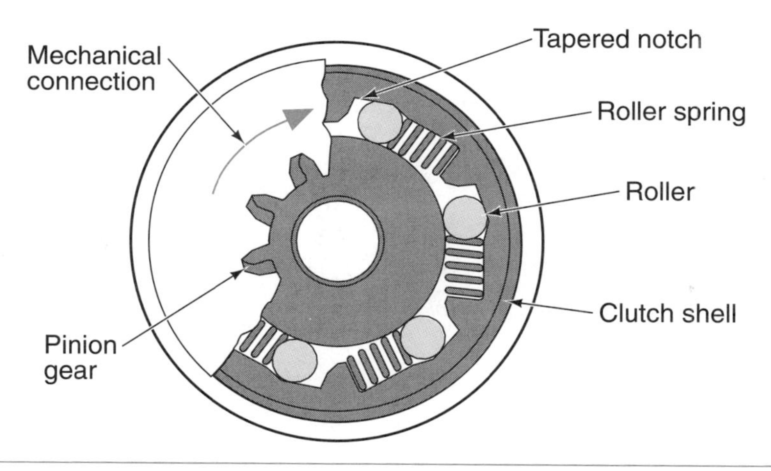

Figure 6-21 When the armature turns, it locks the rollers into the tapered notches. (Courtesy of Robert Bosch Corporation)

Cranking Motor Circuits

The starting system of the vehicle consists of two circuits: the starter control circuit and the motor feed circuit. These circuits are separate but related. The control circuit consists of the starting portion of the ignition switch, the starting safety switch (if applicable), and the wire conductor to connect these components to the relay or solenoid. The motor feed circuit consists of heavy battery cables from the battery to the relay and the starter or directly to the solenoid if the starter is so equipped.

A BIT OF HISTORY

The integrated key starter switch was introduced in 1949 by Chrysler. Before this, the key turned the system on and the driver pushed a starter button.

Starter Control Circuit Components

Shop Manual
Chapter 6, page 166

Magnetic Switches

The starter motor requires large amounts of current (up to 300 amperes) to generate the torque needed to turn the engine. The conductors used to carry this amount of current (battery cables) must be large enough to handle the current with very little voltage drop. It would be impractical to place a conductor of this size into the wiring harness to the ignition switch. To provide control of the high current, all starting systems contain some type of magnetic switch. There are two basic types of magnetic switches used: the solenoid and the relay.

A **solenoid** is an electromagnetic device that uses movement of a plunger to exert a pulling or holding force.

Solenoids. In the solenoid-actuated starter system, the solenoid is mounted directly on top of the starter motor (Figure 6-22). The solenoid switch on a starter motor performs two functions: It closes the circuit between the battery and the starter motor. Then it shifts the starter motor pinion gear into mesh with the ring gear. This is accomplished by a linkage between the solenoid

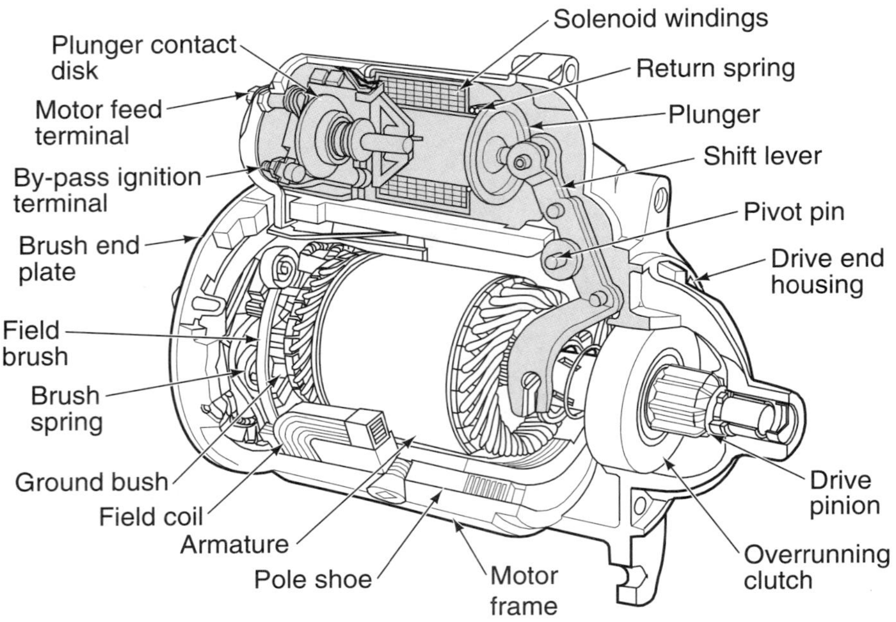

Figure 6-22 Solenoid operated starter has the solenoid mounted directly on top of the motor.

Some manufacturers use a starter relay in conjunction with a solenoid relay. The relay is used to reduce the amount of current flow through the ignition switch.

The two windings of the solenoid are called the **pull-in** and the **hold-in windings.** Their names explain their function.

The amount of current drawn by the two windings can be as high as 50 amperes.

plunger and the shift lever on the starter motor. The most common method of energizing the solenoid is directly from the battery through the ignition switch.

When the circuit is closed and current flows to the solenoid, current from the battery is directed to two separate windings (Figure 6-23). Because it requires more magnetic force to pull the plunger in than to hold it in position, both windings are energized to create a combined magnetic field that pulls the plunger. Once the plunger is moved, the current required to hold the plunger is reduced. This allows the current that was used to pull the plunger in to be used to rotate the starter motor.

When the ignition switch is placed in the START position, voltage is applied to the S terminal of the solenoid (Figure 6-24). The hold-in winding has its own ground to the case of the solenoid. The pull-in winding's ground is through the starter motor. Current will flow through both windings to produce a strong magnetic field. When the plunger is moved into

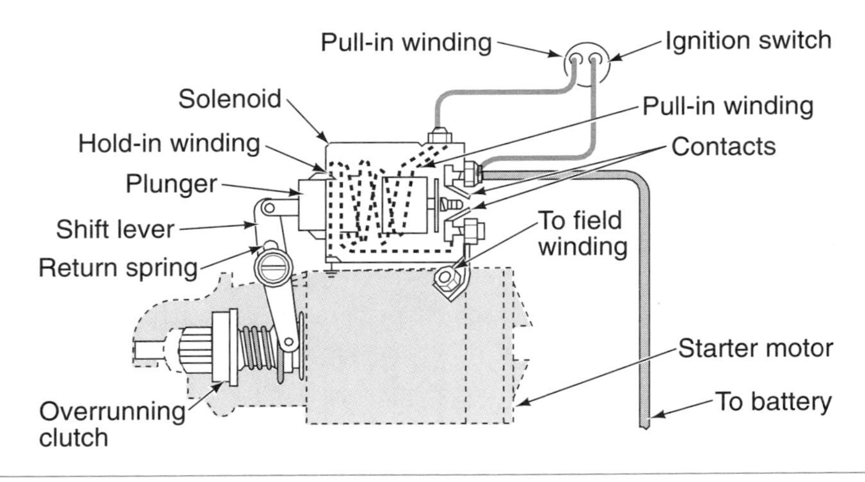

Figure 6-23 A solenoid uses two windings. Both are energized to draw the plunger; then only the hold-in winding is used to hold the plunger in position.

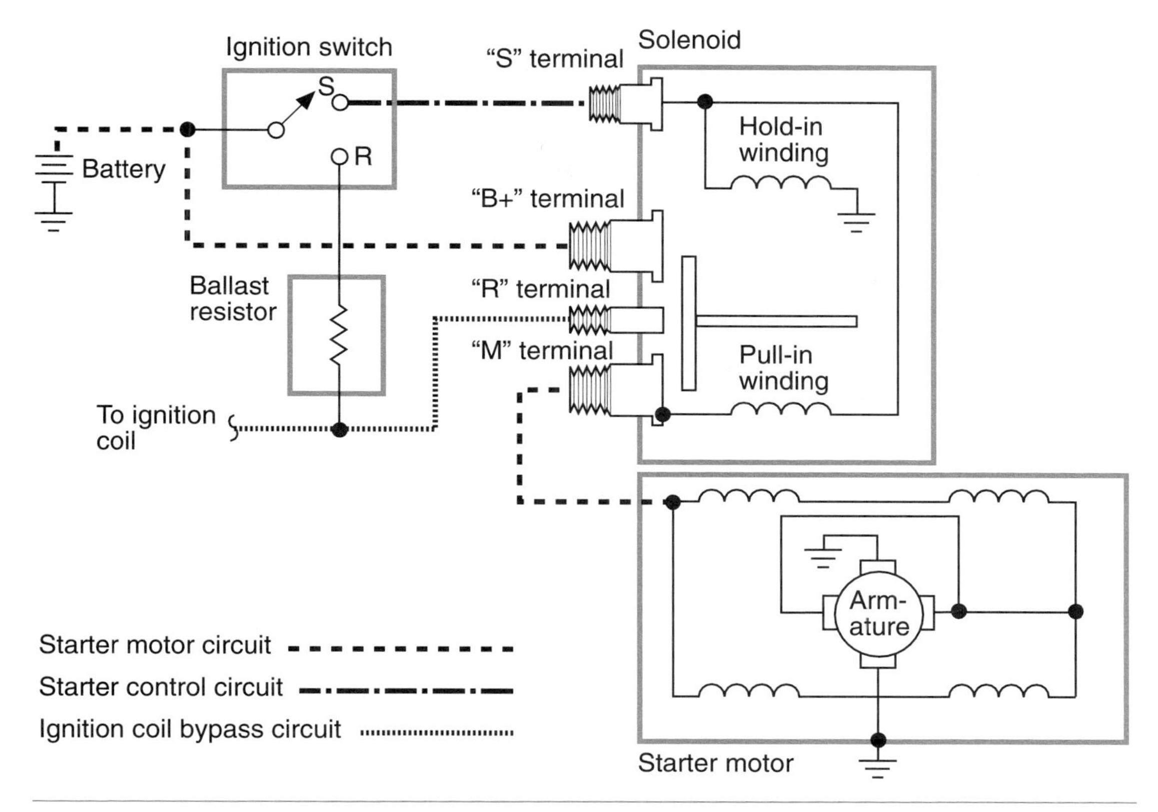

Figure 6-24 Schematic of solenoid-operated starter motor circuit.

contact with the main battery and motor terminals, the pull-in winding is de-energized. The pull-in winding is not energized because the contact places battery voltage on both sides of the coil (Figure 6-25). The current that was directed through the pull-in winding is now sent to the motor.

Because the contact disc does not close the circuit from the battery to the starter motor until the plunger has moved the shift lever, the pinion gear is in full mesh with the flywheel before the armature starts to rotate.

After the engine is started, releasing the key to the RUN position opens the control circuit. Voltage no longer is supplied to the hold-in windings; the return spring causes the plunger to return to its neutral position.

A common problem with the control circuit is that low system voltage or an open in the hold-in windings will cause an oscillating action to occur. The combination of the pull-in winding and the hold-in winding is sufficient to move the plunger. However, once the contacts are closed, there is insufficient magnetic force to hold the plunger in place. This condition is recognizable by a series of clicks when the ignition switch is turned to the START position. Before replacing the solenoid, check the battery condition; a low battery charge will cause the same symptom.

The R terminal provides voltage to the ignition bypass circuit, which is used to provide for full battery voltage to the ignition coil while the engine is cranking. This circuit bypasses the ballast resistor. The bypass circuit is not used on most ignition systems today.

Starter Relays. Many manufacturers use a starter relay (Figure 6-26) instead of or in addition to a solenoid. The relay is usually mounted near the battery on the fender well or radiator support. Unlike the solenoid, the relay does not move the pinion gear into mesh with the flywheel ring gear.

When the ignition switch is turned to the START position, current is supplied through the switch to the relay coil. The coil produces a magnetic field, so it pulls the moveable core into

The **starter relay** is a magnetic switch that uses low current to control the flow of very high current.

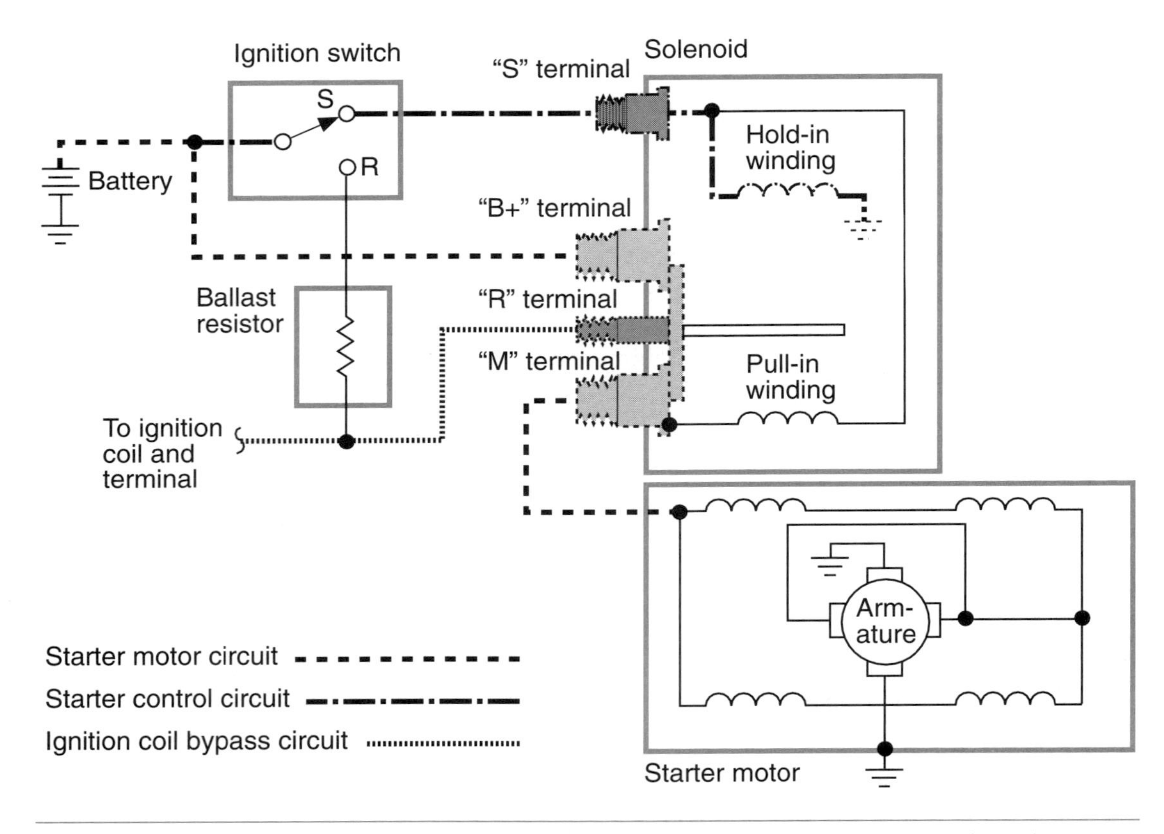

Figure 6-25 Once the contact disc closes the terminals, the hold-in winding is the only one that is energized.

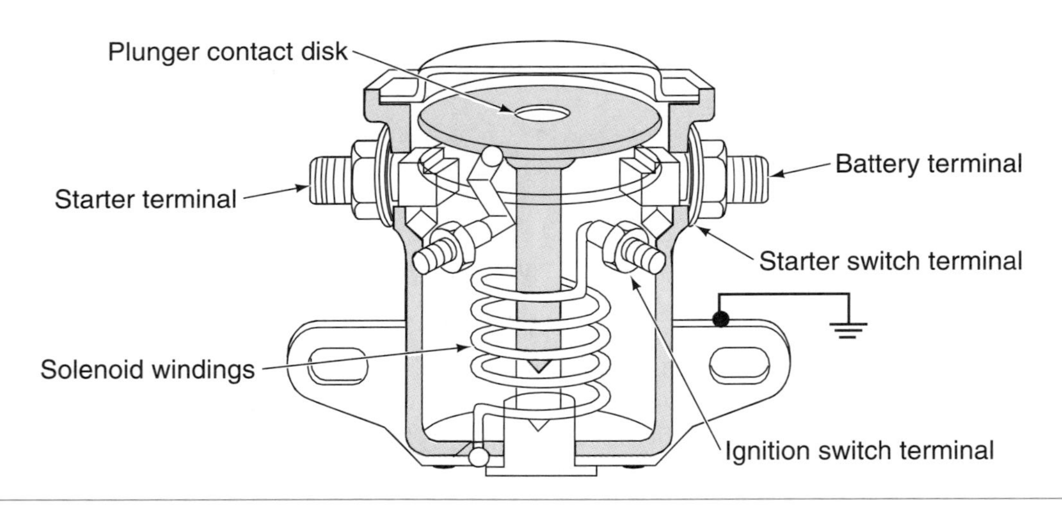

Figure 6-26 A remote starter solenoid, often referred to as the starter relay. (Reprinted with the permission of Ford Motor Company)

contact with the internal contacts of the battery and starter terminals (Figure 6-27). With the contacts closed, full battery current is supplied to the starter motor.

A secondary function of the starter relay is to provide for an alternate path for current to the ignition coil during cranking. This is done by an internal connection that is energized by the relay core when it completes the circuit between the battery and the starter motor.

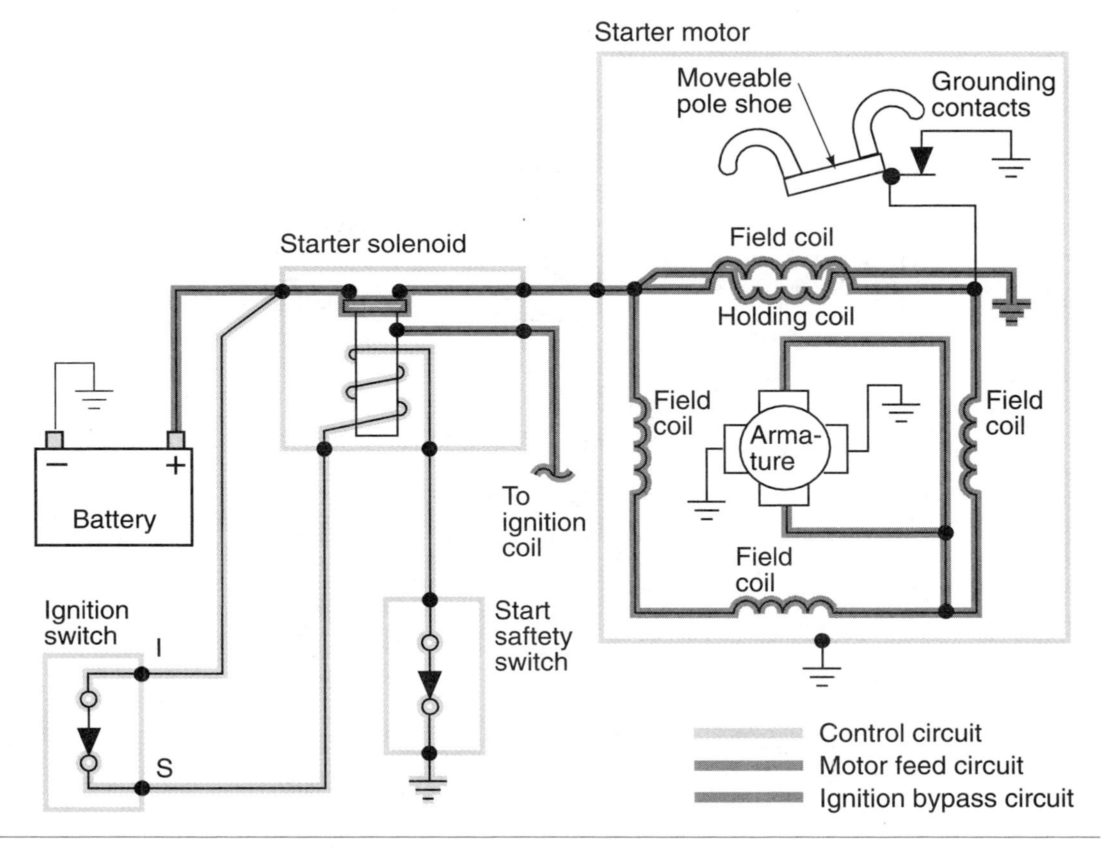

Figure 6-27 Current flow when the starter solenoid is energized.

Ignition Switch

The ignition switch is the power distribution point for most of the vehicle's primary electrical systems (Figure 6-28). Most ignition switches have five positions:

1. ACCESSORIES: Supplies current to the vehicle's electrical accessory circuits. It will not supply current to the engine control circuits, starter control circuit, or the ignition system.
2. LOCK: Mechanically locks the steering wheel and transmission gear selector. All electrical contacts in the ignition switch are open. Most ignition switches must be in this position to insert or remove the key from the cylinder.
3. OFF: All circuits controlled by the ignition switch are opened. The steering wheel and transmission gear selector are unlocked.
4. ON or RUN: The switch provides current to the ignition, engine controls, and all other circuits controlled by the switch. Some systems will power a chime or light with the key in the ignition switch. Other systems power an anti-theft system when the key is removed and turn it off when the key is inserted.
5. START: The switch provides current to the starter control circuit, ignition system, and engine control circuits.

The ignition switch is spring loaded in the START position. This momentary contact automatically moves the contacts to the RUN position when the driver releases the key. All other ignition switch positions are detent positions.

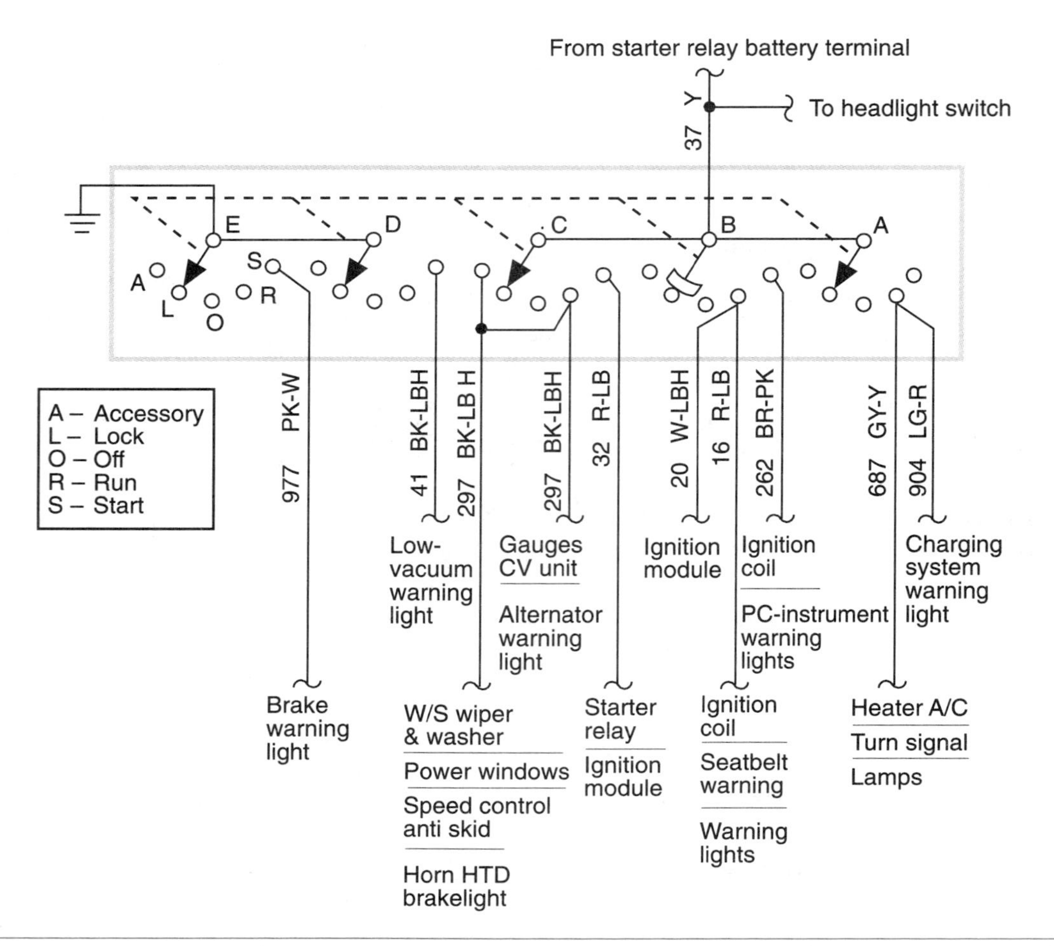

Figure 6-28 Ganged ignition switch. (Reprinted with the permission of Ford Motor Company)

Starting Safety Switch

The **neutral safety switch** is used on vehicles equipped with automatic transmissions. It opens the starter control circuit when the transmission shift selector is in any position except PARK or NEUTRAL.

Actual location of the neutral safety switch depends on the kind of transmission and the location of the shift lever. Some manufacturers place the switch in the transmission (Figure 6-29).

Vehicles equipped with automatic transmissions require a means of preventing the engine from starting while the transmission is in gear. Without this feature the vehicle would lunge forward or backward once it was started, causing personal or property damage. The normally open

Figure 6-29 The neutral safety switch can be combined with the backup light switch and installed in the transmission.

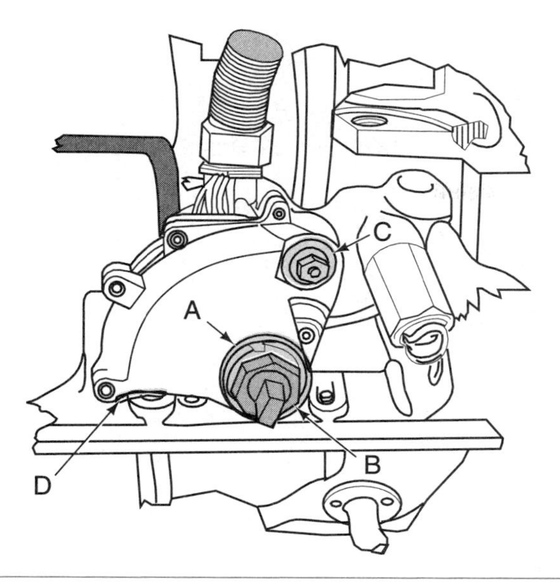

Figure 6-30 Neutral safety switch attached to the shift lever on the console.

neutral safety switch is connected in series in the starting system control circuit and is usually operated by the shift lever (Figure 6-30). When in the PARK or NEUTRAL position, the switch is closed, allowing current to flow to the starter circuit. If the transmission is in a gear position, the switch is opened and current cannot flow to the starter circuit.

Many vehicles equipped with manual transmissions use a similar type of safety switch. The start/clutch interlock switch is usually operated by movement of the clutch pedal (Figure 6-31). When the clutch pedal is pushed downward, the switch closes and current can flow through the starter circuit. If the clutch pedal is left up, the switch is open and current cannot flow.

Some General Motors vehicles use a mechanical linkage that blocks movement of the ignition switch cylinder unless the transmission is in PARK or NEUTRAL (Figure 6-32).

The **start/clutch interlock switch** is used on vehicles equipped with manual transmissions.

Most starter safety switches are adjustable. Sometimes a no-start problem can be corrected by checking and adjusting (or replacing) the starter safety switch.

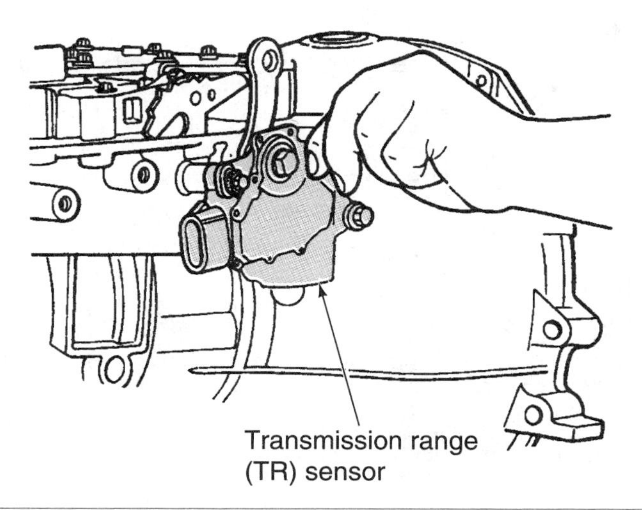

Figure 6-31 A TR sensor mounted to a transmission. The TR sensor is not only a starter safety switch but also tells the PCM the position of the gear shift lever. (Reprinted with the permission of Ford Motor Company)

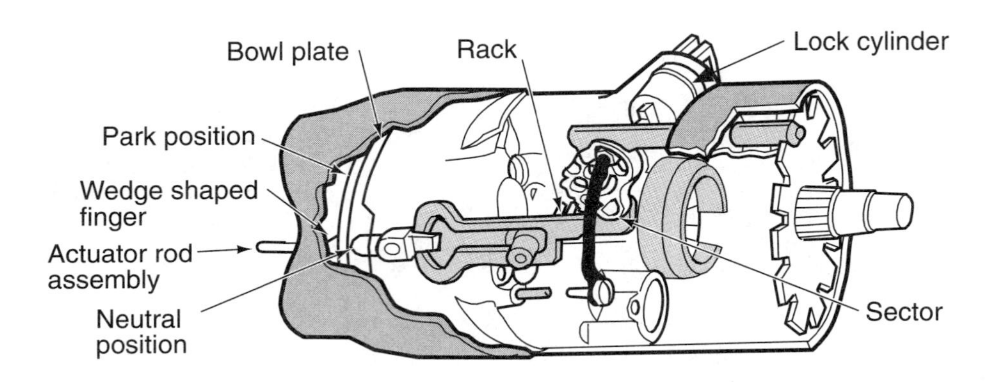

Figure 6-32 Mechanical linkage is used to prevent starting the engine while the transmission is in gear. (Courtesy of Cadillac Motor Division—GMC)

Cranking Motor Designs

The most common type of starter motor used today incorporates the overrunning clutch starter drive instead of the old inertia-engagement bendix drive. There are four basic groups of starter motors:

1. Direct drive
2. Gear reduction
3. Positive-engagement (moveable pole)
4. Permanent magnet

Direct Drive Starters

The direct drive starter motor can be either series-wound or compound motors.

The most common type of starter motor is the solenoid-operated direct drive unit (Figure 6-33). Although there are construction differences between applications, the operating principles are the same for all solenoid-shifted starter motors.

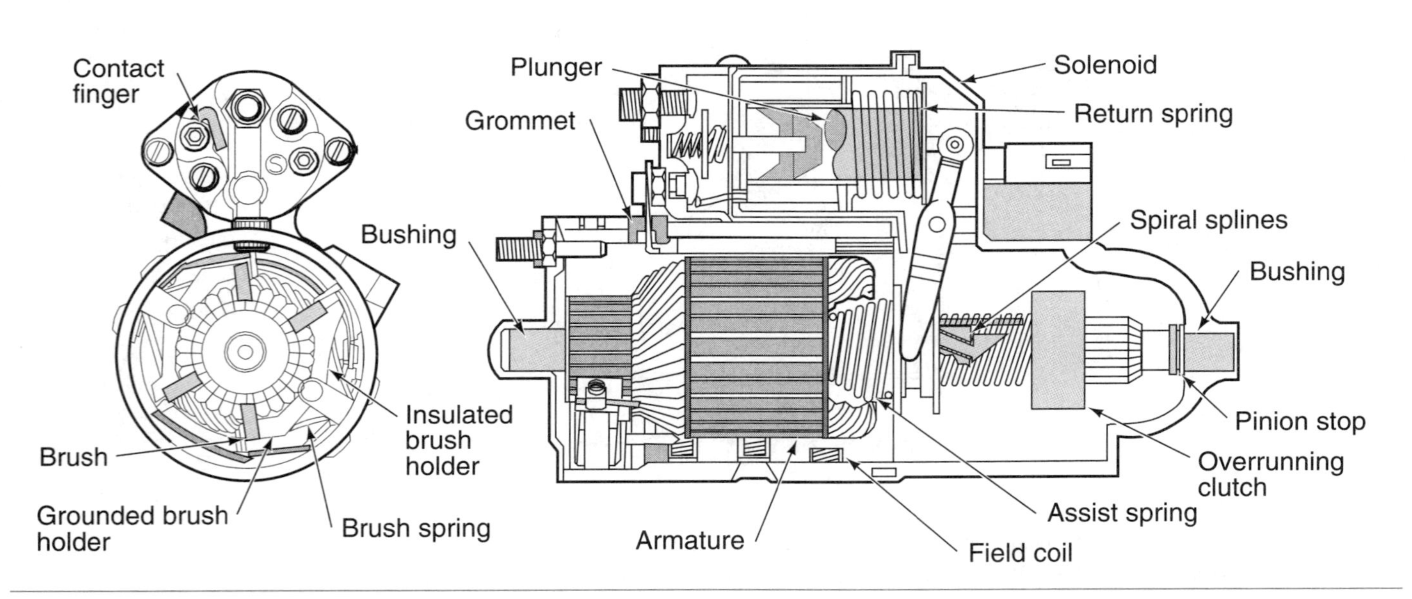

Figure 6-33 Solenoid-operated Delco MT series starter motor. (Courtesy of General Motors Corporation)

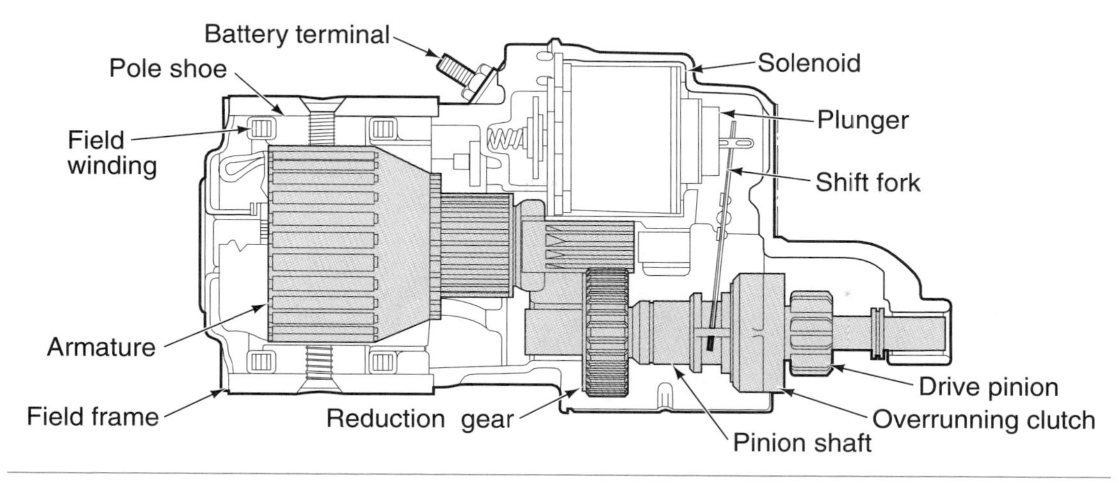

Figure 6-34 Gear reduction starter motor construction. (Reprinted with the permission of Ford Motor Company)

When the ignition switch is placed in the START position, the control circuit energizes the pull-in and hold-in windings of the solenoid. The solenoid plunger moves and pivots the shift lever, which in turn locates the drive pinion gear into mesh with the engine flywheel.

When the solenoid plunger is moved all the way, the contact disc closes the circuit from the battery to the starter motor. Current now flows through the field coils and the armature. This develops the magnetic fields that cause the armature to rotate, thus turning the engine.

Gear Reduction Starters

Some manufacturers use a gear reduction starter to provide increased torque (Figure 6-34). The gear reduction starter differs from most other designs in that the armature does not drive the pinion gear directly. In this design, the armature drives a small gear that is in constant mesh with a larger gear. Depending on the application, the ratio between these two gears is between 2:1 and 3.5:1. The additional reduction allows for a small motor to turn at higher speeds and greater torque with less current draw.

The solenoid operation is similar to that of the solenoid-shifted direct drive starter in that the solenoid moves the plunger, which engages the starter drive.

Some gear reduction starter motors are compound motors.

Most gear reduction starters have the commutator and brushes located in the center of the motor.

Positive-engagement Starters

One of the most commonly used starters on Ford applications is the positive-engagement starter (Figure 6-35). Positive-engagement starters use the shunt coil windings of the starter motor to engage the starter drive. The high starting current is controlled by a starter solenoid mounted close to the battery. When the solenoid contacts are closed, current flows through a drive coil. The drive coil creates an electromagnetic field that attracts a moveable pole shoe. The moveable pole shoe is attached to the starter drive through the plunger lever. When the moveable pole shoe moves, the drive gear engages the engine flywheel.

As soon as the starter drive pinion gear contacts the ring gear, a contact arm on the pole shoe opens a set of normally closed grounding contacts (Figure 6-36). With the return to ground circuit opened, all the starter current flows through the remaining three field coils and through the brushes to the armature. The starter motor then begins to rotate. To prevent the starter drive from disengaging from the ring gear if battery voltage drops while cranking, the moveable-pole shoe is held down by a holding coil. The holding coil is a smaller coil inside the main drive coil and is strong enough to hold the starter pinion gear engaged.

Positive-engagement starters are also called **moveable-pole shoe starters.**

The **drive coil** is a hollowed field coil that is used to attract the moveable pole shoe.

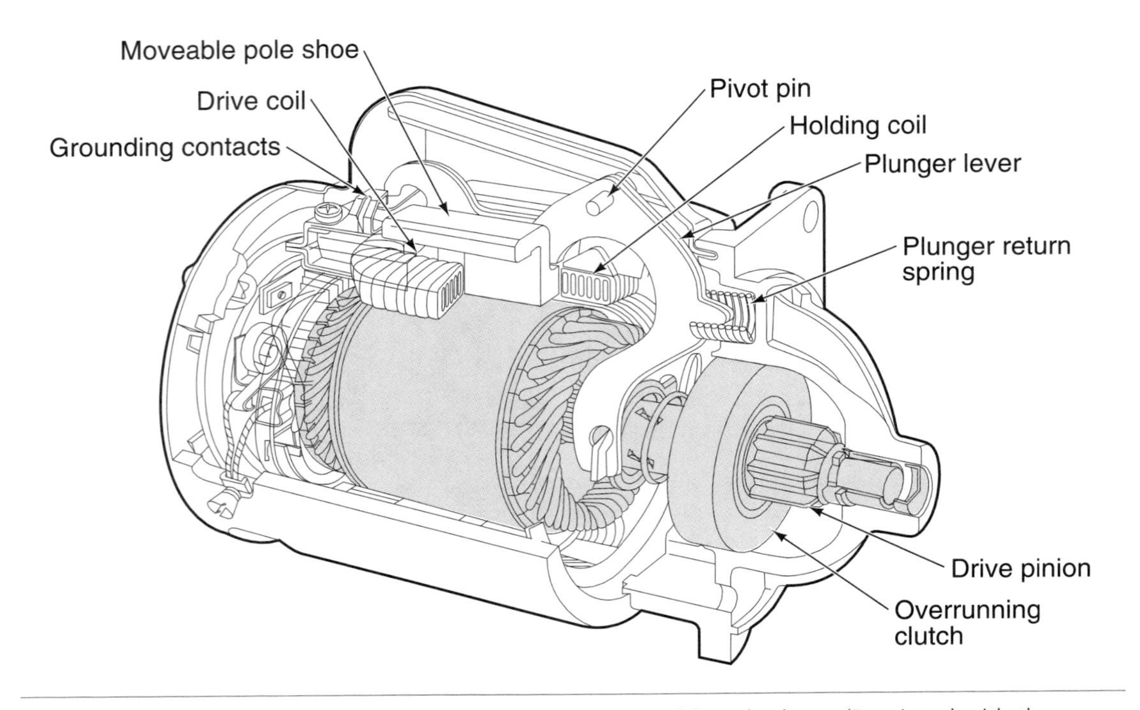

Figure 6-35 Positive-engagement starters use a movable pole shoe. (Reprinted with the permission of Ford Motor Company)

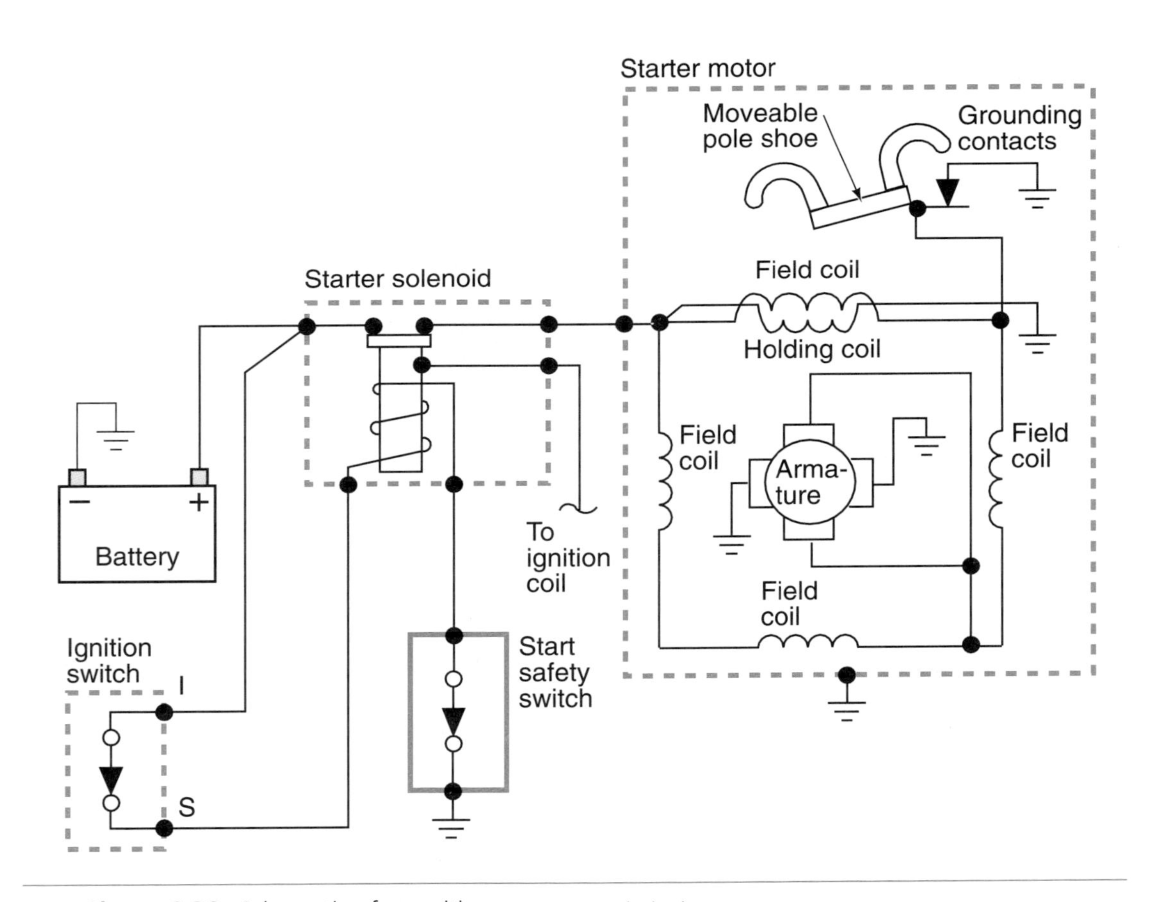

Figure 6-36 Schematic of a positive-engagement starter.

Permanent Magnet Starters

The permanent magnet starter design provides for less weight, simpler construction, and less heat generation as compared to conventional field coil starters (Figure 6-37). Because there are no field coils, current is delivered directly to the armature through the commutator and brushes.

The permanent magnet starter also uses gear reduction through a planetary gear set. The planetary gear train transmits power between the armature and the pinion shaft. This allows the armature to rotate at greater speed and increased torque. The planetary gear assembly consists of a sun gear on the end of the armature, and three planetary carrier gears inside a ring gear. The ring gear is held stationary. When the armature is rotated, the sun gear causes the carrier gears to rotate about the internal teeth of the ring gear. The planetary carrier is attached to the output shaft. The gear reduction provided for by this gear arrangement is 4.5:1. By providing for this additional gear reduction, the demand for high current is lessened.

The electrical operation between the conventional field coil and PMGR starters remains basically the same (Figure 6-38).

The **permanent magnet gear reduction (PMGR)** starter uses four or six permanent magnet field assemblies in place of field coils.

The greatest amount of gear reduction from a planetary gear set is accomplished by holding the ring gear, inputing the sun gear, and outputing the carrier.

CAUTION: Special care must be taken when handling the PMGR starter. The permanent magnets are very brittle and are easily destroyed if the starter is dropped or struck by another object.

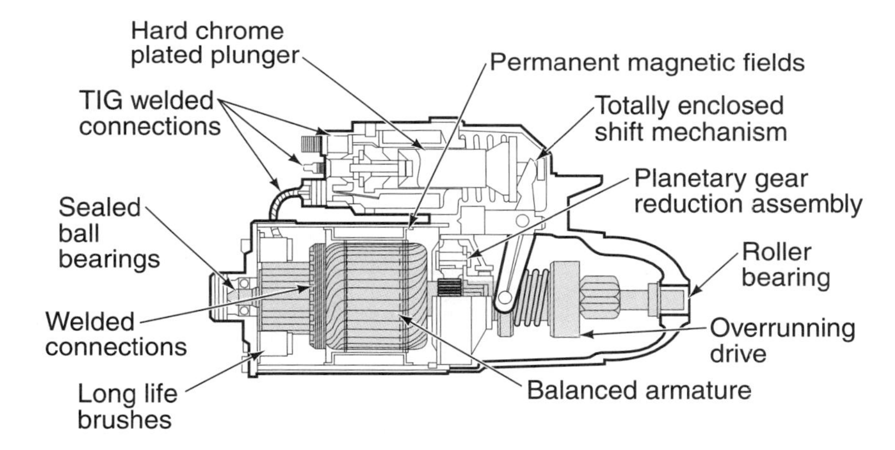

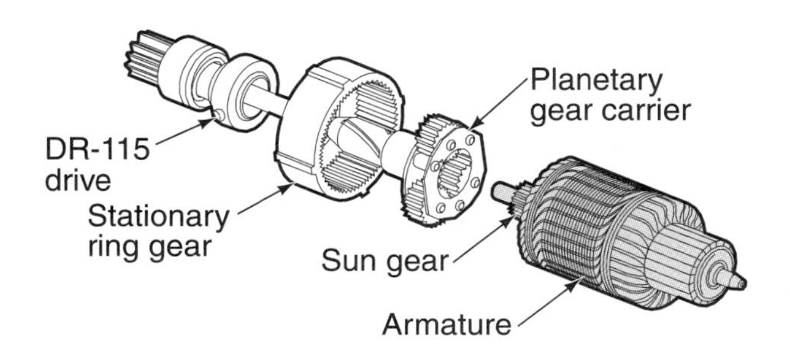

Figure 6-37 The PMGR motor uses a planetary gear set and permanent magnets. (Courtesy of Delco-Remy Division—GMC)

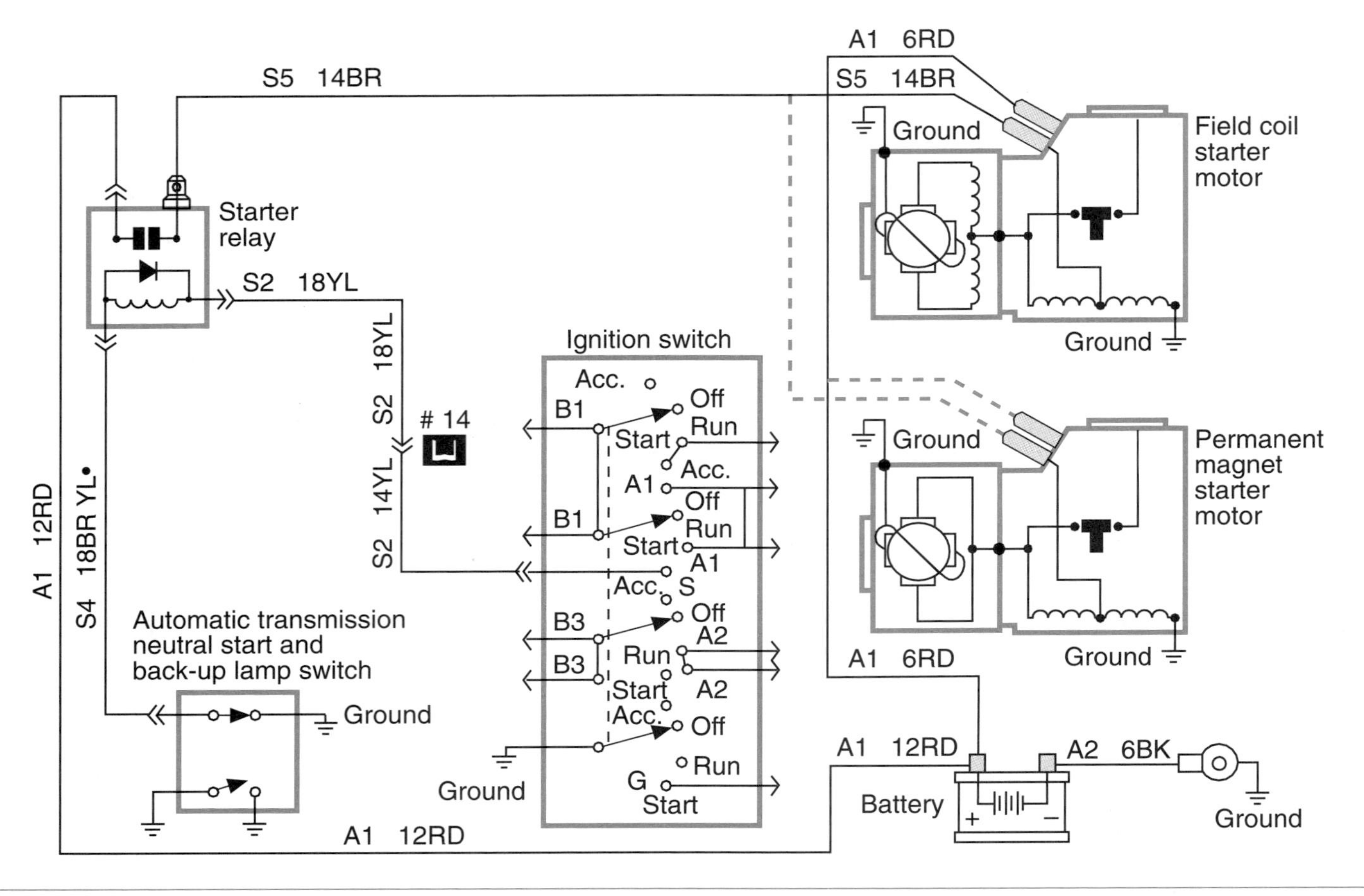

Figure 6-38 Comparison of the electrical circuits used in the field coil and PMGR starters. (Courtesy of Chrysler Corporation)

Summary

Terms to Know

Armature

Bendix drive

Brushes

Commutator

Compound motors

Counter electromotive force (CEMF)

Direct drive

Eddy currents

Gear reduction

Hold-in windings

Inertia

Laminations

Neutral safety switch

- The starting system is a combination of mechanical and electrical parts that work together to start the engine.
- The starting system components include the battery, cable and wires, the ignition switch, the starter solenoid or relay, the starter motor, the starter drive and flywheel ring gear, and the starting safety switch.
- The armature is the moveable component of the motor that consists of a conductor wound around a laminated iron core. It is used to create a magnetic field.
- Pole shoes are made of high magnetic permeability material to help concentrate and direct the lines of force in the field assembly.
- The magnetic forces will cause the armature to turn in the direction of the weaker field.
- Within an electromagnetic style of starter motor, the inside windings are called the armature. The armature rotates within the stationary outside windings, called the field, which has windings coiled around pole shoes.
- The commutator is a series of conducting segments located around one end of the armature.
- A split-ring commutator is in contact with the ends of the armature loops. So, as the brushes pass over one section of the commutator to another, the current flow in the armature is reversed.

- ❑ Two basic winding patterns are used in the armature: lap winding and wave winding.
- ❑ The field coils are electromagnets constructed of wire coils wound around a pole shoe.
- ❑ When current flows through the field coils, strong stationary electromagnetic fields are created.
- ❑ In any DC motor, there are three methods of connecting the field coils to the armature: in series, in parallel (shunt), and a compound connection that uses both series and shunt coils.
- ❑ A starter drive includes a pinion gear set that meshes with the engine flywheel ring gear on the engine.
- ❑ To prevent damage to the pinion gear or the ring gear, the pinion gear must mesh with the ring gear before the starter motor rotates.
- ❑ The bendix drive depends on inertia to provide meshing of the drive pinion with the ring gear.
- ❑ The most common type of starter drive is the overrunning clutch. This is a roller-type clutch that transmits torque in one direction only and freewheels in the other direction.
- ❑ The starting system consists of two circuits called the starter control circuit and the motor feed circuit.
- ❑ The components of the control circuit include the starting portion of the ignition switch, the starting safety switch (if applicable), and the wire conductor to connect these components to the relay or solenoid.
- ❑ The motor feed circuit consists of heavy battery cables from the battery to the relay and the starter or directly to the solenoid if the starter is so equipped.
- ❑ There are four basic groups of starter motors: direct drive, gear reduction, positive engagement (moveable pole), and permanent magnet.

Terms to Know (Continued)

Overrunning clutch drive

Permanent magnet gear reduction (PMGR)

Pinion gear

Pole shoes

Positive-engagement (moveable pole)

Pull-in windings

Ratio

Relay

Series-wound motors

Shunt-wound motors

Solenoid

Starter drive

Start/clutch interlock switch

Static neutral point

Review Questions

Short Answer Essays

1. What is the purpose of the starting system?
2. List and describe the purpose of the major components of the starting system.
3. Explain the principle of operation of the DC motor.
4. Describe the types of magnetic switches used in starting systems.
5. List and describe the operation of the different types of starter drive mechanisms.
6. Describe the differences between the positive-engagement and solenoid shift starter.
7. Explain the operating principles of the permanent magnet starter.
8. Describe the purpose and operation of the armature.
9. Describe the purpose and operation of the field coil.
10. Describe the operation of the two circuits of the starter system.

Fill-in-the-Blanks

1. DC motors use the interaction of magnetic fields to convert the ____________________ energy into ____________________ energy.

2. The ____________________ is the moveable component of the motor, which consists of a conductor wound around a ____________________ iron core and is used to create a ____________ field.

3. Pole shoes are made of high magnetic ____________________ material to help concentrate and direct the ____________________ ____________________ ____________________ in the field assembly.

4. The starter motor electrical connection that permits all of the current that passes through the field coils to also pass through the armature is called the ____________________ motor.

5. ____________________ ____________________ ____________________ is voltage produced in the starter motor itself. This current acts against the supply voltage from the battery.

6. A starter motor that uses the characteristics of a series motor and a shunt motor is called a ____________________ motor.

7. The ____________________ ____________________ is the part of the starter motor that engages the armature to the engine flywheel ring gear.

8. The ____________________ ____________________ is a roller-type clutch that transmits torque in one direction only and freewheels in the other direction.

9. The two circuits of the starting system are called the ____________________ ____________________ circuit and the ____________________ ____________________ circuit.

10. There are two basic types of magnetic switches used in starter systems: the ____________________ and the ____________________ .

ASE Style Review Questions

1. The purpose of the starter system is being discussed:
   Technician A says the starting system is a combination of mechanical and electrical parts that work together to start the engine.
   Technician B says the starting system is designed to change the mechanical energy into electrical energy.
   Who is correct?
   A. A only
   B. B only
   C. Both A and B
   D. Neither A nor B

2. The components of the starting system are being discussed:
   Technician A says the drive belt is part of the starting system.
   Technician B says the starter drive and flywheel ring gear are components of the starting system.
   Who is correct?
   A. A only
   B. B only
   C. Both A and B
   D. Neither A nor B

3. The operation of the DC motor is being discussed:
   Technician A says DC motors use the interaction of magnetic fields.
   Technician B says DC motors use a mechanical connection to the engine that turns the armature.
   Who is correct?
   A. A only
   B. B only
   C. Both A and B
   D. Neither A nor B

4. The starter motor armature is being discussed:
   Technician A says the armature is the stationary component of the motor that consists of a conductor wound around a pole shoe.
   Technician B says the commutator is a series of conducting segments located around one end of the armature.
   Who is correct?
   A. A only
   B. B only
   C. Both A and B
   D. Neither A nor B

5. The starter motor field coils are being discussed:
   Technician A says the field coil is made of wire wound around a non-magnetic pole shoe.
   Technician B says the field coils are always shunt wound to the armature.
   Who is correct?
   A. A only
   B. B only
   C. Both A and B
   D. Neither A nor B

6. The starter solenoid is being discussed:
   Technician A says the solenoid is an electromagnetic device that uses movement of a plunger to exert a pulling or holding force.
   Technician B says the solenoid makes an electrical connection, which allows voltage to the starter motor.
   Who is correct?
   A. A only
   B. B only
   C. Both A and B
   D. Neither A nor B

7. *Technician A* says both windings of the solenoid are energized anytime the ignition switch is in the START position.
   Technician B says the two windings of the solenoid are called the pull-in and the hold-in windings.
   Who is correct?
   A. A only
   B. B only
   C. Both A and B
   D. Neither A nor B

8. Permanent magnet starters are being discussed:
   Technician A says the permanent magnet starter uses four or six permanent magnet field assemblies in place of field coils.
   Technician B says the permanent magnet starter uses a planetary gear set.
   Who is correct?
   A. A only
   B. B only
   C. Both A and B
   D. Neither A nor B

9. *Technician A* says the R terminal of the starter solenoid provides current to the ignition bypass circuit.
Technician B says the ignition switch is a component of the starter feed circuit.
Who is correct?
A. A only
B. B only
C. Both A and B
D. Neither A nor B

10. A customer's no-start complaint is being discussed:
Technician A says the problem may be that the starter safety switch needs adjusting.
Technician B says the battery should be tested before condemning starter system components.
Who is correct?
A. A only
B. B only
C. Both A and B
D. Neither A nor B

CHAPTER 7

Charging Systems

Upon completion and review of this chapter, you should be able to:

- ❑ Explain the purpose of the charging system.
- ❑ Identify the major components of the charging system.
- ❑ Explain the function of the major components of the AC generator.
- ❑ Describe the two styles of stators.
- ❑ Describe how AC current is rectified to DC current in the AC generator.
- ❑ Describe the three principle circuits used in the AC generator.
- ❑ Explain the relationship between regulator resistance and field current.
- ❑ Explain the relationship between field current and AC generator output.
- ❑ Identify the differences between A circuit, B circuit, and isolated circuit.
- ❑ Explain the operation of charge indicators, including lamps, electronic voltage monitors, ammeters, and voltmeters.

Introduction

The automotive storage battery is not capable of supplying the demands of the electrical system for an extended period of time. Every vehicle must be equipped with a means of replacing the current being drawn from the battery. A charging system is used to restore the electrical power to the battery that was used during engine starting. In addition, the charging system must be able to react quickly to high load demands required of the electrical system. It is the vehicle's charging system that generates the current to operate all of the electrical accessories while the engine is running.

Two basic types of charging systems have been used. The first was a DC generator, which was discontinued in the 1960s. Since that time the AC generator has been the predominant charging device. The DC generator and the AC generator both use similar operating principles.

In an attempt to standardize terminology in the industry, the term alternator is being replaced with generator. Often an alternator is referred to as an AC generator.

The purpose of the charging system is to convert the mechanical energy of the engine into electrical energy to recharge the battery and run the electrical accessories. When the engine is first started, the battery supplies all the current required by the starting and ignition systems.

As the battery drain continues, and engine speed increases, the charging system is able to produce more voltage than the battery can deliver. When this occurs, the electrons from the charging device are able to flow in a reverse direction through the battery's positive terminal. The charging device is now supplying the electrical system's load requirements; the reserve electrons build up and recharge the battery (Figure 7-1).

If there is an increase in the electrical demand and a drop in the charging system's output equal to the voltage of the battery, the battery and charging system work together to supply the required current.

As illustrated (Figure 7-2), the entire charging system consists of the following components:

1. Battery
2. AC or DC generator
3. Drive belt
4. Voltage regulator
5. Charge indicator (lamp or gauge)

The ignition switch is considered a part of the charging system because it closes the circuit that supplies current to the indicator lamp and stimulates the field coil.

6. Ignition switch
7. Cables and wiring harness
8. Starter relay (some systems)
9. Fusible link (some systems)

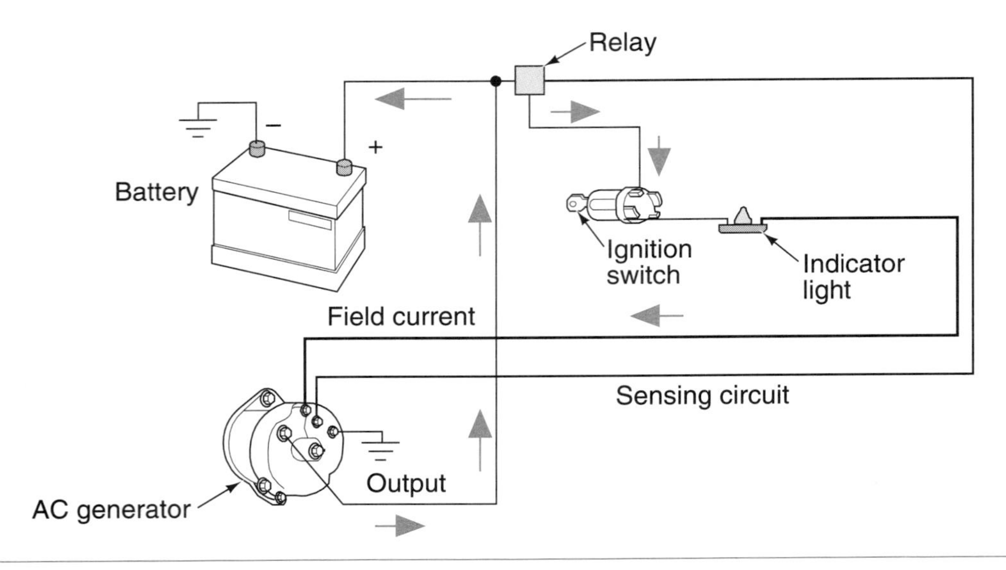

Figure 7-1 Current flow when the AC generator is operating.

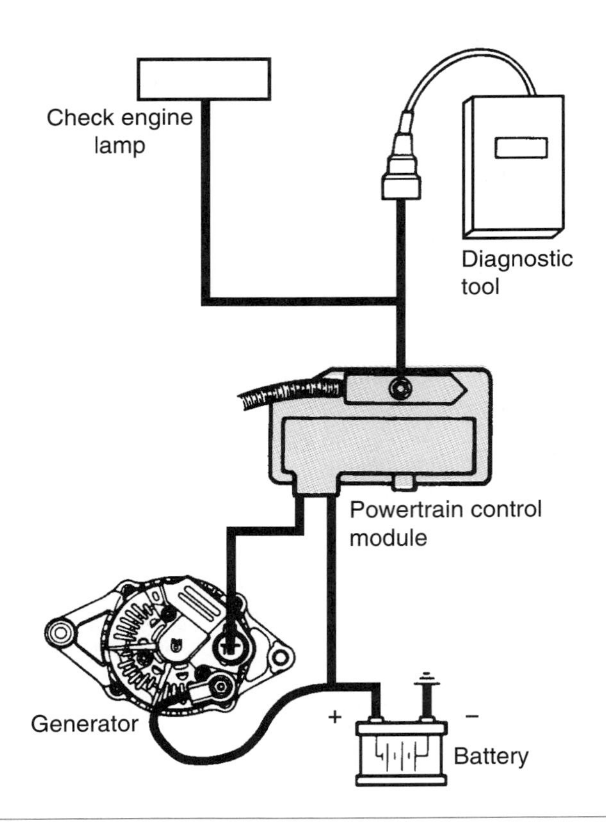

Figure 7-2 Components of the charging system. (Courtesy of Chrysler Corporation)

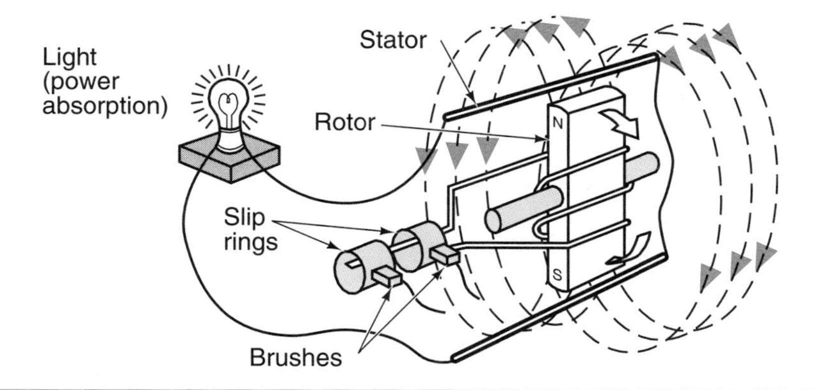

Figure 7-3 Simplified alternator showing electromagnetic induction.

Principle of Operation

Shop Manual
Chapter 7, page 191

All charging systems use the principle of electromagnetic induction to generate electrical power (Figure 7-3). Electromagnetic principle states that a voltage will be produced if motion between a conductor and a magnetic field occurs. The amount of voltage produced is affected by:

1. The speed at which the conductor passes through the magnetic field.
2. The strength of the magnetic field.
3. The number of conductors passing through the magnetic field.

To see how electromagnetic induction produces an AC voltage, refer to the illustration (Figure 7-4). When the conductor is parallel with the magnetic field, the conductor is not cut by any flux lines (Figure 7-4A). At this point in the revolution there is zero voltage and current being produced.

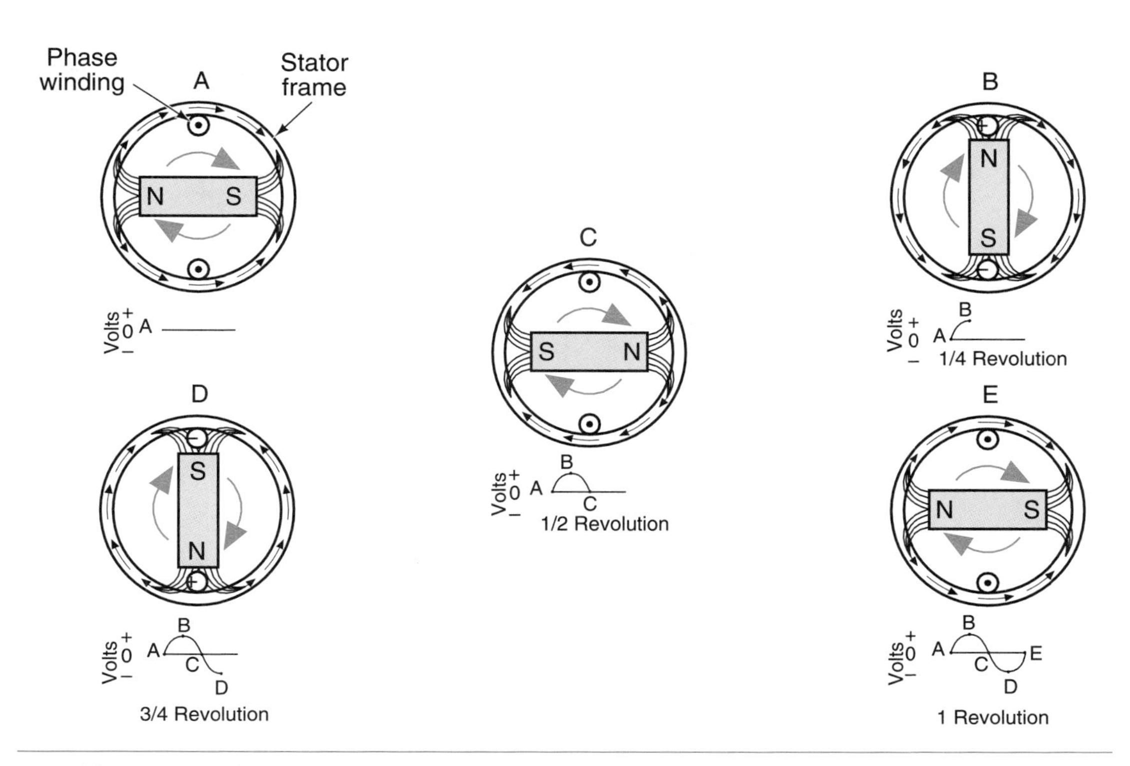

Figure 7-4 Alternating current is produced as the magnetic field is rotated.

As the magnetic field is rotated 90 degrees, the magnetic field is at a right angle to the conductor (Figure 7-4B). At this point in the revolution the maximum number of flux lines cut the conductor at the north pole. With the maximum amount of flux lines cutting the conductor, voltage is at its maximum positive value.

When the magnetic field is rotated an additional 90 degrees, the conductor returns to being parallel with the magnetic field (Figure 7-4C). Once again no flux lines cut the conductor, and voltage drops to zero.

An additional 90-degree revolution of the magnetic field results in the magnetic field being reversed at the top conductor (Figure 7-4D). At this point in the revolution, the maximum number of flux lines cuts the conductor at the south pole. Voltage is now at maximum negative value.

The sine wave produced by a single conductor during one revolution is called **single-phase voltage.**

When the magnetic field completes one full revolution, it returns to a parallel position with the magnetic field. Voltage returns to zero. The **sine wave** is determined by the angle between the magnetic field and the conductor. It is based on the trigonometry sine function of angles. The sine wave shown (Figure 7-5) plots the voltage generated during one revolution.

It is the function of the drive belt to turn the magnetic field. Drive belt tension should be checked periodically to assure proper charging system operation. A loose belt can inhibit charging system efficiency, and a belt that is too tight can cause early bearing failure.

DC Generators

The DC generator is similar to the DC starter motor used to crank the engine. The housing contains two field coils that create a magnetic field. Output voltage is generated in the wire loops of the armature as it rotates inside the magnetic field. This voltage sends current to the battery through the brushes. A DC generator is actually an AC generator whose current is rectified by a commutator.

Polarizing will make sure that the polarity is the same for both the generator and the regulator.

The components must be polarized whenever a replacement DC generator or voltage regulator is installed. To polarize an externally grounded field circuit (A-type field circuit), use a jumper wire and connect between the BAT terminal and the ARM terminal of the voltage regulator. Make this jumper connection for just an instant. Do not hold the jumper wire on the terminals. For an internally grounded field circuit (B-type), jump the F terminal and the BAT terminal.

Most generator terminals are identified as BAT for battery output, F for field, and ARM for armature.

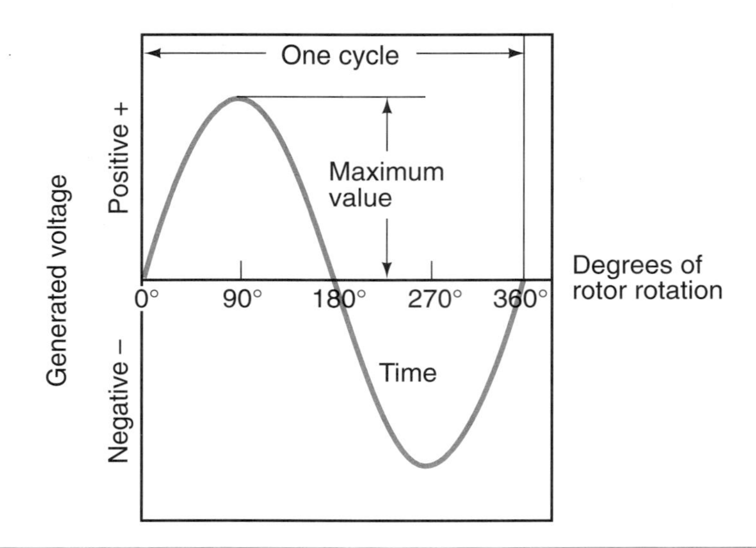

Figure 7-5 Sine wave produced in one revolution of the conductor or magnetic field.

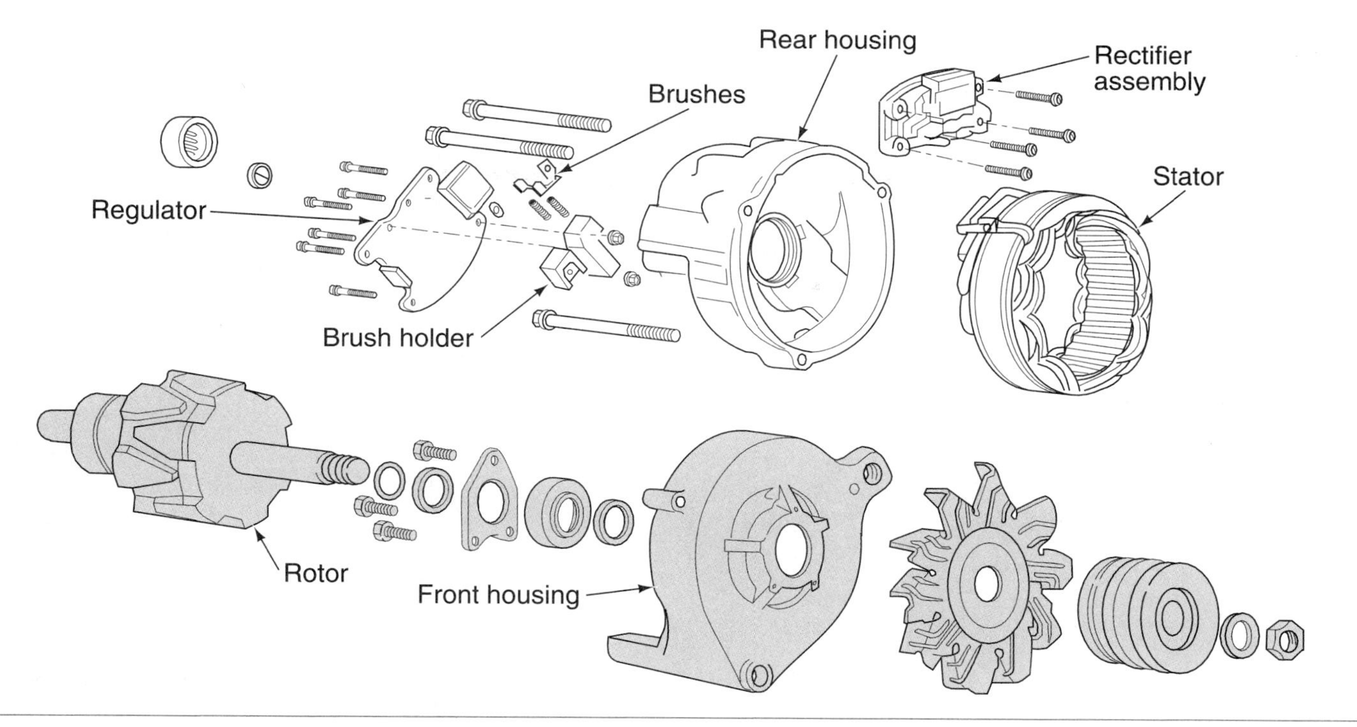

Figure 7-6 Components of an AC generator. (Reprinted with the permission of Ford Motor Company)

AC Generators

The DC generator was unable to produce the sufficient amount of current required when the engine was operating at low speeds. With the addition of more electrical accessories and components, the AC (alternating current) generator, or alternator, replaced the DC generator. The main components of the AC generator are (Figure 7–6):

1. The rotor
2. Brushes
3. The stator
4. The rectifier bridge
5. The housing
6. Cooling fan

Rotors. The rotor is constructed of many turns of copper wire around an iron core. There are metal plates bent over the windings at both ends of the rotor windings (Figure 7-7). The poles do not come into contact with each other, but they are interlaced. When current passes through the coil (1.5 to 3.0 amperes), a magnetic field is produced. The strength of the magnetic field is dependent on the amount of current flowing through the coil and the number of windings.

The poles will take on the polarity (north or south) of the side of the coil they touch. The right-hand rule will show whether a north or south pole magnet is created. When the rotor is assembled, the poles alternate north-south around the rotor (Figure 7- 8). As a result of this alternating arrangement of poles, the magnetic flux lines will move in opposite directions between adjacent poles (Figure 7-9). This arrangement provides for several alternating magnetic fields to intersect the stator as the rotor is turning. These individual magnetic fields produce a voltage by induction in the stationary stator windings.

Shop Manual
Chapter 7, page 217

The **rotor** creates the rotating magnetic field of the AC generator. It is the portion of the AC generator that is rotated by the drive belt.

The metal plates are called **fingers** or **poles**.

The current flow through the coil is referred to as **field current**.

Most rotors have twelve to fourteen poles.

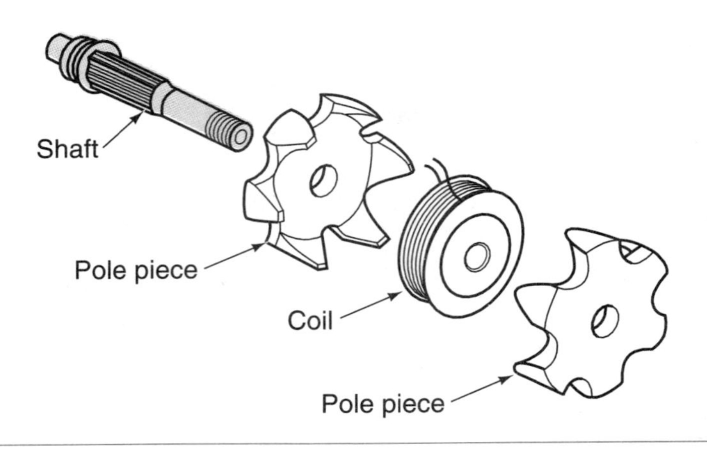

Figure 7-7 Components of a typical AC generator rotor.

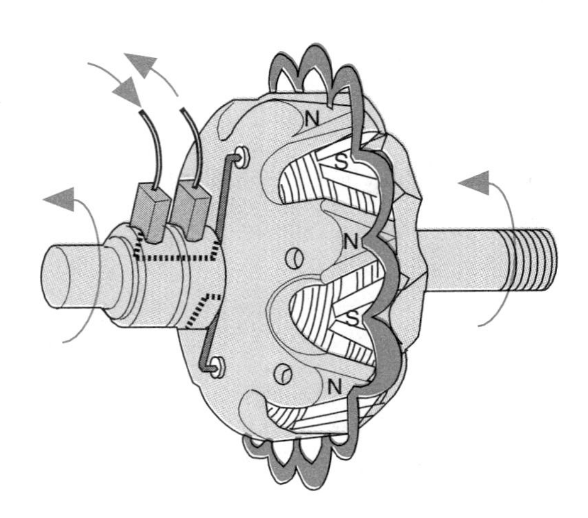

Figure 7-8 The north and south poles of a rotor's field alternate.

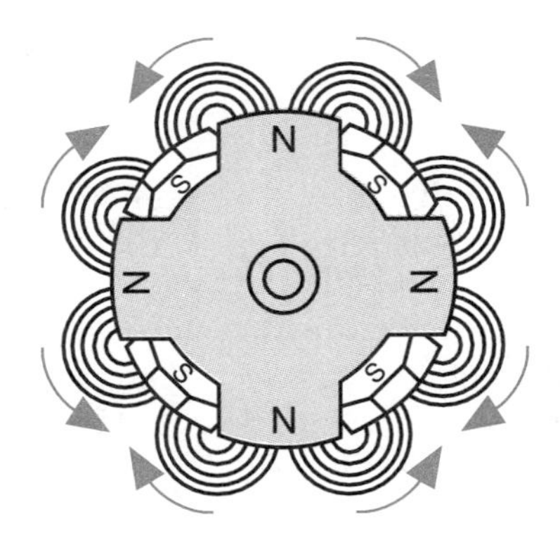

Figure 7-9 Magnetic flux lines move in opposite directions between the rotor poles. (Reprinted with the permission of Ford Motor Company)

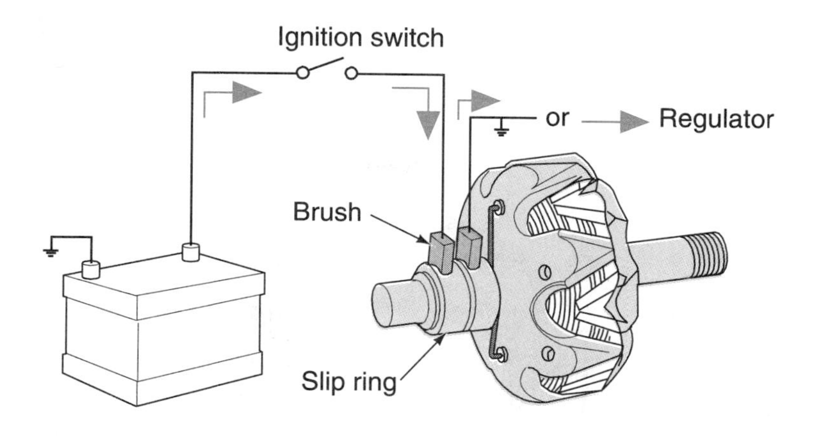

Figure 7-10 The slip rings and brushes provide a current path to the rotor coil.

The wires from the rotor coil are attached to two slip rings that are insulated from the rotor shaft. The insulated stationary carbon brush passes field current into a slip ring, then through the field coil, and back to the other slip ring. Current then passes through a grounded stationary brush (Figure 7-10) or to a voltage regulator.

The **slip rings** function much like the armature commutator in the starter motor, except they are smooth.

Brushes. The field winding of the rotor receives current through a pair of brushes that ride against the slip rings. The brushes and slip rings provide a means of maintaining electrical continuity between stationary and rotating components. The brushes (Figure 7-11) ride the surface of the slip rings on the rotor and are held tight against the slip rings by spring tension provided by the brush holders. The brushes conduct only the field current (2 to 5 amperes). The low current that the brushes must carry contributes to their longer life.

Direct current from the battery is supplied to the rotating field through the field terminal and the insulated brush. The second brush may be the ground brush, which is attached to the AC generator housing or to a voltage regulator.

Shop Manual
Chapter 7, page 217

Brushes are electrically conductive sliding contacts, usually made of copper and carbon.

Stators. The stator contains three main sets of windings wrapped in slots around a laminated, circular iron frame (Figure 7-12). Each of the three windings has the same number of coils as the rotor has pairs of north and south poles. The coils of each winding are evenly spaced around the

The brushes used in the DC generator conduct the output voltage of the generator.

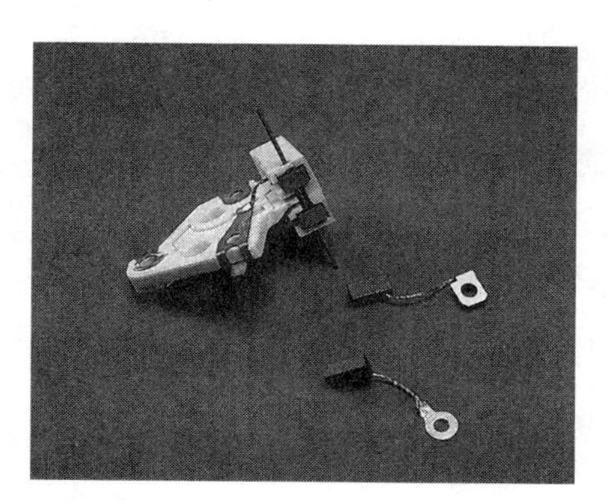

Figure 7-11 Brushes are the stationary electrical contacts to the rotor's slip rings.

Shop Manual
Chapter 7, page 218

The **stator** is the stationary coil in which electricity is produced.

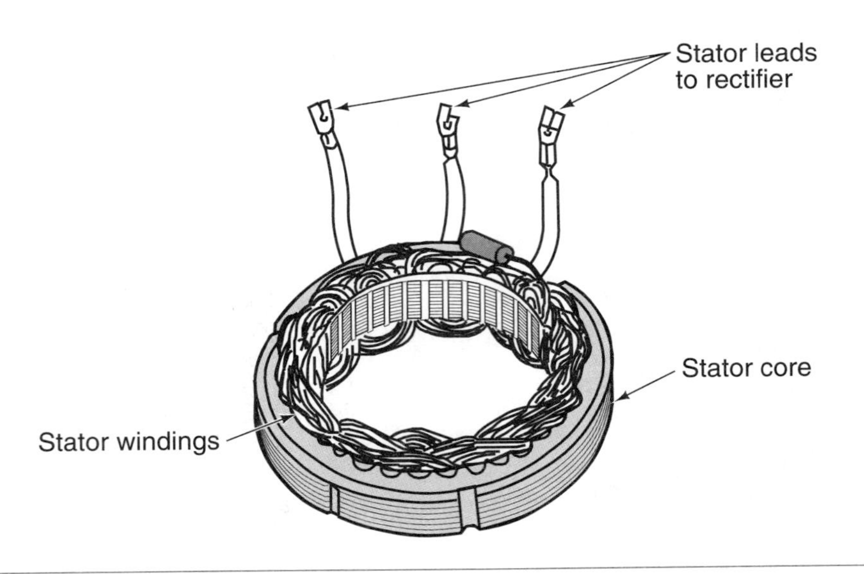

Figure 7-12 Components of a typical stator. (Reprinted with the permission of Ford Motor Company)

core. The three sets of windings alternate and overlap as they pass through the core (Figure 7-13). The overlapping is needed to produce the required phase angles.

The rotor is fitted inside the stator (Figure 7-14). A small air gap (approximately 0.015 inch) is maintained between the rotor and the stator. This gap allows the rotor's magnetic field to energize all of the windings of the stator at the same time and to maximize the magnetic force.

Each group of windings has two leads. The first lead is for the current entering the winding. The second lead is for current leaving. There are two basic means of connecting the leads. In the

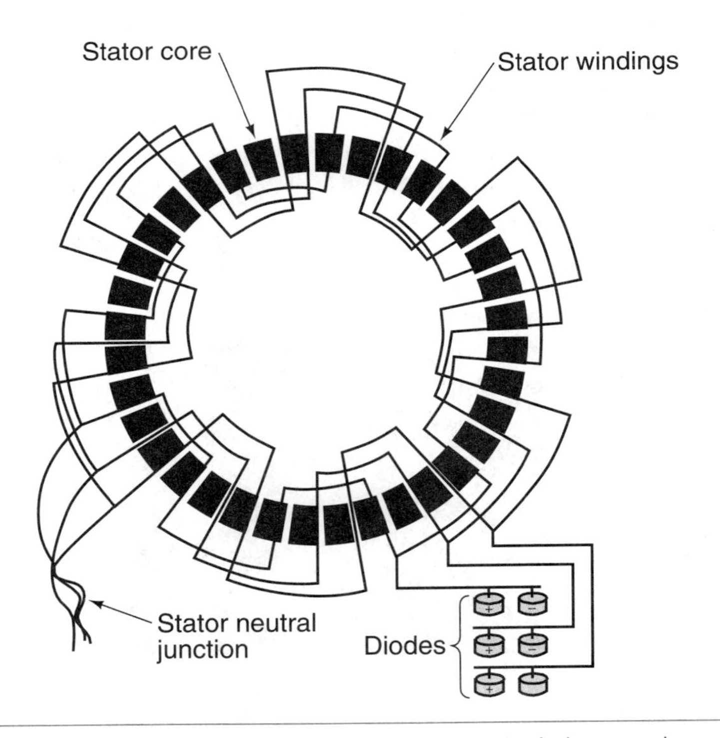

Figure 7-13 Overlapping stator windings produce the required phase angles.

Figure 7-14 A small air gap between the rotor and the stator maximizes the magnetic force.

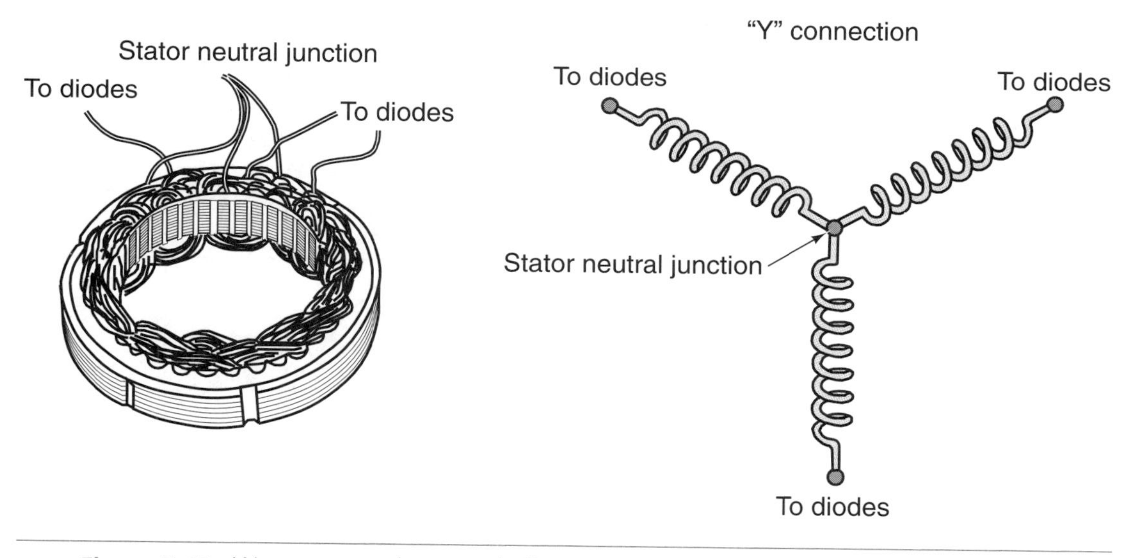

Figure 7-15 Wye-connected stator winding.

The **wye wound connection** receives its name from its resemblance to the letter *Y*.

The common junction in the wye connected winding is called the **stator neutral junction.**

The **delta connection** receives its name from its resemblance to the Greek letter delta (Δ).

past, the most common method was the wye connection (Figure 7-15). In the wye connection one lead from each winding is connected to one common junction. From this junction the other leads branch out in a Y pattern. Today the most common method of connecting the windings is called the delta connection (Figure 7-16). The delta connection connects the lead of one end of the winding to the lead at the other end of the next winding.

Each group of windings occupies one third of the stator, or 120 degrees of the circle. As the rotor revolves in the stator, a voltage is produced in each loop of the stator at different phase angles. The resulting overlap of sine waves that is produced is shown (Figure 7-17). Each of the sine waves is at a different phase of its cycle at any given time. As a result, the output from the stator is divided into three phases.

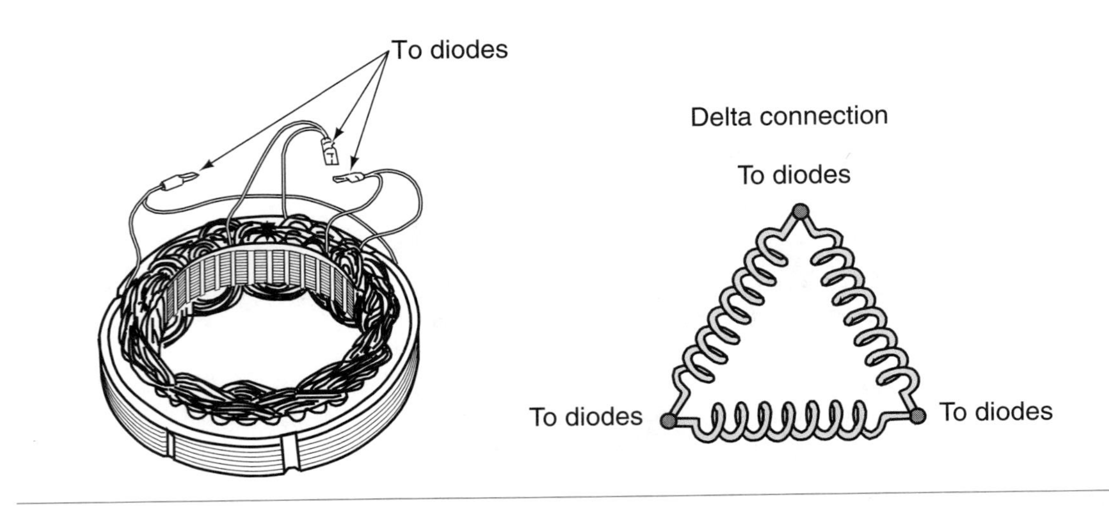

Figure 7-16 Delta-connected stator winding.

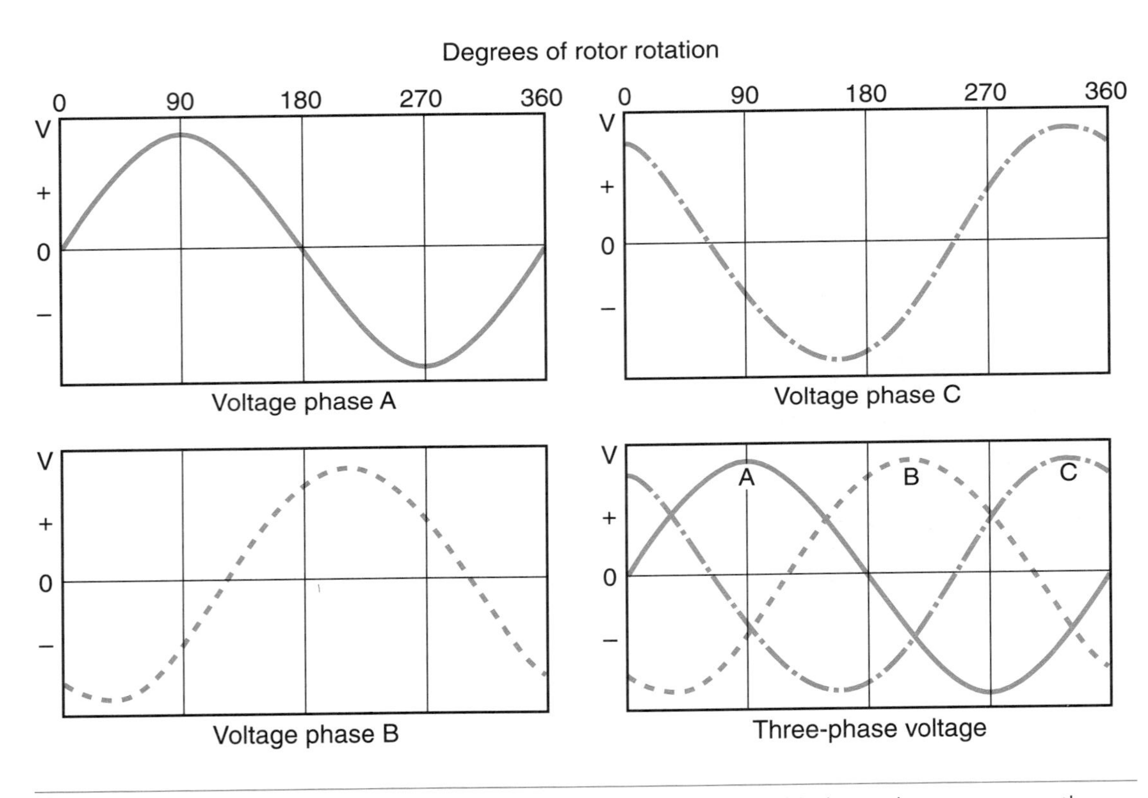

Figure 7-17 The voltage produced in each stator winding is added together to create a three-phase voltage.

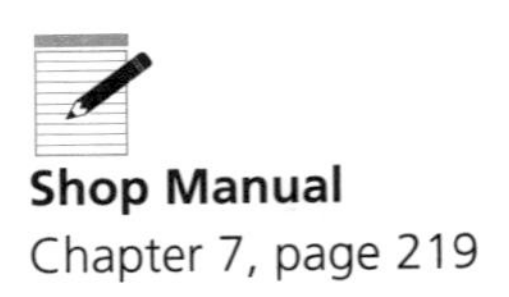

Shop Manual
Chapter 7, page 219

Diode Rectifier Bridge. The battery and the electrical system cannot accept or store AC voltage. For the vehicle's electrical system to be able to use the voltage and current generated in the AC generator, the AC current needs to be converted to DC current. A split-ring commutator cannot be used to rectify AC current to DC current because the stator is stationary in an AC generator. Instead, a diode rectifier bridge is used to change the current in an AC generator (Figure 7-18). Acting as a one-way check valve, the diodes switch the current flow back and forth so that it flows from the AC generator in only one direction.

Figure 7-18 General Motors' rectifier bridge.

When AC current reverses itself, the diode blocks and no current flows. If AC current passes through a positively biased diode, the diode will block off the negative pulse. The result is the scope pattern shown in Figure 7-19. The AC current has been changed to a pulsing DC current.

An AC generator usually uses a pair of diodes for each stator winding, for a total of six diodes (Figure 7-20). Three of the diodes are positive biased and are mounted in a heat sink (Figure 7-21). The three remaining diodes are negative biased and are attached directly to the frame of the AC generator (Figure 7-22). By using a pair of diodes that are reversed biased to each other, rectification of both sides of the AC sine wave is achieved (Figure 7-23). The negative biased diodes allow for conducting current from the negative side of the AC sine wave and putting this current into the circuit. Diode rectification changes the negative current into positive output.

The **diode rectifier bridge** provides reasonably constant DC voltage to the vehicle's electrical system and battery.

The converting of AC current to DC current is called **rectification.**

Diodes are electrical one-way check valves that permit current to flow in one direction only.

The passing of half of the AC pulses is called **half-wave rectification.**

The use of three positive biased and three negative biased diodes provides for **full-wave rectification.**

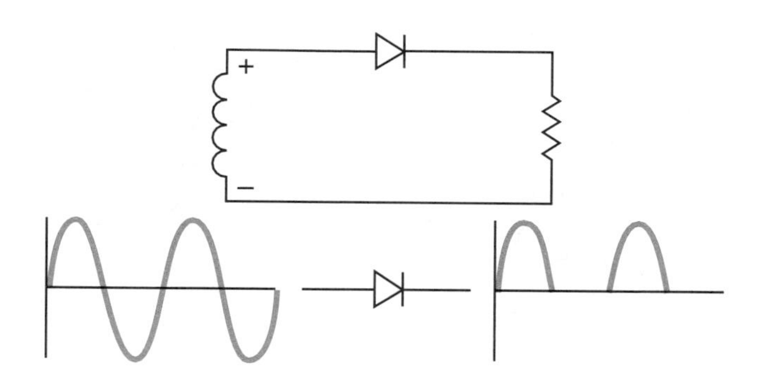

Figure 7-19 AC current rectified to a pulsating DC current after passing through a positive biased diode. This is called half-wave rectification.

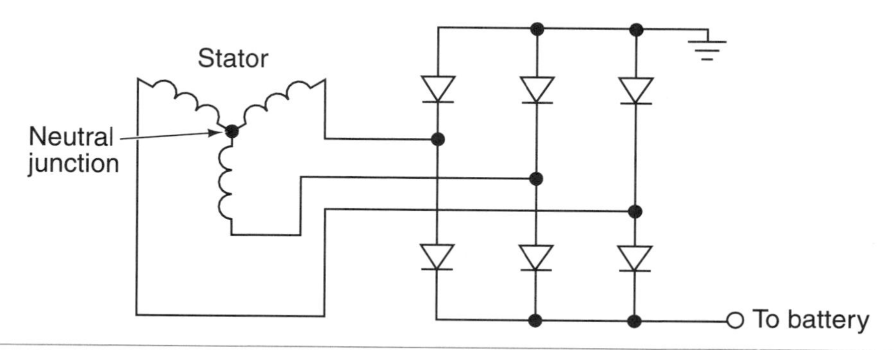

Figure 7-20 A simplified schematic of AC generator windings connected to the diode rectifier bridge.

The diodes of the rectifier bridge produce heat, thus they are mounted in **heat sinks** to dissipate the heat.

Not only do the diodes rectify stator output, but they also block battery drain back when the engine is not running.

The circuit is called **full-wave rectification** because both halves of the sine wave are used.

The rectifier bridge is also known as a **rectifier stack.**

Figure 7-21 The positive-biased diodes are mounted into a heat sink to provide protection.

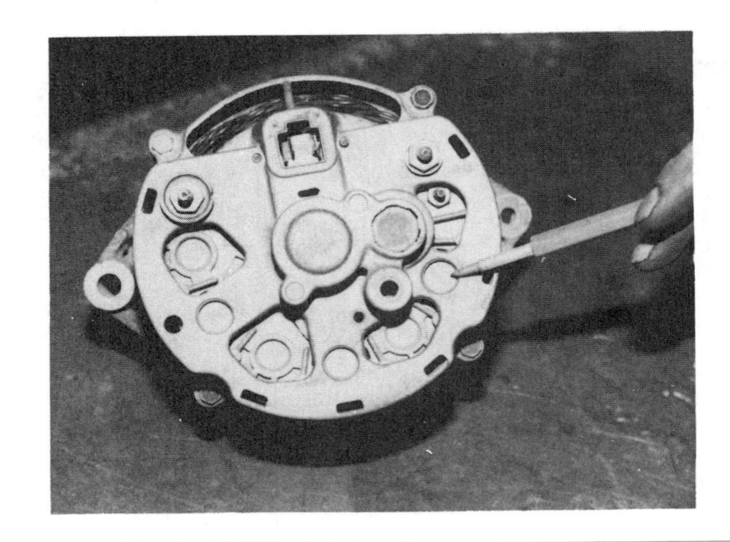

Figure 7-22 Negative-biased diodes pressed into the AC generator housing.

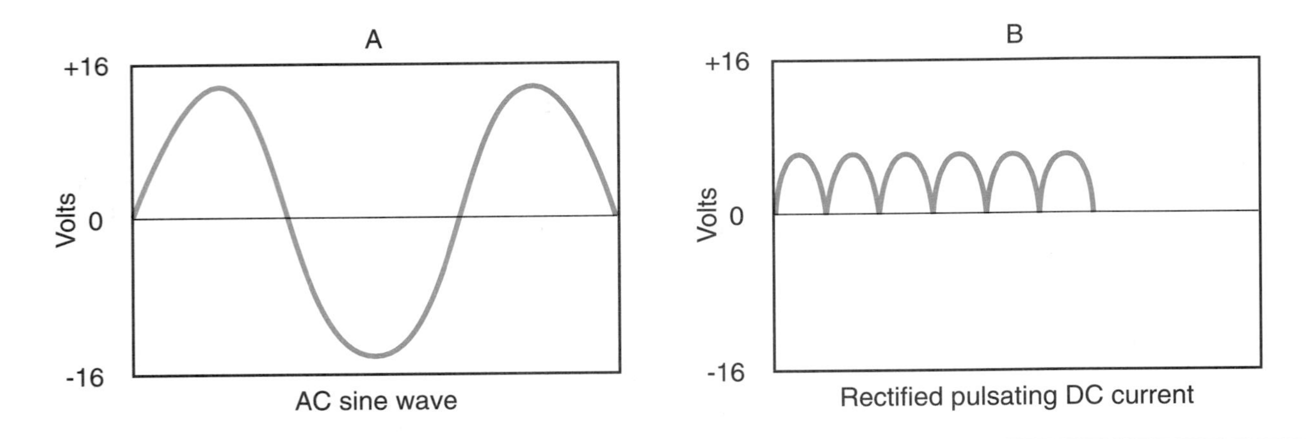

Figure 7-23 Full-wave rectification uses both sides of the AC sine wave to create a pulsating DC current.

With each stator winding connected to a pair of diodes, the resultant waveform of the rectified voltage would be similar to that shown (Figure 7-24). With six peaks per revolution, the voltage will vary only slightly during each cycle.

The examples used so far have been for single-pole rotors in a three-winding stator. Most AC generators use either a twelve- or fourteen-pole rotor. Each pair of poles produces one complete sine wave in each winding per revolution. During one revolution a fourteen-pole rotor will produce seven sine waves. The rotor generates three overlapping sine wave voltage cycles in the stator. The total output of a fourteen-pole rotor per revolution would be twenty-one sine wave cycles (Figure 7-25). With final rectification, the waveform would be similar to the one shown (Figure 7-26).

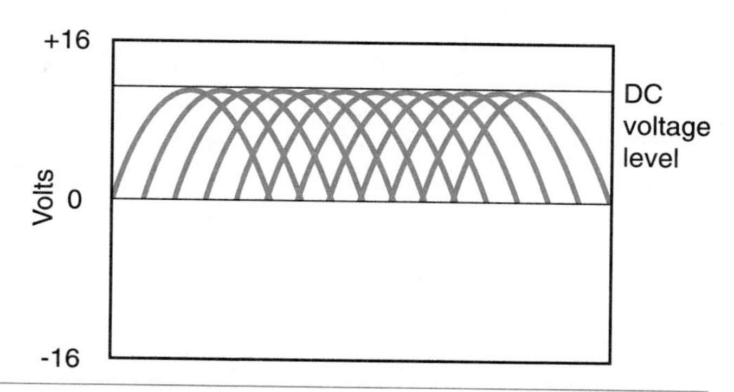

Figure 7-24 With three-phase rectification, the DC voltage level is uniform.

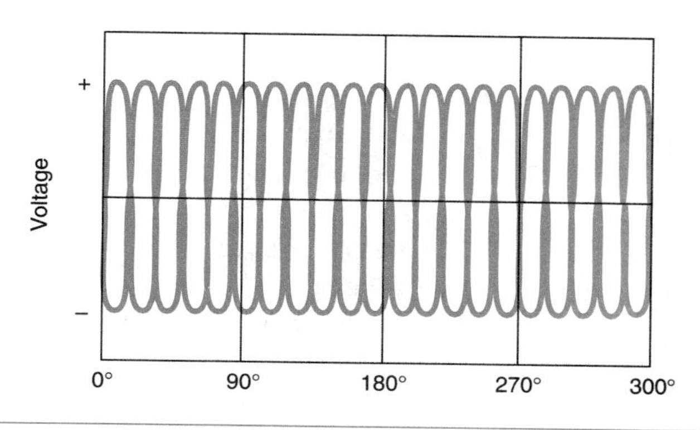

Figure 7-25 Sine wave cycle of a fourteen-pole rotor and three-phase stator.

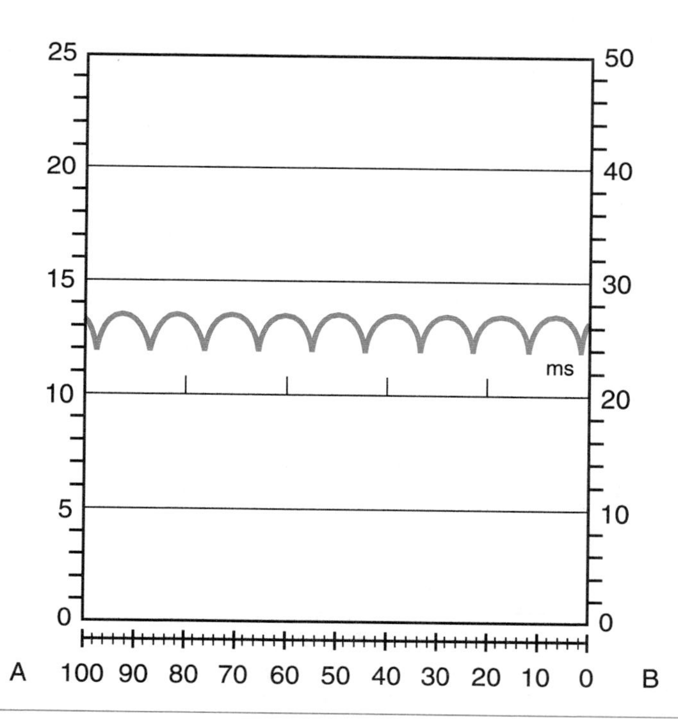

Figure 7-26 Rectified AC output has a ripple that can be shown on an oscilloscope.

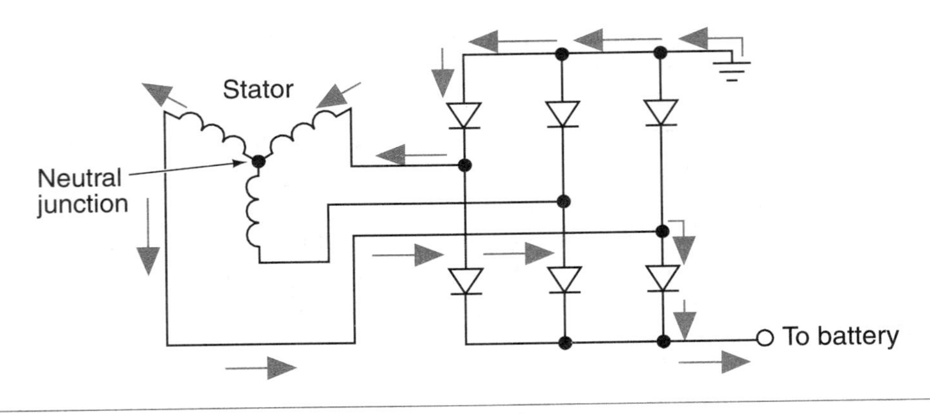

Figure 7-27 Current flow through a wye-wound stator.

Full-wave rectification is desired because using only half-wave rectification wastes the other half of the AC current. Full-wave rectification of the stator output uses the total potential by redirecting the current from the stator windings so that all current is in one direction.

A wye wound stator with each winding connected to a pair of diodes is shown (Figure 7-27). Each pair of diodes has one negative and one positive diode. During rotor movement two stator windings will be in series and the third winding will be neutral. As the rotor revolves it will energize a different set of windings. Also, current flow through the windings is reversed as the rotor passes. Current in any direction through two windings in series will produce DC current.

The action that occurs when the delta wound stator is used is shown (Figure 7-28). Instead of two windings in series, the three windings of the delta stator are in parallel. This makes more current available because the parallel paths allow more current to flow through the diodes.

Shop Manual
Chapter 7, page 214

AC Generator Housing and Cooling Fan. Most AC generator housings are a two-piece construction, made from cast aluminum (Figure 7-29). The two end frames provide support of the rotor and the stator. In addition, the end frames contain the diodes, regulator, heat sinks, terminals, and other components of the AC generator. The two end pieces are referred to as:

1. The drive end housing: This housing holds a bearing to support the front of the rotor shaft. The rotor shaft extends through the drive end housing and holds the drive pulley and cooling fan.
2. The slip ring end housing: This housing also holds a rotor shaft that supports a bearing. In addition, it contains the brushes and has all of the electrical terminals. If the AC generator has an integral regulator, it is also contained in this housing.

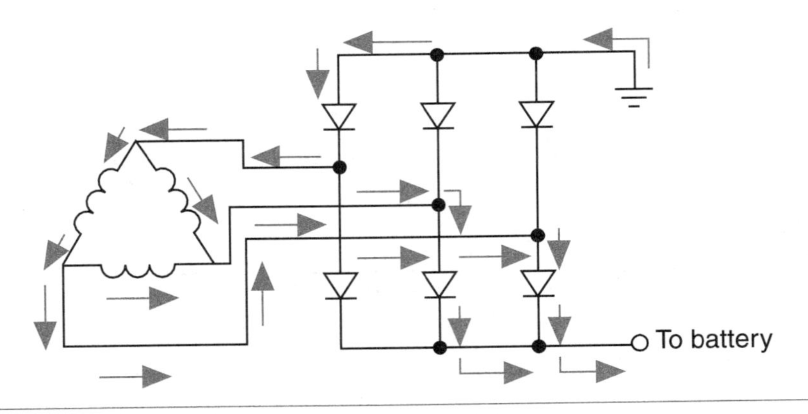

Figure 7-28 Current flow through a delta-wound stator.

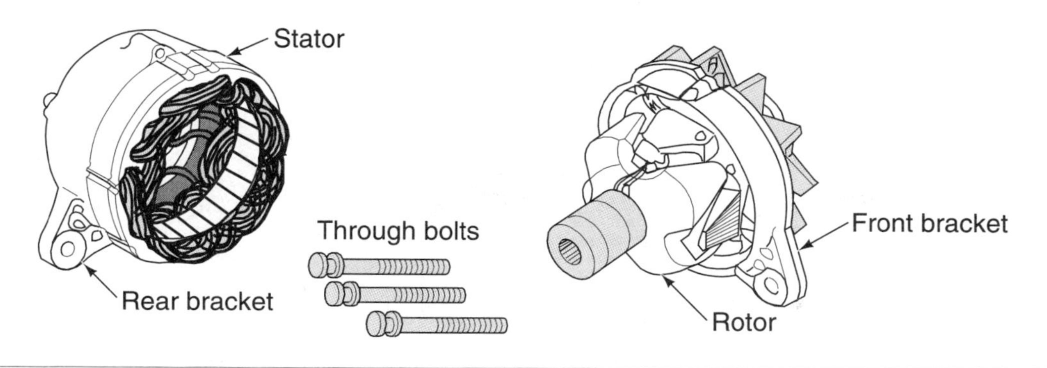

Figure 7-29 Typical two-piece AC generator housing. (Courtesy of Chrysler Corporation)

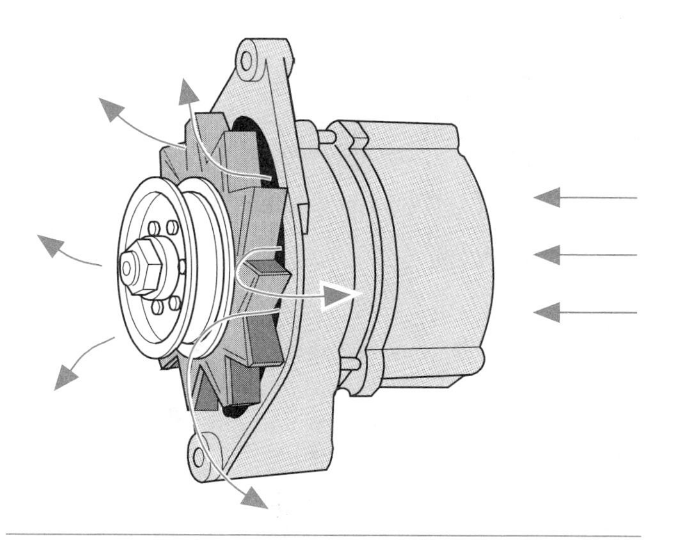

Figure 7-30 The cooling fan draws air in from the rear of the generator to keep the diodes cool. (Courtesy of Robert Bosch Corporation)

Figure 7-31 General Motors' CS series AC generator.

The cooling fan draws air into the housing through the openings at the rear of the housing. The air leaves through openings behind the cooling fan (Figure 7-30).

WARNING: Do not pry on the AC generator housing because any excess force can damage the housing.

General Motors introduced an AC generator called the CS (charging system) series. This generator is smaller than previous designs (Figure 7-31). Additional features include two cooling fans (one external and one internal) and terminals designed to permit connections to an external computer (Figure 7-32).

AC Generator Circuits

Shop Manual
Chapter 7, page 208

There are three principle circuits used in an AC generator:

1. The charging circuit: Consists of the stator windings and rectifier circuits.
2. The excitation circuit: Consists of the rotor field coil and the electrical connections to the coil.
3. The preexcitation circuit: Supplies the initial current for the field coil that starts the buildup of the magnetic field.

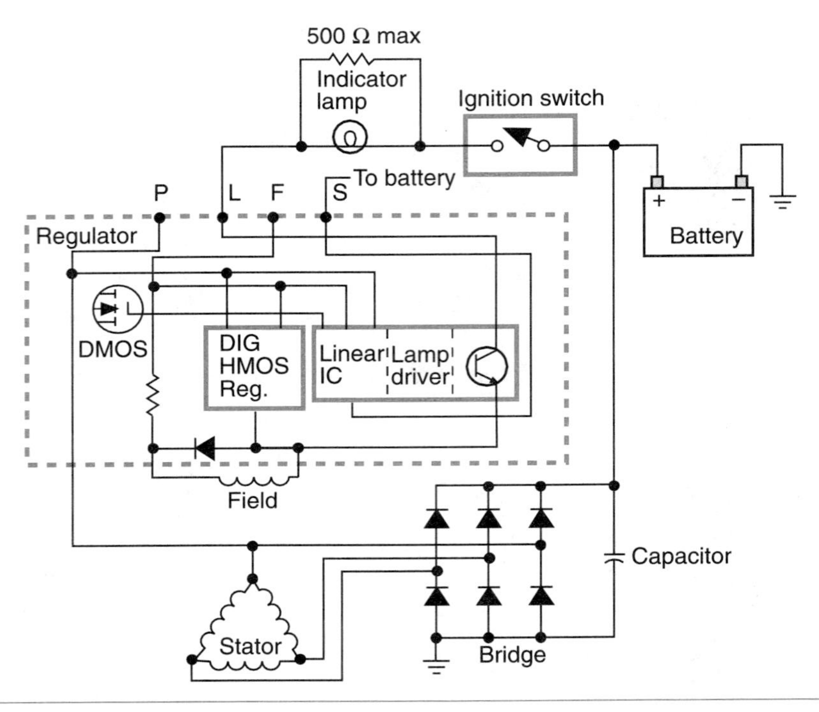

Figure 7-32 CS series AC generator circuit diagram.

If the battery is completely discharged, the vehicle cannot be push started because there is no excitation of the field coil.

The **diode trio** is used by some manufacturers to rectify current from the stator so that it can be used to create the magnetic field in the field coil of the rotor. This eliminates extra wiring.

For the AC generator to produce current, the field coil must develop a magnetic field. The AC generator creates its own field current in addition to its output current.

For excitation of the field to occur, the voltage induced in the stator rises to a point that it overcomes the forward voltage drop of at least two of the rectifier diodes. Before the diode trio can supply field current, the anode side of the diode must be at least 0.6 volt more positive than the cathode side (Figure 7-33). When the ignition switch is turned on, the warning lamp current acts as a small magnetizing current through the field (Figure 7-34). This current preexcites the field, reducing the speed required to start its own supply of field current.

AC Generator Operation Overview

When the engine is running, the drive belt spins the rotor inside the stator windings. This magnetic field inside the rotor generates a voltage in the windings of the stator. Field current flowing through the slip rings to the rotor creates alternating north and south poles on the rotor.

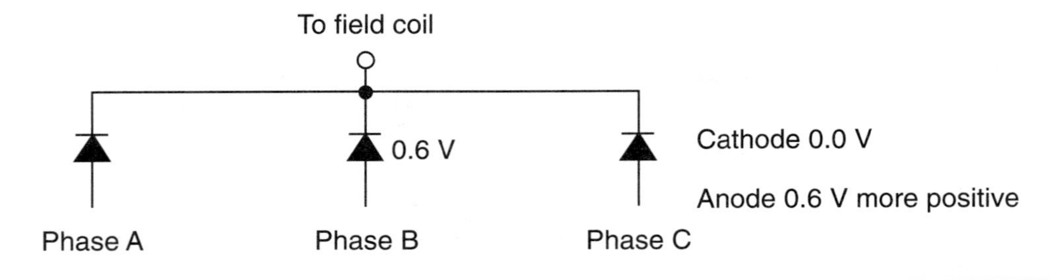

Figure 7-33 The diode trio connects the phase windings to the field. To conduct, there must be 0.6 volts more positive on the anode side of the diodes.

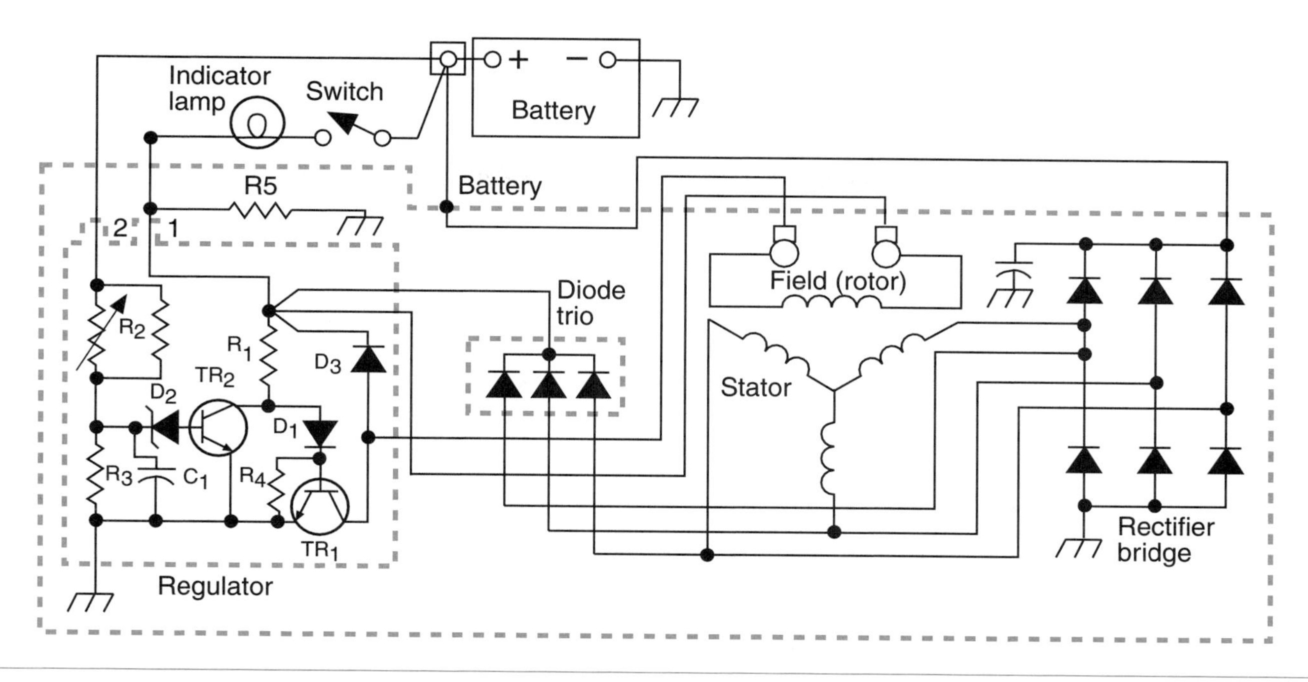

Figure 7-34 Schematic of a charging system.

The induced voltage in the stator is an alternating voltage because the magnetic fields are alternating. As the magnetic field begins to induce voltage in the stator's windings, the induced voltage starts to increase. The amount of voltage will peak when the magnetic field is the strongest. As the magnetic field begins to move away from the stator windings, the amount of voltage will start to decrease. Each of the three windings of the stator generates voltage, so the three combine to form a three-phase voltage output.

In the past, the most common type of stator was the wye connection (Figure 7-35). The output terminals (A, B, and C) apply voltage to the rectifier. Because only two stator windings apply voltage (because the third winding is always connected to diodes that are reverse biased), the voltages come from points A to B, B to C, and C to A.

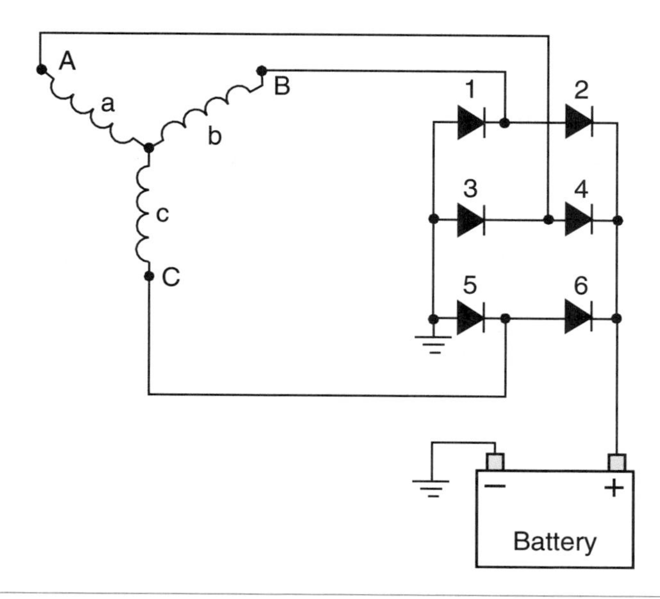

Figure 7-35 Wye-connected stator windings attached to the rectifier bridge.

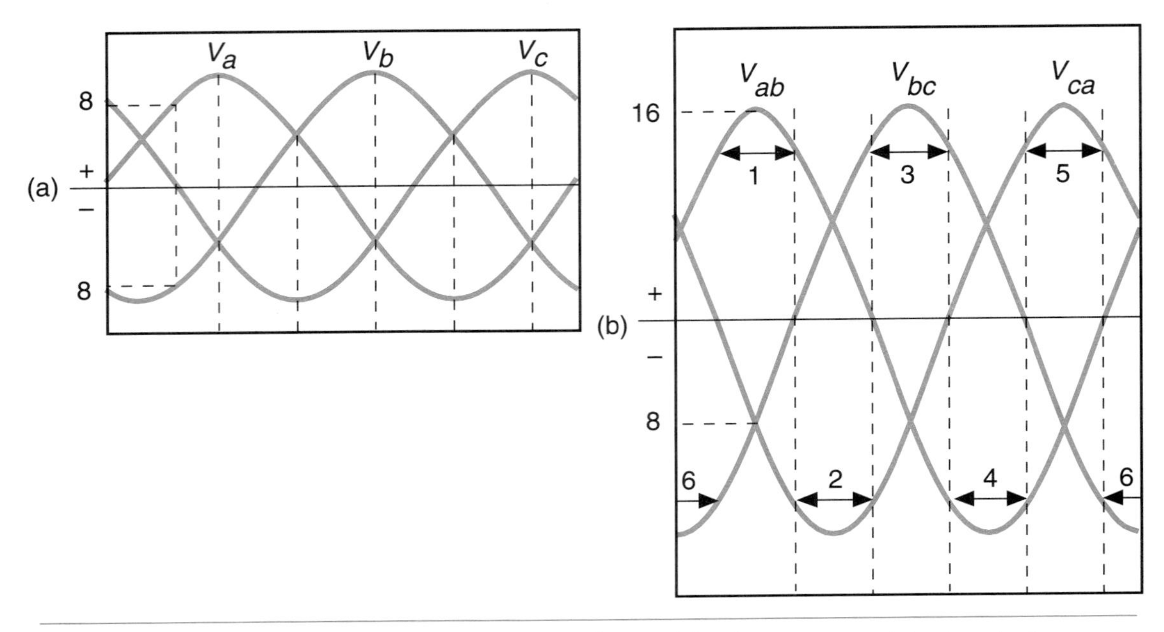

Figure 7-36 (A) Individual stator winding voltages; (B) voltages across stator terminals A, B, and C.

Alternating current is constantly changing, so this formula would have to be performed at several different times.

To determine the amount of voltage produced in the two stator windings, find the difference between the two points. For example, to find the voltage applied from points A and B subtract the voltage at point B from the voltage at point A. If the voltage at point A is 8 volts positive and the voltage at point B is 8 volts negative, the difference is 16 volts. This procedure can be performed for each pair of stator windings at any point in time to get the sine wave patterns (Figure 7-36). The voltages in the windings are designated as VA, VB, and VC. Designations of VAB, VBC, and VCA refer to the voltage difference in the two stator windings. In addition, the numbers refer to the diodes used for the voltages generated in each winding pair.

The current induced in the stator passes through the diode rectifier bridge consisting of three positive and three negative diodes. At this point there are six possible paths for the current to follow. The path that is followed depends on the stator terminal voltages. If the voltage from points A and B is positive (point A is positive in respect to point B), current is supplied to the positive terminal of the battery from terminal A through diode 2 (Figure 7-37). The negative return path is through diode 3 to terminal B.

Both diodes 2 and 3 are forward biased. The stator winding labeled C does not produce current because it is connected to diodes that are reverse biased. The stator current is rectified to DC current to be used for charging the battery and supplying current to the vehicle's electrical system.

When the voltage from terminals C and A is negative (point C is negative in respect to point A), current flow to the battery positive terminal is from terminal A through diode 2 (Figure 7-38). The negative return path is through diode 5 to terminal C.

This procedure is repeated through the four other current paths (Figures 7-39 through 7-42).

Regulation

Shop Manual
Chapter 7, page 208

The battery, and the rest of the electrical system, must be protected from excessive voltages. To prevent early battery and electrical system failure, regulation of the charging system voltage is very important. Also the charging system must supply enough current to run the vehicle's electrical accessories when the engine is running.

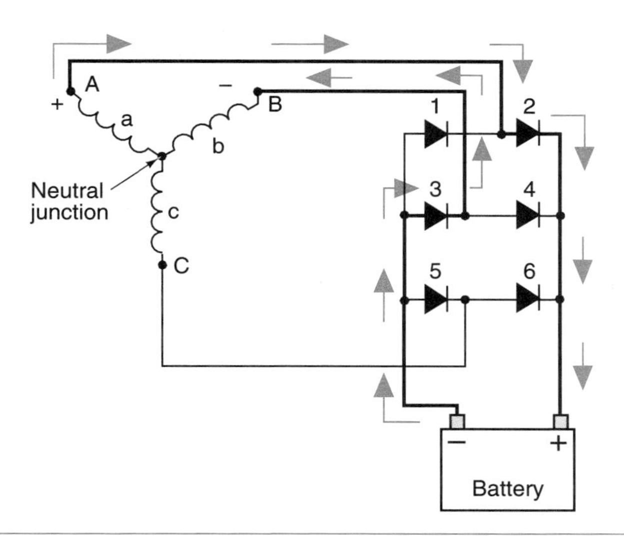

Figure 7-37 Current flow when terminals A and B are positive.

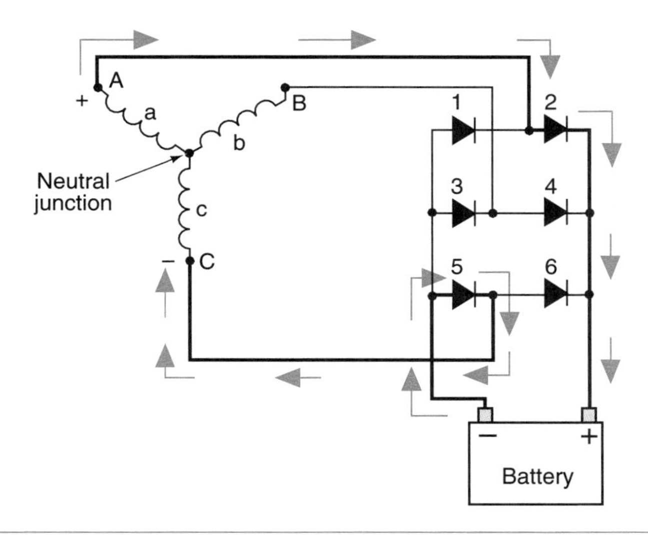

Figure 7-38 Current flow when terminals A and C are negative.

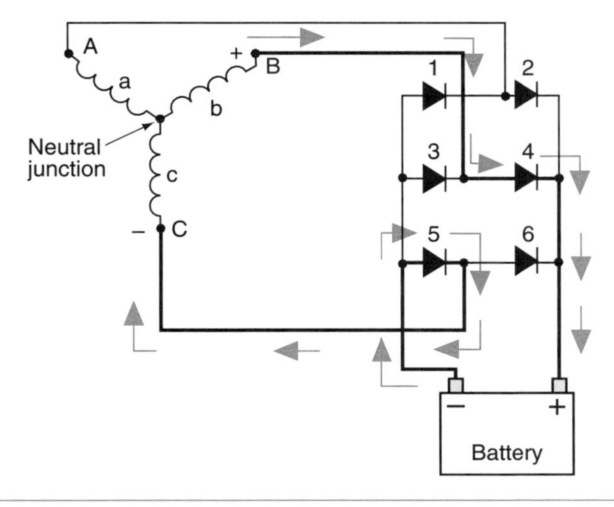

Figure 7-39 Current flow when terminals B and C are positive.

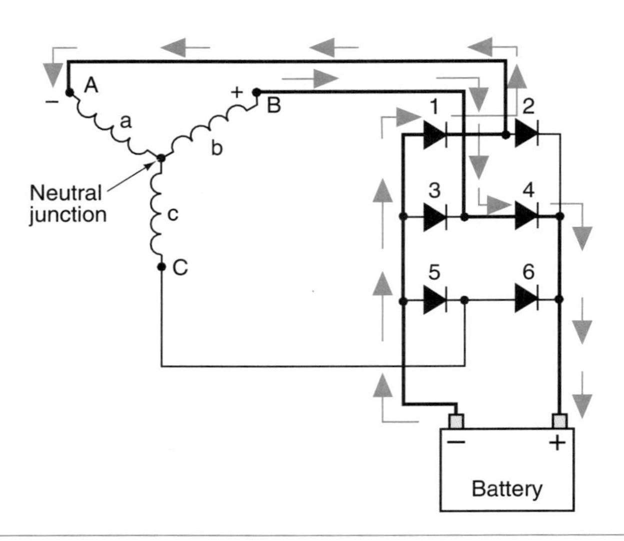

Figure 7-40 Current flow when terminals A and B are negative.

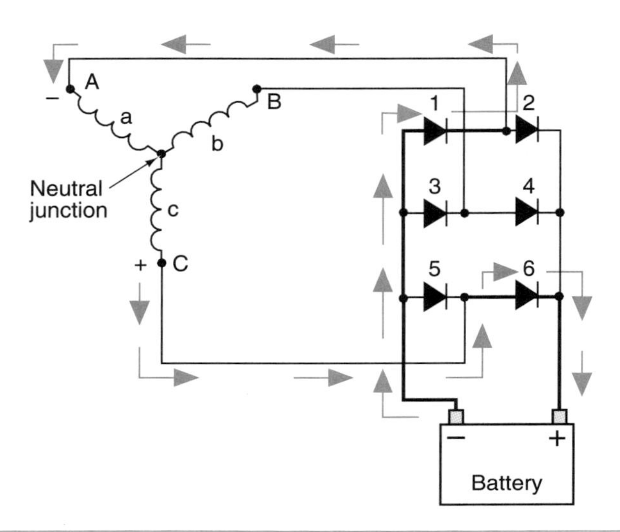

Figure 7-41 Current flow when terminals A and C are positive.

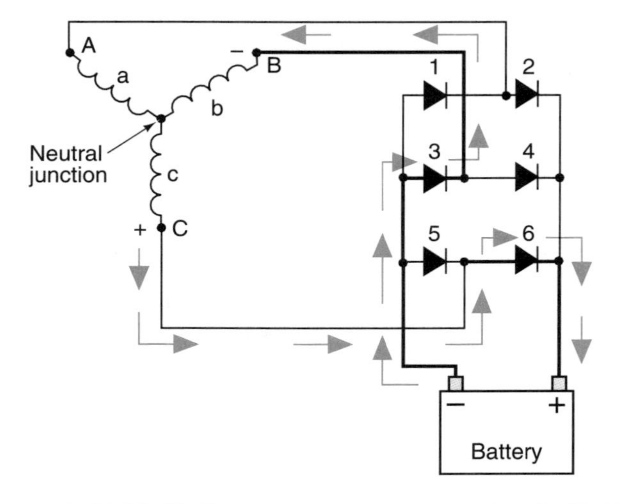

Figure 7-42 Current flow when terminals B and C are negative.

	Volts	
Temperature	Minimum	Maximum
20° F	14.3	15.3
80° F	13.8	14.4
140° F	13.3	14.0
Over 140° F	Less than 13.3	–

Figure 7-43 Chart indicating the relationship between temperature and charge rate.

AC generators do not require current limiters, because of their design they limit their own current output. Current limit is the result of the constantly changing magnetic field because of the induced AC current. As the magnetic field changes, an opposing current is induced in the stator windings. The inductive reactance in the AC generator limits the maximum current that the AC generator can produce. Even though current (amperage) is limited by its operation, voltage is not. The AC generator is capable of producing as high as 250 volts, if it were not controlled.

The **voltage regulator** controls the output voltage of the generator, based on charging system demands, by controlling field current.

The opposing voltage is called **inductive reactance.**

Regulation of voltage is done by varying the amount of field current flowing through the rotor. The higher the field current, the higher the output voltage. By controlling the amount of resistance in series with the field coil, control of the field current and the AC generator output is obtained. To insure a full battery charge, and operation of accessories, most regulators are set for a system voltage between 13.5 and 14.5 volts.

The regulator must have system voltage as an input in order to regulate the output voltage. The input voltage to the AC generator is called **sensing voltage.**

If sensing voltage is below the regulator setting, an increase in charging voltage output results by increasing field current. Higher sensing voltage will result in a decrease in field current and voltage output. A vehicle being driven with no accessories on and a fully charged battery will have a high sensing voltage. The regulator will reduce the charging voltage until it is at a level to run the ignition system while trickle charging the battery. If a heavy load is turned on (such as the headlights) the additional draw will cause a drop in the battery voltage. The regulator will sense this low system voltage and will increase current to the rotor. This will allow more current to the field windings. With the increase of field current, the magnetic field is stronger and AC generator voltage output is increased. When the load is turned off, the regulator senses the rise in system voltage and cuts back the amount of field current and ultimately AC generator voltage output.

Another input that affects regulation is temperature. Because ambient temperatures influence the rate of charge that a battery can accept, regulators are temperature compensated (Figure 7-43). Temperature compensation is required because the battery is more reluctant to accept a charge at lower ambient temperatures. The regulator will increase the system voltage until it is at a higher level so the battery will accept it.

Shop Manual
Chapter 7, page 197

Field Circuits

To properly test and service the charging system, it is important to identify the field circuit being used. Automobile manufacturers use three basic types of field circuits. The first type is called the A circuit. It has the regulator on the ground side of the field coil. The B+ for the field coil is picked up from inside the AC generator (Figure 7-44). By placing the regulator on the ground side of the field coil, the regulator will allow the control of field current by varying the current flow to ground. A regulator can be located anywhere in the series circuit and have the same effect.

The A circuit is called an **external grounded field circuit.**

The second type of field circuit is called the B circuit. In this case, the voltage regulator controls the power side of the field circuit. Also the field coil is grounded from inside the AC generator (Figure 7-45).

Usually the B circuit regulator is mounted externally of the AC generator. The B circuit is an **internally grounded circuit.**

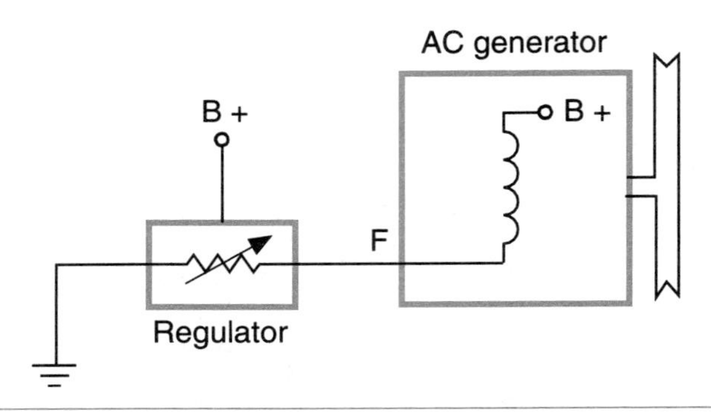

Figure 7-44 Simplified diagram of an A field circuit.

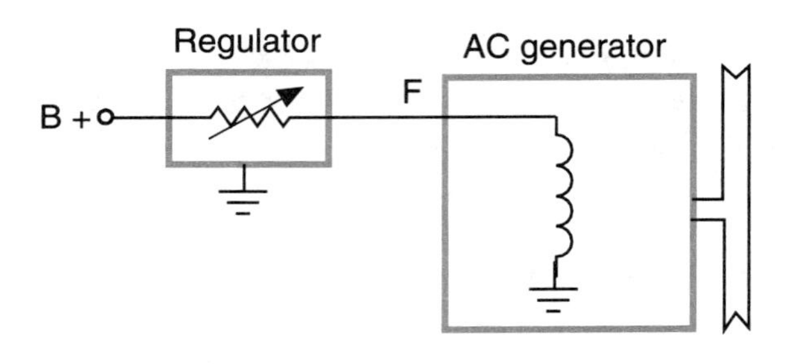

Figure 7-45 Simplified diagram of a B field circuit.

Isolated field AC generators pick up B+ and ground externally.

To remember these circuits: Think of "A" for "After" the field and "B" for "Before" the field.

The third type of field circuit is called the isolated field. The AC generator has two field wires attached to the outside of the case. The voltage regulator can be located on either the ground (A circuit) or on the B+ (B circuit) side (Figure 7-46).

Electromechanical Regulators

Shop Manual
Chapter 7, page 208

There are two basic types of regulators: electromechanical and electronic. Also, on many newer model vehicles regulation is controlled by the computer. Even though the electro-

The electromechanical regulators were discontinued by most manufacturers by 1978.

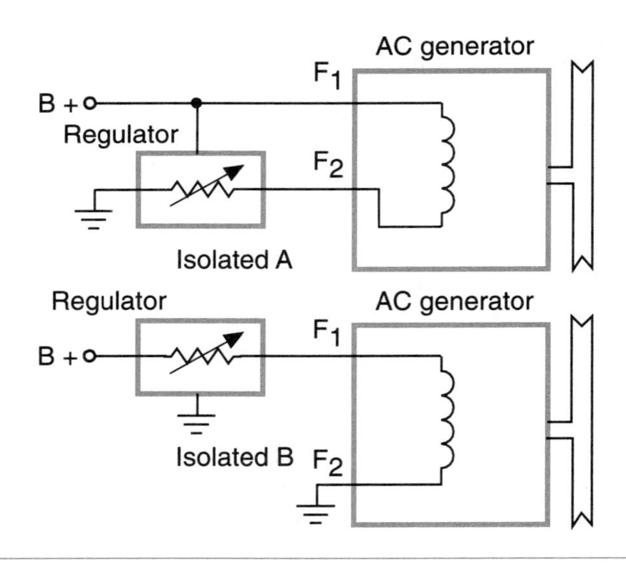

Figure 7-46 In the isolated circuit field AC generator, the regulator can be installed on either side of the field.

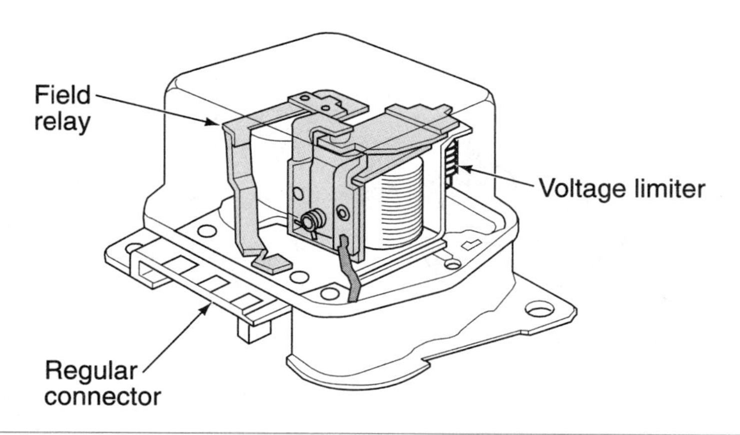

Figure 7-47 An external electromechanical regulator. (Reprinted with the permission of Ford Motor Company)

The **field relay** applies battery voltage to the AC generator field coil after the engine has started.

The **voltage limiter** is connected through the resistor network and determines whether the field will receive high, low, or no voltage. It automatically controls the field voltage for the required amount of charging.

mechanical regulator is obsolete, a study of its operation will help you to understand the more complex systems.

The external electromechanical regulator is a vibrating contact point design (Figure 7-47). The regulator uses electromagnetics to control the opening and closing of the contact points. Inside the regulator are two coils. One coil is the field relay and the second is the voltage regulator. The field relay coil and contact with no current flowing through the coil are shown (Figure 7-48). The contact points are open, preventing current flow. An electromagnetic field develops when current flows through the field relay coil. It pulls the contact arm down (Figure 7-49). Once the contact points close, current will flow.

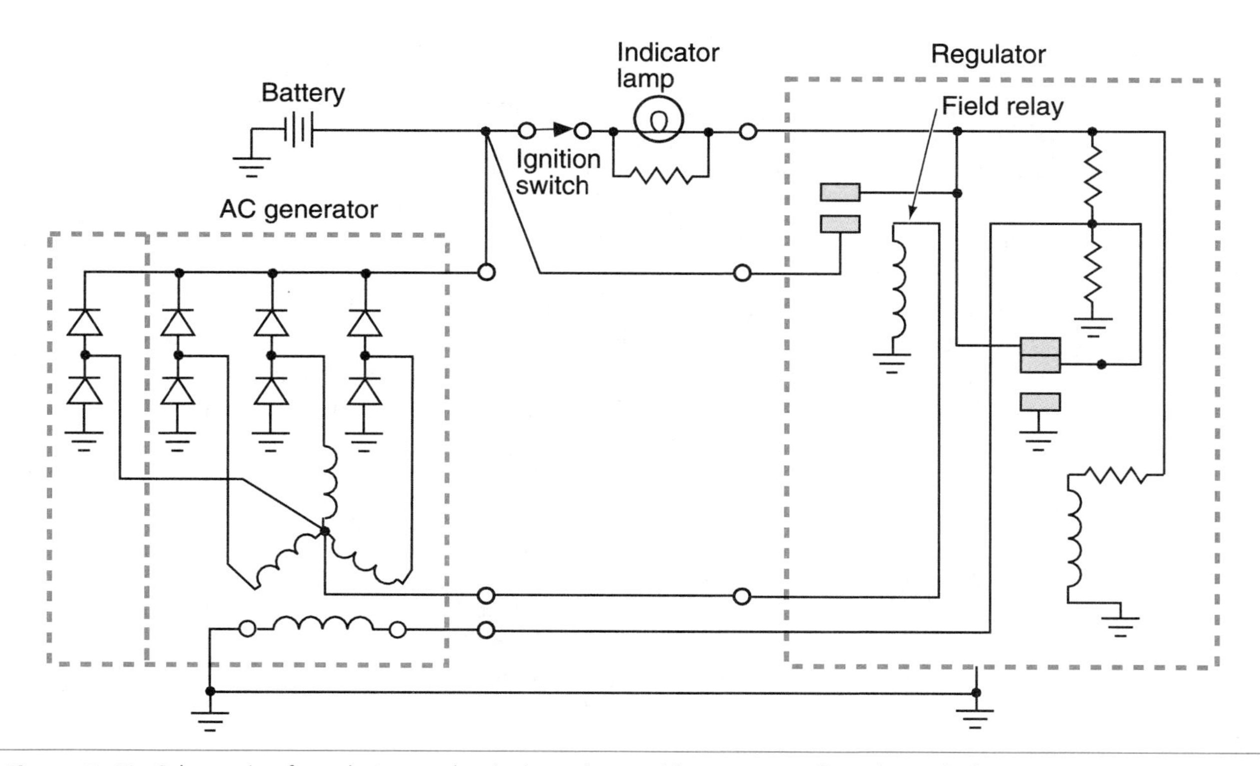

Figure 7-48 Schematic of an electromechanical regulator with no current flow through the coil. (Reprinted with the permission of Ford Motor Company)

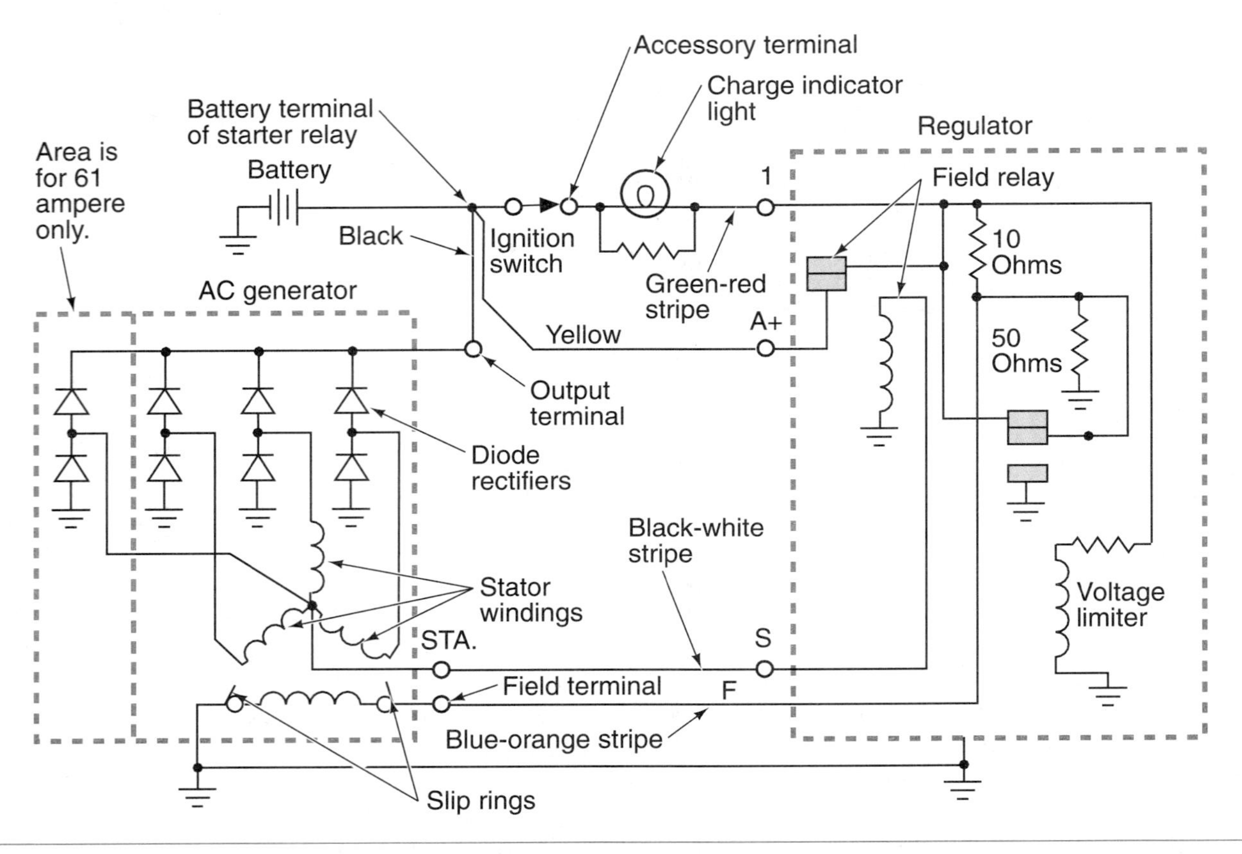

Figure 7-49 Regulator circuit with current flowing through the coil. (Reprinted with the permission of Ford Motor Company)

The voltage regulator coil uses an electromagnetic field to open the contact points. With the points closed, current flows from the battery to the rotor. Current also flows to the regulator coil. As the battery charges, the battery's voltage increases. This increase in voltage strengthens the coil's attraction for the contact points. At a preset voltage level, the coil will overcome the contact point spring tension and open the points. This will stop current from flowing to the rotor. Once the points open, voltage output of the AC generator drops. As the battery voltage decreases, so does the regulator coil's magnetic strength. Spring tension will overcome the magnetic attraction and the points will close again. Once again current flows through the rotor and the AC generator is producing voltage. This action occurs several times per second.

With the ignition switch in RUN, current flow for the field will go through the ignition switch through the resistor and bulb, through the lower contacts of the voltage regulator (closed), and out of the F terminal to the AC generator, through the field coil, and to ground. The indicator lamp bulb will light because the bulb is in series with the field coil.

Once the engine is started, the AC generator will begin to produce voltage. At this time the system voltage may be below 13.5 volts, however, because the rotor is revolving, there is some production of voltage. This allows voltage out of the R terminal to energize the field relay coil. With the relay coil energized, the contact points close and direct battery voltage flows through terminal 3 and out terminal 4. The bulb will go out with battery voltage on both sides of the bulb. Simultaneously, battery voltage flows through the voltage regulator contact points to the field coil in the AC generator. Because the voltage regulator coil is also connected to the battery (above terminal 4), the lower than 13.5 volts is unable to produce a sufficient electromagnetic field to pull the contact points open. This condition will allow maximum AC generator output.

When maximum current is flowing to the rotor it is called **full field current.**

As the battery receives a charge from the AC generator, the battery voltage will increase to over 13.5 volts. The increased voltage will strengthen the electromagnetic field of the voltage regulator coil. The coil will attract the points and cause the lower contacts to open. Current will now flow through the resistor and out terminal F. Because an additional series resistance is added to the field coil circuit of the AC generator, field current is reduced, and AC generator output is also reduced.

When current flows through the series resistor to the field coil, it is called **half field current.**

If the AC generator is producing more voltage than the system requires, both battery and system voltage will increase. As this voltage increases to a level above 14.5 volts, the magnetic strength of the coil increases. This increase in magnetic strength will close the top set of contact points and apply a ground to the F terminal and the AC generator's field coil. Both sides of the field coil are grounded, thus no current will flow through it and there is no output.

Once the engine is shut off, there is no current from the R terminal; the field relay coil is deenergized, allowing the points to open. This prevents battery discharge through the charging system.

Shop Manual
Chapter 7, page 208

Electronic Regulators

An **electronic regulator** uses solid-state circuitry to perform the regulatory functions.

The second type of regulator is the electronic regulator. Electronic regulators can be mounted either externally or internally of the AC generator. There are no moving parts, so it can cycle between 10 and 7,000 times per second. This quick cycling provides more accurate control of the field current through the rotor. Electronic regulation control is through the ground side of the field current (A circuit).

Pulse width modulation controls AC generator output by varying the amount of time the field coil is energized. For example, assume that a vehicle is equipped with a 100-ampere generator. If the electrical demand placed on the charging system requires 50 amperes of current, the regulator would energize the field coil for 50% of the time (Figure 7-50). If the electrical system's demand was increased to 75 amperes, the regulator would energize the field coil 75% of the cycle time (Figure 7-51).

The thermistor changes resistance with temperature.

The electronic regulator uses a zener diode that blocks current flow until a specific voltage is obtained, at which point it allows the current to flow. An electronic regulator is shown (Figure 7-52).

AC generator current from the stator and diodes first goes through a thermistor. Current then flows to the zener diode. When the upper voltage limit (14.5 volts) is reached, the zener diode will conduct current to flow to the base of transistor 2. This turns transistor 2 on and switches off transistor 1. Transistor 1 controls field current to the rotor. If transistor 1 is off, no current can flow through the field coil and the AC generator will not have any output. When no voltage is applied to the zener diode, current flow stops, transistor 2 is turned off, transistor 1 is

The on/off cycling is called **pulse width modulation.** The period of time for each cycle does not change. Only the amount of on time in each cycle changes.

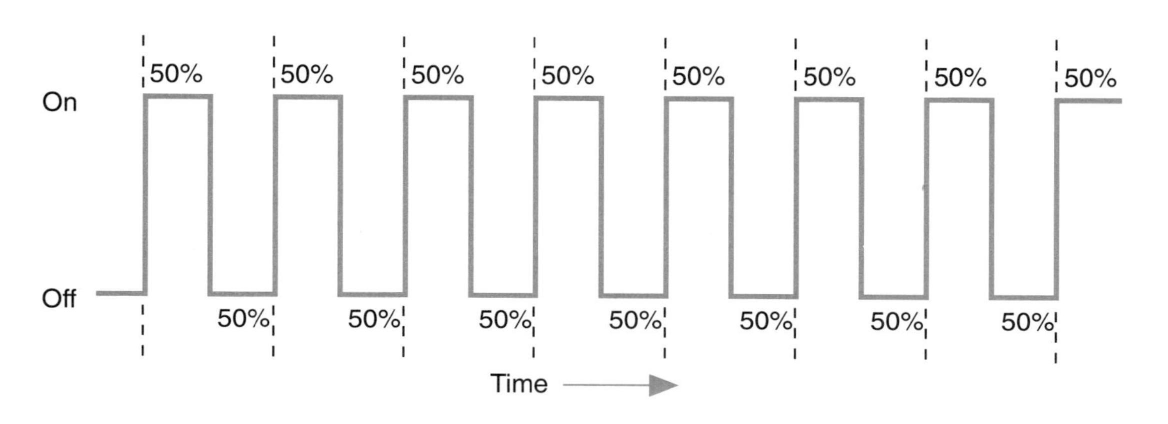

Figure 7-50 Pulse width modulation with 50% on-time.

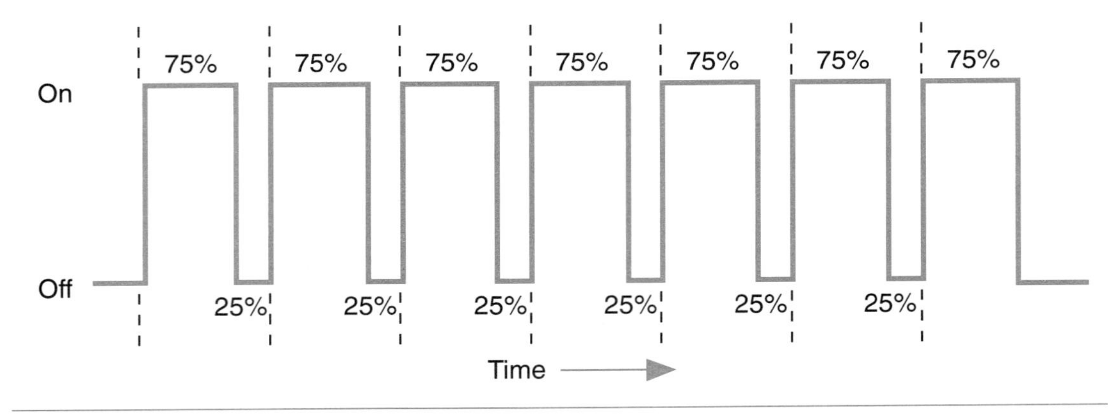

Figure 7-51 Pulse width modulation with 75% on-time.

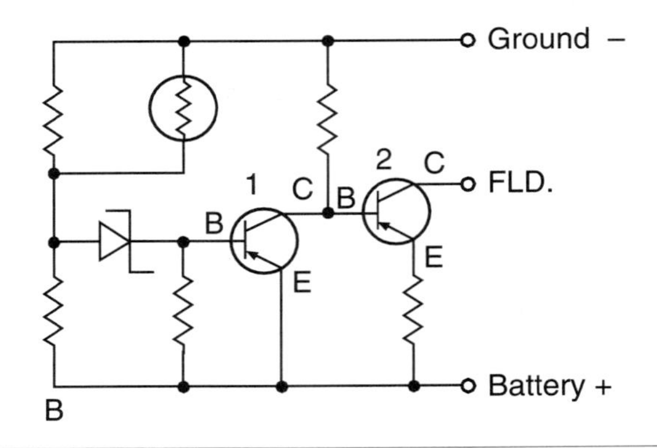

Figure 7-52 A simplified circuit of an electronic regulator utilizing a zener diode.

turned on, and the field circuit is closed. The magnetic field is restored in the rotor, and the AC generator produces output voltage.

Many manufacturers are installing the voltage regulator internally in the AC generator. This eliminates some of the wiring needed for external regulators. The diode trio rectifies AC current from the stator to DC current that is applied to the field windings (Figure 7-53).

The resistor above the indicator lamp is used to ensure that current will flow through terminal 1 if the lamp burns out.

Current flow with the engine off and the ignition switch in the RUN position is illustrated (Figure 7-54). Battery voltage is applied to the field through the common point above R1. TR1 conducts the field current coming from the field coil, producing a weak magnetic field. The indicator lamp lights because TR1 directs current to ground and completes the lamp circuit.

Current flow with the engine running is illustrated (Figure 7-55). When the AC generator starts to produce voltage, the diode trio will conduct and battery voltage is available for the field and terminal 1 at the common connection. Placing voltage on both sides of the lamp gives the same voltage potential at each side; therefore, current doesn't flow and the lamp goes out.

Current flow as the voltage output is being regulated is illustrated (Figure 7-56). The sensing circuit from terminal 2 passes through a thermistor to the zener diode (D2). When the system voltage reaches the upper voltage limit of the zener diode, the zener diode conducts current to TR2. When TR2 is biased it opens the field coil circuit and current stops flowing through the field coil. Regulation of this switching on and off is based on the sensing voltage received through terminal 2. With the circuit to the field coil opened, the sensing voltage decreases and the zener diode stops conducting. TR2 is turned off and the circuit for the field coil is closed.

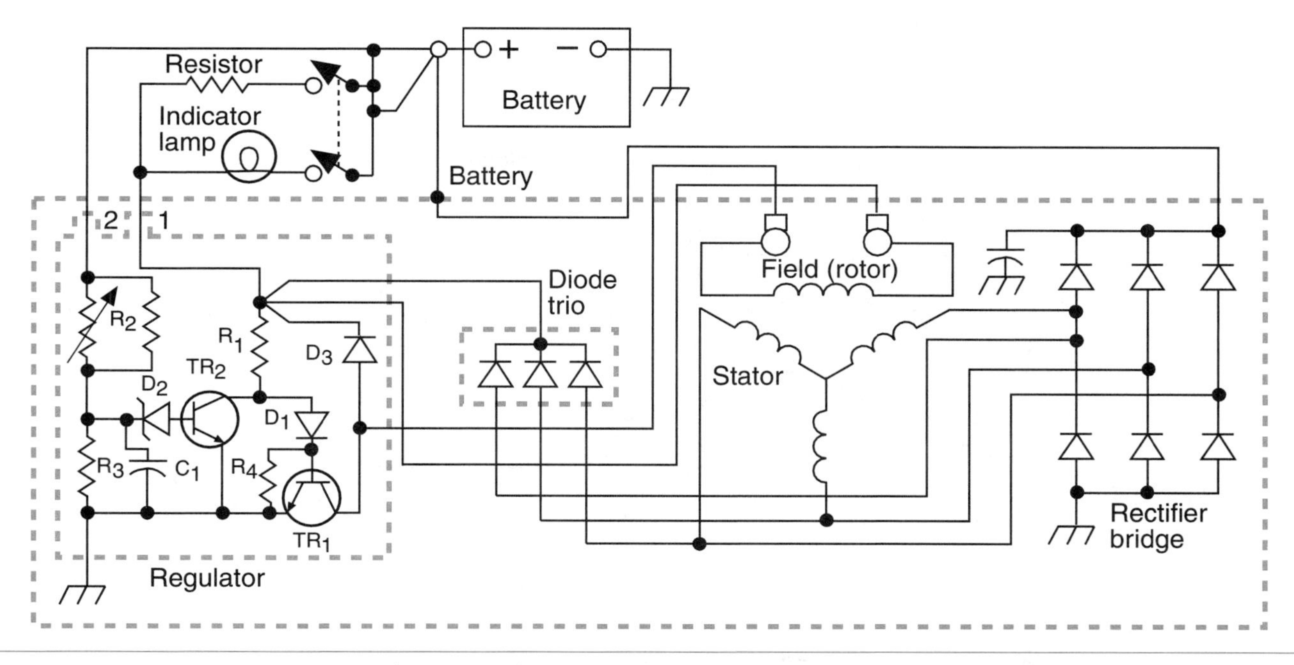

Figure 7-53 AC generator circuit diagram with an internal regulator. Uses a diode trio to rectify the stator current that will be applied to the field coil.

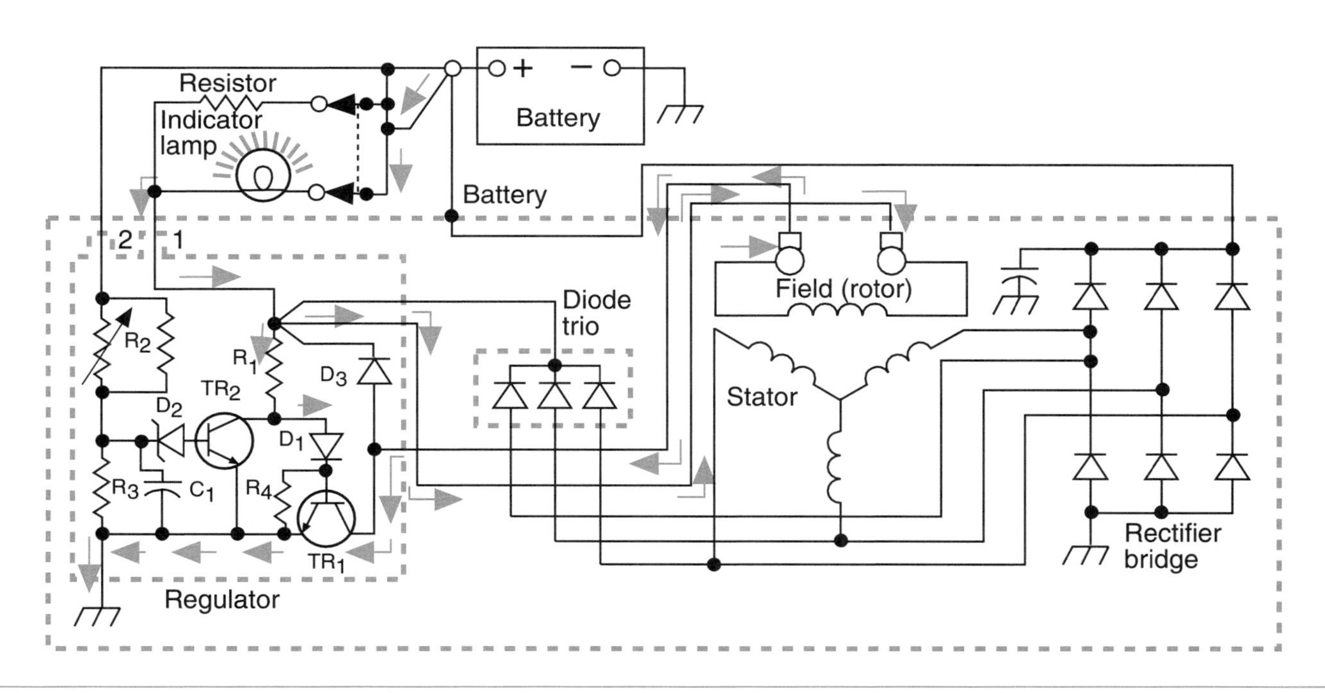

Figure 7-54 Current flow to the rotor with the ignition switch in the RUN position and the engine OFF.

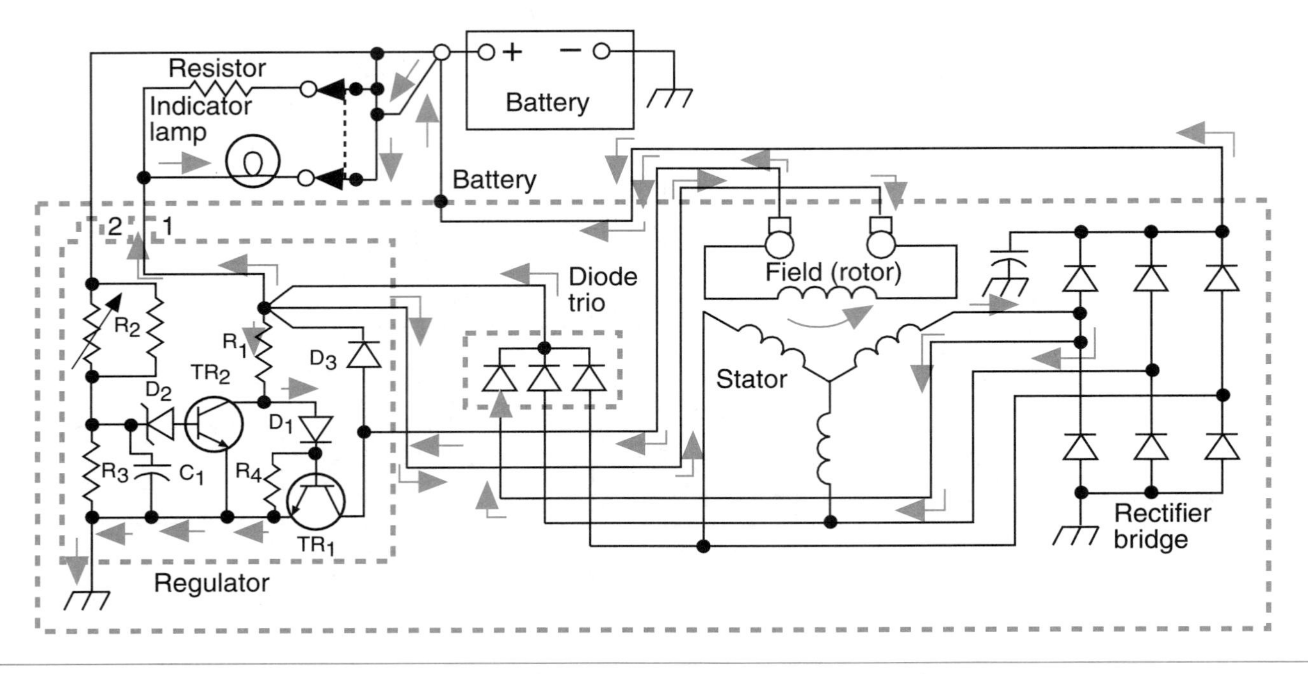

Figure 7-55 Current flow with the engine running and AC generator producing voltage.

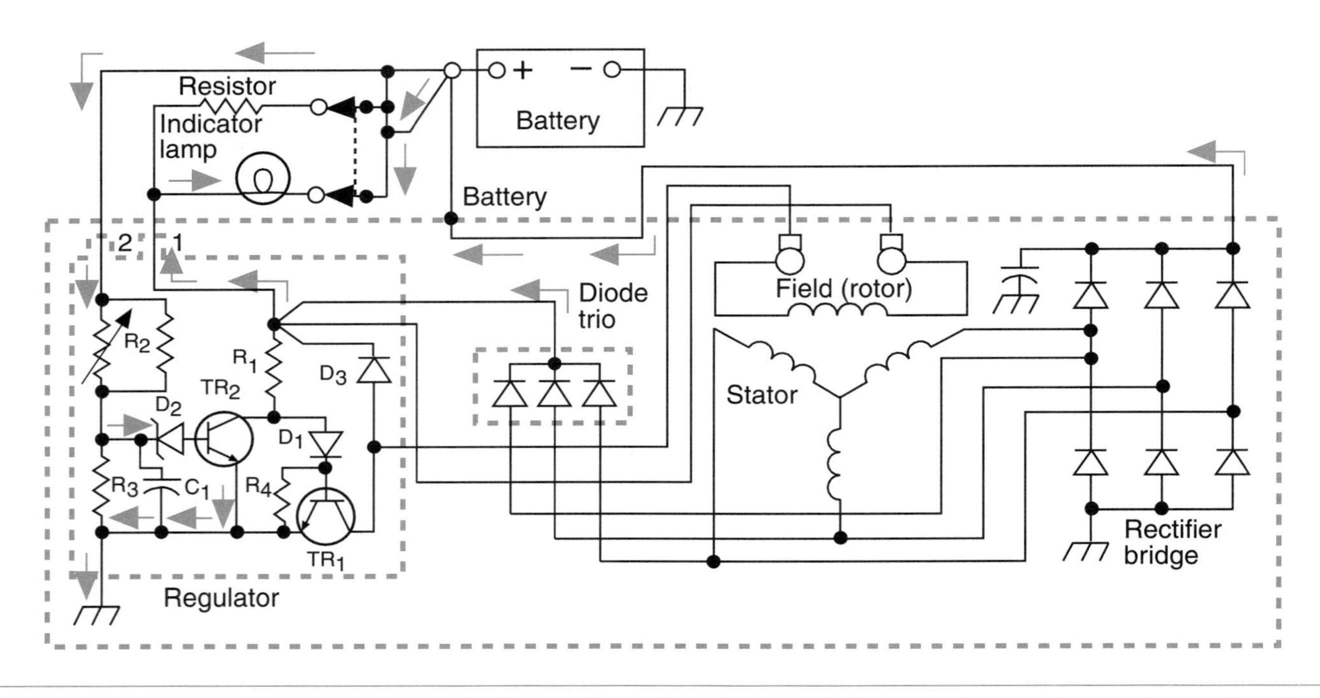

Figure 7-56 When the system voltage is high enough to allow the zener diode to conduct, TR_2 is turned on and TR_1 is shut off, which opens the field circuit.

Shop Manual
Chapter 7, page 201

Computer-Controlled Regulation

On many vehicles after the mid-1980s, the regulator function has been incorporated into the vehicle's engine computer (Figure 7-57). The operation is the same as the internal electronic regulator. Regulation of the field circuit is through the ground (A circuit).

The logic board's decisions, concerning voltage regulation, are based on output voltages and battery temperature. When the desired output voltage is obtained (based on battery temperature)

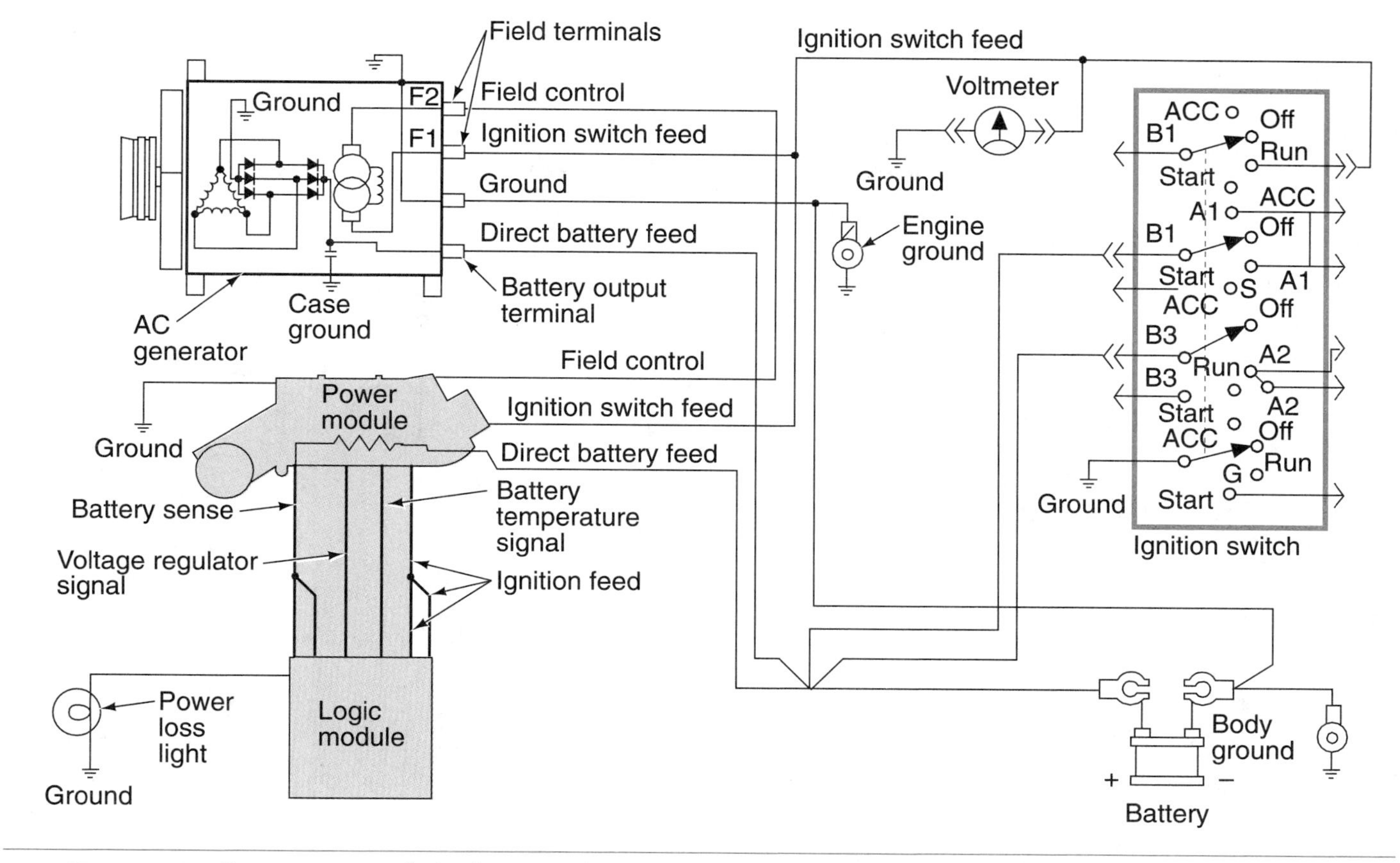

Figure 7-57 Computer-controlled voltage regulator circuit. (Courtesy of Chrysler Corporation)

the logic board duty-cycles a switching transistor. This transistor grounds the AC generator's field to control output voltage.

General Motors' CS series generators may be connected directly to the PCM through terminals L and F at the generator. The voltage regulator portion of the PCM switches the field current on and off at a frequency of about 400 times per second. Varying the on and off time of the field current controls the voltage output of the generator.

Duty-cycle is the percentage of on time to total cycle time.

Charging Indicators

There are four basic methods of informing the driver of the charging system's condition: indicator lamps, electronic voltage monitor, ammeter, and voltmeter.

Indicator Light Operation

As discussed earlier, most indicator lamps operate on the basis of opposing voltages. If the AC generator output is less than battery voltage, there is an electrical potential difference in the lamp circuit and the lamp will light. In the electromechanical regulator system, if the stator is not producing a sufficient amount of current to close the field relay contact points the lamp will light (Figure 7-58). If the voltage at the battery is equal to the output voltage, the two equal voltages on both sides of the lamp result in no electrical potential and the lamp goes out (Figure 7-59)

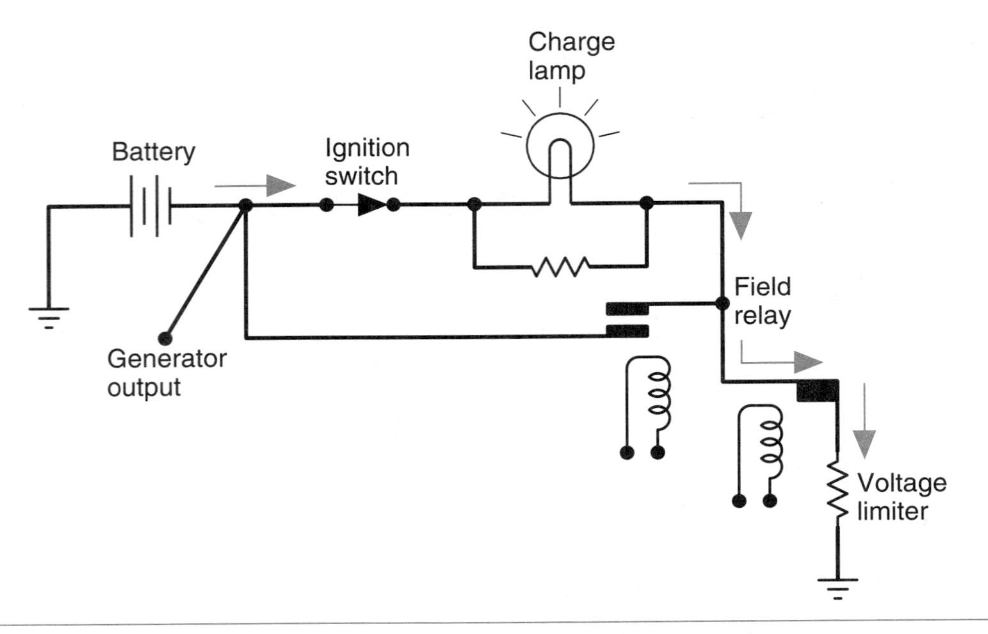

Figure 7-58 Electromechanical regulator with indicator light on and no generator output.

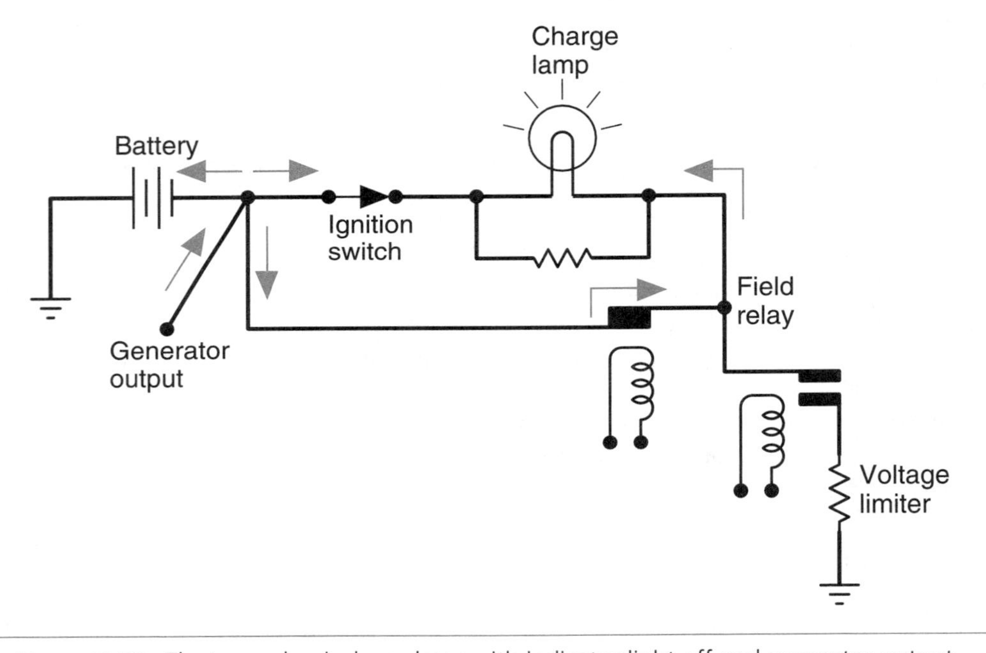

Figure 7-59 Electromechanical regulator with indicator light off and generator output.

Electronic regulators that use an indicator lamp operate on the same principle. If there is no stator output through the diode trio, then the lamp circuit is completed to ground through the rotor field and TR1 (Figure 7-60).

On most systems, the warning lamp will be "proofed" when the ignition switch is in the RUN position before the engine starts. This indicates that the bulb and indicator circuit are operating properly. Proofing the bulb is accomplished because there is no stator output without the rotor turning.

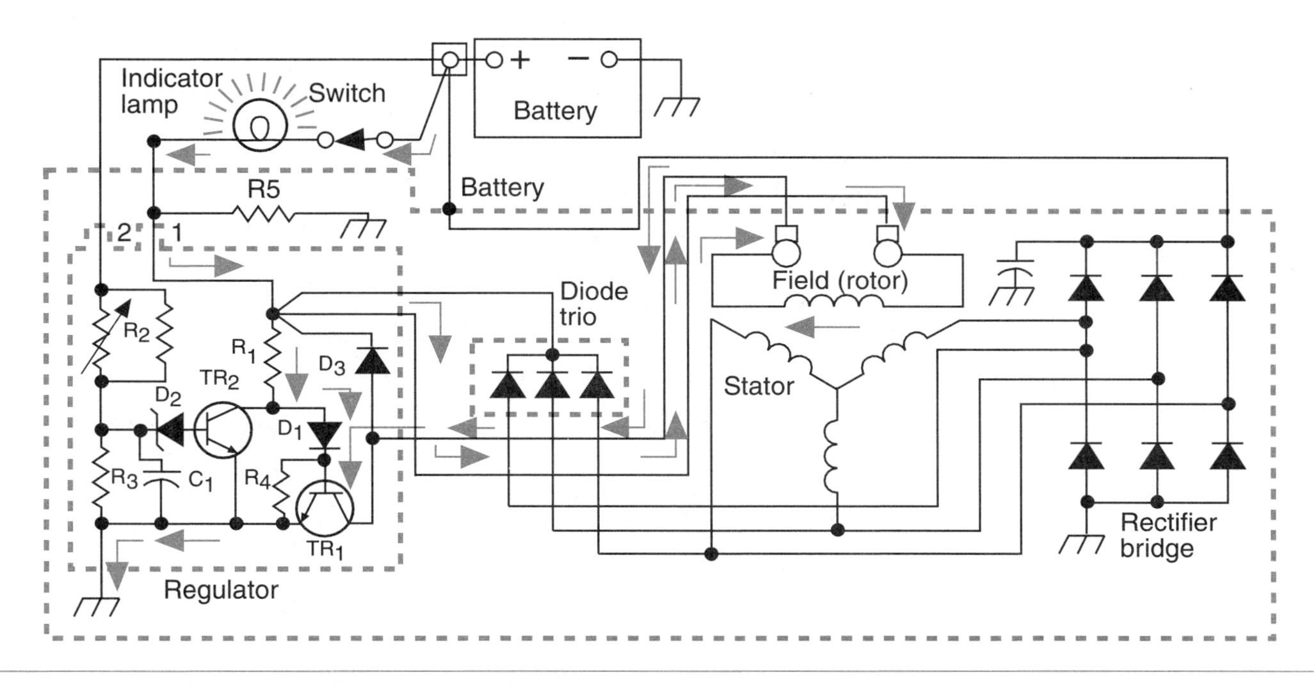

Figure 7-60 Electronic regulator with an indicator light on due to no generator output.

Electronic Voltage Monitor

The electronic voltage monitor module is used to monitor the system voltage (Figure 7-61). The lamp will remain off if the system voltage is above 11.2 volts. Once system voltage drops below 11.2 volts, a transistor amplifier in the module turns on the indicator lamp. This system can use either a lamp or a light-emitting diode.

Ammeter Operation

In place of the indicator light, some manufacturers install an ammeter. The ammeter is wired in series between the AC generator and the battery (Figure 7-62). Most ammeters work on the principle of d'Arsonval movement.

The movement of the ammeter needle under different charging conditions is illustrated (Figure 7-63). If the charging system is operating properly, the ammeter needle will remain within the normal range. If the charging system is not generating sufficient current, the needle will swing toward the discharge side of the gauge. When the charging system is recharging the battery, or is called on to supply high amounts of current, the needle deflects toward the charge side of the gauge.

The **ammeter** measures charging and discharging current in amperes.

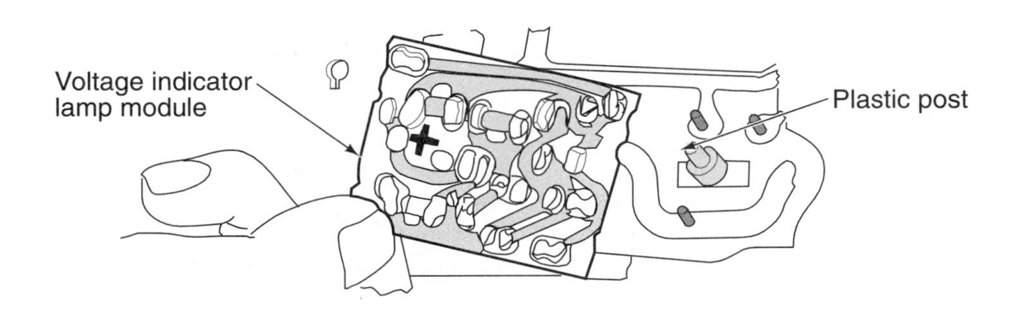

Figure 7-61 Electronic voltage monitor module. (Courtesy of Chrysler Corporation)

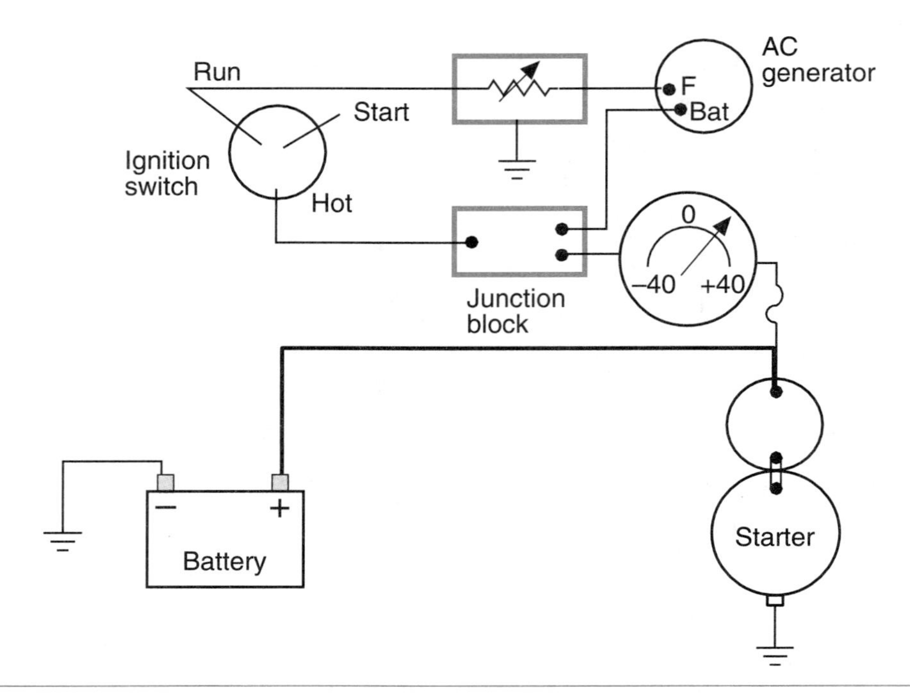

Figure 7-62 Ammeter connected in series to indicate charging system operation.

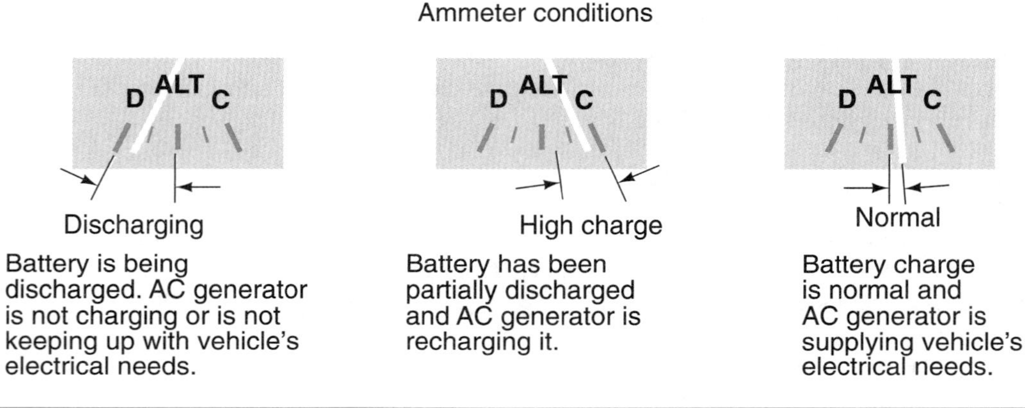

Figure 7-63 Ammeter needle movement indicates charging conditions. (Reprinted with the permission of Ford Motor Company)

It is normal for the gauge to read a high amount of current after initial engine start up. As the battery is recharged, the needle should move more toward the normal range.

Voltmeter Operation

The **voltmeter** indicates the amount of voltage potential between two points of a circuit.

Because the ammeter is a complicated gauge for most people to understand, many manufacturers use a voltmeter to indicate charging system operation. The voltmeter is usually connected between the battery positive and negative terminals (Figure 7-64).

When the engine is started, it is normal for the voltmeter to indicate a reading between 13.2 and 15.2 volts. If the voltmeter indicates a voltage level that is below 13.2, it may mean that the battery is discharging. If the voltmeter indicates a voltage reading that is above 15.2 volts, the

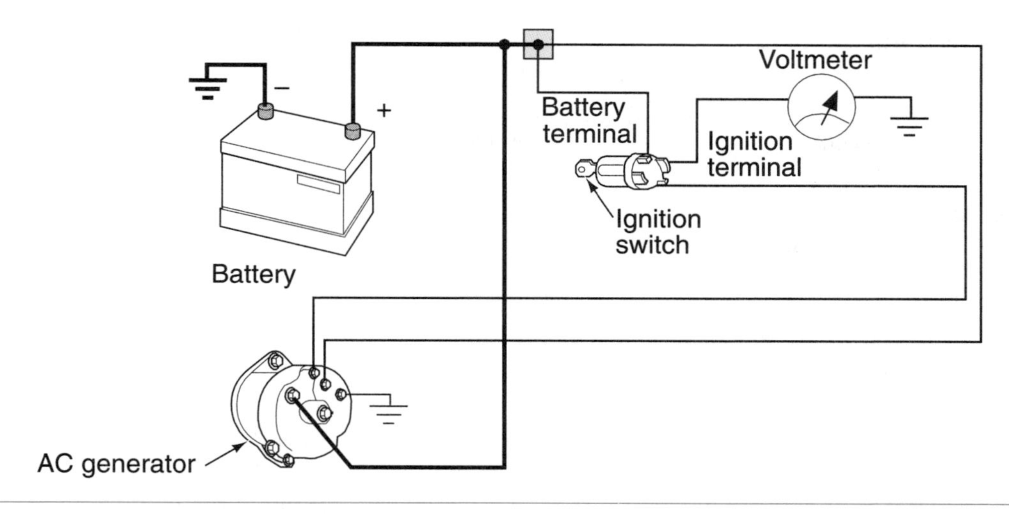

Figure 7-64 Voltmeter connected to the charging circuit to monitor operation.

charging system is overcharging the battery. The battery and electrical circuits can be damaged as a result of higher than normal charging system output.

AC Generator Design Differences

Although all AC generators operate on the same principles, there are differences in the styles and construction.

General Motors 10SI Series

The 10SI series AC generator uses an internal voltage regulator that is mounted to the inside of the slip ring end frame (Figure 7-65). There are three terminals on the rear-end frame of the AC generators:

- **Terminal number 1:** Connects to the field through one brush and slip ring and to the output of the diode trio. In addition, this terminal is connected to a portion of the regulator and warning light circuitry.

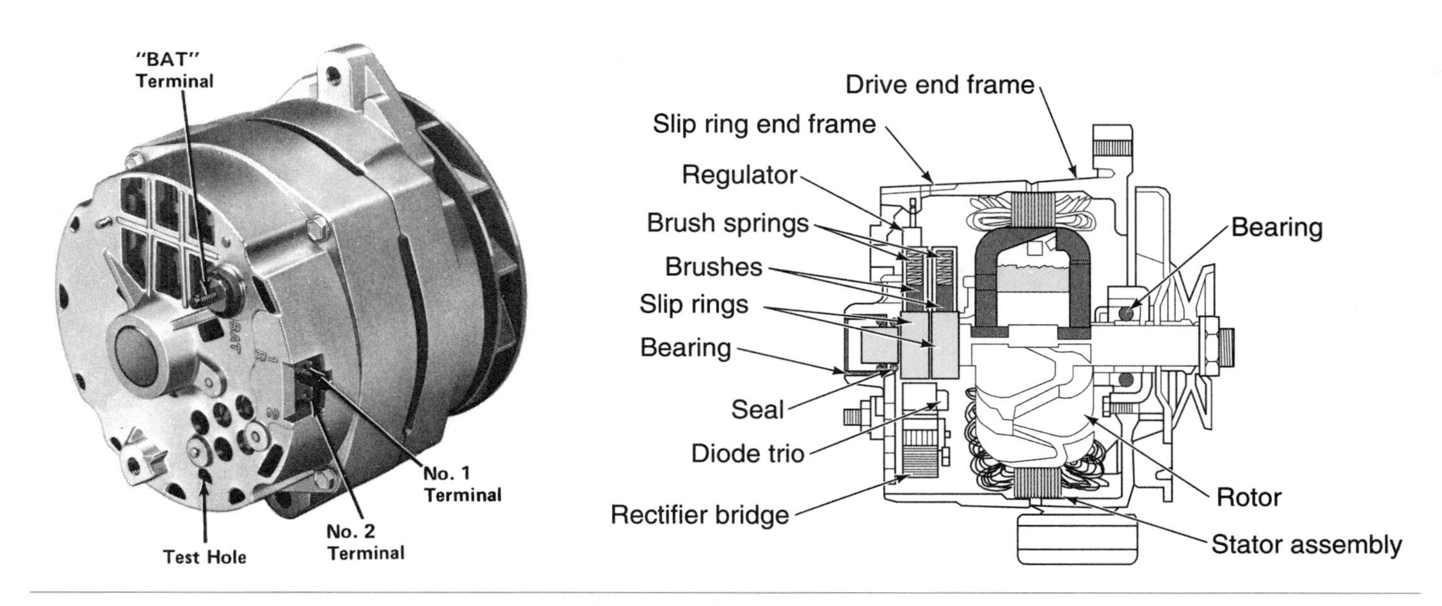

Figure 7-65 10SI AC generator. (Courtesy of General Motors)

❑ **Terminal number 2:** Connects to the regulator to supply battery voltage to a portion of the regulator circuitry that senses system voltage.

❑ **BAT terminal:** Connects to the output of the stator windings and supplies the battery with charging voltage.

Ford AC Generators

Shop Manual
Chapter 7, page 198

There are four different designs of AC generators that Ford used (Figure 7-66). For many years Ford used the common rear or side terminal AC generator. The rear terminal AC generators used two different types of rectifiers. One rectifier has a single plate that contains all six of the diodes. The diodes are built into the plate (Figure 7-67).

This style of rectifier is called **flat-type rectifier.**

The second type of rectifier uses two plates that are stacked on top of each other (Figure 7-68). One plate contains the positive diodes and the other contains the negative diodes.

This type of rectifier is called a **stacked-type rectifier.**

On the rear-terminal AC generator there are four terminals on the end frame:

❑ **BAT terminal:** Stator output connection that supplies charging voltage to the battery.

❑ **FLD terminal:** Connects one side of the field winding through the insulated brush and the slip ring.

❑ **STA terminal:** Connects to the neutral stator junction.

❑ **GRD terminal:** Provides a connection for the ground wire from the regulator.

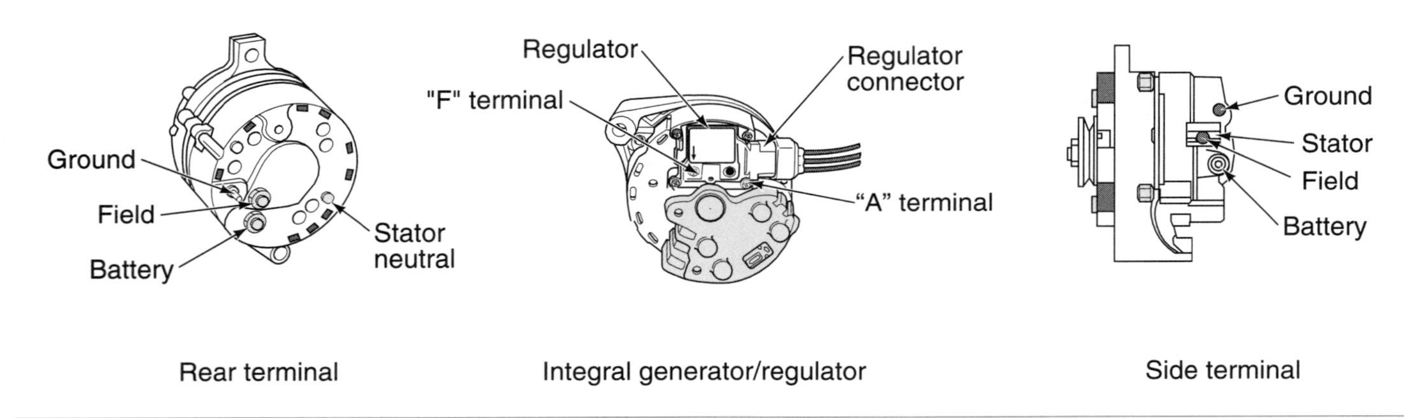

Figure 7-66 Three designs of AC generators used by Ford. (Reprinted with the permission of Ford Motor Company)

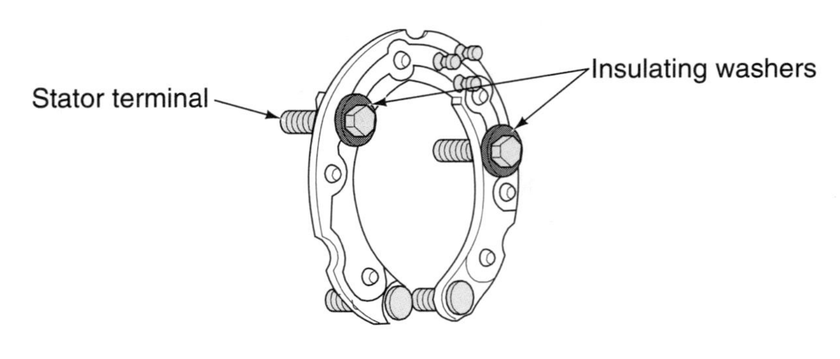

Figure 7-67 Flat-type rectifier has a single plate containing all six diodes.

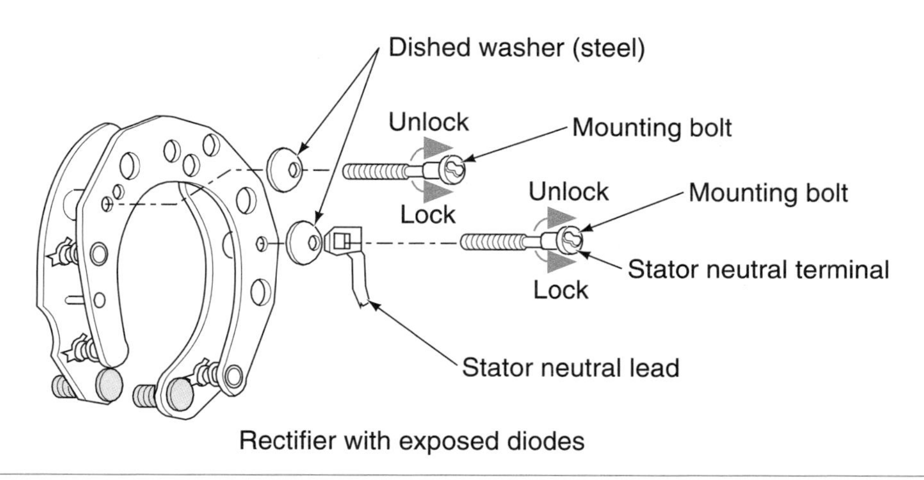

Figure 7-68 Stacked-type rectifier with two plates.

The Ford side-terminal AC generator is larger and has higher output capacities. The same four terminals are used, however, they are arranged differently.

In 1984, Ford introduced an integral alternator regulator (IAR) AC generator. The regulator is mounted on the exterior of the rear-end frame, which simplifies testing and replacement of the regulator. The F and A terminals are used to test the charging system. Additional modifications include the brushes being a part of the regulator.

When the ignition switch is in the RUN position, voltage is sent to the I terminal of the regulator (Figure 7-69). Regulator terminal A senses system voltage. Field current is drawn through this terminal also.

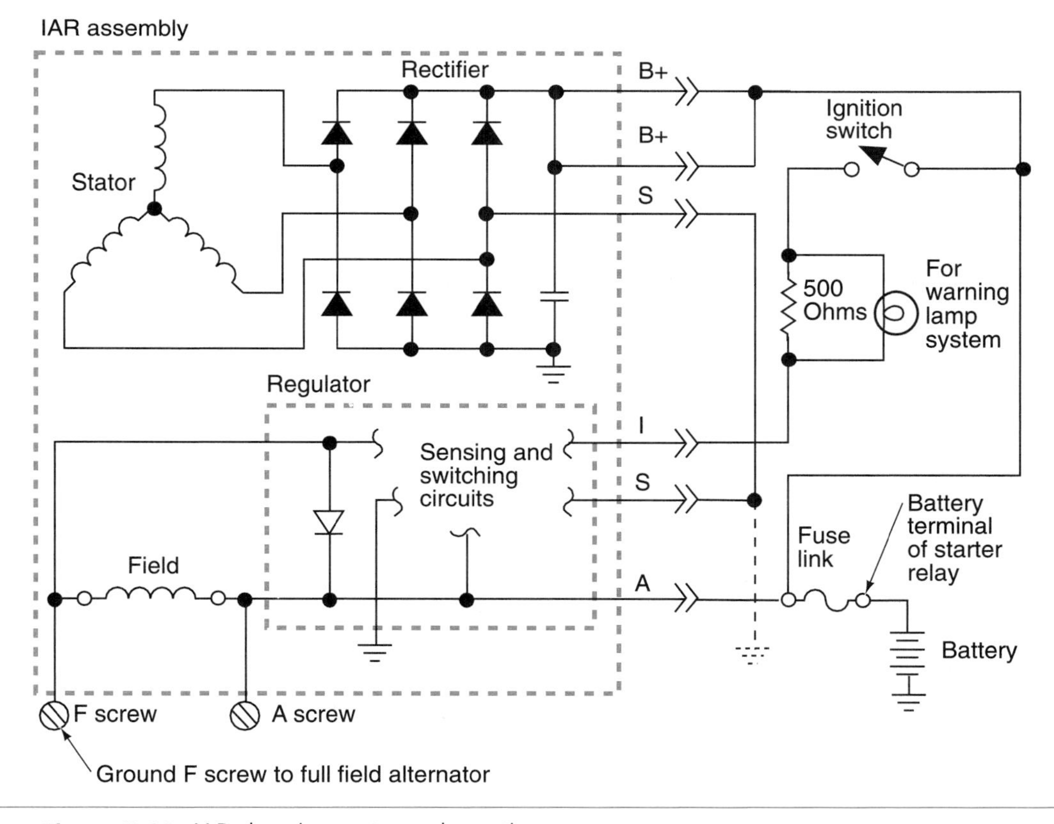

Figure 7-69 IAR charging system schematic.

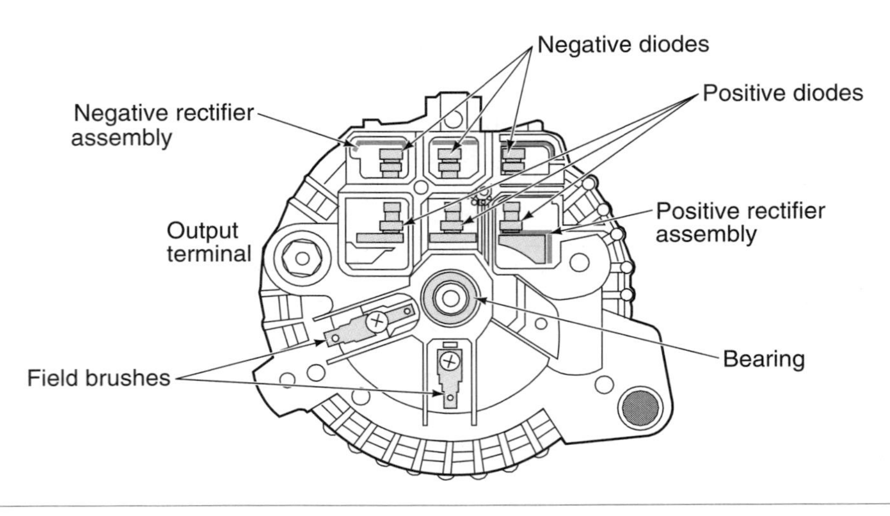

Figure 7-70 Rear end-frame of a Chrysler AC generator.

EVR stands for External Voltage Regulator.

Ford's EVR charging system uses a solid-state external regulator. This style is built either with rear or side terminals. The side-terminal design provides higher output by using delta wound stators.

Chrysler AC Generators

The Chrysler AC generator (Figure 7-70) uses separate heat sinks for the positive and negative diodes. Both heat sinks are attached to the rear-end frame. Also, the brushes are attached to the exterior of the end frame. This allows for brush replacement without having to disassemble the AC generator (in fact the brushes can usually be replaced without having to remove the AC generator from the vehicle).

CAUTION: Do not attempt to replace the brushes without first disconnecting the battery negative terminal. Failure to disconnect the battery may result in damage to the charging system or injury to the technician.

The three terminals on the rear-end frame are connected as follows:

- **BAT terminal:** Connects the stator output to the battery to supply charging voltage.
- **FLD terminal:** There are two field terminals. Battery voltage is applied to one of the field terminals; the regulator connects to the second field terminal.

In recent years Chrysler combined the logic and power module into a single unit.

In 1985 Chrysler introduced a delta-wound, dual-output, computer-controlled charging system (Figure 7-71). This system has some unique capabilities:

1. The system is capable of varying charging system output based on the ambient temperature and the system's voltage needs.
2. The computer monitors the charging system and is capable of self-diagnosis.

A BIT OF HISTORY

Chrysler equipped its vehicles with AC generators in the late 1950s making it the first manufacturer to use an AC generator. Chrysler introduced the dual-output AC generator (40 or 90 amperes) with computer control in 1985.

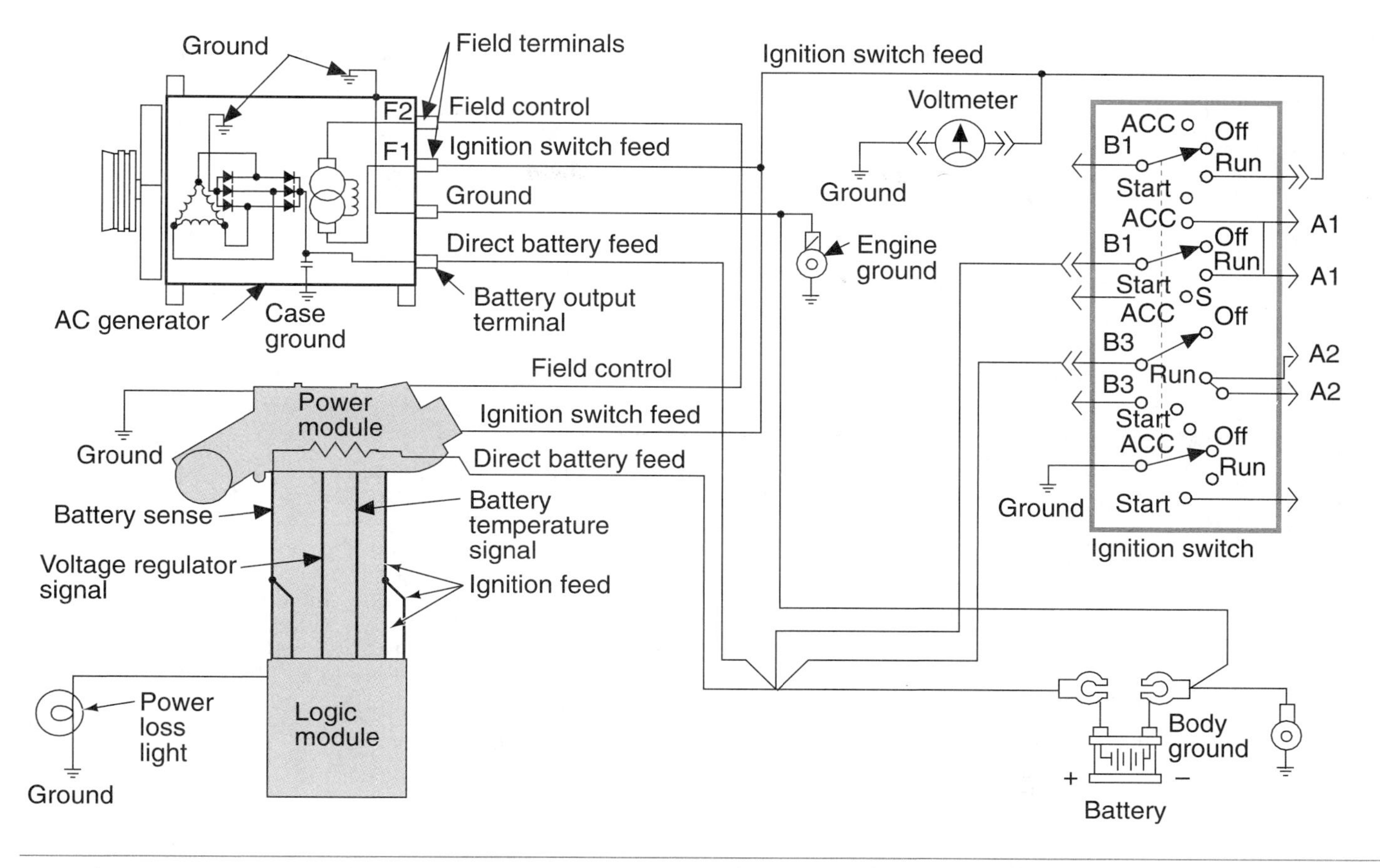

Figure 7-71 Computer-controlled dual output charging system. (Courtesy of Chrysler Corporation)

When the ignition switch is in the RUN position, the logic module checks the ambient temperature and signals the power module to turn on the field circuit. Based on inputs to the logic module relating to temperature and system requirements, it signals the power module when current output adjustments are required.

Mitsubishi AC Generator

The Mitsubishi AC generator uses an internal voltage regulator (Figure 7-72). It also has two separate wye connected stator windings (Figure 7-73). Each of the stator windings have their own set of six diodes for rectification.

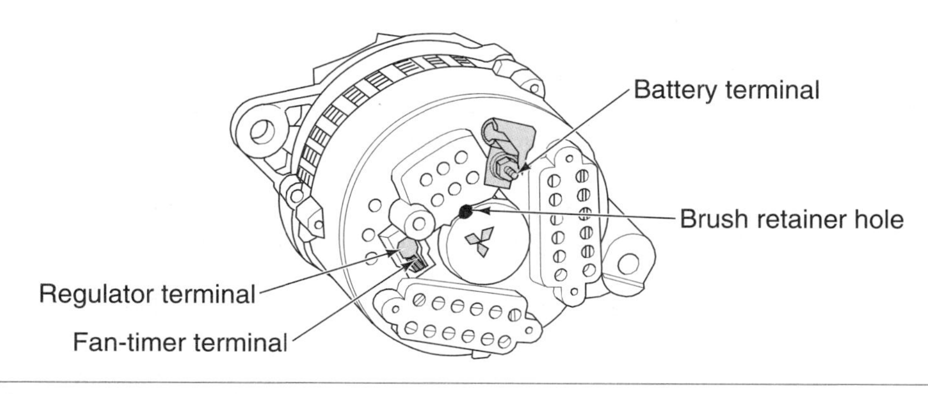

Figure 7-72 Mitsubishi AC generator terminals. (Courtesy of Chrysler Corporation)

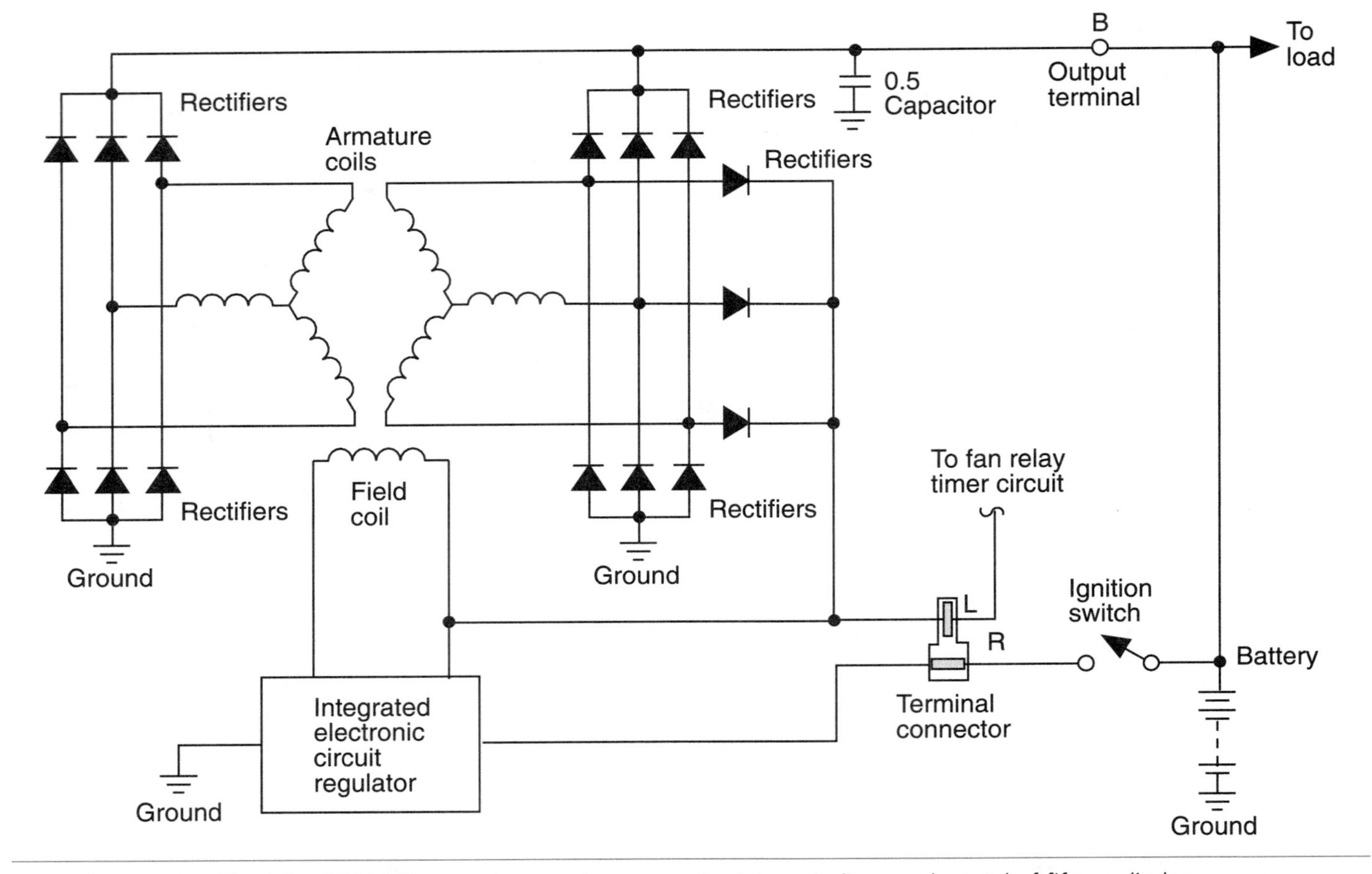

Figure 7-73 The Mitsubishi AC generator uses two separate stator windings and a total of fifteen diodes.

This AC generator also uses a diode trio to rectify stator voltage to be used in the field winding. The three terminals are connected as follows:

- **B terminal:** Connects the output of both stator windings to the battery, supplying charging voltage.
- **R terminal:** Supplies 12 volts to the regulator.
- **L terminal:** Connects to the output of the diode trio to provide rectified stator voltage to designated circuits.

Brushless AC Generators

Brushless AC generators are normally used in heavy-duty trucks.

Some manufacturers have developed AC generators that do not require the use of brushes or slip rings. In these AC generators, the field winding and the stator winding are stationary (Figure 7-74). A screw terminal is used to make the electrical connection. The rotor contains the pole pieces and is fitted between the field winding and the stator winding.

The magnetic field is produced when current is applied to the field winding. The air gaps in the magnetic path contain a nonmetallic ring to divert the lines of force into the stator winding.

The pole pieces on the rotor concentrate the magnetic field into alternating north and south poles. When the rotor is spinning, the north and south poles alternate as they pass the stator winding. The moving magnetic field produces an electrical current in the stator winding. The alternating current is rectified in the same manner as in conventional AC generators.

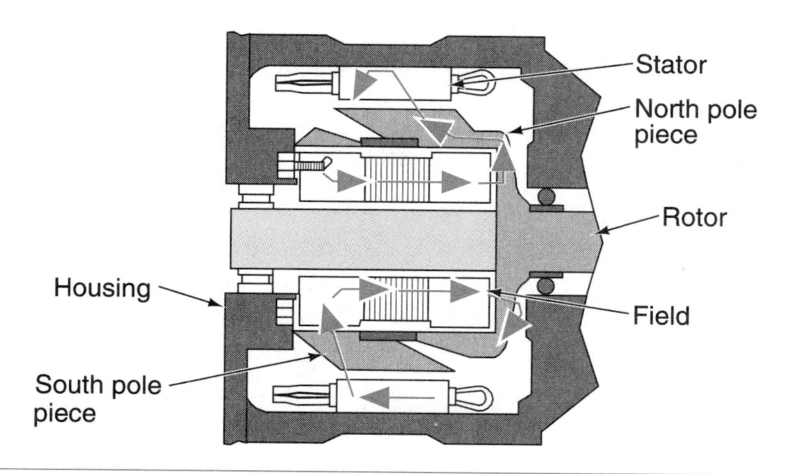

Figure 7-74 Brushless AC generator with stationary field and stator windings.

Summary

- The purpose of the charging system is to convert the mechanical energy of the engine into electrical energy to recharge the battery and run the electrical accessories.
- The charging system consists of the battery, AC or DC generator, drive belt, voltage regulator, charge indicator (lamp or gauge), ignition switch, cables and wiring harness, starter relay (some systems), and fusible links (some systems).
- All charging systems use the principle of electromagnetic induction to generate the electrical power.
- Electromagnetic principle states that a voltage will be produced if motion between a conductor and a magnetic field occurs. The amount of current produced is affected by the speed that the conductor passes through the magnetic field, the strength of the magnetic field, and the number of conductors passing through the magnetic field.
- The main components of the AC generator are the rotor, the brushes, the stator, the rectifier bridge, the housing, and the cooling fan.
- The rotor creates the rotating magnetic field of the AC generator. It is the portion of the AC generator that is rotated by the drive belt.
- The insulated stationary carbon brush passes field current into a slip ring, then through the field coil, and back to the other slip ring. Current then passes through to the grounded stationary brush or voltage regulator.
- The stator is the stationary coil in which current is produced.
- The stator contains three main sets of windings wrapped in slots around a laminated, circular iron frame.
- The most common method of stator connection is called the wye connection. In the wye connection, one lead from each winding is connected to one common junction. From this junction the other leads branch out in a Y pattern.
- Another method of stator connection is called the delta connection. The delta connection connects the lead of one end of the winding to the lead at the other end of the next winding.
- The diode rectifier bridge provides reasonably constant DC voltage to the vehicle's electrical system and battery. The diode rectifier bridge is used to change the current in an AC generator.

Terms to Know

A circuit
B circuit
Brushes
Delta connection
Diode rectifier bridge
Diode trio
Duty-cycle
External grounded field circuit
Field current
Field relay
Full field current
Full-wave rectification
Half field current
Half-wave rectification
Heat sink
Inductive reactance
Internally grounded circuit
Polarizing
Pulse width modulation
Rectification

Terms to Know (Continued)

Rotor

Sensing voltage

Sine wave

Single-phase voltage

Stator

Stator neutral junction

Voltage limiter

Voltage regulator

Wye connection

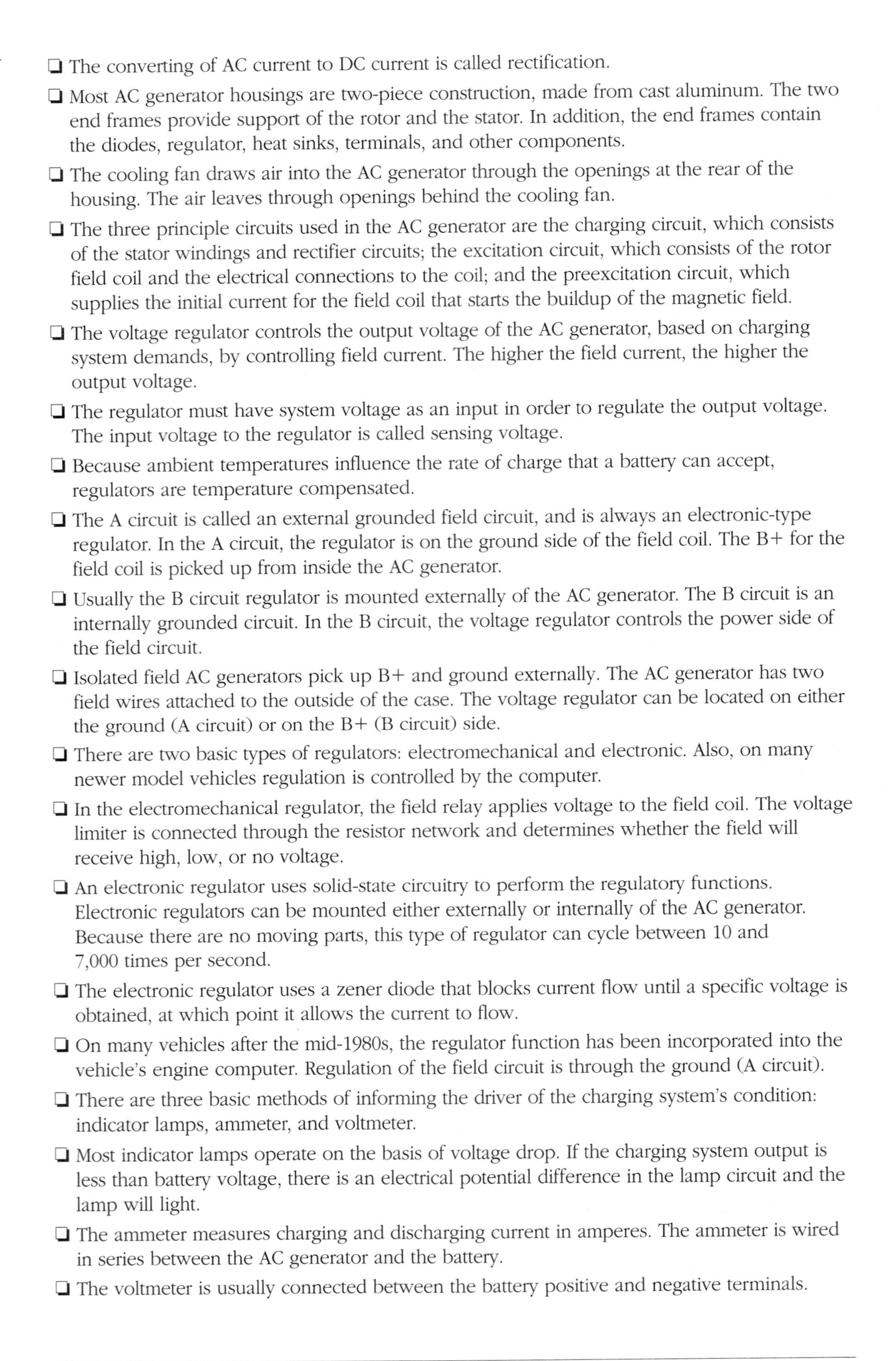

- ❑ The converting of AC current to DC current is called rectification.
- ❑ Most AC generator housings are two-piece construction, made from cast aluminum. The two end frames provide support of the rotor and the stator. In addition, the end frames contain the diodes, regulator, heat sinks, terminals, and other components.
- ❑ The cooling fan draws air into the AC generator through the openings at the rear of the housing. The air leaves through openings behind the cooling fan.
- ❑ The three principle circuits used in the AC generator are the charging circuit, which consists of the stator windings and rectifier circuits; the excitation circuit, which consists of the rotor field coil and the electrical connections to the coil; and the preexcitation circuit, which supplies the initial current for the field coil that starts the buildup of the magnetic field.
- ❑ The voltage regulator controls the output voltage of the AC generator, based on charging system demands, by controlling field current. The higher the field current, the higher the output voltage.
- ❑ The regulator must have system voltage as an input in order to regulate the output voltage. The input voltage to the regulator is called sensing voltage.
- ❑ Because ambient temperatures influence the rate of charge that a battery can accept, regulators are temperature compensated.
- ❑ The A circuit is called an external grounded field circuit, and is always an electronic-type regulator. In the A circuit, the regulator is on the ground side of the field coil. The B+ for the field coil is picked up from inside the AC generator.
- ❑ Usually the B circuit regulator is mounted externally of the AC generator. The B circuit is an internally grounded circuit. In the B circuit, the voltage regulator controls the power side of the field circuit.
- ❑ Isolated field AC generators pick up B+ and ground externally. The AC generator has two field wires attached to the outside of the case. The voltage regulator can be located on either the ground (A circuit) or on the B+ (B circuit) side.
- ❑ There are two basic types of regulators: electromechanical and electronic. Also, on many newer model vehicles regulation is controlled by the computer.
- ❑ In the electromechanical regulator, the field relay applies voltage to the field coil. The voltage limiter is connected through the resistor network and determines whether the field will receive high, low, or no voltage.
- ❑ An electronic regulator uses solid-state circuitry to perform the regulatory functions. Electronic regulators can be mounted either externally or internally of the AC generator. Because there are no moving parts, this type of regulator can cycle between 10 and 7,000 times per second.
- ❑ The electronic regulator uses a zener diode that blocks current flow until a specific voltage is obtained, at which point it allows the current to flow.
- ❑ On many vehicles after the mid-1980s, the regulator function has been incorporated into the vehicle's engine computer. Regulation of the field circuit is through the ground (A circuit).
- ❑ There are three basic methods of informing the driver of the charging system's condition: indicator lamps, ammeter, and voltmeter.
- ❑ Most indicator lamps operate on the basis of voltage drop. If the charging system output is less than battery voltage, there is an electrical potential difference in the lamp circuit and the lamp will light.
- ❑ The ammeter measures charging and discharging current in amperes. The ammeter is wired in series between the AC generator and the battery.
- ❑ The voltmeter is usually connected between the battery positive and negative terminals.

Review Questions

Short Answer Essays

1. List the major components of the charging system.
2. List and explain the function of the major components of the AC generator.
3. How does the regulator control the charging system's output?
4. What is the relationship between field current and AC generator output?
5. Identify the differences between A, B, and isolated circuits.
6. Explain the operation of charge indicator lamps.
7. Describe the two styles of stators.
8. What is the difference between half-wave and full-wave rectification?
9. Describe how AC voltage is rectified to DC voltage in the AC generator.
10. What is the purpose of the charging system?

Fill-in-the-Blanks

1. The charging system converts the ____________________ energy of the engine into ____________________ energy to recharge the battery and run the electrical accessories.

2. All charging systems use the principle of ____________________ ____________________ to generate the electrical power.

3. The ____________________ creates the rotating magnetic field of the AC generator.

4. ____________________ are electrically conductive sliding contacts, usually made of copper and carbon.

5. In the ____________________ connection stator, one lead from each winding is connected to one common junction.

6. The ____________________ ____________________ controls the output voltage of the AC generator, based on charging system demands, by controlling ____________________ current.

7. In an electronic regulator, ____________________ ____________________ ____________________ controls AC generator output by varying the amount of time the field coil is energized.

8. In most electronic regulators that use an indicator lamp, if there is no ____________ ____________, then the lamp circuit is completed to ground.

9. Full-wave rectification in the AC generator requires ____________ pair of diodes.

10. The ____________ is the stationary coil that current is produced in the AC generator.

ASE Style Review Questions

1. The purpose of the charging system is being discussed:
Technician A says the charging system is used to restore the electrical power to the battery that was used during engine starting.
Technician B says the charging system must be able to react quickly to high load demands required of the electrical system.
Who is correct?
A. A only
B. B only
C. Both A and B
D. Neither A nor B

2. The operation of the charging system is being discussed:
Technician A says battery condition has no affect on the charging system operation.
Technician B says the stator is the rotating component that creates the magnetic field.
Who is correct?
A. A only
B. B only
C. Both A and B
D. Neither A nor B

3. Rectification is being discussed:
Technician A says the AC generator uses a segmented commutator to rectify AC current.
Technician B says the DC generator uses a pair of diodes to rectify AC current.
Who is correct?
A. A only
B. B only
C. Both A and B
D. Neither A nor B

4. Rotor construction is being discussed:
Technician A says the poles will take on the polarity of the side of the coil that they touch.
Technician B says the magnetic flux lines will move in opposite directions between adjacent poles.
Who is correct?
A. A only
B. B only
C. Both A and B
D. Neither A nor B

5. Stator construction is being discussed:
Technician A says the wye connection is the most common.
Technician B says the wye connection connects the lead of one end of the winding to the lead at the other end of the next winding.
Who is correct?
A. A only
B. B only
C. Both A and B
D. Neither A nor B

6. Charging system regulation is being discussed:
Technician A says DC generators require current limiters.
Technician B says typically the higher the field current, the lower the output voltage.
Who is correct?
A. A only
B. B only
C. Both A and B
D. Neither A nor B

7. Indicator lamp operation is being discussed:
Technician A says in a system with an electronic regulator, the lamp will light if there is no stator output through the diode trio.
Technician B says when there is stator output the lamp circuit has voltage applied to both sides and the lamp will not light.
Who is correct?
A. A only
B. B only
C. Both A and B
D. Neither A nor B

8. AC generator differences are being discussed:
Technician A says the Mitsubishi AC generator uses two separate wye connected stator windings.
Technician B says Ford rear-terminal AC generators use two different types of rectifiers.
Who is correct?
A. A only
B. B only
C. Both A and B
D. Neither A nor B

9. *Technician A* says only two stator windings apply voltage because the third winding is always connected to diodes that are reverse biased.
Technician B says AC generators that use half-wave rectification are the most efficient.
Who is correct?
A. A only
B. B only
C. Both A and B
D. Neither A nor B

10. The operation of the electromechanical regulator is being discussed:
Technician A says the field relay applies battery voltage to the field coil after the engine has started.
Technician B says the voltage limiter is connected through the resistor network and determines whether the field will receive high, low, or no voltage.
Who is correct?
A. A only
B. B only
C. Both A and B
D. Neither A nor B

Ignition Systems

Upon completion and review of this chapter, you should be able to:

- ❑ Describe the three major functions of an ignition system.
- ❑ Name the operating conditions of an engine that affect ignition timing.
- ❑ Name the two major electrical circuits used in ignition systems and their common components.
- ❑ Describe the operation of ignition coils, spark plugs, and ignition cables.
- ❑ Describe the various types of spark timing systems, including electronic switching systems and their related engine position sensors.
- ❑ Describe the operation of distributor-based ignition systems.
- ❑ Describe the operation of distributorless ignition systems.

Introduction

One of the requirements for an efficient running engine is the correct amount of heat delivered into the cylinders at the right time. This requirement is the responsibility of the ignition system. The ignition system supplies properly timed, high-voltage surges to the spark plugs. These voltage surges cause an arc across the electrodes of a spark plug, and this heat begins the combustion process inside the cylinder. For each cylinder in an engine, the ignition system has three main jobs. First, it must generate an electrical spark that has enough heat to ignite the air/fuel mixture in the combustion chamber. Second, it must maintain that spark long enough to allow for the combustion of all the air and fuel in the cylinders. Third, it must deliver the spark to each cylinder so that combustion can begin at the right time during the compression stroke of each cylinder.

When the combustion process is completed, a very high pressure is exerted against the top of the piston. This pressure pushes the piston down on its power stroke. This pressure is the force that gives the engine power. In order for an engine to produce the maximum amount of power that it can, the maximum pressure from combustion should be present when the piston is at 10° to 23° after top dead center (**ATDC**). Because combustion of the air/fuel mixture within a cylinder takes a short period of time, usually measured in thousandths of a second (milliseconds), the combustion process must begin before the piston is on its power stroke. Therefore, the delivery of the spark must be timed to arrive at some point before the piston reaches top dead center.

In order to have an efficient running engine there must be the correct amount of air mixed with the correct amount of fuel, in a sealed container, shocked by the correct amount of heat at the right time.

Determining how much before TDC the spark should begin gets complicated by the fact that as the speed of the piston increases as it moves from its compression stroke to its power stroke, the time needed for combustion stays about the same. This means that the spark should be delivered earlier as the engine's speed increases (Figure 8-1). However, as the engine has to provide more power to do more work, the load on the crankshaft tends to slow down the acceleration of the piston and the spark needs to be somewhat delayed.

TDC stands for top dead center or the uppermost position of the piston in its cylinder.

Calculating when the spark should begin gets more complicated with the fact that the rate of combustion varies according to certain factors. Higher compression pressures tend to speed up combustion. Higher octane gasolines ignite less easily and require more burning time. Increased vaporization and turbulence tend to decrease combustion times. Other factors, including intake air temperature, humidity, and barometric pressure, also affect combustion. Because of all of these complications, delivering the spark at the right time is a difficult task.

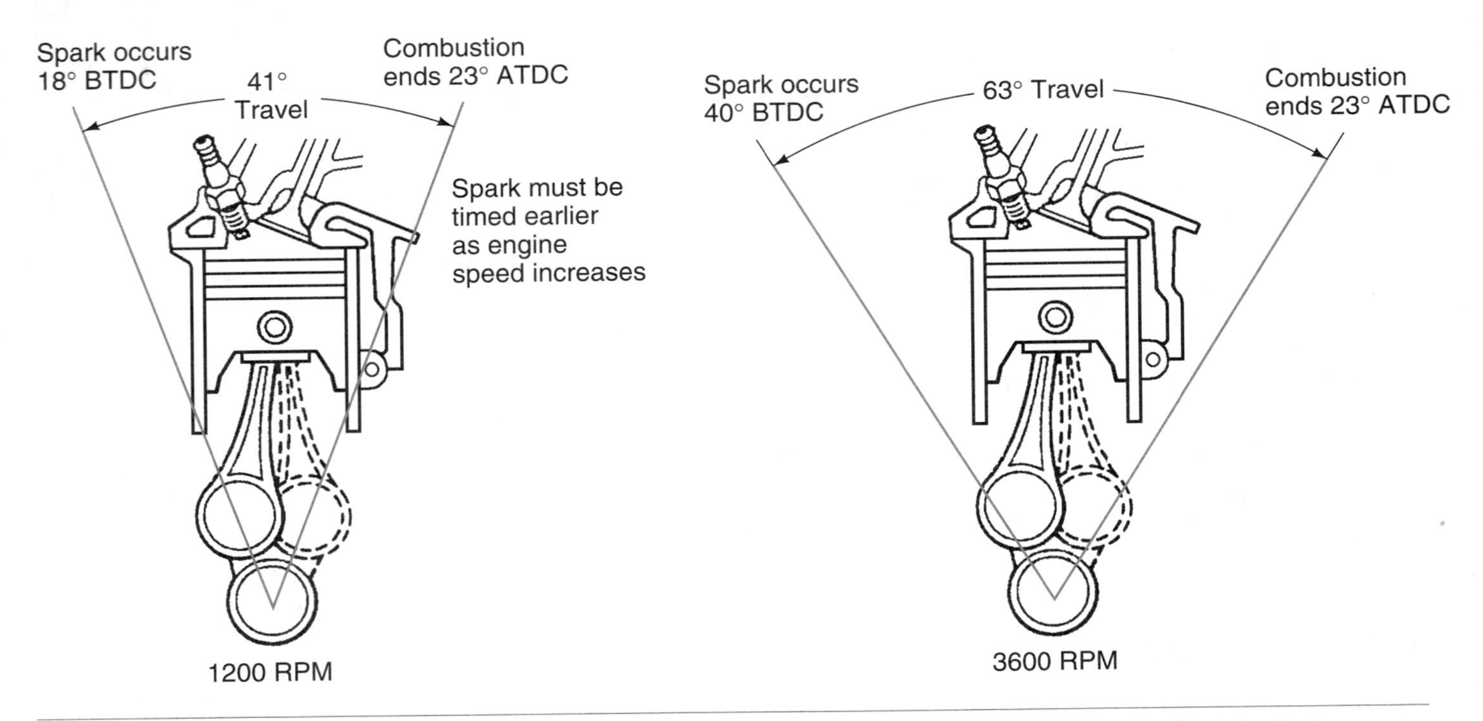

Figure 8-1 With an increase in speed, ignition must begin earlier to end by 23 degrees ATDC. (Reprinted with the permission of Ford Motor Company)

Ignition Timing

Shop Manual
Chapter 8, page 270

Ignition timing refers to the precise time that spark occurs. Ignition timing is specified by referring to the position of the number one piston in relationship to crankshaft rotation. Ignition timing reference marks can be located on a pulley or flywheel to indicate the position of the number one piston (Figure 8-2). Vehicle manufacturers specify initial, or **basic** ignition **timing**.

When the marks are aligned at TDC, or 0, the piston in cylinder number one is at TDC of its compression stroke. Additional numbers on a scale indicate the number of degrees of crankshaft rotation before TDC (**BTDC**) or after TDC (**ATDC**). In a majority of engines, the initial timing is specified at a point between TDC and 20 degrees BTDC.

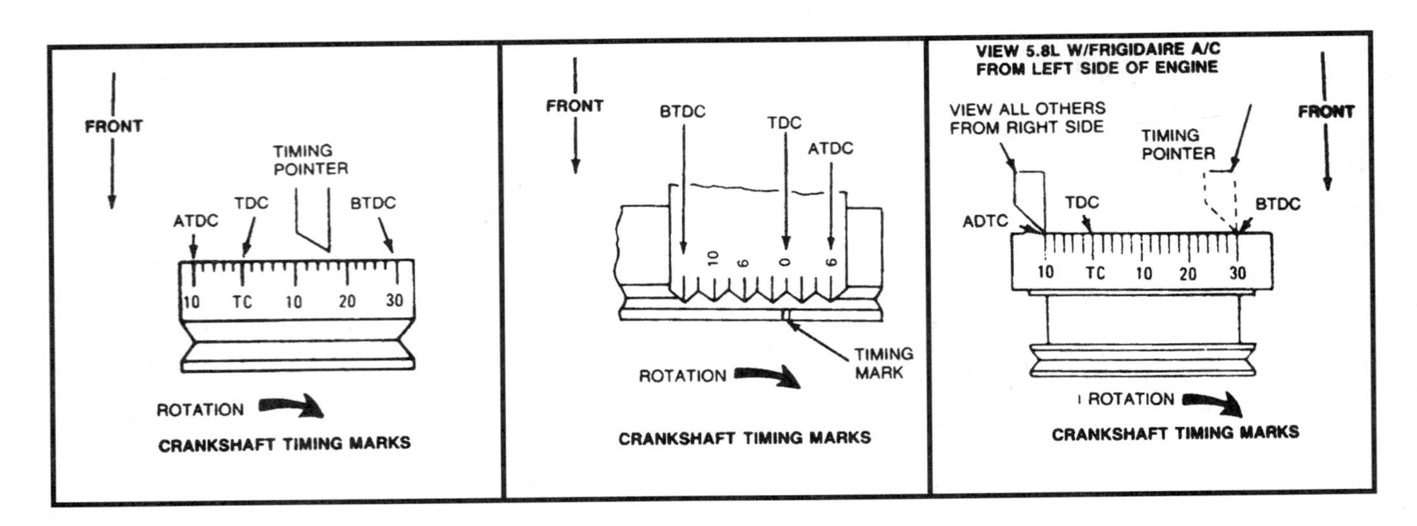

Figure 8-2 Various crankshaft pulleys with timing marks. (Reprinted with the permission of Ford Motor Company)

If optimum engine performance is to be maintained, the ignition timing of the engine must change as the operating conditions of the engine change. All of the different operating conditions affect the speed of the engine and the load on the engine. All ignition timing changes are made in response to these primary factors.

Engine Speed

At higher engine speeds, the crankshaft turns through more degrees in a given period of time. If combustion is to be completed by 10 degrees ATDC, ignition timing must occur sooner or be advanced.

However, air/fuel mixture turbulence increases with rpm. This causes the mixture, inside the cylinder, to burn faster. Increased turbulence requires that ignition must occur slightly later or be slightly retarded.

These two factors must be balanced for best engine performance. Therefore, while the ignition timing must be advanced as engine speed increases, the amount of advance must be decreased some to compensate for the increased turbulence.

Engine Load

The load on an engine is related to the work it must do. Driving up hills or pulling extra weight increases engine load. Under load, the pistons accelerate more slowly and the engine runs less efficiently. A good indication of engine load is the amount of vacuum formed during the intake stroke.

Engine vacuum is formed during the intake stroke of the cylinder. Engine vacuum is any pressure lower than atmospheric pressure.

Under light loads and with the throttle plate(s) partially opened, a high vacuum exists in the intake manifold. The amount of air/fuel mixture drawn into the manifold and cylinders is small. This means the air and fuel molecules are relatively far apart. On compression, this thin mixture produces less combustion pressure and combustion time is slow. To complete combustion by 10 degrees ATDC, ignition timing must be advanced to allow for a longer burning time.

Under heavy loads, when the throttle is opened fully, a larger mass of the air/fuel mixture can be drawn in, and the vacuum in the manifold is low. Combustion is fast because the air and fuel molecules are close together. High combustion pressure and rapid burning results. In such a case, the ignition timing must be advanced less or retarded to prevent complete burning from occurring before 10 degrees ATDC.

Firing Order

Up to this point, the primary focus of discussion has been ignition timing as it relates to any one cylinder. However, the function of the ignition system extends beyond timing the arrival of a spark to a single cylinder. It must perform this task for each cylinder of the engine in a specific sequence.

Each cylinder of an engine produces power once every 720 degrees of crankshaft rotation. Each cylinder must have a power stroke at its own appropriate time during the rotation. To make this possible, the pistons and rods are arranged in a precise fashion. This is called the engine's **firing order**. The firing order is arranged to reduce rocking and imbalance problems. Because the potential for this rocking is determined by the design and construction of the engine, the firing order varies from engine to engine. Vehicle manufacturers simplify identifying each cylinder by numbering them (Figure 8-3). Regardless of the particular firing order used, the number one cylinder always starts the firing order, with the rest of the cylinders following in a fixed sequence.

Shop Manual
Chapter 8, page 243

The ignition system must be able to monitor the rotation of the crankshaft and the relative position of each piston in order to determine which piston is on its compression stroke. It must also be able to deliver a high-voltage surge to each cylinder at the proper time during its compression stroke. How the ignition system does these things depends on the design of the system.

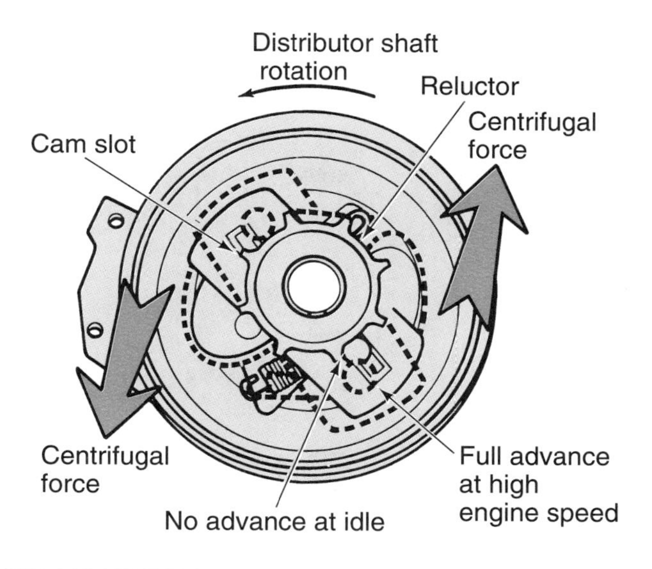

Figure 8-27 Action of a centrifugal advance assembly. (Courtesy of Chrysler Corporation)

During part-throttle engine operation, high vacuum is present in the intake manifold. To get the most power and the best fuel economy from the engine, the plugs must fire even earlier during the compression stroke than is provided by a centrifugal advance mechanism.

The heart of the vacuum advance mechanism (Figure 8-28) is the spring-loaded diaphragm, which fits inside a metal housing and connects to a movable plate on which the pickup coil is mounted. Vacuum is applied to one side of the diaphragm in the housing chamber while the other side of the diaphragm is open to the atmosphere. Any increase in vacuum allows atmospheric pressure to push the diaphragm. In turn, this causes the movable plate to rotate. The more vacuum present on one side of the diaphragm, the more atmospheric pressure is able to cause a change in timing. The rotation of the movable plate moves the pickup coil so the armature develops a signal earlier. These units are also equipped with a spring that allows the timing to return to only centrifugal timing advance as vacuum decreases.

Shop Manual
Chapter 8, page 272

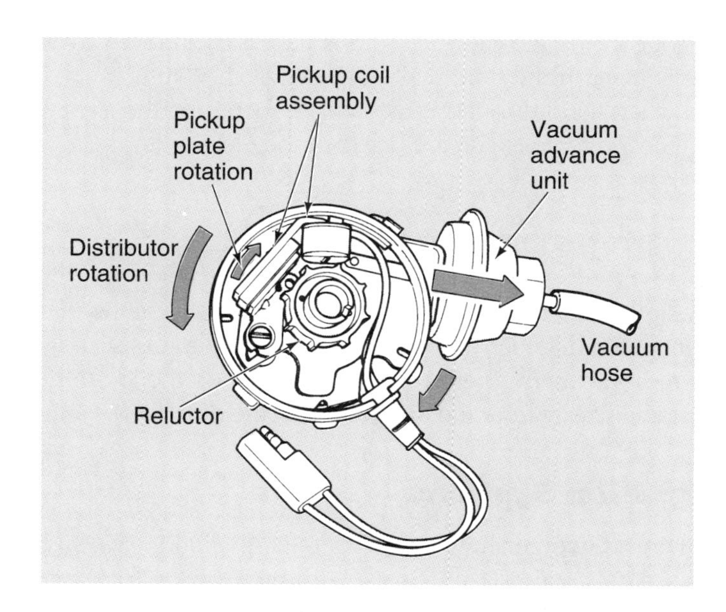

Figure 8-28 Action of a vacuum advance assembly. (Courtesy of Chrysler Corporation)

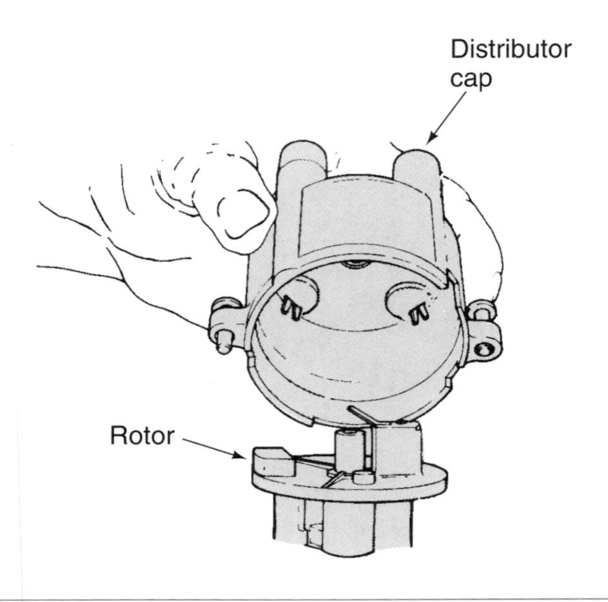

Figure 8-29 A typical distributor cap. (Courtesy of Chrysler Corporation)

Spark Distribution

The distributor cap and rotor receive the high voltage from the secondary winding via a high-tension wire. The voltage enters the distributor cap through the coil tower, or center terminal. The rotor then sends the voltage from the coil tower to the spark plug wire electrodes inside the distributor cap. The rotor mounts on the upper portion of the distributor shaft and rotates with it.

The distributor cap (Figure 8-29) is made from low carbon fiber, reinforced plastic, silicone plastic, or similar material that offers protection from chemical attack and resists electrical arcing and carbon tracking. It is attached to the distributor housing with screws or spring-loaded clips. The coil tower contains a carbon insert that carries the voltage from the high-tension coil lead to the raised portion of the electrode on the rotor. Spaced evenly around the coil tower are the spark plug wire electrodes and towers for each spark plug wire.

An air gap of a few thousandths of an inch exists between the tip of the rotor electrode and the spark plug electrode inside the cap. This gap is necessary in order to prevent the two electrodes from making contact. If they did make contact, both would wear out rapidly. This gap cannot be measured when the distributor is assembled; therefore, the gap is usually described in terms of the voltage needed to create an arc between the electrodes.

Common DI Systems

Through the years there have been many different designs of DI systems; all operate basically the same way but are configured differently. The systems described in this section represent the different designs used by manufacturers. These designs are based on the location of the ECU. Systems either mounted the module away from the distributor, on the distributor, or in the distributor.

Dura-Spark Ignition Systems

Ford Motor Company used two generations of Dura-Spark ignition systems. The second design (Dura-Spark II) is based on the first (Dura-Spark I) and will be the focus of this discussion. The Dura-Spark II had the ECU mounted away from the distributor, typically on a fender wall. The distributor is fitted with a centrifugal advance assembly and a vacuum advance unit. An armature is mounted to the distributor shaft with a roll pin. The armature has a high point for each cylinder. The pickup coil is mounted to the breaker plate (Figure 8-30).

In this system, a resistance wire is connected between the ignition switch and the positive primary coil terminal. A bypass wire is also connected from the ignition switch to the positive coil terminal (Figure 8-31).

When the engine is running, 5 to 7 volts are supplied through the resistance wire to the coil. While the engine is cranking, full battery voltage is supplied to the coil through the bypass wire. This increases primary current flow and allows the coil to provide high voltage to ensure good starting.

The negative terminal of an ignition coil is commonly referred to as the tachometer (tach) terminal.

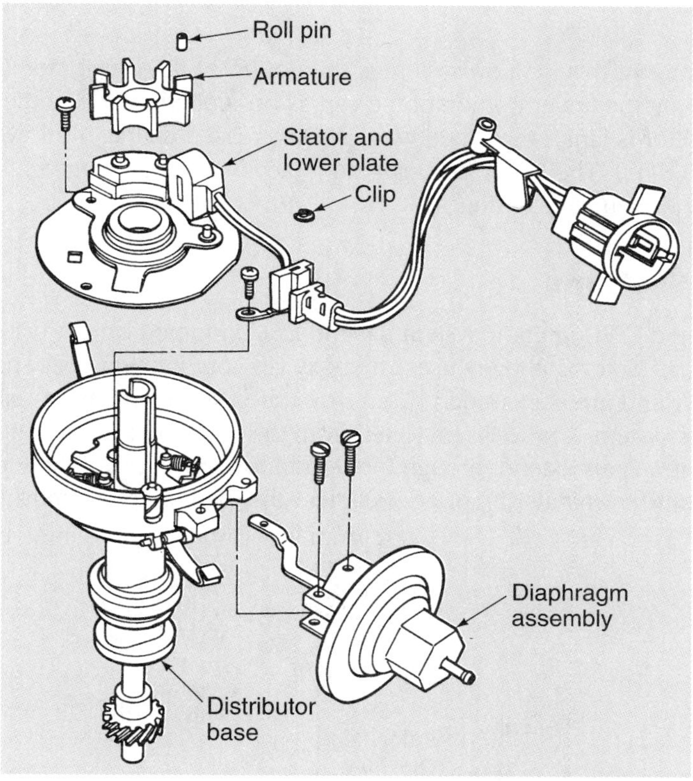

Figure 8-30 A Dura-Spark II distributor assembly. (Reprinted with the permission of Ford Motor Company)

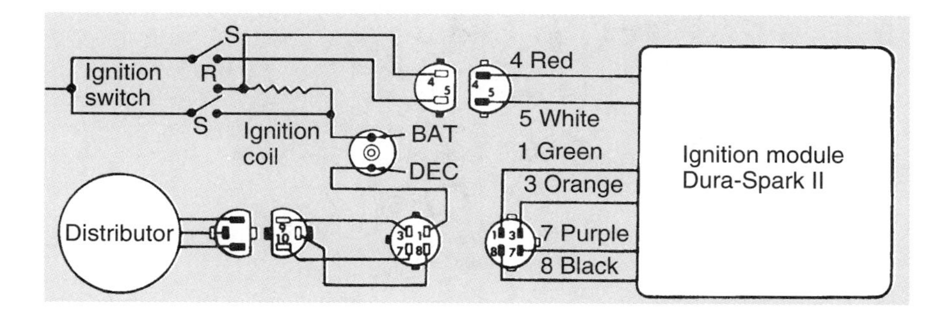

Figure 8-31 The circuitry for a Dura-Spark II ignition system. (Reprinted with the permission of Ford Motor Company)

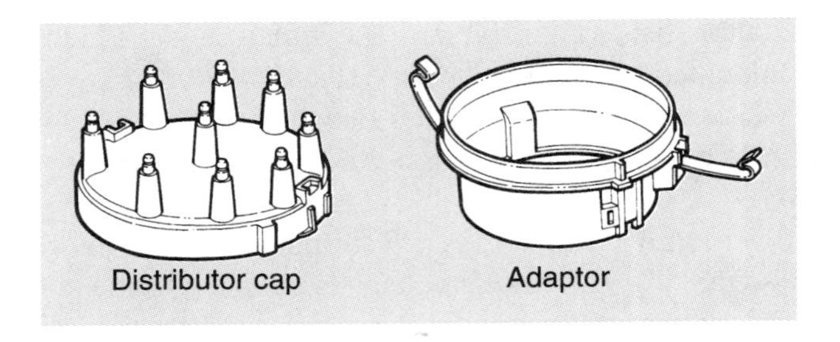

Figure 8-32 The distributor cap assembly for a Dura-Spark II distributor. (Reprinted with the permission of Ford Motor Company)

A unique feature of this ignition system is the design of the distributor. The cap is a two-piece unit. An adaptor, the lower portion of the cap, is positioned on top of the distributor housing. Its upper diameter is larger than the lower. This increased diameter allows for a larger distributor cap (Figure 8-32). The larger diameter cap places the spark plug wire terminals further apart. This helps to prevent cross-firing.

TFI Ignition Systems

A thick film integrated (TFI) ignition system has the ECU mounted on the distributor. This ignition system from Ford Motor Company uses an epoxy (E) core ignition coil. The windings of the coil are set in epoxy and are surrounded by an iron core. Neither a ballast resistor nor resistor wire are used in this system. A wire is connected from the ignition switch to the TFI module. The module-to-pick-up terminals extend through the distributor housing, and three pickup lead wires are connected to these terminals (Figure 8-33). Heat dissipating grease must be placed on the

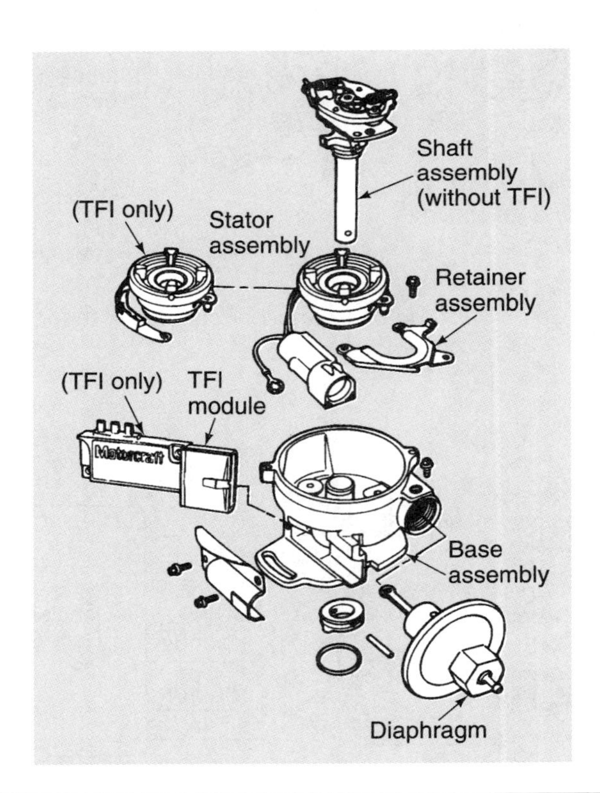

Figure 8-33 A thick film integrated (TFI) distributor assembly. (Reprinted with the permission of Ford Motor Company)

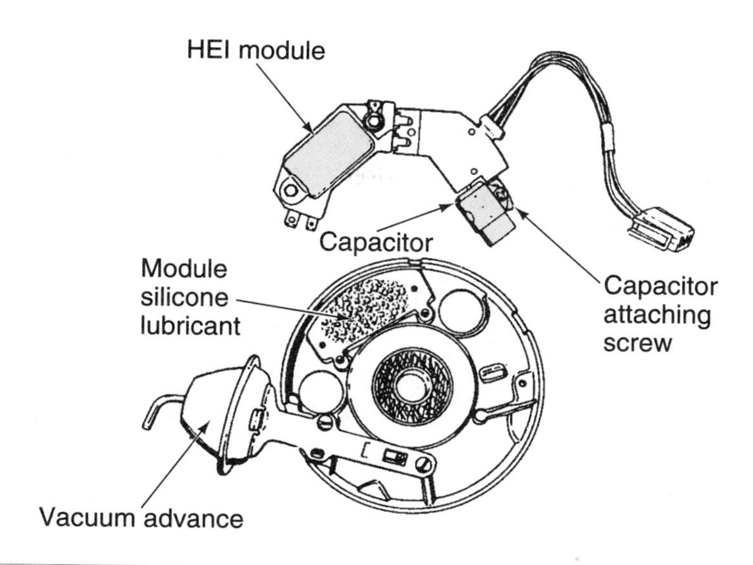

Figure 8-34 The mounting of an ECU in an HEI distributor. (Courtesy of Oldsmobile Division—GMC)

back of the module to prevent overheating of the module. The distributor is also equipped with centrifugal and vacuum advance assemblies.

HEI Systems

The high energy ignition (HEI) systems from General Motors are characterized by the mounting of the ECU inside the distributor (Figure 8-34). Some HEI units also contain the ignition coil; others have the coil remotely mounted away from the distributor. Some HEI designs have centrifugal and vacuum advance units, while others utilize electronic spark timing.

The pickup coil surrounds the distributor shaft, and a flat magnetic plate is bolted between the pickup coil and the pole piece. A timer core with one high point for each cylinder is attached to the distributor shaft. The pole piece has the same number of high points as the timer core. This allows the timer core high points to be aligned with the pole piece teeth at the same time. The module is mounted to the breaker plate and is set in heat dissipating grease. A capacitor is connected from the module voltage supply terminal to ground.

In some HEI designs, the coil is mounted in the top of the distributor cap; other designs have externally mounted coils. The coil battery terminal is connected directly to the ignition switch, and the negative coil terminal is connected to the module (Figure 8-35). HEI coils are basically E-core coils. An E-core coil is an ignition coil assembly that is not contained in a housing. Rather, the windings are looped around an E-shaped core. The core is made of steel plates. Rather than use oil to cool the coil, E-core coils allow the heat to dissipate into the air.

A wire also extends from the coil's battery terminal to the module. In systems with an internal ignition coil, a ground wire is connected between the frame of the coil and the distributor housing. This lead is used to dissipate any voltage induced in the coil's frame (Figure 8-36).

Computer-Controlled Ignition System Operation

Computer-controlled ignition systems (Figure 8-37) control the primary circuit and distribute the firing voltages in the same manner as other types of electronic ignition systems. The main difference between the systems is the elimination of any mechanical or vacuum advance devices from

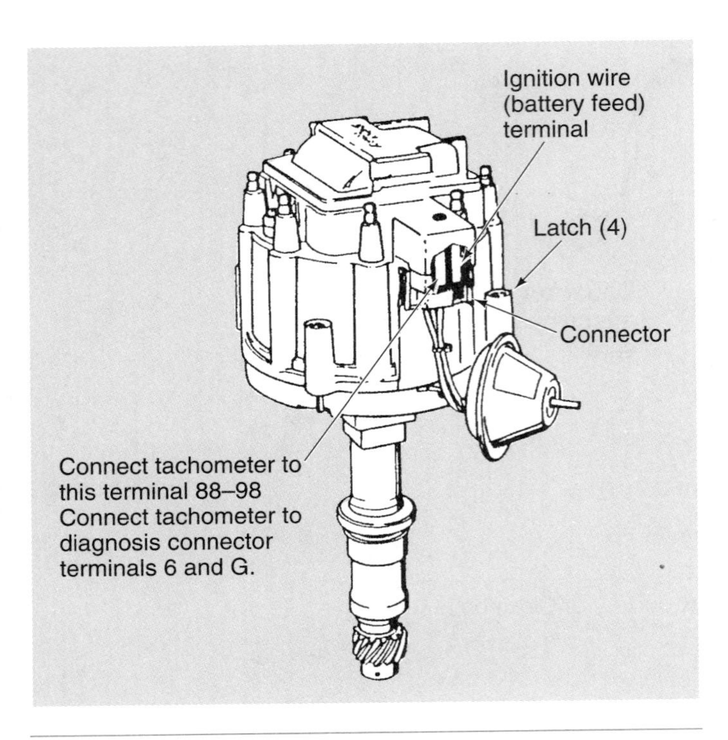

Figure 8-35 The external hook-ups for an HEI distributor. (Courtesy of Oldsmobile Division—GMC)

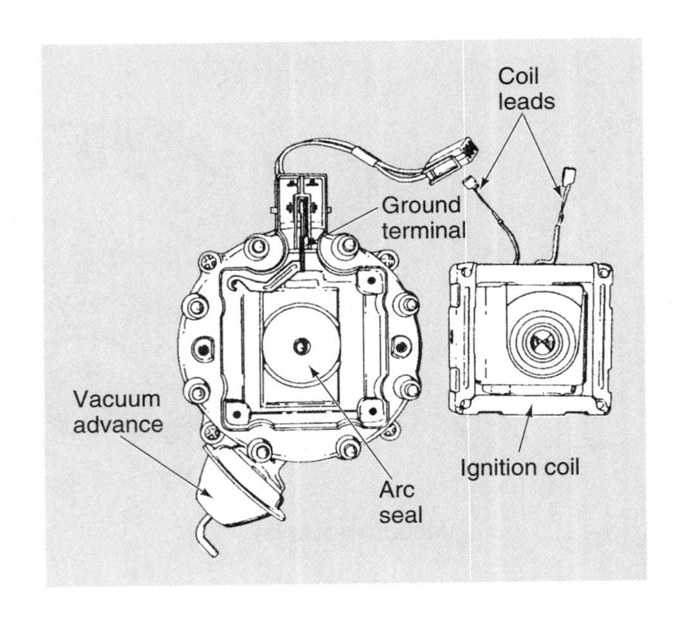

Figure 8-36 Mounting and electrical connections for an internal ignition coil of an HEI distributor. (Courtesy of Oldsmobile Division—GMC)

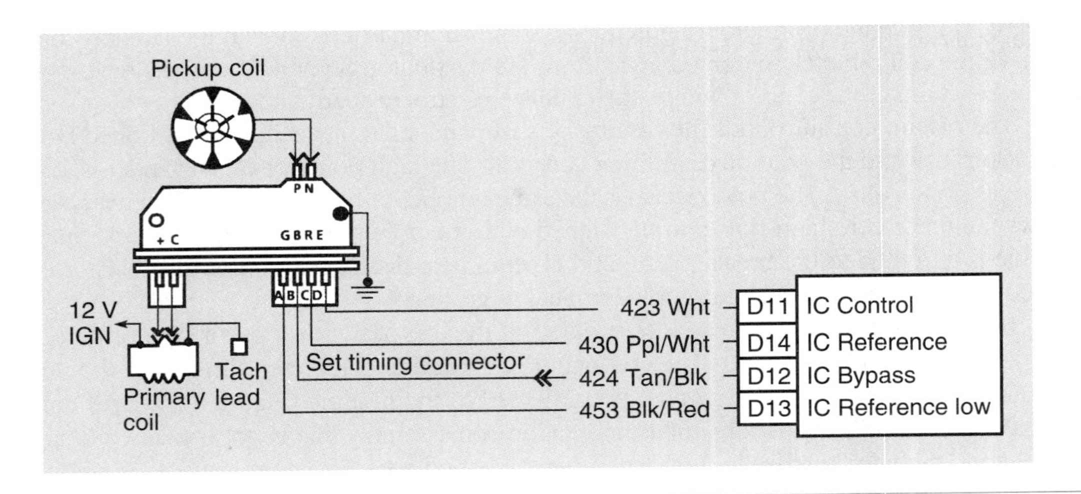

Figure 8-37 Typical computer-controlled ignition system. (Courtesy of Chevrolet Motor Division—GMC)

the distributor in the computer-controlled systems. In these systems, the distributor's sole purpose is to generate the primary circuit's switching signal and distribute the secondary voltage to the spark plugs. Timing advance is controlled by a microprocessor, or computer. In fact, some of these systems have even removed the primary switching function from the distributor by using a crankshaft position sensor. The distributor's only job is to distribute secondary voltage to the spark plugs.

Spark timing on these systems is controlled by a computer that continuously varies ignition timing to obtain optimum air/fuel combustion. The computer monitors the engine operating

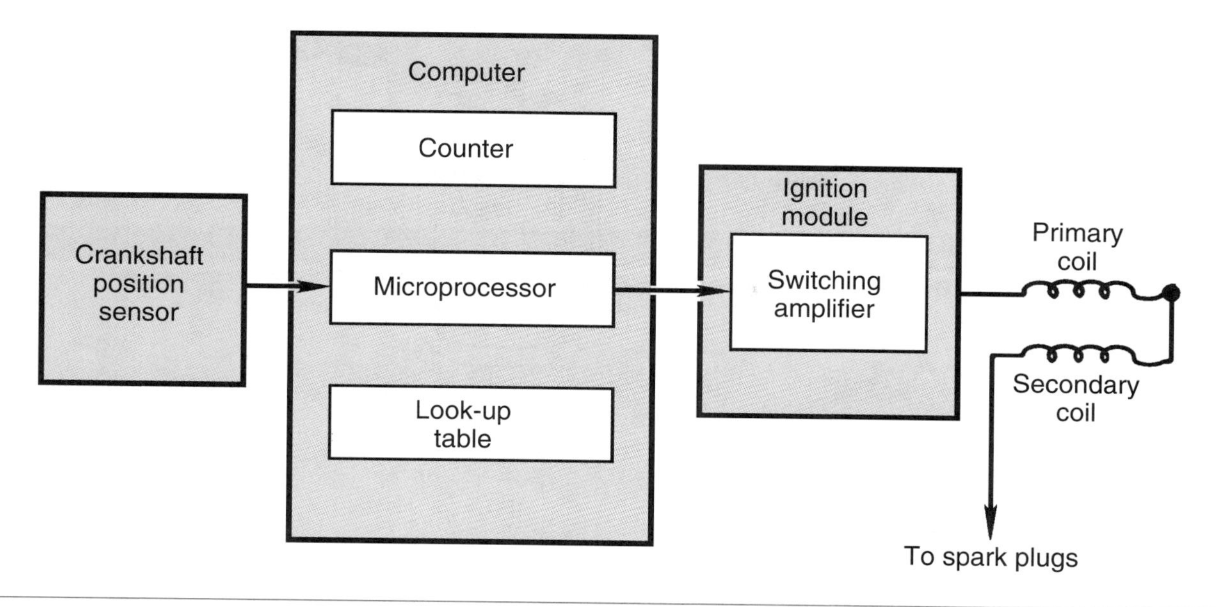

Figure 8-38 Simple diagram of a computer-controlled ignition system.

parameters with sensors. Based on this input, the computer signals an ignition module to collapse the primary circuit, allowing the secondary circuit to fire the spark plugs (Figure 8-38).

Timing control is selected by the computer's program. During engine starting, computer control is bypassed and the mechanical setting of the distributor controls spark timing. Once the engine is started and running, spark timing is controlled by the computer. This scheme or **strategy** allows the engine to start regardless of whether the electronic control system is functioning properly or not.

The goal of computerized spark timing is to produce maximum engine power, top fuel efficiency, and minimum emissions levels during all types of operating conditions. The computer does this by continuously adjusting ignition timing. The computer determines the best spark timing based on certain engine operating conditions such as crankshaft position, engine speed, throttle position, engine coolant temperature, and initial and operating manifold or barometric pressure. Once the computer receives input from these and other sensors, it compares the existing operating conditions to information permanently stored or programmed into its memory. The computer matches the existing conditions to a set of conditions stored in its memory, determines proper timing setting, and sends a signal to the ignition module to fire the plugs.

The computer continuously monitors existing conditions, adjusting timing to match what its memory tells it is the ideal setting for those conditions. It can do this very quickly, making thousands of decisions in a single second. The control computer typically has the following types of information permanently programmed into it:

- Speed-related spark advance. As engine speed increases to a particular point, there is a need for more advanced timing. As the engine slows, the timing should be retarded or have less advance. The computer bases speed related spark advance decisions on engine speed and signals from the TP sensor.
- Load-related spark advance. This is used to improve power and fuel economy during acceleration and heavy load conditions. The computer defines the load and the ideal spark advance by processing information from the TP sensor, MAP, and engine speed sensors. Typically, the more load on an engine, the less spark advance is needed.
- Warm-up spark advance. This is used when the engine is cold, since a greater amount of advance is required while the engine warms up.

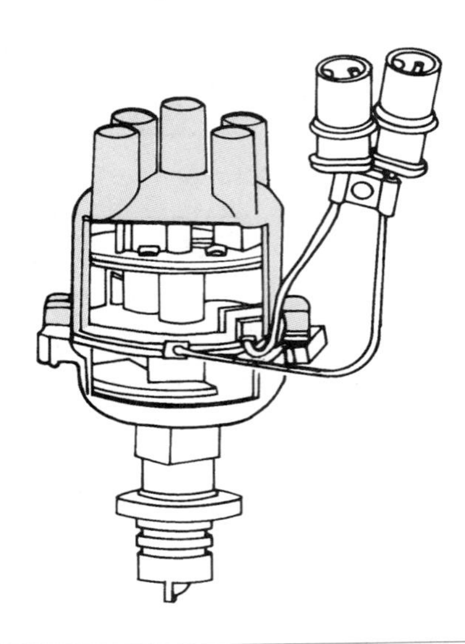

Figure 8-39 A Chrysler distributor with dual Hall-effect switches, one located above the breaker plate and one below. (Courtesy of Chrysler Corporation)

- ❑ Special spark advance. This is used to improve fuel economy during steady driving conditions. During constant speed and load conditions, the engine will be more efficient with much advance timing.
- ❑ Spark advance due to barometric pressure. This is used when barometric pressure exceeds a preset calibrated value.
- ❑ Spark advance is used to control engine idle speed.

All of this information is looked at by the computer to determine the ideal spark timing for all conditions. The calibrated or programmed information in the computer is contained in what is called software **look-up tables**.

Ignition timing can also work in conjunction with the electronic fuel control system to provide emission control, optimum fuel economy, and improved drivability. They are all dependent on spark advance. An example of this type of system is used by Chrysler.

Chrysler's system has two Hall-effect switches in the distributor when the engine is equipped with port fuel injection (Figure 8-39). In some units, the pickup unit used for ignition triggering is located above the pickup plate in the distributor and is referred to as the reference pickup. The second pickup unit is positioned below the plate. A ring with two notches is attached to the distributor shaft and rotates through the lower pickup unit. This lower pickup is called the synchronizer (SYNC) pickup.

In other designs, the two pickup units are mounted below the pickup plate, and one set of blades rotates through both Hall-effect units (Figure 8-40). The shutter blade representing the number one cylinder has a large opening in the center of the blade. When this blade rotates through the SYNC pickup, a different signal is produced compared to the other blades. This number one blade signal informs the power train control module (PCM) when to activate the injectors.

EI System Operation

Electronic ignition (EI) systems (Figure 8-41) electronically perform the functions of a distributor. They control spark timing and advance in the same manner as the computer-controlled ignition systems. Yet the EI is a step beyond the computer-controlled system because it also distributes spark electronically instead of mechanically. The distributor is completely eliminated from these systems.

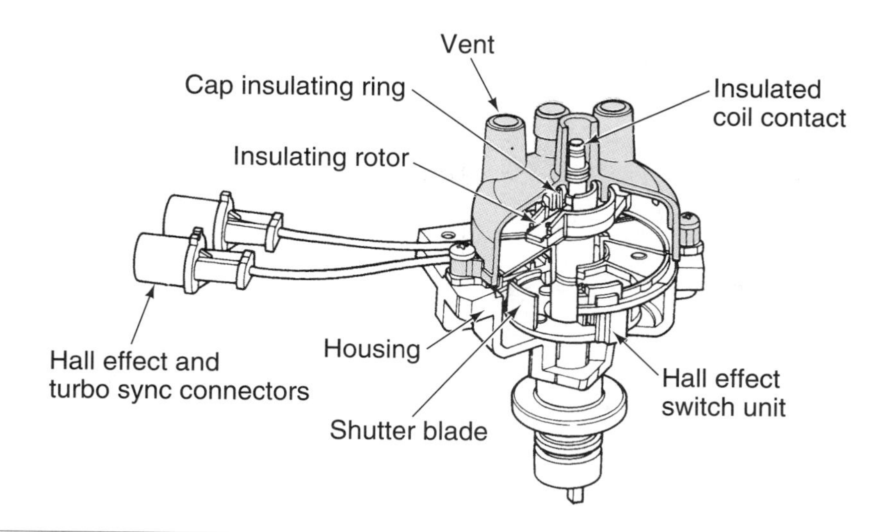

Figure 8-40 A Chrysler distributor with dual Hall-effect switches, both located below the breaker plate. (Courtesy of Chrysler Corporation)

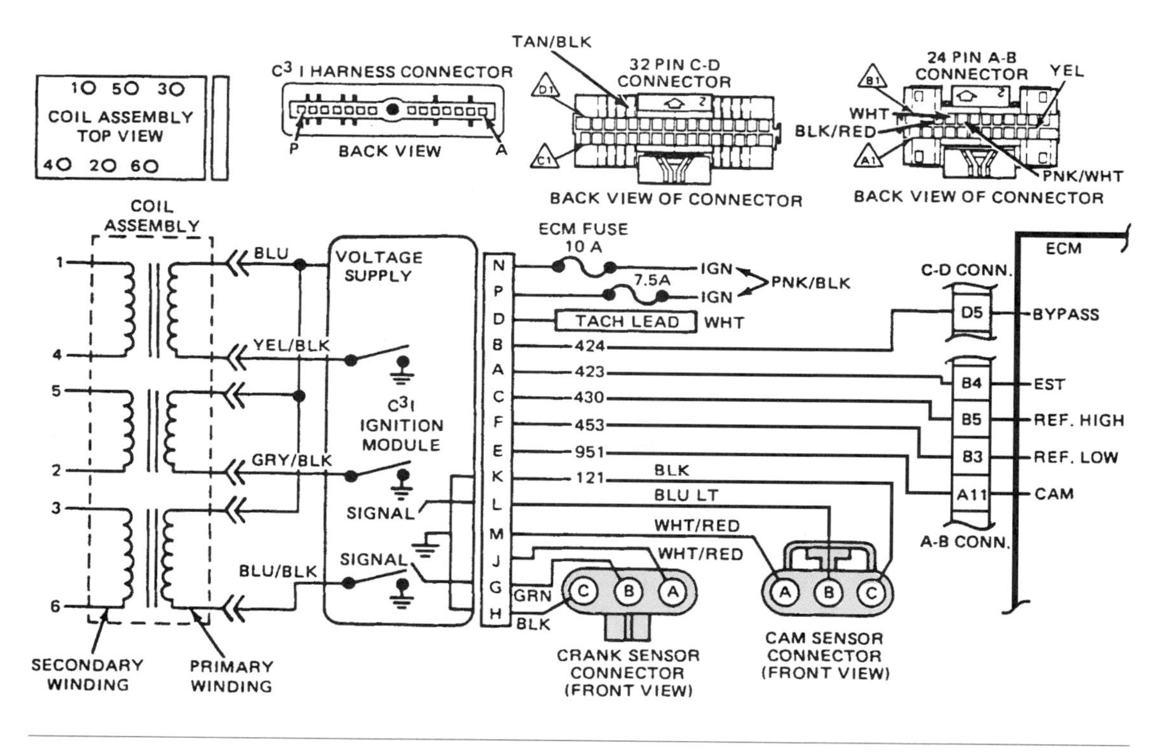

Figure 8-41 A typical electronic ignition (EI) system. (Courtesy of Buick Motor Divison—GMC)

The computer, ignition module, and position sensors combine to control spark timing and advance. The computer collects and processes information to determine the ideal amount of spark advance for the operating conditions. The ignition module uses crank/cam sensor data to control the timing of the primary circuit in the coils. Remember that there is more than one coil in a distributorless ignition system. The ignition module synchronizes the coils' firing sequence in relation to crankshaft position and firing order of the engine. Therefore, the ignition module takes the place of the distributor.

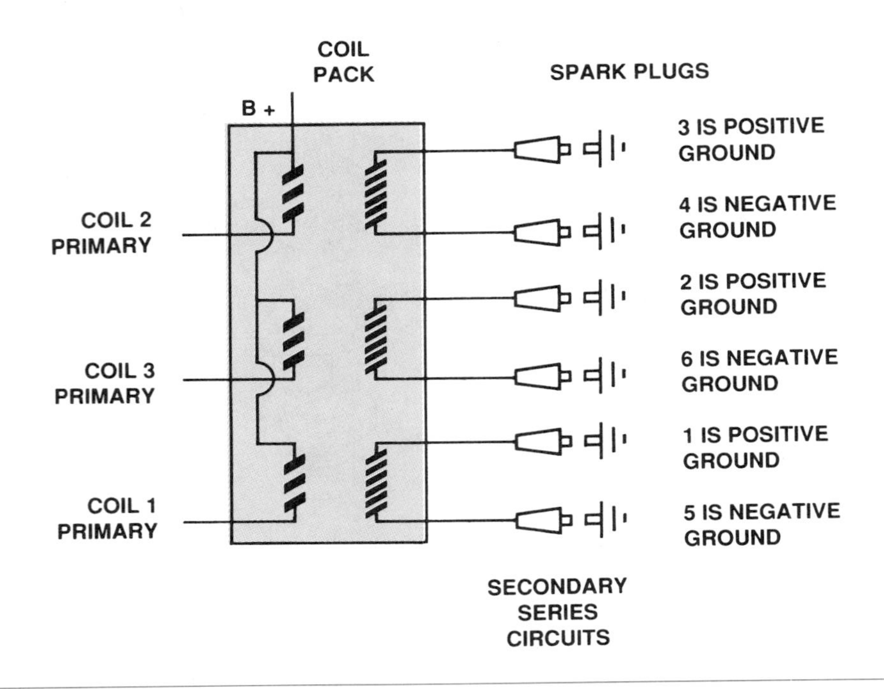

Figure 8-42 Spark plug firing for a six-cylinder engine with EI. (Reprinted with the permission of Ford Motor Company)

The ignition module also adjusts spark timing below 400 rpm (for starting) and when the vehicle's control computer bypass circuit becomes open or grounded. Depending on the exact EI system, the ignition coils can be serviced as a complete unit or separately. The coil assembly is typically called a **coil pack** and consists of two or more individual coils.

On those EI systems that use one coil per spark plug, the electronic ignition module determines when each spark plug should fire and controls the on/off time of each plug's coil.

The systems with a coil for every two spark plugs also use an electronic ignition module, but they use the **waste spark** method of spark distribution. Each end of the coil's secondary winding is attached to a spark plug. Each coil is connected to a pair of spark plugs in cylinders whose pistons rise and fall together. When the field collapses in the coil, voltage is sent to both spark plugs that are attached to the coil. In all V-6s, the paired cylinders are 1 and 4, 2 and 5, and 3 and 6 (or 4 and 1 and 3 and 2 on 4-cylinder engines). With this arrangement, one cylinder of each pair is on its compression stroke while the other is on the exhaust stroke. Both cylinders get spark simultaneously, but only one spark generates power while the other is wasted out the exhaust. During the next revolution, the roles are reversed.

Due to the way the secondary coils are wired, when the induced voltage cuts across the primary and secondary windings of the coil, one plug fires in the normal direction—positive center electrode to negative side electrode—and the other plug fires just the reverse—side to center electrode (Figure 8-42). As shown in Figure 8-43, both plugs fire simultaneously, completing the series circuit. Each plug always fires the same way on both the exhaust and compression strokes.

The coil is able to overcome the increased voltage requirements caused by reversed polarity and still fire two plugs simultaneously because each coil is capable of producing up to 100,000 volts. There is very little resistance across the plug gap on exhaust, so the plug requires very little voltage to fire, thereby providing its mate (the plug that is on compression) with plenty of available voltage.

Figure 8-44 shows a waste spark system in which the coils are mounted directly over the spark plugs so that no wiring between the coils and plugs is necessary. This type system operates

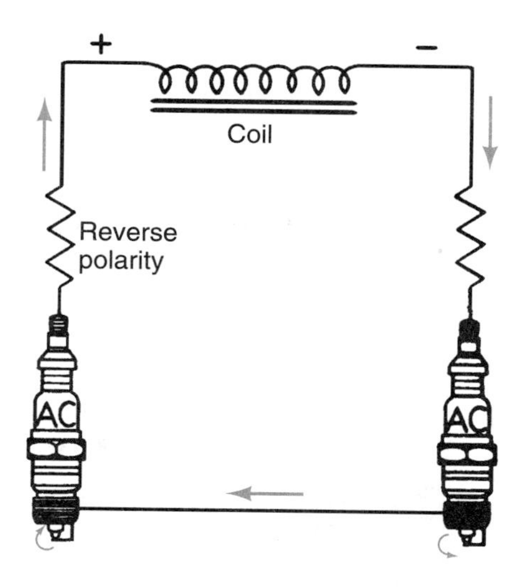

Figure 8-43 Complete circuit for spark plug firing in an EI ignition system.

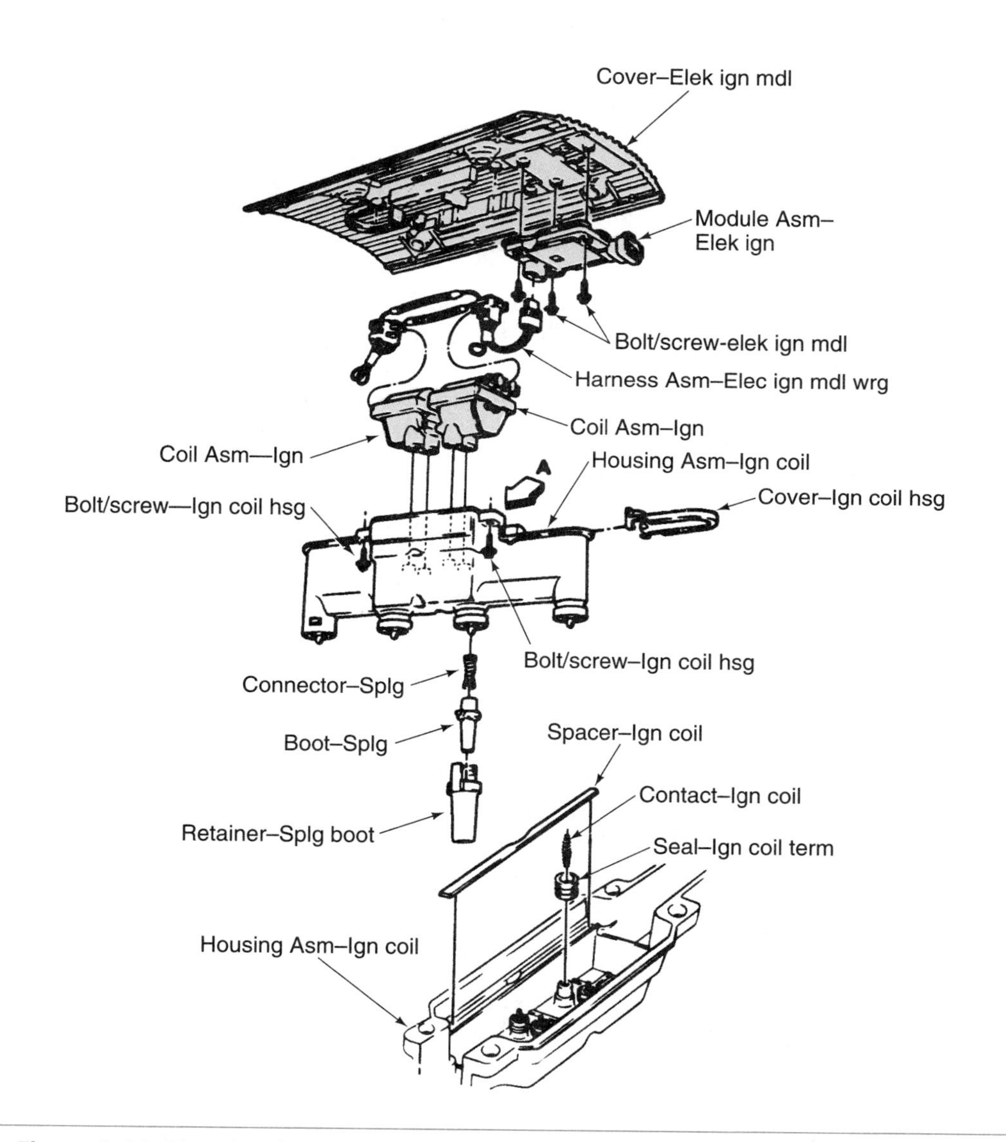

Figure 8-44 GM's Quad 4 with the ignition coils mounted directly over the spark plugs. (Courtesy of Oldsmobile Division—GMC)

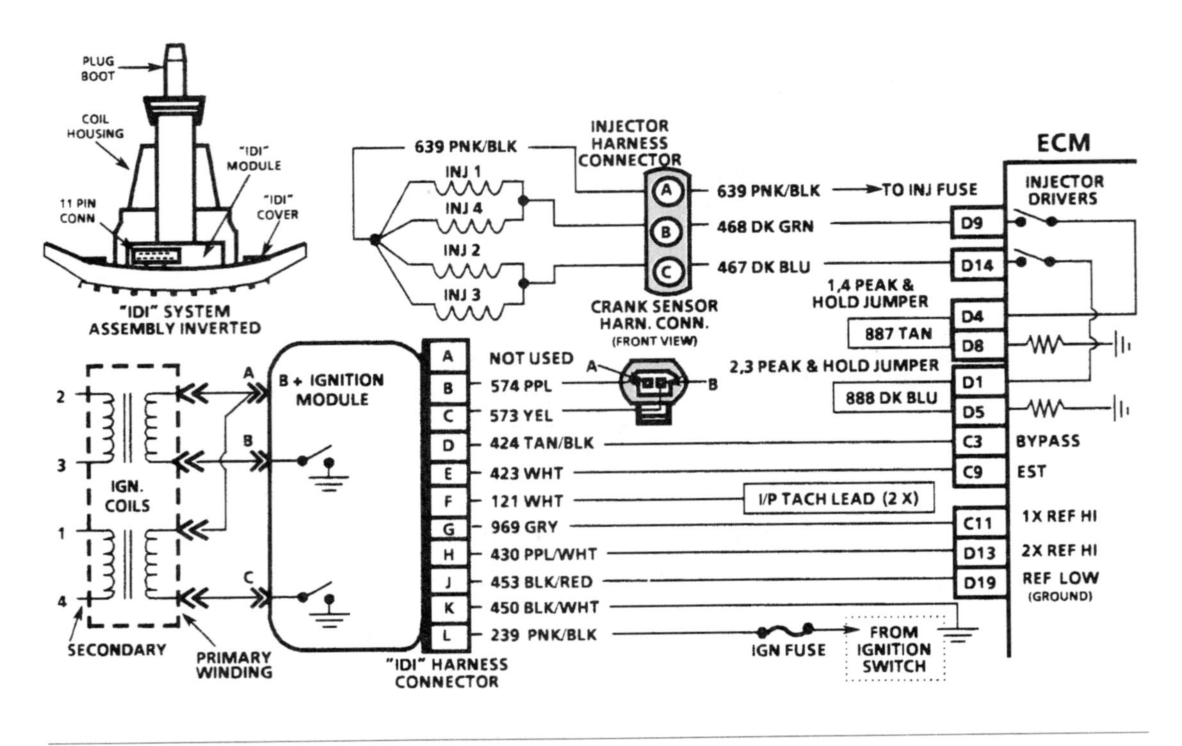

Figure 8-45 GM's Quad 4 ignition circuit. (Courtesy of Oldsmobile Division—GMC)

in the same way as other EI systems (Figure 8-45). On other systems, the coil packs are mounted remote from the spark plugs. High-tension secondary wires carry high-voltage current from the coils to the plugs (Figures 8-46 and 8-47).

A few EI systems have one coil per cylinder with two spark plugs per cylinder. During starting only one plug is fired. Once the engine is running the other plug also fires. One spark plug is located on the intake side of the combustion chamber, while the other is located on the exhaust

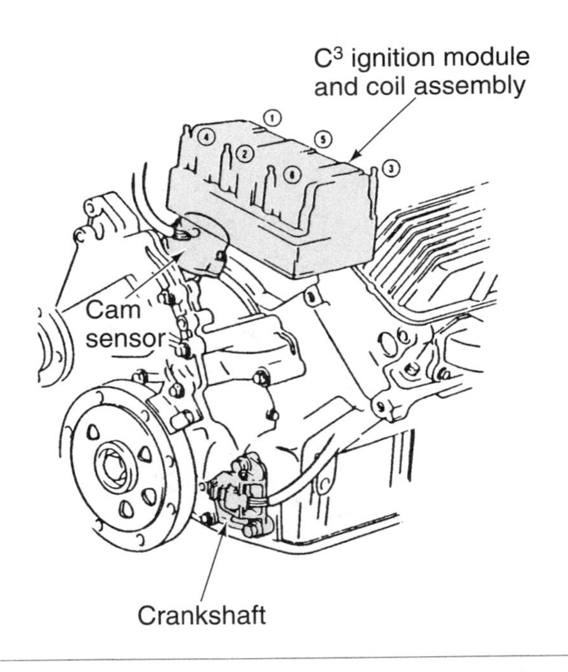

Figure 8-46 A coil pack. (Courtesy of Oldsmobile Division—GMC)

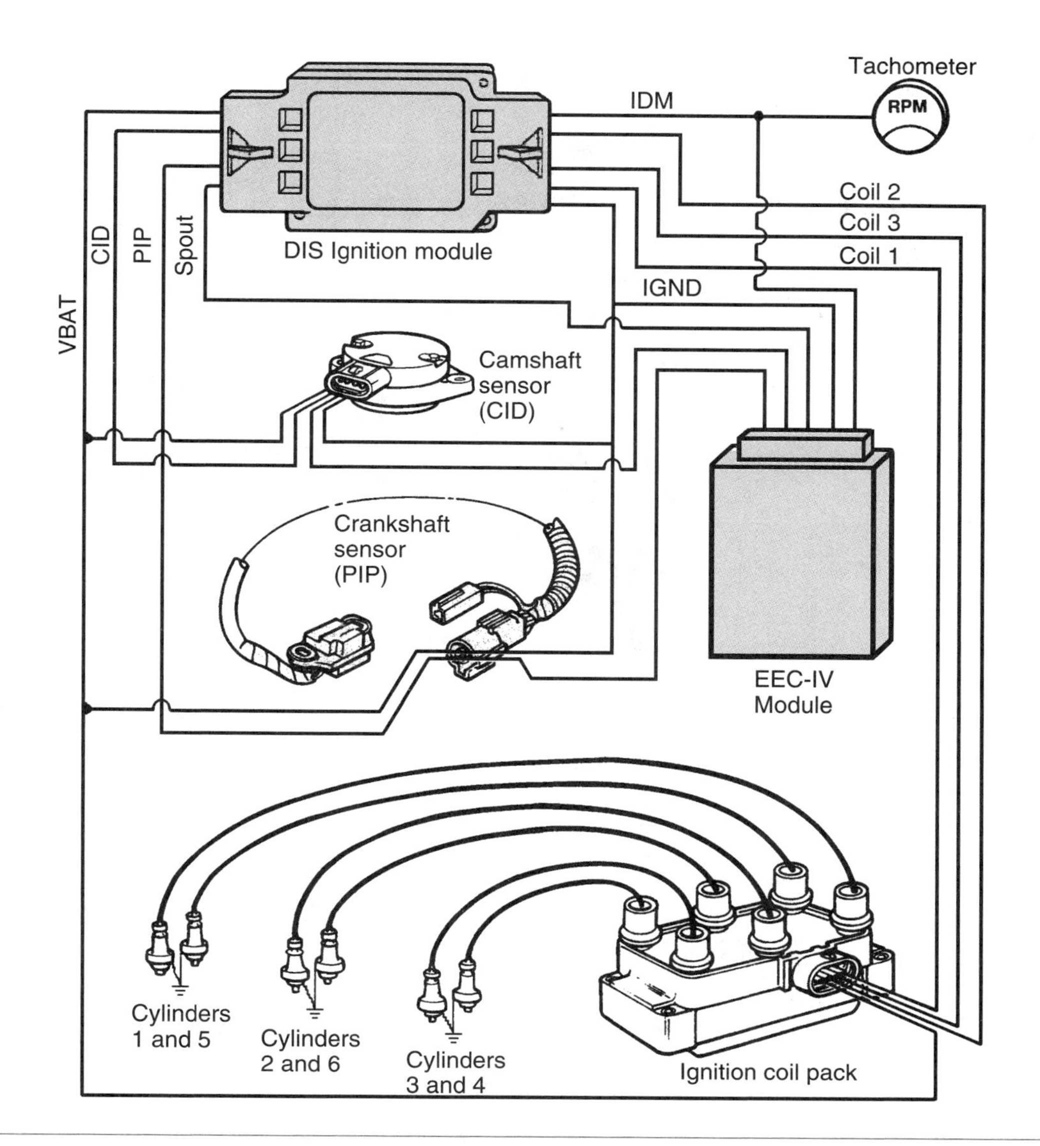

Figure 8-47 A coil pack and the wiring to and from it. (Reprinted with the permission of Ford Motor Company)

side. Two coil packs are used, one for the plugs on the intake side and the other for the plugs on the exhaust side. These systems are called **dual plug** systems (Figure 8-48). During dual plug operation, the two coil packs are synchronized so that each cylinder's two plugs fire at the same time. The coils fire two spark plugs at the same time. Therefore on a four-cylinder engine, four spark plugs are fired at a time, two during the compression stroke of the cylinder and two during the exhaust stroke of another cylinder.

Timing References

From a general operating standpoint, most distributorless ignition systems are similar. However, there are variations in the way different distributorless systems obtain a timing reference in regard to crankshaft and camshaft position.

Some engines use separate Hall-effect sensors to monitor crankshaft (Figure 8-49) and camshaft position for the control of ignition and fuel injection firing orders. The crankshaft pulley

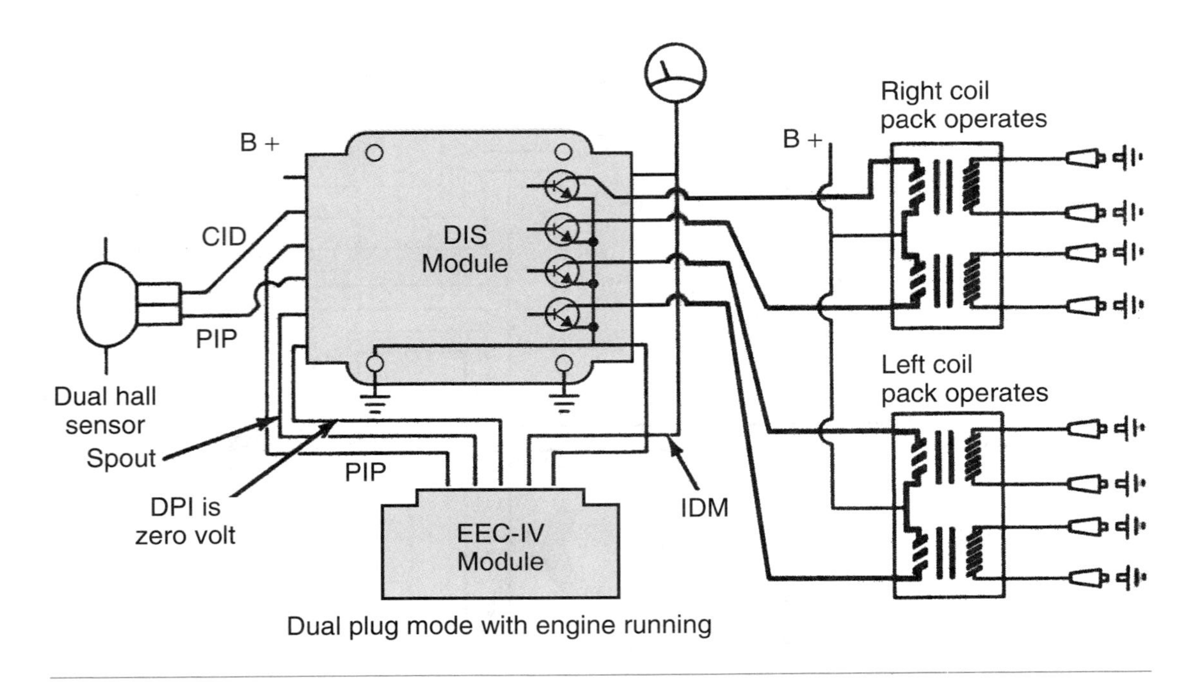

Figure 8-48 A dual plug EI system for a 4-cylinder engine. (Reprinted with the permission of Ford Motor Company)

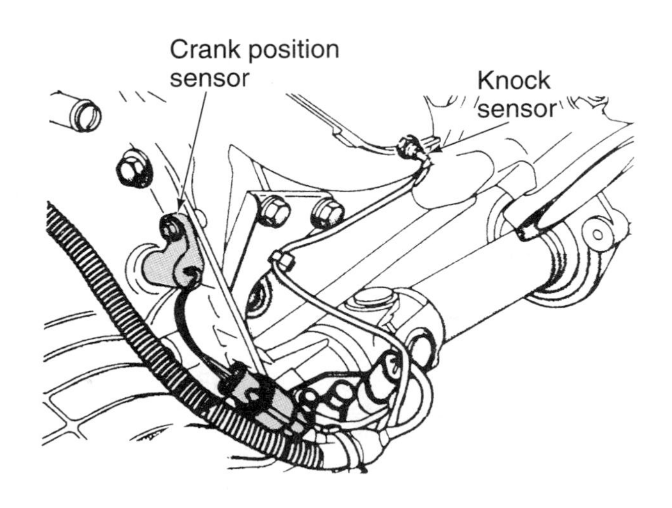

Figure 8-49 A crank position sensor. (Courtesy of Chrysler Corporation)

has interrupter rings that are equal in number to half of the cylinders of the engine (Figure 8-50). The resultant signal informs the PCM as to when to fire the plugs. The camshaft sensor helps the computer determine when the number one piston is at TDC on the compression stroke.

Other systems use a dual Hall-effect sensor. One sensor generates three signals per crankshaft rotation, at 120 degree intervals. The other sensor generates one signal per revolution, which tells the computer when the number one cylinder is on TDC. From these signals the computer can calculate the position of the camshaft, as well as know the position of the crankshaft.

Defining the different types of EI systems used by manufacturers focuses on the location and type of sensors used. There are other differences, such as the construction of the coil pack, wherein some are a sealed assembly and others have individually mounted ignition coils. Some EI systems have a camshaft sensor mounted in the opening where the distributor was mounted.

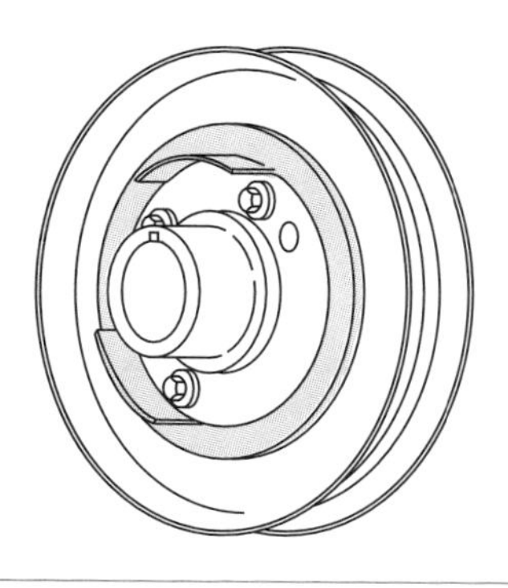

Figure 8-50 A crankshaft pulley with three blades. (Courtesy of Buick Motor Division—GMC)

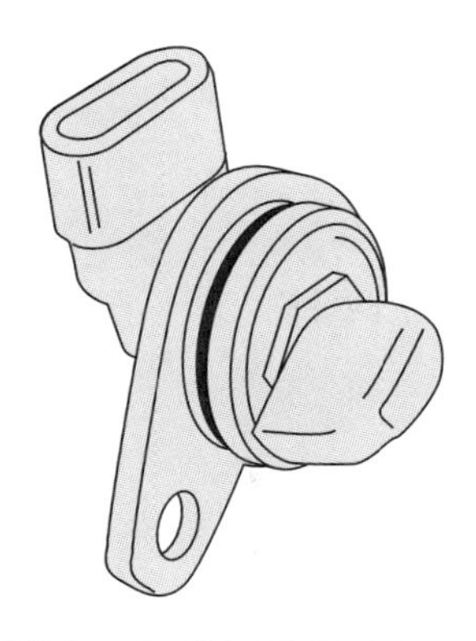

Figure 8-51 A camshaft position sensor. (Courtesy of Buick Motor Division—GMC)

The camshaft sensor ring has one notch and produces a leading edge and trailing edge signal once per camshaft revolution. These systems also use a crankshaft sensor. Both the camshaft and crankshaft sensors are Hall-effect sensors.

Some systems have the camshaft sensor mounted in the front of the timing chain cover (Figure 8-51). A magnet on the camshaft gear rotates past the inner end of the camshaft sensor and produces a signal for each camshaft revolution.

Other systems use a dual crankshaft sensor located behind the crankshaft pulley. When this type of sensor is used, there are two interrupter rings on the back of the pulley that rotate through the Hall-effect switches at the dual crankshaft sensor (Figure 8-52). The inner ring with three equally spaced blades rotates through the inner Hall-effect switch, whereas the outer ring with one opening rotates through the outer Hall-effect (Figure 8-53).

In this dual sensor, the inner sensor provides three leading edge signals, and the outer sensor produces one leading edge during one complete revolution of the crankshaft (Figure 8-54). The outer sensor is the SYNC sensor.

The examples given so far depend on two revolutions of the crankshaft to inform the PCM as to when the number one cylinder is ready. These systems are referred to as slow-start systems because the engine must crank through two crankshaft revolutions before ignition begins.

While starting the engine, slow-start EI systems may require up to two crankshaft revolutions before the ignition system begins firing the spark plugs.

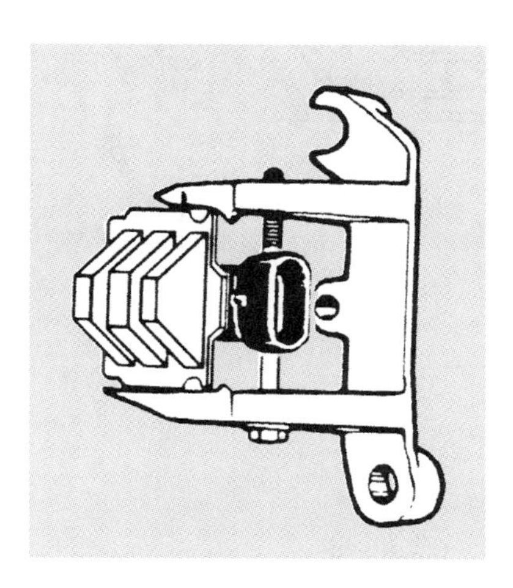

Figure 8-52 A dual crankshaft position sensor. (Courtesy of Oldsmobile Division—GMC)

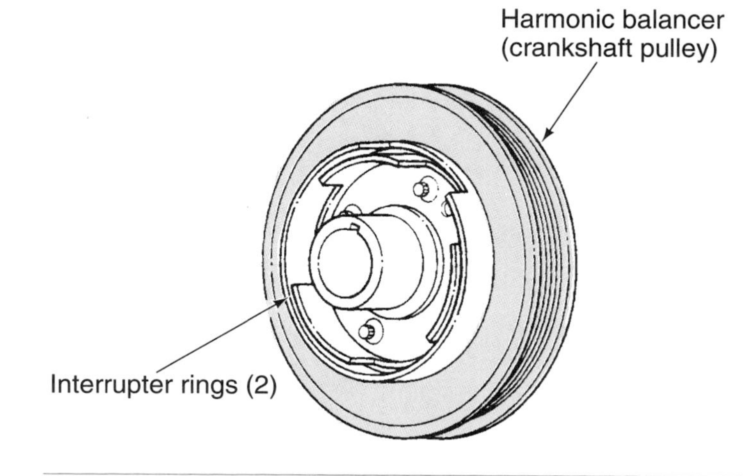

Figure 8-53 A crankshaft pulley fitted with two sets of interrupter rings for a dual sensor. (Courtesy of Oldsmobile Division—GMC)

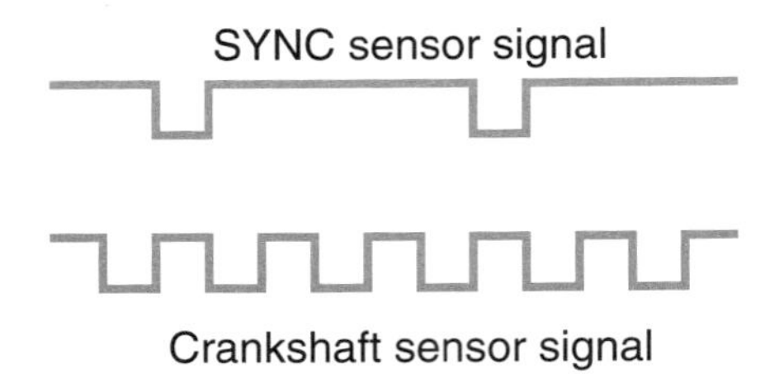

Figure 8-54 The signals generated by a dual crankshaft position sensor. (Courtesy of Oldsmobile Divison—GMC)

While starting the engine, fast-start EI systems start firing the spark plugs in 180 degrees or less of crankshaft rotation.

The **fast-start** electronic ignition system used in GM's Northstar system uses two crankshaft position sensors. A reluctor ring with 24 evenly spaced notches and 8 unevenly spaced notches is cast onto the center of the crankshaft (Figure 8-55).

When the reluctor ring rotates past the magnetic-type sensors, each sensor produces 32 high and low voltage signals per crankshaft revolution. The A sensor is positioned in the upper

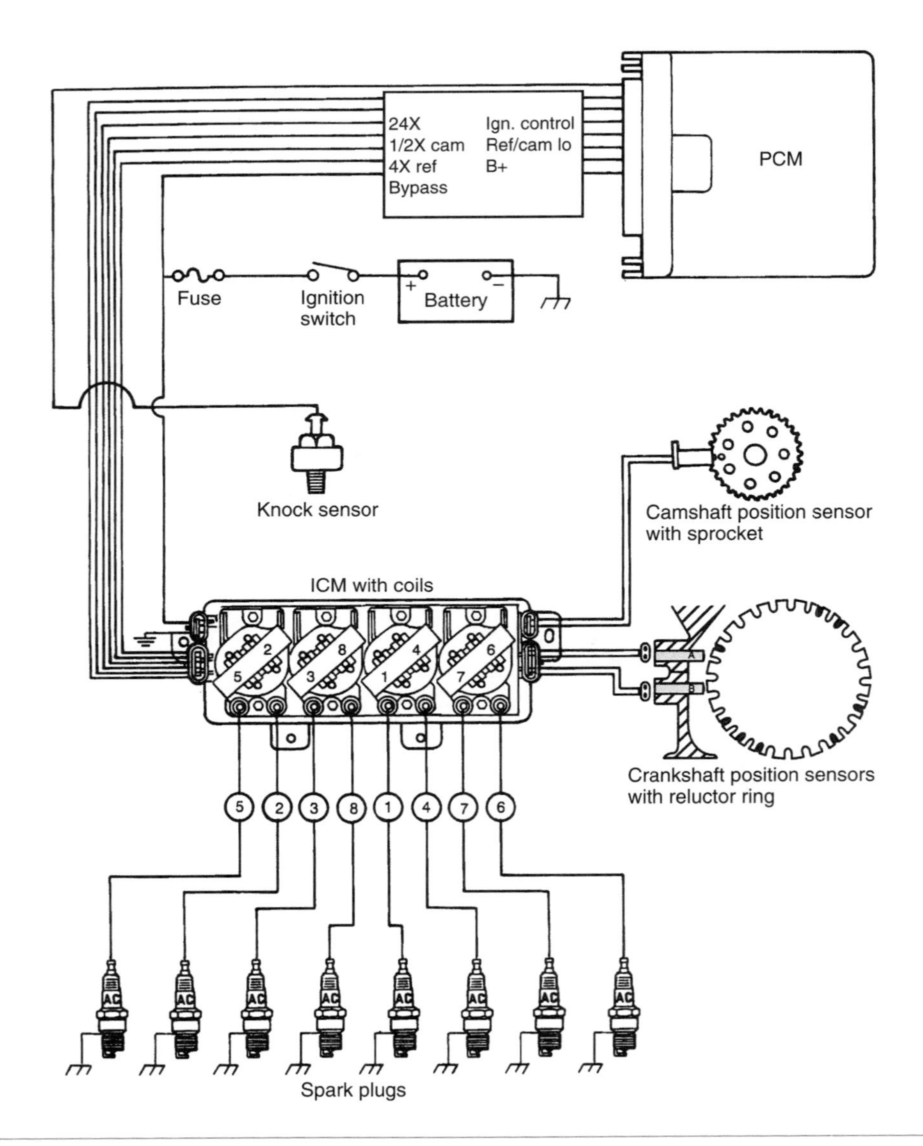

Figure 8-55 GM's Northstar fast-start EI system. (Courtesy of Cadillac Motor Division—GMC)

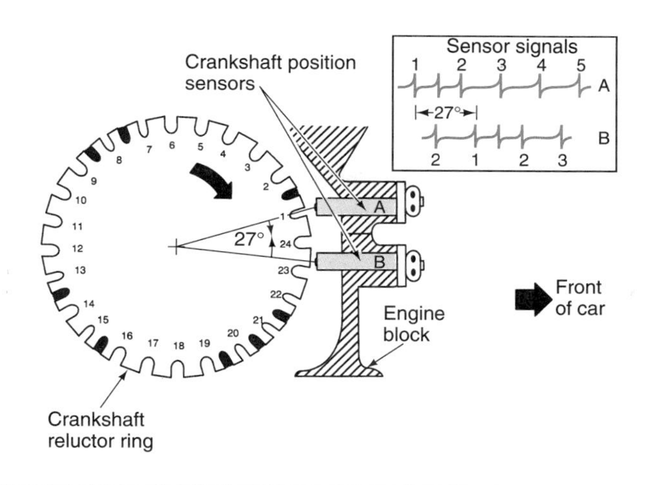

Figure 8-56 GM's Northstar dual position sensors and the crankshaft reluctor. (Courtesy of Cadillac Motor Division—GMC)

crankcase, and the B sensor is positioned in the lower crankcase. Since the A sensor is above the B sensor, the signal from the A sensor occurs 27 degrees before the B sensor signal (Figure 8-56).

The signals from the two sensors are sent to the ignition control module. This module counts the number of B sensor signals between the A sensor signals to sequence the ignition coils properly. There can be zero, one, or two B sensor signals between the A signals. When starting the engine, the module begins counting B sensor signals between the A signals as soon as the module senses zero B signals between sensor signals. After the module senses four B signals, the module sequences the coils properly. This allows the ignition system to begin firing the spark plugs within 180 degrees of crankshaft rotation while starting the engine. This system allows for much quicker starting than other EI systems that require the crankshaft to rotate one or two times before the coils are sequenced.

Finally, some engines use a magnetic pulse generator. The timing wheel is cast on the crankshaft and has machined slots on it. If the engine is a six-cylinder, there will be seven slots, six of which are spaced exactly 60 degrees apart and the seventh notch is located 10 degrees from the number six notch and is used to synchronize the coil firing sequence in relation to crankshaft position (Figure 8-57). The same triggering wheel can be and is used on four-cylinder engines. The computer only needs to be programmed to interpret the signals differently than on a six-cylinder.

The magnetic sensor, which protrudes into the side of the block to within 0.050 in. (±0.020 in.) of the crankshaft reluctor, generates a small AC voltage each time one of the machined slots passes by.

By counting the time between pulses, the ignition module picks out the unevenly spaced seventh slot, which starts the calculation of the ignition coil sequencing. Once its counting is synchronized with the crankshaft, the module is programmed to accept the AC voltage signals of the select notches for firing purposes.

When the system is working properly, there is no base timing to adjust and there are no moving parts to wear.

The development and spreading popularity of EI is the result of reduced emissions, improved fuel economy, and increased component reliability brought about by these systems.

EI offers advantages in production costs and maintenance considerations. By removing the distributor, the manufacturers realize a substantial savings in ignition parts and related machining costs. By eliminating the distributor, they also do away with cracked caps, eroded carbon buttons, burned-through rotors, moisture misfiring, base timing adjustments, and the like.

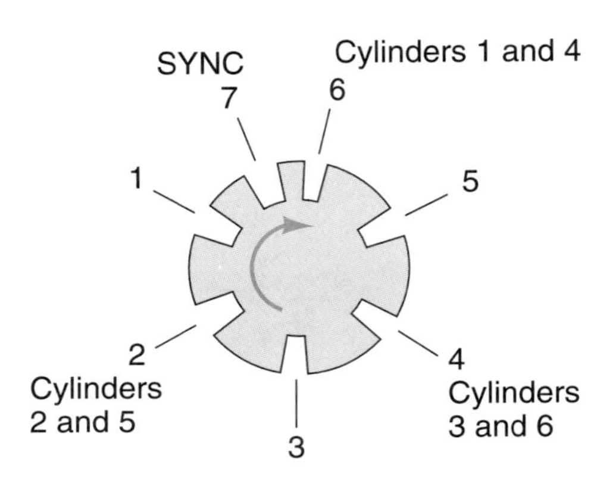

Figure 8-57 The timing wheel for a 6-cylinder engine equipped with a magnetic pulse generator. The seventh slot is the synchronizing slot.

Summary

Terms to Know

ATDC
Ballast resistor
Barometric pressure switch
Basic timing
Breaker plate
Breaker point
BTDC
Centrifugal advance
Contact points
Control module
DIS
Distributor
Fast start
Firing order
Heat range
Ignition timing
Inductive reluctance
Look-up tables
Photoelectric sensor
Primary circuit
Pulse transformer
Rotor
Secondary circuit
Waste spark

- The ignition system supplies high voltage to the spark plugs to ignite the air/fuel mixture in the combustion chambers.
- The arrival of the spark is timed to coincide with the compression stroke of the piston. This basic timing can be advanced or retarded under certain conditions, such as high engine rpm or extremely light or heavy engine loads.
- The ignition system has two interconnected electrical circuits: a primary circuit and a secondary circuit.
- The primary circuit supplies low voltage to the primary winding of the ignition coil. This creates a magnetic field in the coil.
- A switching device interrupts primary current flow, collapsing the magnetic field, and creating a high-voltage surge in the ignition coil secondary winding.
- The switching device used in electronic systems is an NPN transistor. Old ignitions use mechanical breaker point switching.
- The secondary circuit carries high-voltage surges to the spark plugs. On some systems, the circuit runs from the ignition coil, through a distributor, to the spark plugs.
- The distributor may house the switching device plus centrifugal or vacuum timing advance mechanisms. Some systems locate the switching device outside the distributor housing.
- Ignition timing is directly related to the position of the crankshaft. Magnetic pulse generators and Hall-effect sensors are the most widely used engine position sensors. They generate an electrical signal at certain times during crankshaft rotation. This signal triggers the electronic switching device to control ignition timing.
- Direct ignition systems eliminate the distributor. Each spark plug, or in some cases, pair of spark plugs, has its own ignition coil. Primary circuit switching and timing control are done using a special ignition module tied into the vehicle control computer.
- Computer-controlled ignition eliminates centrifugal and vacuum timing mechanisms. The computer receives input from numerous sensors. Based on this data, the computer determines the optimum firing time and signals an ignition module to activate the secondary circuit at the precise time needed.

Review Questions

Short Answer Essay

1. Under what condition is the ballast resistor in an ignition system's primary circuit bypassed?
2. Under light loads, what must be done to complete air/fuel combustion in the combustion chamber by the time the piston reaches 10 degrees ATDC?
3. At high engine rpm, what must be done to complete air/fuel combustion in the combustion chamber by the time the piston reaches 10 degrees ATDC?
4. How do EI ignition systems differ from conventional electronic ignition systems?
5. Explain the components and operation of a magnetic pulse generator.
6. Name the engine operating conditions that affect ignition timing requirements.
7. Name the three major functions of an ignition system.
8. Explain why a distributorless ignition system has more than one ignition coil.
9. What is the basic difference between the primary and secondary ignition circuits?
10. What primary role does a rotor have in the ignition system?

Fill-in-the Blanks

1. Modern ignition cables contain fiber cores that act as a ________________ in the secondary circuit to cut down on radio and television interference and reduce spark plug wear.
2. The magnetic field surrounding the pickup coil in a magnetic pulse generator moves when the ________________ .
3. The arrival of the spark is timed to coincide with the ________________ stroke of the piston.
4. Basic ignition timing is typically ________________ with increases in engine speed and ________________ with increases of engine load.
5. The ignition system has two interconnected electrical circuits: a ________________ circuit and a ________________ circuit.
6. A ________________ device interrupts primary current flow, collapsing the magnetic field, and creating a high-voltage surge in the ignition coil secondary winding.
7. The switching device used in electronic systems is a(n) ________________ .
8. ________________ ________________ ________________ and ________________ - ________________ sensors are the most widely used engine position sensors.

9. With EI, primary circuit switching and timing control is done using a special ignition ____________________ tied into the vehicle control computer.

10. Computer-controlled ignition systems rely on the inputs from various ____________________ to control ignition timing.

ASE Style Review Questions

1. *Technician A* says EI systems typically use one coil for two spark plugs.
   Technician B says some EI systems rely on a waste spark system to fire the spark plug.
   Who is correct?
   A. A only
   B. B only
   C. Both A and B
   D. Neither A nor B

2. *Technician A* says the switching transistor in an EI ignition system turns off current flow to the ignition coil primary winding whenever the pickup coil is generating a voltage signal.
   Technician B says the transistor turns off current flow when the pickup coil is not generating a voltage signal.
   Who is correct?
   A. A only
   B. B only
   C. Both A and B
   D. Neither A nor B

3. *Technician A* says ignition systems equipped with Hall-effect sensors do not require a ballast resistor or resistance wire to regulate primary circuit current.
   Technician B says these systems do require resistance control in their primary circuit.
   Who is correct?
   A. A only
   B. B only
   C. Both A and B
   D. Neither A nor B

4. In EI systems using one ignition coil for every two cylinders,
   Technician A says two plugs fire at the same time, with one wasting the spark on the exhaust stroke.
   Technician B says one plug fires in the normal direction (center to side electrode) and the other in reversed polarity (side to center electrode).
   Who is correct?
   A. A only
   B. B only
   C. Both A and B
   D. Neither A nor B

5. While discussing what happens when the low-voltage current flow in the coil primary winding is interrupted by the switching device:
   Technician A says the magnetic field collapses.
   Technician B says a high-voltage surge is induced in the coil secondary winding.
   Who is correct?
   A. A only
   B. B only
   C. Both A and B
   D. Neither A nor B

6. *Technician A* says an ignition system must generate sufficient voltage to force a spark across the spark plug gap.
   Technician B says the ignition system must time the arrival of the spark to coincide with the movement of the engine's pistons and vary it according to the operating conditions of the engine.
   Who is correct?
   A. A only
   B. B only
   C. Both A and B
   D. Neither A nor B

7. While discussing electronic ignition systems:
Technician A says a transistor actually controls primary current flow through the coil.
Technician B says a reluctor controls the primary coil current.
Who is correct?
A. A only
B. B only
C. Both A and B
D. Neither A nor B

8. *Technician A* says a magnetic pulse generator is equipped with a permanent magnet.
Technician B says a Hall-effect switch is equipped with a permanent magnet.
Who is correct?
A. A only
B. B only
C. Both A and B
D. Neither A nor B

9. While discussing PM generators:
Technician A says the pickup coil does not produce a voltage signal when a reluctor tooth approaches the coil.
Technician B says the pickup coil does not produce a voltage signal when a reluctor tooth moves away from the coil.
Who is correct?
A. A only
B. B only
C. Both A and B
D. Neither A nor B

10. While discussing engine position sensors:
Technician A says a metal detection sensor needs to have its voltage signal amplified, inverted, and shaped into a clean square wave.
Technician B says a Hall-effect sensor needs to have its voltage signal amplified, inverted, and shaped into a clean square wave signal.
Who is correct?
A. A only
B. B only
C. Both A and B
D. Neither A nor B

Lighting Circuits

Upon completion and review of this chapter, you should be able to:

- ❑ Describe the operation and construction of automotive lamps.
- ❑ Describe the differences between conventional sealed-beam, halogen, and composite headlight lamps.
- ❑ Describe the operation and controlled circuits of the headlight switch.
- ❑ Describe the operation of the dimmer switch.
- ❑ Explain the operation of the most common styles of concealed headlight systems.
- ❑ Describe the operation of the various exterior light systems, including parking, tail, brake, turn, side, clearance, and hazard warning lights.
- ❑ Explain the operating principles of the turn signal and hazard light flashers.
- ❑ Describe the operation of the various interior light systems, including courtesy and instrument panel lights.

Introduction

Today's technician is required to understand the operation and purpose of the various lighting circuits on the vehicle. The addition of computers and their many sensors and actuators (some that interlink to the lighting circuits) make it impossible for mechanics to just bypass part of the circuit and rewire the system to their own standards. If a lighting circuit is not operating properly the safety of the driver, the passengers, people in other vehicles, and pedestrians are in jeopardy. When today's technician performs repairs on the lighting systems, the repairs must meet at least two requirements: They must assure vehicle safety and meet all applicable laws.

A **lamp** is a device that produces light as a result of current flow through a filament. The filament is enclosed within a glass envelope and is a type of resistance wire that is generally made from tungsten (Figure 9-1).

The lighting circuits of today's vehicles can consist of more than 50 light bulbs and hundreds of feet of wiring. Incorporated within these circuits are circuit protectors, relays, switches, lamps, and connectors. In addition, more sophisticated lighting systems use computers and sensors. The lighting circuits consist of an array of interior and exterior lights, including headlights, taillights, parking lights, stoplights, marker lights, dash instrument lights, courtesy lights, and so on.

The process of changing energy forms to produce light is called **incandescence.**

The lighting circuits are largely regulated by federal laws, so the systems are similar between the various manufacturers. However, there are variations. Before attempting to do any repairs on an unfamiliar circuit, the technician should always refer to the manufacturer's service manuals. This chapter provides information about the types of lamps used, describes the headlight circuit, discusses different types of concealed headlight systems, and explores the various exterior and interior light circuits individually.

Many lamps are designed to execute more than one function. A **double filament lamp** (Figure 9-2) can be used in the stop light circuit, taillight circuit, and the turn signal circuit combined.

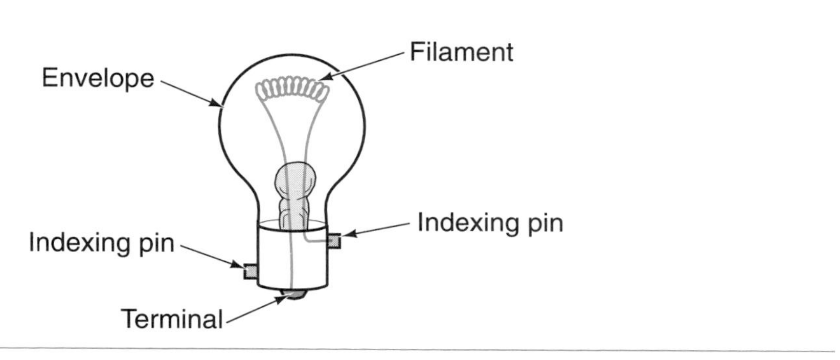

Figure 9-1 A single filament bulb.

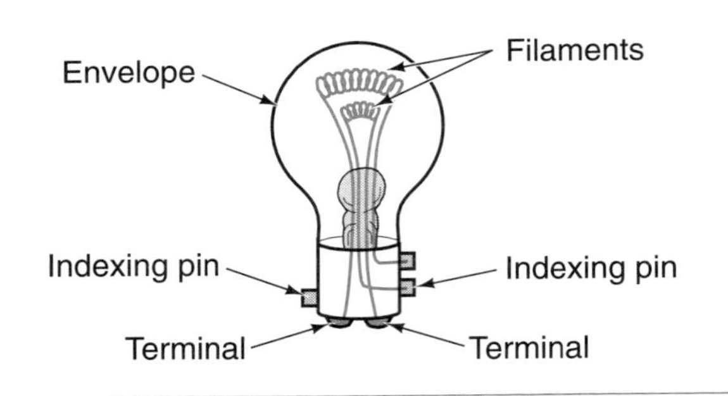

Figure 9-2 A double filament bulb.

Lamps

A lamp generates light as current flows through the filament. This causes it to get very hot. The changing of electrical energy to heat energy in the resistive wire filament is so intense that the filament starts to glow and emits light. The lamp must have a vacuum surrounding the filament to prevent it from burning so hot that the filament burns in two. The glass envelope that encloses the filament maintains the presence of vacuum. When the lamp is manufactured, all the air is removed and the glass envelope seals out the air. If air is allowed to enter the lamp, the oxygen would cause the filament to oxidize and burn up.

It is important that any burned-out lamp be replaced with the correct lamp. The technician can determine what lamp to use by checking the lamp's standard trade number (Figures 9-3).

TYPICAL AUTOMOTIVE LIGHT BULBS

Trade Number	Design Volts	Design Amperes	Watts: P = I x E
37	14.0	0.09	1.3
37E	14.0	0.09	1.3
51	7.5	0.22	1.7
53	14.4	0.12	1.7
55	7.0	0.41	2.9
57	14.0	0.24	3.4
57X	14.0	0.24	3.4
63	7.0	0.63	4.4
67	13.5	0.59	8.0
68	13.5	0.59	8.0
70	14.0	0.15	2.1
73	14.0	0.08	1.1
74	14.0	0.10	1.4
81	6.5	1.02	6.6
88	13.0	0.58	7.5
89	13.0	0.58	7.5
90	13.0	0.58	7.5
93	12.8	1.04	13.3
94	12.8	1.04	13.3
158	14.0	0.24	3.4
161	14.0	0.19	2.7
168	14.0	0.35	4.9

Figure 9-3 Chart of typical automotive light bulbs.

TYPICAL AUTOMOTIVE LIGHT BULBS (continued)

Trade Number	Design Volts	Design Amperes	Watts: P = I x E
192	13.0	0.33	4.3
194	14.0	0.27	3.8
194E-1	14.0	0.27	3.8
194NA	14.0	0.27	3.8
209	6.5	1.78	11.6
211-2	12.8	0.97	12.4
212-2	13.5	0.74	10.0
214-2	13.5	0.52	7.0
561	12.8	0.97	12.4
562	13.5	0.74	10.0
563	13.5	0.52	7.0
631	14.0	0.63	8.8
880	12.8	2.10	27.0
881	12.8	2.10	27.0
906	13.0	0.69	9.0
912	12.8	1.00	12.8
1003	12.8	0.94	12.0
1004	12.8	0.94	12.0
1034	12.8	1.80/0.59	23.0/7.6
1073	12.8	1.80	23.0
1076	12.8	1.80	23.0
1129	6.4	2.63	16.8
1133	6.2	3.91	24.2
1141	12.8	1.44	18.4
1142	12.8	1.44	18.4
1154	6.4	2.63/.75	16.8/4.5
1156	12.8	2.10	26.9
1157	12.8	2.10/0.59	26.9/7.6
1157A	12.8	2.10/0.59	26.9/7.6
1157NA	12.8	2.10/0.59	26.9/7.6
1176	12.8	1.34/0.59	17.2/7.6
1195	12.5	3.00	37.5
1196	12.5	3.00	37.5
1445	14.4	0.135	1.9
1816	13.0	0.33	4.3
1889	14.0	0.27	3.8
1891	14.0	0.24	3.4
1892	14.4	0.12	1.7
1893	14.0	0.33	4.6
1895	14.0	0.27	3.8
2057	12.8	2.10/0.48	26.9/6.1
2057NA	12.8	2.10/0.48	26.9/6.1
P25-1	13.5	1.86	25.1
P25-2	13.5	1.86	25.1
R19/5	13.5	0.37	
R19/10	13.5	0.74	
W10/3	13.5	0.25	

Figure 9-3 (continued)

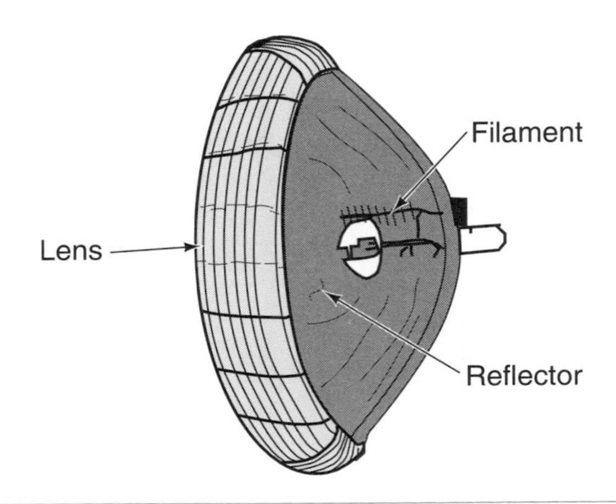

Figure 9-4 Sealed-beam headlight construction.

Headlights

There are three basic types of headlights used on automobiles today: (1) standard sealed beam, (2) halogen sealed beam, and (3) composite.

Sealed-Beam Headlights

Shop Manual
Chapter 9, page 300

The **sealed-beam** headlight is a self-contained glass unit made up of a filament, an inner reflector, and an outer glass lens.

From 1939 to about 1975 the headlights used on vehicles remained virtually unchanged. During this time the headlight was a round lamp. The introduction of the rectangle headlight in 1975 enabled the vehicle manufacturers to lower the hood line of their vehicles. Both the round and rectangle headlights were sealed-beam construction (Figure 9-4). The standard sealed-beam headlight does not surround the filament with its own glass envelope (bulb). The glass lens is fused to the parabolic reflector that is sprayed with vaporized aluminum. The inside of the lamp is filled with argon gas. All oxygen must be removed from the standard sealed-beam headlight to prevent the filament from becoming oxidized. The reflector intensifies the light produced by the filament, and the lens directs the light to form the required light beam pattern.

The vaporized aluminum gives a reflecting surface that is comparable to silver.

The lens is designed to produce a broad flat beam. The light from the reflector is passed through concave prisms in the glass lens (Figure 9-5). The illustration (Figure 9-6) shows the horizontal spreading and the vertical control of the light beam to prevent upward glaring.

Lens prisms redirect the light beam and create a broad, flat beam.

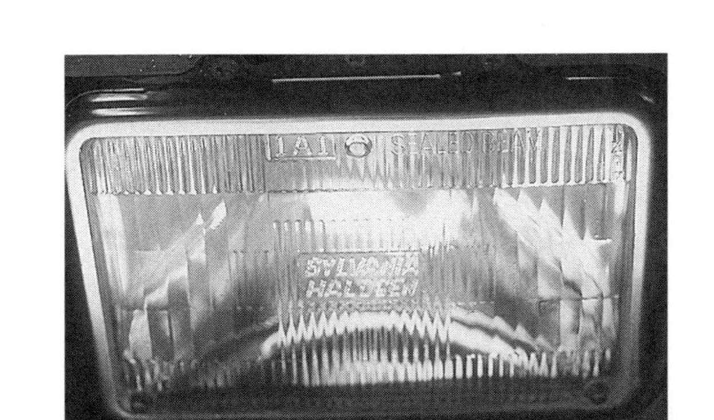

Figure 9-5 The lens uses prisms to redirect the light.

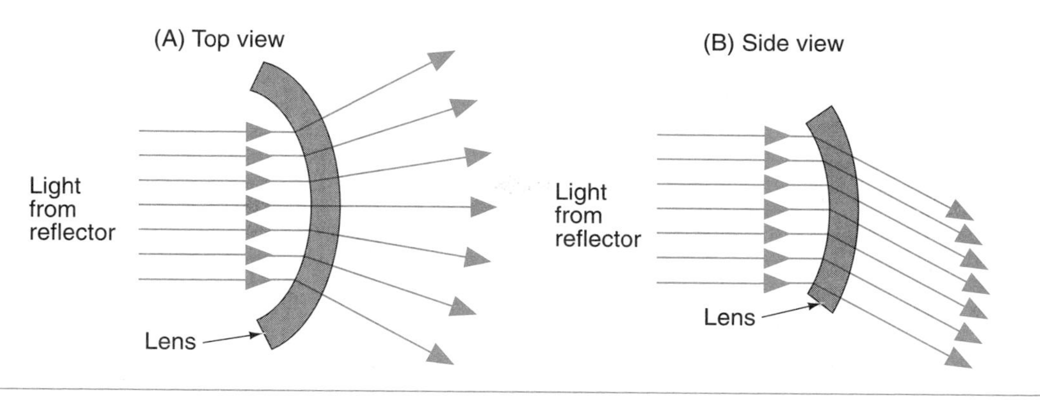

Figure 9-6 The prism directs the beam into (A) a flat horizontal pattern and (B) downward.

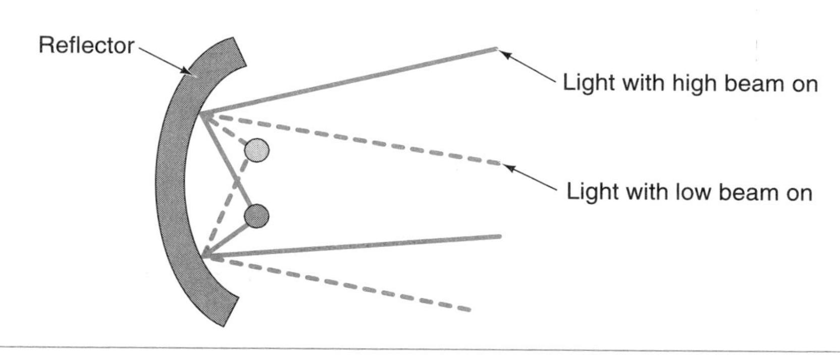

Figure 9-7 Filament placement controls the projection of the light beam.

Halogen is the term used to identify a group of chemically related nonmetallic elements. These elements include chlorine, fluorine, and iodine.

Because the filament is contained in its own bulb, cracking or breaking of the lens does not prevent halogen headlight operation. As long as the filament envelope has not been broken, the filament will continue to operate. However, a broken lens will result in poor light quality and should be replaced.

WARNING: Because of the construction and placement of prisms in the lens, it is important that the technician install the headlight in its proper position. The lens is usually marked "TOP" to indicate the proper installation position.

By placing the filament in different locations on the reflector, the direction of the light beam is controlled (Figure 9-7). In a dual filament lamp, the lower filament is used for the high beam and the upper filament is used for the low beam.

A BIT OF HISTORY

Improved sealed-beam headlamps were introduced in 1955.

Halogen Headlights

The halogen lamp most commonly used in automotive applications consists of a small bulb filled with iodine vapor. The bulb is made of a high temperature resistant glass or plastic that surrounds a tungsten filament. This inner bulb is installed in a sealed glass housing (Figure 9-8). With the halogen added to the bulb, the tungsten filament is capable of withstanding higher temperatures than that of conventional sealed-beam lamps. The halogen lamp can withstand higher temperatures and thus is able to burn brighter.

In a conventional sealed-beam headlight, the heating of the filament causes atoms of tungsten to be released from the surface of the filament. These released atoms deposit on the glass

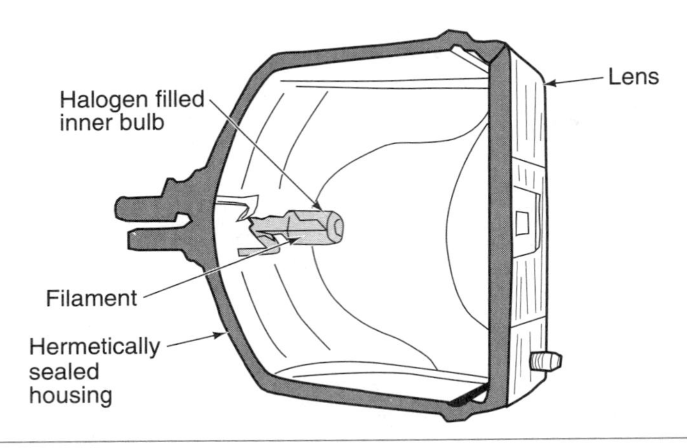

Figure 9-8 A halogen sealed-beam headlight with iodine vapor bulb.

envelope and create black spots that affect the light output of the lamp. In a halogen lamp, the iodine vapor causes the released tungsten atoms to be redeposited back onto the filament. This virtually eliminates any black spots. It also allows for increased high beam output of 25% over conventional lamps.

WARNING: It is not recommended that halogen sealed-beam and standard sealed-beam headlights be mixed on the vehicle. Also, if the vehicle was originally equipped with halogen headlights, do not replace these with standard sealed beams. Doing so may result in poor light quality.

Shop Manual
Chapter 9, page 306

Many of today's vehicles have a halogen headlight system that uses a replaceable bulb. This system is called **composite headlights.**

Composite Headlights

By using the composite headlight system, vehicle manufacturers are able to produce any style of headlight lens they desire (Figure 9-9). This improves the aerodynamics, fuel economy, and styling of the vehicle.

Many manufacturers vent the composite headlight housing because of the increased amount of heat developed by these bulbs. Because the housings are vented, condensation may develop inside of the lens assembly. This condensation is not harmful to the bulb and does not affect headlight operation. When the headlights are turned on, the heat generated from the halogen bulbs

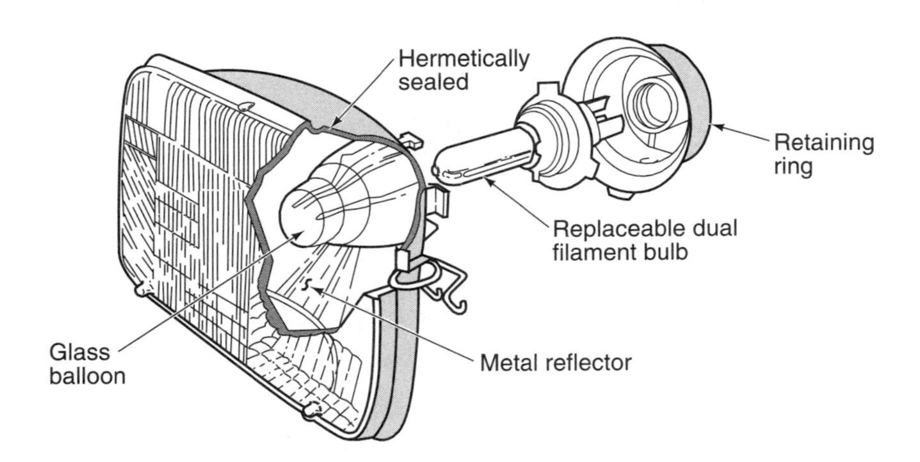

Figure 9-9 A composite headlight system with a replaceable halogen bulb.

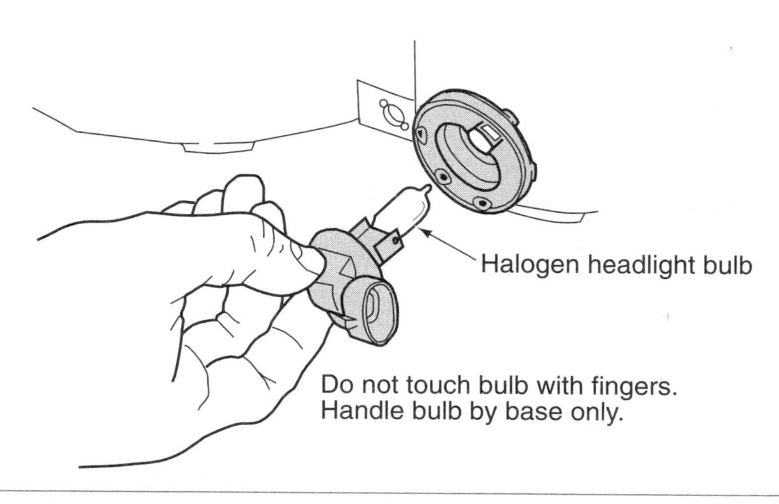

Figure 9-10 The correct method of handling the composite bulb during replacement. (Reprinted with the permission of Ford Motor Company)

will dissipate the condensation quickly. Ford uses integrated nonvented composite headlights. On these vehicles condensation is not considered normal. The assembly should be replaced.

WARNING: Whenever technicians replace a composite lamp, care must be taken not to touch the envelope with the fingers. Staining the bulb with normal skin oil can substantially shorten the life of the bulb. Handle the lamp only by its base (Figure 9-10). Also dispose of the lamp properly.

HID Headlamps

High intensity discharge (HID) headlamps are the latest headlight development. These headlamps (Figure 9-11) put out three times more light and twice the light spread on the road than conventional halogen headlamps. They also use about two-thirds less power to operate and will last two to three times as long. HID lamps produce light in both ultraviolet and visible wavelengths. This advantage allows highway signs and other reflective materials to glow. This type lamp first appeared on select models from BMW in 1993, Ford in 1995, and Porsche in 1996.

Figure 9-11 A Lincoln equipped with HID headlamps; note the reduced size of the headlamp assemblies. (Reprinted with the permission of Ford Motor Company)

These lamps do not rely on a glowing filament for light. Rather light is provided as a high voltage bridges an air gap between two electrodes. The presence of an inert gas amplifies the light given off by the arcing. More than 15,000 volts are used to jump the gap between the electrodes. To provide this voltage, a voltage booster and controller is required. Once the gap is bridged by the high voltage, only about 80 volts is required to keep current flow across the gap.

The great light output of these lamps allows the headlamp assembly to be smaller and lighter. These advantages allow designers more flexibility in body designs as they attempt to make their vehicles more aerodynamic and efficient.

Shop Manual
Chapter 9, page 315

The headlight switch may be located either on the dash by the instrument panel or on the steering column (Figure 9-12).

Headlight Switches

The headlight switch controls most of the vehicle's lighting systems (Figure 9-13). The most common style of headlight switch is the three-position type with OFF, PARK, and HEADLIGHT positions. The headlight switch will generally receive direct battery voltage to two terminals of the switch. This allows the light circuits to be operated without having the ignition switch in the RUN or ACC (accessory) position.

When the headlight switch is in the OFF position, the open contacts prevent battery voltage from continuing to the lamps (Figure 9-14). When the switch is in the PARK position, battery volt-

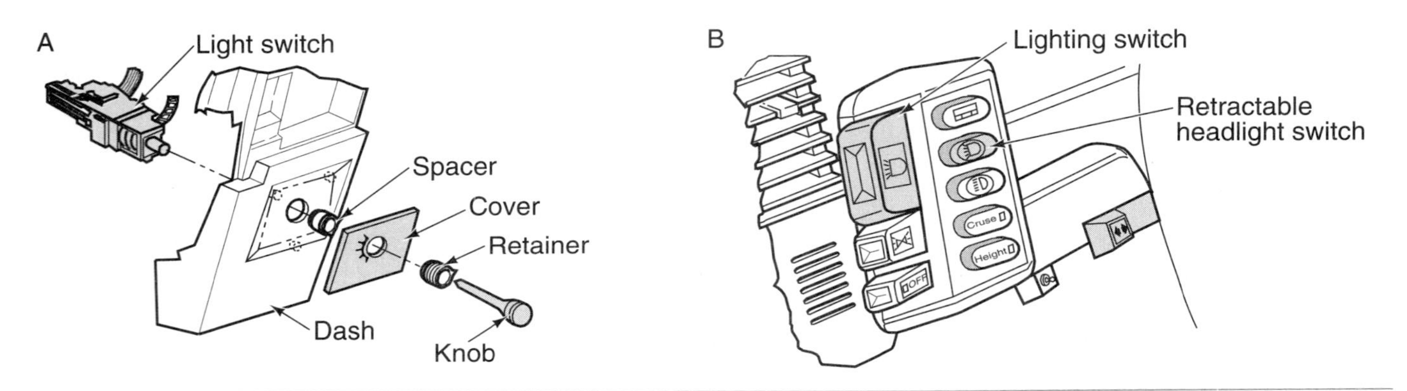

Figure 9-12 (A) Instrument panel-mounted headlight switch. (Reprinted with the permission of Ford Motor Company) (B) Steering column-mounted headlight switch. (Courtesy of Toyota Motor Corporation)

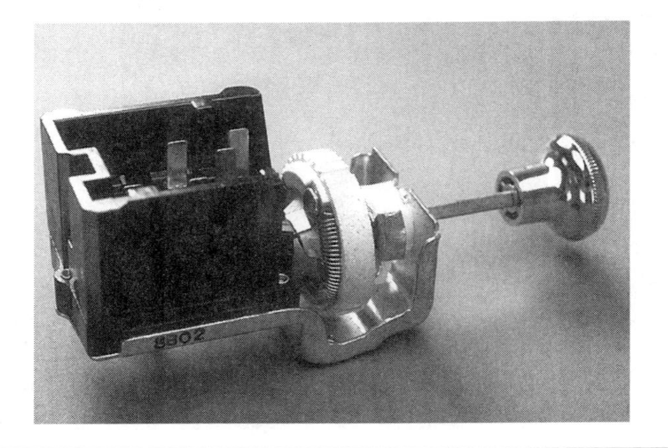

Figure 9-13 Typical dash-mounted headlight switch. Pushing in on the release button allows for removal of the knob.

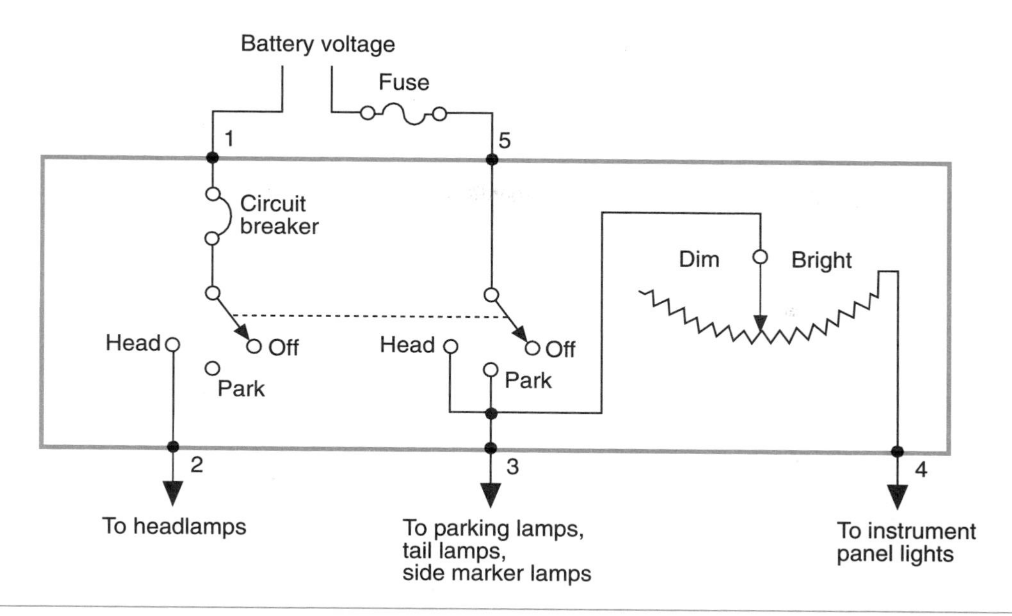

Figure 9-14 A headlight switch in the OFF position.

age that is present at terminal 5 is able to be applied through the closed contacts to the side marker, taillight, license plate, and instrument cluster lights (Figure 9-15). This circuit is usually protected by a 15- to 20-ampere fuse that is separate from the headlight circuit.

A BIT OF HISTORY

The Model T Ford used a headlight system that had a replaceable bulb. The owner's manual warned against touching the reflector except with a soft cloth.

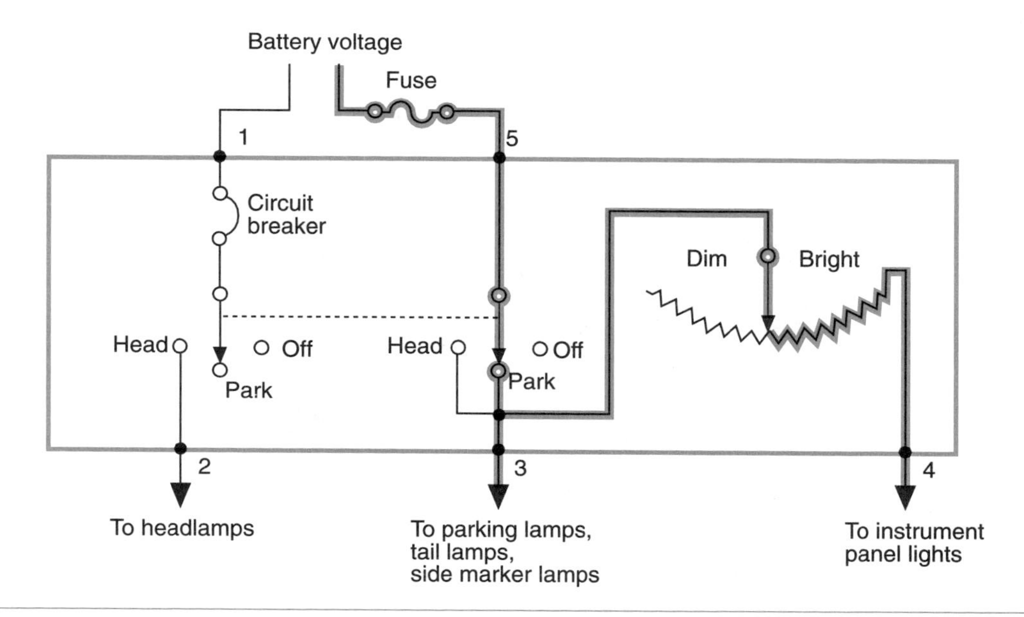

Figure 9-15 A headlight switch in the PARK position.

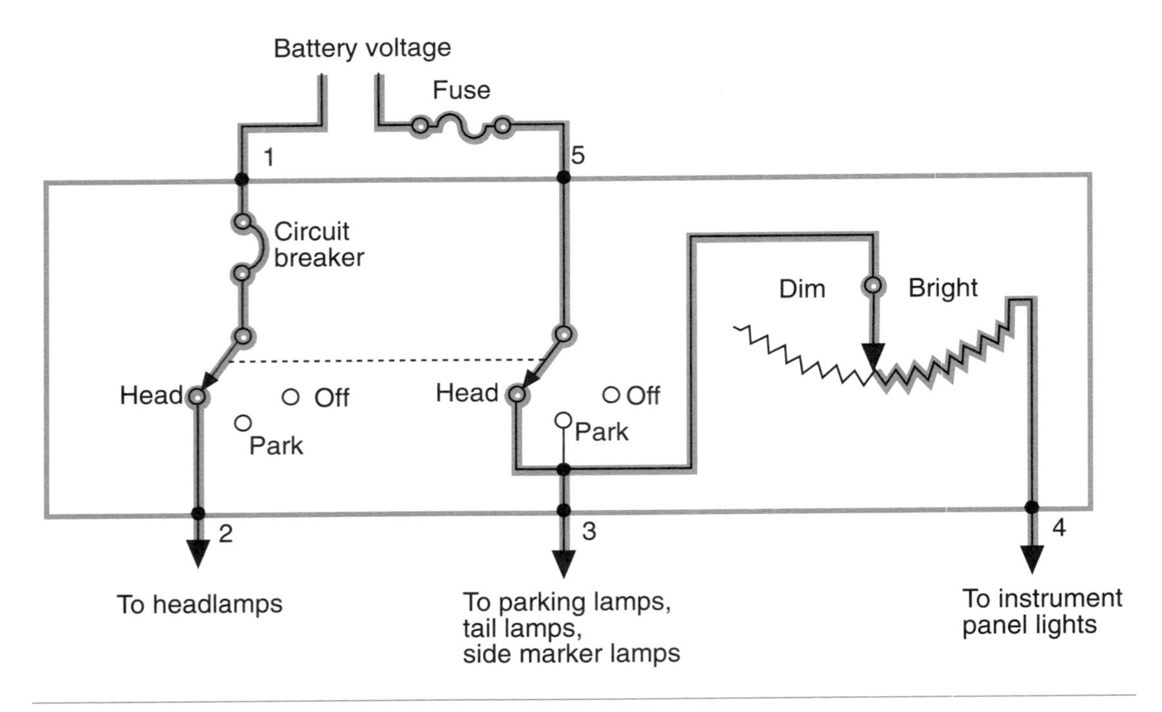

Figure 9-16 A headlight switch in the HEADLIGHT position.

In vehicles that have the headlight switch located in the steering column, the rheostat may be a separate unit located on the dash near the instrument panel (Figure 9-17).

Shop Manual
Chapter 9, page 324

When the switch is located in the HEADLIGHT position, battery voltage that is present at terminal 1 is able to be applied through the circuit breaker and the closed contacts to light the headlights. Battery voltage from terminal 5 continues to light the lights that were on in the PARK position (Figure 9-16). The circuit breaker is used to prevent temporary overloads to the system from totally disabling the headlights.

The rheostat is a variable resistor that the driver uses to control the instrument cluster illumination lamp brightness. As the driver turns the light switch knob, the resistance in the rheostat is changed. The greater the resistance, the dimmer the instrument panel illumination lights glow.

Dimmer Switches

In the past, the most common location of the dimmer switch was on the floor board next to the left kick panel. This switch is operated by the driver pressing on it with a foot. Positioning the

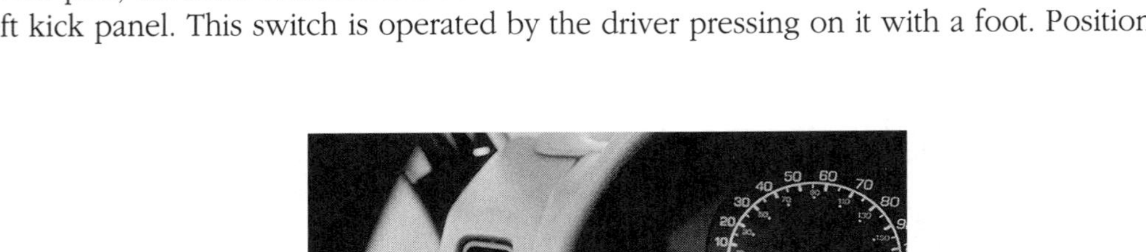

Figure 9-17 A headlight switch mounted on the steering column with a separate control on the dash to control the brightness of the instrument panel lights.

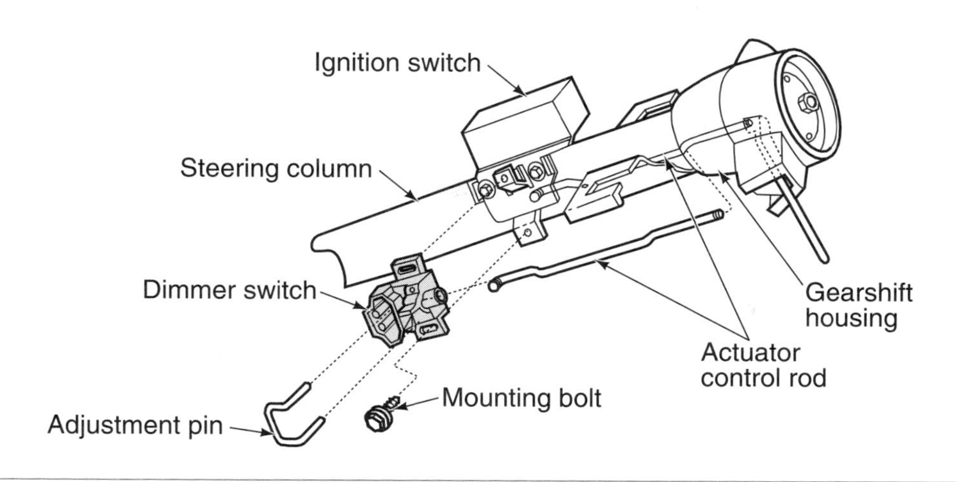

Figure 9-18 A steering column-mounted dimmer switch. (Reprinted with the permission of Ford Motor Company)

The **dimmer switch** provides the means for the driver to select either high or low beam operation, and to switch between the two. The dimmer switch is connected in series within the headlight circuit and controls the current path for high and low beams.

switch on the floor board made the switch subject to damage because of rust, dirt, and so forth. Most newer vehicles locate the dimmer switch on the steering column to prevent early failure and to increase driver accessibility (Figure 9-18).

A BIT OF HISTORY

Foot operated dimmer switches became standard equipment in 1923.

Headlight Circuits

Shop Manual
Chapter 9, page 311

The complete headlight circuit consists of the headlight switch, dimmer switch, high beam indicator, and the headlights. When the headlight switch is pulled to the HEADLIGHT position, current flows to the dimmer switch through the closed contacts (Figure 9-19). If the dimmer switch is in the LOW position, current flows through the low beam filament of the headlights. When the dimmer switch is placed on the HIGH position, current flows through the high beam filaments of the headlights (Figure 9-20).

The headlight circuits just discussed are designed with insulated side switches and grounded bulbs. In this system, battery voltage is present to the headlight switch. The switch must be closed in order for current to flow through the filaments and to ground. The circuit is complete because the headlights are grounded to the vehicle body or chassis. Many import manufacturers use a system design that has insulated bulbs and ground side switches (Figure 9-21). In this system, when the headlight switch is located in the HEADLIGHT position the contacts are closed to complete the circuit path to ground. The headlight switch is located after the headlight lamps in the circuit. Battery voltage is applied directly to the headlights when the relays are closed. But the headlights will not light until the switch completes the ground side of the relay circuits. In this system, both the headlight and dimmer switches complete the circuits to ground.

No matter if the headlights use insulated side switches or ground side switches, each system is wired in parallel to each other. This prevents total headlight failure if one filament burns out.

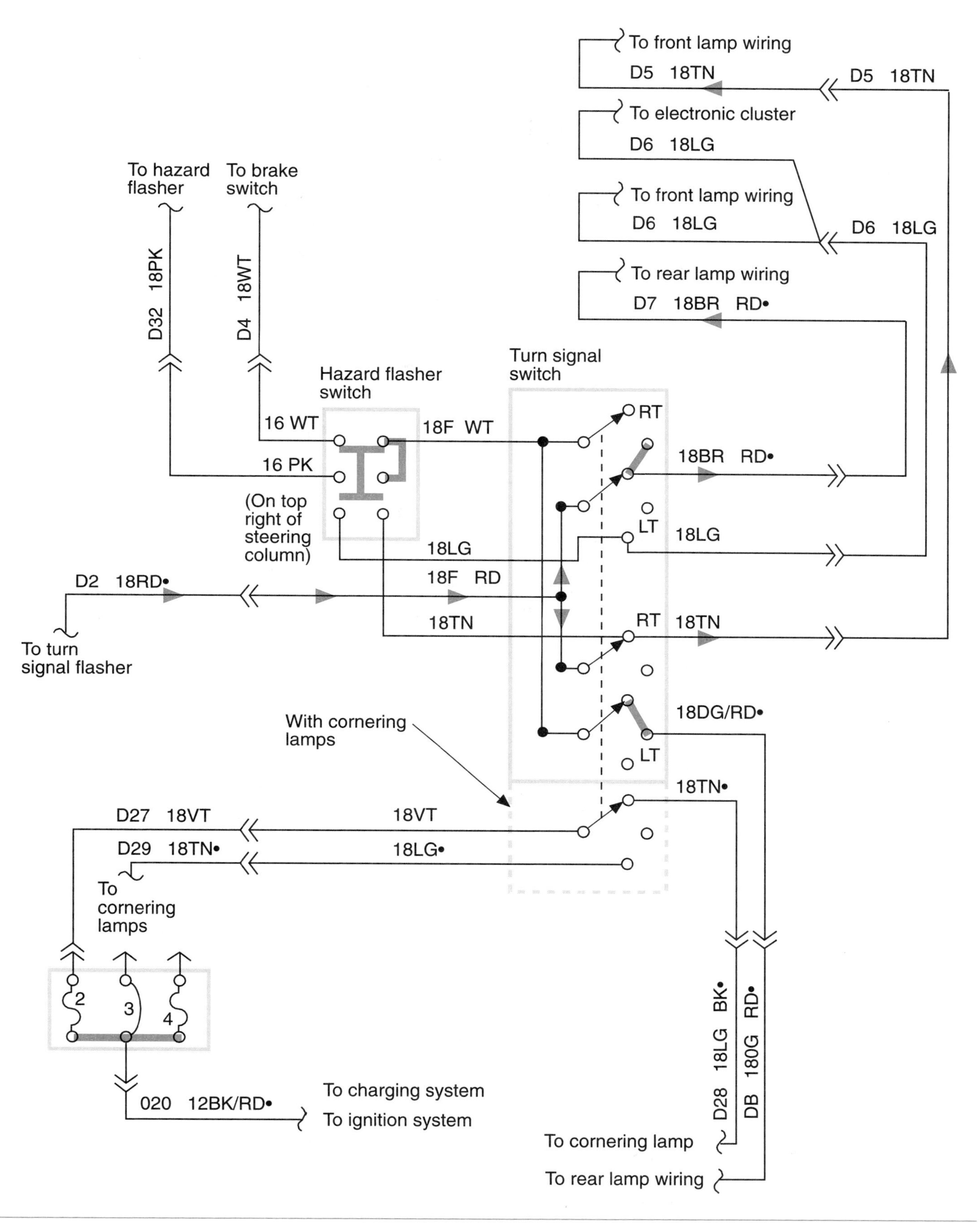

Figure 9-44 Turn signal circuit with the switch in the right turn position. (Courtesy of Chrysler Corporation)

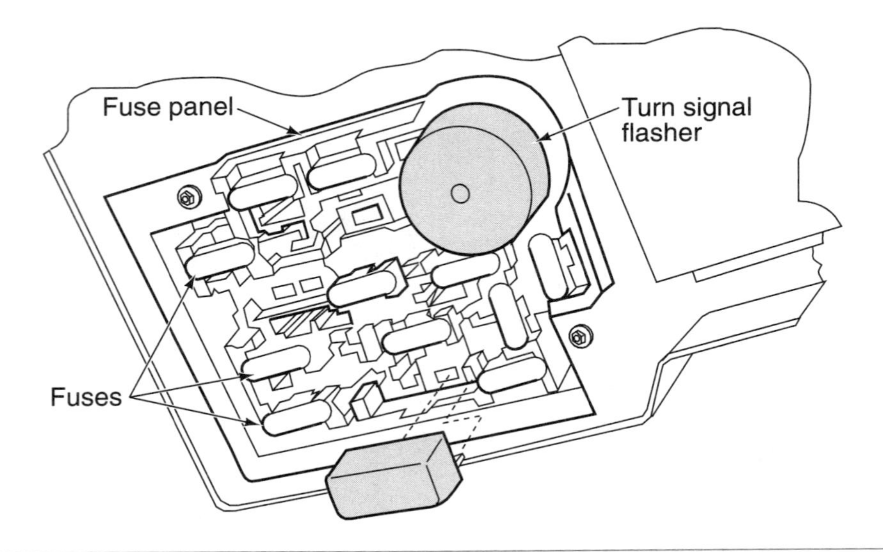

Figure 9-45 A turn signal flasher located in the fuse box. (Reprinted with the permission of Ford Motor Company)

SERVICE TIP: If the turn signals operate properly in one direction but do not flash in the other, the problem is not in the flasher unit. A burned-out lamp filament will not cause enough current to flow to sufficiently heat the bimetallic strip to cause it to open. Thus, the lights do not flash. Locate the faulty bulb and replace it.

With the contacts closed, power flows from the flasher through the turn signal switch to the lamps. The flasher (Figure 9-45) consists of a set of normally closed contacts, a bimetallic strip, and a coil heating element (Figure 9-46). These three components are wired in series. As current flows through the heater element, it increases in temperature, which heats the bimetallic strip. The strip then bends and opens the contact points. Once the points are open, current flow stops. The bimetallic strip cools and the contacts close again. With current flowing again the process is repeated. Because the flasher is in series with the turn signal switch, this action causes the turn signal lights to turn on and off.

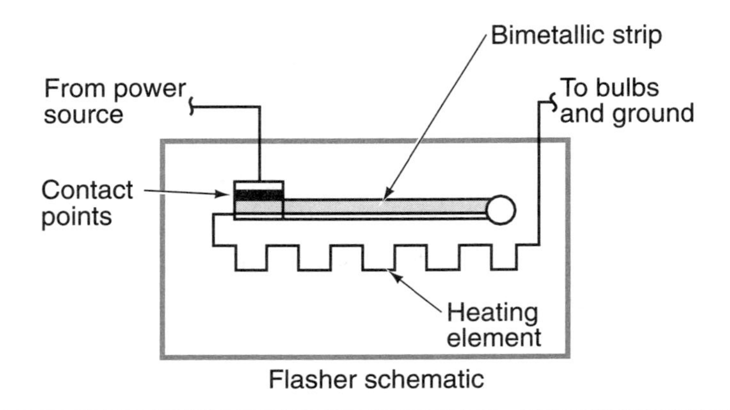

Figure 9-46 The flasher uses a bimetallic strip and a heating coil to flash the turn signal lights. (Reprinted with the permission of Ford Motor Company)

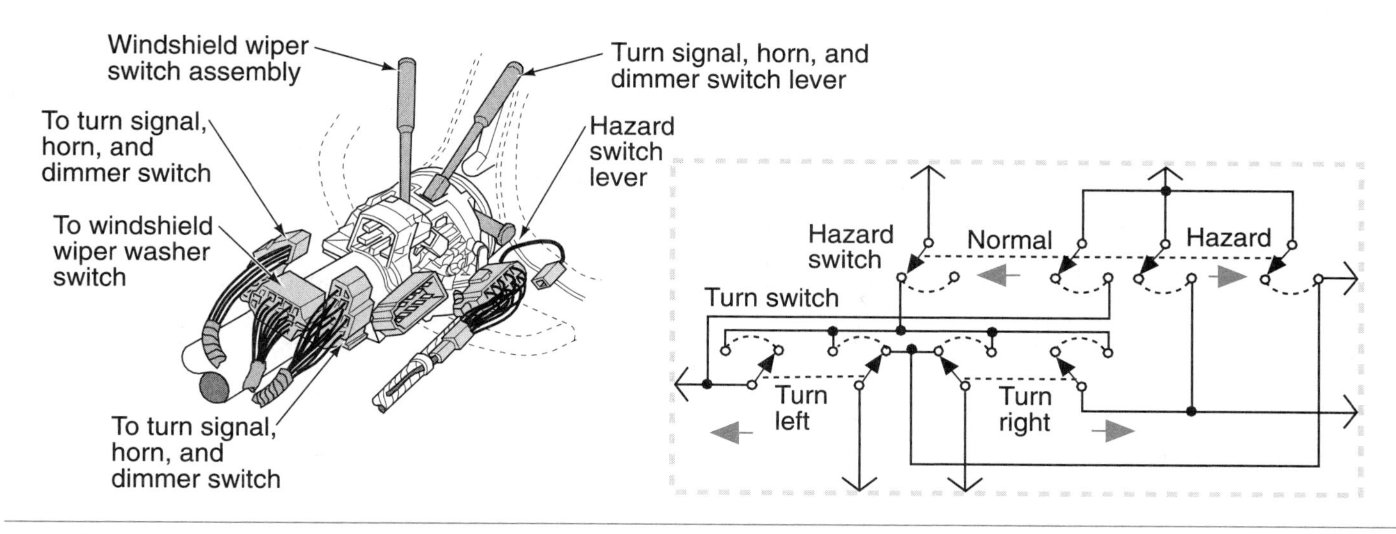

Figure 9-47 The hazard warning system is incorporated into the turn signal system. (Reprinted with the permission of Ford Motor Company)

WARNING: The incorrect selection of the turn signal flasher may result in fast, slow, or no turn signal operation.

The hazard warning system is part of the turn signal system (Figure 9-47). It has been included on all vehicles sold in North America since 1967. All four turn signal lamps flash when the hazard warning switch is turned on. Depending on the manufacturer, a separate flasher can be used for the hazard lights than the one used for the turn signal lights. The operation of the hazard flasher is identical to that of the turn signal. The only difference is that the hazard flasher is capable of carrying the additional current drawn by all four turn signals. And, it receives its power source directly from the battery. Figure 9-48 shows the current flow through the hazard warning system.

Neon Third Brake Lights

In 1995 Ford Motor Company began equipping select models with rear high-mount brake lights that use neon lights. These lights are more energy efficient and turn on more quickly than the regular lights. Behind the third brake light lens is a single neon bulb. Since neon bulbs have no filament, the neon bulb should last much longer than a regular bulb.

The neon bulbs turn on within 3 milliseconds after being activated. Halogen lamps require 300 milliseconds. The importance of this quickness is that it gives the driver behind the vehicle an earlier warning to stop. This early warning can give the approaching driver 19 more feet for stopping when driving at 60 miles per hour.

Cornering lights are lamps that illuminate when the turn signals are activated. They burn steady when the turn signal switch is in a turn position to provide additional illumination of the road in the direction of the turn (Figure 9-49).

Cornering Lights

Vehicles equipped with cornering lights have an additional set of contacts in the turn signal switch. These contacts operate the cornering light circuit only. The contacts can receive voltage from either the ignition switch or the headlight switch. If the ignition switch provides the power, the cornering lights will be activated any time the turn signals are used (Figure 9-50). If the contacts receive the voltage from the headlight switch, the cornering lights do not operate unless the headlight switch is in the PARK or HEADLIGHT position (Figure 9-51).

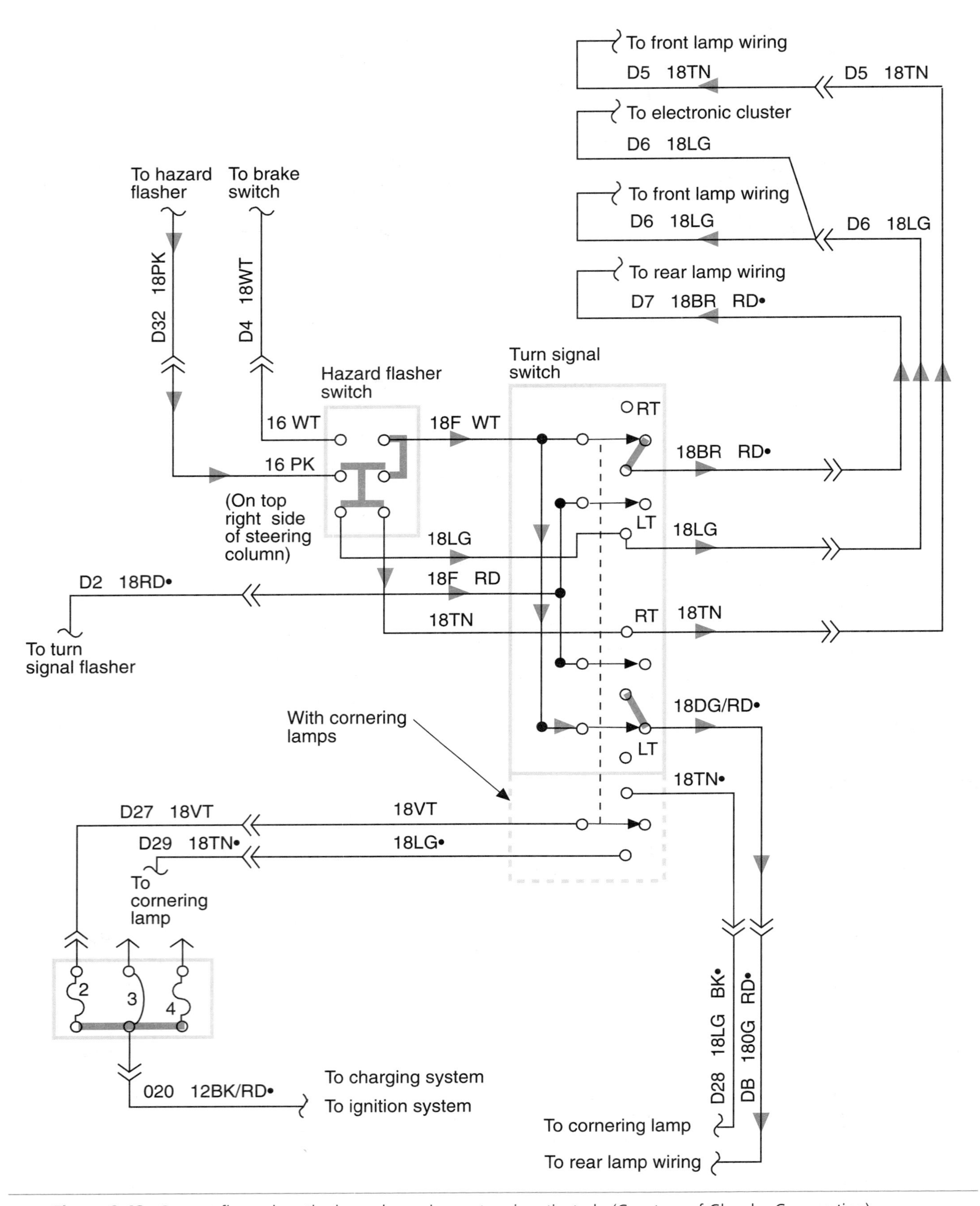

Figure 9-48 Current flow when the hazard warning system is activated. (Courtesy of Chrysler Corporation)

Figure 9-49 Cornering lights are used to provide additional illumination during turns.

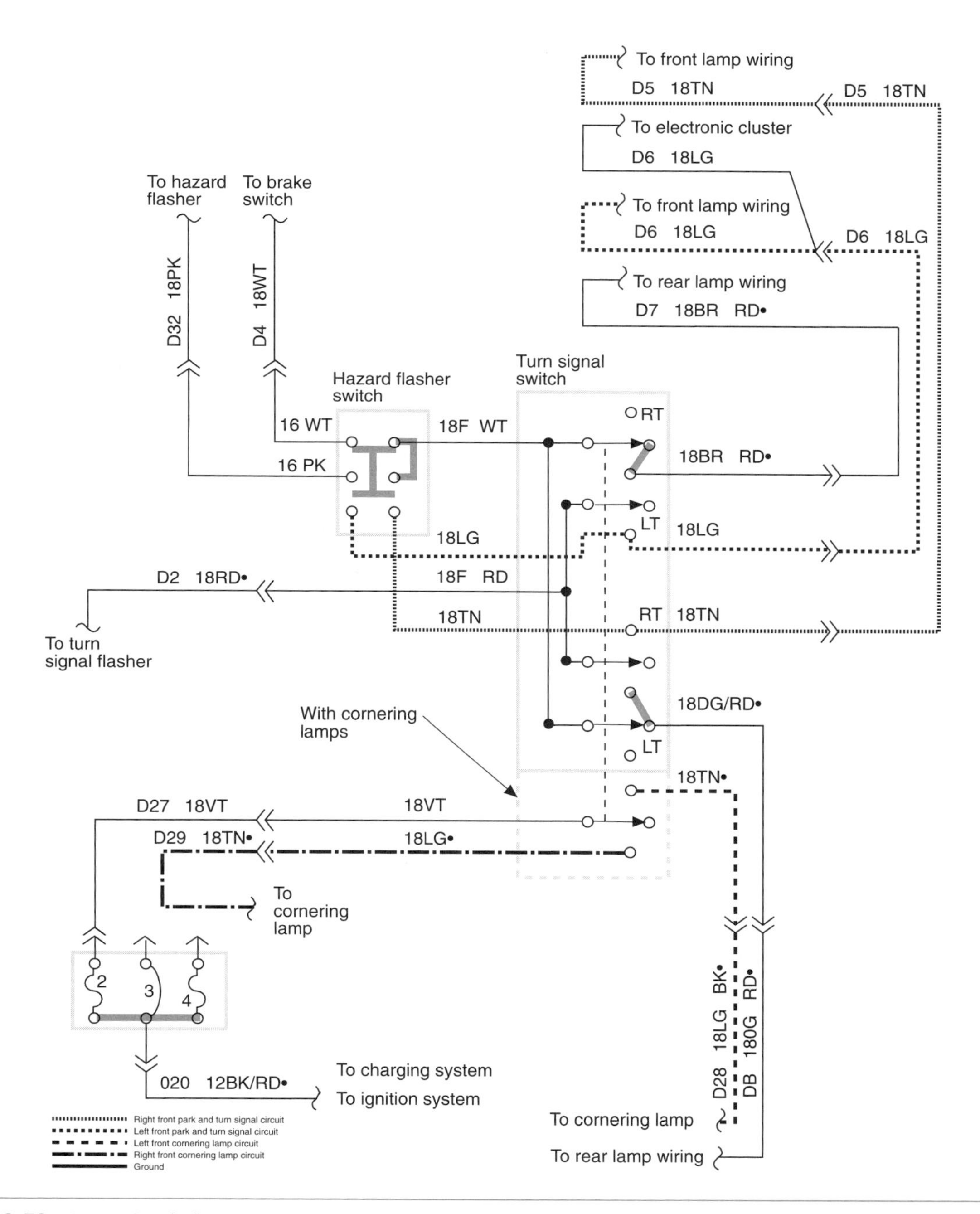

Figure 9-50 Cornering light circuit powered through the ignition switch. (Courtesy of Chrysler Corporation)

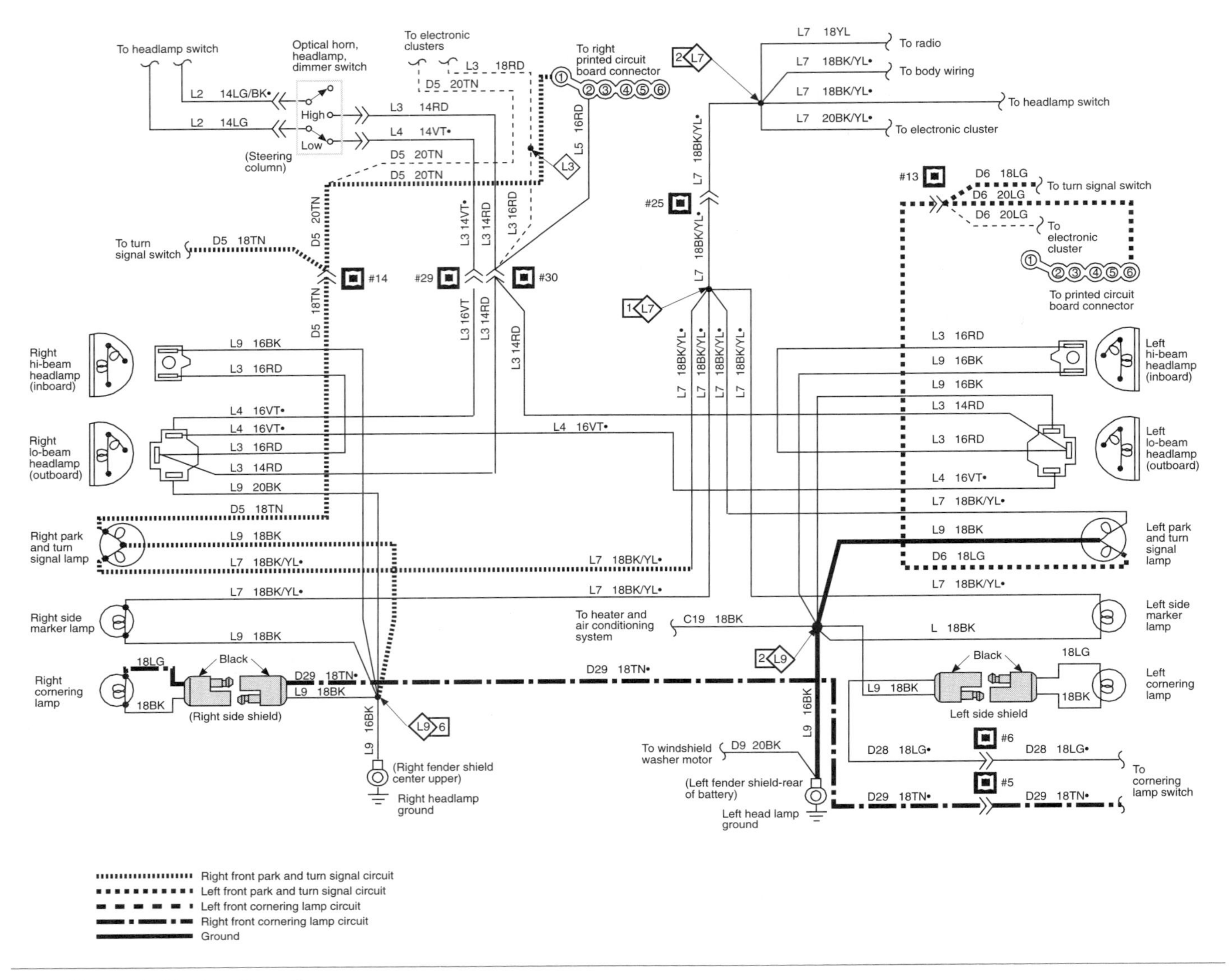

Figure 9-50 (continued)

Backup Lights

All vehicles sold in North America after 1971 are required to have backup lights. Backup lights illuminate the road behind the vehicle and warn other drivers and pedestrians of the driver's intention to back up. Figure 9-52 illustrates a backup light circuit. Power is supplied through the ignition switch when it is in the RUN position. When the driver shifts the transmission into reverse, the backup light switch contacts are closed and the circuit is completed.

Many automatic transmission equipped vehicles incorporate the backup light switch into the neutral safety switch. Most manual transmissions are equipped with a separate switch. Either style of switch can be located on the steering column, on the floor console, or on the transmission (Figure 9-53). Depending on the type of switch used, there may be a means of adjusting the switch to assure that the lights are not on when the vehicle is in a forward gear selection.

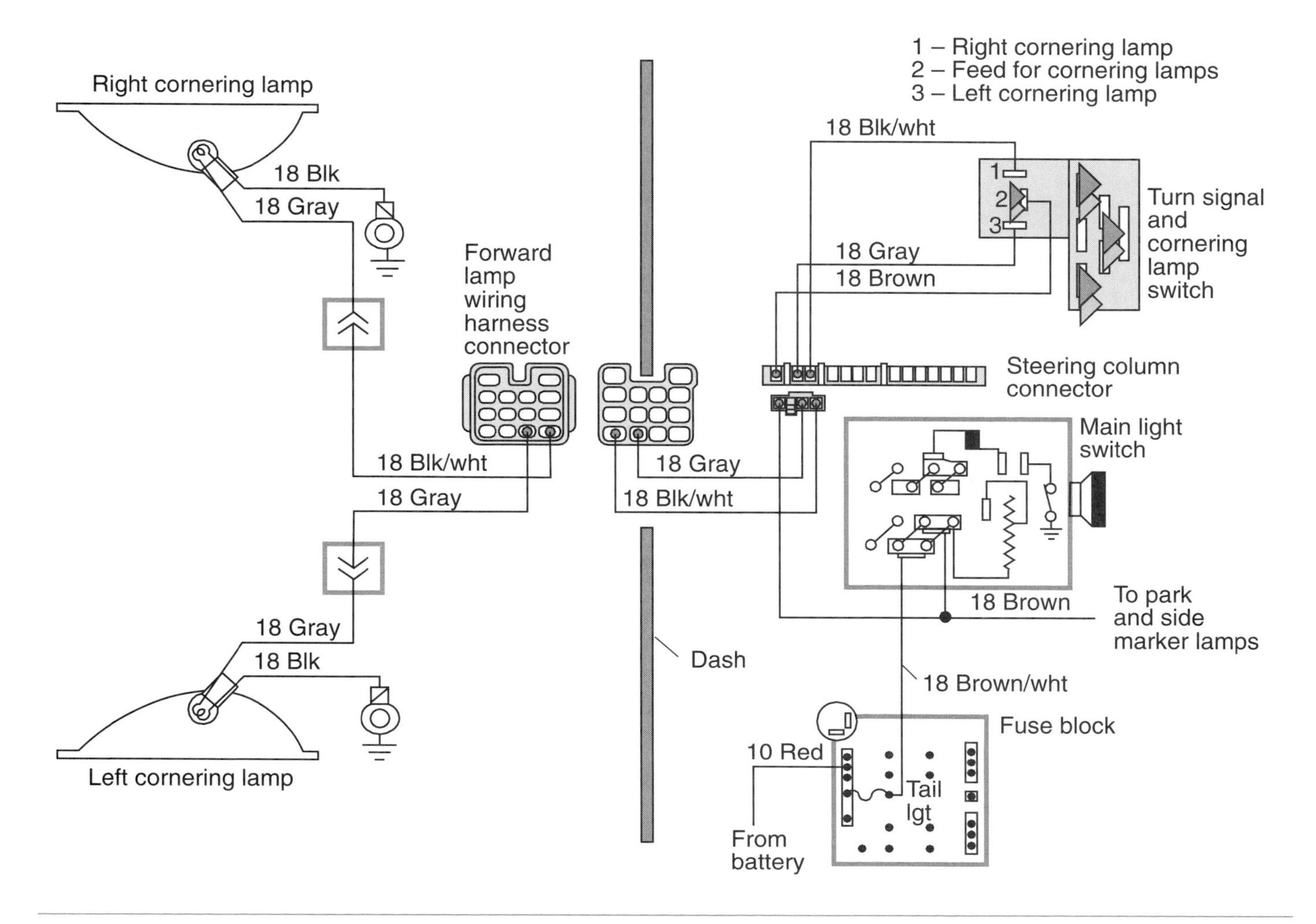

Figure 9-51 Cornering light circuit powered when the taillights are on.

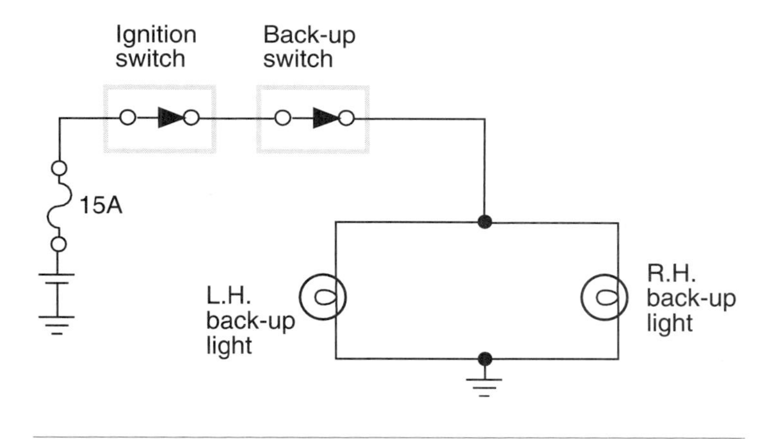

Figure 9-52 Backup light circuit.

Figure 9-53 A combination backup and neutral safety switch installed on an automatic transmission.

Figure 9-54 Wrap-around headlights serve as side marker lights.

A BIT OF HISTORY

The 1921 Wills-St. Claire was the first car to display a backup lamp.

Side Marker and Clearance Lights

Side marker lights are installed on all vehicles sold in North America since 1969. These lamps permit the vehicle to be seen when entering a roadway from the side. This also provides a means for other drivers to determine vehicle length. The front side marker light lens must be amber and the rear lens must be red. Vehicles that use wrap-around headlight and taillight assemblies also use this lens for the side marker lights (Figure 9-54). Vehicles that surpass certain length and height limits are also required to have clearance lights that face both to the front and rear of the vehicle.

The common method of wiring the side marker lights is in parallel with the parking lights. Wired in this manner, the side marker lights would only illuminate when the headlight switch was in the PARK or HEADLIGHT position.

Many vehicle manufacturers use a method of wiring in the side marker lights so that they flash when the turn signals are activated. The side marker light is wired across the parking light and turn signal light (Figure 9-55). If the parking lights are on, voltage is applied to the side marker light from the parking light circuit. Ground for the side marker light is provided through the turn signal filament. Because of the large voltage drop across the side marker lamp, the turn signal bulb will barely illuminate. In this condition the side marker light stays on constantly (Figure 9-56).

If the parking lights are off and the turn signal is activated, the side marker light receives its voltage source from the turn signal circuit. Ground for the side marker light is provided through the parking light filament. The voltage drop over the side marker light is so high that the parking light will not illuminate. The side marker light will flash with the turn signal light (Figure 9-57).

If the turn signal is activated while the parking lights are illuminated, the side marker light will flash alternately with the turn signal light. When both the turn signal light and the parking light are on, there is equal voltage applied to both sides of the side marker light. There is no voltage potential across the bulb, so the light does not illuminate (Figure 9-58). The turn signal light

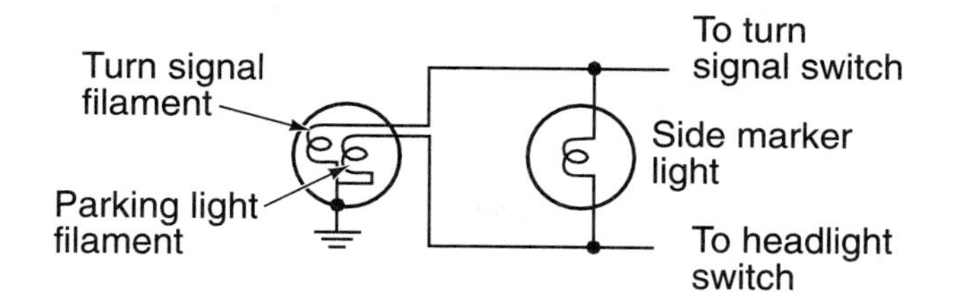

Figure 9-55 A side marker light wired across two circuits.

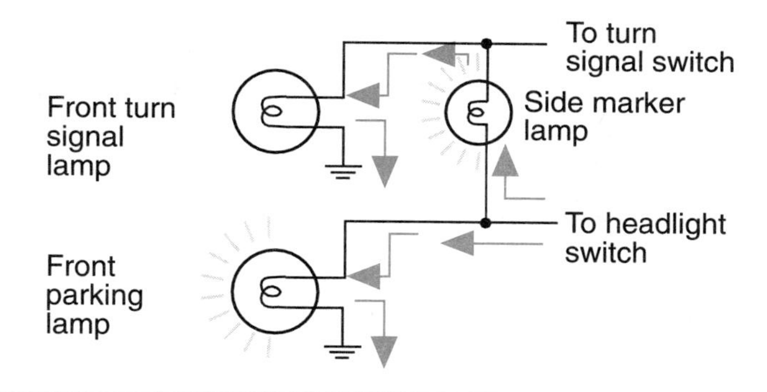

Figure 9-56 Current flow to the side marker light with the parking light on.

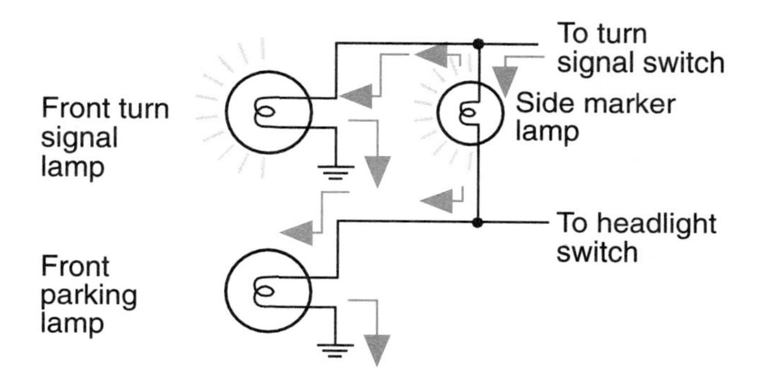

Figure 9-57 Side marker operation with the turn signal switch activated.

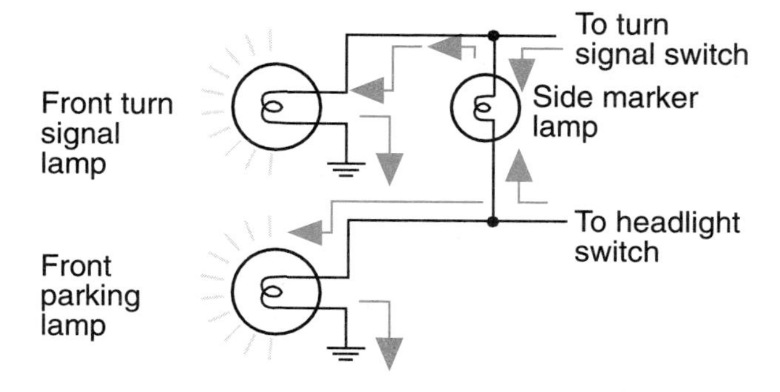

Figure 9-58 Side marker light with the turn signal and parking light on.

turns off as a result of the flasher opening. Then the turn signal light filament provides a ground path and the side marker light comes on. The side marker light will stay on until the flasher contacts close to turn on the turn signal light again.

Interior Lights

Shop Manual
Chapter 9, page 332

Interior lighting includes courtesy lights, map lights, and instrument panel lights.

Courtesy Lights

Courtesy lights illuminate the vehicle interior when the doors are open.

Courtesy lights operate from the headlight and door switches and receive their power source directly from a fused battery connection. The switches can be either ground switch circuit (Figure 9-59) or insulated switch circuit design (Figure 9-60). In the insulated switch circuit, the switch is used as the power relay to the lights. In the grounded switch circuit, the switch controls the grounding portion of the circuit for the lights. The courtesy lights may also be activated by

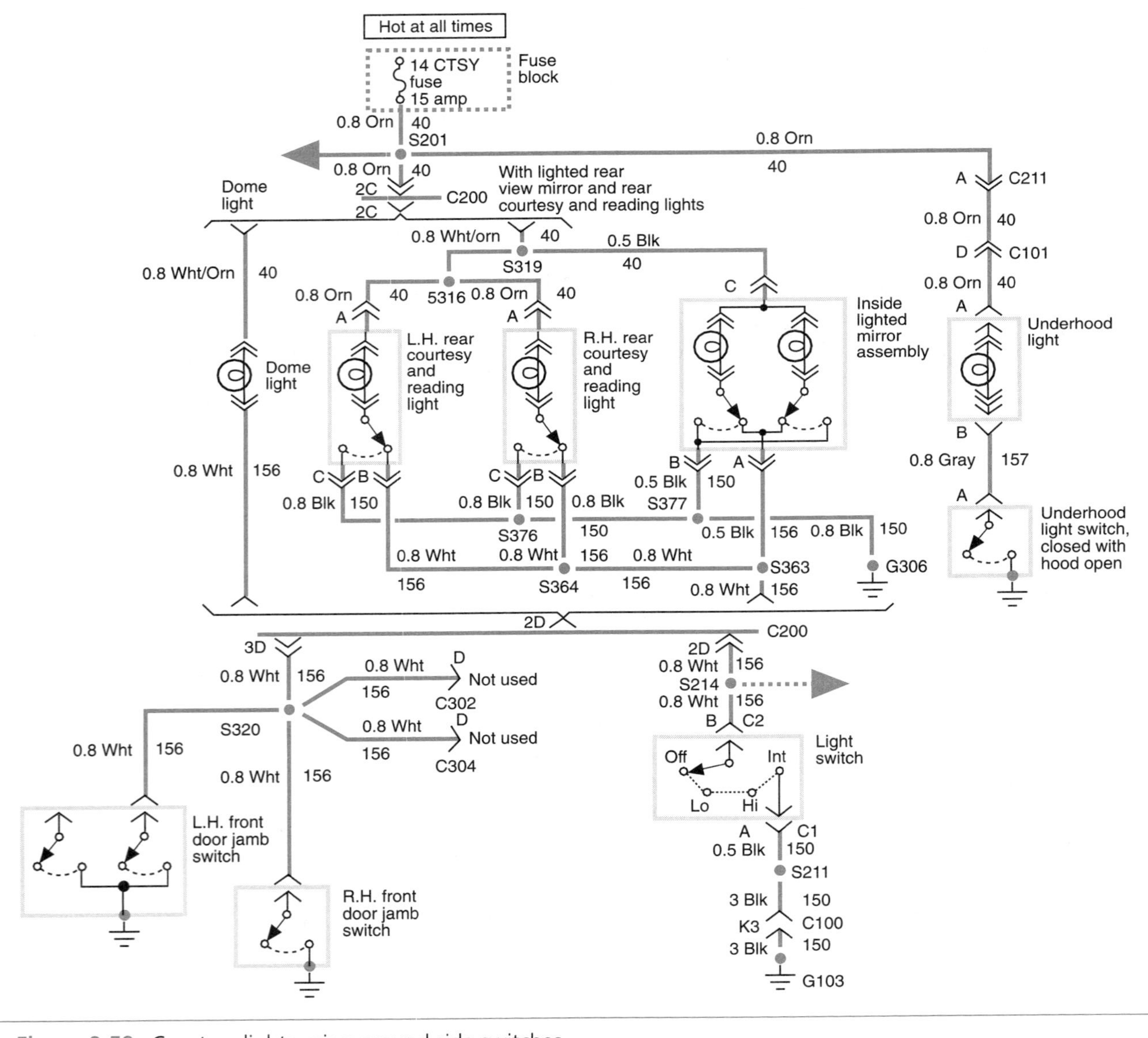

Figure 9-59 Courtesy lights using ground side switches.

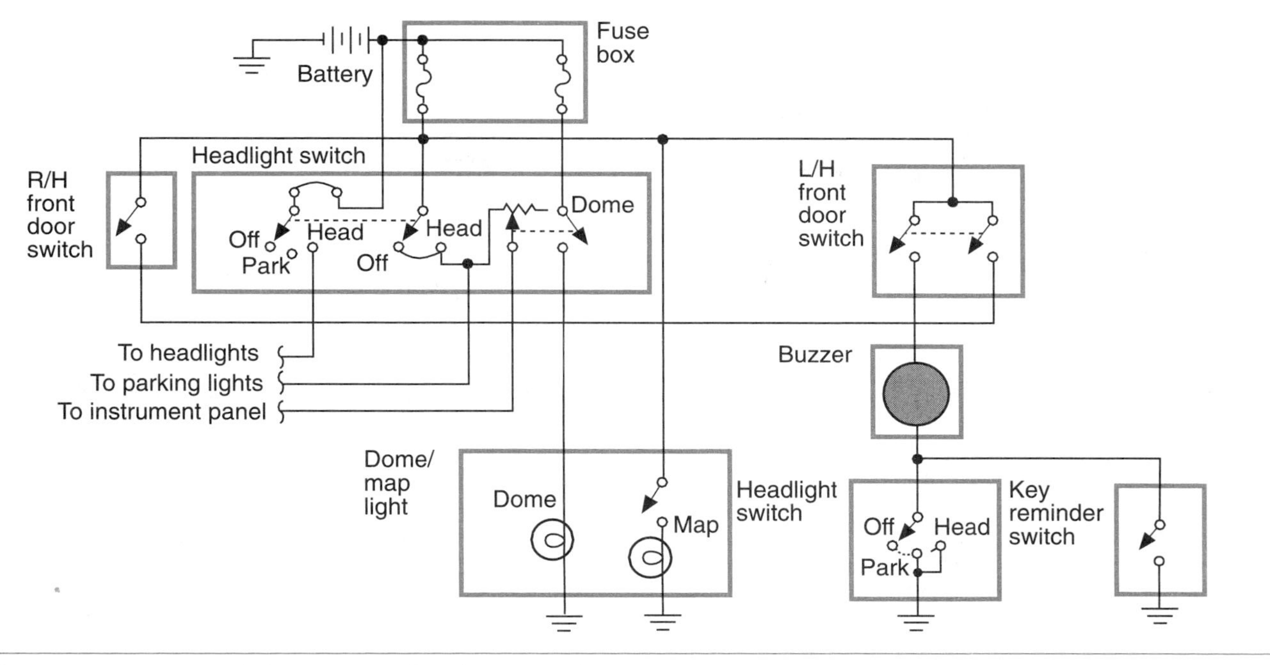

Figure 9-60 Courtesy lights using insulated side switches.

the headlight switch. When the headlight switch knob is turned to the extreme counterclockwise position, the contacts in the switch close and complete the circuit.

A BIT OF HISTORY

The 1913 Spaulding touring car had such luxuries as four seats with folding backs, air mattresses, and electric reading lamps.

Instrument Cluster and Panel Lights

Consider the following three types of lighting circuits within the instrument cluster:

1. **Warning lights** alert the driver to potentially dangerous conditions such as brake failure or low oil pressure.
2. **Indicator lights** include turn signal indicators.
3. **Illumination lights** provide indirect lighting to illuminate the instrument gauges, speedometer, heater controls, clock, ashtray, radio, and other controls.

The power source for the instrument panel lights is provided through the headlight switch. The contacts are closed when the headlight switch is located in the PARK or HEADLIGHT position. The current must flow through a variable resistor (rheostat) that is either a part of the headlight switch or a separate dial on the dash. The resistance of the rheostat is varied by turning the knob. By varying the resistance, changes in the current flow to the lamps control the brightness of the lights (Figure 9-61).

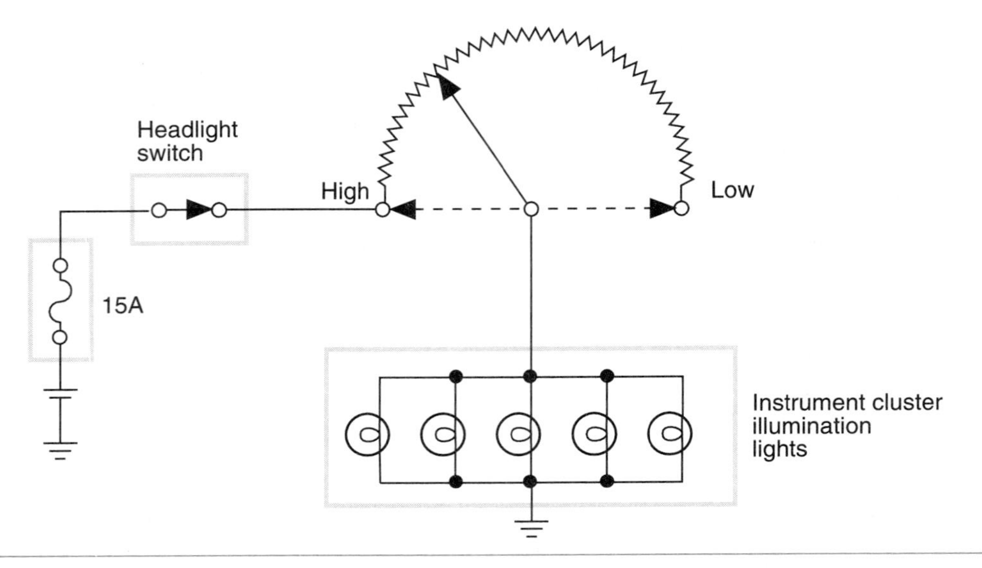

Figure 9-61 A rheostat controls the brightness of the instrument panel lights.

SERVICE TIP: On some vehicles the instrument cluster illumination light fuse is located after the headlight switch. On these circuits the headlight switch must be turned on when using a test light or digital volt-ohm meter (DVOM) to check the fuse.

Feedback occurs when electricity seeks a path of lower resistance. This alternate path operates another component than the one intended. Feedback can be classified as a short.

Lighting System Complexity

Today's vehicles have a sophisticated lighting system and electrical interconnections. It is possible to have problems with lights and accessories that cause them to operate when they are not supposed to. This is through a condition called feedback. If there is an open in the circuit, electricity will seek another path to follow. This may cause any lights or accessories in that path to turn on.

Examples of feedback and how it may cause undesired operation are illustrated in Figures 9-62 through 9-69. The illustration (Figure 9-62) shows a system that has the dome light, tail light,

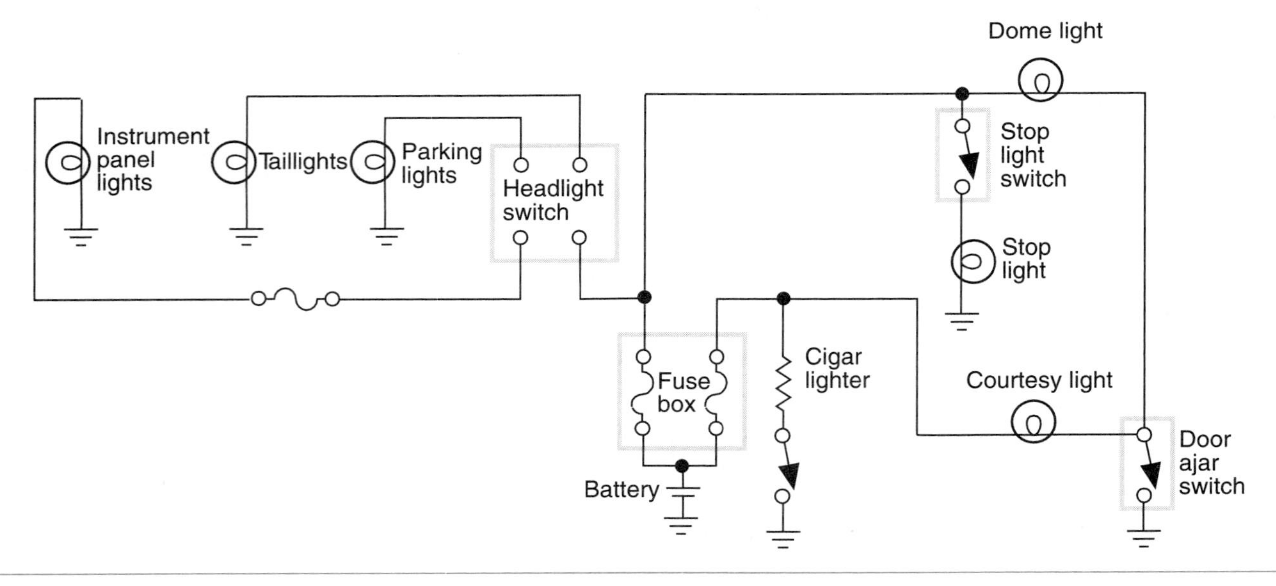

Figure 9-62 A normally operating light circuit.

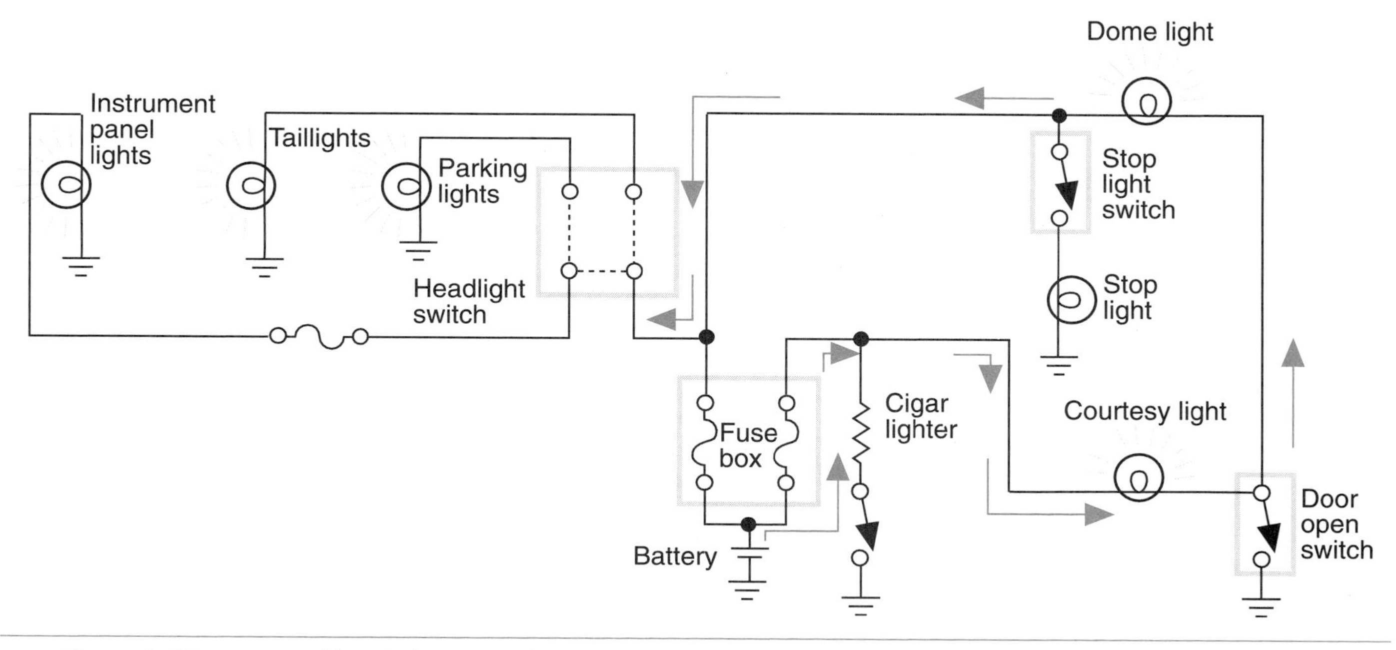

Figure 9-63 An open (blown) dome light fuse can cause feedback to other circuits when the headlights are turned on.

and brake light circuits on one fuse; the cigar lighter circuit has its own fuse. The two fuses are located in the main fuse block and share a common bus bar on the power side.

If the dome light fuse blew, and the headlight switch was in the PARK or HEADLIGHT position, the courtesy lights, dome light, taillight, parking lights, and instrument lights would all be very dimly lit (Figure 9-63). Current would flow through the cigar lighter fuse to the courtesy light and on to the door light switch. Current will then continue through the dome light to the headlight switch. Because the headlight switch is now closed, the instrument panel lights are also in the circuit. The lights are dim because all the bulbs are now connected in series.

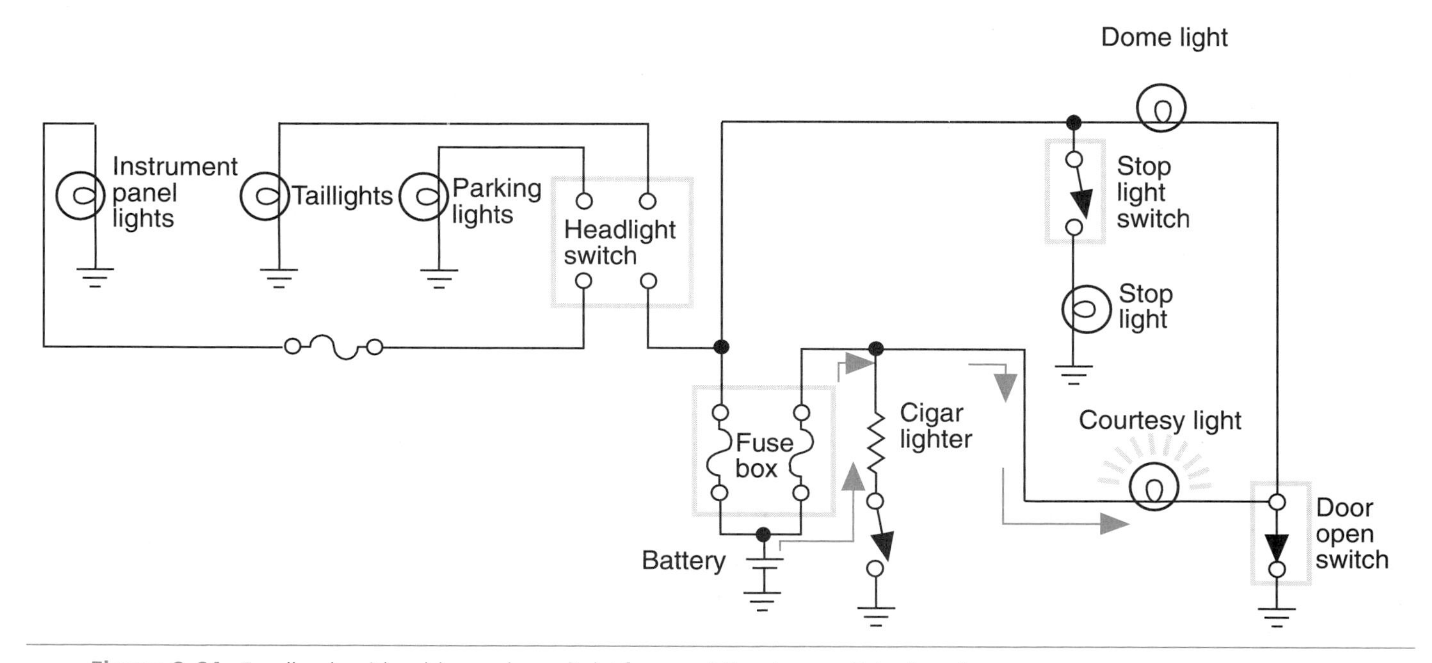

Figure 9-64 Feedback with a blown dome light fuse and the door switch closed.

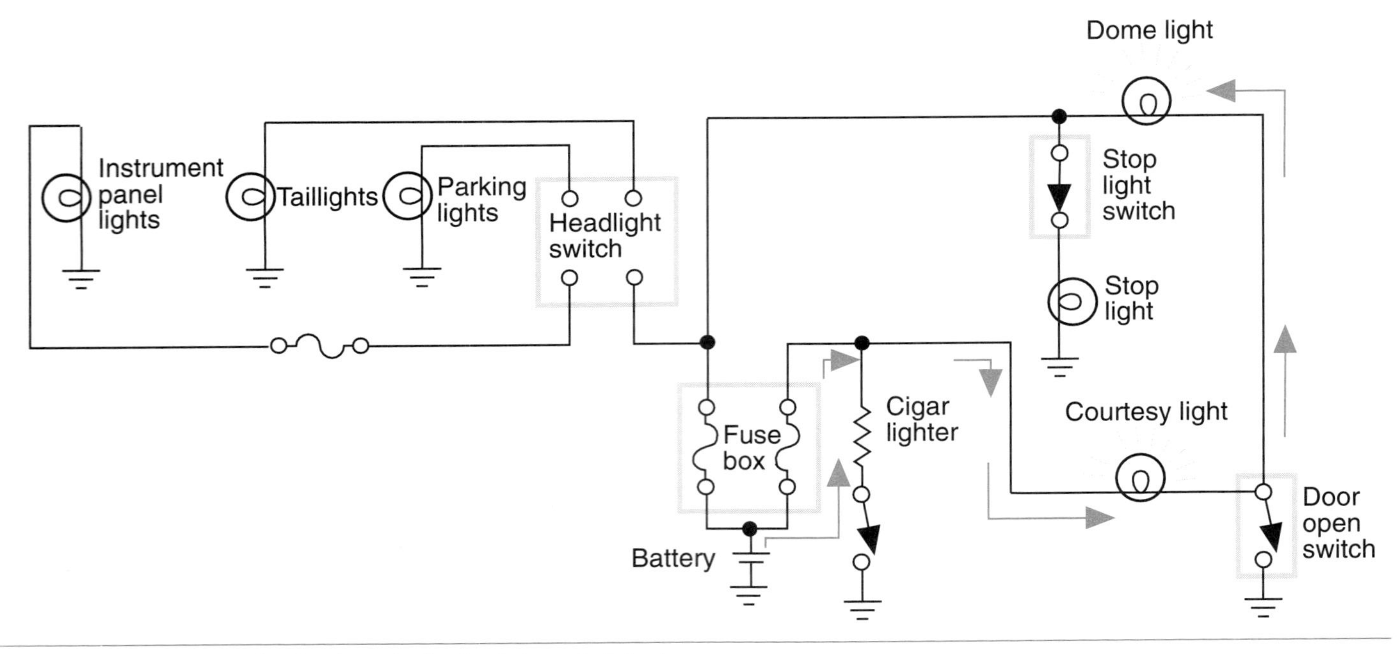

Figure 9-65 Feedback when the brake light switch is closed.

If the dome light fuse is blown and the headlight switch is in the OFF position, all lights will turn off. However, if the door is opened, the courtesy lights will come on but the dome light will not (Figure 9-64).

With the same blown fuse and the brake light switch closed, the dome light, courtesy light, and brake light will all illuminate dimly because the loads are in series (Figure 9-65).

In this example, if the dome light and courtesy lights come on dimly when the cigar lighter is pushed in, the problem can be caused by a blown cigar lighter fuse (Figure 9-66). With the cigar lighter pushed in, a path to ground is completed. The lights and cigar lighter are now in

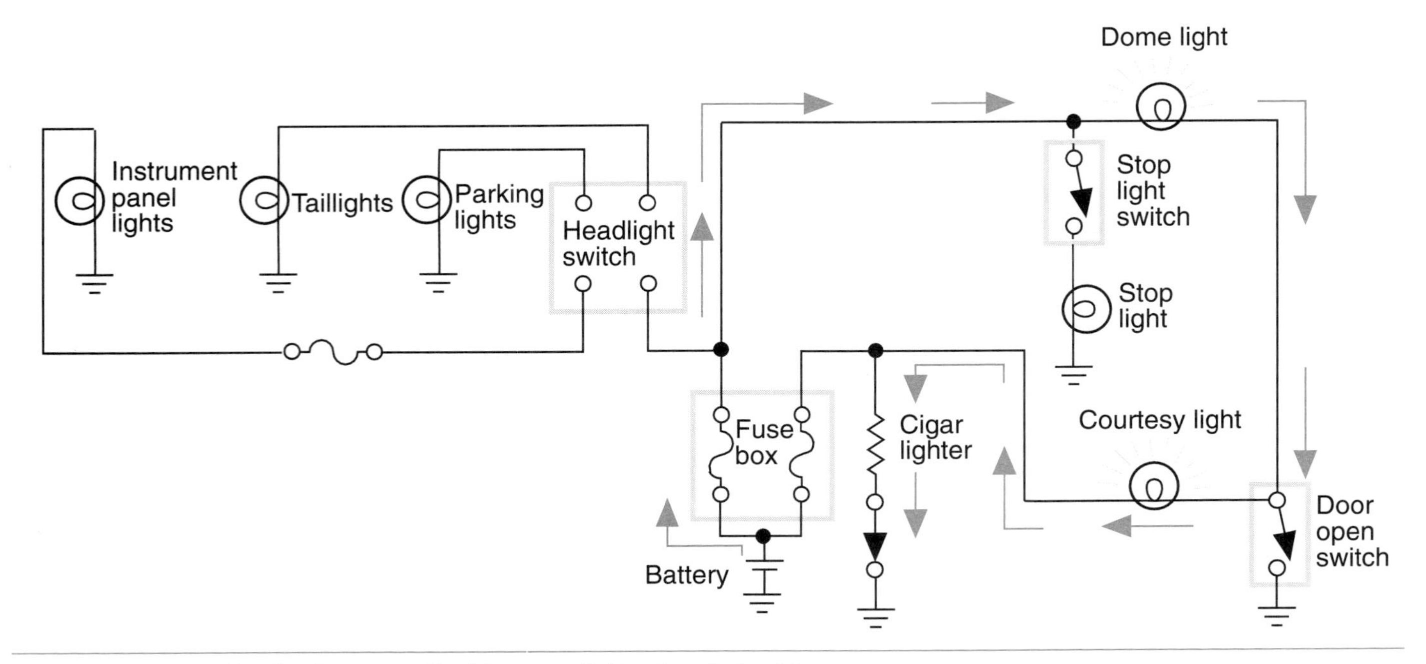

Figure 9-66 Feedback as a result of the cigar lighter fuse being blown.

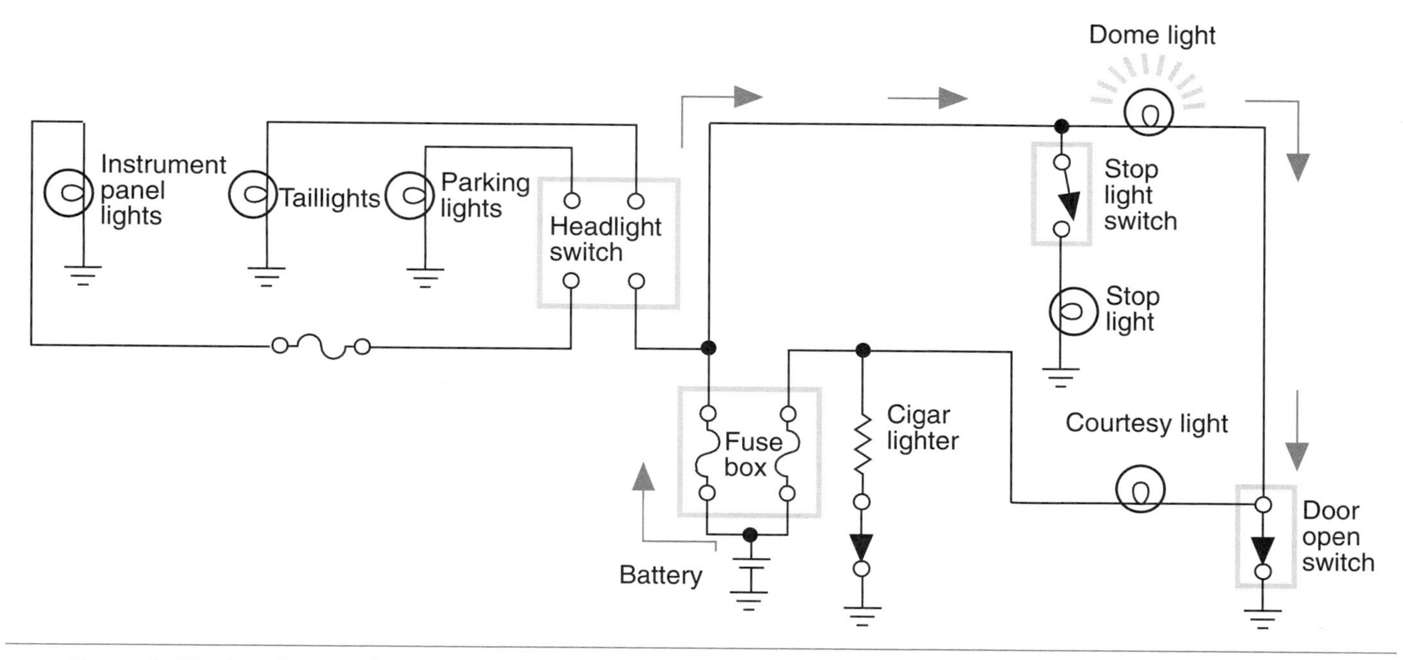

Figure 9-67 Opening the door will cause the courtesy light to go out and the dome light to get brighter.

series, thus the lights are dim and there is not enough current to heat and release the cigar lighter. If the cigar lighter was left in this position the battery would eventually drain down.

A blown cigar lighter fuse will also cause the dome light to get brighter when the doors are open, and the courtesy lights will go out (Figure 9-67). Also, if the lighter is pushed in and the brake light switch is closed, the dome and courtesy lights will go out (Figure 9-68).

Feedback can also be the result of a conductive corrosion that is developed at a connection. If the corrosion allows for current flow from one conductor to an adjacent conductor in the

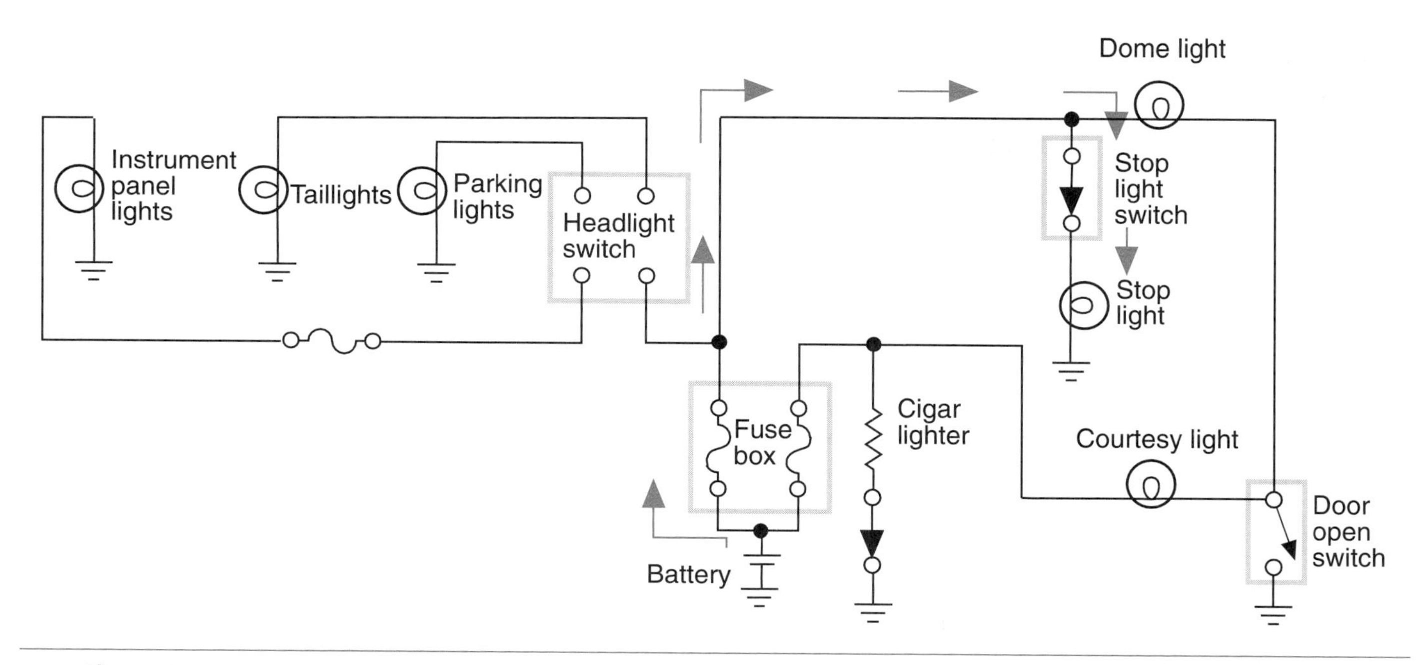

Figure 9-68 Dome and courtesy lights go out when the brakes are applied.

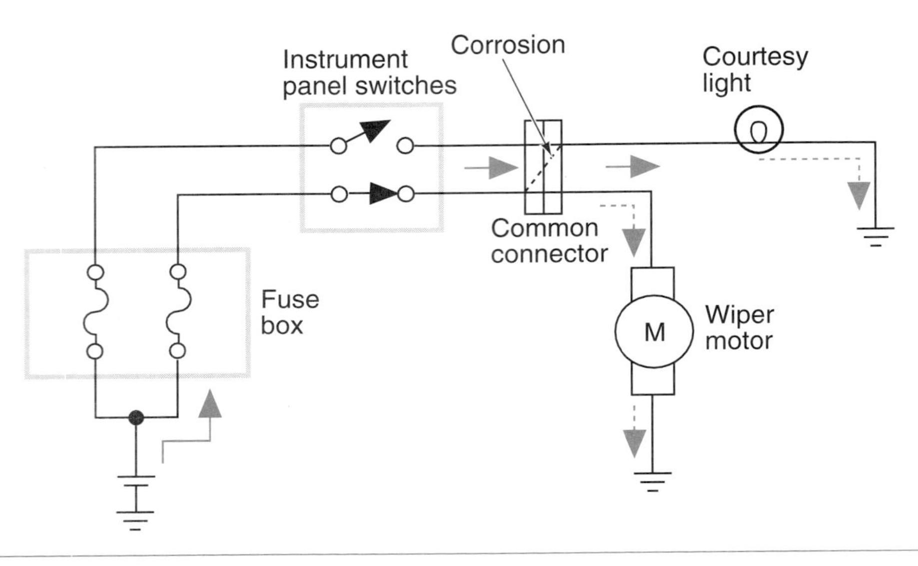

Figure 9-69 A corroded common connector can cause feedback.

Dual filament light bulb sockets are subjectable to corrosion that can cause feedback. The most common indicator of this problem is parking lights that illuminate when the brake pedal is applied.

connection, the other circuit will also be activated. The illustration (Figure 9-69) shows how corrosion in a common connector can cause the dome light to illuminate when the wiper motor is turned on. Because the wiper motor has a greater resistance than the light bulb, more voltage will flow through the bulb than through the motor. The bulb will light brightly, but the motor will turn very slowly or not at all. The same effect will result if the courtesy light switch is turned on with the motor switch off.

Summary

Terms to Know

CHMSL
Composite bulb
Cornering lights
Courtesy lights
Dimmer switch
Double filament lamp
Feedback
Filament
Halogen
HID
Illumination lights
Incandescence
Indicator lights
Lamp
Limit switches
Neon lights
Optical horn

- The most commonly thought of light circuit is the headlights. But there are many lighting systems in the vehicle.
- Different types of lamps are used to provide illumination for the systems. The lamp may be either a single filament bulb that performs a single function, or a double filament that performs several functions.
- The headlight lamps can be one of three designs: standard sealed beam, halogen sealed beam, or composite.
- The headlight filament is located on a reflector that intensifies the light, which is then directed through the lens. The lens is designed to change the circular light pattern into a broad, flat light beam. Placement of the filament in the reflector provides for low and high beam light patterns.
- Some manufacturers use concealed headlights to improve the aerodynamics of the vehicle. The concealed headlight doors can operate from vacuum or by electrically controlled motors. Some systems incorporate the use of IC chips into the concealed headlight door control.
- In addition to the headlight system, the lighting systems include:

  Stop lights
  Turn signals
  Hazard lights
  Parking lights
  Taillights

Backup lights

Side marker lights

Courtesy lights

Instrument panel lights

- ❑ The headlight switch can be used as the control of many of these lighting systems. Most headlight switches have a circuit breaker that is an integral part of the switch. The circuit breaker provides protection of the headlight system without totally disabling the headlight operation if a circuit overload is present.
- ❑ A rheostat is used in conjunction with the headlight switch to control the brightness of the instrument panel illumination lights.

Terms to Know (Continued)

Prism lens

Reflector

Sealed beam

Vacuum distribution valve

Warning lights

Review Questions

Short Answer Essays

1. What lighting systems are controlled by the headlight switch?
2. How is the brightness of the instrument cluster lamps controlled?
3. What is CHMSL?
4. What is the purpose of the lamp filament?
5. What three lighting circuits are incorporated within the instrument cluster?
6. List and describe the three types of headlight lamps used.
7. Describe the influence that the turn signal switch has on the operation of the brake lights in a two-bulb taillight assembly.
8. What is the purpose of the diodes on some CHMSL circuits?
9. How are most cornering light circuits wired to allow the cornering light to be on steady when the turn signal switch is activated?
10. Describe what the term *feedback* means and how it can affect the operation of the electrical system.

Fill-in-the-Blanks

1. When today's technician performs repairs on the lighting systems, the repairs must meet at least two requirements: They must assure vehicle ________________ and meet all ________________ ________________ .
2. A ________________ is a device that produces light as a result of current flow through a ________________ .
3. ________________ ________________ redirect the light beam and create a broad, flat beam.
4. The ________________ ________________ controls most of the vehicle's lighting systems.

5. The ________________ ________________ provides the means for the driver to select either high or low beam operation.

6. The complete headlight circuit consists of the ________________ ________________, ________________ ________________, ________________ ________________ ________________, and the ________________.

7. The headlight doors of a concealed system can be controlled by either ________________ ________________ or by ________________.

8. On vehicles equipped with cornering lights, the turn signal switch has an additional set of ________________ that operate the cornering light circuit only.

9. Most limit switches operate off of a ________________ on the motor.

10. In most automatic transmission equipped vehicles, the backup light switch is part of the ________________ ________________ ________________. Most manual transmissions are equipped with a ________________ switch.

ASE Style Review Questions

1. Two-bulb taillight assemblies are being discussed:
   Technician A says the brake lights use the high intensity filament of the taillight bulb.
   Technician B says the current from the brake lights flows through the turn signal switch before going to the brake lights.
   Who is correct?
   A. A only
   B. B only
   C. Both A and B
   D. Neither A nor B

2. Composite headlights are being discussed:
   Technician A says composite headlights have replaceable bulbs.
   Technician B says a cracked or broken lens will prevent the operation of the composite headlight.
   Who is correct?
   A. A only
   B. B only
   C. Both A and B
   D. Neither A nor B

3. The brake light circuit is being discussed:
   Technician A says the brake light switch receives current from the ignition switch.
   Technician B says the brake light switch receives current from the headlight switch.
   Who is correct?
   A. A only
   B. B only
   C. Both A and B
   D. Neither A nor B

4. The turn signal circuit is being discussed:
   Technician A says the dimmer switch is a part of the circuit.
   Technician B says some flashers use a bimetallic strip to open and close the circuit.
   Who is correct?
   A. A only
   B. B only
   C. Both A and B
   D. Neither A nor B

5. The turn signals work properly in one direction, but in the other direction the indicator light stays on steadily:
Technician A says a burned out light bulb may be the fault.
Technician B says the flasher is at fault.
Who is correct?
A. A only
B. B only
C. Both A and B
D. Neither A nor B

6. The concealed headlight system is being discussed:
Technician A says the system can use vacuum to operate the doors.
Technician B says the system can use electric motors to operate the doors.
Who is correct?
A. A only
B. B only
C. Both A and B
D. Neither A nor B

7. The CHMSL circuit is being discussed:
Technician A says the diodes are used to assure proper turn signal operation.
Technician B says the diodes are used to prevent radio static when the brake light is activated.
Who is correct?
A. A only
B. B only
C. Both A and B
D. Neither A nor B

8. The exterior lights of a vehicle are being discussed:
Technician A says the cornering lights use an additional set of contacts in the turn signal switch.
Technician B says the cornering light circuit can receive its voltage from the ignition switch.
Who is correct?
A. A only
B. B only
C. Both A and B
D. Neither A nor B

9. *Technician A* says the side marker lights can be wired to flash with the turn signals.
Technician B says wrap-around lenses can be used for side marker lights.
Who is correct?
A. A only
B. B only
C. Both A and B
D. Neither A nor B

10. *Technician A* says feedback is normal during the operation of the electrical system.
Technician B says feedback is a form of a short.
Who is correct?
A. A only
B. B only
C. Both A and B
D. Neither A nor B

Conventional Analog Instrumentation, Indicator Lights, and Warning Devices

Upon completion and review of this chapter, you should be able to:

- ❑ Describe the operation of mechanical speedometers and odometers.
- ❑ Describe the function and operation of the tachometer.
- ❑ Explain the purpose of an instrument voltage regulator.
- ❑ Describe the operation of bimetallic gauges.
- ❑ Describe the operation of electromagnetic gauges, including d'Arsonval, three coil, two coil, and air core.
- ❑ Explain the function and operation of the various gauge sending units, including thermistors, piezoresistive, and mechanical variable resistors.
- ❑ Explain the operation of various warning lamp circuits.
- ❑ Explain the operation of various audible warning systems.

Introduction

Analog instrument gauges and indicator lights monitor the various vehicle operating systems. They provide information to the driver about their current operation (Figure 10-1). Warning devices also provide information to the driver, however, these are usually associated with an audible signal. Some vehicles use a voice module to alert the driver to certain conditions.

This chapter details the operation of analog electrical gauges, indicator lights, and warning devices.

Indicator/Warning Light Key

1. Turn signals
2. Charging system (AMP)
3. Door ajar
4. Malfunction of anti-lock brake system
5. High beam
6. Brake
7. Engine temp. gauge
8. Fuel gauge
9. Seat belt
10. Check engine
11. Speedometer
12. Tachometer
13. Boost gauge
14. Low fuel *
15. Low oil pressure *
16. Trac-control active

* Low oil pressure/hi coolant temp/low fuel (below 1/8) redundant with gauges

Figure 10-1 Instrument panel. (Courtesy of Chrysler Corporation)

The **speedometer** is used to indicate the speed of the vehicle.

The aluminum drum is not attracted to the magnetic field. However, it is a conductor of electric current.

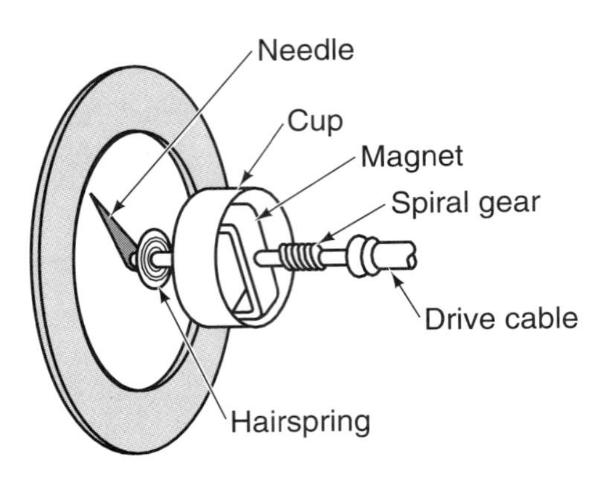

Figure 10-2 The main elements of a conventional-type speedometer.

Shop Manual
Chapter 10, page 353

Eddy currents are small induced currents

Speedometers

A conventional speedometer (Figure 10-2) uses a cable that is connected to the output shaft of the transmission (or transfer case if four-wheel drive). The rotation of the output shaft causes the speedometer cable to rotate within its housing (Figure 10-3). The cable then transfers rotation to the speedometer assembly. In the speedometer assembly, the cable is connected to a permanent magnet that is surrounded by an aluminum drum. The speedometer needle is attached to the drum. There is no solid or mechanical connection between the speedometer needle and the cable.

The rotating permanent magnet produces a rotating magnetic field around the drum. The rotating magnetic field generates circulating currents in the drum. These eddy currents produce a small magnetic field that interacts with the field of the rotating magnet. This interaction of the two magnetic fields pulls the drum and needle around with the rotating magnet. A fine hairspring holds back the drum to prevent the needle from bouncing. As the drum rotates, the needle moves across the speedometer scale (Figure 10-4).

Shop Manual
Chapter 10, page 358

Some manufacturers use an odometer with seven wheels to display mileages and kilometers over 100,000.

Odometers

The odometer is driven by the speedometer cable through a worm gear (Figure 10-5). The odometer usually has six wheels; each wheel is numbered 0 through 9. The wheels are geared together so that as the right wheel makes one full revolution, the next wheel to the left moves one position. This action is repeated for all of the wheels.

A BIT OF HISTORY

One of the early styles of speedometers used a regulated amount of air pressure to turn a speed dial. The air pressure was generated in a chamber containing two intermeshing gears. The gears were driven by a flexible shaft that was connected to a front wheel or the driveshaft. The air was applied against a vane inside of the speed dial. The amount of air applied was proportional to the speed of the vehicle.

WARNING: Federal and state laws prohibit tampering with the correct mileage as indicated on the odometer. If the odometer must be replaced, it must be set to the reading of the original odometer. Or, a door sticker must be installed indicating the reading of the odometer when it was replaced.

The **odometer** is a mechanical counter in the speedometer unit that indicates total miles accumulated on the vehicle. Many vehicles also have a second odometer that can be reset to zero. It is referred to as a **trip odometer.**

A **worm gear** is a spiral cut gear on a shaft.

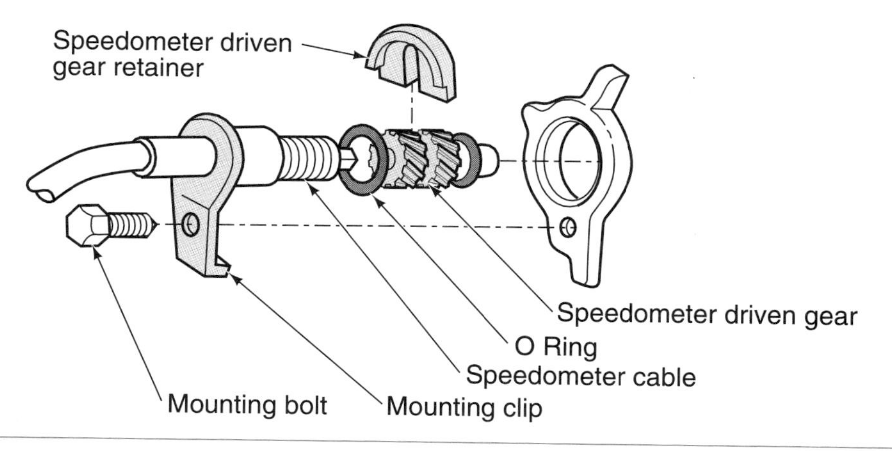

Figure 10-3 Speedometer cable connection to the transmission.

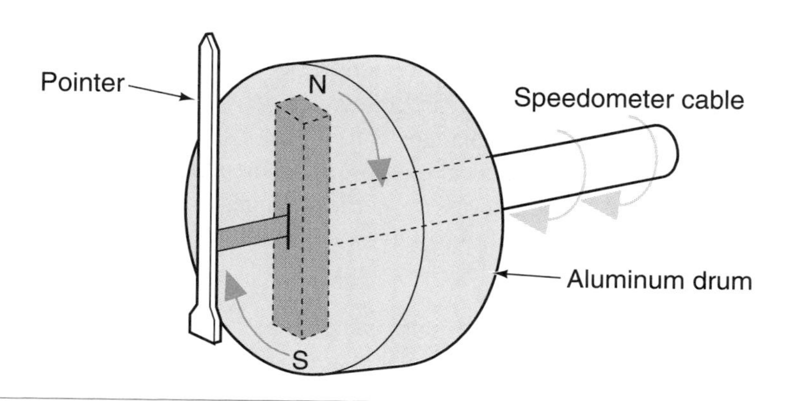

Figure 10-4 Eddy currents cause the needle to move across the scale.

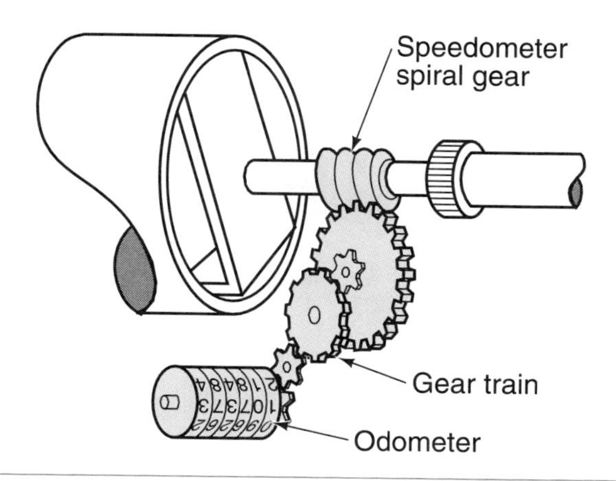

Figure 10-5 Odometer gears and wheels.

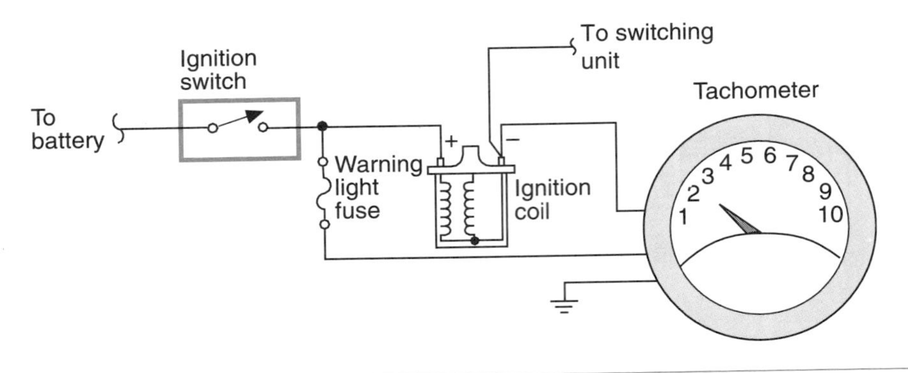

Figure 10-6 Electrical tachometer wired into the ignition system.

A **tachometer** is an instrument that measures the speed of the engine in revolutions per minute (rpm).

Tachometers

Shop Manual
Chapter 10, page 358

An electrical tachometer receives voltage pulses from the ignition system—usually the coil (Figure 10-6). Each of the voltage pulses represents the generation of one spark at the spark plug. The rate of spark plug firing is in direct relationship to the speed of the engine. A circuit within the tachometer converts the ignition pulse signal into a varying voltage. The voltage is applied to a voltmeter that serves as the engine speed indicator.

Gauges

Shop Manual
Chapter 10, page 359

A **gauge** is a device that displays the measurement of a monitored system by the use of a needle or pointer that moves along a calibrated scale.

Most instrument panels will include at least one gauge. The electromechanical gauge acts as an ammeter because the gauge reading changes with variations in the sender unit resistance. There are two basic types of electromechanical gauges: the bimetallic gauge and the electromagnetic gauge. The basic principles of operation of electrical gauges are very simple. Once these principles are mastered, the technician should have no difficulty in locating and repairing a faulty gauge circuit.

WARNING: These gauges are called analog because they use needle movement to indicate current levels. However, many newer instrument panels use computer driven analog gauges that operate under different principles. It is important that the technician follows the manufacturer's procedures for testing the gauges or gauge damage will result.

A BIT OF HISTORY

In 1914, Studebaker introduced a dash-mounted fuel gauge. Before that gas levels were measured with a wooden dipstick.

Instrument Voltage Regulator

Shop Manual
Chapter 10, page 362

Some gauge systems require the use of an instrument voltage regulator (IVR) that provides a constant voltage to the gauge so it will read accurately under all charging conditions. The IVR consists of a set of normally closed contacts and a heating coil wrapped around a bimetallic strip (Figure 10-7). When voltage is applied to the IVR, current flows through the contacts and the heating coil. Because the heating coil is connected to a ground, the flowing current heats the coil and

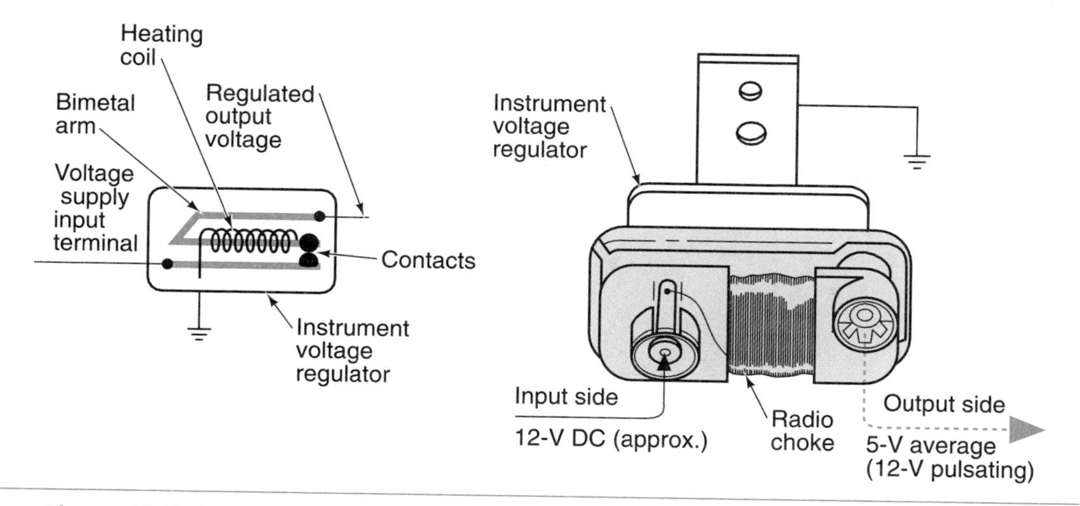

Figure 10-7 Instrument voltage regulator (IVR).

The gauge is called an **electro-mechanical device** because it is operated electrically. However, its movement is mechanical.

The **sender unit** is the sensor for the gauge. It is a variable resistor that changes resistance values with changing monitored conditions.

The **instrument voltage regulator (IVR)** provides a constant voltage to the gauge regardless of the voltage output of the charging system.

Opened is the term used to mean that current flow is stopped. The path for electron flow is broken by opening the circuit. **Closed circuit** means there are no breaks in the path and current will flow.

the bimetallic strip bends. When the strip bends, the contacts open and current flows through the IVR and the heater coil stops. With no current flowing through the coil, the bimetallic strip cools and straightens. This closes the contact points and the process is repeated. This constant opening and closing of the contact points causes a pulsating DC current with a varying potential equivalent in heating effects to approximately 5 volts. The equivalent voltage is constant, and is applied to the gauge. The IVR output is approximately 5 volts, regardless of the generator output.

Bimetallic Gauges

Bimetallic gauges (or thermoelectric gauges) are simple dial and needle indicators that transform the heating effect of electricity into mechanical movement. The movement of the needle is slow and smooth, so the needle avoids responding to sudden changes that could cause a flickering needle. This prevents the gauges from constantly moving as a result of vehicle movement (as would be the case with a fuel gauge for example).

The construction of the bimetallic gauge features an indicating needle that is linked to the free arm of a U-shaped bimetallic strip (Figure 10-8). The free arm has a heater coil that is connected to the gauge terminal posts. When current flows through the heater coil, it heats the bimetallic arm. This causes the arm to bend and moves the needle across the gauge dial.

The amount that the bimetallic strip bends is proportional to the heat produced in the heater coil. When the current flow through the heater coil is small, less heat is created and needle movement is slight. If the current through the heater coil is increased, more heat is created

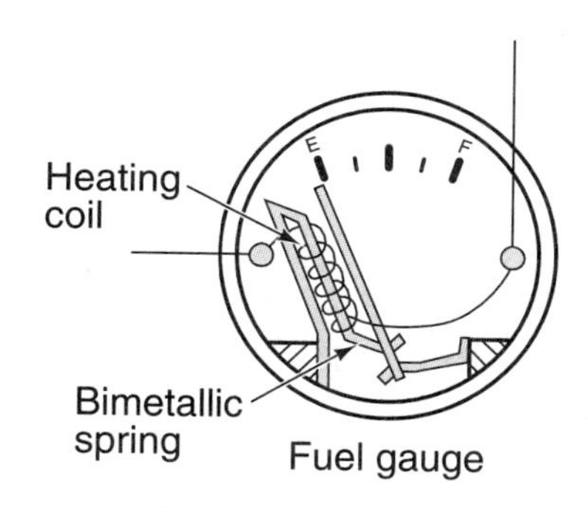

Figure 10-8 Bimetallic gauge construction. (Reprinted with the permission of Ford Motor Company)

A **bimetallic strip** consists of two different types of metals. Heating one of the metals more rapidly than the other causes the strip to flex in proportion to the amount of heat.

Opening and closing of the contact points creates induced voltage (like that in an ignition coil). This voltage is directed through a **radio choke** that absorbs it and prevents static in the vehicle's radio.

The bimetallic gauge has the advantage that it is not affected by changes in the charging system because gauge input is controlled by the IVR.

A **variable resistor** provides for an infinite number of resistance values within a range.

Electromagnetic gauges produce needle movement by magnetic forces instead of heat.

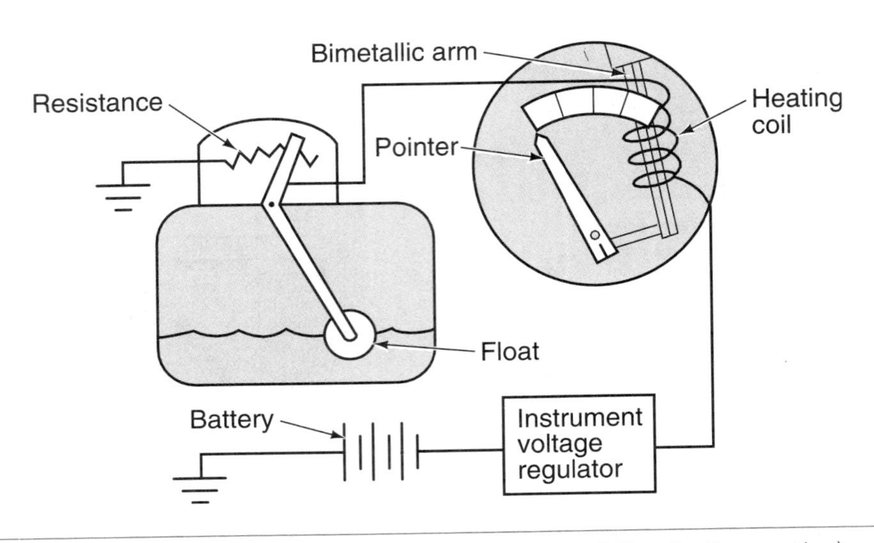

Figure 10-9 Typical bimetallic fuel gauge circuit. (Courtesy of Chrysler Corporation)

and the needle will move more. The variable resistance used to control the current flow through the heater coil is in the sending unit (Figure 10-9).

The bimetallic gauge uses heat to move the needle. Thus it is influenced by changes as a result of high or low ambient temperatures. Change in ambient temperatures is compensated for by the U-shape of the bimetallic element. When ambient temperature bends the free arm, it also affects the fixed arm. The effect on the fixed arm is equal to the free arm but in the opposite direction. This action then cancels the effects of outside temperatures.

Electromagnetic Gauges

There are four types of electromagnetic gauges: the d'Arsonval, the three coil, the two coil, and the air core.

The d'Arsonval Gauge. The d'Arsonval gauge consists of a permanent horseshoe-type magnet that surrounds a moveable electromagnet (armature) that is attached to a needle (Figure 10-10). When current flows through the armature it becomes an electromagnet and is repelled by the permanent magnet. When current flow through the armature is low, the strength of the electromagnet is weak and needle movement is small. When the current flow is increased, the magnetic

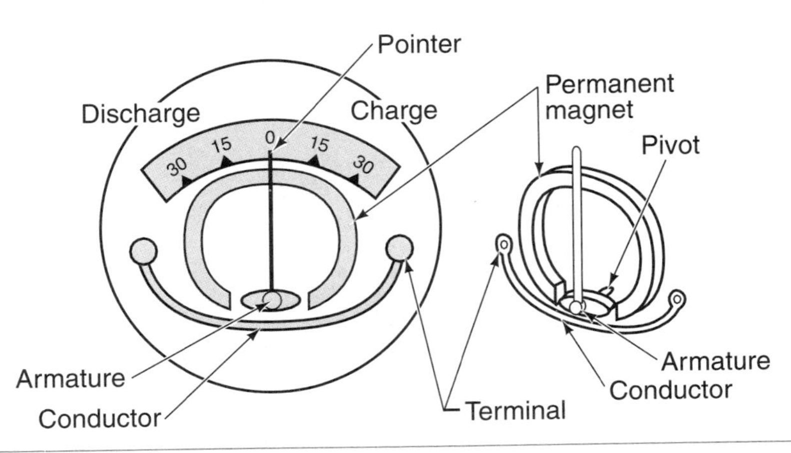

Figure 10-10 d'Arsonval gauge needle movement. (Reprinted with the permission of Ford Motor Company)

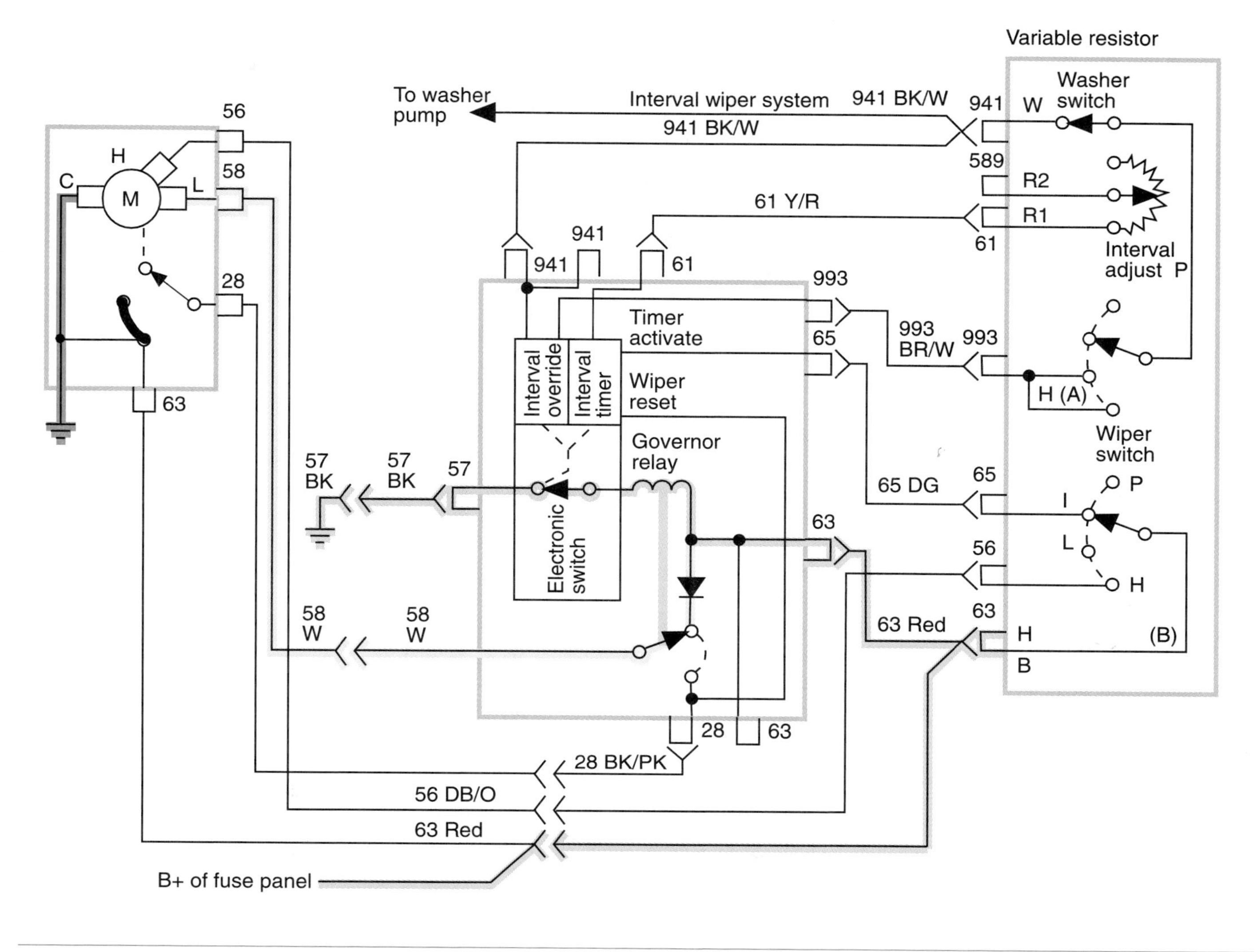

Figure 11-29 Current flow when intermittent wiper mode is initiated. (Reprinted with the permission of Ford Motor Company)

intermittent wiper mode is initiated when the wipers are in their parked position, the park switch is in the ground position. Current is sent to the solid-state module to the timer activate terminal (Figure 11-28). The internal timer unit "triggers" the electronic switch to close the circuit for the governor relay, which then closes the circuit to the low speed brush (Figure 11-29). The wiper will operate until the park switch swings back to the PARK position.

The delay between wiper sweeps is determined by the amount of resistance the driver puts into the potentiometer control. By rotating the intermittent control knob, the resistance value is altered. The module contains a capacitor that is charged through the potentiometer. Once the capacitor is saturated, the electronic switch is "triggered" to send current to the wiper motor. The capacitor discharge is long enough to start the wiper operation and the park switch is returned to the RUN position. The wiper will continue to run until one sweep is completed and the park switch opens. The amount of time between sweeps is based on the length of time required to saturate the capacitor. As more resistance is added to the potentiometer, it takes longer to saturate the capacitor.

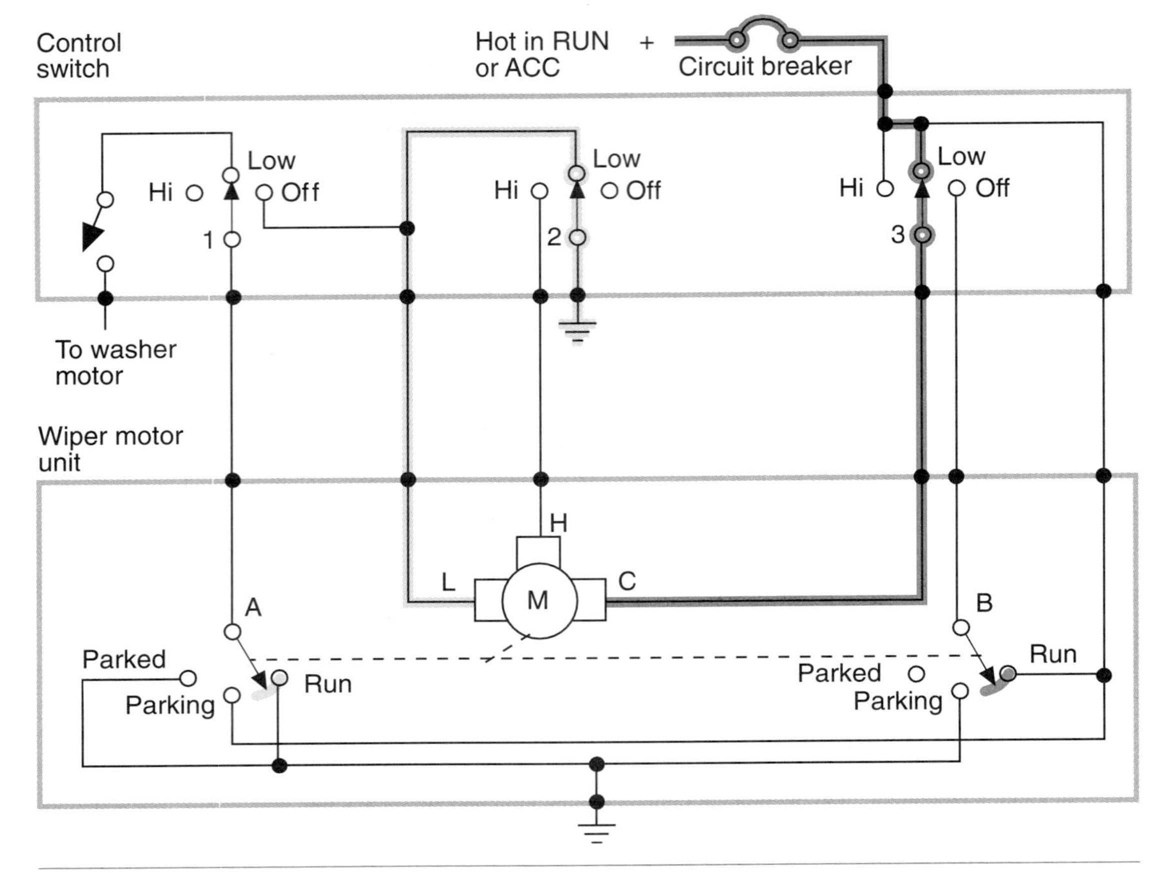

Figure 11-30 Depressed park wiper system in LOW SPEED position.

Depressed Wiper Systems

Systems that have a depressed-park feature use a second set of contacts with the park switch. These contacts are used to reverse the rotation of the motor for about 15 degrees after the wipers have reached the normal park position. The circuitry of the depressed circuit is different from that of standard wiper motors.

Note the difference in operation between the depressed-park wiper system and the one shown in Figure 11-16.

The operation of a depressed-park wiper system in the LOW SPEED position is shown (Figure 11-30). Current flows through the number 3 wiper to the common brush. Ground is provided through the low speed brush and switch wiper 2.

When the switch is placed in the OFF position, current is supplied through the park switch wiper B and switch wiper 3 (Figure 11-31). Ground is supplied through the low speed brush and switch wiper 1, then to park switch wiper A.

When the wipers reach their park position, the park switch swings to the PARKING position (Figure 11-32). Current flow is through the park switch wiper A, to switch wiper 1. Wiper 1 directs the current to the low speed brush. The ground path is through the common brush, switch wiper 3, and park switch wiper B. This reversed current flow is continued until the wipers reach the depressed-park position, when park switch wiper A swings to the PARKED position.

Shop Manual
Chapter 11, page 396

Washer Pumps

Windshield washers spray a washer fluid solution onto the windshield and work in conjunction with the wiper blades to clean the windshield of dirt. Some vehicles that have composite headlights

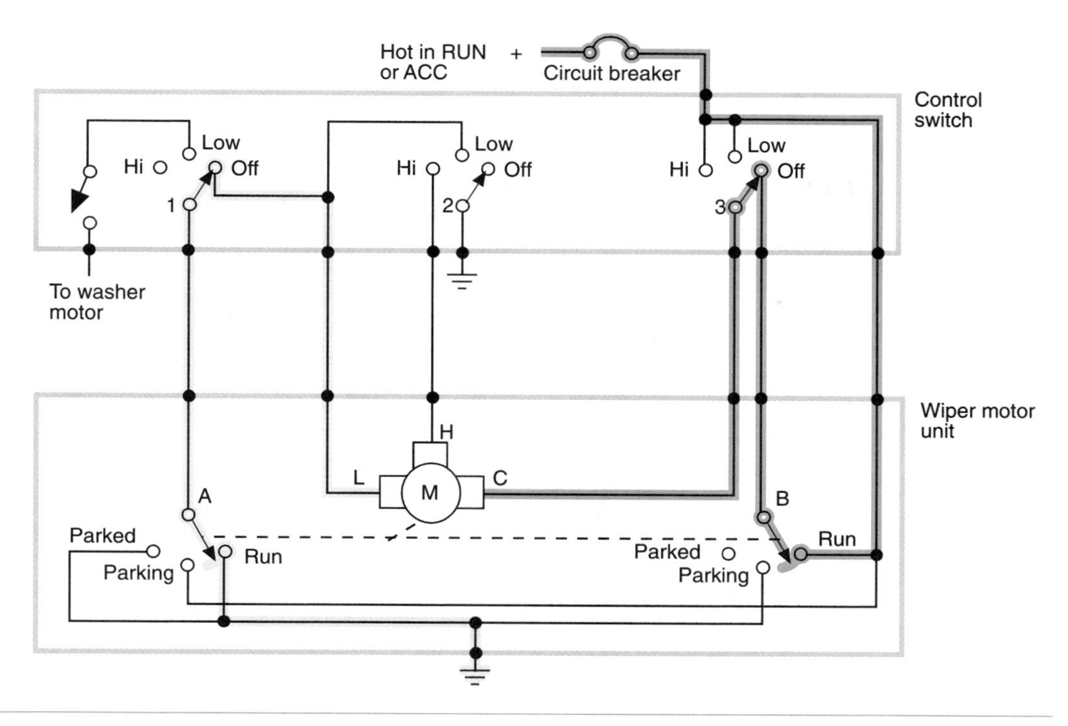

Figure 11-31 Current flow when the switch is turned OFF after operation (parking).

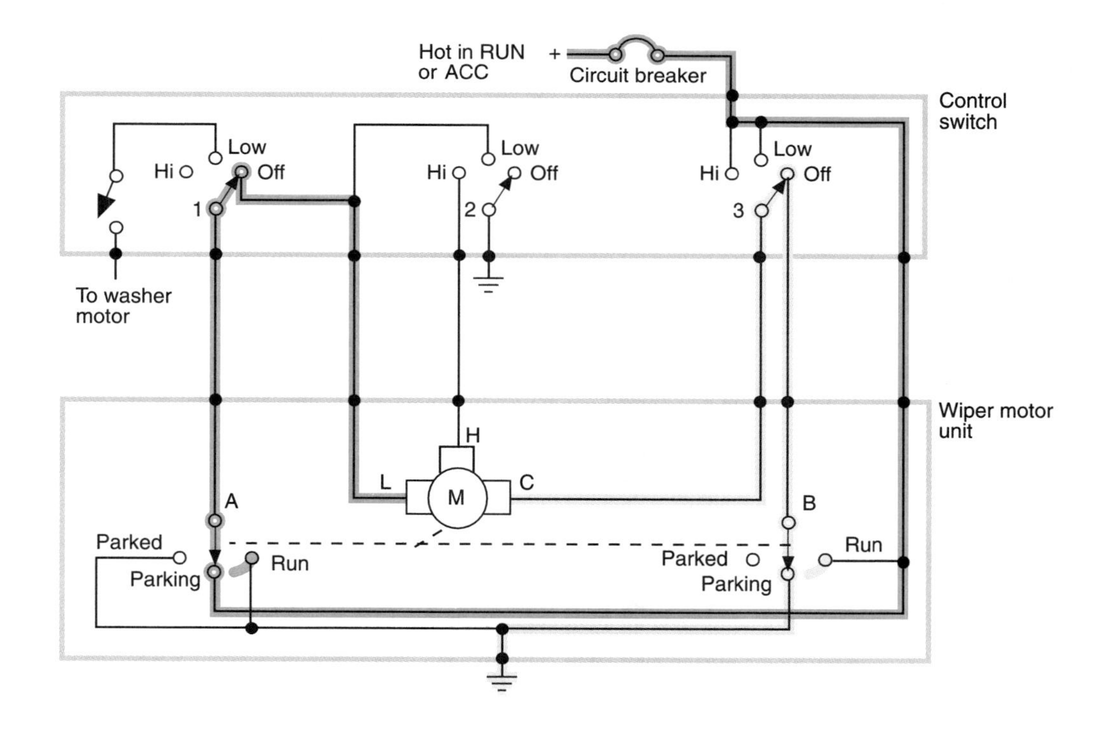

Figure 11-32 Current flow when wipers are parking into the depressed position.

The washer system includes plastic or rubber hoses to direct fluid flow to the nozzles and produce the spray pattern.

incorporate a headlight washing system along with the windshield washer (Figure 11-33). Most systems have the washer pump motor installed into the reservoir (Figure 11-34). General Motors uses a pulse-type washer pump that operates off the wiper motor (Figure 11-35).

The system is activated by holding the washer switch (Figure 11-36). If the wiper/washer system also has an intermittent control module, a signal is sent to the module when the washer switch is activated (Figure 11-37). An override circuit in the module operates the wipers on low speed for a programmed length of time. The wipers will either return to the parked position or will operate in intermittent mode, depending on system design.

Figure 11-33 Headlight washer system may operate with the windshield washer or have a separate switch.

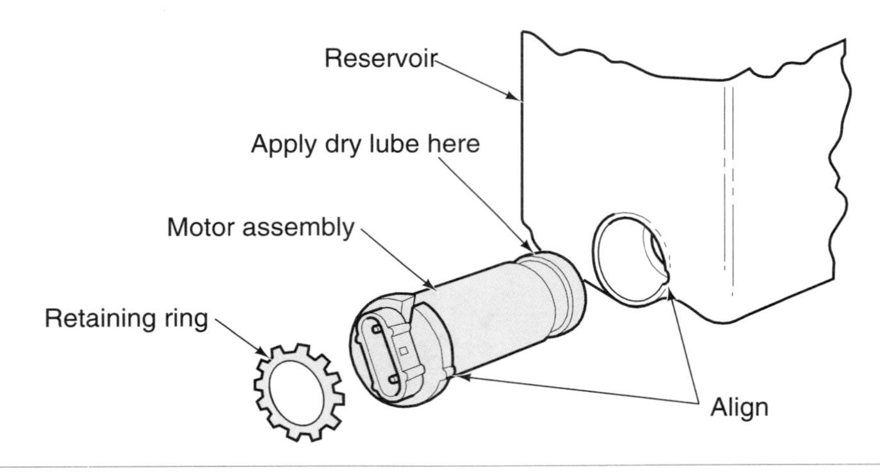

Figure 11-34 Washer motor installed into the reservoir. (Reprinted with the permission of Ford Motor Company)

Figure 11-35 General Motors' pulse-type washer system incorporates the washer motor into the wiper motor.

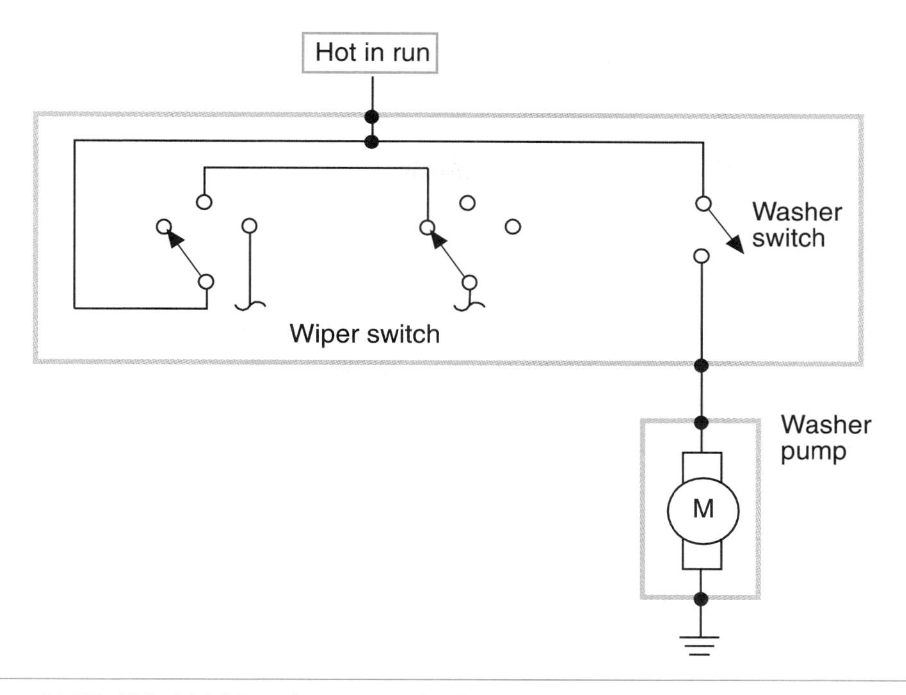

Figure 11-36 Windshield washer motor circuit.

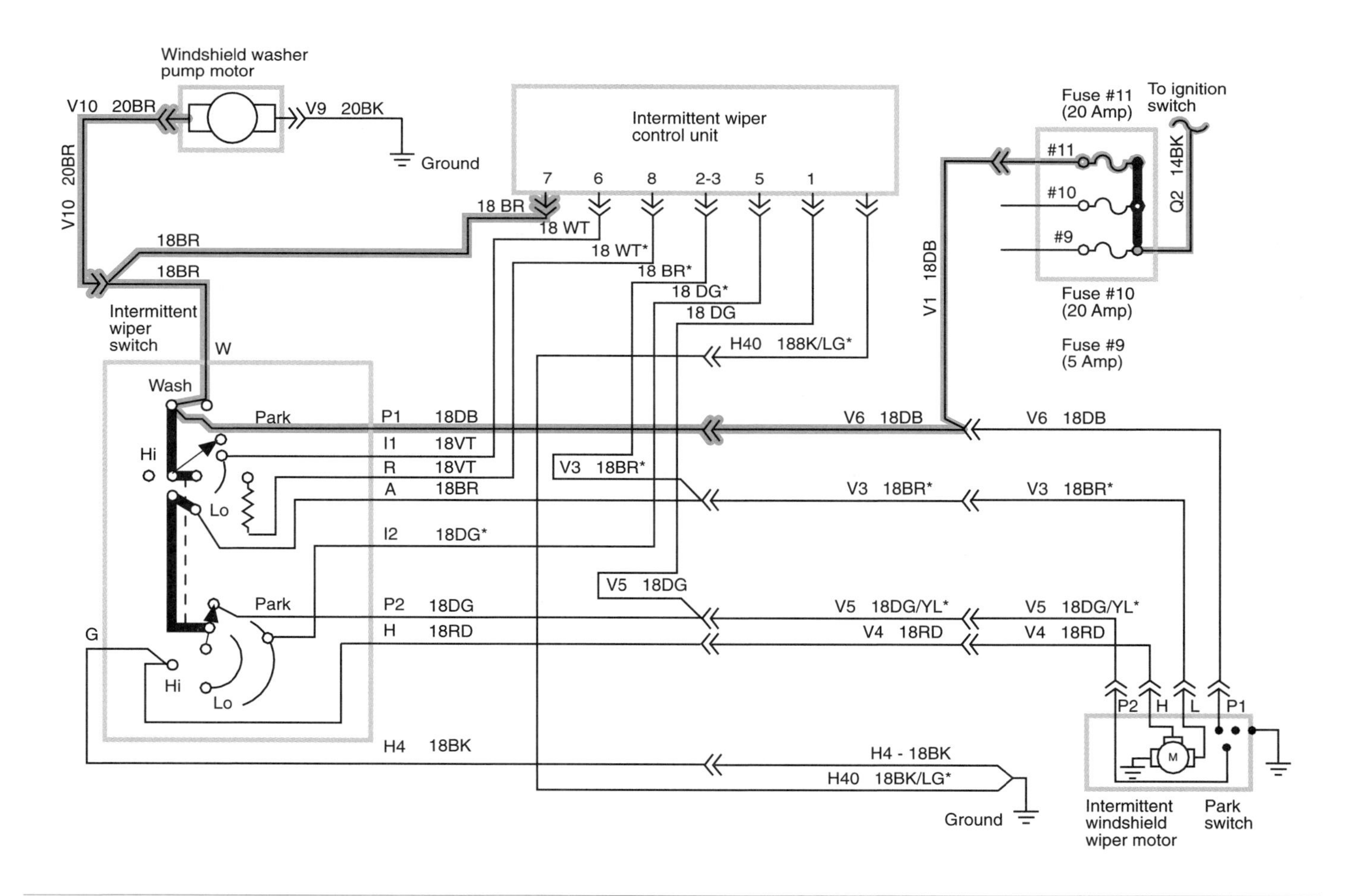

Figure 11-37 Input signal alerts the module that the washers are activated. (Courtesy of Chrysler Corporation)

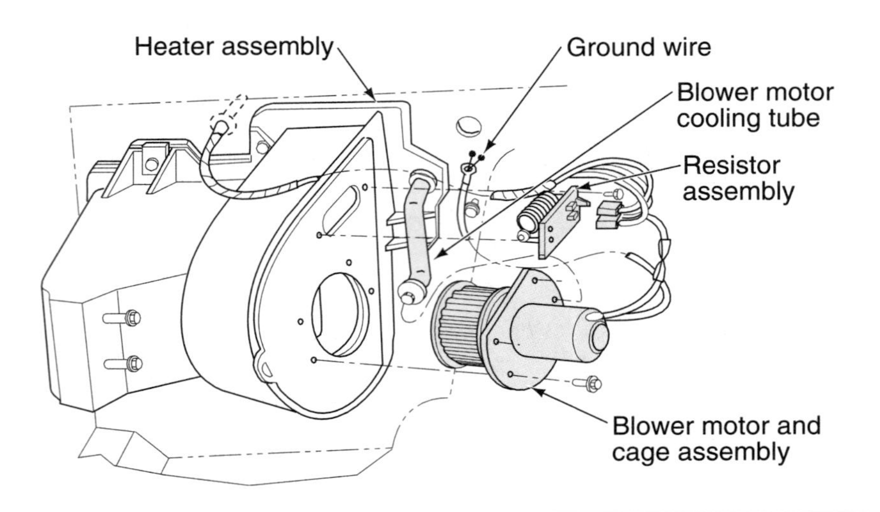

Figure 11-38 The blower motor is usually installed into the heater assembly. Mode doors control if vent, heater, or A/C cooled air is blown by the motor cage.

Shop Manual
Chapter 11,
page 398

The **resistor block** consists of two or three helically wound wire resistors wired in series.

Blower Motor Circuit

The blower motor is used to move air inside the vehicle for air conditioning, heating, defrost, and ventilation. The motor is usually a permanent magnet, single-speed motor and is located in the heater housing assembly (Figure 11-38). A blower motor switch mounted on the dash controls the fan speed (Figure 11-39). The switch position directs current flow to a resistor block that is wired in series between the switch and the motor (Figure 11-40).

The blower motor circuit includes the control assembly, blower switch, resistor block, and the blower motor (Figure 11-41). This system uses an insulated side switch and a grounded motor. Battery voltage is applied to the control head when the ignition switch is in the RUN or ACC positions. The current can flow from the control head to the blower switch and resistor block in any control head position except OFF.

When the blower switch is in the LOW position, the blower switch wiper opens the circuit. Current can only flow to the resistor block directly through the control head. The current must

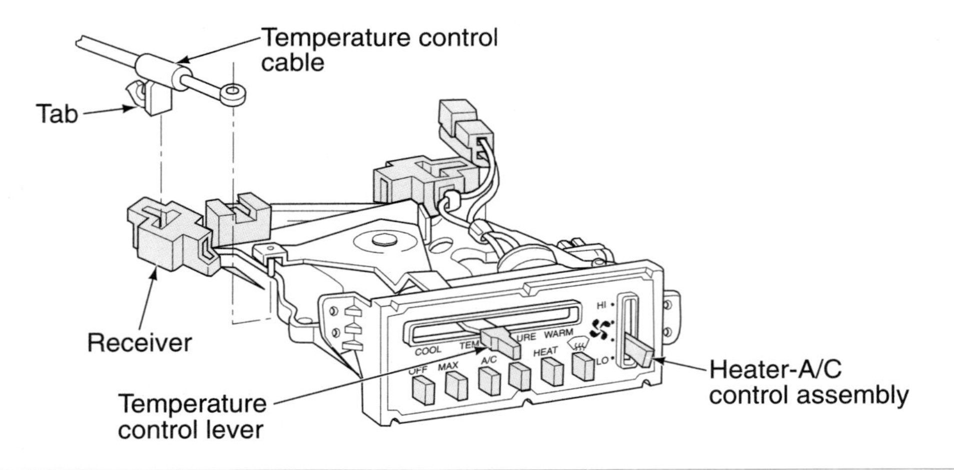

Figure 11-39 Comfort control assembly. (Courtesy of Chrysler Corporation)

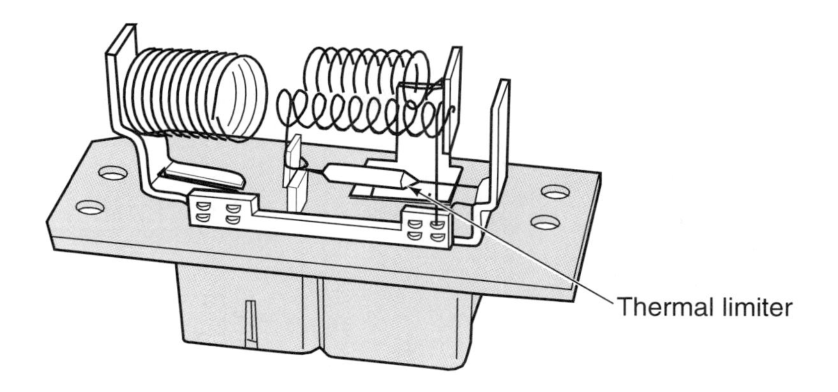

Figure 11-40 Fan motor resistor block. (Reprinted with the permission of Ford Motor Company)

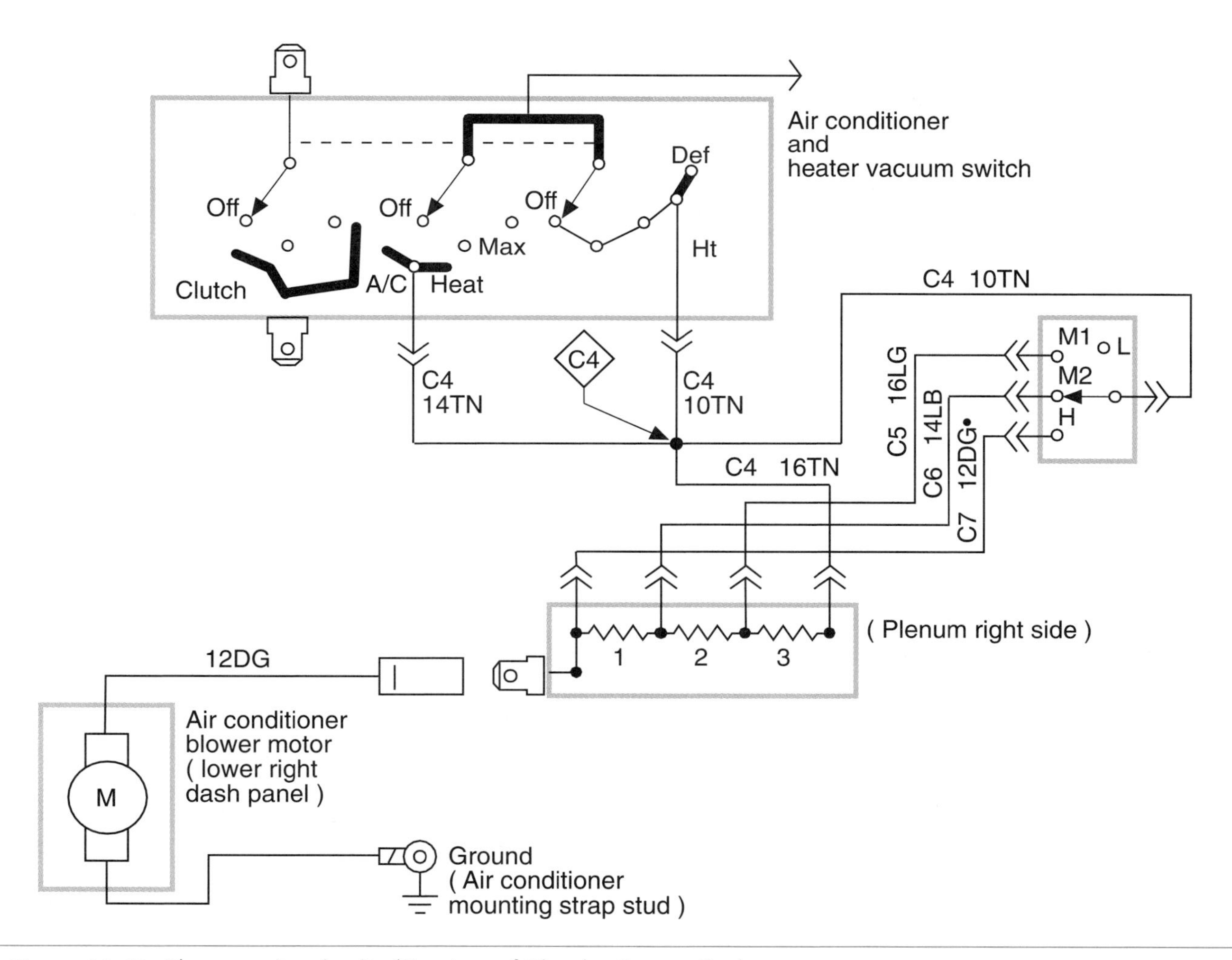

Figure 11-41 Blower motor circuit. (Courtesy of Chrysler Corporation)

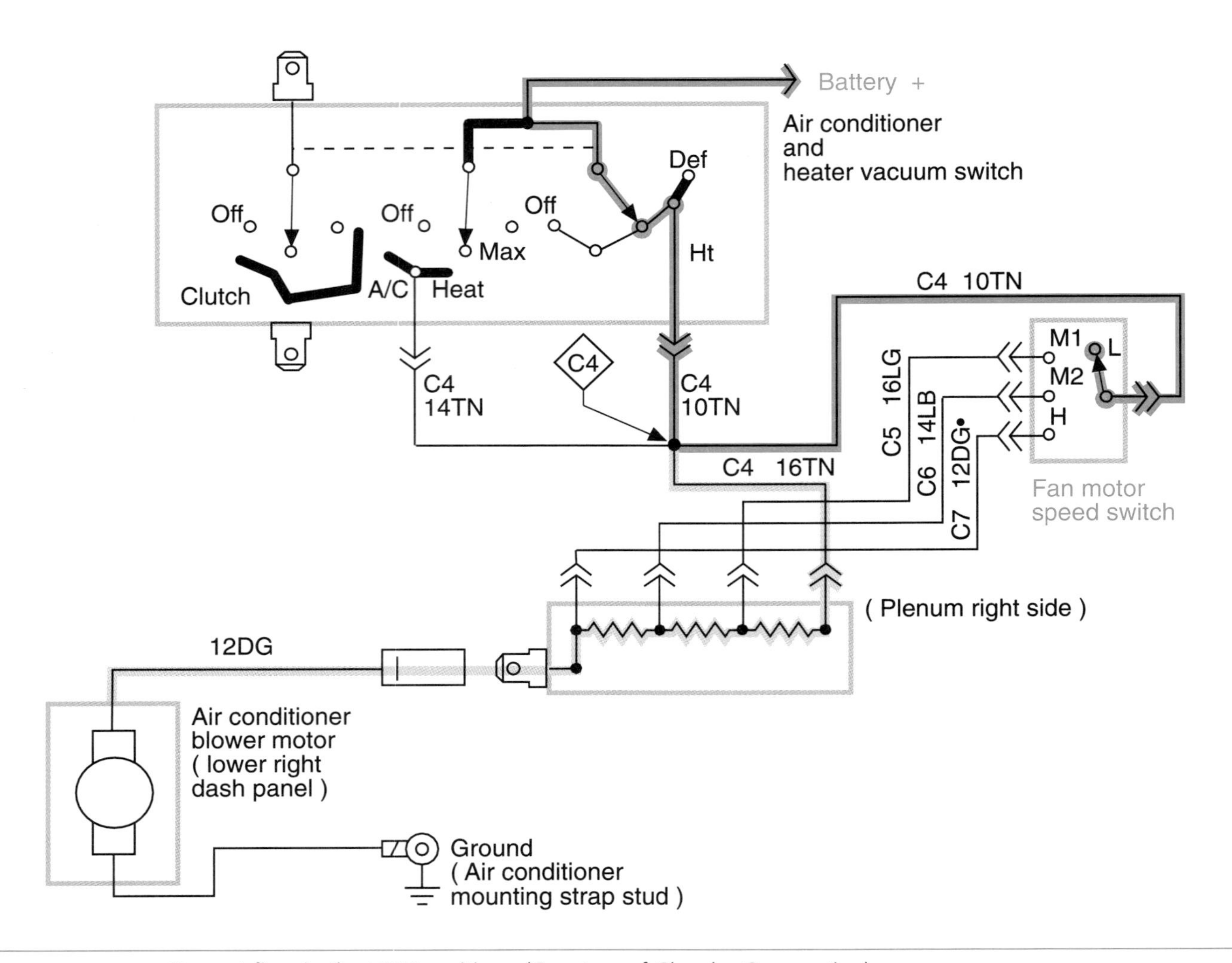

Figure 11-42 Current flow in the LOW position. (Courtesy of Chrysler Corporation)

Many of today's vehicles are using the body computer to control fan speed by pulse width modulation.

pass through all the resistors before reaching the motor. With the voltage dropped over the resistors, the motor speed is slowed (Figure 11-42).

When the blower switch is placed in the MED 1, MED 2, or HIGH position, the current flows through the blower switch to the resistor block. Depending on the speed selection, the current must pass through one, two, or none of the resistors (Figure 11-43). With more applied voltage to the motor, the fan speed is increased as the amount of resistance decreases.

Current through the circuit will remain constant; varying the amount of resistance changes the voltage applied to the motor. Because the motor is a single-speed motor, it obtains its fastest rotational speed with full battery voltage. The resistors drop the amount of voltage to the motor, resulting in slower speeds.

Some manufacturers use ground side switching with an insulated motor (Figure 11-44). The switch completes the circuit to ground. Depending on wiper position, current flow is directed through the resistor block. The operating principles are identical to that of the insulated switch already discussed.

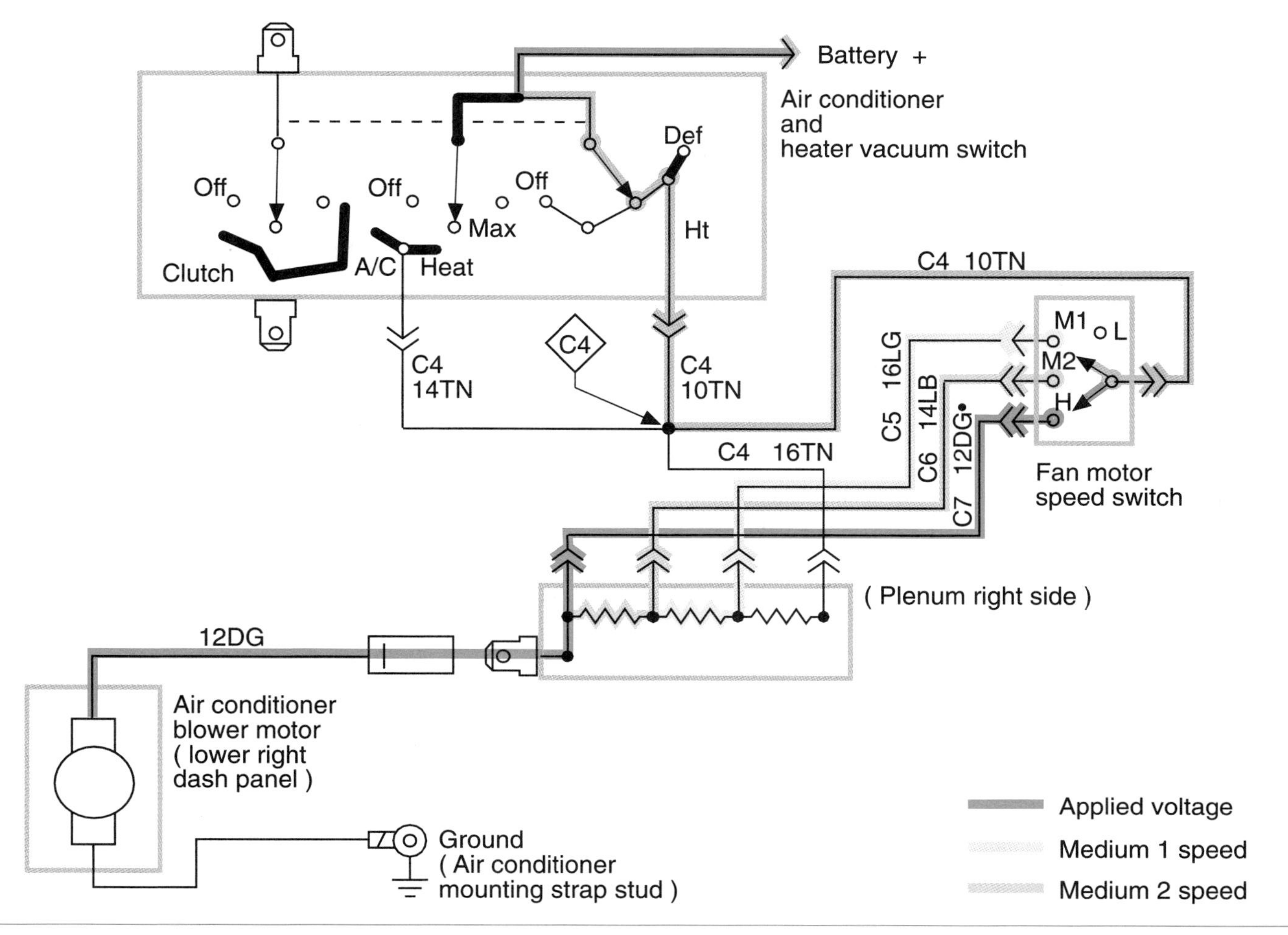

Figure 11-43 Current flow in the different speed selections. (Courtesy of Chrysler Corporation)

A BIT OF HISTORY

Automotive electric heaters were introduced at the 1917 National Auto Show. Hot water in-car heaters were introduced in 1926.

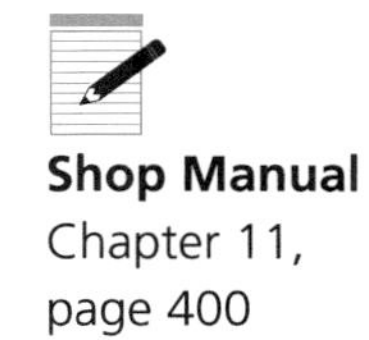

Shop Manual
Chapter 11, page 400

Electric Defoggers

Electric defoggers heat the rear window to remove ice and/or condensation. Some vehicles use the same circuit to heat the outside driver's side mirror.

When electrons are forced to flow through a resistance, heat is generated. Rear window defoggers use this principle of controlled resistance to heat the glass. The resistance is through a grid that is baked on the inside of the glass (Figure 11-45). The terminals are soldered to the vertical bus bars. One terminal supplies the current from the switch; the other provides the ground (Figure 11-46).

The system may incorporate a timer circuit that controls the relay (Figure 11-47). The timer is used due to the high amount of current required to operate the system (approximately 30 amperes). If this drain were allowed to continue for extended periods of time, battery and charging system failure could result. Because of the high current draw, most vehicles equipped with a rear window defogger use a high output AC generator.

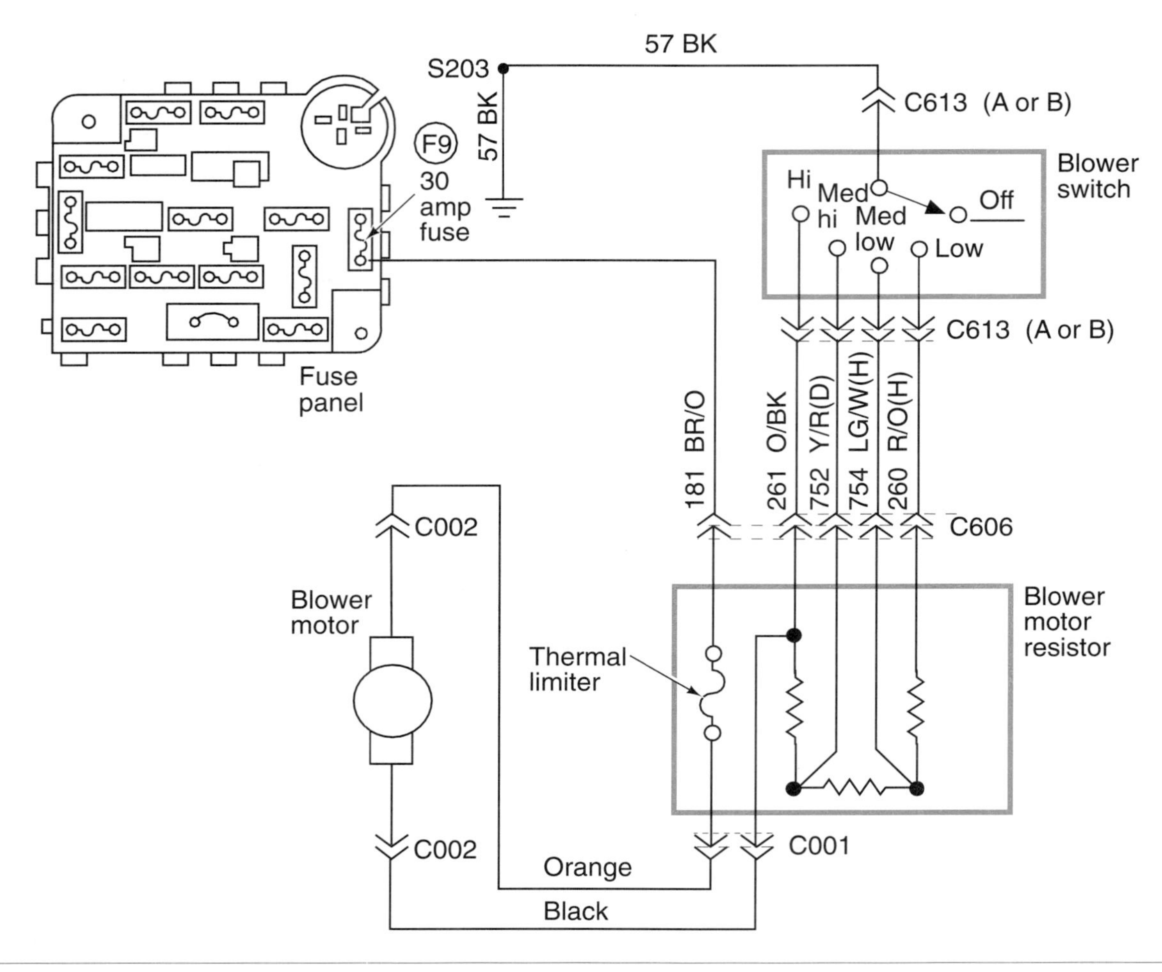

Figure 11-44 Ground side switch to control the blower motor system. (Reprinted with the permission of Ford Motor Company)

The rear window defogger **grid** is a series of horizontal ceramic silver-compounded lines baked into the surface of the window.

Figure 11-45 Rear window defogger grid.

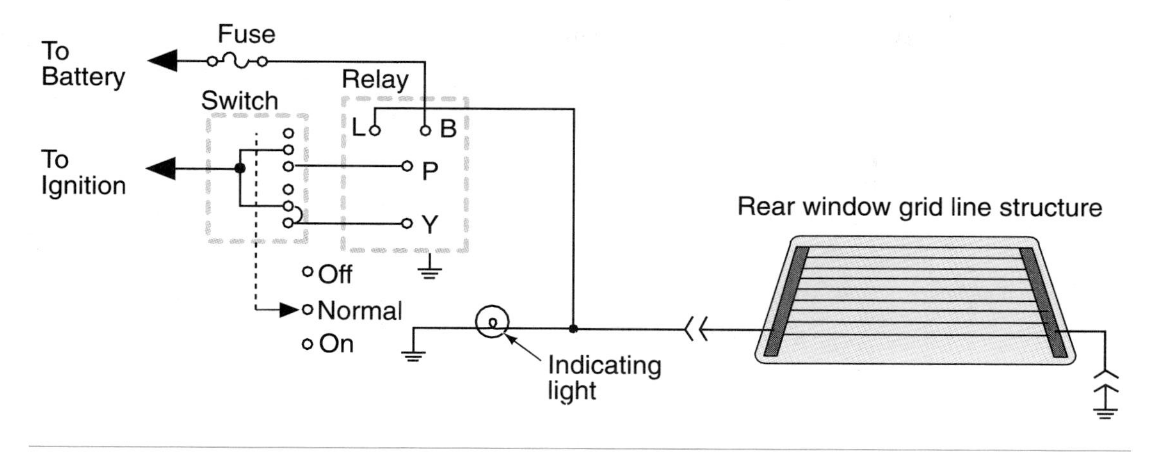

Figure 11-46 Rear window defogger circuit schematic.

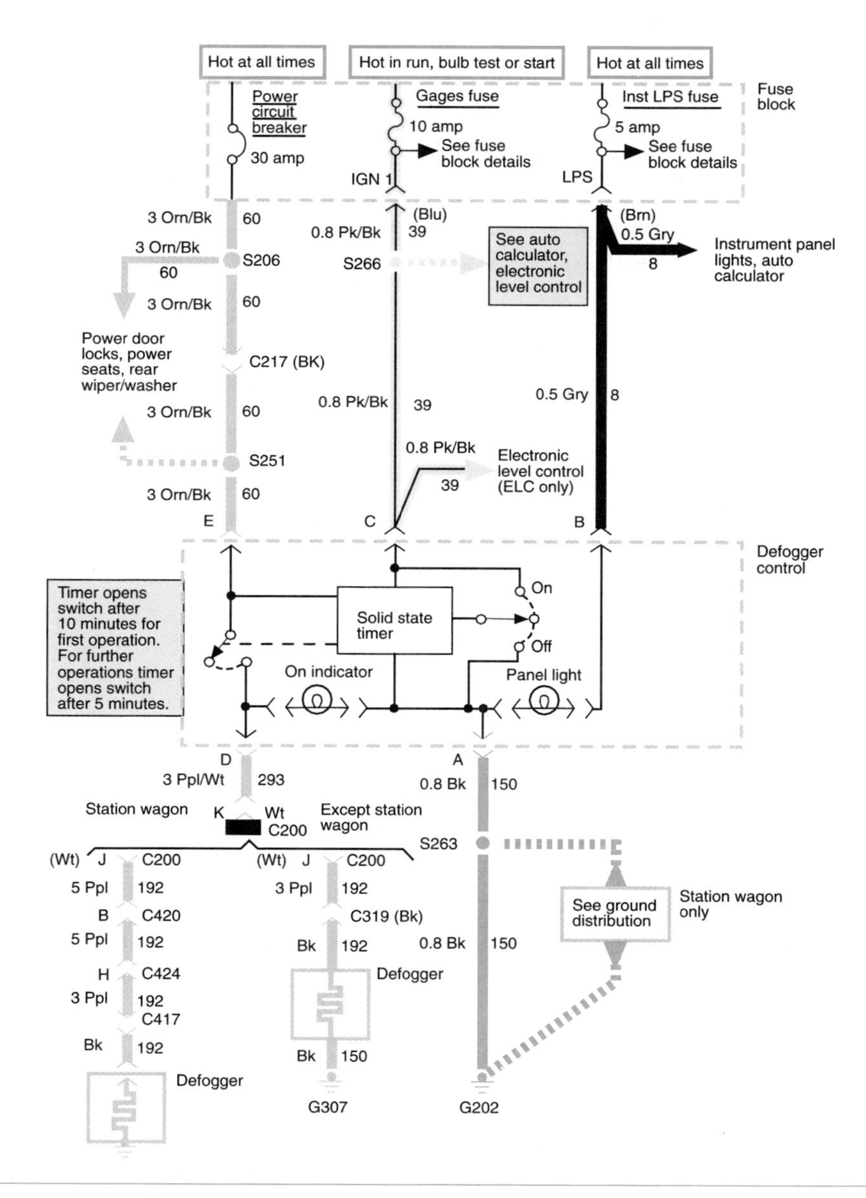

Figure 11-47 Defogger circuit using a solid-state timer.

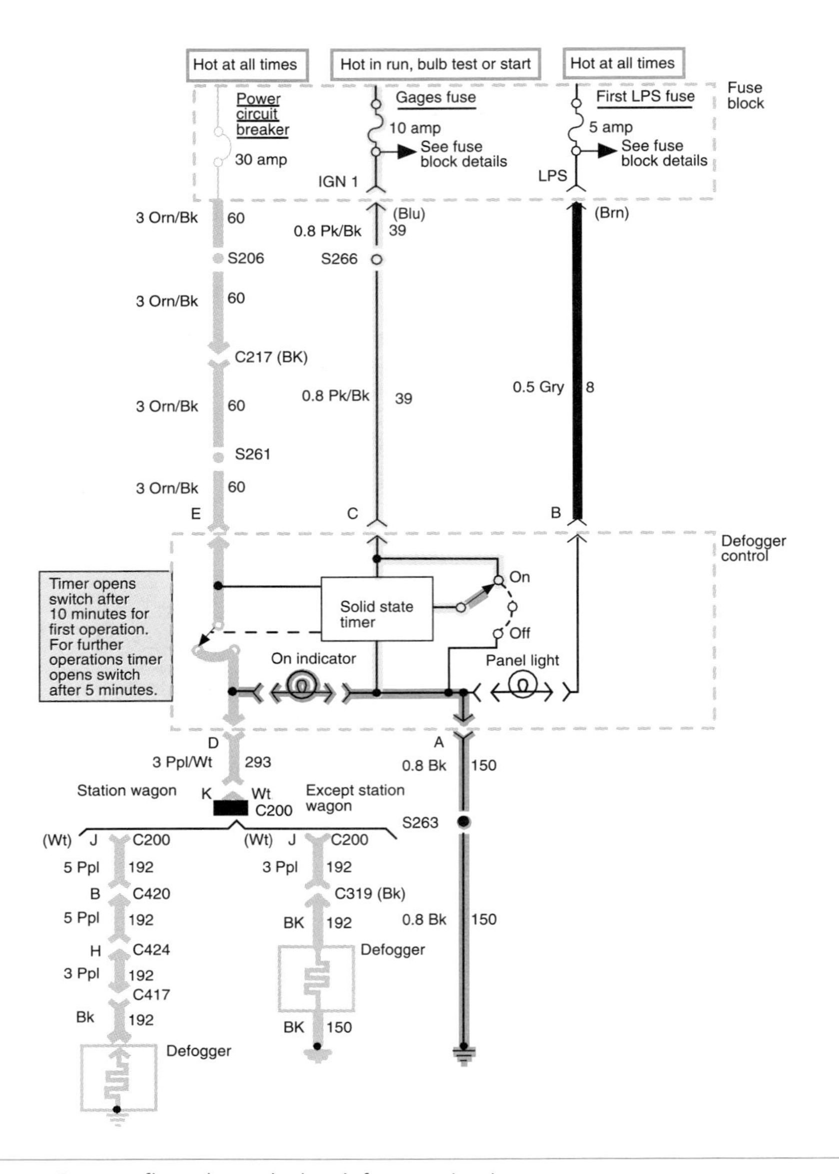

Figure 11-48 Current flow through the defogger circuit.

The ON indicator can be either a bulb or light-emitting diode (LED).

The control switch is a three-position, spring-loaded switch that returns to the center position after making momentary contact to the ON or OFF terminals. Activation of the switch energizes the electronic timing circuit, which energizes the relay coil. With the relay contacts closed, direct battery voltage is sent to the heater grid (Figure 11-48). At the same time, voltage is applied to the ON indicator. The timer is activated for 10 minutes. At the completion of the timed cycle, the relay is de-energized and the circuit to the grid and indicator light is broken. If the switch is activated again, the timer will energize the relay for 5 minutes.

The timer sequence can be aborted by moving the switch to the OFF position or by turning off the ignition switch. If the ignition switch is turned off while the timer circuit is activated, the rear window defogger switch will have to be returned to the ON position to activate the system again.

Ambient temperatures have an effect on electrical resistance, thus the amount of current flow through the grid depends on the temperature of the grid. As the ambient temperature

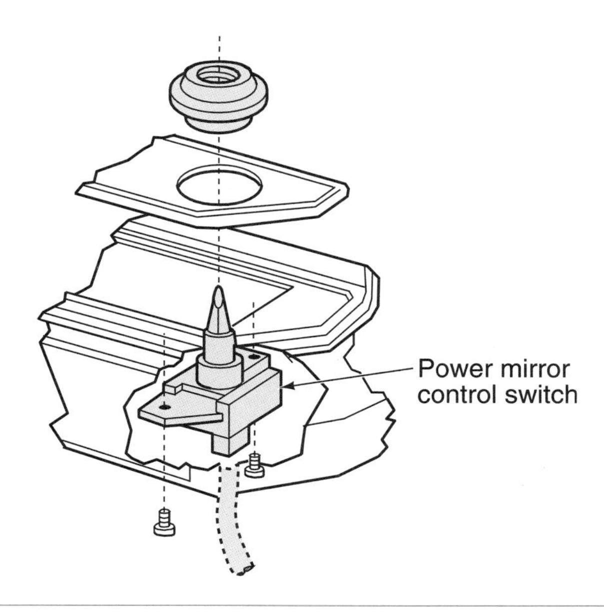

Figure 11-49 Control switch for power mirrors. (Reprinted with the permission of Ford Motor Company)

decreases, the resistance value of the grid also decreases. A decrease in resistance increases the current flow and results in quick warming of the window. The defogger system tends to be self-regulated to match the requirements for defogging.

Power Mirrors

The electrically controlled mirror allows the driver to position the outside mirrors by use of a switch (Figure 11-49). The mirror assembly will use built-in dual drive, reversible permanent magnet (PM) motors (Figure 11-50).

A single switch for controlling both the left and right side mirrors is used. On many systems, selection of the mirror to be adjusted is by rotating the knob counterclockwise for the left mirror

Power mirrors are outside mirrors that are electrically positioned from the inside of the driver compartment.

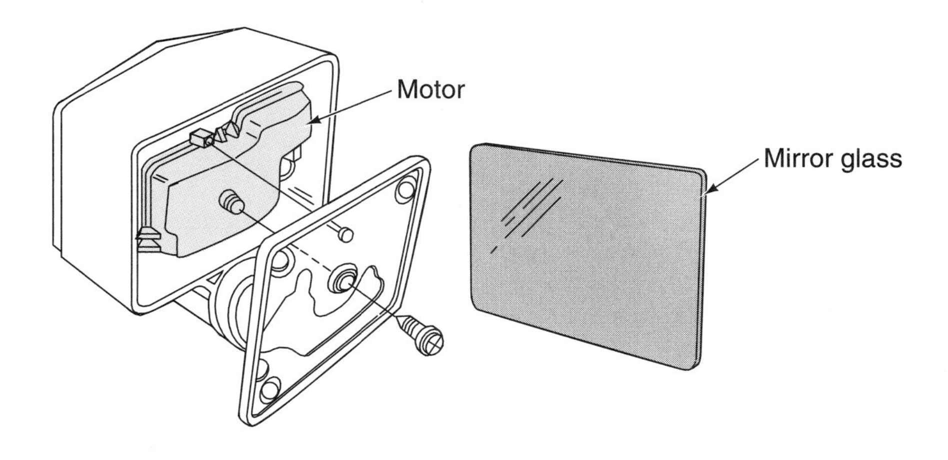

Figure 11-50 Power mirror motor.

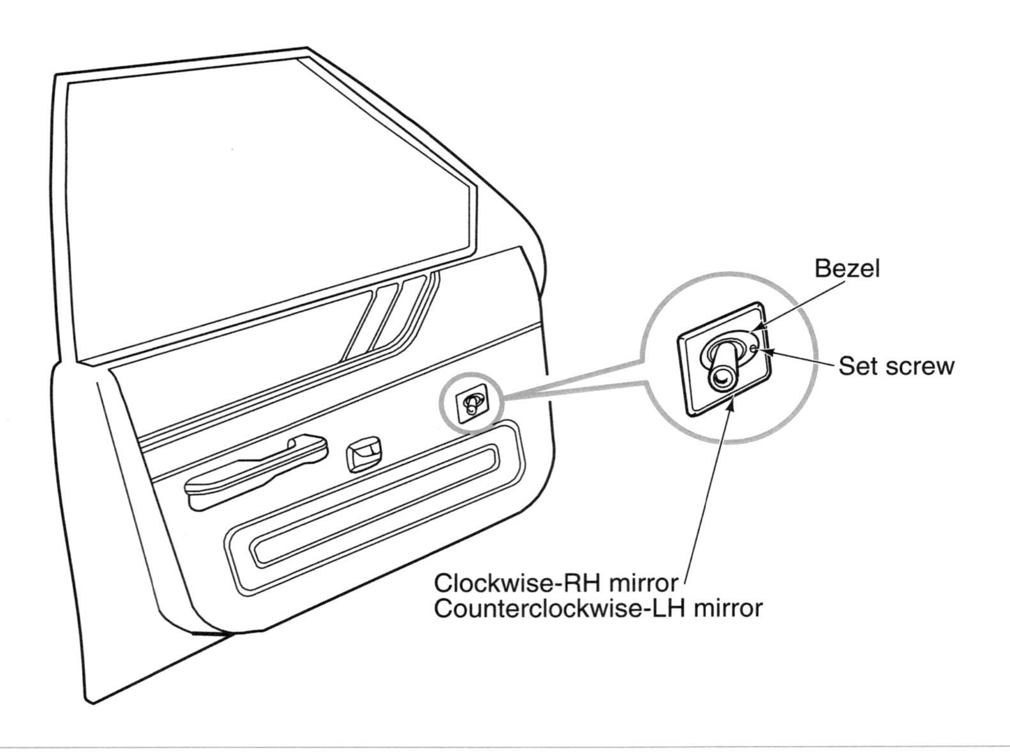

Figure 11-51 Mirror selection from the control switch. (Reprinted with the permission of Ford Motor Company)

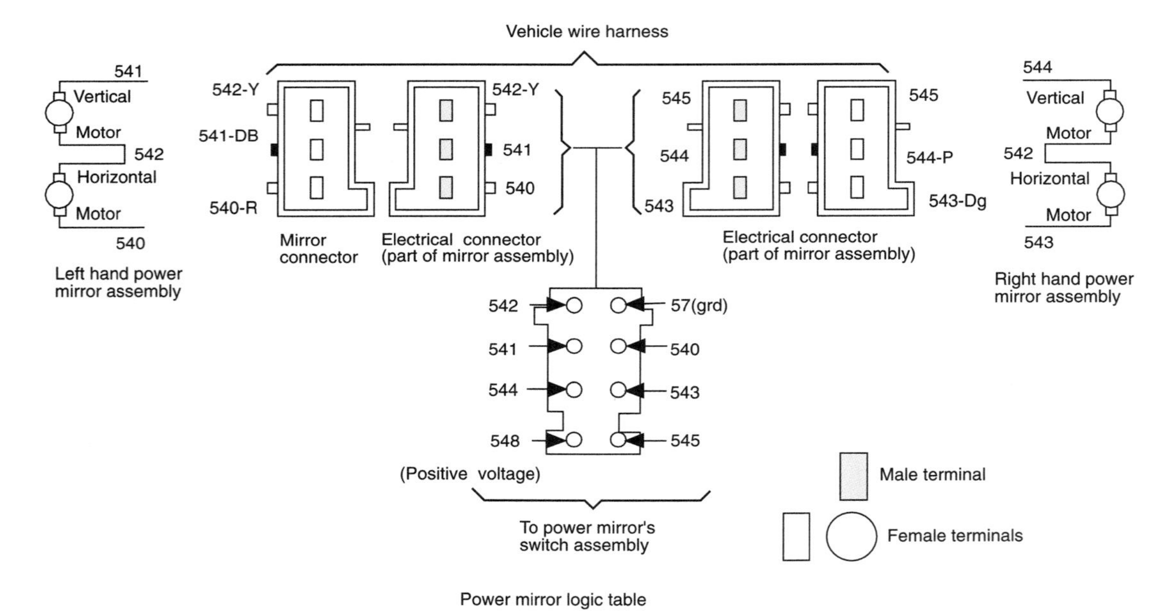

Mirror operational mode	Directional movement	Left hand mirror		Right hand mirror	
		Positive voltage (+)	Ground (-)	Positive voltage (+)	Ground (-)
Vertical	Up	541	542	544	565
Vertical	Down	542	541	545	544
Horizontal	Left	540	542	543	545
Horizontal	Right	542	540	565 545 Exp	543

Figure 11-52 Power mirror logic table. (Reprinted with the permission of Ford Motor Company)

and clockwise for the right mirror (Figure 11-51). After the mirror is selected, movement of the joystick (up, down, left, or right) moves the mirror in the corresponding direction. The illustration (Figure 11-52) shows a logic table for the mirror switch and motors.

Automatic Rear View Mirror

Some manufacturers have developed interior rear view mirrors that automatically tilt when the intensity of light that strikes the mirror is sufficient enough to cause discomfort to the driver.

The system has two photocells mounted in the mirror housing. One of the photocells is used to measure the intensity of light inside the vehicle. The second is used to measure the intensity of light being received by the mirror. When the intensity of the light striking the mirror is greater than that of ambient light, by a predetermined amount, a solenoid is activated that tilts the mirror.

Electrochromic Mirrors

Electrochromic mirrors are available that automatically adjust the amount of reflectance based on the intensity of glare (Figure 11-53). The process is similar to that used by photo-gray or photochromic sunglasses. If the glare is heavy, the mirror darkens to about 6% reflectivity. The electrochromic mirror has the advantage that it provides a comfort zone where the mirror will provide 20% to 30% reflectivity. When no glare is present, the mirror changes to the daytime reflectivity rating of up to 85%. The reduction of the glare by darkening of the mirror does not impair visibility.

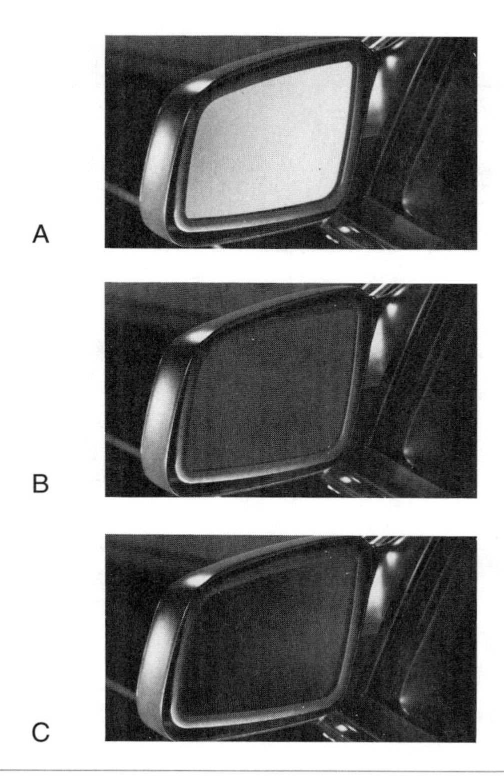

Figure 11-53 Electrochromic mirror operation: (A) day time, (B) mild glare, and (C) high glare. (Courtesy of Gentex Corporation)

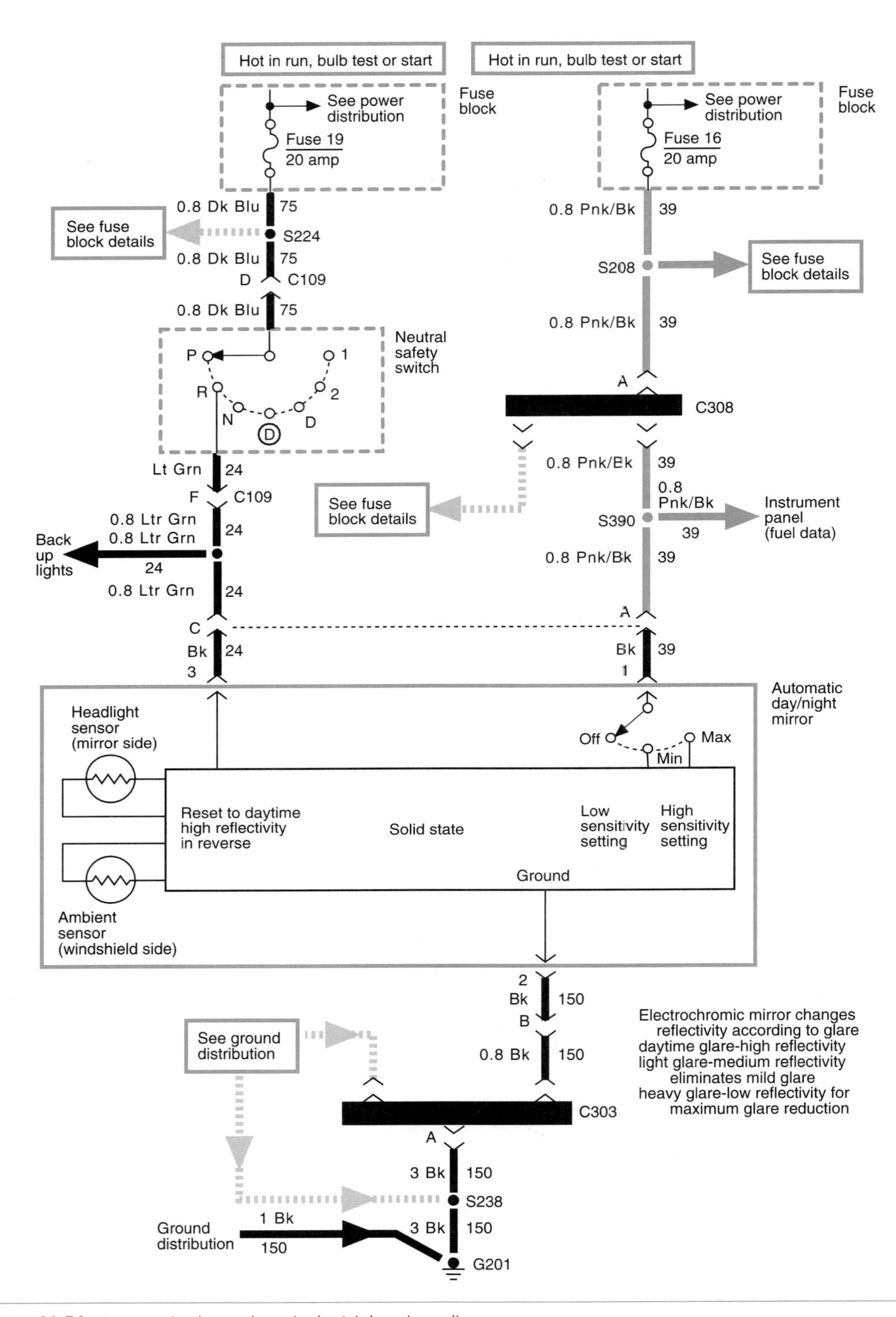

Figure 11-54 Automatic electrochromic day/night mirror diagram.

Electrochromic mirrors are also installed as inside mirrors. The mirror is constructed of a thin layer of electrochromic material that is placed between two plates of conductive glass. There are two photocell sensors that measure light intensity in front and in back of the mirror. During night driving, the headlight beam striking the mirror causes the mirror to gradually become darker as the light intensity increases. The darker mirror absorbs the glare. Sensitivity of the mirror can be adjusted by the driver through a three-position switch (Figure 11-54).

When the ignition switch is placed in the RUN position, battery voltage is applied to the three-position switch. If the switch is in the MIN position battery voltage is applied to the solid-state unit, and sets the sensitivity to a low level. The MAX setting causes the mirror to darken more at a lower glare level. When the transmission is placed in reverse, the reset circuit is activated. This returns the mirror to daytime setting for clearer viewing to back-up.

The MIN position is used for city driving.

Power Windows

Shop Manual
Chapter 11, page 402

Many luxury vehicles replace the conventional window crank with electric motors that operate the side windows. In addition, most station wagon models are equipped with electric rear tailgate windows. The motor used in the power window system is a reversible PM or two-field winding motor.

Power windows are raised and lowered by use of electrical motors.

The power window system usually consists of the following components:

1. Master control switch
2. Individual control switches
3. Individual window drive motors
4. Lock-out or disable switch.

Another design is to use rack-and-pinion gears. The rack is a flexible strip of gear teeth with one end attached to the window (Figure 11-55).

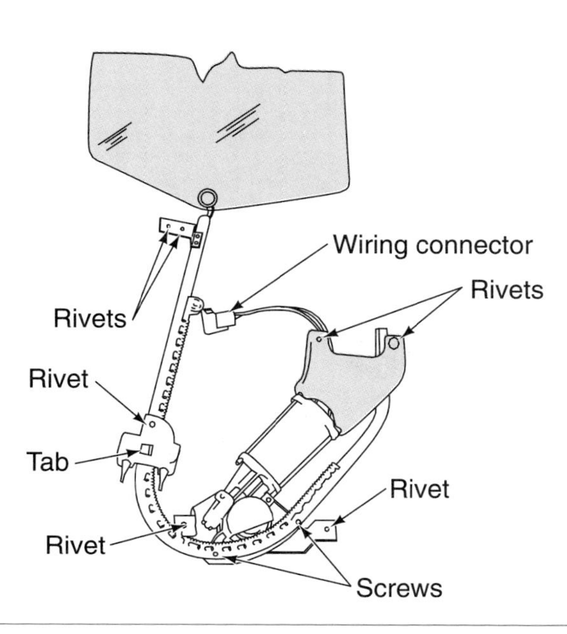

Figure 11-55 Rack-and-pinion style power mirror and regulator. (Courtesy of Chrysler Corporation)

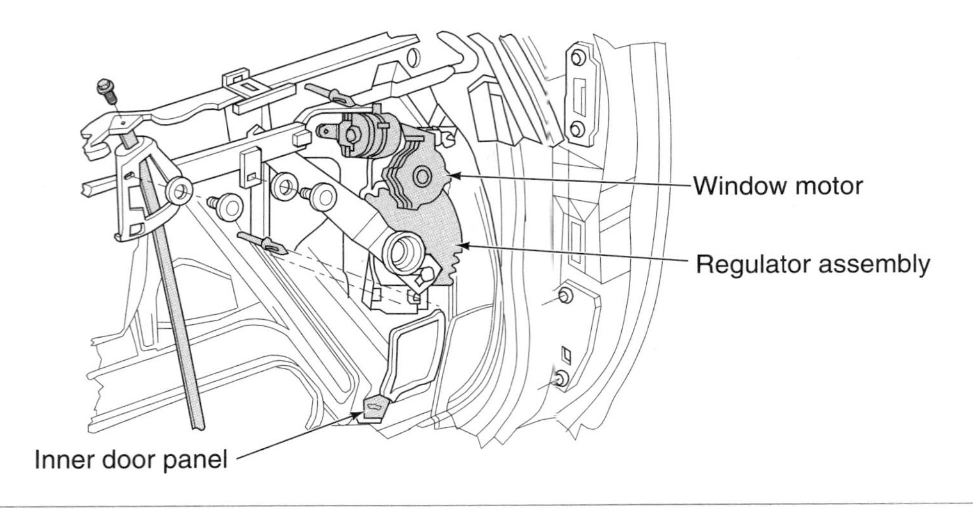

Figure 11-56 Window regulator. (Courtesy of Chrysler Corporation)

A BIT OF HISTORY

Power windows were introduced in 1939.

A **window regulator** converts the rotary motion of the motor into the vertical movement of the window.

The **sector gear** is the section of gear teeth on the regulator.

The motor operates the window regulator either through a cable or directly. On direct drive motors, the motor pinion gear meshes with the sector gear (Figure 11-56). As the window is lowered, the spiral spring is wound. The spring unwinds as the window is raised to assist in raising the window. The spring reduces the amount of current that would be required to raise the window by the motor itself.

The master control switch provides the overall control of the system (Figure 11-57). Power to the individual switches is provided through the master switch. The master switch

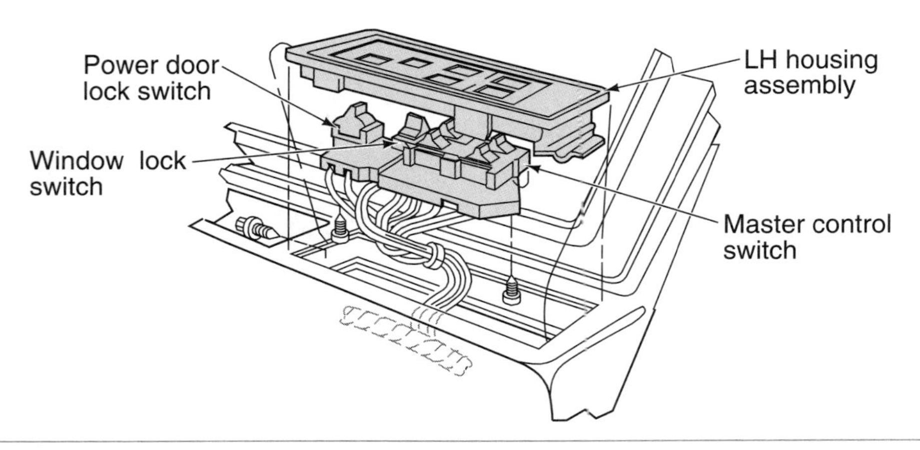

Figure 11-57 Power window master control switch. (Reprinted with the permission of Ford Motor Company)

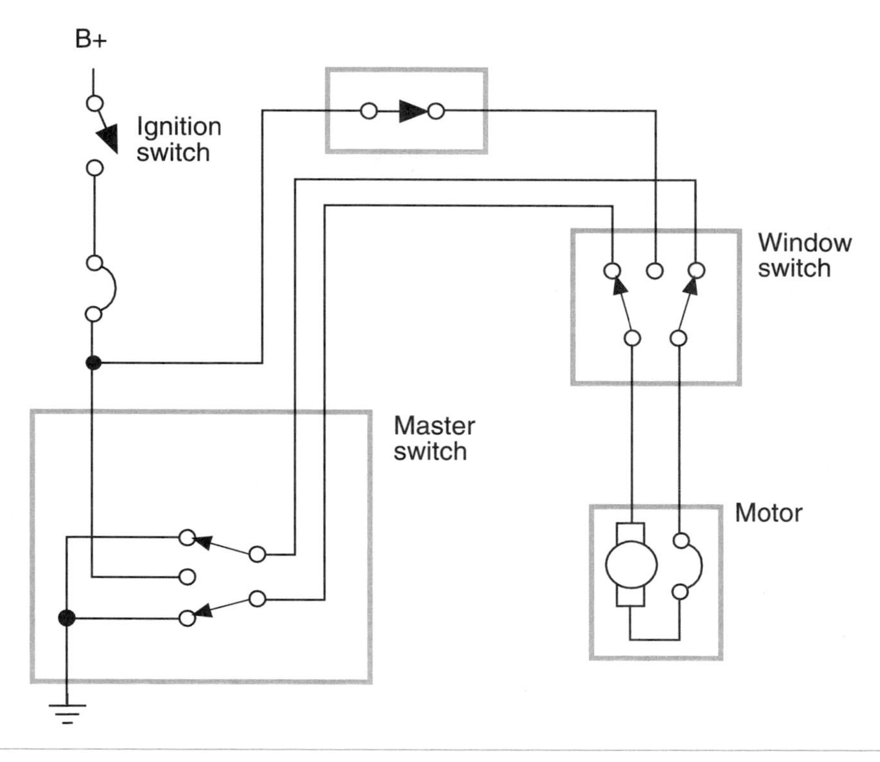

Figure 11-58 Power window wiring diagram.

may also have a safety lock switch to prevent operation of the windows by the individual switches. When the safety switch is activated, it opens the circuit to the other switches and control is only by the master switch. As an additional safety feature, some systems prevent operation of the individual switches unless the ignition switch is in the RUN or ACC position (Figure 11-58).

Wiring circuits depend on motor design. Most PM-type motors are insulated, with ground provided through the master switch (Figure 11-59). When the master control switch is placed in the UP position, current flow is from the battery, through the master switch wiper, through the individual switch wiper, to the top brush of the motor. Ground is through the bottom brush and circuit breaker to the individual switch wiper, to the master switch wiper and ground (Figure 11-60).

When the window is raised from the individual switch, battery voltage is supplied directly to the switch and wiper from the ignition switch. The ground path is through the master control switch (Figure 11-61).

When the window is lowered from the master control switch, the current path is reversed (Figure 11-62). In the illustration shown (Figure 11-63) current flows through the individual switch to lower the window.

Some manufacturers use a two-field coil motor that is grounded with insulated side switches. The two field coils are wired in opposite directions, and only one coil is energized at a time. Direction of the motor is determined by which coil is activated (Figure 11-64).

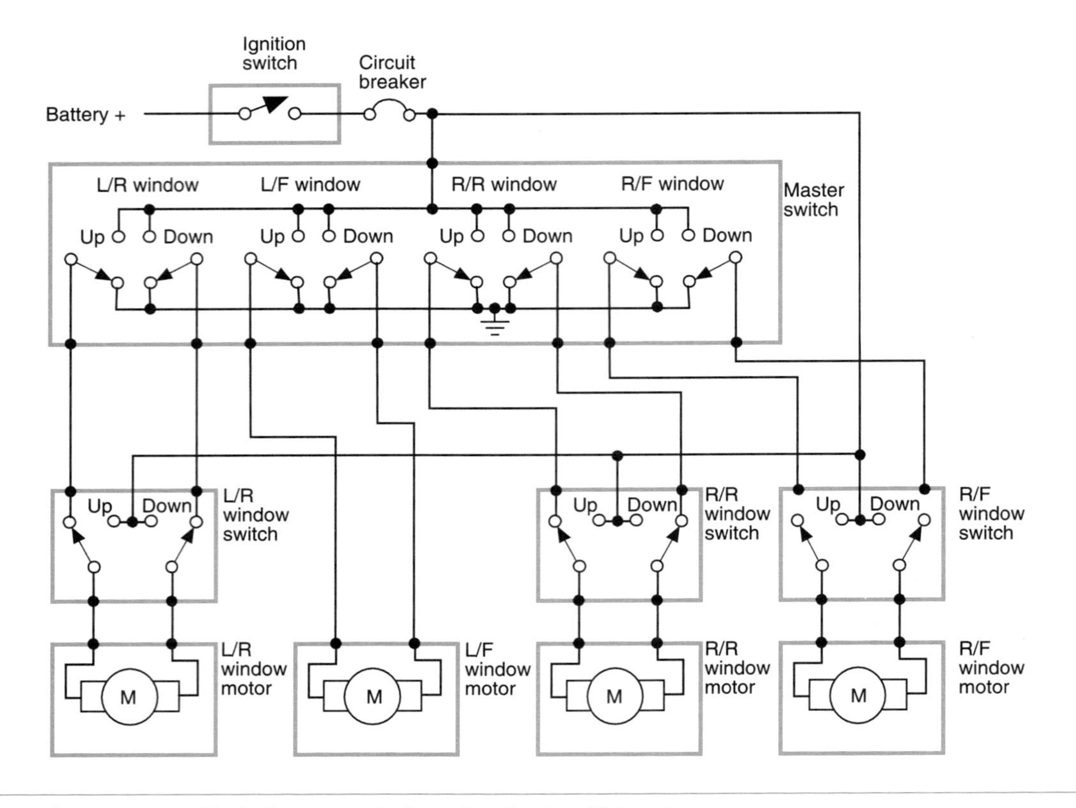

Figure 11-59 Typical power window circuit using PM motors.

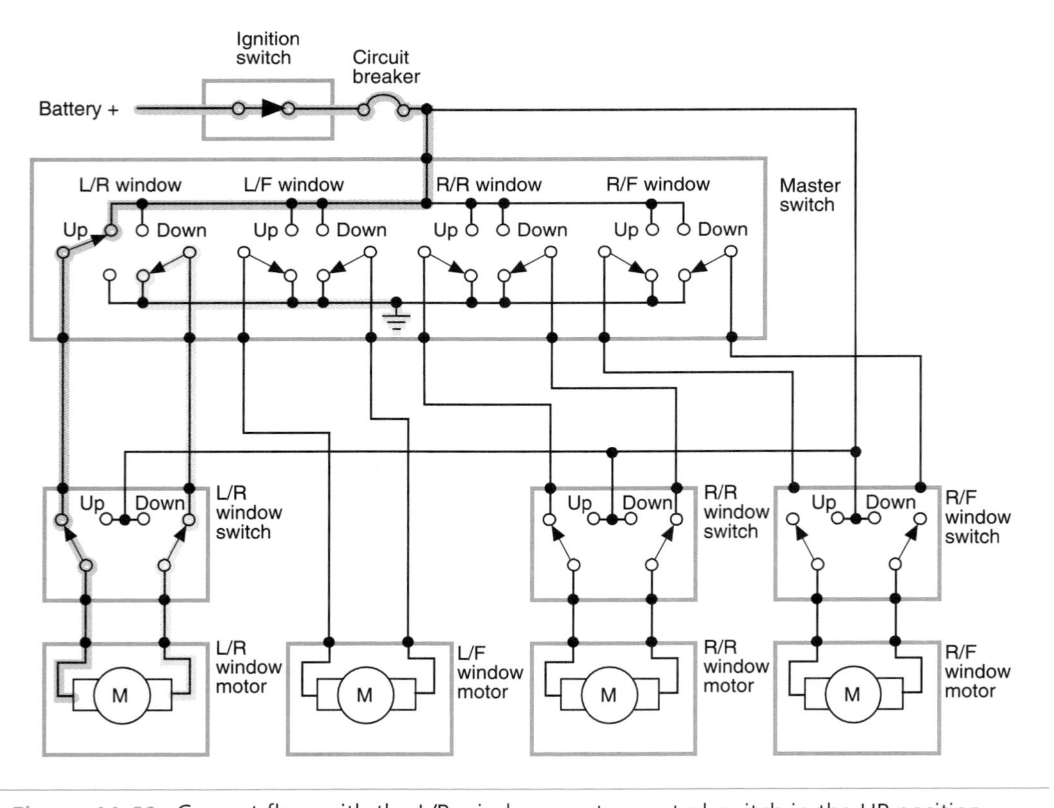

Figure 11-60 Current flow with the L/R window master control switch in the UP position.

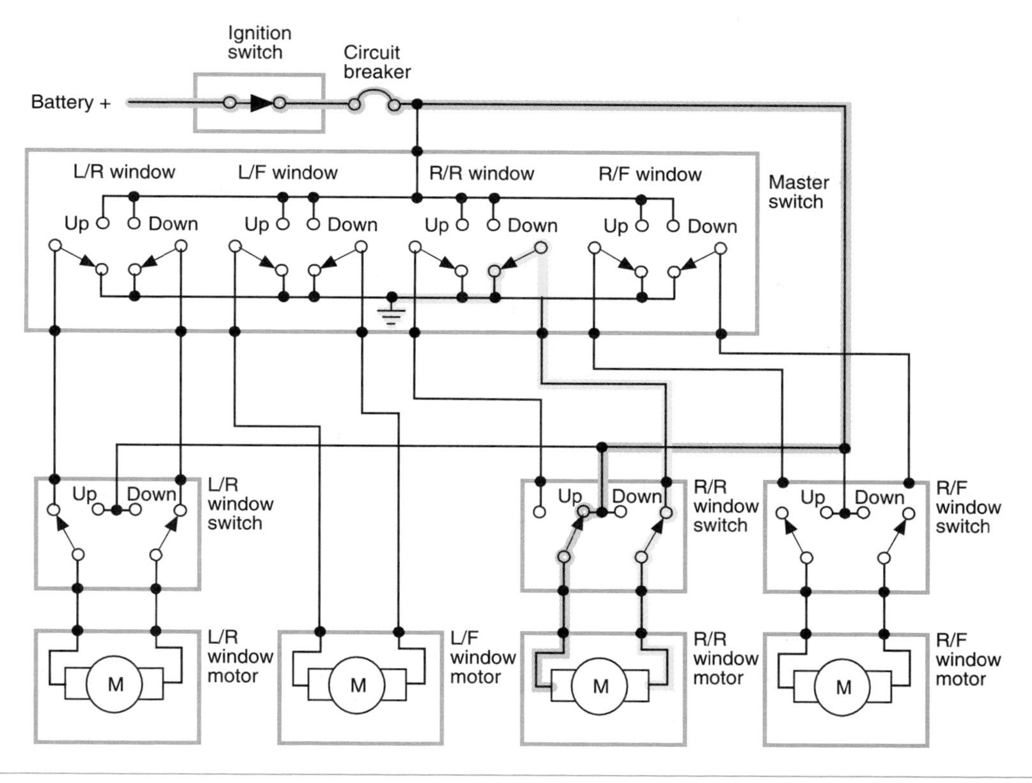

Figure 11-61 Current flow when the R/R window switch is in the UP position.

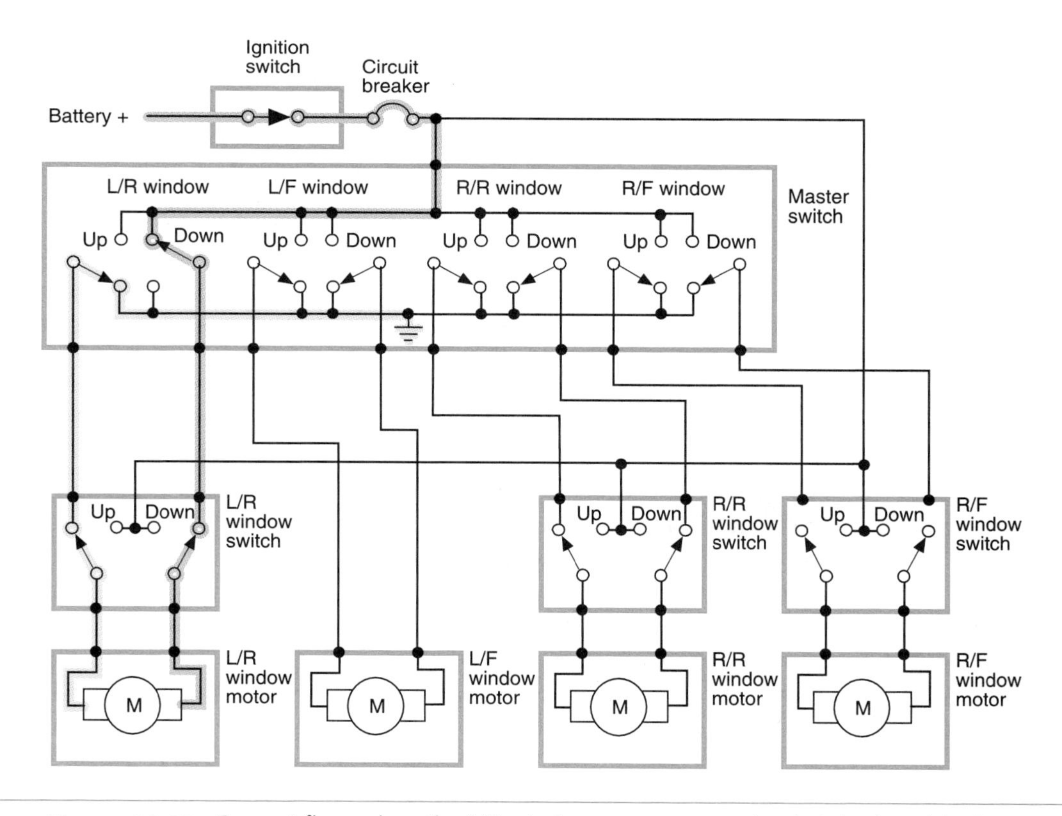

Figure 11-62 Current flow when the L/R window master control switch is placed in the DOWN position.

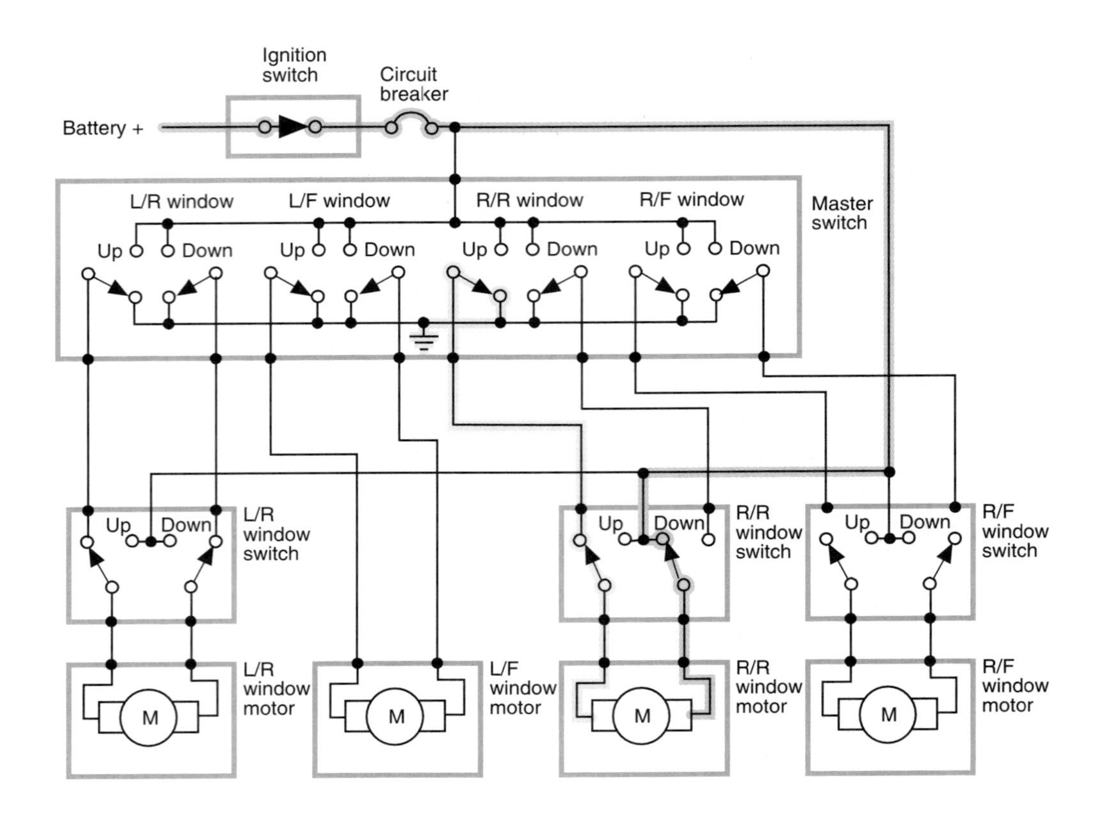

Figure 11-63 Current flow through the individual switch to lower the R/R window.

Shop Manual
Chapter 11,
page 407

The three armature motor is called a **trimotor**.

Early General Motors six-way power seats used a single motor. Solenoids were used to connect the motor to one of three transmissions that would move the seat.

Some seat back latches use a solenoid to lock the seat unless the door is open. The solenoid is controlled by the door jam switch.

Power Seats

The power seat system is classified by the number of ways in which the seat is moved. The most common classifications are:

1. Two-way: Moves the seat forward and backward.
2. Four-way: Moves the seat forward, backward, up, and down.
3. Six-way: Moves the seat forward, backward, up, down, front tilt, and rear tilt.

All modern six-way power seats use a reversible, permanent magnet trimotor (Figure 11-65). The motor may transfer rotation to a rack-and-pinion or to a worm gear drive transmission. A typical control switch consists of a four-position knob and a set of two position switches (Figure 11-66). The four-position knob controls the forward, rearward, up, and down movement of the seat. The separate two-position switches are used to control the front tilt and rear tilt of the seat.

Current direction through the motor determines the rotation direction of the motor. The switch wipers control the direction of current flow (Figure 11-67). If the driver pushes the four-way switch into the down position, the entire seat lowers (Figure 11-68). Switch wipers 3 and 4 are swung to the left and battery voltage is sent through wiper 4 to wipers 6 and 8. These wipers direct the current to the front and rear height motors. The ground circuit is provided through wipers 5 and 7, to wiper 3 and ground.

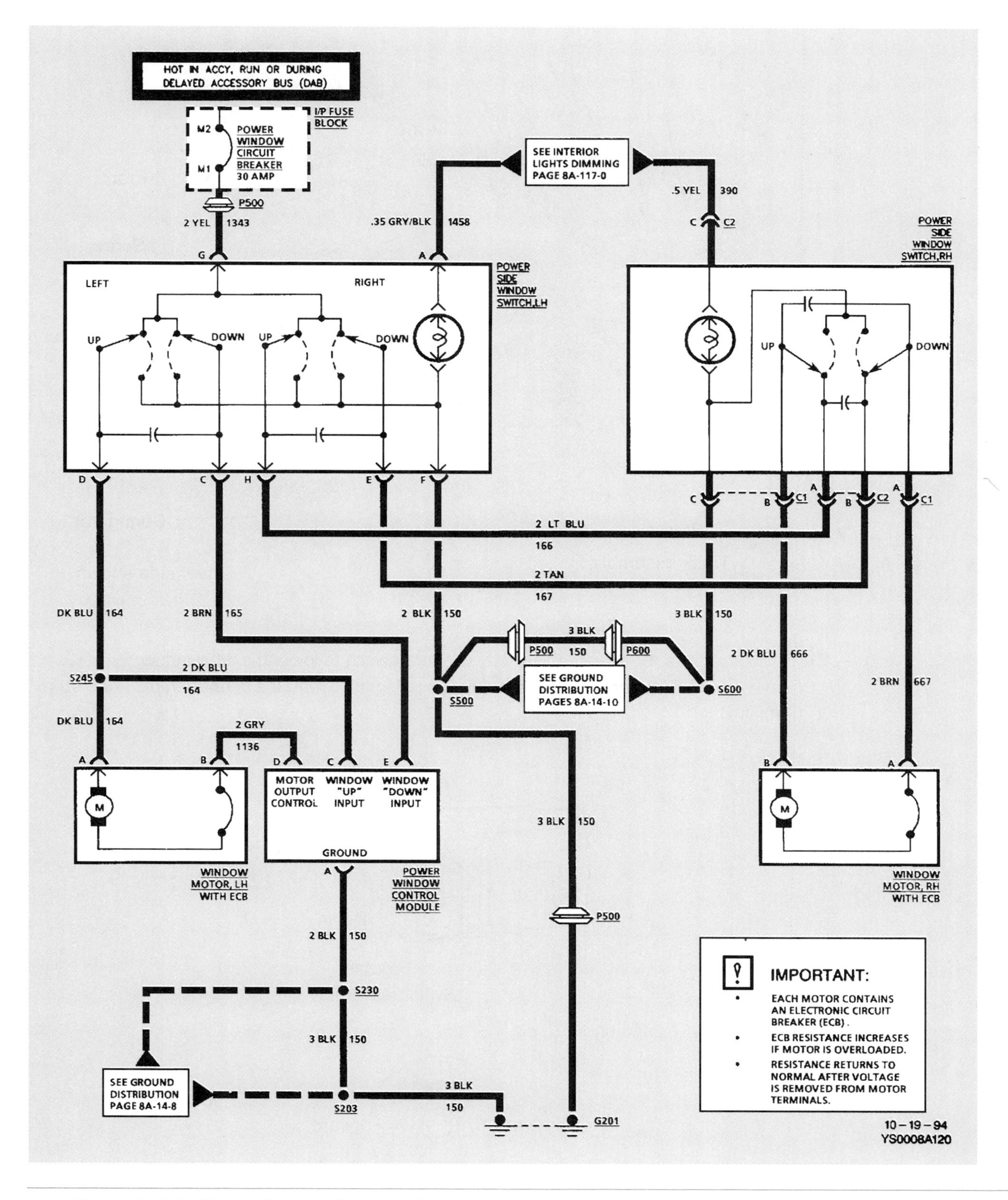

Figure 11-64 Wiring diagram of power window circuit using electromagnetic motors and electronic circuit breakers. (Courtesy of Chevrolet Motor Division—GMC)

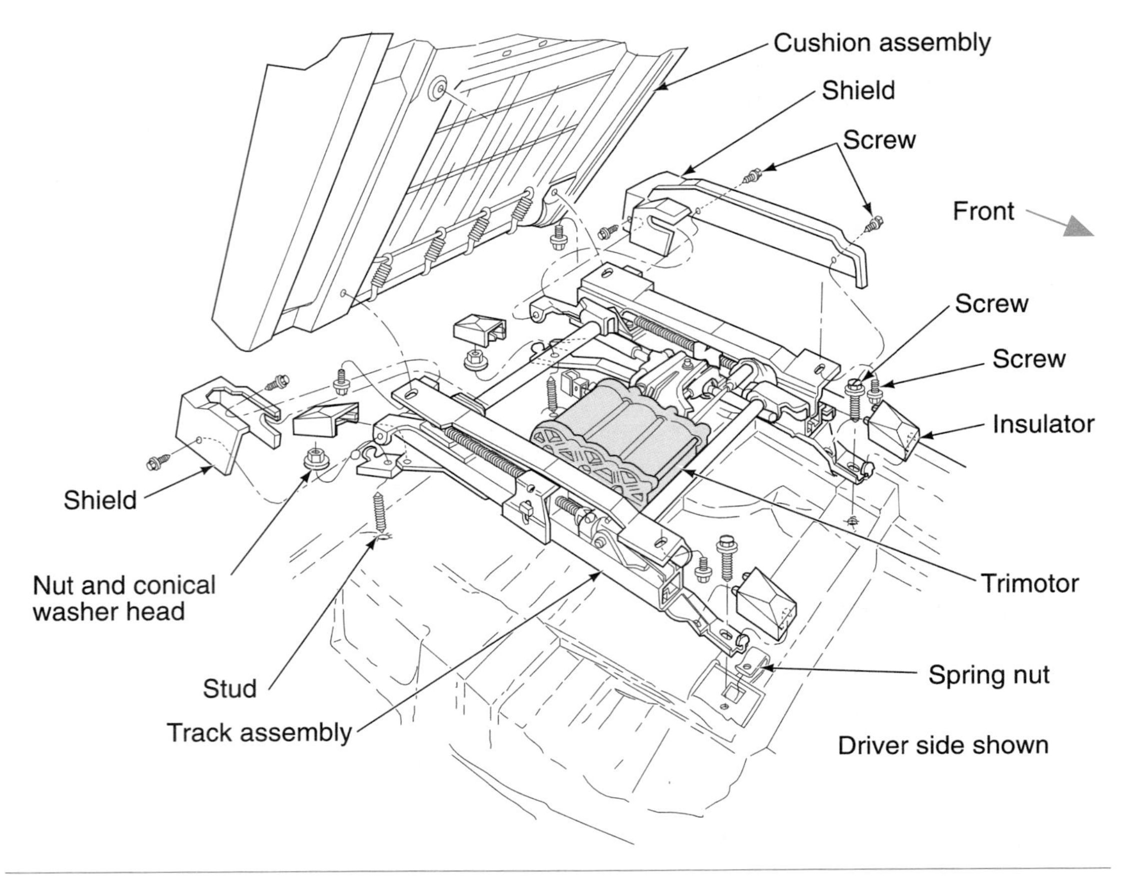

Figure 11-65 Trimotor power seat installation. (Reprinted with the permission of Ford Motor Company)

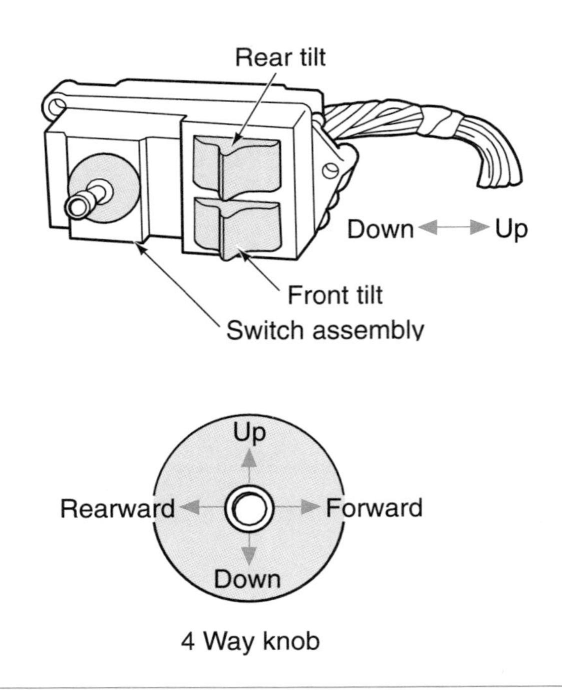

Figure 11-66 Power seat control switch. (Reprinted with the permission of Ford Motor Company)

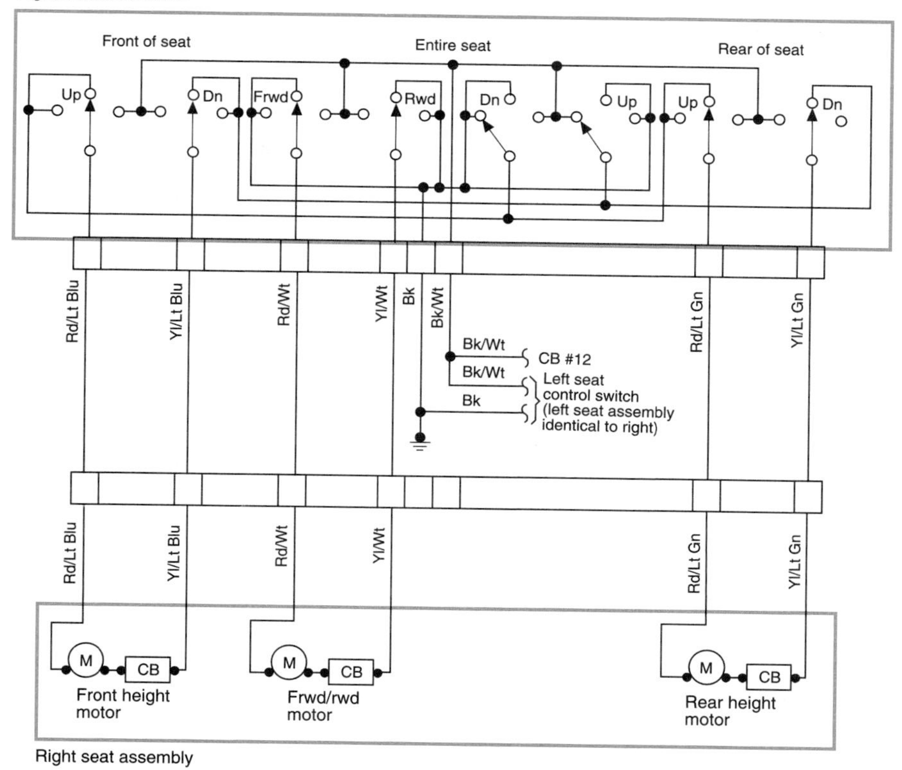

Figure 11-67 Six-way power seat circuit diagram.

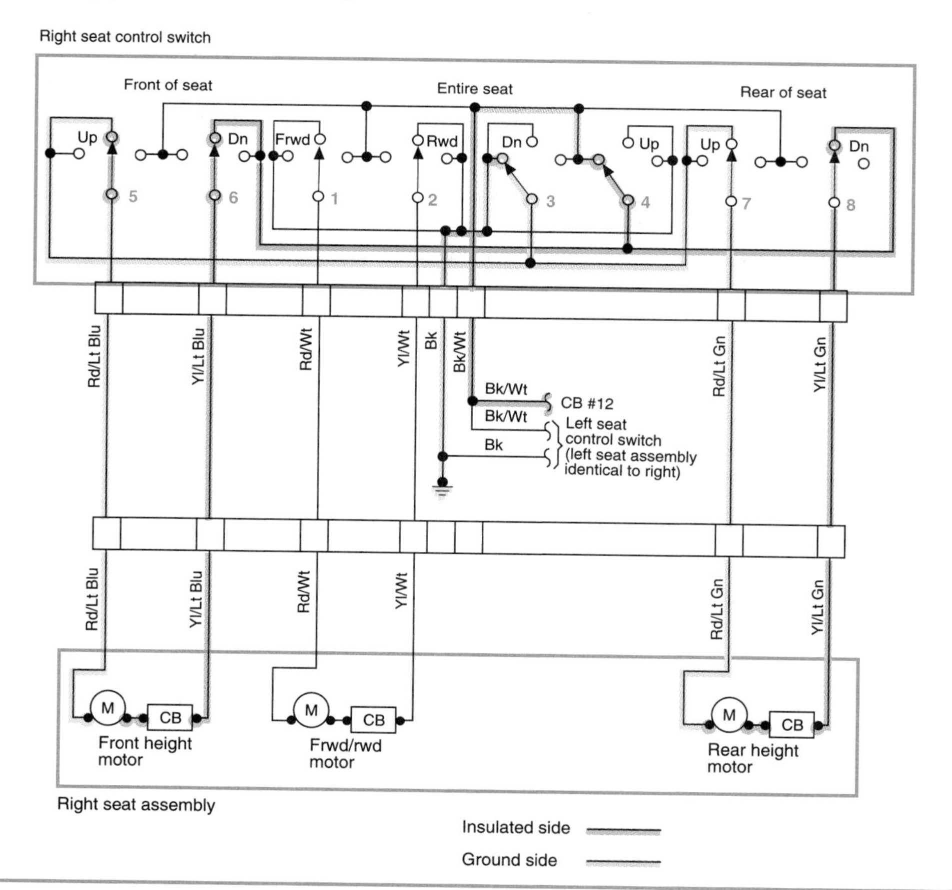

Figure 11-68 Current flow in the seat LOWER position.

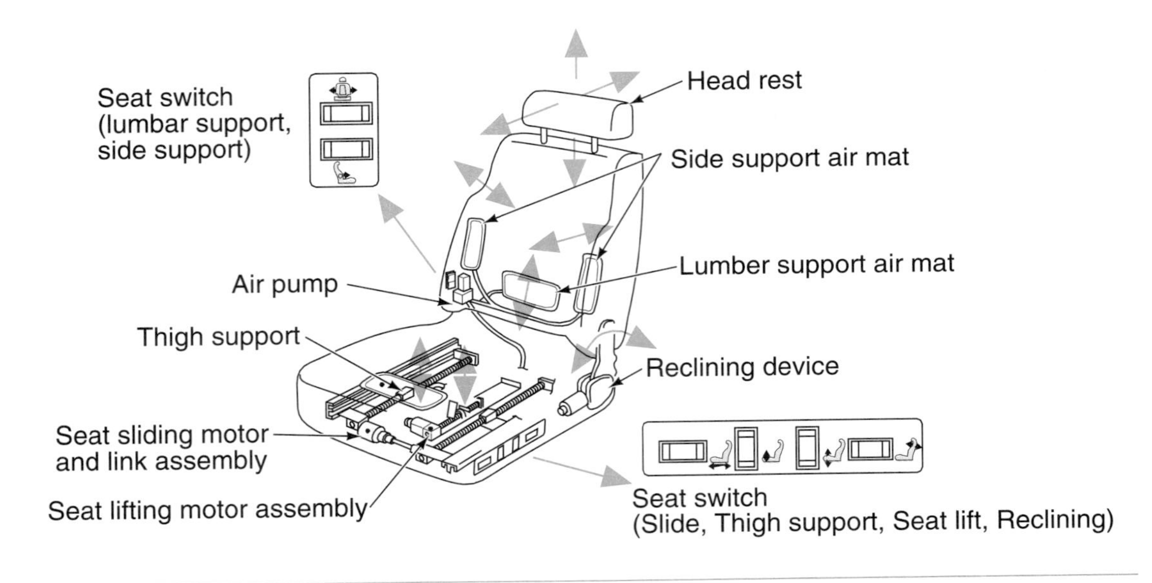

Figure 11-69 Adjustable seat cushions increase driver comfort and safety. (Courtesy of Nissan Motor Corporation, USA)

Some manufacturers equip their seats with adjustable support mats that shape the seat to fit the driver (Figure 11-69). The lumbar support mat provides the driver with additional comfort by supporting the back curvature. The system uses air that is pumped into the mats.

Power Door Locks

Shop Manual
Chapter 11,
page 408

Some manufacturers use a **child safety latch** in the door lock system that prevents the door from being opened from the inside, regardless of the position of the door lock knob. The child safety latch is activated by a switch designed into the latch bellcrank (Figure 11-70).

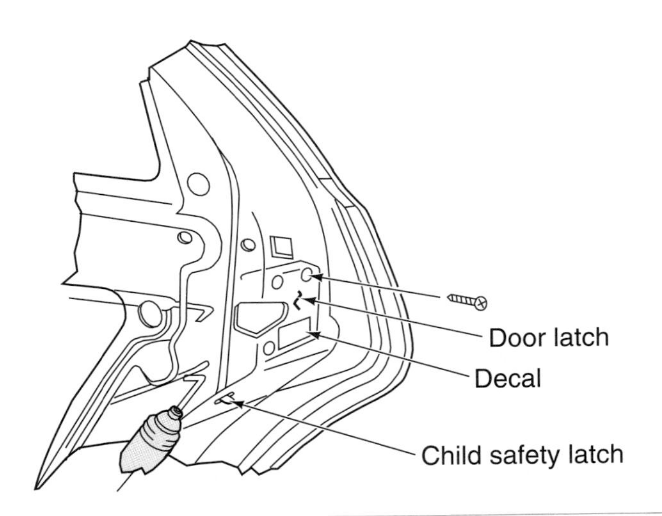

Figure 11-70 Child safety latch. (Reprinted with the permission of Ford Motor Company)

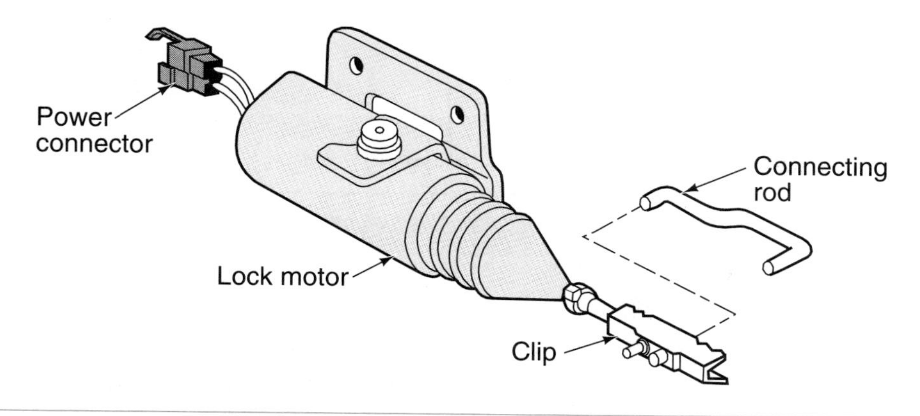

Figure 11-71 PM power door lock motor. (Courtesy of Chrysler Corporation)

Electric power locks use either a solenoid or a permanent magnet reversible motor. Due to the high current demands of solenoids, most modern vehicles use PM motors (Figure 11-71). Depending on circuit design, the system may incorporate a relay (Figure 11-72). The relay has two coils and two sets of contacts to control current direction. In this system the door lock switch energizes one of the door lock relay coils to send battery voltage to the motor. If the door lock

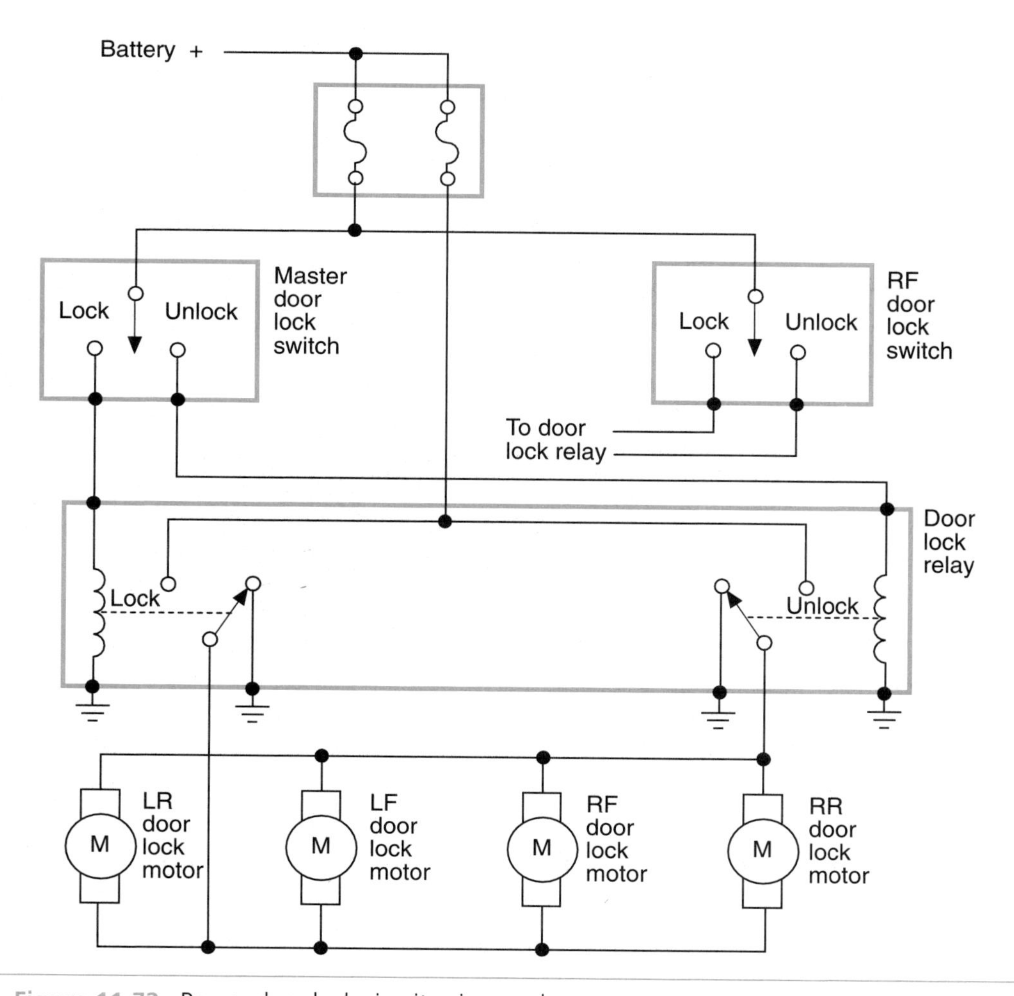

Figure 11-72 Power door lock circuit using a relay.

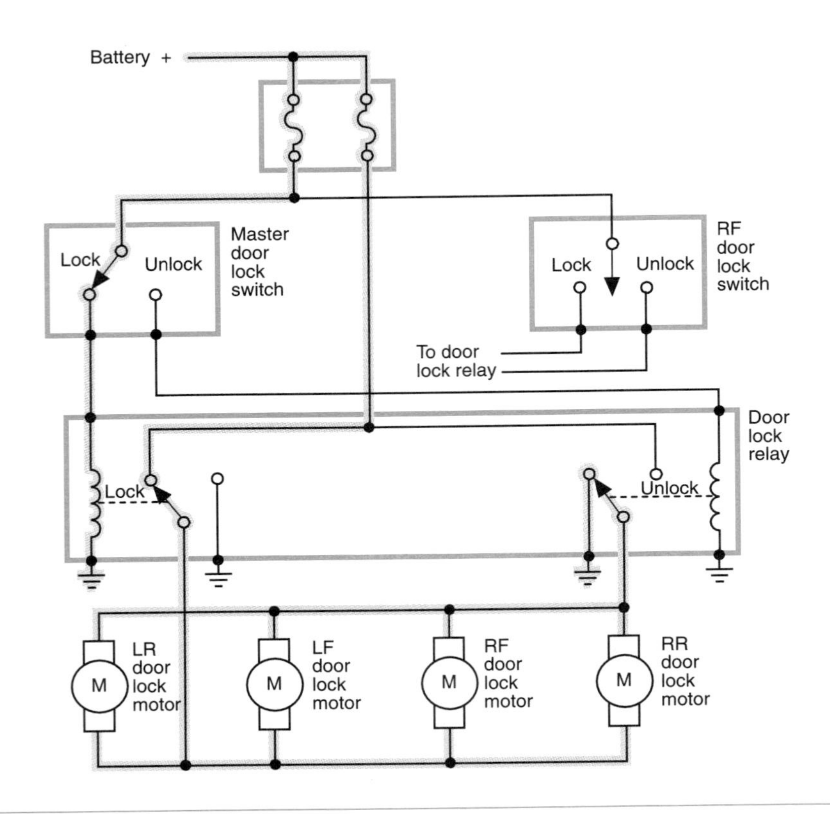

Figure 11-73 Current flow in the LOCK position.

switch is placed in the LOCK position, current flow is that shown (Figure 11-73). In the illustration (Figure 11-74), current flows when the door lock switch is placed in the UNLOCK position.

A system that does not use relays is shown (Figure 11-75). The switch provides control of current flow in the same manner as power seats or windows.

Many vehicles are equipped with automatic door locks that are activated when the gear shift lever is placed in the DRIVE position. The doors unlock when the selector is returned to the PARK position.

Summary

Terms to Know

Child safety latch

Depressed park

Diaphragm

Electric defoggers

Grid

High-note horn

Horn

- ❑ Automotive electrical horns operate on an electromagnetic principle that vibrates a diaphragm to produce a warning signal.
- ❑ Horn switches are either installed in the steering wheel or as a part of the multifunction switch. Most horn switches are normally open switches.
- ❑ Horn switches that are mounted on the steering wheel require the use of sliding contacts. The contacts provide continuity for the horn control in all steering wheel positions.
- ❑ The most common type of horn circuit control is to use a relay.
- ❑ Most two-speed windshield wiper motors use permanent magnet fields whereby the motor speed is controlled by the placement of the brushes on the commutator.

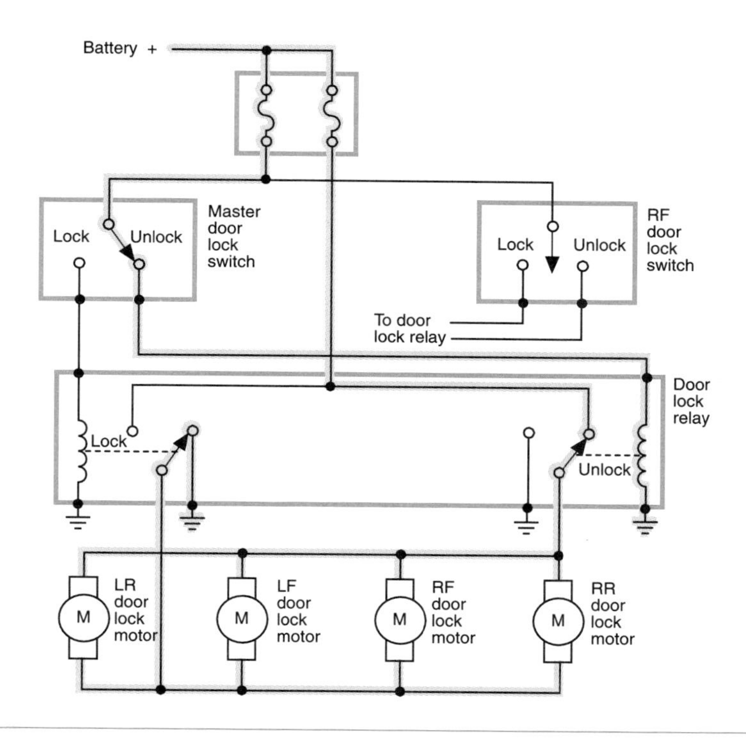

Figure 11-74 Current flow in the UNLOCK position.

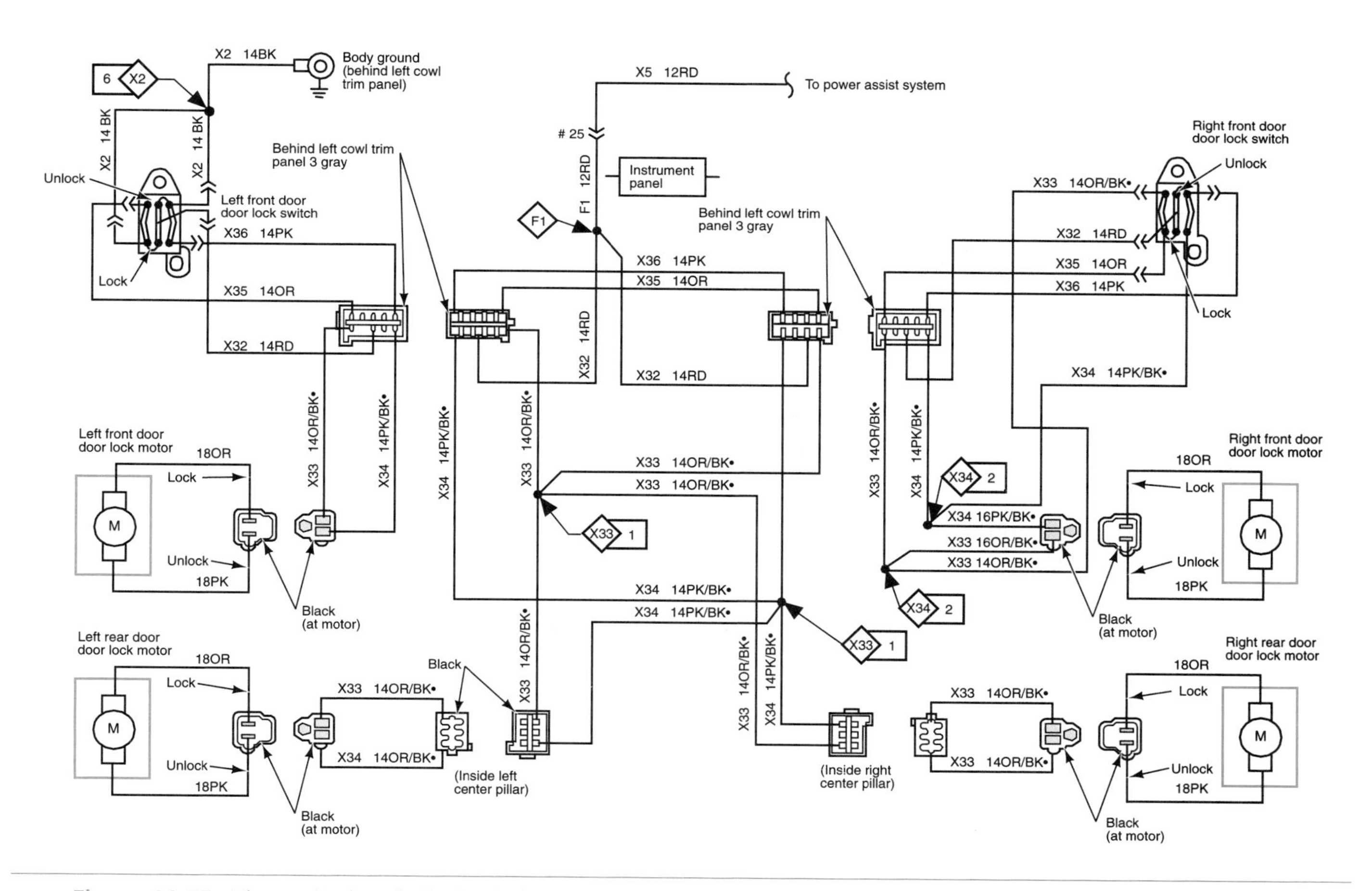

Figure 11-75 Electronic door lock circuit diagram. (Courtesy of Chrysler Corporation)

Terms to Know (Continued)

Intermittent
Low-note horn
Park contacts
Power mirrors
Power windows
Resistor block
Sector gear
Trimotor
Window regulator
Windshield wipers

- ❑ Some two-speed, and all three-speed wiper motors use two electromagnetic field windings: series field and shunt field. The two field coils are wound in opposite directions so that their magnetic fields will oppose each other. The strength of the total magnetic field will determine at what speed the motor will operate.
- ❑ Park contacts are located inside the wiper motor assembly, and supply current to the motor after the switch has been turned to the PARK position. This allows the motor to continue operating until the wipers have reached the PARK position.
- ❑ Intermittent wiper mode provides a variable interval between wiper sweeps and is controlled by a solid-state module.
- ❑ Systems that have a depressed-park feature use a second set of contacts with the park switch, which are used to reverse the rotation of the motor for about 15 degrees after the wipers have reached the normal park position.
- ❑ Blower fan motors use a resistor block that consists of two or three helically wound wire resistors that are connected in series to control fan speed.
- ❑ The blower motor circuit includes the control assembly, blower switch, resistor block, and the blower motor.
- ❑ Electric defoggers heat the rear window by means of a resistor grid.
- ❑ Electric defoggers may incorporate a timer circuit to prevent the high current required to operate the system from damaging the battery or charging system.
- ❑ The electrically controlled mirror allows the driver to position the outside mirrors by use of a switch that controls dual drive, reversible PM motors.
- ❑ Power windows, seats, and door locks usually use reversible PM motors, whereby motor rotational direction is determined by the direction of current flow through the switch wipers.

Review Questions

Short Answer Essays

1. Describe the operation of a relay-controlled horn circuit.
2. Explain how brush placement determines the speed of a two-speed, permanent magnet motor.
3. How do wiper motor systems that use a three-speed, electromagnetic motor control wiper speed?
4. What is the purpose of the capacitor in some intermittent wiper systems?
5. Describe the method used to control blower fan motor speeds.
6. Describe the operation of electric defoggers.
7. Briefly explain the principles of operation for power windows.
8. Describe the operation of a six-way, trimotor power seat.
9. Explain how depressed-park wipers operate.
10. What is the advantage of PM motors over solenoids in the power lock system?

Fill-in-the-Blanks

1. Electrical accessories provide for additional ________________ and ________________ .

2. The ________________ is a thin, flexible, circular plate that is held around its outer edge by the horn housing, allowing the middle to flex.

3. Horn switches that are mounted on the steering wheel require the use of ________________ ________________ to provide continuity in all steering wheel positions.

4. In systems equipped with ________________ ________________ , the blades drop down below the lower windshield molding to hide them.

5. PM windshield wiper motors use ________________ brushes.

6. The operational speed of electromagnetic field winding motors used in wiper systems depends on the strength of the ________________ ________________ ________________ .

7. Most blower motor fan speeds are controlled through a ________________ ________________ that is wired in series.

8. Electric defoggers operate on the principle that when electrons are forced to flow through a ________________ , heat is generated.

9. A window ________________ converts the rotary motion of the motor into the vertical movement of the window.

10. The three armature motor is called a ________________ .

ASE Style Review Questions

1. The sound generation from the horn is being discussed:
   Technician A says electrical horns operate by generating heat to vibrate a diaphragm.
   Technician B says the vibration of the column of air in the horn produces the sound.
   Who is correct?
   A. A only
   B. B only
   C. Both A and B
   D. Neither A nor B

2. The horn circuit is being discussed:
   Technician A says if the circuit does not use a relay, the horn switch carries the total current requirements of the horns.
   Technician B says most systems that use a relay have battery voltage present to the lower contact plate of the horn switch and the switch closes the path for the relay coil.
   Who is correct?
   A. A only
   B. B only
   C. Both A and B
   D. Neither A nor B

3. Permanent magnet wiper motors are being discussed:
Technician A says motor speed is controlled by the placement of the brushes on the commutator.
Technician B says the motor operates by the attraction and repelling of two electromagnetic fields.
Who is correct?
A. A only
B. B only
C. Both A and B
D. Neither A nor B

4. *Technician A* says if there are more armature windings connected between the common and high speed brushes, there is less magnetism in the armature and a lower counterelectromotive force, thus the motor turns faster.
Technician B says with less CEMF in the armature, the greater the armature current will be.
Who is correct?
A. A only
B. B only
C. Both A and B
D. Neither A nor B

5. Electromagnetic field wiper motors are being discussed:
Technician A says the two field coils are wound in opposite directions so that their magnetic fields will oppose each other.
Technician B says the strength of the total magnetic field will determine the speed of the motor.
Who is correct?
A. A only
B. B only
C. Both A and B
D. Neither A nor B

6. Intermittent wiper systems are being discussed:
Technician A says the system uses a solid-state module or is incorporated into the body computer.
Technician B says the capacitor operates the wiper motor directly until it is fully discharged.
Who is correct?
A. A only
B. B only
C. Both A and B
D. Neither A nor B

7. The blower motor circuit is being discussed:
Technician A says the motor is usually a permanent magnet, single-speed motor.
Technician B says the switch position directs current flow to one of the three different brushes in the motor.
Who is correct?
A. A only
B. B only
C. Both A and B
D. Neither A nor B

8. *Technician A* says a resistor block that is wired in series between the switch and the motor controls blower fan speed.
Technician B says a thermal limiter acts as a circuit breaker to protect the circuit if the resistor block gets too hot.
Who is correct?
A. A only
B. B only
C. Both A and B
D. Neither A nor B

9. The electric rear window defogger is being discussed:
Technician A says the grid is a series of controlled voltage amplifiers.
Technician B says the system may incorporate a timer circuit to protect the vehicle's electrical system.
Who is correct?
A. A only
B. B only
C. Both A and B
D. Neither A nor B

10. *Technician A* says the master control switch for power windows provides the overall control of the system.
Technician B says current direction through the power seat motor determines the rotation direction of the motor.
Who is correct?
A. A only
B. B only
C. Both A and B
D. Neither A nor B

Introduction to the Body Computer

Upon completion and review of this chapter, you should be able to:

- ❑ Describe the principle of analog and digital voltage signals.
- ❑ Explain the principle of computer communications.
- ❑ Describe the basics of logic gate operation.
- ❑ Describe the basic function of the central processing unit (CPU).
- ❑ Explain the basic method by which the CPU is able to make determinations.
- ❑ List and describe the differences in memory types.
- ❑ List and describe the functions of the various sensors used by the computer.
- ❑ List and describe the operation of output actuators.
- ❑ Explain the principle of multiplexing.

Introduction

This chapter introduces the basic theory and operation of the digital computer used to control many of the vehicle's accessories. The aura of mystery surrounding automotive computers is so great that some technicians are afraid to work on them. Knowledge is key to dispelling this myth. Although it is not necessary to understand all of the concepts of computer operation to service the systems they control, knowledge of the digital computer will make you feel more comfortable when working on these systems.

A **computer** is an electronic device that stores and processes data. It is capable of controlling other devices.

The use of computers on automobiles has expanded to include control and operation of several functions, including climate control, lighting circuits, cruise control, antilock braking, electronic suspension systems, and electronic shift transmissions. Some of these are functions of what is known as a body computer module (BCM). Some body computer-controlled systems include direction lights, rear window defoggers, illuminated entry, intermittent wipers, and other systems once thought of as basic.

A computer processes the physical conditions that represent information (data). The operation of the computer is divided into four basic functions:

1. *Input:* A voltage signal sent from an input device. This device can be a sensor or a switch activated by the driver or technician.
2. *Processing:* The computer uses the input information and compares it to programmed instructions. The logic circuits process the input signals into output demands.
3. *Storage:* The program instructions are stored in an electronic memory. Some of the input signals are also stored for later processing.
4. *Output:* After the computer has processed the sensor input and checked its programmed instructions, it will put out control commands to various output devices. These output devices may be the instrument panel display or a system actuator. The output of one computer can also be used as an input to another computer.

Understanding these four functions will help today's technician organize the troubleshooting process. When a system is tested, the technician will be attempting to isolate the problem to one of these functions.

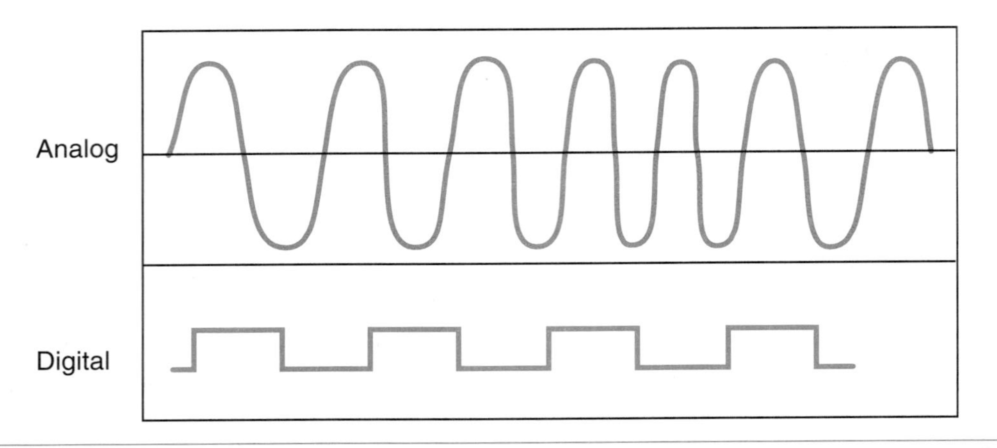

Figure 12-1 Analog voltage signals are constantly variable. Digital voltage patterns are either on or off, high or low; digital signals are referred to as square waves.

Analog and Digital Principles

Voltage does not flow through a conductor; current flows and voltage is the pressure that "pushes" the current. However, voltage can be used as a signal, for example, difference in voltage levels, frequency of change, or switching from positive to negative values can be used as a signal.

A **program** is a set of instructions the computer must follow to achieve desired results.

The computer is capable of reading only voltage signals. The programs used by the computer are "burned" into IC chips using a series of numbers. These numbers represent various combinations of voltages that the computer can understand. The voltage signals to the computer can be either analog or digital. Most of the inputs from the sensors are analog variables. For example, ambient temperature sensors do not change abruptly. The temperature varies in infinite steps from low to high. The same is true for several other inputs such as engine speed, vehicle speed, fuel flow, and so on.

Analog means a voltage signal is infinitely variable, or can be changed, within a given range.

Digital means a voltage signal that is one of two states either on-off, yes-no, or high-low.

Compared to an analog voltage representation, digital voltage patterns are square-shaped because the transition from one voltage level to another is very abrupt (Figure 12-1). A digital signal is produced by an on/off or high/low voltage. The simplest generator of a digital signal is a switch (Figure 12-2). If 5 volts is applied to the circuit, the voltage sensor will read close to 5 volts (a high voltage value) when the switch is open. Closing the switch will result in the voltage sensor reading close to 0 volts. This measuring of voltage drops sends a digital signal to the computer. The voltage values are represented by a series of digits, which create a binary code.

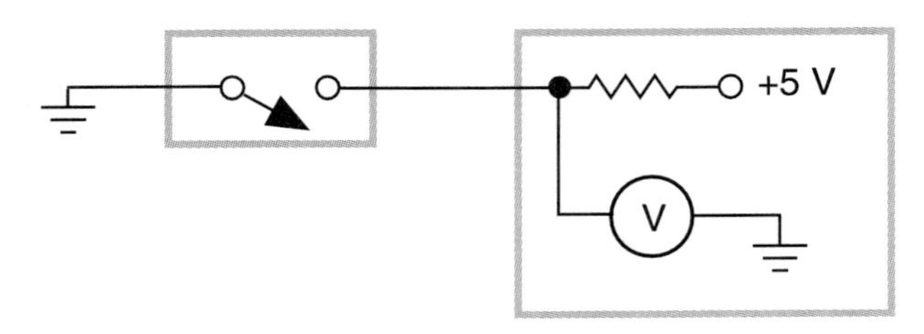

Figure 12-2 Simplified voltage sensing circuit that indicates whether the switch is opened or closed.

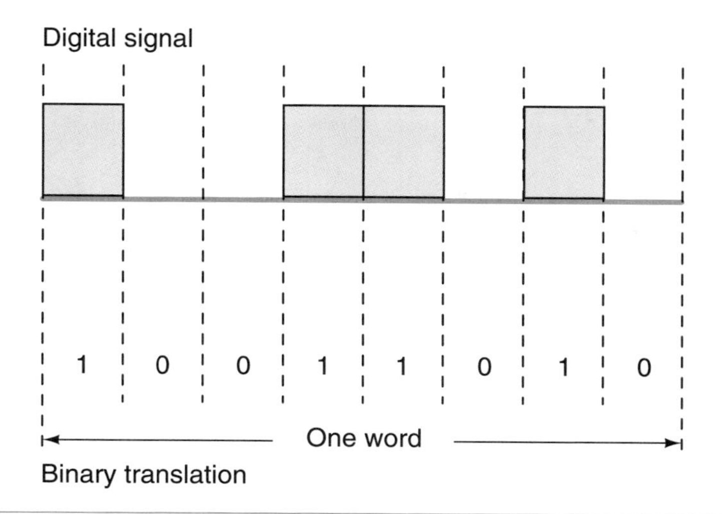

Figure 12-3 Each binary 1 and 0 is one bit of information. Eight bits equal one byte. (Courtesy of General Motors Corporation)

Binary Numbers

A transistor that operates as a relay is the basis of the digital computer. As the input signal switches from off to on, the transistor output switches from cutoff to saturation. The on and off output signals represent the binary digits 1 and 0.

The computer converts the digital signal into binary code by translating voltages above a given value to 1, and voltages below a given value to 0. As shown (Figure 12-3), when the switch is open and 5 volts is sensed, the voltage value is translated into a 1 (high voltage). When the switch is closed, lower voltage is sensed and the voltage value is translated into a 0. Each 1 or 0 represents one bit of information.

Binary numbers are represented by the numbers 1 and 0. Any number and word can be translated into a combination of binary 1's and 0's.

In the binary system, whole numbers are grouped from right to left. Because the system uses only two digits, the first portion must equal a 1 or a 0. To write the value of 2, the second position must be used. In binary, the value of 2 would be represented by 10 (one two and zero ones). To continue, a 3 would be represented by 11 (one two and one one). Figure 12-4 illustrates the conversion of binary numbers to digital base ten numbers. If a thermistor is sensing

Decimal number	Binary number code 8 4 2 1	Binary to decimal conversion
0	0000	= 0 + 0 = 0
1	0001	= 0 + 1 = 1
2	0010	= 2 + 0 = 2
3	0011	= 2 + 1 = 3
4	0100	= 4 + 0 = 4
5	0101	= 4 + 1 = 5
6	0110	= 4 + 2 = 6
7	0111	= 4 + 2 + 1 = 7
8	1000	= 8 + 0 = 8

Figure 12-4 Binary number code conversion to base ten numbers.

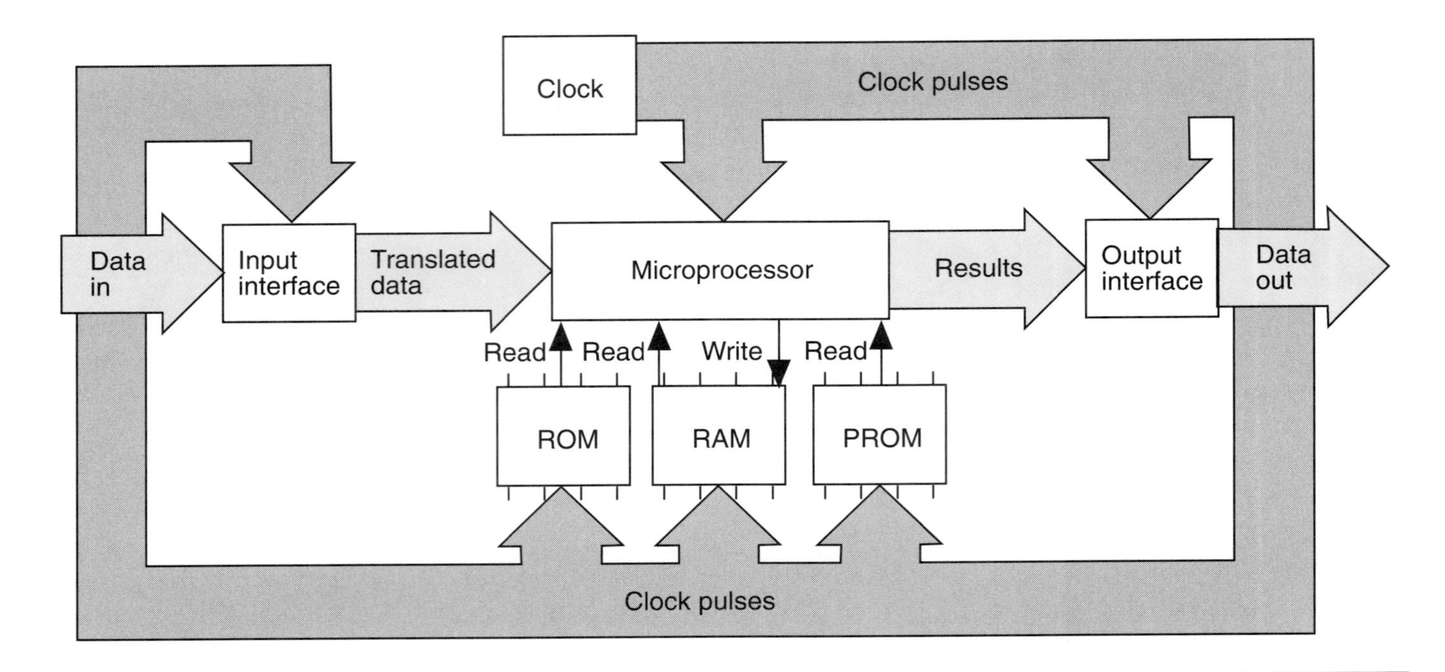

Figure 12-5 Interaction of the main components of the computer. All of the components monitor clock pulses. (Courtesy of General Motors Corporation)

150 degrees, the binary code would be 10010110. If the temperature increases to 151 degrees, the binary code changes to 10010111.

The **clock** is a crystal that electrically vibrates when subjected to current at certain voltage levels. As a result, the chip produces very regular series of voltage pulses.

The computer contains a crystal oscillator or clock that delivers a constant time pulse The clock maintains an orderly flow of information through the computer circuits by transmitting one bit of binary code for each pulse (Figure 12-5). In this manner the computer is capable of distinguishing between the binary codes such as 101 and 1001.

Signal Conversion

An **interface** is used to protect the computer from excessive voltage levels and to translate input and output signals.

The input interface is used as an analog to digital (A/D) convertor.

For the computer to receive information from the sensor, and to give commands to actuators, it requires an interface. The computer will have two interface circuits: input and output. The digital computer cannot accept analog signals from the sensors and requires an input interface to convert the analog signal to digital. Also, some of the controlled actuators may require an analog signal. In this instance, an output digital to analog (D/A) converter is used (Figure 12-6).

Logic Gates

Logic gates are the thousands of field effect transistors (FET) incorporated into the computer circuitry. The FETs use the incoming voltage patterns to determine the pattern of pulses leaving the gate. The following are some of the most common logic gates and their operation. The symbols represent functions and not electronic construction:

1. **NOT gate:** A NOT gate simply reverses binary 1's to 0's and vice versa (Figure 12-7). A high input results in a low output and a low input results in a high output.

Computer housing
Sensor voltage weak and varied
Amplifier increases voltage signal strength
Microprocessor uses digital or number signal
Clock reference
Clock
Analog-to-digital converter
On
Off
Conditioner or interface
Digital signal to microprocessor
Brain
Digital-to-analog converter
Conditioner or interface
Reference voltage to sensor
Voltage regulator
Power transistor or driver
Memory
Digital signal compared to known signal to determine correct output
Sensor
Condition or operating input
Solenoid uses computer output
Varying analog voltage output
Mechanical movement output

Figure 12-6 The interface converts analog inputs into digital signals so the computer can use the voltage signal. Some output actuators require analog signals; the D/A convertor converts the digital commands of the computer into analog voltages. (Courtesy of General Motors Corporation)

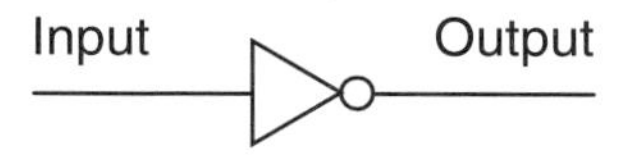

Truth table	
Input	Output
0	1
1	0

Figure 12-7 The NOT gate symbol and a truth table. The NOT gate inverts the input signal.

A truth table shows all of the possible outputs from the gate for all possible inputs.

The NOT gate is also called an **inverter**.

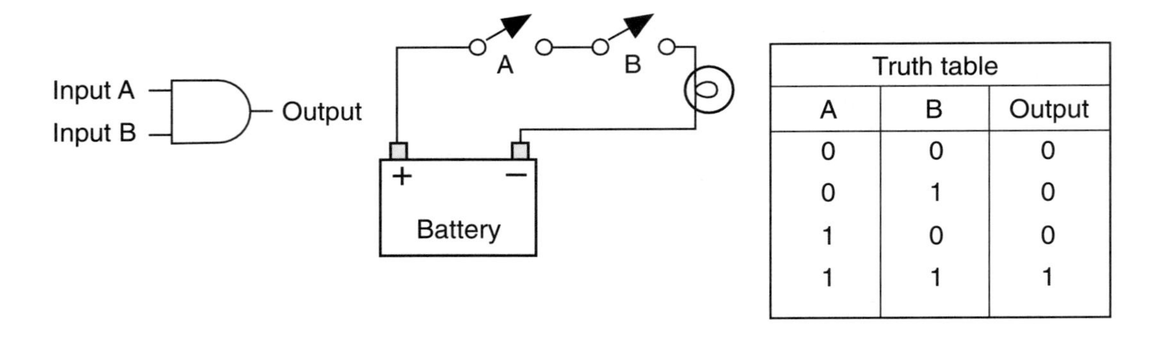

Truth table		
A	B	Output
0	0	0
0	1	0
1	0	0
1	1	1

Figure 12-8 The AND gate symbol and a truth table. The AND gate operates similarly to switches in series.

These circuits are called **logic gates** because they act as gates to output voltage signals depending on different combinations of input signals.

2. **AND gate:** The AND gate will have at least two inputs and one output. The operation of the AND gate is similar to two switches in series to a load (Figure 12-8). The only way the light will turn on is if switches A *and* B are closed. The output of the gate will be high only if both inputs are high. Before current can be present at the output of the gate, current must be present at the base of both transistors (Figure 12-9).

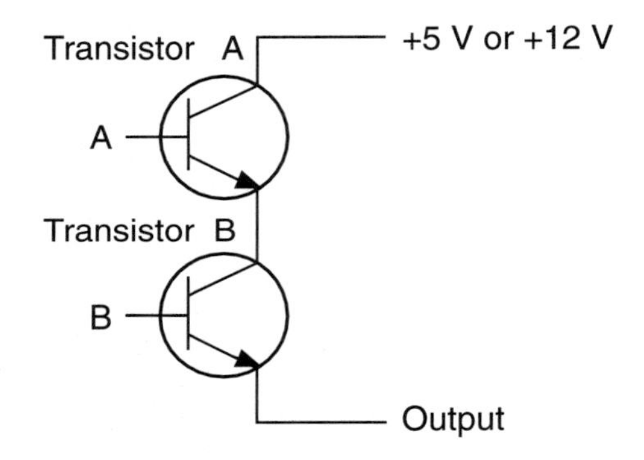

Figure 12-9 The AND gate circuit.

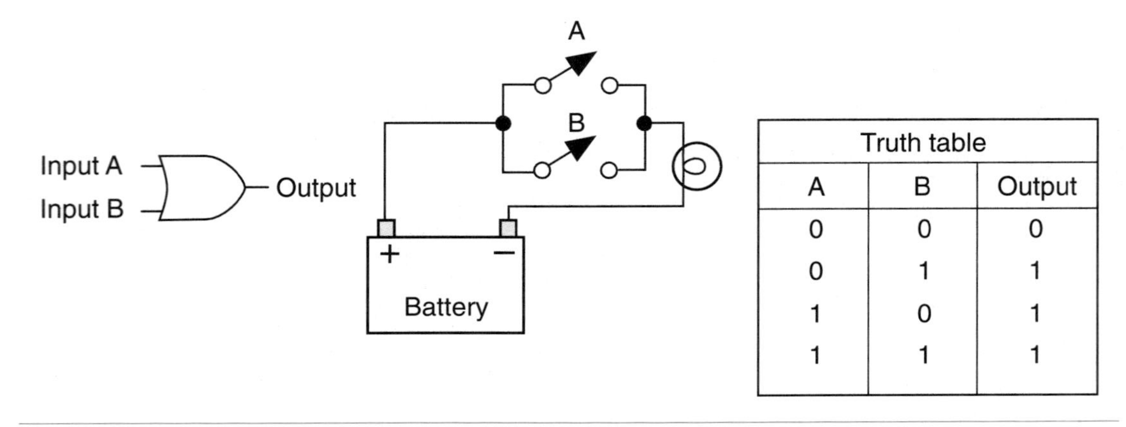

Truth table		
A	B	Output
0	0	0
0	1	1
1	0	1
1	1	1

Figure 12-10 OR gate symbol and a truth table. The OR gate is similar to parallel switches.

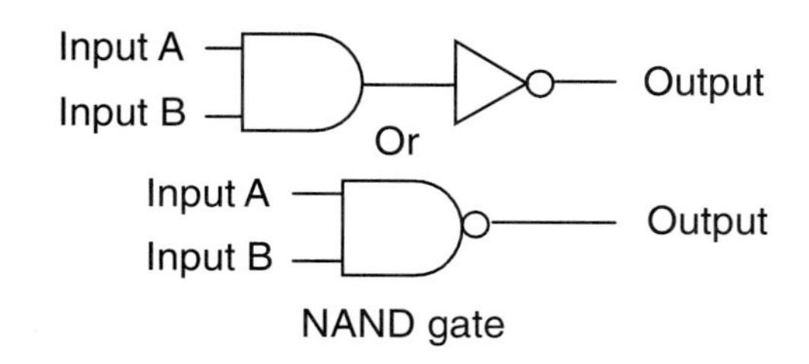

Truth table		
A	B	Output
0	0	1
0	1	1
1	0	1
1	1	0

Truth table		
A	B	Output
0	0	1
0	1	0
1	0	0
1	1	0

Figure 12-11 Symbols and truth tables for NAND and NOR gates. The small circle represents an inverted output on any logic gate symbol.

3. **OR gate:** The OR gate operates similarly to two switches that are wired in parallel to a light (Figure 12-10). If switch A *or* B is closed, the light will turn on. A high signal to either input will result in a high output.
4. **NAND** and **NOR gates:** A NOT gate placed behind an OR or AND gate inverts the output signal (Figure 12-11).
5. **Exclusive-OR (XOR) gate:** A combination of gates that will produce a high output signal only if the inputs are different (Figure 12-12).

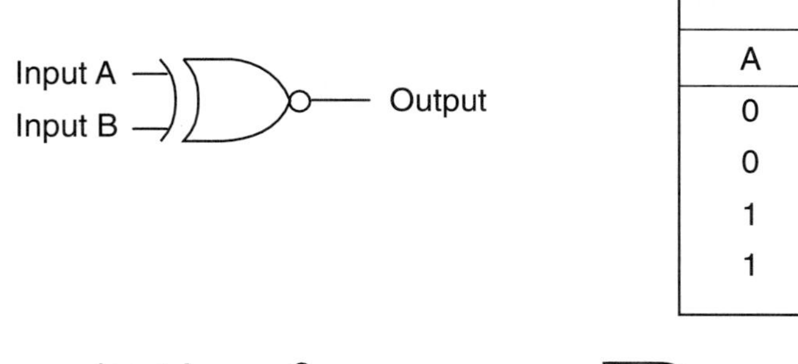

Truth table		
A	B	Output
0	0	0
0	1	1
1	0	1
1	1	0

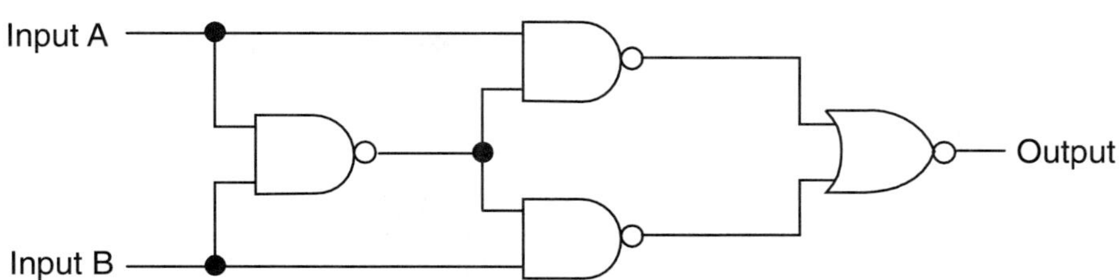

Figure 12-12 XOR gate symbol and truth table. An XOR gate is a combination of NAND and NOR gates.

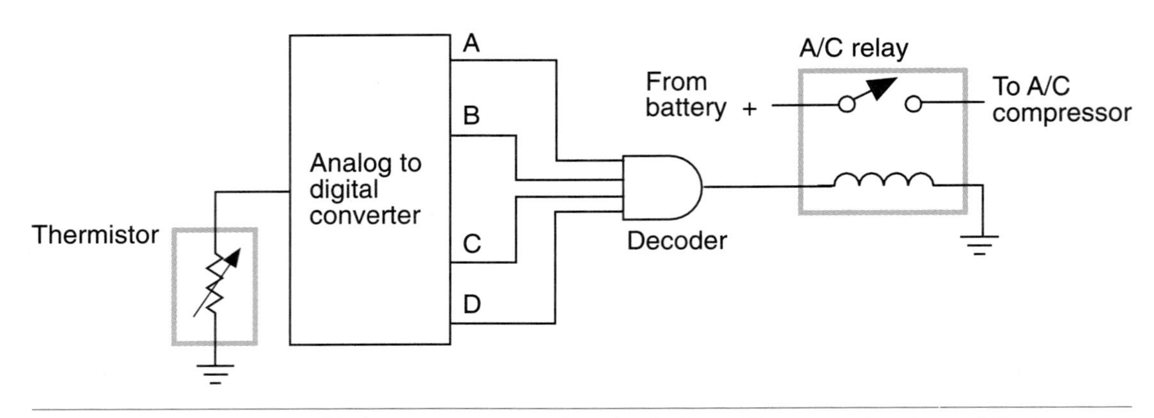

Figure 12-13 Simplified temperature sensing circuit that will turn on the air-conditioning compressor when inside temperature reaches a predetermined value.

These different gates are combined to perform the processing function. The following are some of the most common combinations:

1. **Decoder circuit:** A combination of AND gates used to provide a certain output based on a given combination of inputs (Figure 12-13). When the correct bit pattern is received by the decoder, it will produce the high voltage signal to activate the relay coil.
2. **Multiplexer (MUX):** The basic computer is not capable of looking at all of the inputs at the same time. A multiplexer is used to examine one of many inputs depending on a programmed priority rating (Figure 12-14).
3. **Demultiplexer (DEMUX):** Operates similar to the MUX except that it controls the order of the outputs (Figure 12-15).
4. **RS and clocked RS flip-flop circuits:** Logic circuits that remember previous inputs and do not change their outputs until they receive new input signals. The illustration (Figure 12-16) shows a basic RS flip-flop circuit. The clocked flip-flop circuit has an inverted clock signal as an input so that circuit operations occur in the proper order (Figure 12-17).
5. **Driver circuits:** A *driver* is a term used to describe a transistor device that controls the large current in the output circuit. Drivers are controlled by a computer to operate such things as fuel injectors, ignition coils, and many other high current circuits. The high currents handled by a driver are not really that high; they are just more than what is typically handled by a transistor. Several types of driver circuits are used on automobiles, such as Quad, Discrete, Peak and Hold, and Saturated Switch driver circuits.

The process that the MUX and DEMUX operate on is called **sequential sampling.** This means the computer will deal with all of the sensors and actuators one at a time.

Flip-flop circuits are called **sequential logic circuits** because the output is determined by the sequence of inputs. A given input affects the output produced by the next input.

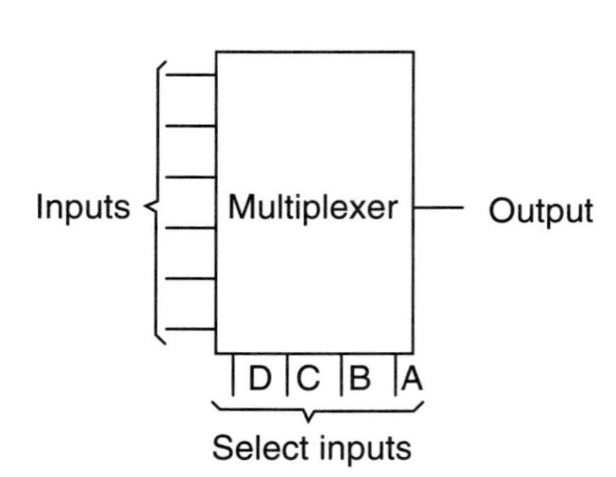

Figure 12-14 Selection at inputs D, C, B, and A will determine which data input will be processed.

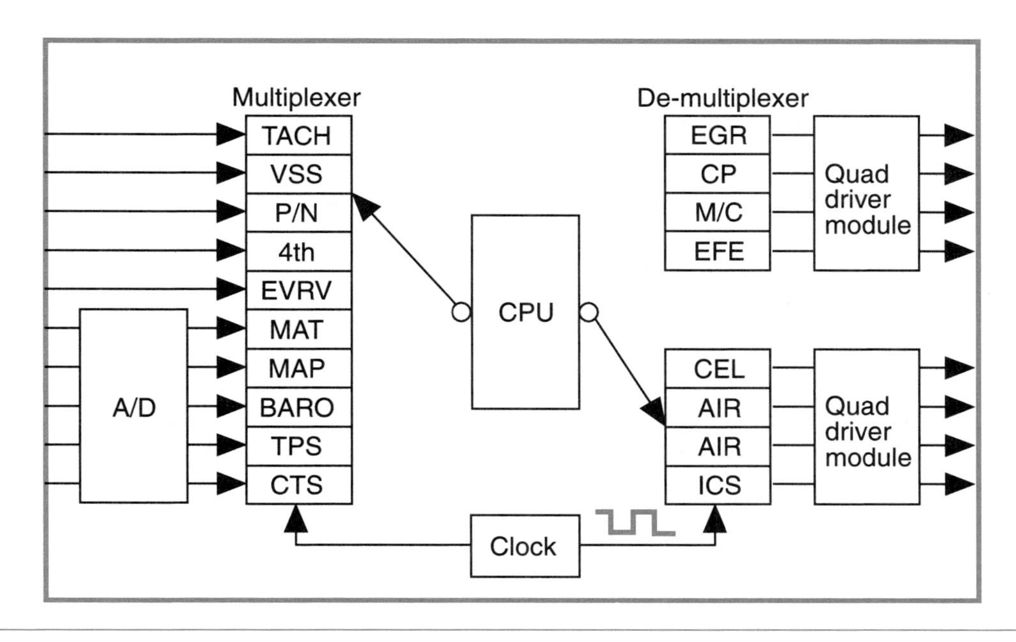

Figure 12-15 Block diagram representation of the MUX and DEMUX circuit.

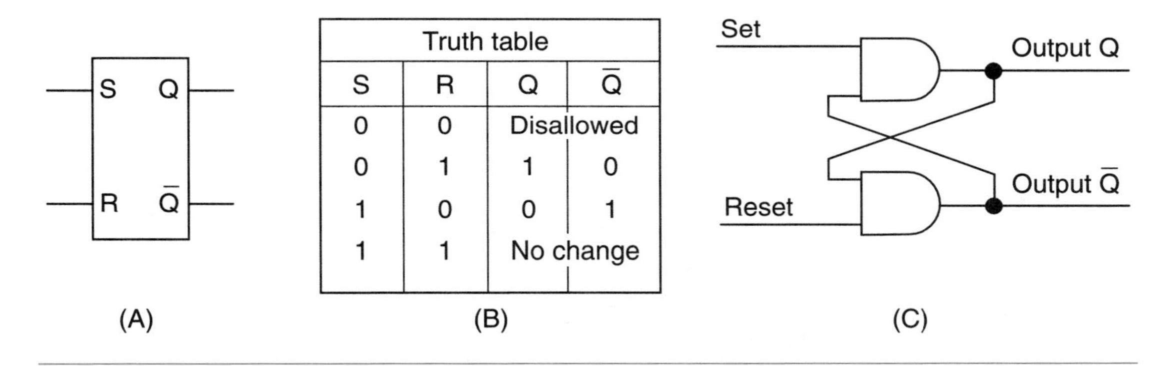

Truth table			
S	R	Q	$\bar{Q}$
0	0	Disallowed	
0	1	1	0
1	0	0	1
1	1	No change	

Figure 12-16 (A) RS flip-flop symbol. (B) Truth table. (C) Logic diagram. Variations of the circuit may include NOT gates at the inputs. If used, the truth table outputs would be reversed.

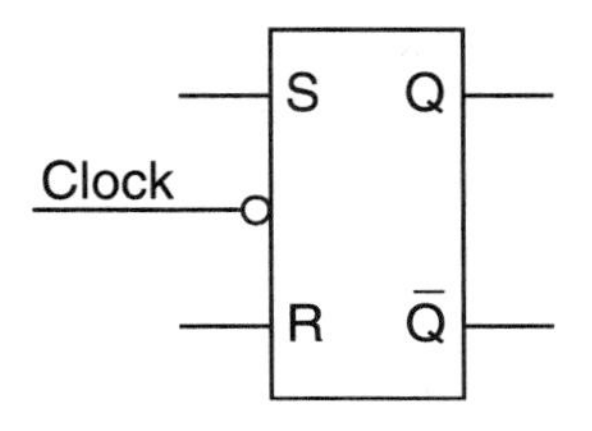

Figure 12-17 Clocked RS flip-flop symbol.

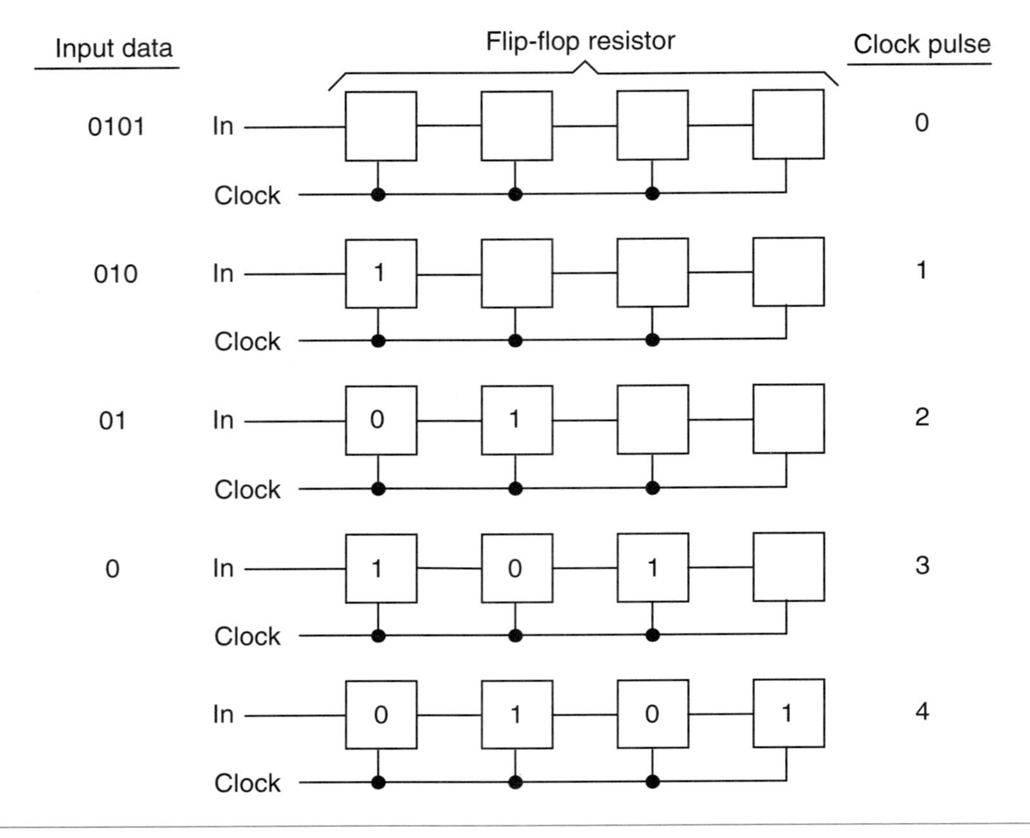

Figure 12-18 It takes four clock pulses to load four bits into the register.

6. **Registers:** Used in the computer to temporarily store information. A register is a combination of flip-flops that transfer bits from one to another every time a clock pulse occurs (Figure 12-18).
7. **Accumulators:** Registers designed to store the results of logic operations that can become inputs to other modules.

Central Processing Unit

The terms microprocessor and central processing unit are basically interchangeable.

The central processing unit (CPU) is the brain of the computer. The CPU is constructed of thousands of transistors that are placed on a small chip. The CPU brings information into and out of the computer's memory. The input information is processed in the CPU and checked against the program in memory. The CPU also checks memory for any other information regarding programmed parameters. The information obtained by the CPU can be altered according to the program instructions. The program may have the CPU apply logic decisions to the information. Once all calculations are made, the CPU will deliver commands to make the required corrections or adjustments to the operation of the controlled system.

The CPU has several main components (Figure 12-19). The registers used include the accumulator, the data counter, the program counter, and the instruction register. The control unit implements the instructions located in the instruction register. The arithmetic logic unit (ALU) performs the arithmetic and logic functions.

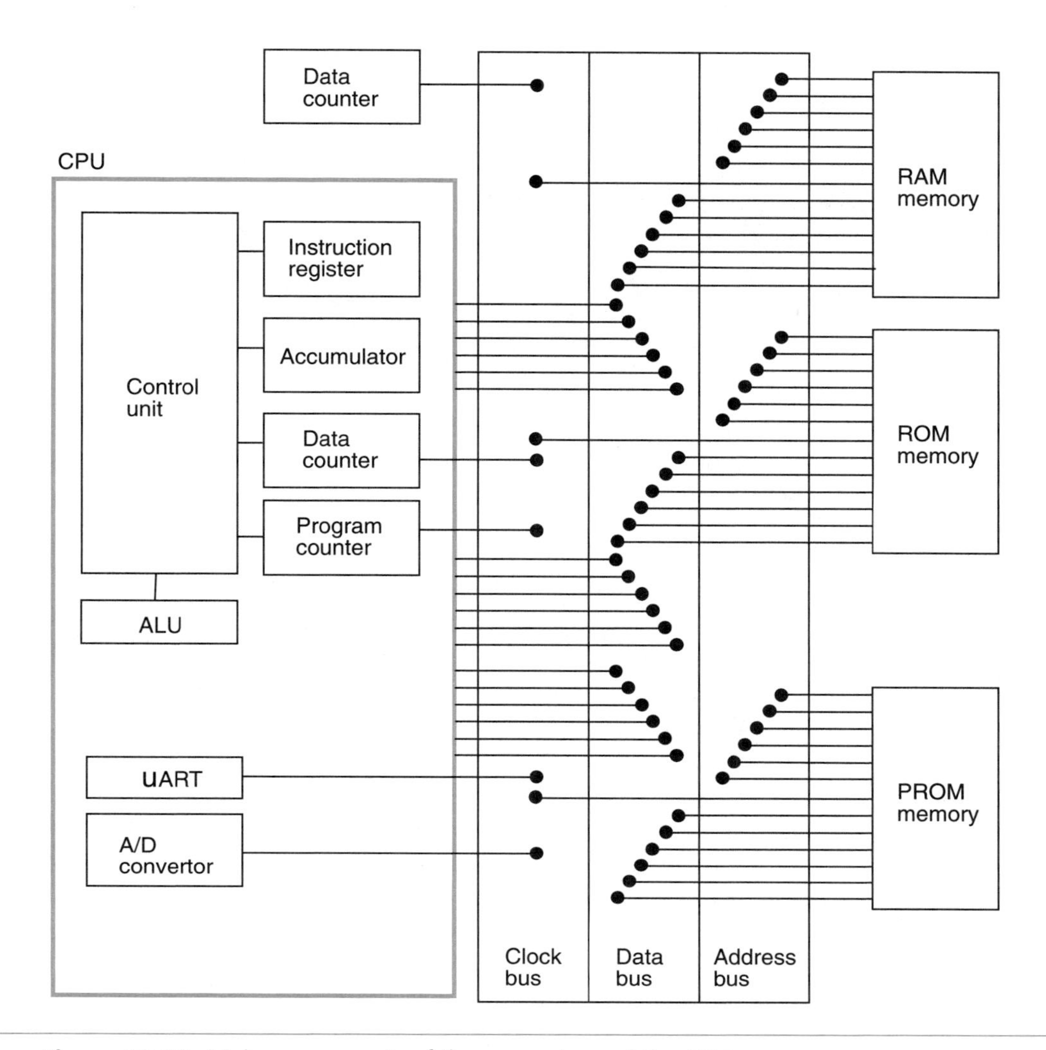

Figure 12-19 Main components of the computer and the CPU.

Computer Memory

The computer requires a means of storing both permanent and temporary memory. Several types of memory chips are used in the body computer:

1. **Read only memory (ROM)** contains a fixed pattern of 1's and 0's that represent permanent stored information. This information is used to instruct the computer on what to do in response to input data. The CPU reads the information contained in ROM, but it cannot write to it or change it. ROM is permanent memory that is programmed in. This memory is not lost when power to the computer is lost. ROM contains formulas, calibrations, ignition-timing tables, and so on. ROM contains the basic operating perimeters for the vehicle.
2. **Random access memory (RAM)** is constructed from flip-flop circuits formed into the chip. The RAM will store temporary information that can be read from or written to by the CPU. RAM stores information that is waiting to be acted upon, and it stores output signals that are waiting to be sent to an output device. RAM can be designed as volatile or nonvolatile. In volatile RAM, the data will be retained as long as current flows through the memory. RAM that is connected to the battery through the ignition switch

The terms ROM, RAM, and PROM are used fairly consistently in the computer industry. However, the names vary between automobile manufacturers.

Figure 12-20 Assortment of PROM chips. Many manufacturers design PROMS that are removable by the use of a special tool.

will lose its data when the switch is turned off. A variation of RAM is keep alive memory (KAM). This is connected directly to the battery through circuit protection devices. KAM will be lost when the battery is disconnected, if the battery drains too low, or if the circuit opens.

3. **Programmable read only memory (PROM)** contains specific data that pertains to the exact vehicle in which the computer is installed. This information may be used to inform the CPU of the accessories that are equipped on the vehicle. The information stored in the PROM is the basis for all computer logic. The information in PROM is used to define or adjust the operating perimeters held in ROM. In many instances, the computer is interchangeable between models of the same manufacturers; however, the PROM is not. Consequently, the PROM may be replaceable and plug into the computer (Figure 12-20).
4. **Erasable PROM (EPROM)** is similar to PROM except that its contents can be erased to allow new data to be installed. A piece of Mylar tape covers a window. If the tape is removed, the microcircuit is exposed to ultraviolet light that erases its memory (Figure 12-21).
5. **Electrically erasable PROM (EEPROM)** allows changing the information electrically one bit at a time. Some manufacturers use this type of memory to store information concerning mileage, vehicle identification number, and options.
6. **Nonvolatile RAM (NVRAM)** is a combination of RAM and EEPROM into the same chip. During normal operation, data is written to and read from the RAM portion of the chip. If the power is removed from the chip, or at programmed timed intervals, the data is transferred from RAM to the EEPROM portion of the chip. When the power is restored to the chip, the EEPROM will write the data back to the RAM.

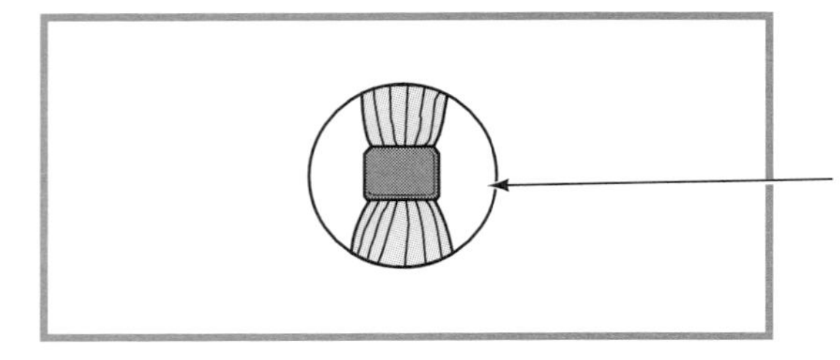

Figure 12-21 EPROM memory is erased when ultraviolet rays contact the microcircuitry.

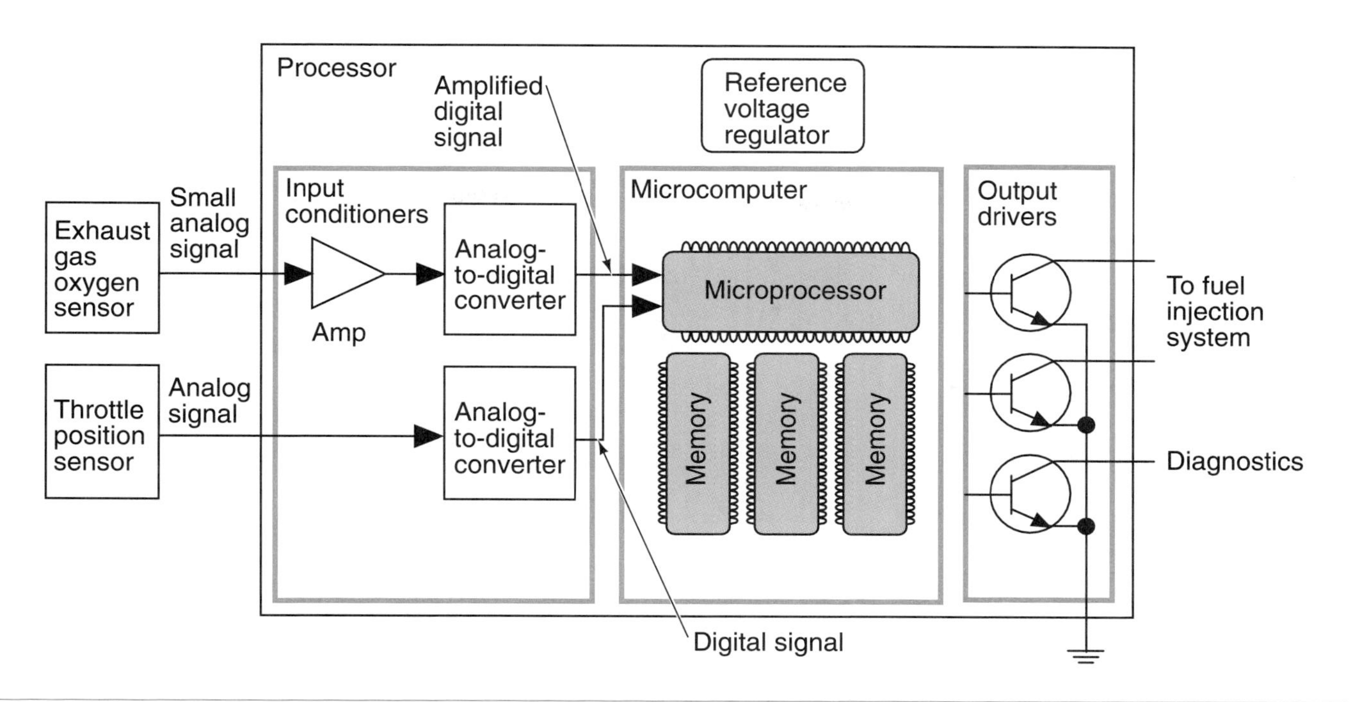

Figure 12-22 The input signals are processed in the microprocessor. The microprocessor directs the output drivers to activate actuators as instructed by the program.

WARNING: Installing a PROM chip backward will immediately destroy the chip. In addition, electrostatic discharge (ESD) will destroy the chip. Static straps should be used when working on a computer to prevent ESD while you are working on the unit.

Inputs

As discussed earlier, the CPU receives inputs that it checks with programmed values. Depending on the input, the computer will control the actuator(s) until the programmed results are obtained (Figure 12-22). The inputs can come from other computers, the driver, the technician, or through a variety of sensors.

Driver input signals are usually provided by momentarily applying a ground through a switch. The computer receives this signal and performs the desired function. For example, if drivers wish to reset the trip odometer on a digital instrument panel, they would push the reset switch. This switch will provide a momentary ground that the computer receives as an input and sets the trip odometer to zero (Figure 12-23).

Switches can be used as an input for any operation that only requires a yes-no, or on-off, condition. Other inputs include those supplied by means of a sensor and those signals returned to the computer in the form of feedback.

Sensors

There are many different designs of sensors. Some are nothing more than a switch that completes the circuit. Others are complex chemical reaction devices that generate their own voltage under different conditions. Repeatability, accuracy, operating range, and linearity are all requirements of a sensor.

Shop Manual
Chapter 12, page 440

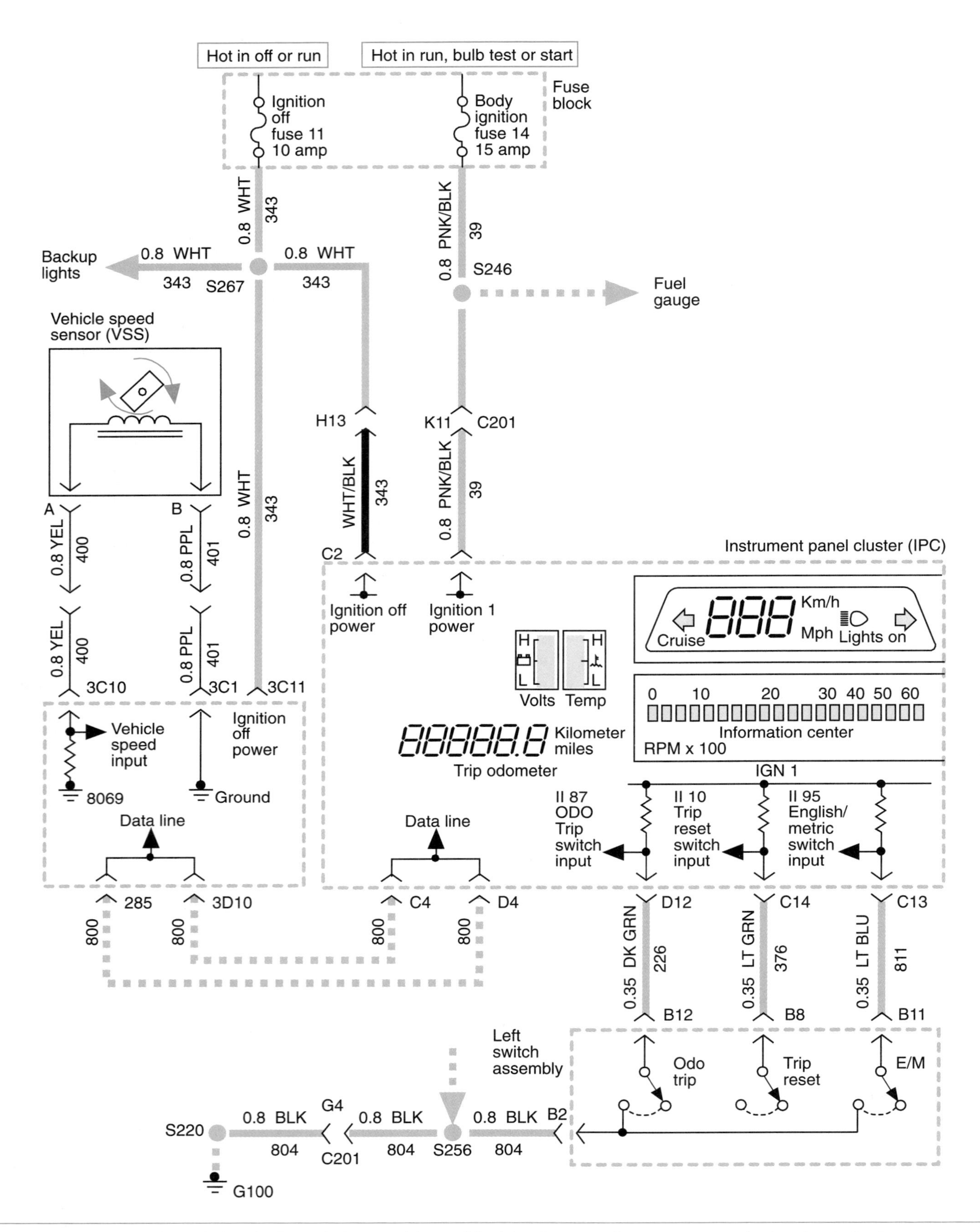

Figure 12-23 The switches in the switch assembly provide driver input to the computer to display the desired information.

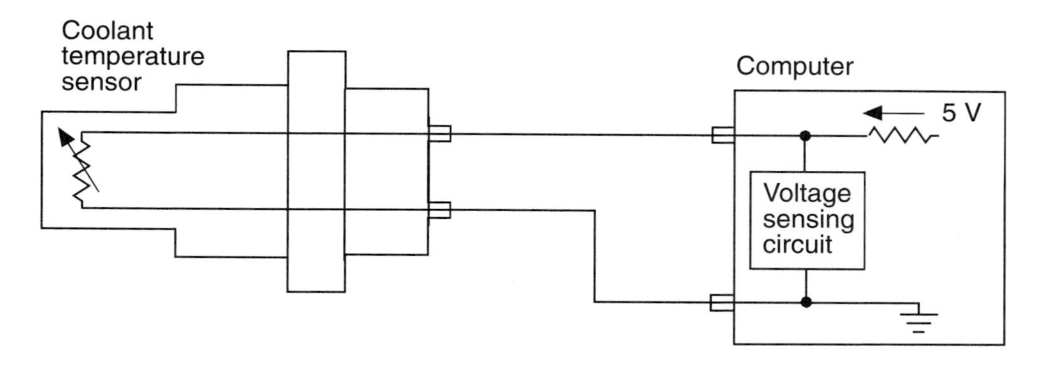

Figure 12-24 A thermistor is used to measure temperature. The sensing unit measures the resistance change and translates the data into temperature values.

Sensors convert some measurement of vehicle operation into an electrical signal.

Linearity refers to the sensor signal being as constantly proportional to the measured value as possible. It is an expression of the sensor's accuracy.

Thermistors. A thermistor is used to sense engine coolant or ambient temperatures. By monitoring the thermistor's resistance value, the computer is capable of observing very small changes in temperature. The computer sends a reference voltage to the thermistor (usually 5 volts) through a fixed resistor. As the current flows through the thermistor resistance to ground, a voltage sensing circuit measures the voltage drop (Figure 12-24). Using its programmed values, the computer is able to translate the voltage drop into a temperature value.

Shop Manual
Chapter 12,
page 440

Wheatstone Bridges. The illustration (Figure 12-25) shows the construction of the Wheatstone bridge. Usually three of the resistors are kept at exactly the same value and the fourth is the sensing resistor. When all four resistors have the same value, the bridge is balanced and the voltage sensor will indicate a value of 0 volts. If there is a change in the resistance value of the sense resistor, a change will occur in the circuit's balance. The sensing circuit will receive a voltage reading that is proportional to the amount of resistance change. If the Wheatstone bridge is used to measure temperature, temperature changes will be indicated as a change in voltage by the sensing circuit. Wheatstone bridges are also used to measure pressure (piezoresistive) and mechanical strain. A Wheatstone bridge is nothing more than two simple series circuits connected in parallel across a power supply. A common use of a Wheatstone bridge is the hot wire sensor in a mass air flow (MAF) sensor. The sensor consists of a hot wire circuit, a cold wire circuit, and an electronic signal processing area. The hot and cold wire circuits form the Wheatstone

A **thermistor** is a solid-state variable resistor made from a semiconductor material that changes resistance in relation to temperature changes.

Negative temperature coefficient (NTC) thermistors reduce their resistance as the temperature increases. **Positive temperature coefficient (PTC)** thermistors increase their resistance as the temperature increases. The NTC is most commonly used.

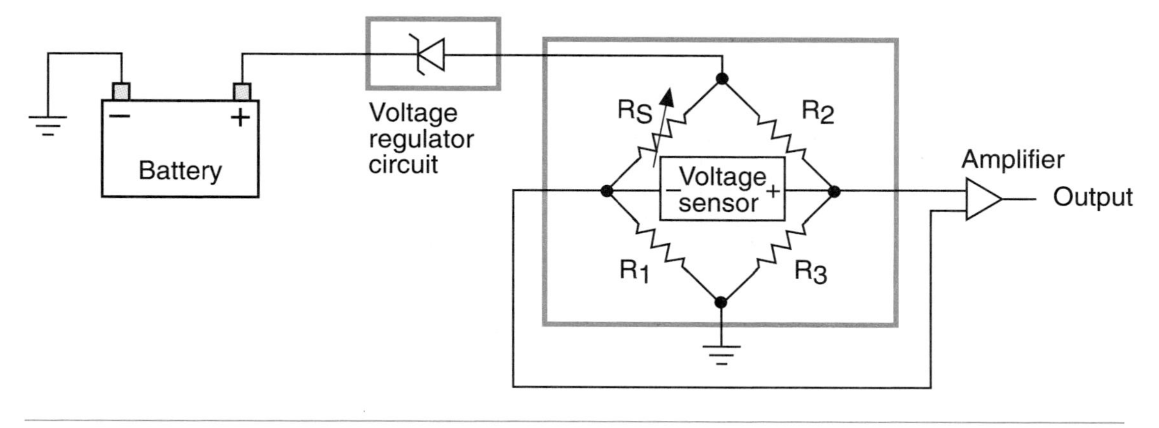

Figure 12-25 Wheatstone bridge.

The **Wheatstone bridge** is a series-parallel arrangement of resistors between an input terminal and ground.

An MAF is used to measure the amount of air flowing into the engine.

bridge. The cold wire circuit is made of a fixed resistor and a thermistor. The amount of voltage dropped across the two resistors is determined by the temperature of the thermistor.

The hot wire circuit is made up of a fixed resistor and a variable resistance heat element (hot wire). The heat element generates heat in proportion to the amount of current flowing through it. This heat, in turn, changes its resistance. As air flows past the hot wire, it moves a small amount of heat from the element. This cooling of the element causes the voltage drop across it to change. The voltage drop across the hot wire is compared to the voltage drop across the fixed resistor in the cold wire circuit, and air flow is determined.

The word "Piezoelectric" comes from the Greek word "piezo," which means pressure.

Piezoelectric Devices. Piezoelectric devices are used to measure fluid and air pressures. The most commonly found piezoelectric device is the engine knock sensor. The knock sensor measures engine knock, or vibration, and converts the vibration into a voltage signal.

The sensor is a voltage generator and has a resistor connected in series with it. The resistor protects the sensor from excessive current flow in case the circuit becomes shorted. The voltage generator is a thin ceramic disc attached to a metal diaphragm. When engine knock occurs, the vibration of the noise puts pressure on the diaphragm. This puts pressure on the piezoelectric crystals in the ceramic disc. The disc generates a voltage that is proportional to the amount of pressure. The voltage generated ranges from zero to one or more volts. Each time the engine knocks, a voltage spike is generated by the sensor.

Piezoresistive Devices: In construction, a piezoresistive device is similar to a piezoelectric one. However, it operates differently. The sensor acts like a variable resistor. Its resistance value changes as the pressure applied to the crystal changes. A voltage regulator supplies a constant voltage to the sensor. Since the amount of voltage dropped by the sensor will change with the change in resistance, the control module can determine the amount of pressure on the crystals by measuring the voltage drop across the sensor. Piezoresistive sensors are commonly used as gauge sending units.

Shop Manual
Chapter 12, page 440

A **potentiometer** is a variable resistor that provides accurate voltage drop readings to the computer.

Potentiometers. The potentiometer usually consists of a wire wound resistor with a moveable center wiper (Figure 12-26). A constant voltage value (usually 5 volts) is applied to terminal A. If the wiper (which is connected to the shaft or moveable component of the unit that is being monitored) is located close to this terminal, there will be low voltage drop represented by high voltage signal back to the computer through terminal B. As the wiper is moved toward the C terminal, the sensor signal voltage to terminal B decreases. The computer interprets the different voltage value into different shaft positions. The potentiometer can measure linear or rotary movement. As the wiper is moved across the resistor, the position of the unit can be tracked by the computer.

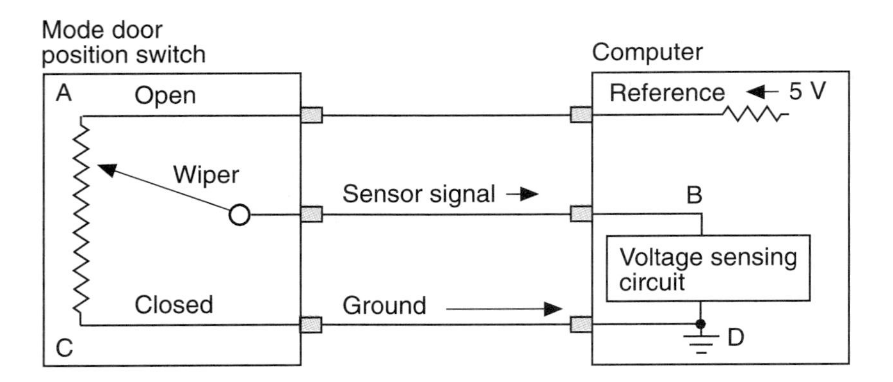

Figure 12-26 A potentiometer sensor circuit measures the amount of voltage drop to determine position.

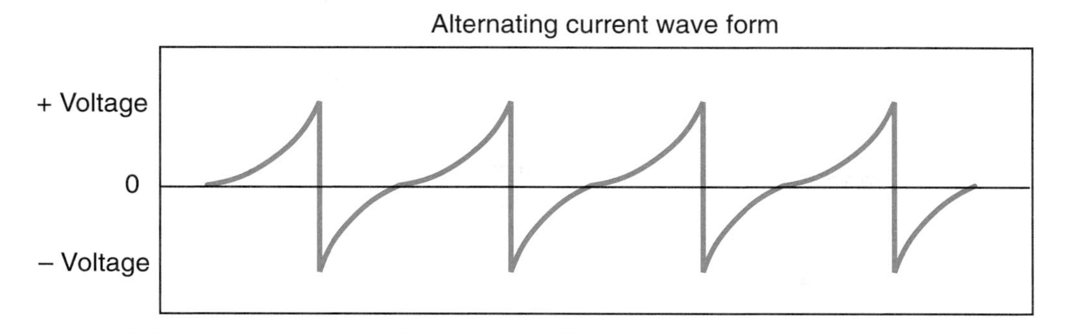

Figure 12-27 Pulse signal wave form.

Magnetic Pulse Generators. Magnetic pulse generators are commonly used to send data to the computer about the speed of the monitored component. This data provides information about vehicle speed and individual wheel speed. The signals from the speed sensors are used for computer-driven instrumentation, cruise control, antilock braking, speed sensitive steering, and automatic ride control systems. The magnetic pulse generator is also used to inform the computer of the position of a monitored component. This is common in engine controls where the computer needs to know the position of the crankshaft in relation to rotational degrees.

The components of the pulse generator are:

1. A timing disc that is attached to the rotating shaft or cable. The number of teeth on the timing disc is determined by the manufacturer and depends on application. The teeth will cause a voltage generation that is constant per revolution of the shaft. For example, a vehicle speed sensor may be designed to deliver 4,000 pulses per mile. The number of pulses per mile remains constant regardless of speed. The computer calculates how fast the vehicle is going based on the frequency of the signal.
2. A pick-up coil consists of a permanent magnet that is wound around by fine wire.

An air gap is maintained between the timing disc and the pick-up coil. As the timing disc rotates in front of the pick-up coil, the generator sends a pulse signal (Figure 12-27). As a tooth on the timing disc aligns with the core of the pick-up coil, it repels the magnetic field. The magnetic field is forced to flow through the coil and pick-up core (Figure 12-28). When the tooth

Shop Manual
Chapter 12,
page 442

Magnetic pulse generators use the principle of magnetic induction to produce a voltage signal.

The magnetic pulse generator is also called a **permanent magnet (PM) generator.**

The **timing disc** is known as an armature, reluctor, trigger wheel, pulse wheel, or timing core. It is used to conduct lines of magnetic force.

The **pick-up coil** is also known as a stator, sensor, or pole piece. It remains stationary while the timing disc rotates in front of it. The changes of magnetic lines of force generate a small voltage signal in the coil.

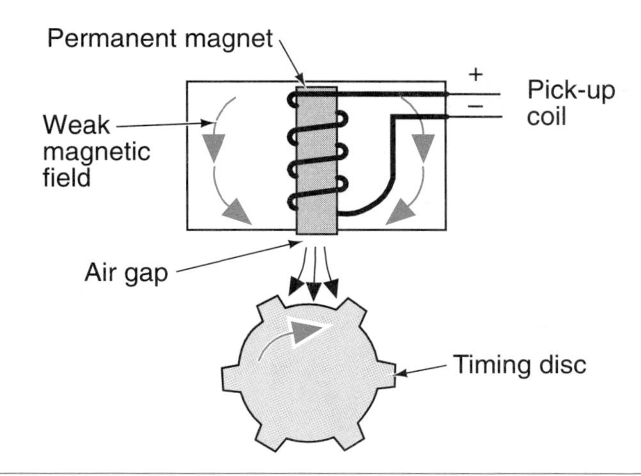

Figure 12-28 A strong magnetic field is produced in the pick-up coil as the teeth align with the core.

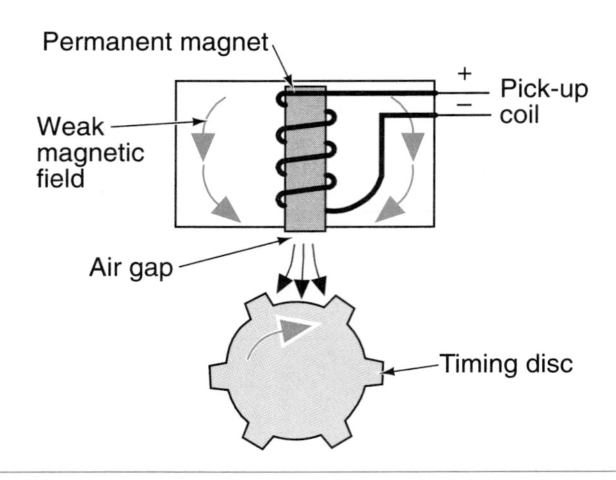

Figure 12-29 The magnetic field expands (weakens) as the teeth pass the core.

passes the core, the magnetic field is able to expand (Figure 12-29). This action is repeated every time a tooth passes the core. The moving lines of magnetic force cut across the coil windings and induce a voltage signal.

When a tooth approaches the core, a positive current is produced as the magnetic field begins to concentrate around the coil (Figure 12-30). When the tooth and core align, there is no more expansion or contraction of the magnetic field (thus no movement) and the voltage drops to zero (Figure 12-31). When the tooth passes the core, the magnetic field expands and a negative current is produced (Figure 12-32). The resulting pulse signal is amplified, digitalized, and sent to the microprocessor.

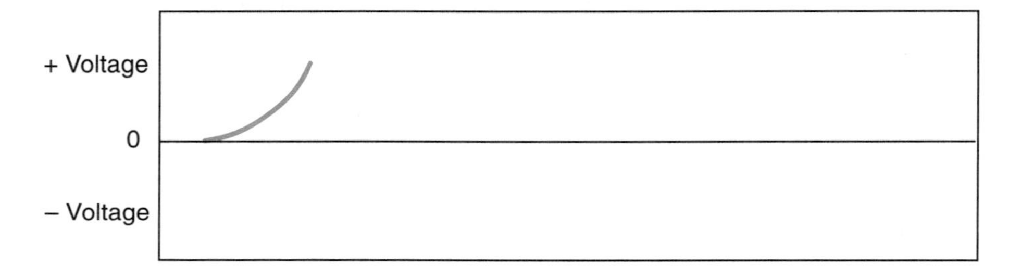

Figure 12-30 A positive voltage swing is produced as the tooth approaches the core.

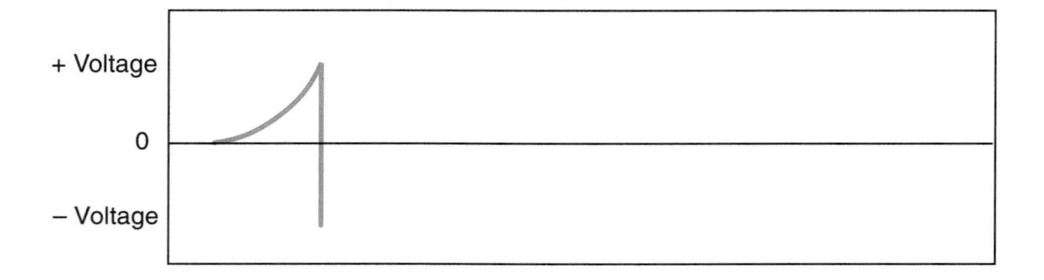

Figure 12-31 There is no magnetic movement and no voltage when the tooth aligns with the core.

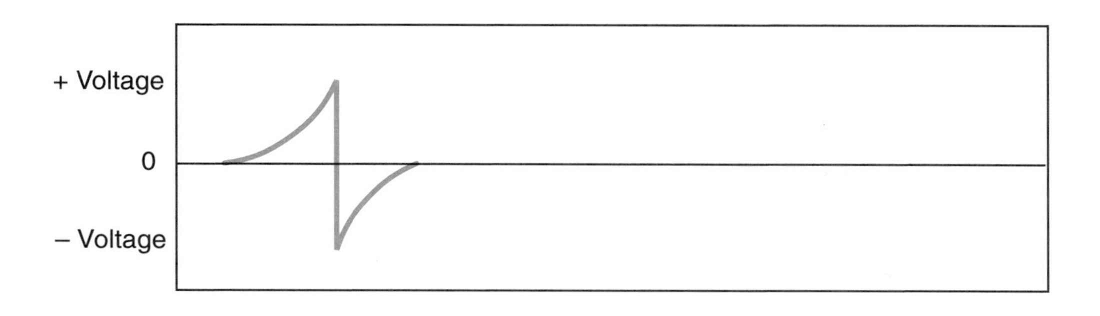

Figure 12-32 A negative waveform is created as the tooth passes the core.

Hall-effect switches operate on the principle that if a current is allowed to flow through thin conducting material that is exposed to a magnetic field, another voltage is produced (Figure 12-33).

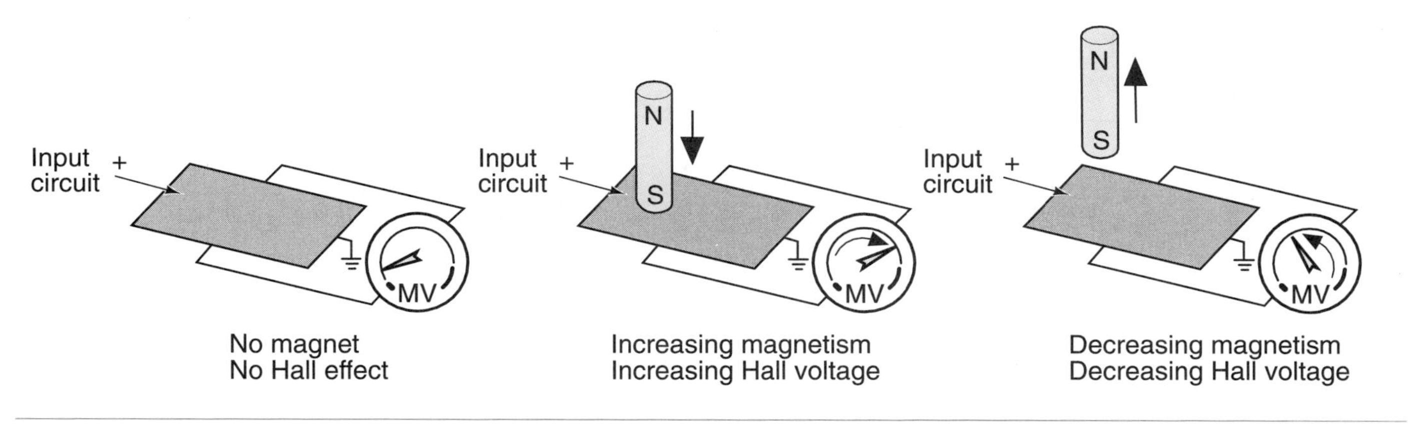

Figure 12-33 Hall-effect principles of voltage induction.

Hall-Effect Switches. The Hall-effect switch performs the same functions as the magnetic pulse generator. It contains a permanent magnet and a thin semiconductor layer made of gallium arsenate crystal (Hall layer), and a shutter wheel (Figure 12-34). The Hall layer has a negative and a positive terminal connected to it. Two additional terminals located on either side of the Hall layer are used for the output circuit.

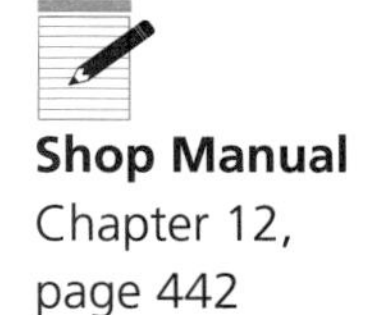

Shop Manual
Chapter 12,
page 442

The permanent magnet is located directly across from the Hall layer so that its lines of flux will bisect at right angles to the current flow. The permanent magnet is mounted so that a small air gap is between it and the Hall layer.

A steady current is applied to the crystal of the Hall layer. This produces a signal voltage that is perpendicular to the direction of current flow and magnetic flux. The signal voltage produced is a result of the effect the magnetic field has on the electrons. When the magnetic field bisects the supply current flow, the electrons are deflected toward the Hall layer negative terminal (Figure 12-35). This results in a weak voltage potential being produced in the Hall switch.

The **shutter wheel** consists of a series of alternating windows and vanes. It creates a magnetic shunt that changes the strength of the magnetic field from the permanent magnet.

A shutter wheel is attached to a rotational component. As the wheel rotates, the shutters (vanes) will pass in this air gap. When a shutter vane enters the gap, it intercepts the magnetic field and shields the Hall layer from its lines of force. The electrons in the supply current are no longer disrupted and return to a normal state. This results in low voltage potential in the signal circuit of the Hall switch.

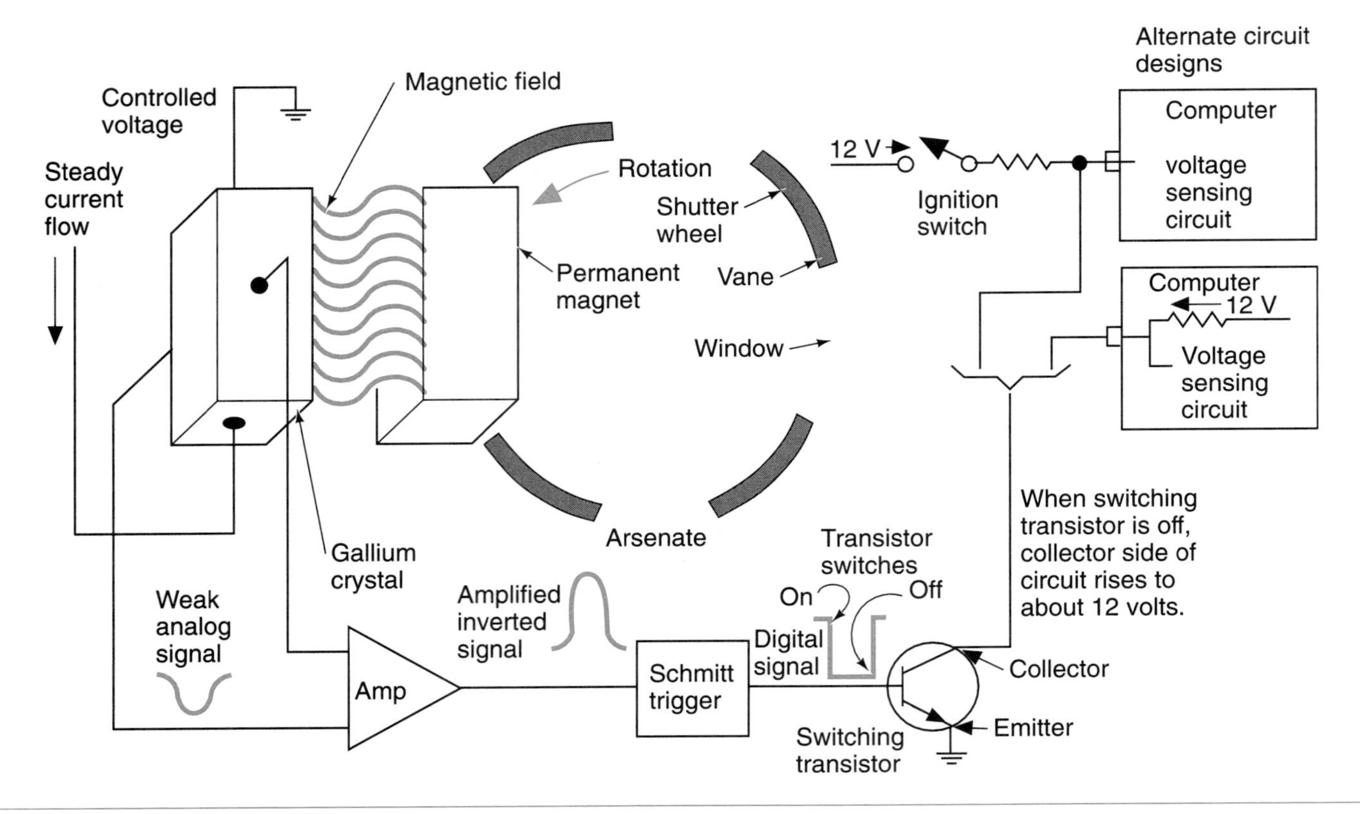

Figure 12-34 Typical circuit of a Hall-effect switch.

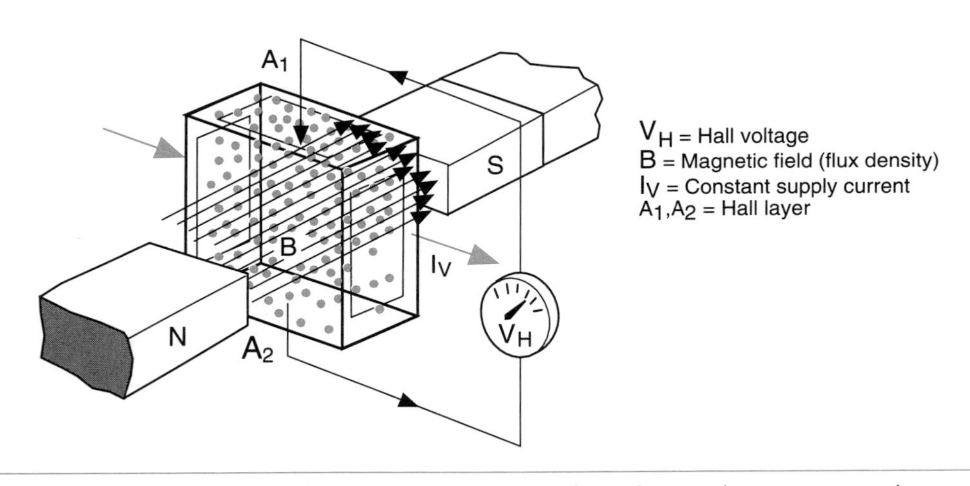

Figure 12-35 The magnetic field causes the electrons from the supply current to gather at the Hall layer negative terminal. This creates a voltage potential.

A **Schmitt trigger** is an A/D convertor.

The signal voltage leaves the Hall layer as a weak analog signal. To be used by the computer, the signal must be conditioned. It is first amplified because it is too weak to produce a desirable result. The signal is also inverted so that a low input signal is converted into a high output signal. It is then sent through a Schmitt trigger where it is digitized and conditioned into a clean square wave signal. The signal is finally sent to a switching transistor. The computer senses the turning on and off of the switching transistor to determine the frequency of the signals and calculates speed.

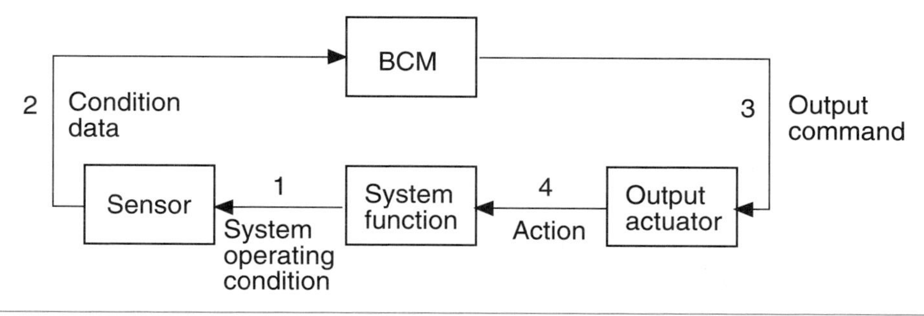

Figure 12-36 Principle of feedback signals.

Feedback Signals

If the computer sends a command signal to open a blend door in an automatic climate control system, a feedback signal may be sent back from the actuator to inform the computer the task was performed. The feedback signal will confirm both the door position and actuator operation (Figure 12-36). Another form of feedback is for the computer to monitor voltage as a switch, relay, or other actuator is activated. Changing states of the actuator will result in a predictable change in the computer's voltage sensing circuit. The computer may set a diagnostic code if it does not receive the correct feedback signal.

Feedback means that data concerning the effects of the computer's commands are fed back to the computer as an input signal.

Oxygen Sensors: One of the most commonly used feedback sensors is the O_2 sensor. Although this is a principle input for engine control systems, its basic operation needs to be described here. The O_2 sensor is mounted in the exhaust gas stream and provides PCM with a measurement of the oxygen in the engine's exhaust. The sensor is contructed of a zirconium dioxide ceramic thimble covered with a thin layer of platinum.

An oxygen sensor is commonly referred to as an O_2 sensor.

When the thimble is filled with oxygen-rich outside air and the outer surface of the thimble is exposed to oxygen-depleted exhaust gases, a chemical reaction in the sensor produces a voltage. The generation of voltage is similar to the same activity that takes place in a battery, except at much lower voltages. The voltage output varies with the level of oxygen present in the exhaust. As oxygen in the exhaust decreases, the voltage output increases. Likewise, as the oxygen level in the exhaust increases, the output voltage decreases.

Outputs

Once the computer's programming instructs that a correction or adjustment must be made in the controlled system, an output signal is sent to an actuator. This involves translating the electronic signals into mechanical motion.

An output driver is used within the computer to control the actuators. The circuit driver usually applies the ground circuit of the actuator. The ground can be applied steadily if the actuator must be activated for a selected amount of time. For example, if the BCM inputs indicate that the automatic door locks are to be activated, the actuator is energized steadily until the locks are latched. Then the ground is removed.

Other systems require the actuator to either be turned on and off very rapidly or for a set amount of cycles per second. It is duty cycled if it is turned on and off a set amount of cycles per second. Most duty cycled actuators cycle ten times per second. To complete a cycle it must go from off to on to off again. If the cycle rate is ten times per second, one actuator cycle is completed in one tenth of a second. If the actuator is turned on for 30% of each tenth of a second and off for 70%, it is referred to as a 30% duty cycle (Figure 12-37).

Duty cycle is the percentage of on-time to total cycle time.

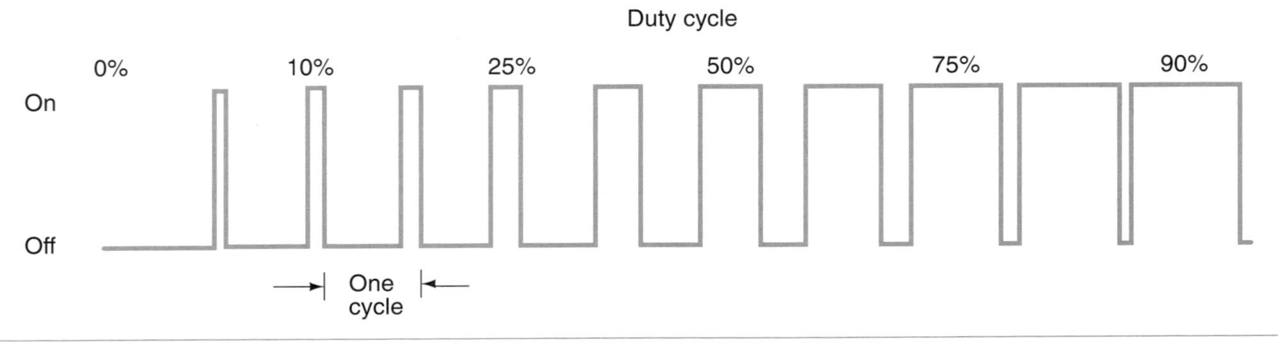

Figure 12-37 Duty cycle is the percentage of on-time per cycle. Duty cycle can be changed; however, the total cycle time remains constant.

Pulse width is the length of time in milliseconds that an actuator is energized.

If the actuator is cycled on and off very rapidly, the pulse width can be varied to provide the programmed results. For example, the computer program will select an illumination level of the digital instrument panel based on the intensity of the ambient light in the vehicle. The illumination level is achieved through pulse width modulation of the lights. If the lights need to be bright, the pulse width is increased, which increases the length of on-time. As the light intensity needs to be reduced, the pulse width is decreased (Figure 12-38).

Shop Manual
Chapter 12, page 438

Actuators perform the actual work commanded by the computer. They can be in the form of a motor, relay, switch, or solenoid.

Actuators

Most computer-controlled actuators are electromechanical devices that convert the output commands from the computer into mechanical action. These actuators are used to open and close switches, control vacuum flow to other components, and operate doors or valves depending on the requirements of the system.

Although they do not fall into the strict definition of an actuator, the BCM can also control lights, gauges, and driver circuits.

Relays. A relay allows control of a high current draw circuit by a very low current draw circuit. The computer usually controls the relay by providing the ground for the relay coil (Figure 12-39). The use of relays protects the computer by keeping the high current from passing through it. For example, the motors used for power door locks require a high current draw to operate them. Instead of having the computer operate the motor directly, it will energize the relay. With the relay energized, a direct circuit from the battery to the motor is completed (Figure 12-40).

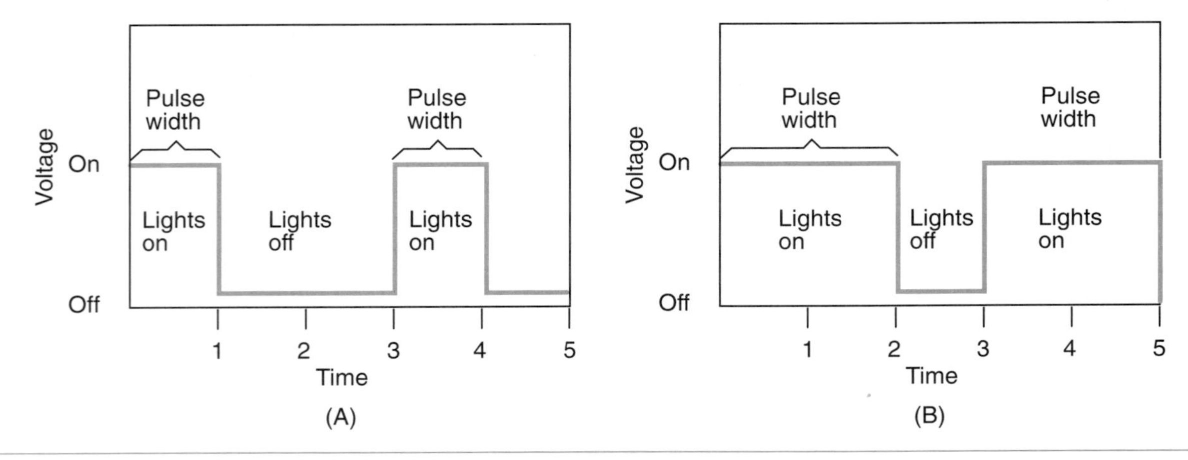

Figure 12-38 Pulse width is the duration of on-time: (A) pulse width modulation to achieve dimmer dash lights, (B) pulse width modulation to achieve brighter dash lights.

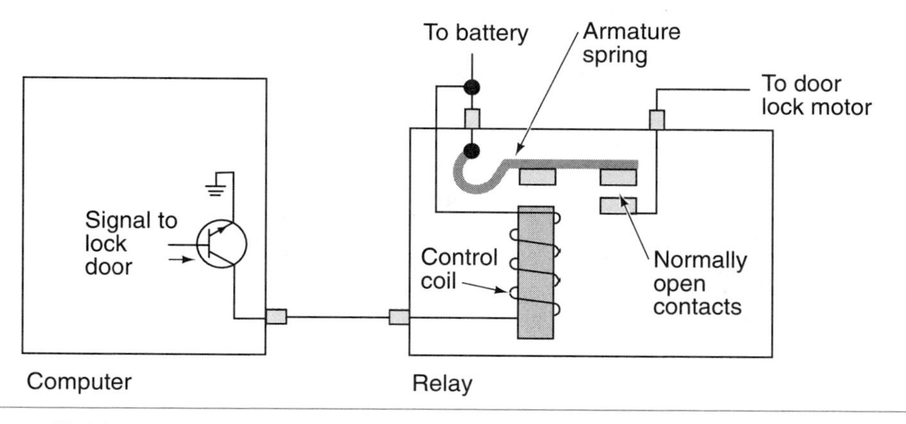

Figure 12-39 The computer's output driver applies the ground for the relay coil.

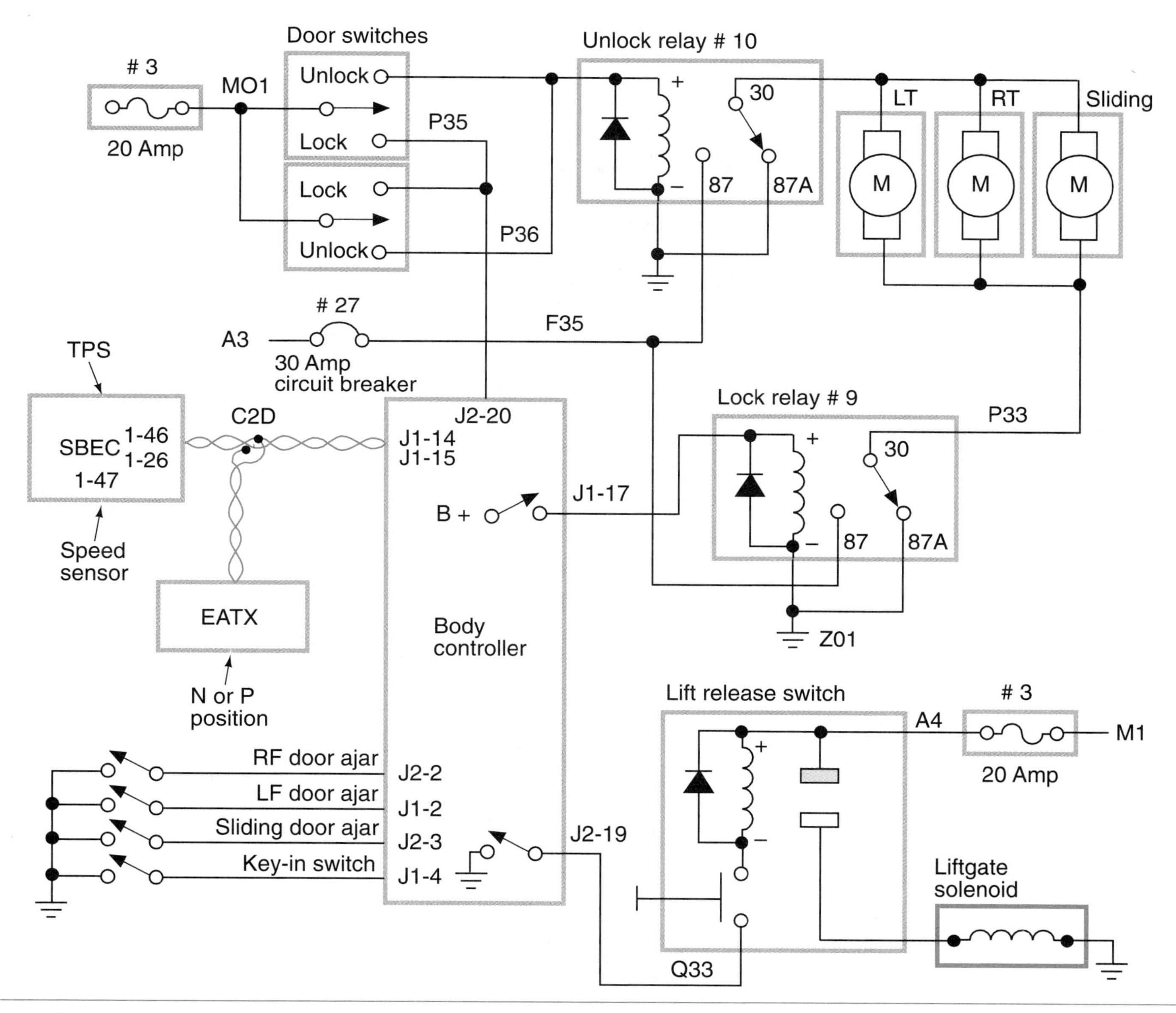

Figure 12-40 The computer controls the operation of the door lock motors by controlling the relays. (Courtesy of Chrysler Corporation)

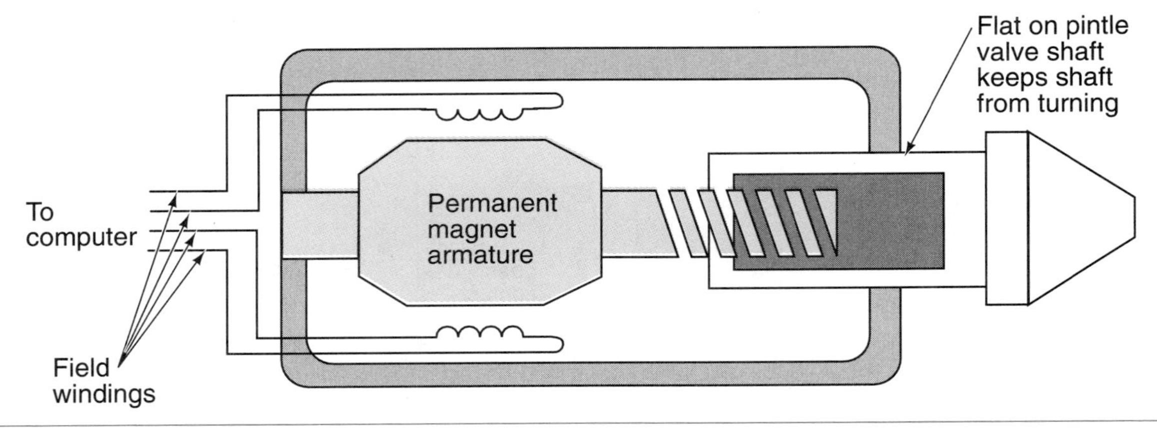

Figure 12-41 Typical stepper motor.

A **stepper motor** contains a permanent magnet armature with two, four, or more field coils (Figure 12-41).

A **servomotor** produces rotation of less than a full turn. A feedback mechanism is used to position itself to the exact degree of rotation required.

Shop Manual Chapter 12, page 362

The common acronym for multiplexing is **MUX**.

Solenoids. Computer control of the solenoid is usually provided by applying the ground through the output driver. A solenoid is commonly used as an actuator because it operates well under duty cycling conditions.

One of the most common uses of the solenoid is to control vacuum to other components. Many automatic climate control systems use vacuum motors to move the blend doors. The computer can control the operation of the doors by controlling the solenoid.

Motors. Many computer-controlled systems use a stepper motor to move the controlled device to whatever location is desired. By applying voltage pulses to selected coils of the motor, the armature will turn a specific number of degrees. When the same voltage pulses are applied to the opposite coils, the armature will rotate the same number of degrees in the opposite direction.

Some applications require the use of a permanent magnet field motor (Figure 12-42). The polarity of the voltage applied to the armature windings determines the direction the motor rotates. The computer can apply a continuous voltage to the armature until the desired result is obtained.

Multiplexing

Multiplexing is the ability to use a single cable to distribute data to different control modules throughout the vehicle. Because the data is transmitted through a single wire, bulky wiring harnesses are eliminated. In a multiplexed system, all sensors are connected to the same wire, which is connected to a control module.

A MUX wiring system uses bus data links that connect each module. Each module can transmit and receive digital codes over the bus data links (Figure 12-43). This allows the modules to share their information. The signal sent from a sensor can go to any of the modules and can be used by the other modules (Figure 12-44). By sharing this data, the need for separate conductors from the sensor to each module is eliminated.

A chip is used to prevent the digital codes from overlapping by allowing only one code to be transmitted at a time. Each digital message is proceeded by an identification code that establishes its priority. If two modules attempt to send a message at the same time, the message with the higher priority code is transmitted first.

The major difference between a multiplexed system and a nonmultiplexed system is the way data is gathered and processed. In nonmultiplexed systems, the signal from a sensor is sent

To reduce the effects of electromagnetic interference, the multiplex wiring system uses a twisted pair design or shielded wires.

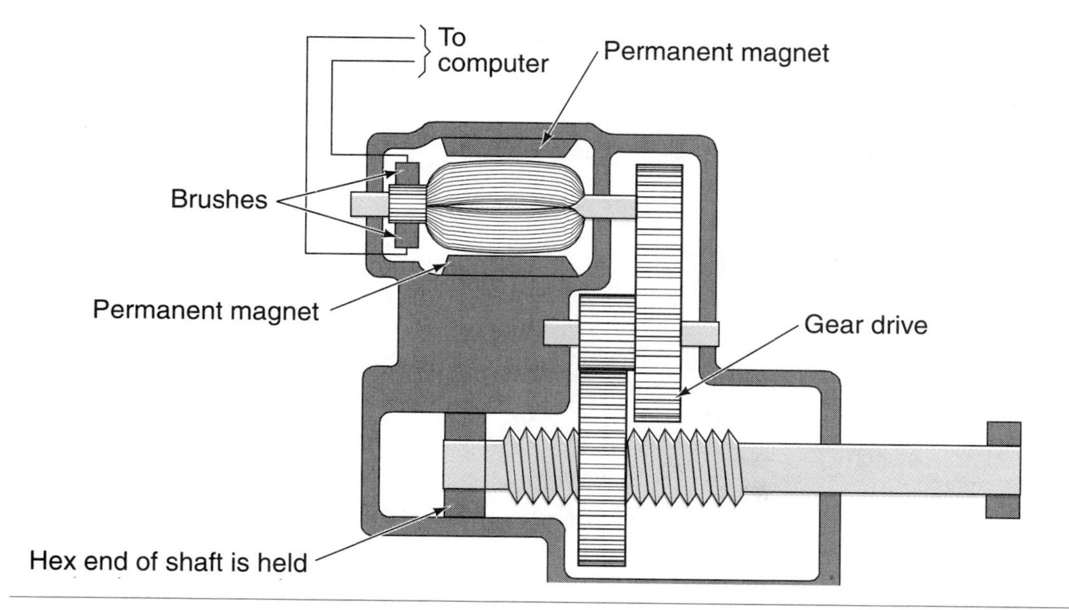

Figure 12-42 Reversible permanent magnet motor.

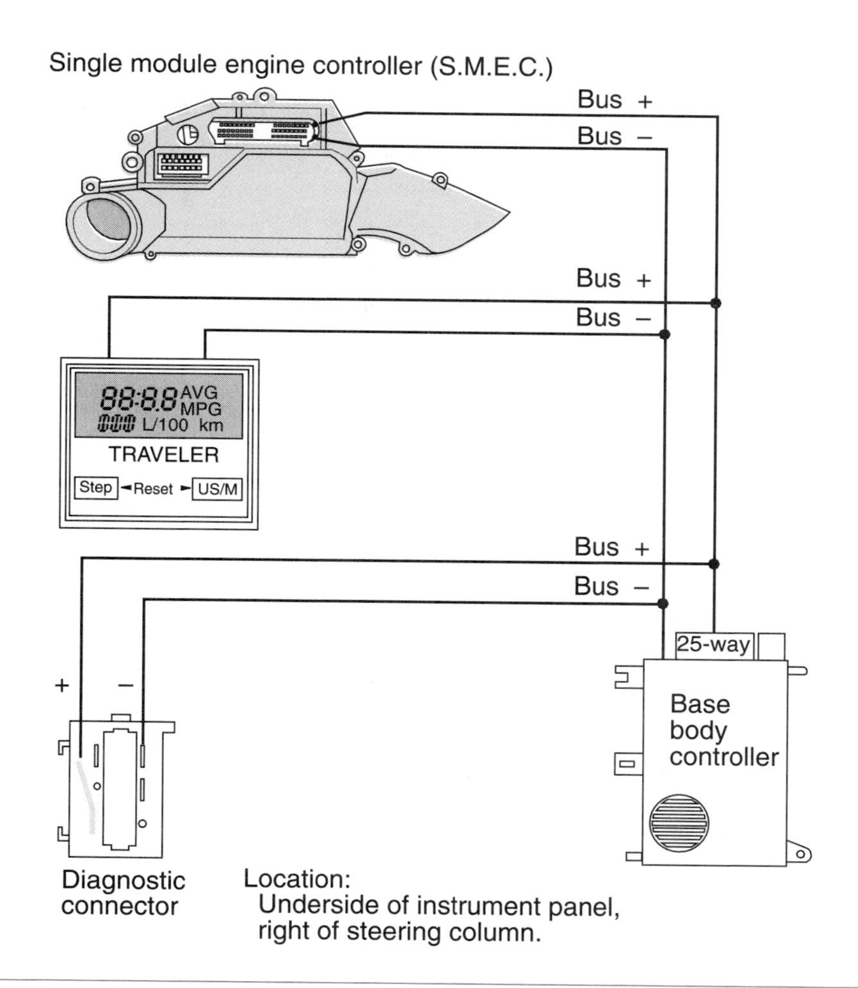

Figure 12-43 Computers use multiplexing to reduce the number of conductors that would be required. (Courtesy of Chrysler Corporation)

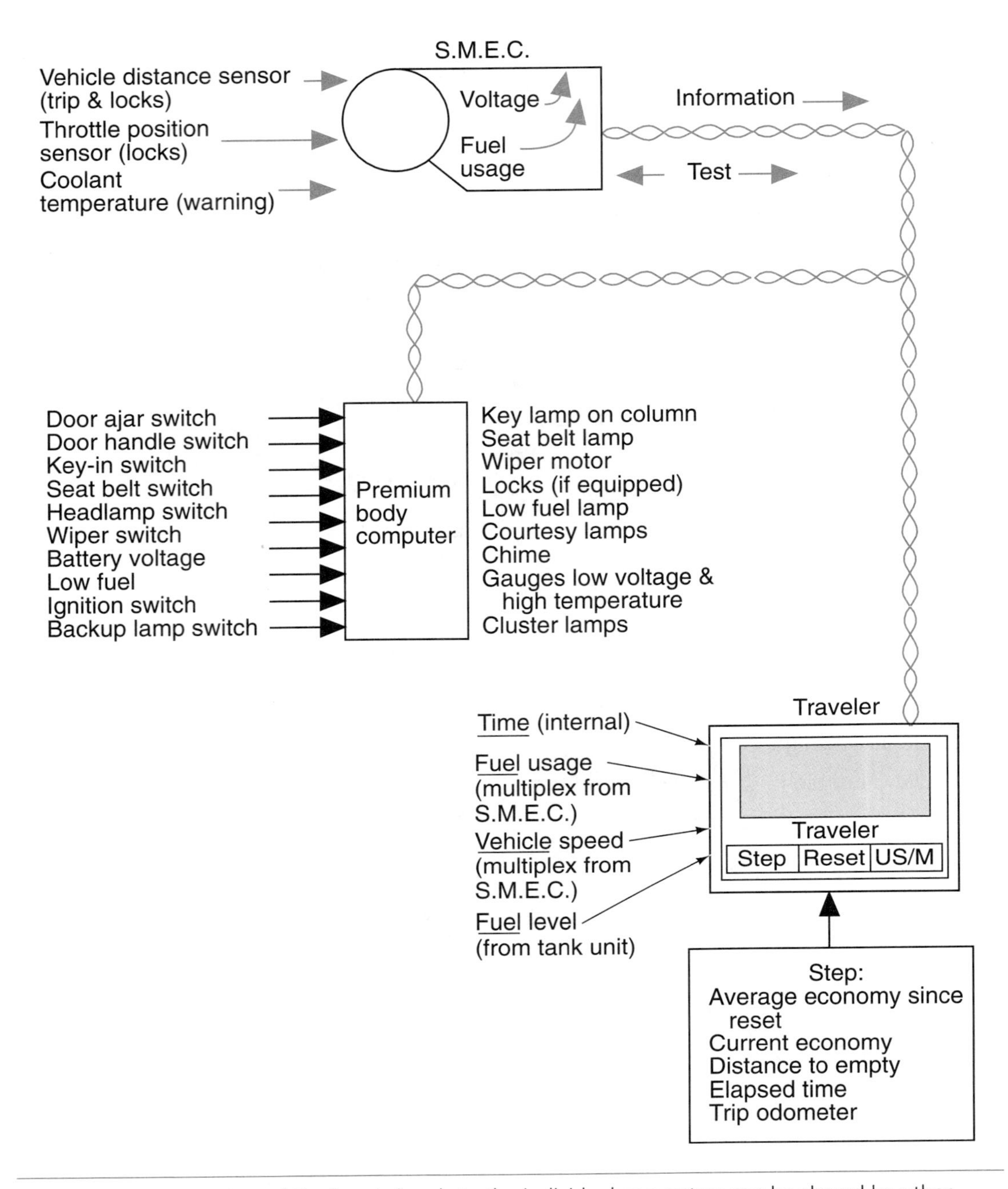

Figure 12-44 Any of the input signals to the individual computers can be shared by other computers that may require the information. (Courtesy of Chrysler Corporation)

as an analog signal through a dedicated wire to the computer or computers. At the computer, the signal is changed from an analog to a digital signal. Because each sensor requires its own dedicated signal wire, the number of wires required to feed data from all of the sensors and transmit control signals to all of the output devices is great.

In a MUX system, the sensors process the information and send a digital signal to the computer. Since the computer or control module of any system can only processs one input at a time, it calls for input signals as it needs them. By timing the transmission of data from the sensors to the control module, a single data wire can be used. Between each transmission of data to the control module, the sensor is electronically disconnected from the control module.

Summary

- ❑ A computer is an electronic device that stores and processes data and is capable of operating other devices.
- ❑ The operation of the computer is divided into four basic functions: input, processing, storage, and output.
- ❑ Binary numbers are represented by the numbers 1 and 0. A transistor that operates as a relay is the basis of the digital computer. As the input signal switches from off to on, the transistor output switches from cutoff to saturation. The on and off output signals represent the binary digits 1 and 0.
- ❑ Logic gates are the thousands of field effect transistors that are incorporated into the computer circuitry. The FETs use the incoming voltage patterns to determine the pattern of pulses that leave the gate. The most common logic gates are NOT, AND, OR, NAND, NOR, and XOR gates.
- ❑ There are several types of memory chips used in the body computer; ROM, RAM, and PROM are the most common types.
- ❑ ROM (read only memory) contains a fixed pattern of 1's and 0's representing permanent stored information used to instruct the computer on what to do in response to input data.
- ❑ RAM (random access memory) will store temporary information that can be read from or written to by the CPU.
- ❑ PROM (programmable read only memory) contains specific data that pertains to the exact vehicle in which the computer is installed.
- ❑ EPROM (Erasable PROM) is similar to PROM except its contents can be erased to allow new data to be installed.
- ❑ EEPROM (Electrically Erasable PROM) allows changing the information electrically one bit at a time.
- ❑ NVRAM (Nonvolatile RAM) is a combination of RAM and EEPROM into the same chip.
- ❑ Inputs provide the computer with system operation information or driver requests.
- ❑ Driver input signals are usually provided by momentarily applying a ground through a switch.
- ❑ Switches can be used as an input for any operation that only requires a yes-no, or on-off, condition.
- ❑ Sensors convert some measurement of vehicle operation into an electrical signal. There are many different designs of sensors: thermistors, Wheatstone bridge, potentiometers, magnetic pulse generator, and Hall-effect switches.
- ❑ A thermistor is a solid-state variable resistor made from a semiconductor material that changes resistance in realation to temperature changes. Negative temperature coefficient (NTC) thermistors reduce their resistance as the temperature increases. Positive temperature coefficient (PTC) thermistors increase their resistance as the temperature increases.
- ❑ The Wheatstone bridge is a series-parallel arrangement of resistors between an input terminal and ground. Usually three of the resistors are kept at exactly the same value while the fourth is the sensing resistor. When all four resistors have the same value, the bridge is balanced and the voltage sensor will indicate a value of 0 volts. If there is a change in the resistance value of the sense resistor, a change will occur in the circuit's balance. The sensing circuit will receive a voltage reading that is proportional to the amount of resistance change.

Terms to Know

Actuator
Analog
Binary numbers
Body Computer Module (BCM)
Central Processing Unit (CPU)
Computer
Digital
Duty cycle
Feedback
Hall-effect switch
Interface
Linearity
Logic gate
Magnetic pulse generator
Multiplexing
Negative Temperature Coefficient (NTC)
Nonvolatile
O_2 sensor
Pick-up coil
Piezoelectric
Piezoresistive
Positive Temperature Coefficient (PTC)
Program
PROM
Pulse width
RAM
ROM
Schmitt trigger
Sensor
Servomotor
Stepper motor
Thermistor
Volatile

- ❑ A potentiometer is a variable resistor that usually consists of a wire-wound resistor with a moveable center wiper.
- ❑ Magnetic pulse generators use the principle of magnetic induction to produce a voltage signal and are commonly used to send data concerning the speed of the monitored component to the computer.
- ❑ Hall-effect switches operate on the principle that if a current is allowed to flow through thin conducting material that is exposed to a magnetic field, another voltage is produced.
- ❑ Actuators are devices that perform the actual work commanded by the computer. They can be in the form of a motor, relay, switch, or solenoid.
- ❑ A servomotor produces rotation of less than a full turn. A feedback mechanism is used to position itself to the exact degree of rotation required.
- ❑ A stepper motor contains a permanent magnet armature with two, four, or more field coils. It is used to move the controlled device to whatever location is desired by applying voltage pulses to selected coils of the motor.
- ❑ Multiplexing is a system in which electrical signals are transmitted by a peripheral serial bus instead of conventional wires. This allows several devices to share signals on a common conductor.

Review Questions

Short Answer Essays

1. What is binary code?
2. Describe the basics of NOT, AND, and OR logic gate operation.
3. List and describe the four basic functions of the computer.
4. What is the difference between ROM, RAM, and PROM?
5. Explain the principle of multiplexing.
6. How does the Hall-effect switch generate a voltage signal?
7. Describe the basic function of a stepper motor.
8. What is meant by feedback as it relates to computer control?
9. What is the difference between duty cycle and pulse width?
10. What are the purposes of the interface?

Fill-in-the-Blanks

1. In binary code, the number 4 is represented by ____________________ .
2. The ____________________ is a crystal that electrically vibrates when subjected to current at certain voltage levels.

3. ____________________ are registers designed to store the results of logic operations.

4. The ____________________ ____________________ ____________________ is the heart of the computer.

5. ____________________ contains specific data that pertains to the exact vehicle in which the computer is installed.

6. ____________________ convert some measurement of vehicle operation into an electrical signal.

7. Negative temperature coefficient (NTC) thermistors ____________________ their resistance as the temperature increases.

8. ____________________ switches operate on the principle that if a current is allowed to flow through thin conducting material exposed to a magnetic field, another voltage is produced.

9. Magnetic pulse generators use the principle of ____________________ ____________________ to produce a voltage signal.

10. ____________________ means that data concerning the effects of the computer's commands are fed back to the computer as an input signal.

ASE Style Review Questions

1. *Technician A* says during the processing function the computer uses input information and compares it to programmed instructions. *Technician B* says during the output function the computer will put out control commands to various output devices. Who is correct?
   A. A only
   B. B only
   C. Both A and B
   D. Neither A nor B

2. *Technician A* says analog means the voltage signal is either on-off, yes-no, or high-low. *Technician B* says digital means the voltage signal is infinitely variable within a given range. Who is correct?
   A. A only
   B. B only
   C. Both A and B
   D. Neither A nor B

3. Logic gates are being discussed:
Technician A says NOT gate operation is similar to that of two switches in series to a load.
Technician B says an AND gate simply reverses binary 1's to 0's and vice versa.
Who is correct?
A. A only
B. B only
C. Both A and B
D. Neither A nor B

4. Computer memory is being discussed:
Technician A says ROM can be written to by the CPU.
Technician B says RAM will store temporary information that can be read from or written to by the CPU.
Who is correct?
A. A only
B. B only
C. Both A and B
D. Neither A nor B

5. *Technician A* says volatile RAM is erased when it is disconnected from its power source.
Technician B says nonvolatile RAM will retain its memory if removed from its power source.
Who is correct?
A. A only
B. B only
C. Both A and B
D. Neither A nor B

6. *Technician A* says EPROM memory is erased if the tape is removed and the microcircuit is exposed to ultraviolet light.
Technician B says electrostatic discharge will destroy the memory chip.
Who is correct?
A. A only
B. B only
C. Both A and B
D. Neither A nor B

7. *Technician A* says negative temperature coefficient thermistors reduce their resistance as the temperature decreases.
Technician B says positive temperature coefficient thermistors increase their resistance as the temperature increases.
Who is correct?
A. A only
B. B only
C. Both A and B
D. Neither A nor B

8. *Technician A* says magnetic pulse generators are commonly used to send data to the computer concerning the speed of the monitored component.
Technician B says an on-off switch sends a digital signal to the computer.
Who is correct?
A. A only
B. B only
C. Both A and B
D. Neither A nor B

9. While discussing speed sensors:
Technician A says the timing disc is stationary and the pick-up coil rotates in front of it.
Technician B says the number of pulses produced per mile increases as rotational speed increases.
Who is correct?
A. A only
B. B only
C. Both A and B
D. Neither A nor B

10. *Technician A* says a Hall-effect switch uses a steady supply current to generate a signal.
Technician B says a Hall-effect switch consists of a permanent magnet wound with a wire coil.
Who is correct?
A. A only
B. B only
C. Both A and B
D. Neither A nor B

Advanced Lighting Circuits and Electronic Instrumentation

Upon completion and review of this chapter, you should be able to:

- ❑ Explain the operation of the most common types of automatic headlight dimming systems.
- ❑ Describe the operation of twilight systems that are body computer controlled.
- ❑ Explain the use and function of fiber optics.
- ❑ Describe the different methods used to provide lamp outage indicators.
- ❑ Describe the operating principles of the digital speedometer.
- ❑ Explain the operation of IC chip and stepper motor odometers.
- ❑ Describe the operation of electronic fuel, temperature, oil, and voltmeter gauges.
- ❑ Explain the use and operation of light emitting diodes, liquid crystal, vacuum fluorescent, and CRT displays in electronic instrument clusters.
- ❑ Describe the operation of quartz analog instrumentation.
- ❑ Explain the operation of body computer-controlled instrument panel illumination light dimming.

Advanced Lighting Circuits Introduction

With the addition of solid-state circuitry in the automobile, manufacturers have been able to incorporate several different lighting circuits or modify the existing ones. Some of the refinements that were made to the lighting system include automatic headlight washers, automatic headlight dimming, automatic on/off with timed delay headlights, and illuminated entry systems. Some of these systems use sophisticated body computer-controlled circuitry and fiber optics.

Some manufacturers have included such basic circuits as turn signals into their body computer to provide for pulse width dimming in place of a flasher unit. The body computer can also be used to control instrument panel lighting based on inputs, including if the side marker lights are on or off. By using the body computer to control many of the lighting circuits, the amount of wiring has been reduced. In addition, the use of computer control has provided a means of self-diagnosis in some applications.

Computer-Controlled Concealed Headlights

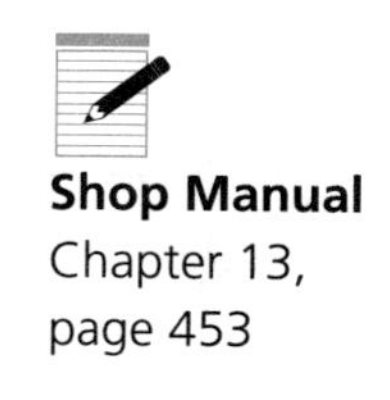

Shop Manual
Chapter 13, page 453

The body computer has been utilized by some manufacturers to operate the concealed headlight system. The body computer will receive inputs from the headlight and flash-to-pass switches (Figure 13-1). When the headlight switch is turned on, the body computer receives a signal that the headlights are being activated. To open the headlight doors, the computer energizes the open door relay. The contacts of the open relay are closed and battery voltage is applied to the door motor (Figure 13-2). Battery voltage to operate the door motor is supplied from the 30-ampere circuit breaker. The computer energizes the door open relay through circuit L50, which moves the normally open relay contact arm. Ground is provided through the door close relay contact.

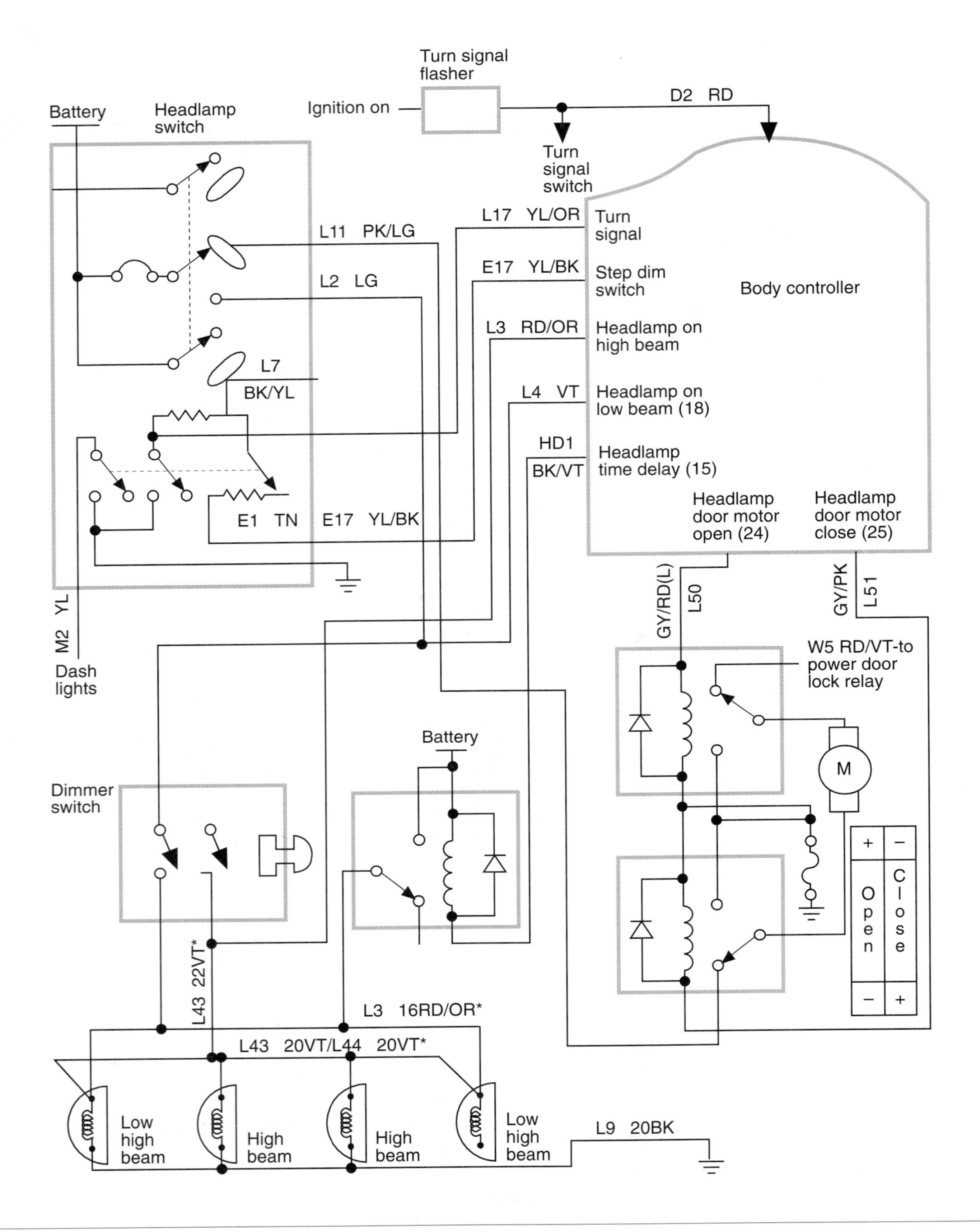

Figure 13-1 Computer-controlled concealed headlight door circuit. (Courtesy of Chrysler Corporation)

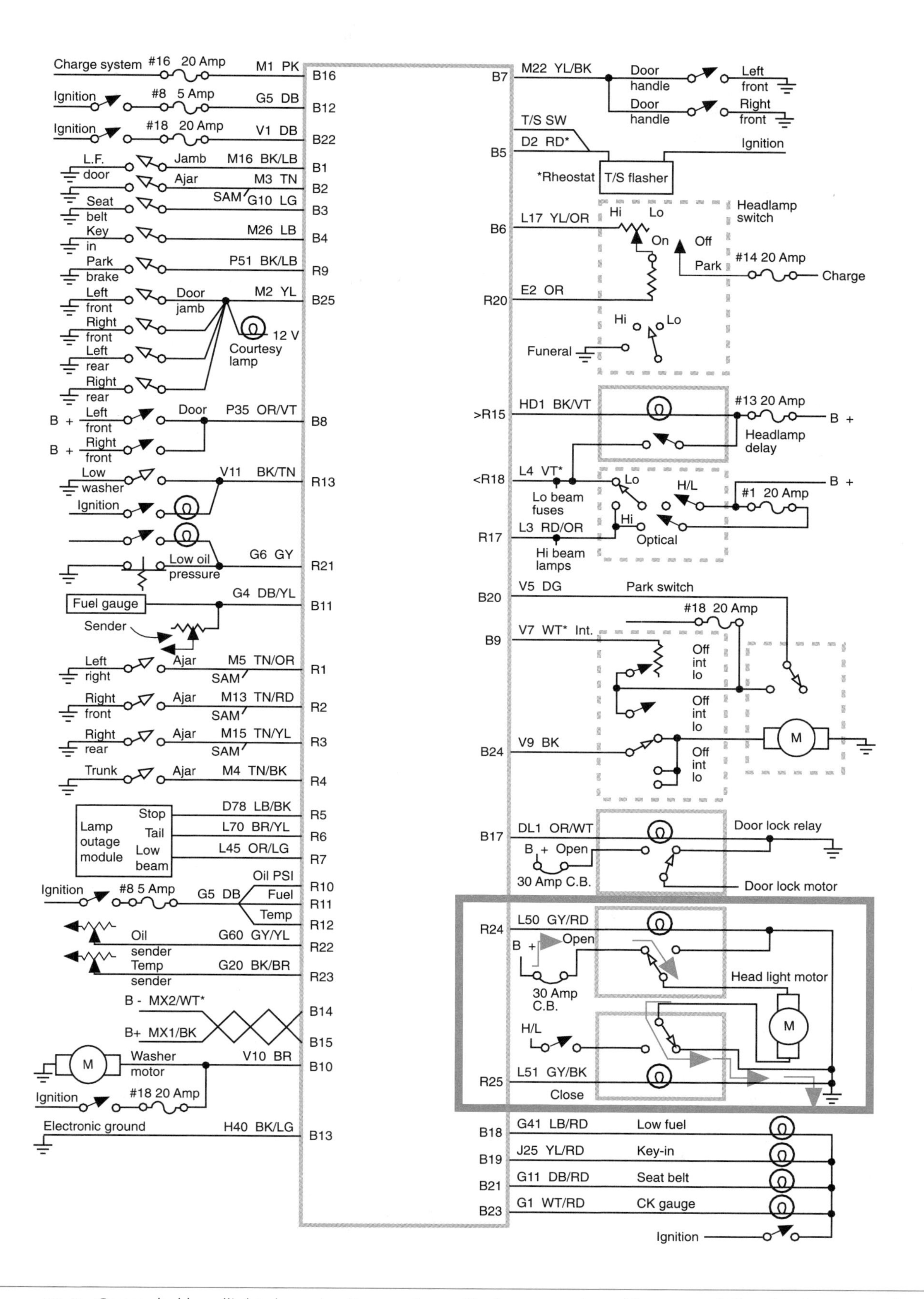

Figure 13-2 Concealed headlight door circuit operation with doors opening. (Courtesy of Chrysler Corporation)

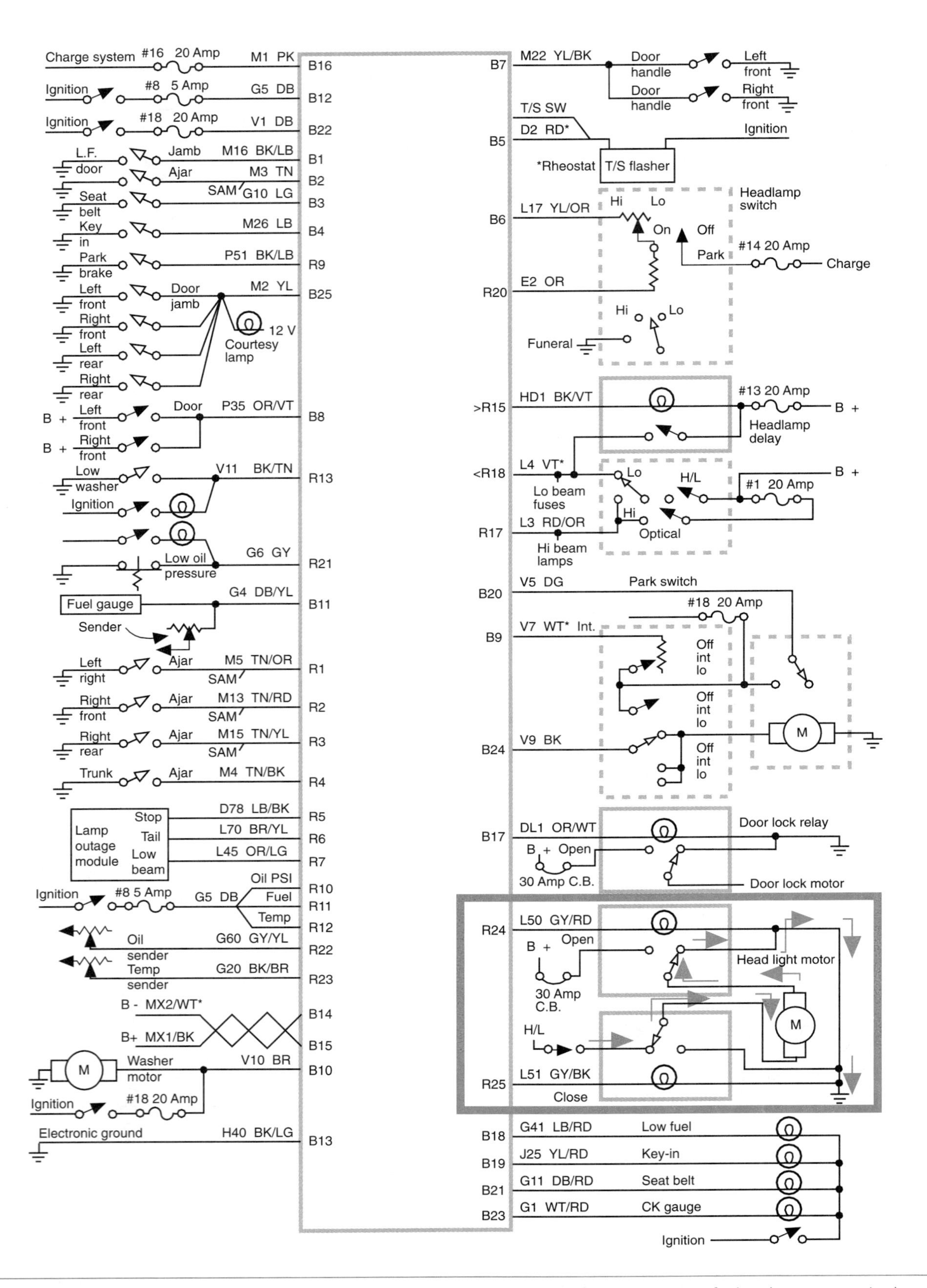

Figure 13-3 Concealed headlight door operation circuit with doors closing. (Courtesy of Chrysler Corporation)

When the headlight switch is turned off, the computer energizes the close door relay through circuit L51. With the door close relay energized, the contacts provide battery voltage to the door motor. The ground is supplied through the door open relay (Figure 13-3). Reversing the polarity through the door motor closes the door.

If the flash-to-pass option is activated, the body computer receives a high (on) signal and energizes the door open relay. When the switch is released, the computer receives a low (off) signal and activates the door close relay. The computer delays the activation of the door close relay for 3 seconds.

Automatic Headlight Dimming

Modern automatic headlight dimming systems use solid-state circuitry and electromagnetic relays to control the beam switching. Most systems consist of the following major components:

1. Light sensitive photocell and amplifier unit
2. High-low beam relay
3. Sensitivity control
4. Dimmer switch
5. Flash-to-pass relay
6. Wiring harness

Automatic headlight dimming automatically switches the headlights from high beams to low beams under two different conditions: when light from oncoming vehicles strikes the photocell-amplifier, or light from the taillights of a vehicle being passed strikes the photocell-amplifier.

The photocell-amplifier is usually mounted behind the front grill, but ahead of the radiator. The sensitivity control sets the intensity level at which the photocell-amplifier will energize. This control is set by the driver and is located next to, or is a part of, the headlight switch assembly (Figure 13-4). The driver is able to adjust the sensitivity level of the system by rotating the control knob. An increase in the sensitivity level will make the headlights switch to the low beams sooner (approaching vehicle is further away). A decrease in the sensitivity level will switch the headlights to low beams when the approaching vehicle is closer. If the knob is rotated to the full counterclockwise position, the system goes into manual override.

The **photocell** is like a variable resistor that uses light to change resistance.

The **sensitivity control** is a potentiometer that allows the driver to adjust the sensitivity of the automatic dimmer system to surrounding ambient light conditions.

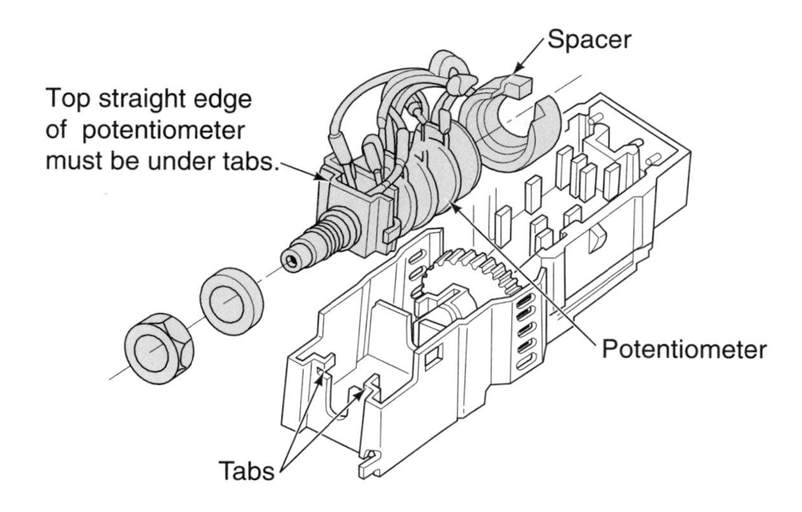

Figure 13-4 The drive sets the sensitivity of the automatic headlight dimmer system by rotating the potentiometer to change resistance values. (Courtesy of Chilton Book Company)

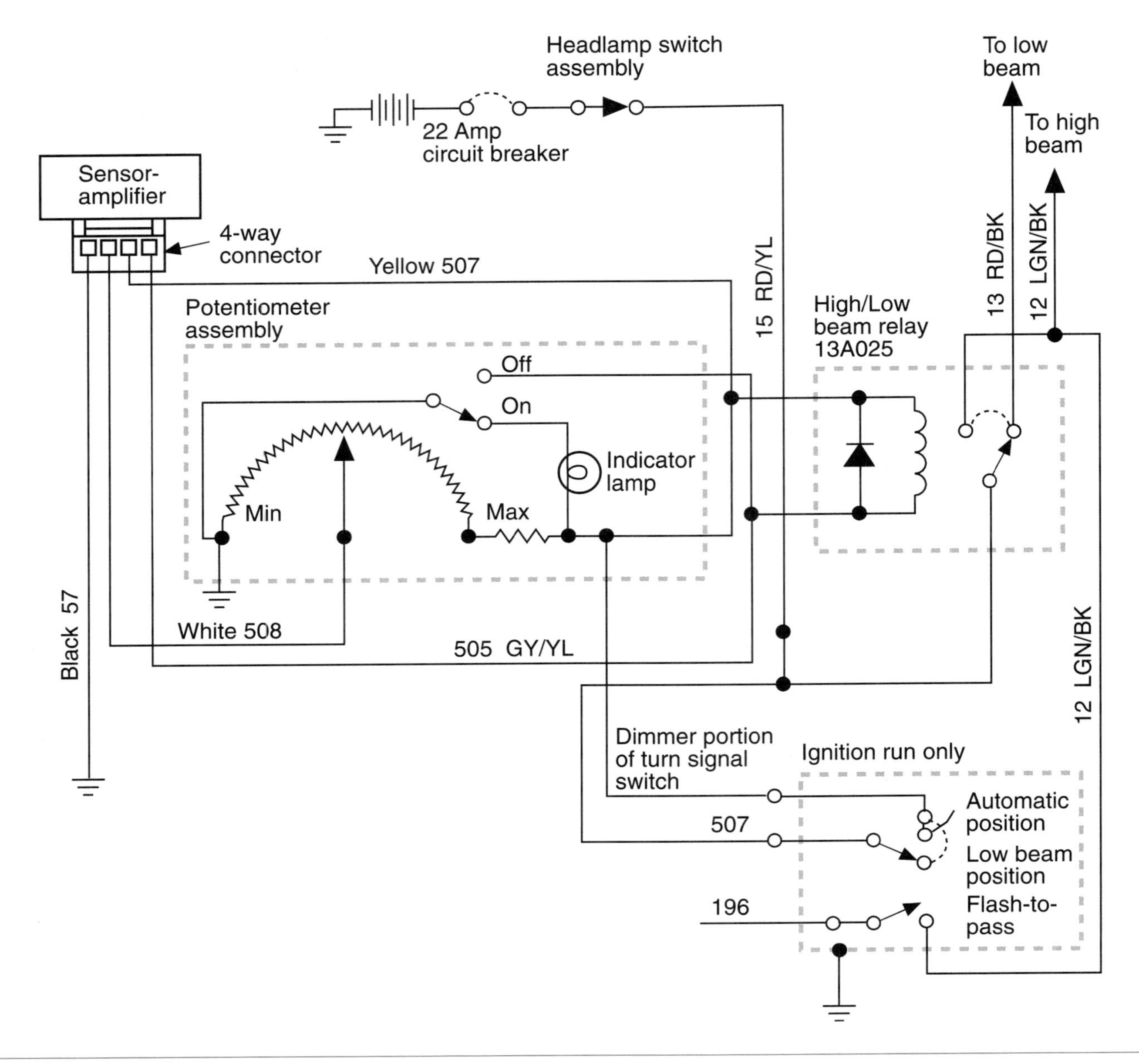

Figure 13-5 Automatic headlight dimming circuit uses a high-low beam relay to switch beam settings. (Reprinted with the permission of Ford Motor Company)

Many vehicle manufacturers install the sensor-amplifier in the rearview mirror support.

The aiming screw is used to adjust the vertical position of the sensor.

The high-low relay is a single-pole, double-throw unit that provides the switching of the headlight beams (Figure 13-5). The relay also contains a clamping diode for electrical transient damping to protect the photocell and amplifier assembly.

The dimmer switch is usually a flash-to-pass design. If the turn signal lever is pulled partway up, the flash-to-pass relay is energized. The high beams will stay on as long as the lever is held in this position, even if the headlights are off. In addition, the driver can select either low beams or automatic operation through the dimmer switch.

Although the components are similar in most systems, there are differences in system operations. Systems differ in how the manufacturer uses the relay to do the switching from high beams to low beams. The system can use either an energized relay to activate the high beams or an energized relay to activate the low beams. If the system uses an energized relay to activate the high beams, the relay control circuit is opened when the dimmer switch is placed in the low

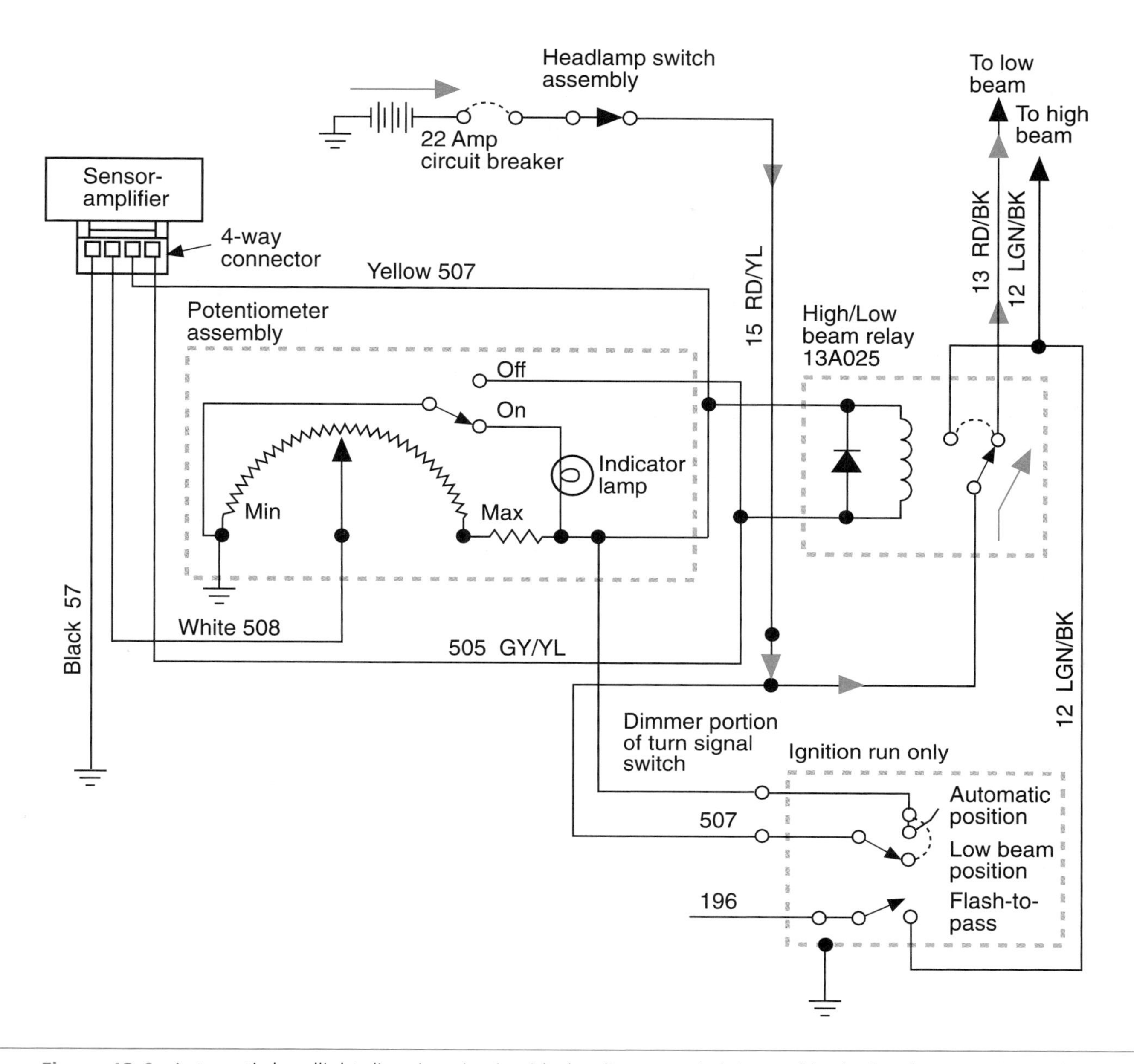

Figure 13-6 Automatic headlight dimming circuit with the dimmer switch located in the low beam position. (Reprinted with the permission of Ford Motor Company)

beam position or the driver manually overrides the system (Figure 13-6). With the headlight switch in the ON position and the dimmer switch in the low beam position, battery voltage is applied to the relay coil through circuit 221. The relay coil is not energized because there is no ground provided. The automatic feature is bypassed when the dimmer switch is in the low beam position.

With the dimmer switch in the automatic position, ground is provided for the relay coil through the sensor-amplifier (Figure 13-7). The energized coil closes the relay contacts to the high beams and battery voltage is applied to the headlamps. When the photocell sensor receives enough light to overcome the sensitivity setting, the amplifier opens the relay's circuit to ground. This de-energizes the relay coil and switches battery voltage from high beam to low beam position.

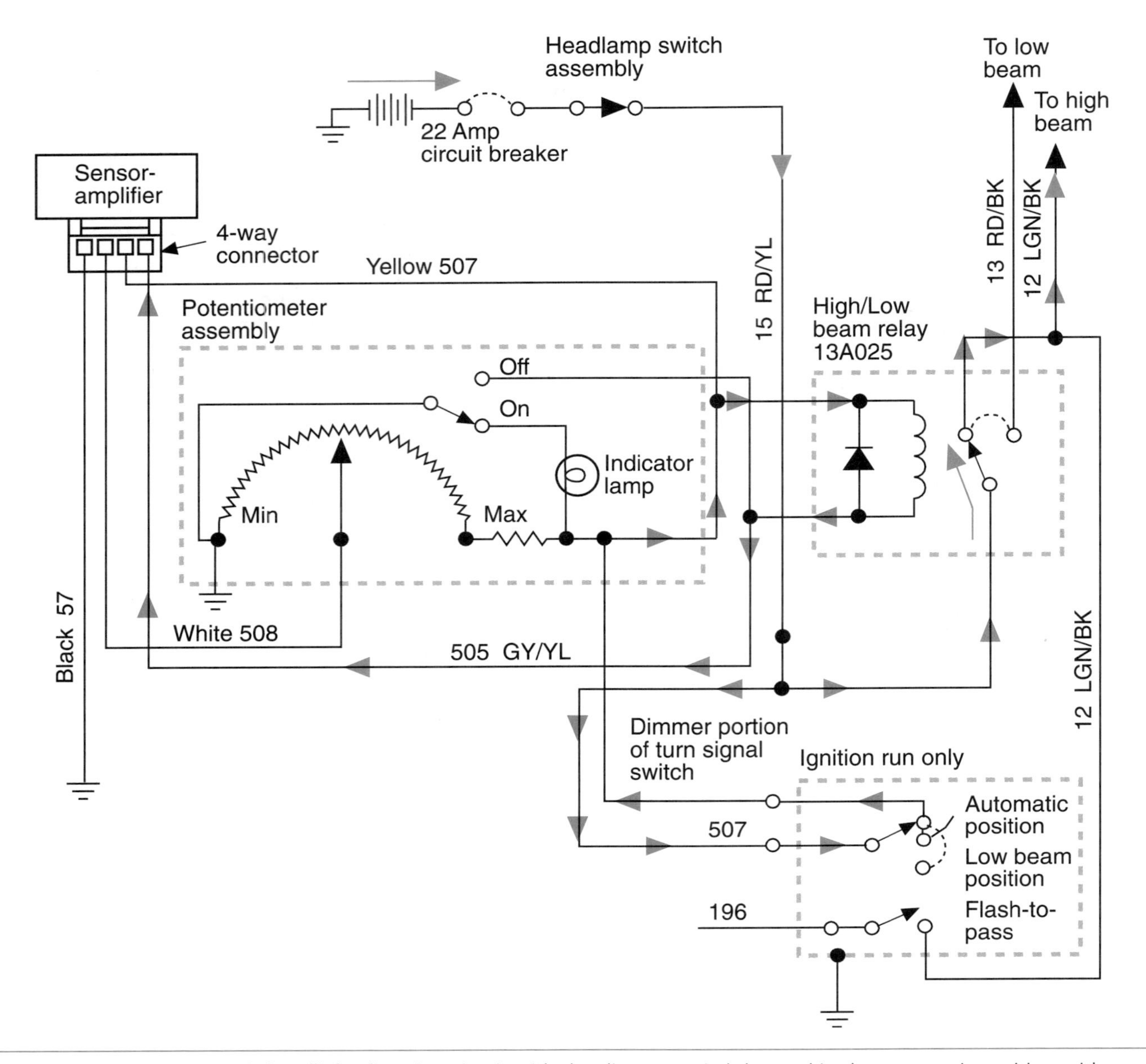

Figure 13-7 Automatic headlight dimming circuit with the dimmer switch located in the automatic position with no oncoming light sensed. (Reprinted with the permission of Ford Motor Company)

If the system uses an energized relay to switch to low beams, placing the dimmer switch in the low beam position will energize the relay (Figure 13-8). With the headlights turned on and the dimmer switch in the automatic position, battery voltage is applied to the photocell-amplifier, one terminal of the high-low control, and through the relay contacts to the high beams. The voltage drop through the high-low control is an input to the photocell-amplifier. When enough light strikes the photocell-amplifier to overcome the sensitivity setting, the amplifier allows battery current to flow through the high-low relay closing the contact points to the low beams. Once the light has passed, the photocell-amplifier opens battery voltage to the relay coil and the contacts close to the high beams.

When flash-to-pass is activated, the switch closes to ground. This bypasses the sensitivity control and de-energizes the relay to switch from low beams to high beams.

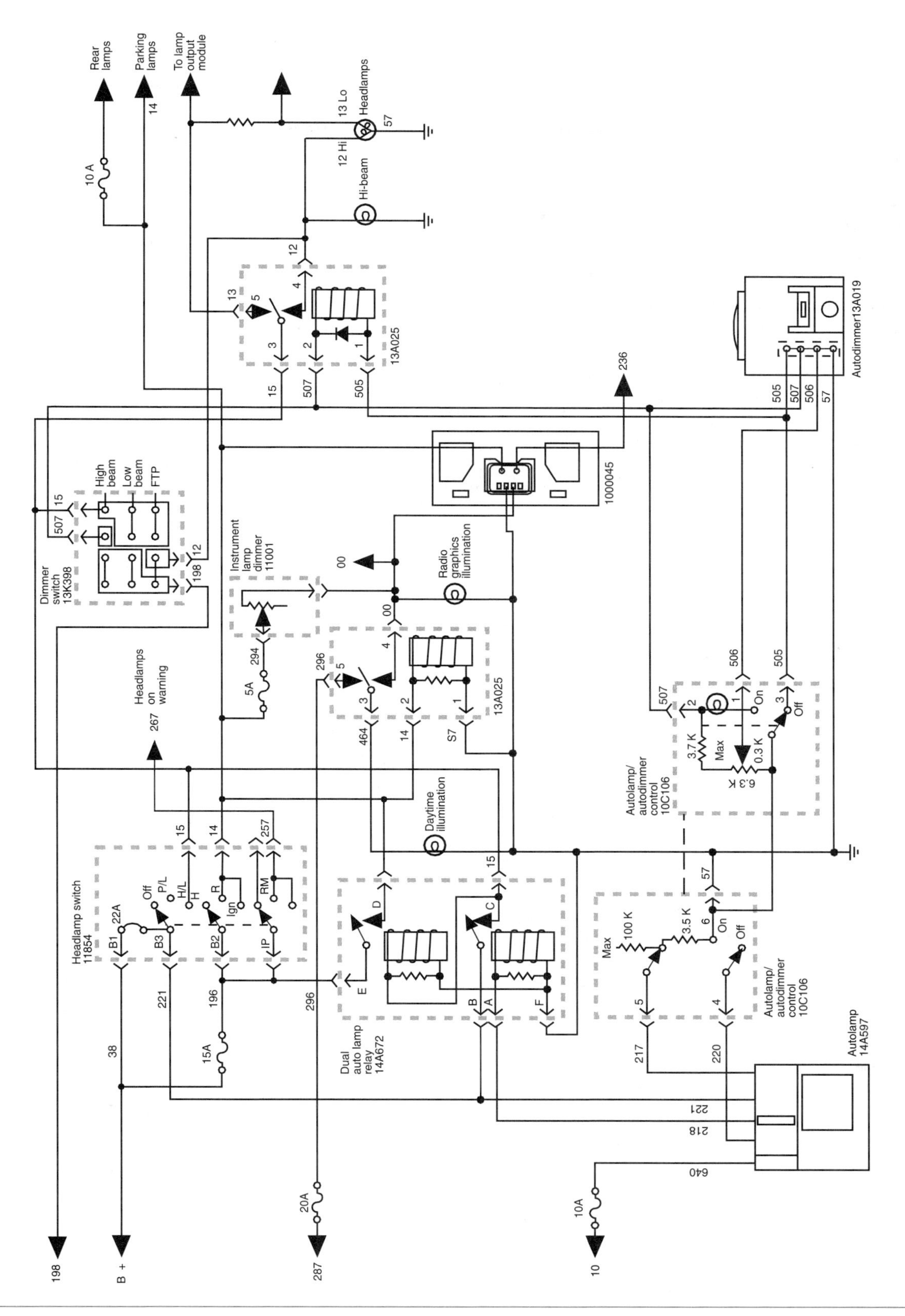

Figure 13-8 System that uses an energized relay to switch to low beams. (Reprinted with the permission of Ford Motor Company)

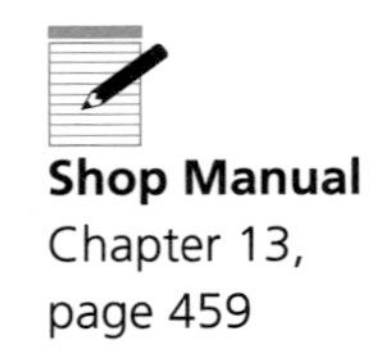

Shop Manual
Chapter 13,
page 459

This system is given several different names by the various manufacturers. Some of the more common names include: Twilight Sentinel, Auto-lamp/Delayed Exit, and Safeguard Sentinel.

Automatic On/Off with Time Delay

The common components of the automatic on/off with time delay include:

1. Photocell and amplifier
2. Power relay
3. Timer control

In a typical system, a photocell is located inside the vehicle's dash to sense outside light (Figure 13-9). In most systems, the headlight switch must be in the OFF position to activate the automatic mode (Figure 13-10). Battery voltage is applied to the normally open headlight contacts of the relay through the headlight switch. Battery voltage is also supplied to the normally open exterior light contacts through the fuse panel (Figure 13-11).

To activate the automatic on/off feature, the photocell and amplifier must receive voltage from the ignition switch (circuit 640). As the ambient light level decreases, the internal resistance of the photocell increases. When the resistance value reaches a predetermined value, the photocell and amplifier trigger the sensor-amplifier module. The sensor-amplifier module energizes the relay, turning on the headlights and exterior parking lights (Figure 13-12).

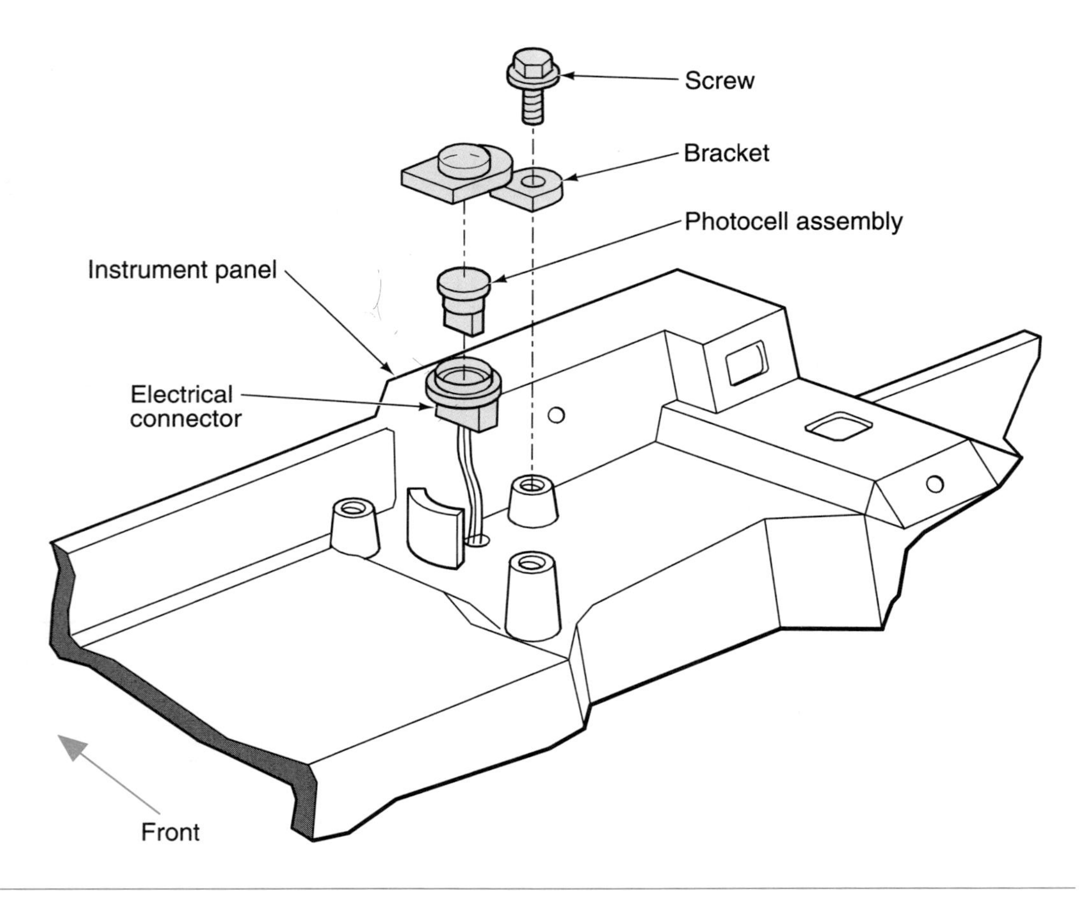

Figure 13-9 Most automatic on/off headlight systems have the photocell located in the dash to sense incoming light levels. (Courtesy of General Motors Corporation)

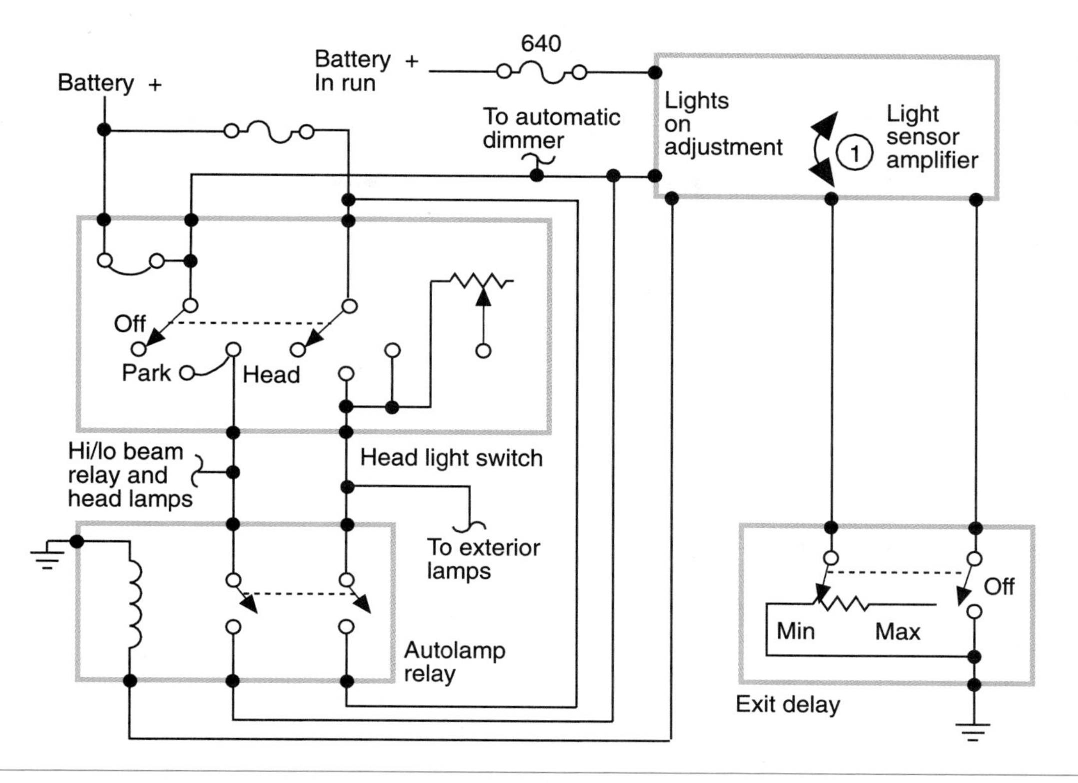

Figure 13-10 Schematic of automatic headlight on/off with time delay system.

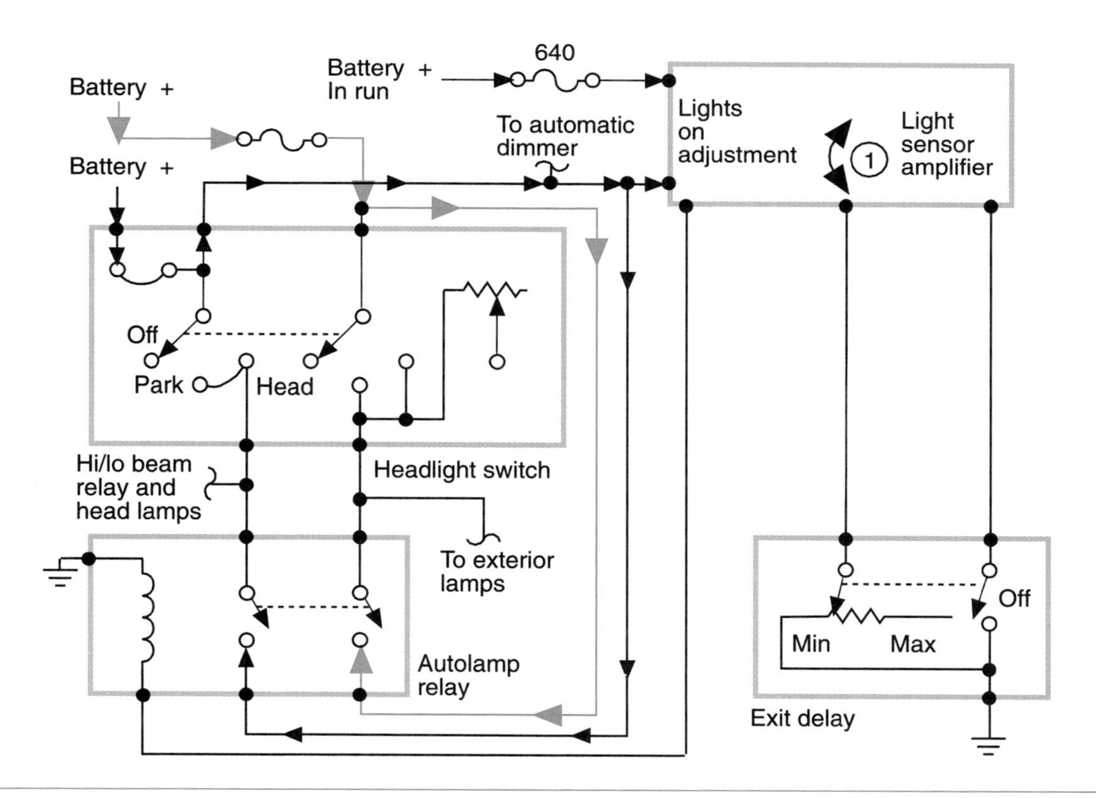

Figure 13-11 Current flow in automatic mode with the headlights off.

The **automatic on/off with time delay** has two functions: to turn on the headlights automatically when ambient light decreases to a predetermined level and to allow the headlights to remain on for a certain amount of time after the vehicle has been turned off. This system can be used in combination with automatic dimming systems.

The **timer control** is a potentiometer that is a part of the headlight switch in most systems. The timer control unit controls the automatic operation of the system and the length of time that the headlights stay on after the ignition switch is turned off.

The Twilight Sentinel System is overridden when the delay control switch is in the OFF position. The delay control switch is located on the left switch panel.

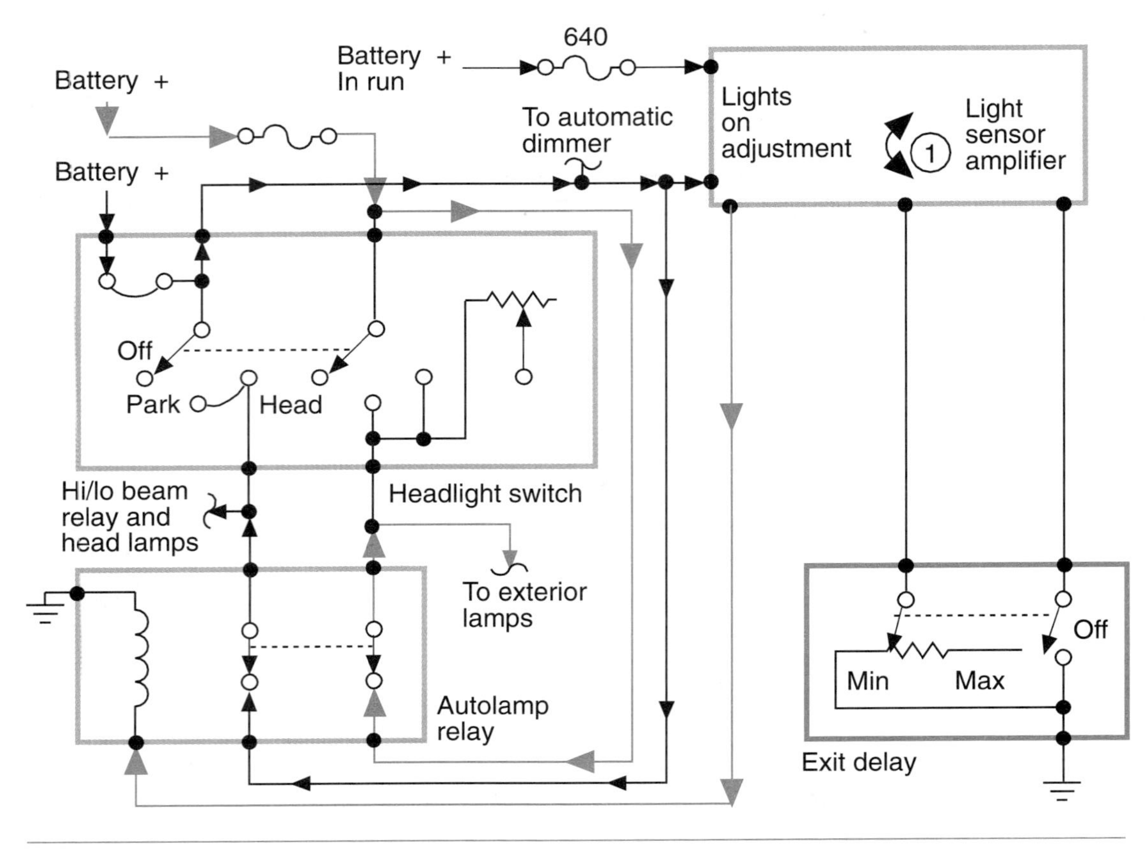

Figure 13-12 Current flow in the automatic mode with the headlights on.

Some systems provide a time delay feature that allows drivers to set a timer circuit to control how long the headlights remain on after they leave the vehicle. The control is a potentiometer that signals the sensor-amplifier module to energize the relay for the requested amount of time.

If the headlights are on when the ignition switch is turned off, the photocell and amplifier's voltage from circuit 640 is opened. This activates a timer circuit in the amplifier. The amplifier still receives battery voltage from the headlight switch, and uses this voltage to keep the relay energized for the requested time interval. When the preset length of time has passed, the amplifier module removes power to the relay and the headlights (and exterior lights) turn off.

The driver can override the automatic on/off feature by placing the headlight switch in the ON position. This bypasses the relay and sends battery voltage directly to the headlight circuit (Figure 13-13).

Depending on model application, General Motors' Twilight Sentinel System can use the body computer to control system operation (Figure 13-14). The body control module (BCM) senses the voltage drop across the photocell and the delay control switch. The delay control switch resistance is wired in series with the photocell. If the ambient light level drops below a specific value, the BCM grounds the headlamp and parklamp relay coils. The BCM also keeps the headlights on for a specific length of time after the ignition switch is turned off.

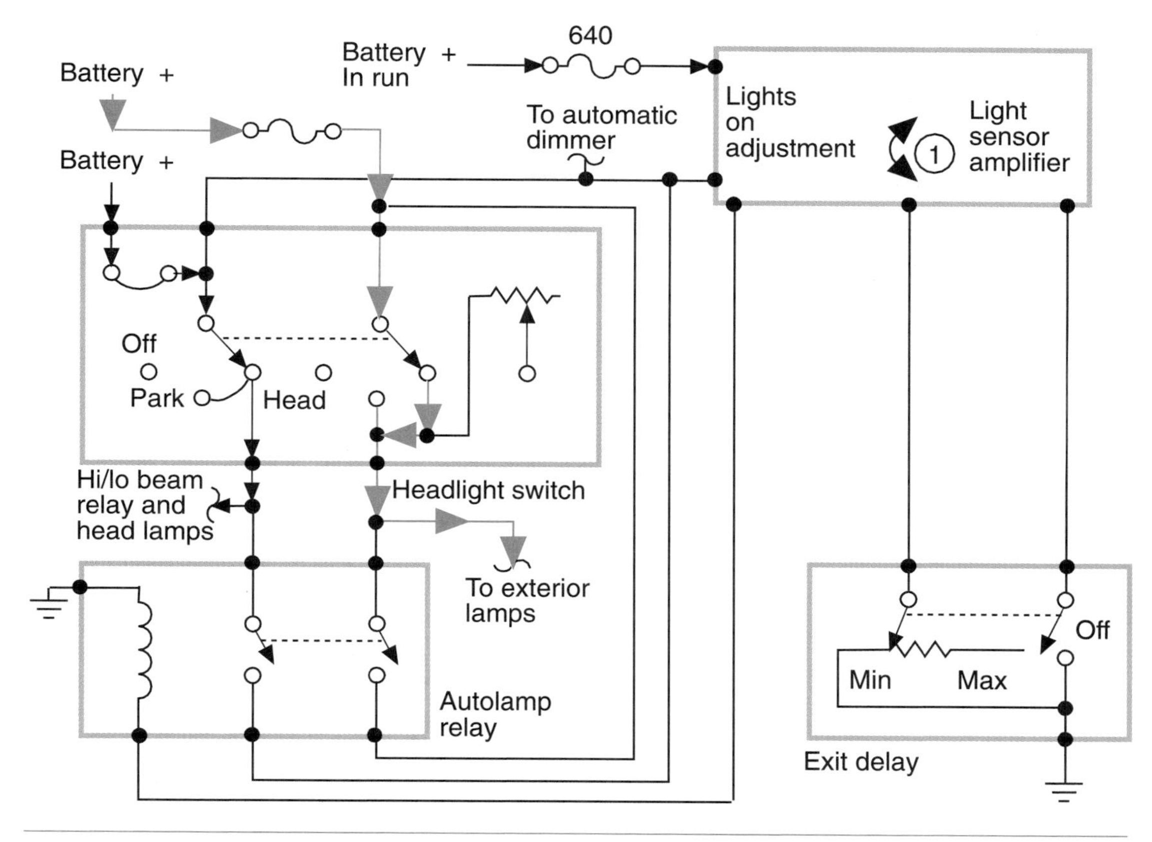

Figure 13-13 Current flow with the automatic headlight control bypassed.

Daytime Running Lamps

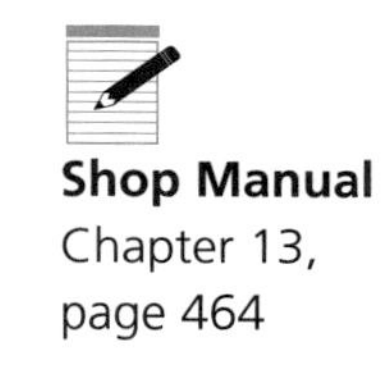

Shop Manual
Chapter 13,
page 464

All late-model Canadian vehicles and newer GM vehicles are equipped with daytime running lamps. The basic idea behind daytime running lights is dimly lit headlamps during the day. This allows other drivers to see the vehicle from a distance. Manufacturers have taken many different approaches to achieve this lighting. Most have a control module or relay (Figure 13-15) that turns the lights on when the engine is running and allows normal headlamp operation when the driver turns on the headlights.

The dimmer headlights can result from headlight current passing through a resistor during daylight hours (Figure 13-16). The resistor reduces the voltage available and the current flowing through the circuit to the headlights. The resistor is bypassed during normal headlamp operation.

Other systems use a control module (Figure 13-17), which sends out short bursts of electricity to the high-beam lamps. These short bursts of power cause the lamps to light with reduced illumination and without triggering the high-beam indicator in the instrument panel.

GM's daytime running lamp (DRL) system includes a solid state control module assembly, a relay, and an ambient light sensor assembly. The system lights the low beam headlights at a reduced intensity when the ignition switch is in the RUN position during daylight. The daytime running lamp system is designed to light the low beam headlamps at full intensity when low light conditions exist.

As the intensity of the light reaching the ambient light sensor increases, the electrical resistance of the sensor assembly decreases. When the DRL control module assembly senses the low resistance, the module allows voltage to be applied to the DRL diode assembly and then to the low beam headlamps. Because of the voltage drop across the diode assembly, the low beam headlamps are on with a low intensity.

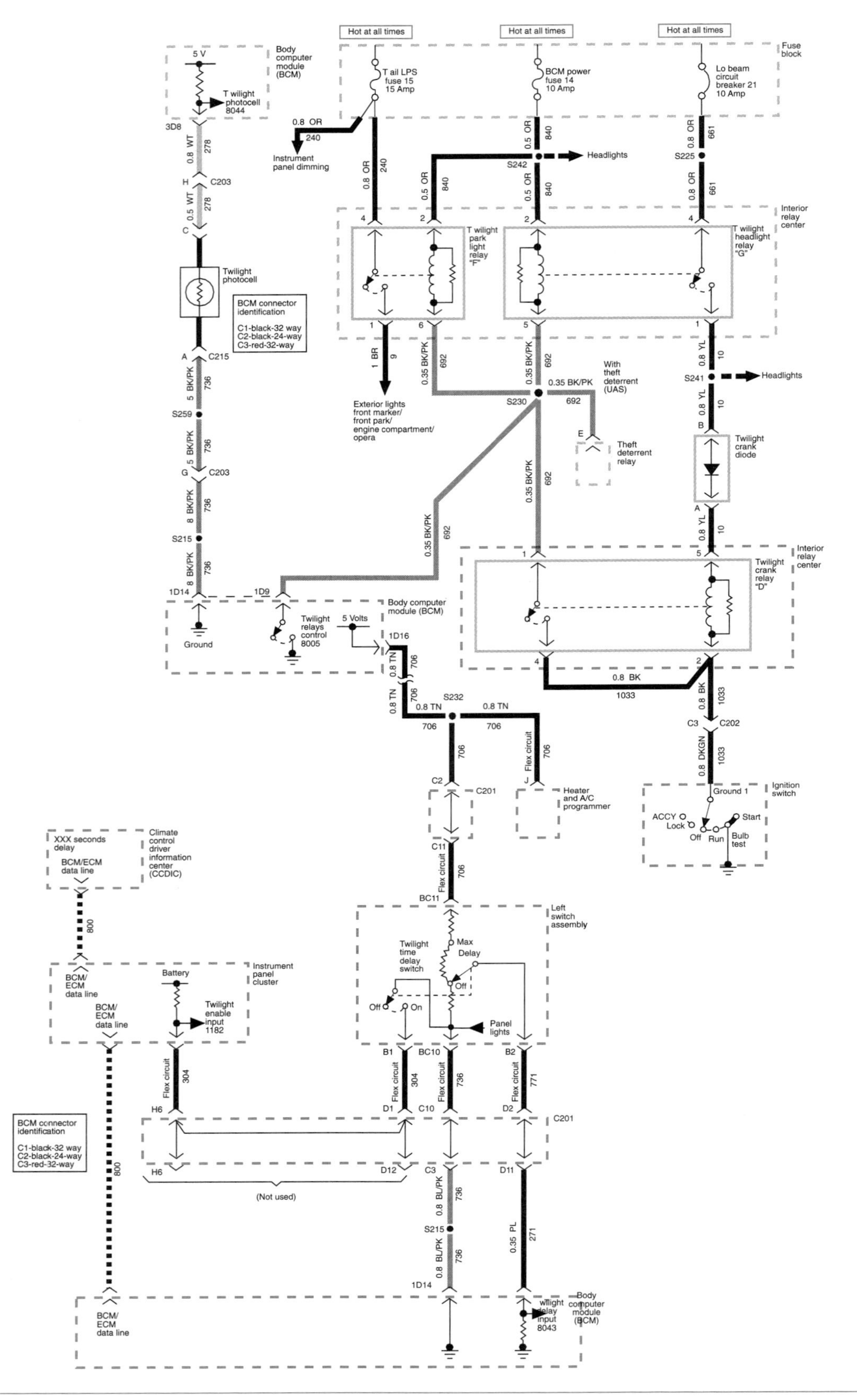

Figure 13-14 Some General Motors' Twilight Sentinel systems use the BCM to sense inputs from the photocell and delay control switch. (Courtesy of Chilton Book Company)

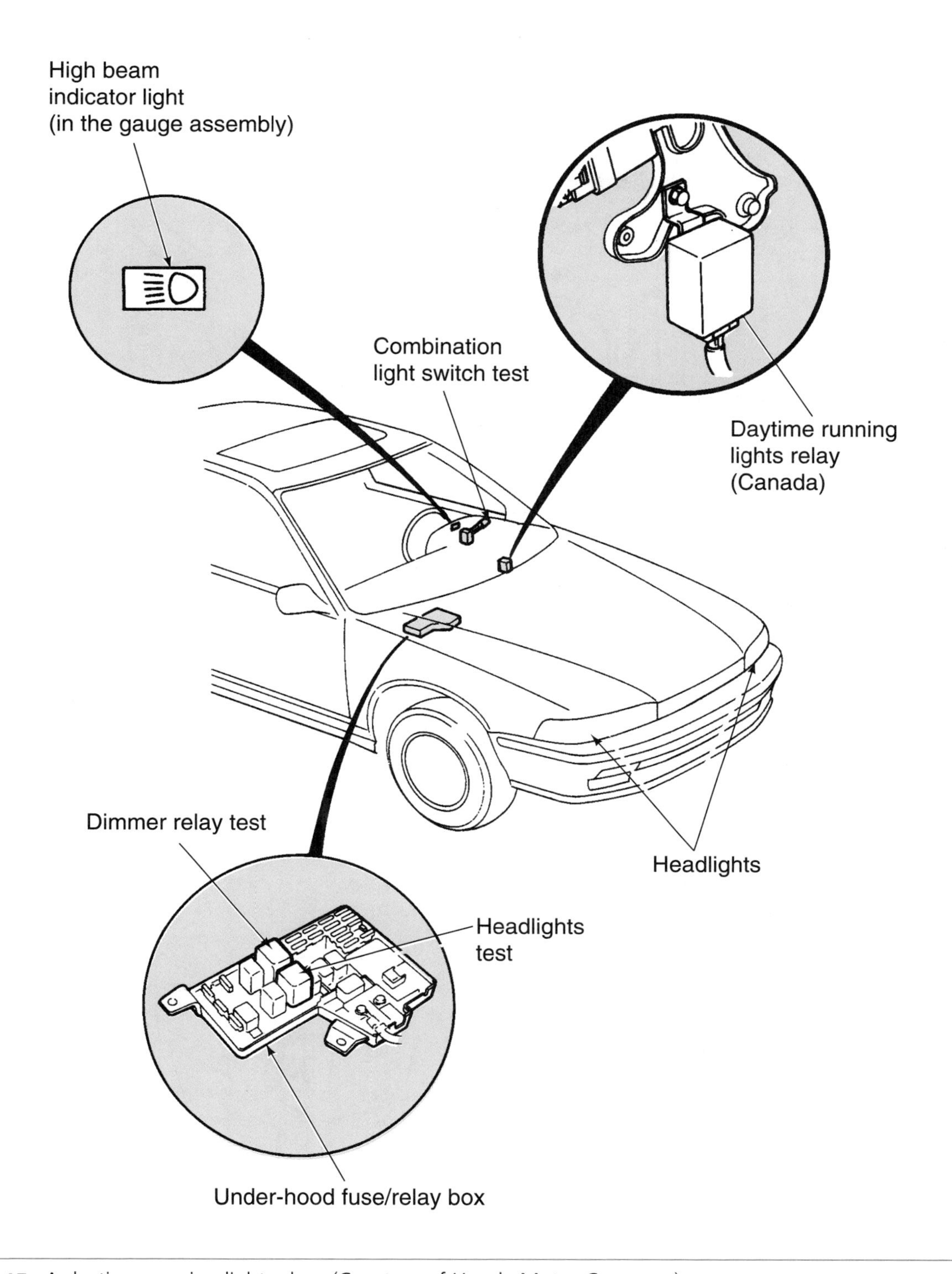

Figure 13-15 A daytime running light relay. (Courtesy of Honda Motor Company)

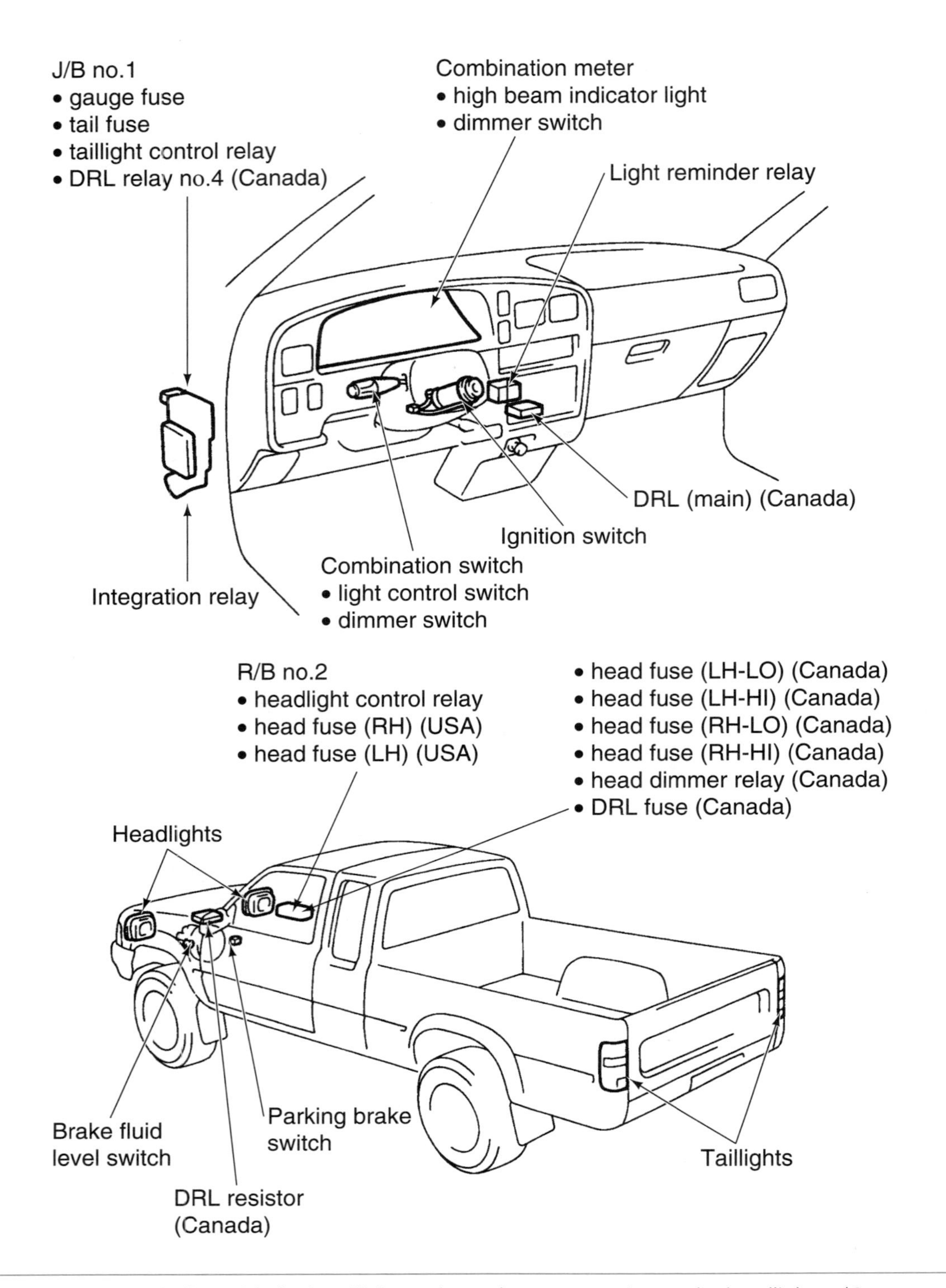

Figure 13-16 A resistor in-line with the headlights reduces the current going to the headlights. (Courtesy of Toyota Motor Corporation)

As the intensity of the light reaching the ambient light sensor decreases, the electrical resistance of the sensors increases. When the DRL module assembly senses high resistance in the sensor, the module closes an internal relay that allows the low beam headlamps to illuminate with full intensity.

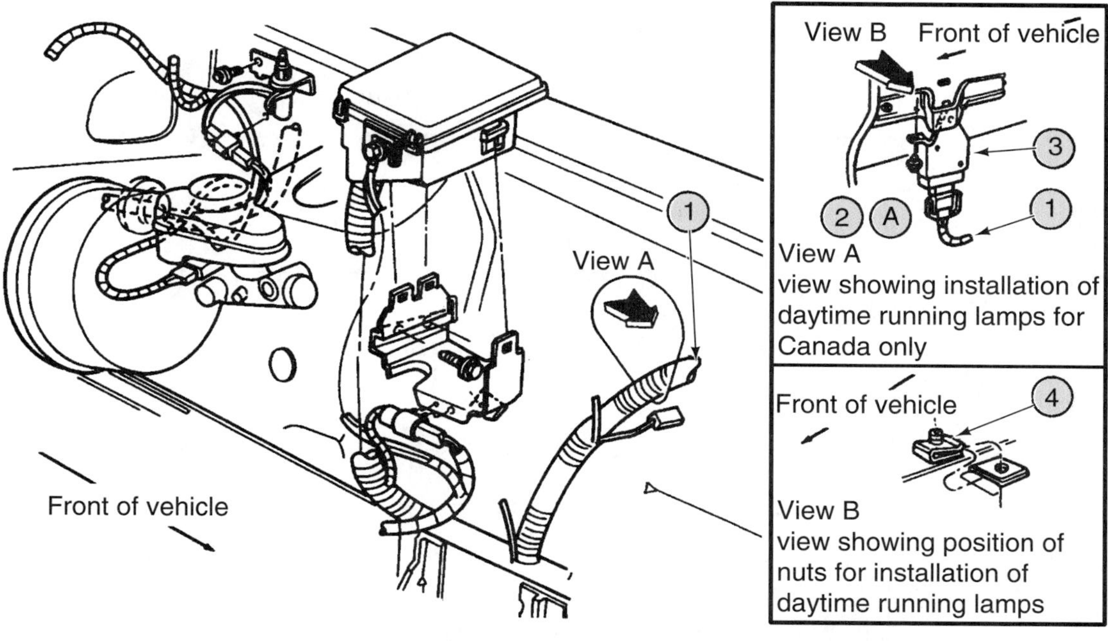

Item	Part Number	Description
1	14290	Headlamp Dash Panel Junction Wire
2	N605892-S36	Screw
3	15A272	Daytime Running Lamp Control Module
4	N800854-S100	Nut
A	—	Tighten to 9-14 N·m (7-10 Lb-Ft)

Figure 13-17 A DRL system that uses a control module to pulse current to the high-beam headlights. This results in reduced illumination. (Reprinted with the permission of Ford Motor Company)

Illuminated Entry Systems

Shop Manual
Chapter 13, page 465

The **illuminated entry system** turns on the courtesy lights before opening the doors.

Most modern illuminated entry systems incorporate solid-state circuitry that includes an illuminated entry actuator and side door switches in the door handles. Illumination of the door lock tumblers can be provided by the use of fiber optics or light emitting diodes.

When either of the front door handles are lifted a switch in the handle will close the ground path from the actuator. This signals the logic module to energize the relay (Figure 13-18). With the relay energized, the contacts close and the interior and door lock lights come on. A timer circuit is incorporated that will turn off the lights after 25 to 30 seconds. If the ignition switch is placed in the RUN position before the timer circuit turns off the interior lights, the timer sequence is interrupted and the interior lights turn off.

Some manufacturers have incorporated the illuminated entry actuator into their body computer (Figure 13-20). Activation of the system is identical as discussed. The signal from the door handle switch can also be used as a "wake-up" signal to the body computer.

A **wake-up signal** is used to notify the body computer that an engine start and operation of accessories is going to be initiated soon. This signal is used to warm up the circuits that will be processing information (Figure 13-19).

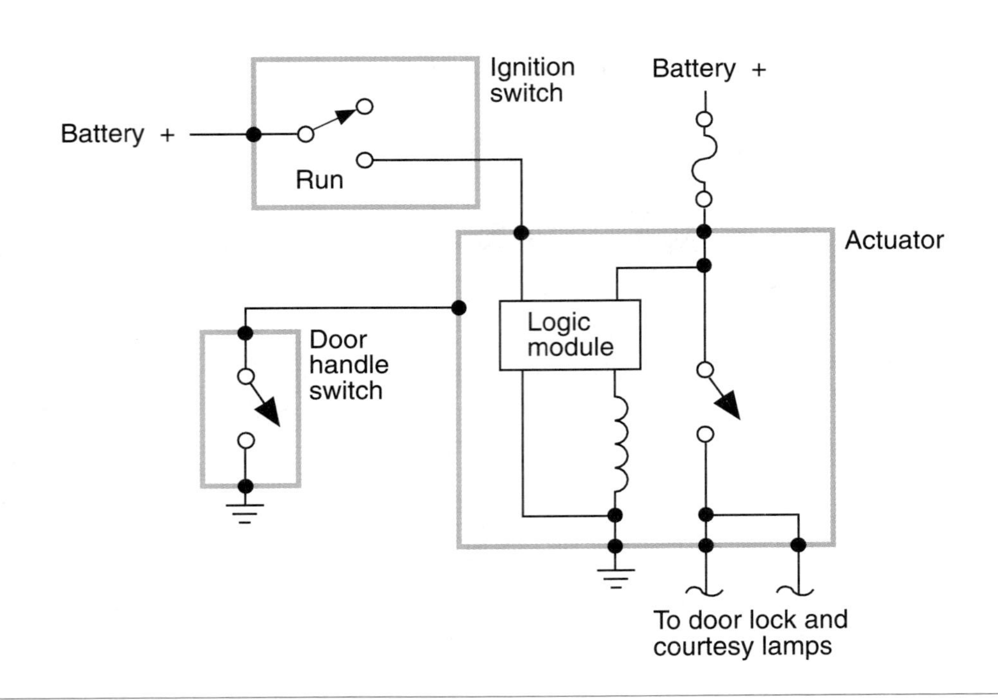

Figure 13-18 Illuminated entry actuator circuit.

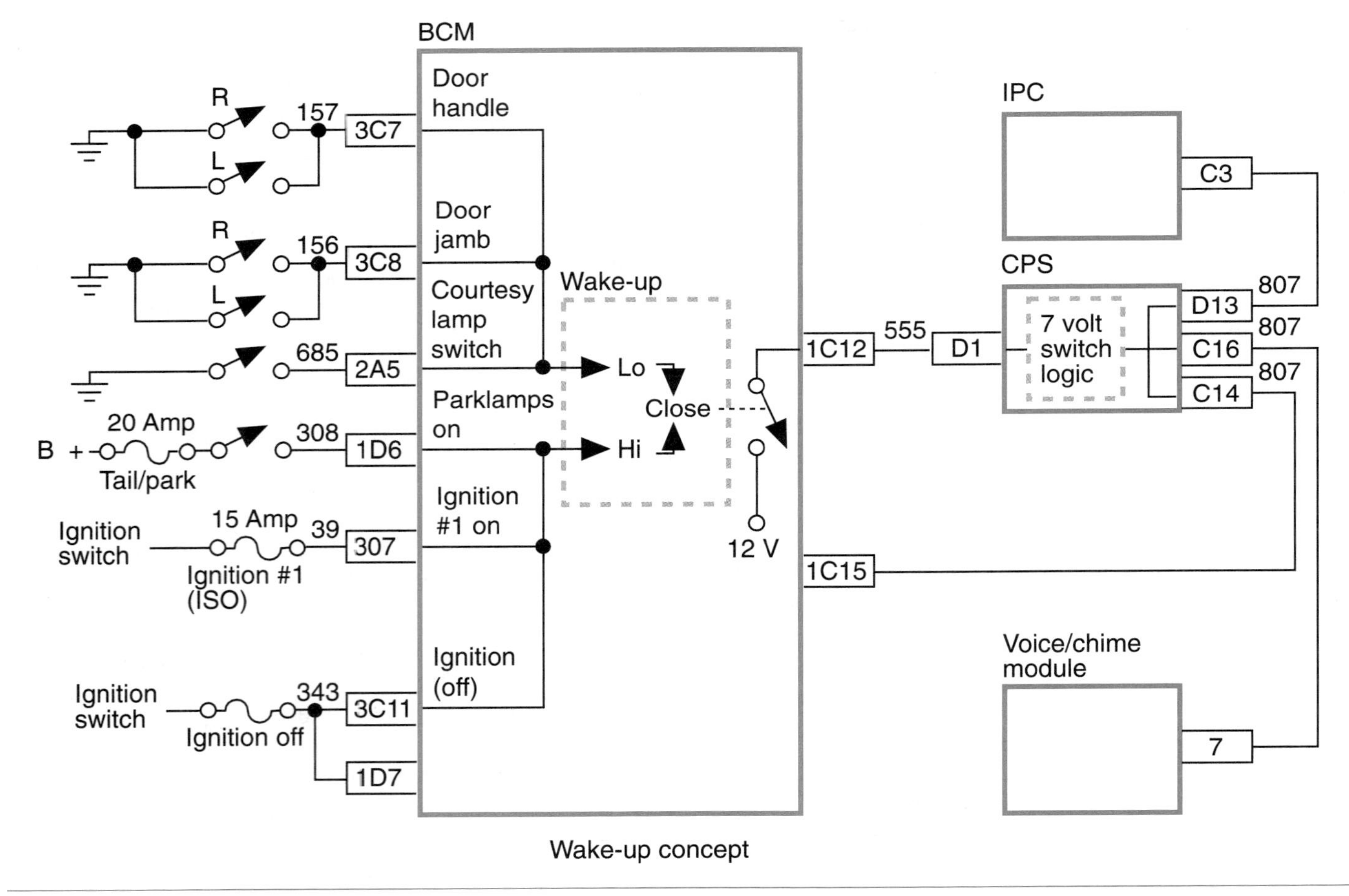

Figure 13-19 The illuminated entry system can also send a wake-up signal to the body computer. (Courtesy of Chilton Book Company)

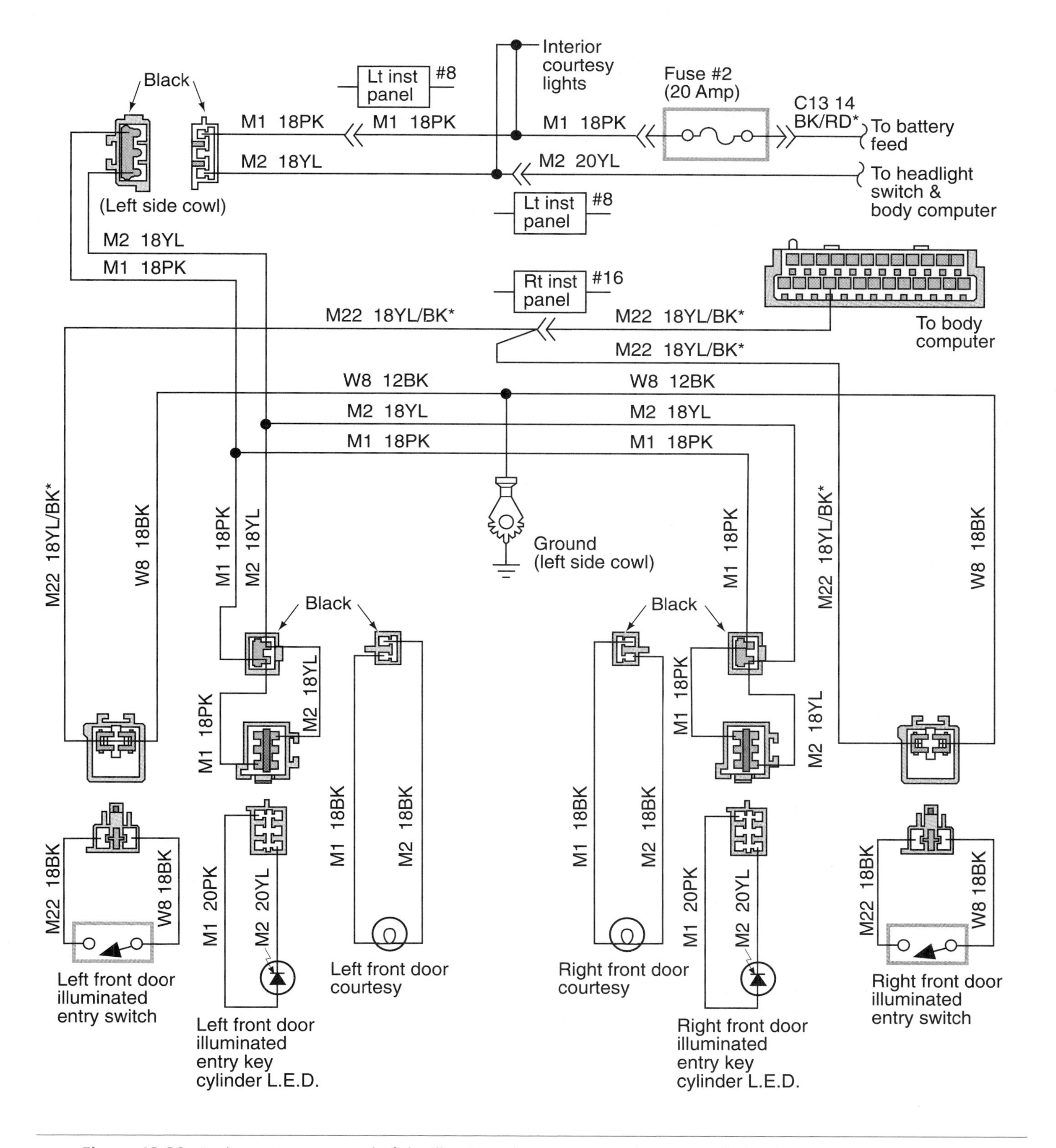

Figure 13-20 Body computer control of the illuminated entry system. (Courtesy of Chrysler Corporation)

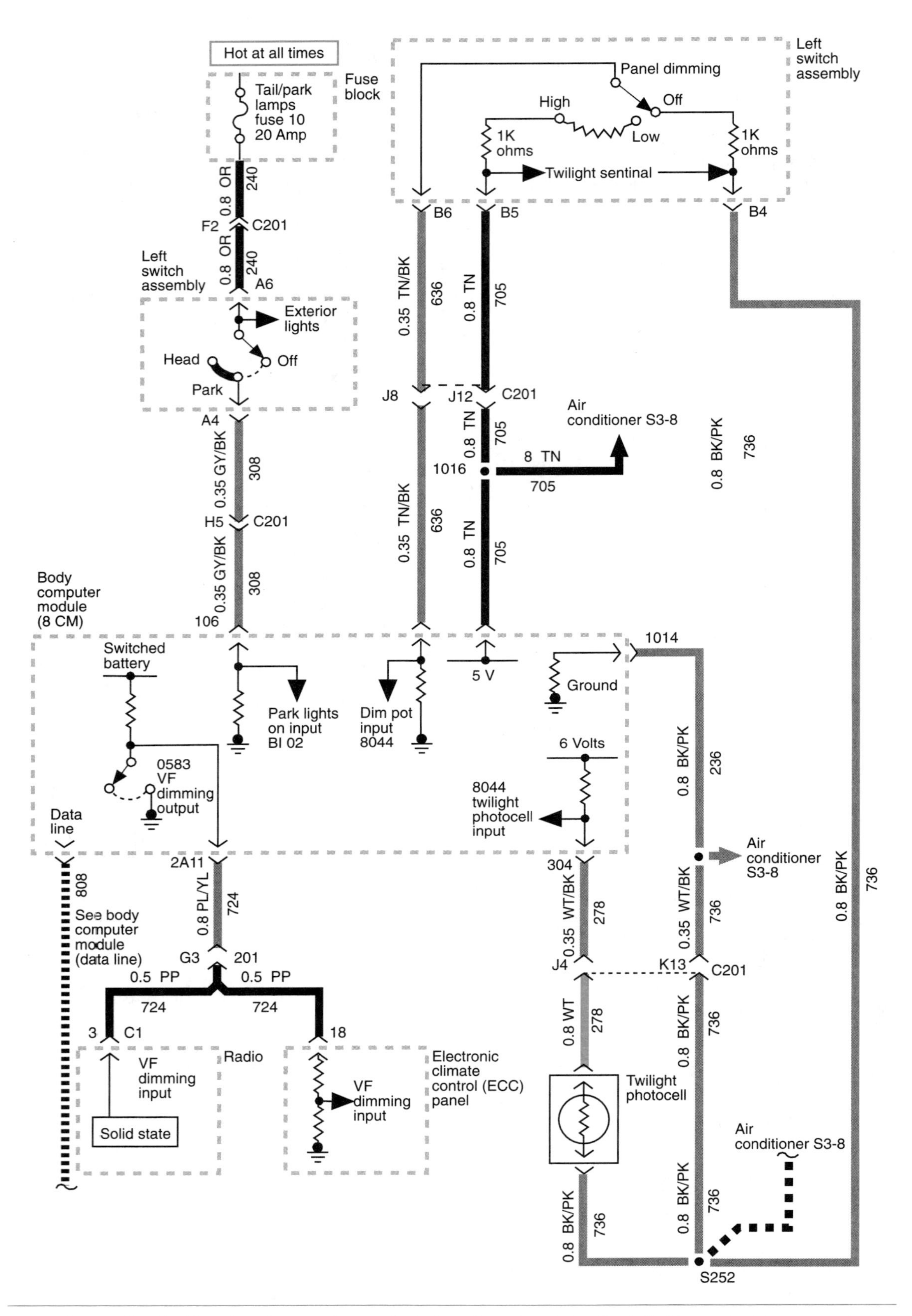

Figure 13-21 The dimming control and photocell are inputs to the BCM to control instrument panel dimming. (Courtesy of General Motors Corporation)

The signal from the door handle switch informs the body computer to activate the courtesy light relay control circuit. Some systems use a pair of door jam switches that signal the body computer to keep the courtesy lights on when the door is open. When the door is closed and the ignition switch is in the RUN position, the lights are turned off.

Some manufacturers use the twilight photocell to inform the body computer of ambient light conditions. If the ambient light is bright, the photocell signals the body computer that courtesy lights are not required.

In the computer-controlled **instrument panel dimming** system, the headlight switch dimming control is used as an input to the computer instead of having direct control of the illumination lights.

Instrument Panel Dimming

The body computer uses inputs from the panel dimming control and photocell to determine the illumination level of the instrument panel lights (Figure 13-21). With the ignition switch in the RUN position, a 5-volt signal is supplied to the panel dimming control potentiometer. The wiper of the potentiometer returns the signal to the body computer.

When the dimmer control is moved toward the dimmer positions, the increased resistance results in a decreased voltage signal to the body computer. By measuring the voltage that is returned, the body computer is able to determine the resistance value of the potentiometer. The body computer controls the intensity level of the illumination lamps by pulse width modulation.

Shop Manual
Chapter 13,
page 470

Fiber Optics

The invention of fiber optics has provided a means of providing illumination of several objects by a single light source (Figure 13-22). Plastic fiber optic strands are used to transmit light from the source to the object to be illuminated. The strands of plastic are sheathed by a polymer that insulates the light rays as they travel within the strands. The light rays travel through the strands by means of internal reflections.

Fiber optics is the transmission of light through polymethyl-methacrylate plastic that keeps the light rays parallel even if extreme bends are in the plastic.

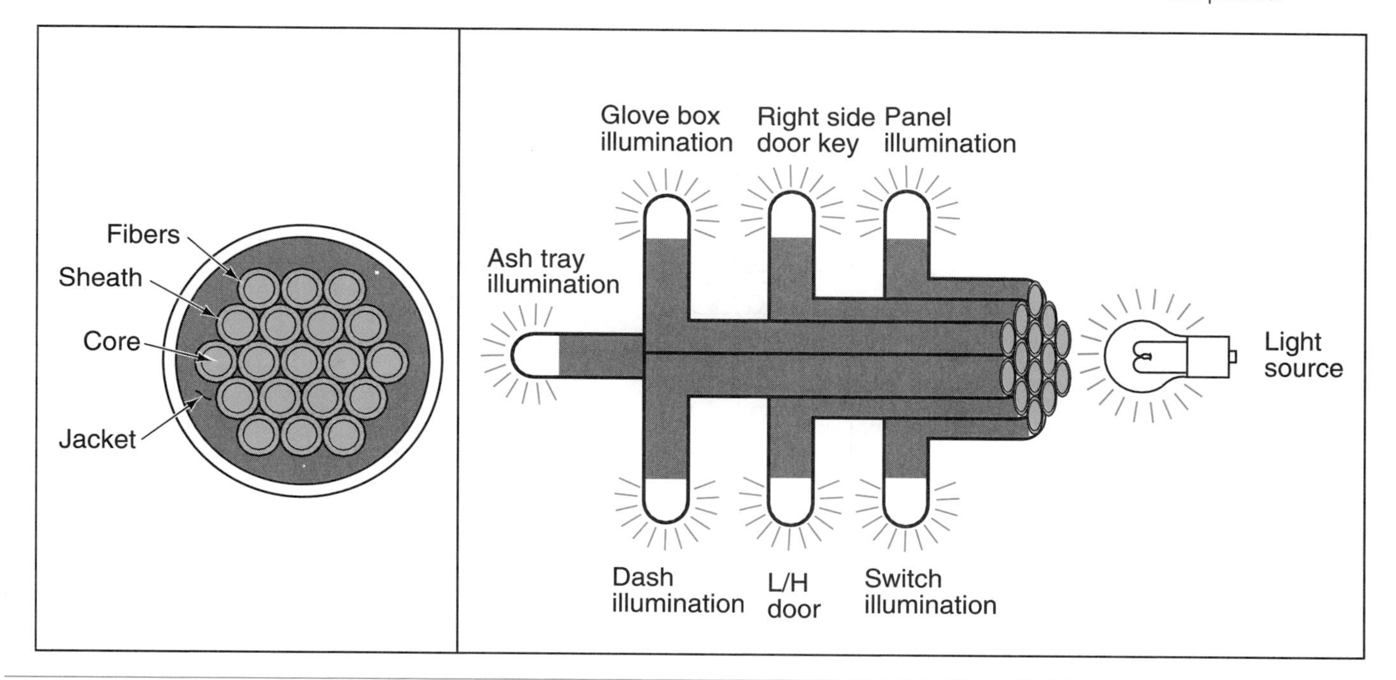

Figure 13-22 One light source can illuminate several areas by using fiber optics.

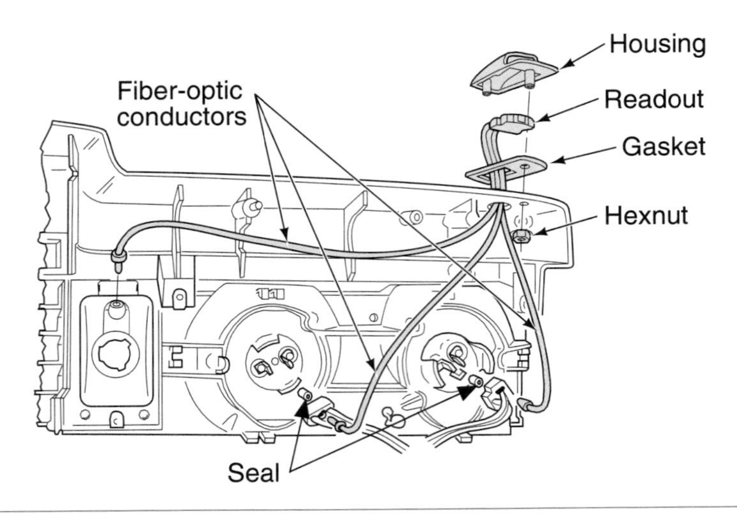

Figure 13-23 Fiber optics can be used to indicate the operation of exterior lights to the driver. (Courtesy of General Motors Corporation)

Fiber optics are commonly used as indicator lights to show the driver that certain lights are functioning. Many vehicles with fender-mounted turn signal indicators use fiber optics from the turn signal light to the indicator (Figure 13-23). The indicator will only show light if the turn signal light is on and working properly.

Some manufacturers send battery voltage to the dimming rheostat, which uses the voltage to the BCM as an input for illumination level. An input of approximately 1 volt indicates a request for high illumination level; higher voltage level inputs indicate less intensity.

Some manufacturers use fiber optics to provide illumination of the lock cylinder "halo" during illuminated entry operation (Figure 13-24). When the illuminated entry system is activated, the light collector provides the source light to the fiber optics and the halo lens receives the light from the fiber optic cable.

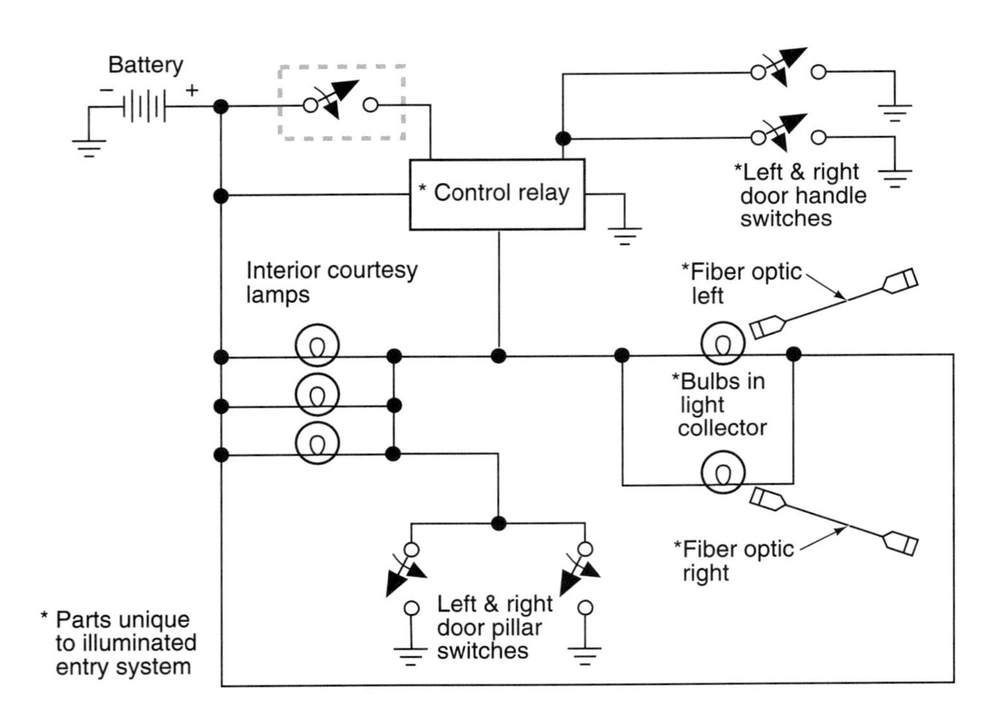

Figure 13-24 Fiber optics used in an illuminated entry system. (Courtesy of Chrysler Corporation)

The advantage of fiber optics is it can be used to provide light in areas where bulbs would be inaccessible for service. Other uses of fiber optics include:

- Lighting ash trays
- Illuminating instrument panels
- Dash lighting over switches

Lamp Outage Indicators

The most common lamp outage indicator uses a translucent drawing of the vehicle (Figure 13-25). If one of the monitored systems fails, or is in need of driver attention, the graphic display illuminates a light to indicate the location of the problem.

The basic lamp outage indicator system is used to monitor the stop light circuit. This system consists of a reed switch and opposing electromagnetic coils (Figure 13-26). When the ignition switch is turned to the RUN position, battery voltage is applied to the normally open reed switch. When the brake light switch is closed, current flows through the coils on the way to the stop light bulbs. If both bulbs are operating properly, the coils create opposing magnetic fields that keep the reed switch in the open position. If one of the stop light bulbs burns out, current will only flow through one of the coils which attracts the reed switch contacts and closes them. This completes the stop light warning circuit and illuminates the warning light on the dash. The warning light will remain on as long as the stop light switch is closed.

Opposing magnetic fields are created because the coils are wound in opposite directions.

The lamp outage module can be used alone or in conjunction with the vehicle's body computer. If the module is a "stand-alone" unit, it will operate the warning light directly. The module monitors the voltage drop of the resistors. If the circuits are operating properly, there is a 0.5-volt input signal to the module. If one of the monitored bulbs burns out, the voltage input signal drops to about 0.25 volt. The module completes the ground circuit to the warning light to alert the driver that a bulb has burned out. The module is capable of monitoring several different light circuits.

A **lamp outage module** is a current measuring sensor that contains a set of resistors, wired in series with the power supply to the headlights, taillights, and stop lights.

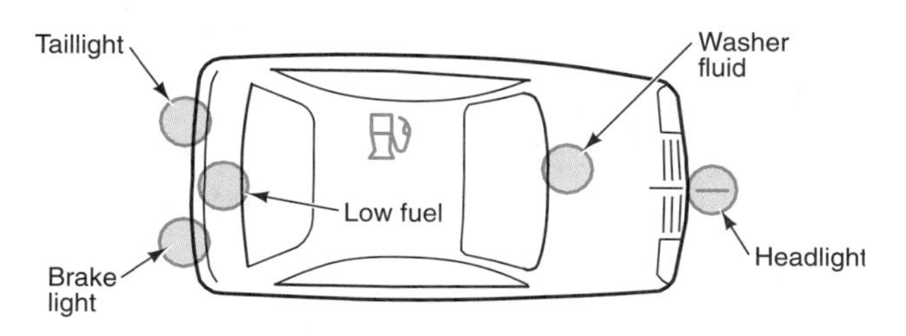

Figure 13-25 Most vehicle information systems may use a graphic display to indicate warning areas to the driver. (Courtesy of Chrysler Corporation)

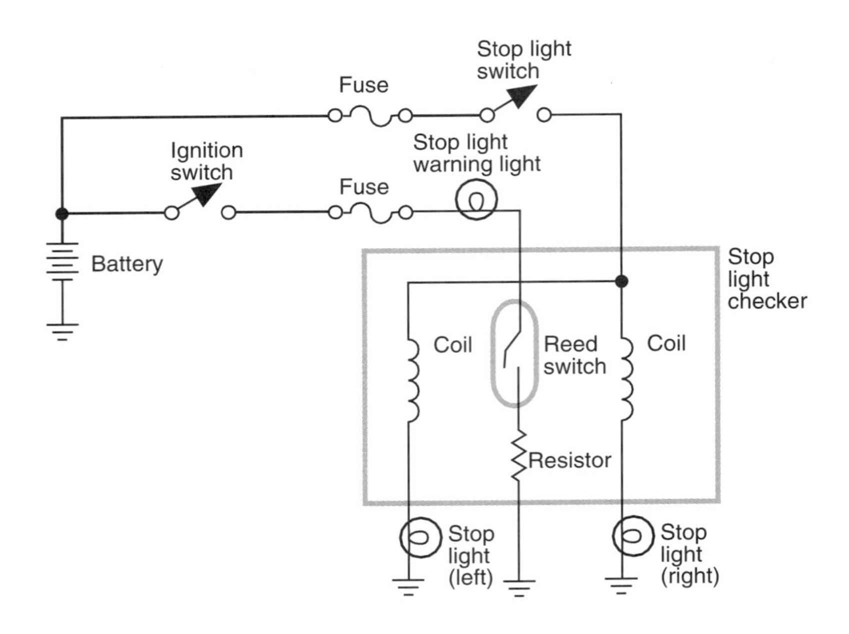

Figure 13-26 Stop light lamp outage indicator circuit. (Courtesy of General Motors Corporation)

Many vehicles today use a computer-driven information center to keep the driver informed of the condition of monitored circuits (Figure 13-27). The vehicle information center usually receives its signals from the vehicle's body computer circuit (Figure 13-28). In this system the lamp outage module is used to send signals to the body computer. The body computer will either illuminate a warning light, give a digital message, or activate a voice warning device to alert the driver that a light bulb is burned out.

The bulbs are monitored only when current is supplied to them.

A burned out light bulb means there is a loss of current flow in one of the resistors of the lamp outage module. A monitoring chip in the module compares the voltage drop across the resistor (Figure 13-29). If there is no voltage drop across the resistor, there is an open in the circuit

Figure 13-27 The computer-driven vehicle information center keeps the driver aware of the condition of monitored systems. (Courtesy of General Motors Corporation)

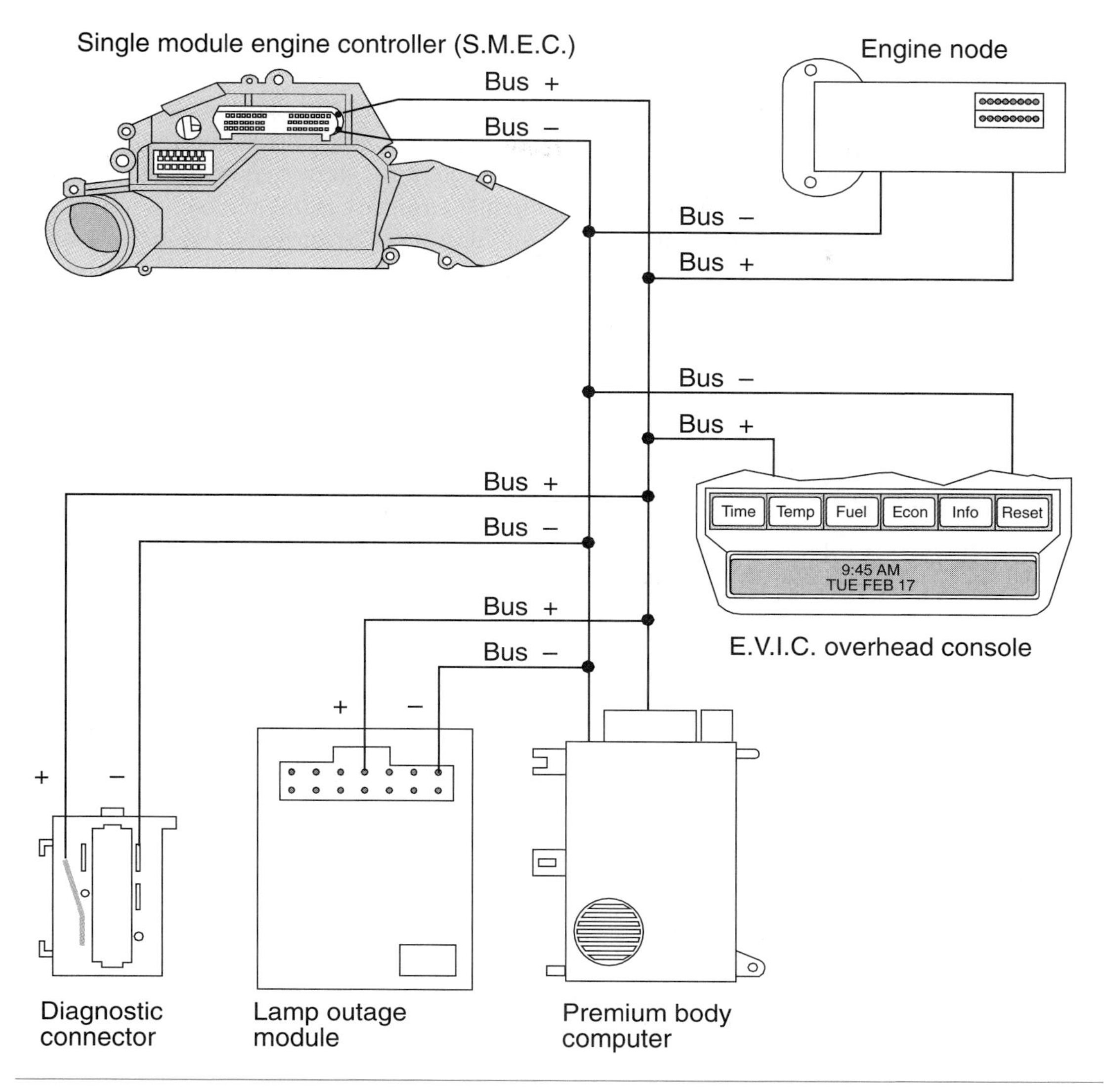

Figure 13-28 The body computer can be used to receive signals from various inputs and to give signals to control the information center. (Courtesy of Chrysler Corporation)

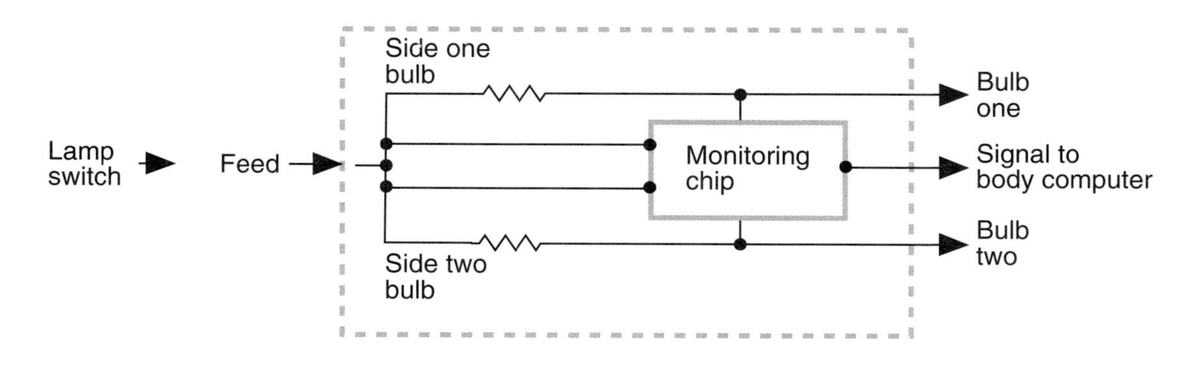

Figure 13-29 One section of the lamp outage module circuit. The monitoring chip compares voltage on both sides of the resistors to measure voltage drop changes. (Courtesy of Chrysler Corporation)

(burned out light bulb). When the chip measures no voltage drop across the resistor, it signals the body computer, which then gives the necessary message to the vehicle information center (Figure 13-30).

General Motors uses the lamp monitor module to connect the light circuits to ground (Figure 13-31). When the circuits are operating properly, the ground connection in the module causes a low circuit voltage. Input from the lamp circuits are through two equal resistance wires. The module output to the bulbs is from the same module terminals as the inputs.

If a bulb burns out, the voltage at the lamp monitor module terminal will increase. The module will open the appropriate circuit from the BCM, signaling the BCM to send a communication to the IPC computer, which displays the message in the information center.

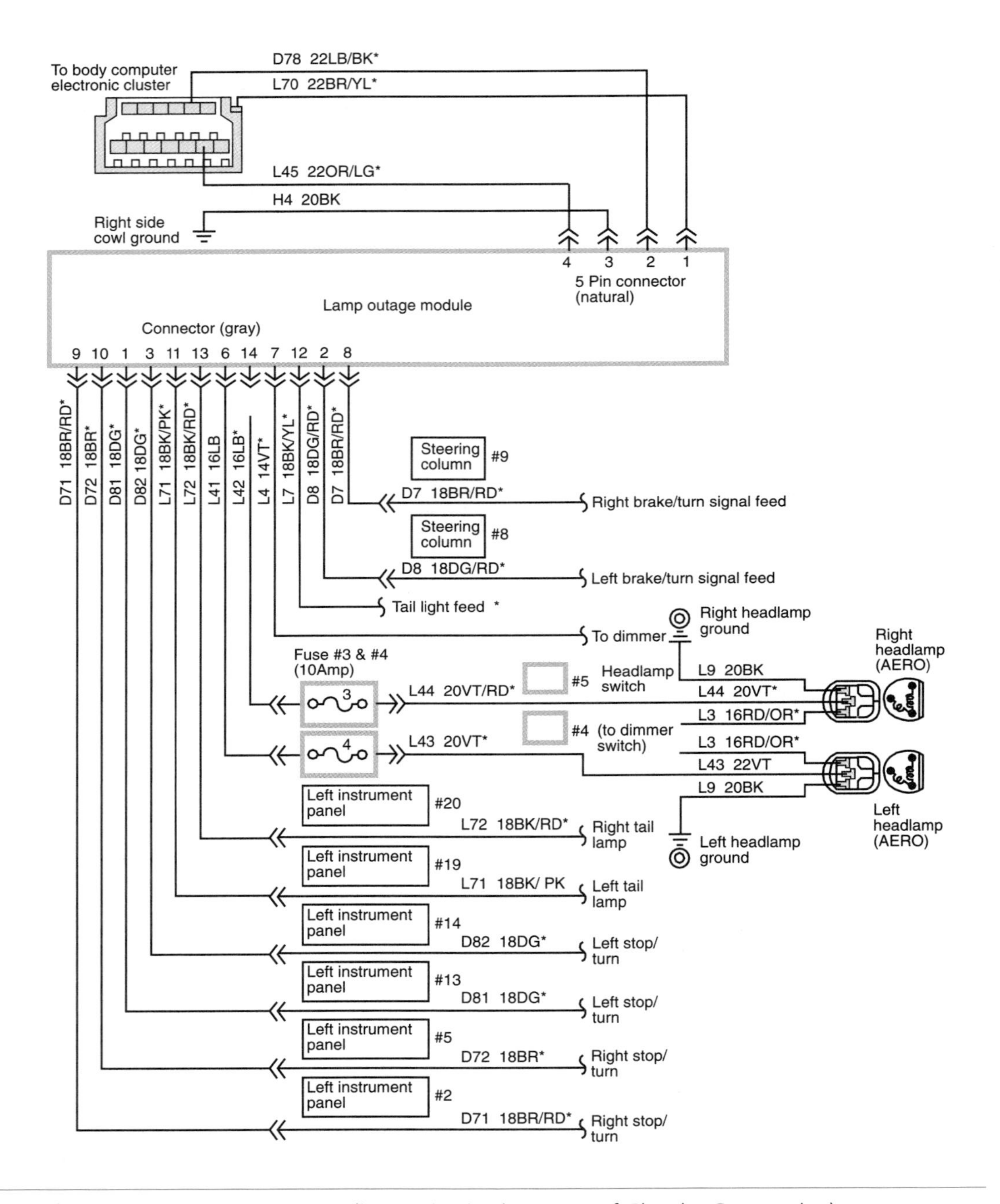

Figure 13-30 Lamp outage indicator circuit. (Courtesy of Chrysler Corporation)

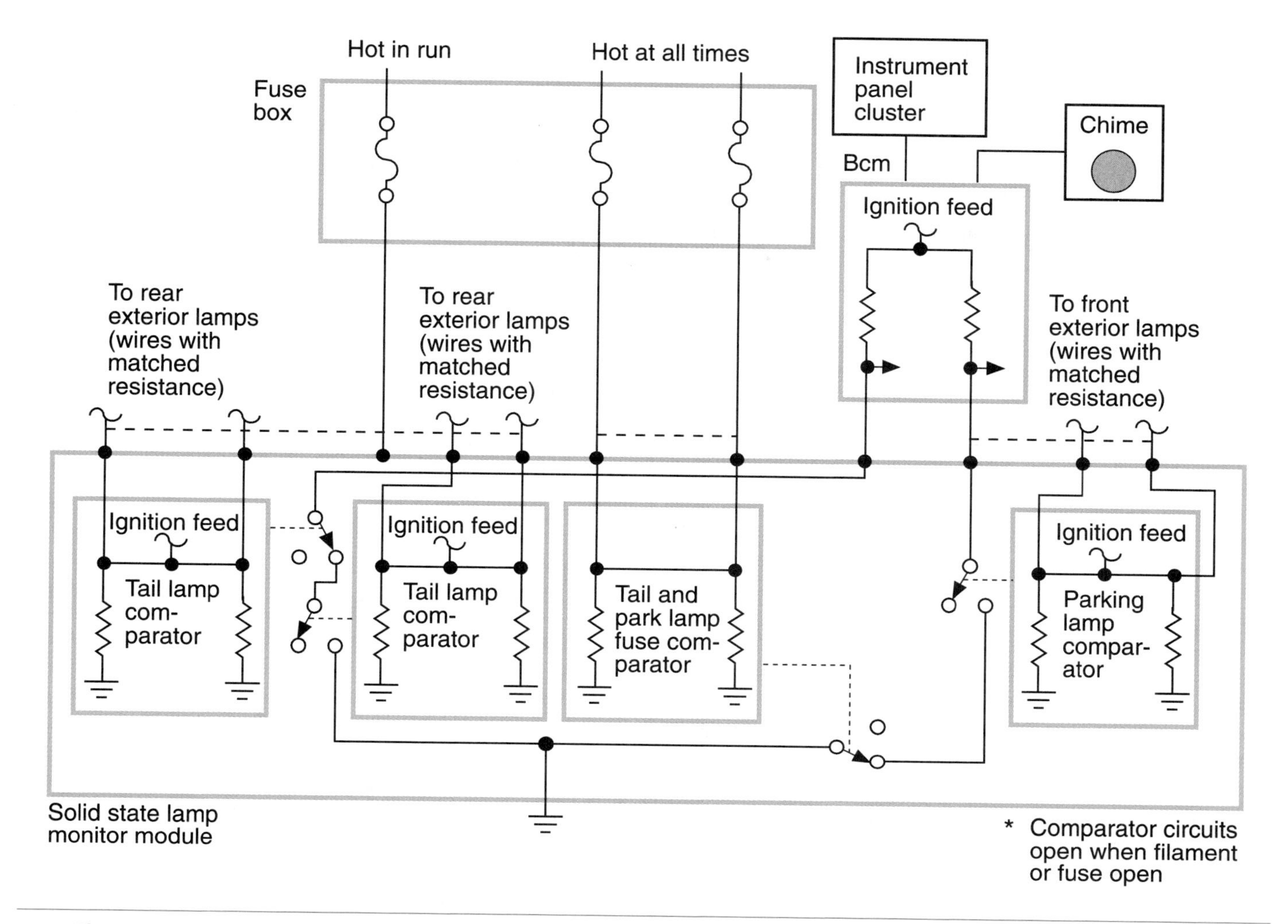

Figure 13-31 Lamp monitor module circuit.

Electronic Instrumentation Introduction

Computer-driven instruments are becoming increasingly popular on today's vehicle. These instruments provide far more accurate readings than their conventional analog counterparts. Today's technician will be required to service these systems on a more frequent basis as they grow in popularity. This section introduces you to the most commonly used computer-driven instrumentation systems. These systems include the speedometer, odometer, fuel, oil, and temperature gauges.

The computer-driven instrument panel uses a microprocessor to process information from various sensors and to control the gauge display. Depending on the manufacturer, the microprocessor can be a separate computer that receives direct information from the sensors and makes the calculations, or it may use the body computer to perform all functions. The illustration (Figure 13-32) shows the different inputs and outputs to the computer-driven instrument panel. The computer may control a digital instrument or an analog cluster.

In addition, there are many types of information systems used today. These systems are used to keep the driver informed of a variety of monitored conditions, including vehicle maintenance, trip information, and navigation.

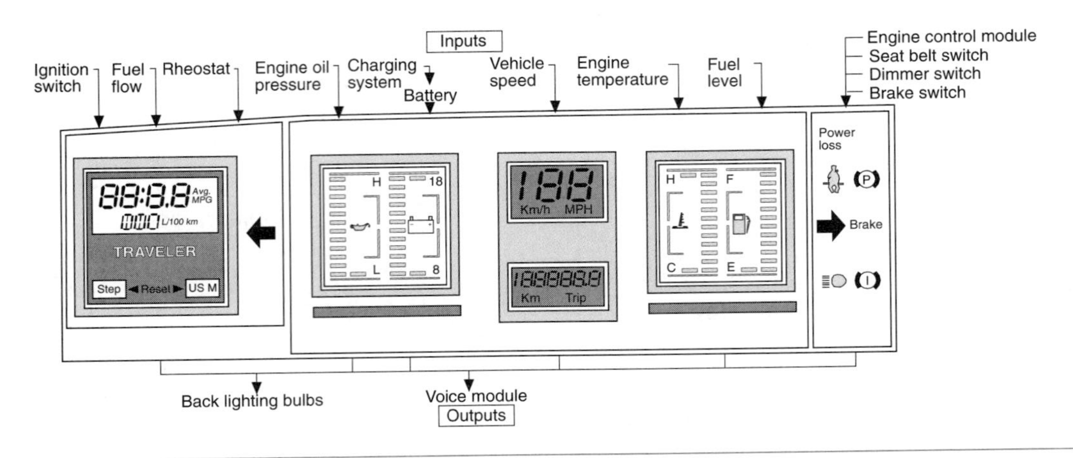

Figure 13-32 Inputs and outputs of the electronic instrument panel. (Courtesy of Chrysler Corporation)

Shop Manual
Chapter 13,
page 470

Digital instrument clusters use digital and linear displays to notify the drive of monitored system conditions (Figure 13-33).

Digital Instrumentation

Digital instrumentation is far more precise than conventional analog gauges. Analog gauges display an average of the readings received from the sensor; a digital display will present exact readings. In some systems the information to the gauge is updated as often as 16 times per second.

Most digital instrument panels provide for display in English or metric values. Also, many gauges are a part of a multigauge system. Drivers select which gauges they wish to have displayed. Most of these systems will automatically display the gauge to indicate a potentially dangerous situation. For example, if the driver has chosen the oil pressure gauge to be displayed and the engine temperature increases above set limits, the temperature gauge will automatically be displayed to warn the driver. A warning light and/or a chime will also activate to get the driver's attention.

Most electronic instrument panels have self-diagnostic capablities. The tests are initiated through a scan tool or by pushing selected buttons on the instrument panel. The instrument panel cluster also initiates a self-test every time the ignition switch is turned to ACC or RUN. Usually the entire dash is illuminated and every segment of the display is lighted. ISO symbols generally flash during this test. At the completion of the test, all gauges will display current readings. A code is displayed to alert the driver if a fault is found.

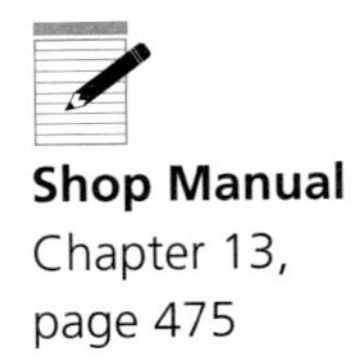

Shop Manual
Chapter 13,
page 475

Speedometers

Ford, GM, and Toyota have used optical vehicle speed sensors. The Ford and Toyota optical sensors are operated from the conventional speedometer cable. The cable rotates a slotted wheel

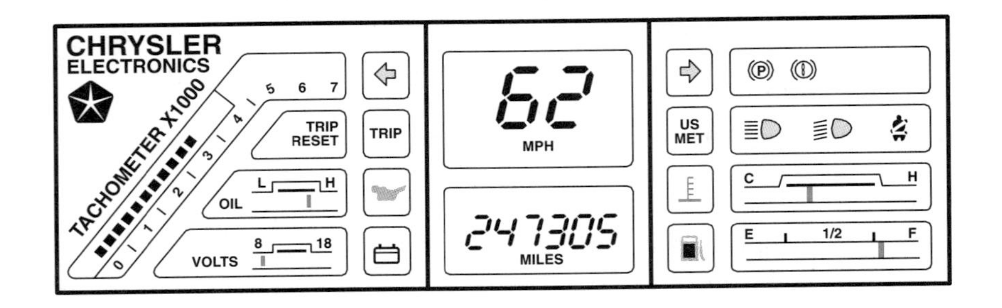

Figure 13-33 Digital instrument cluster. (Courtesy of Chrysler Corporation)

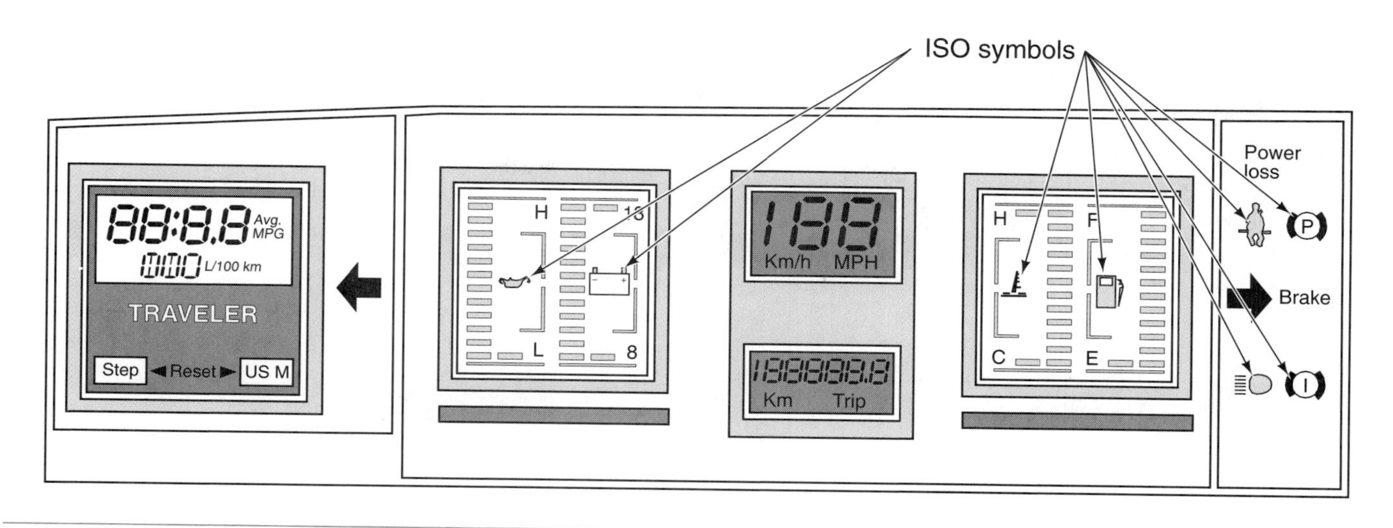

Figure 13-34 A few of the ISO symbols used to identify the gauge. (Courtesy of Chrysler Corporation)

International Standards Organization (ISO) symbols are used to represent the gauge function (Figure 13-34).

between an LED and a phototransistor (Figure 13-35). As the slots in the wheel break the light, the transistor conducts an electronic pulse signal to the speedometer. An integrated circuit rectifies the analog input signal from the optical sensor and counts the pulses per second. The value is calculated into mph and displayed in the digital readout. The display is updated every 1/2 second. If the driver selected the readout to be in kilometers per hour, the computer makes an additional calculation to convert the readout. These systems may use a conventional gear-driven odometer.

The early style of GM speed sensor also operated from the conventional speedometer cable. The LED directs its light onto the back of the speedometer cup. The cup is painted black, and the drive magnet has a reflective surface applied to it. As the drive magnet rotates in front of the LED, its light is reflected back to a phototransistor. A small voltage is created every time the phototransistor is hit with the reflective light.

The electronic speedometer receives voltage signals from the vehicle speed sensor (VSS). This sensor can be a PM generator, Hall-effect switch, or an optical sensor.

The illustration (Figure 13-36) is a schematic of an instrument panel cluster that uses a PM generator for the VSS. As the PM generator is rotated, it causes a small AC voltage to be induced in its coil. This AC voltage signal is sent to the engine control module (ECM) and is shared with the body control module (BCM). The signal is rectified into a digital signal that is used to control the output to the instrument panel cluster (IPC) module. The body computer calculates the vehicle speed and provides this information to the IPC module through the serial data link. The IPC module turns on the proper display.

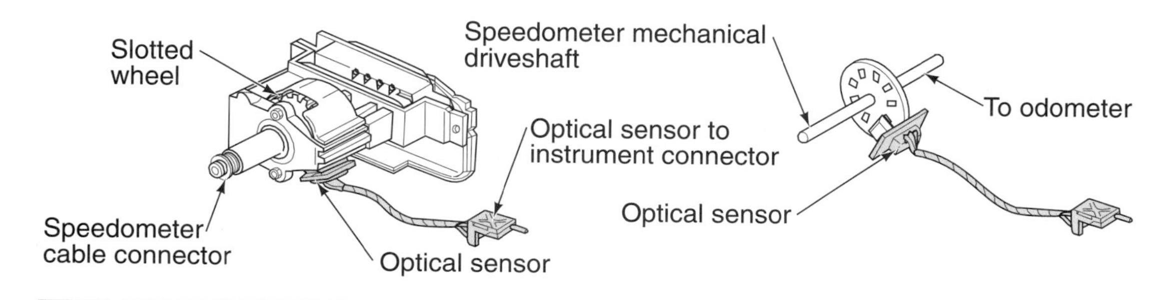

Figure 13-35 Optical speed sensor. (Courtesy of Chilton Book Company)

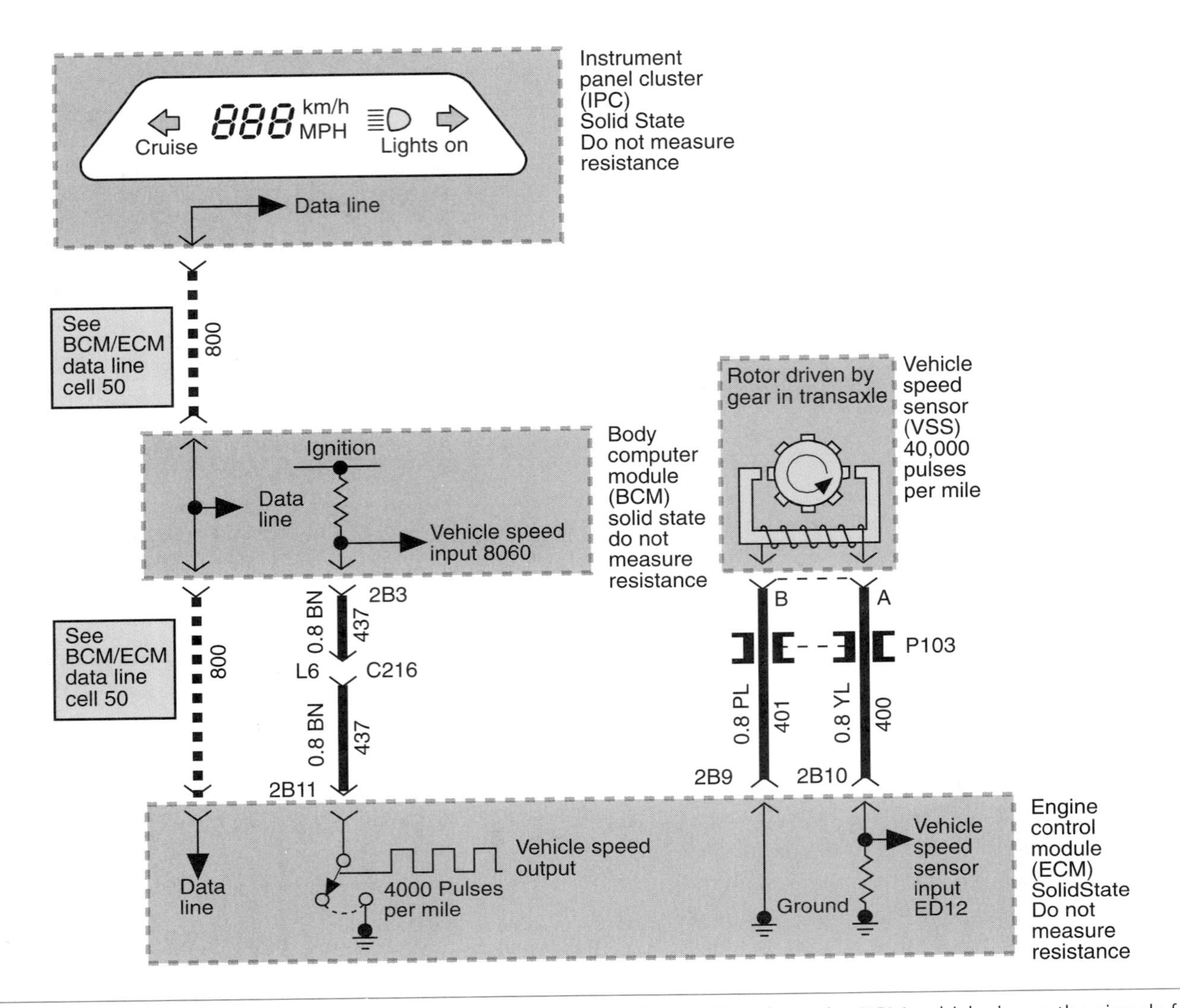

Figure 13-36 The instrument panel cluster module receives its instructions from the BCM, which shares the signals from the VSS with the ECM. (Courtesy of General Motors Corporation)

The microprocessor will initiate a self-check of the electronic instrument cluster anytime the ignition switch is placed in the ACC or RUN position. The self-check usually runs for about 3 seconds. The most common sequence for the self-check is as follows:

1. All display segments are illuminated (Figure 13-37).
2. All displays go blank.
3. 0 mph or 0 km/h is displayed (Figure 13-38).

Shop Manual
Chapter 13,
page 475

A **stepper motor** moves in incremental steps between no voltage to full voltage positions.

Odometers

If the speedometer uses an optical sensor, the odometer may be of conventional design. Two other types of odometer are used with electronic displays: the electromechanical type with a stepper motor, and the electronic design using an IC chip.

Stepper Motor. The electromechanical odometer uses a DC stepper motor that receives control signals from the speedometer circuit (Figure 13-39). The digital signal impulses from the speedometer are processed through a circuit that will halve the signal. The stepper motor

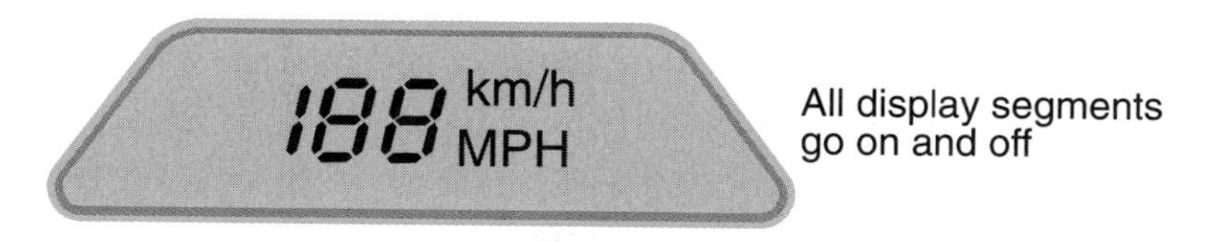

Figure 13-37 During the first portion of the self-test, all segments of the speedometer display are lighted.

Figure 13-38 At the completion of the self-test, the current speed is indicated.

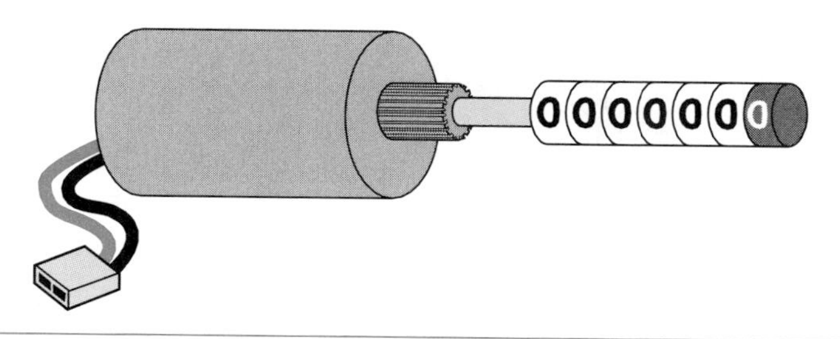

Figure 13-39 A stepper motor is used to rotate the odometer dial.

receives one-half of the VSS signals sent to the instrument panel cluster. As the stepper motor is activated, the rollers are rotated to accurately display accumulated mileage.

General Motors controls the stepper motor through the same impulses that are sent to the speedometer (Figure 13-40). The stepper motor uses these signals to turn the odometer drive IC on and off. An H-gate arrangement of transistors is used to drive the stepper motor by alternately activating a pair of its coils (Figure 13-41). The H-gate is constantly reversing system polarity, causing the permanent magnet poles to rotate in the same direction.

An **H-gate** is a set of four transistors that reverse current to the motor windings to keep the motor rotating in one direction.

IC Chip. The IC chip-type odometer uses a nonvolatile RAM that receives distance information from the speedometer circuit or from the engine controller. The controller can update the odometer display every 1/2 second.

In most systems distance is updated to the RAM every 10 miles and whenever the ignition switch is turned off.

Many instrument panel clusters cannot display both trip mileage and odometer readings at the same time. Drivers must select which function they wish to have displayed (Figure 13-42). By depressing the trip reset button, a ground is applied as an input to the microprocessor. The microprocessor clears the trip odometer readings from memory and returns the display to zero. The trip odometer will continue to store trip mileage even if this function is not selected for display.

If the IC chip fails, some manufacturers provide for replacement of the chip. Depending on the manufacturer, the new chip may be programmed to display the last odometer reading. Most replacement chips will display an X, S, or * to indicate the odometer has been changed. It is

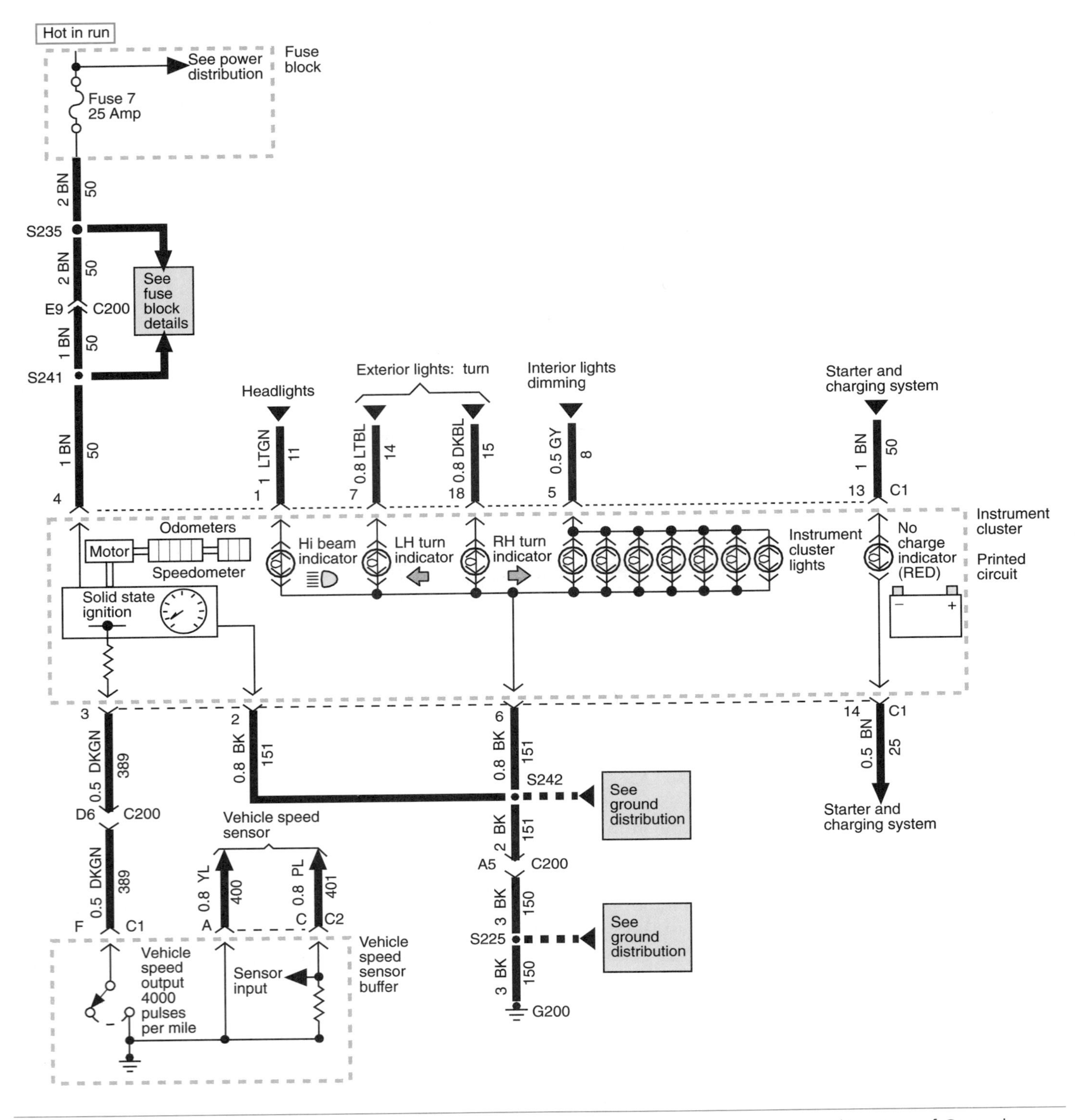

Figure 13-40 The stepper motor receives its control signals through the speedometer circuit. (Courtesy of General Motors Corporation)

impossible to "turn back" an IC odometer. If the odometer IC chip cannot be programmed to display correct accumulated mileage, a door sticker must be installed to indicate the odometer has been replaced. IC odometer reading corrections or changes can only be performed in the first 10 miles of operation.

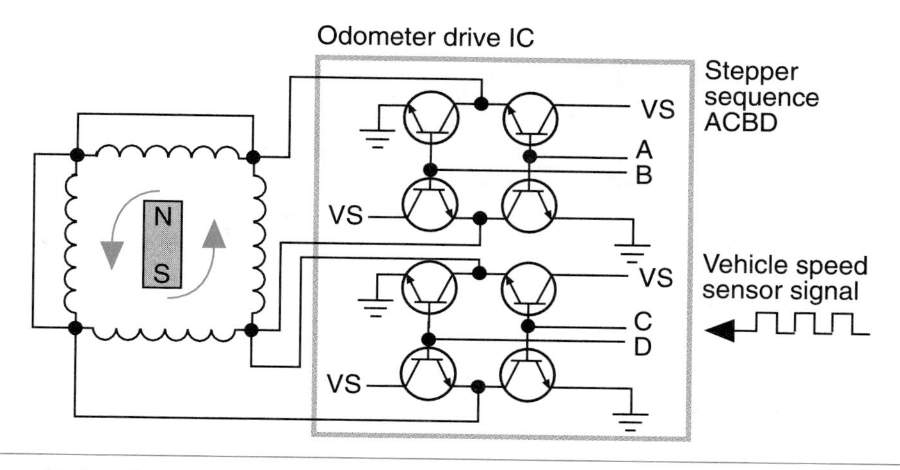

Figure 13-41 The **H-gate** energizes two coils at a time and constantly reverses system polarity.

Federal Motor Vehicle Safety Standards require the odometer be capable of storing up to 500,000 miles in nonvolatile memory. Most odometer readouts are up to 199,999.9 miles.

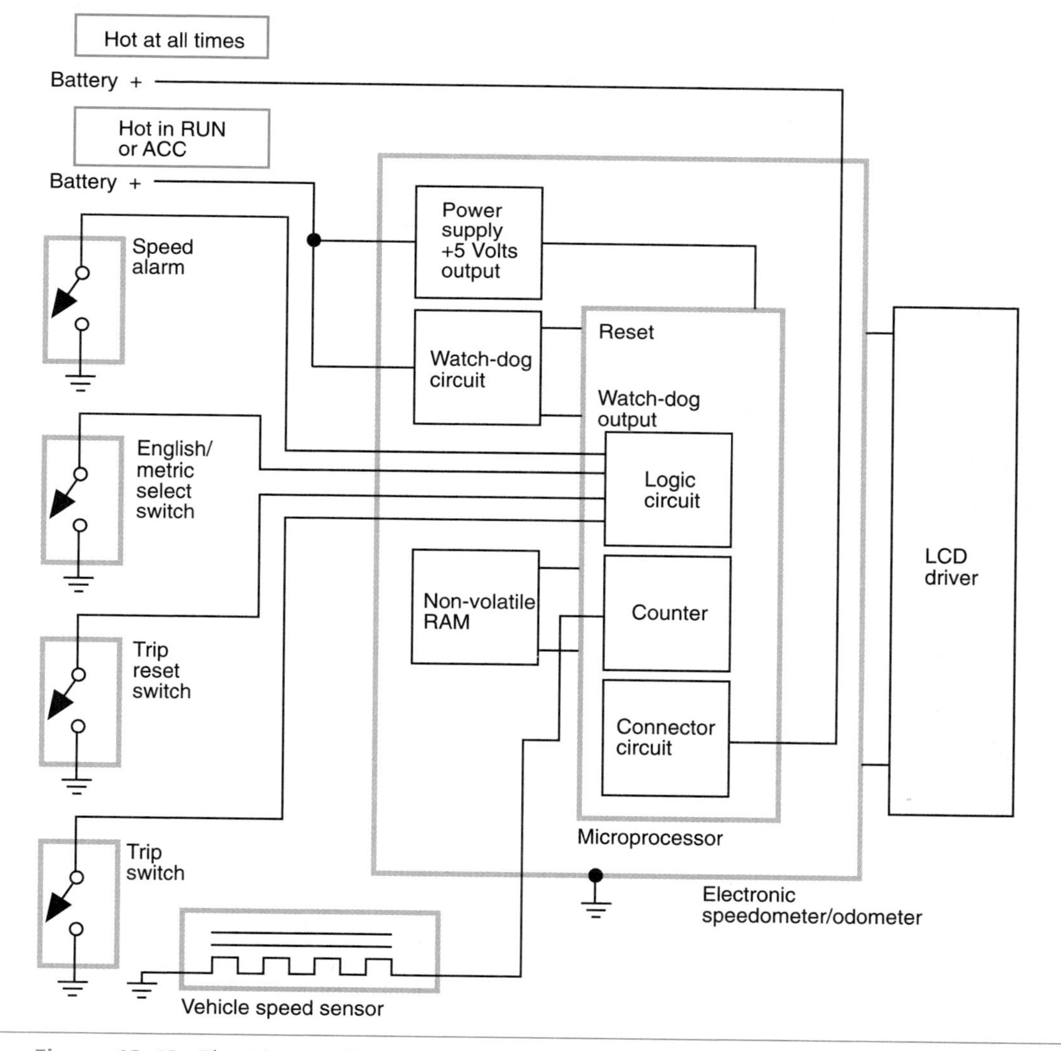

Figure 13-42 The trip reset button provides a ground signal to the logic circuit, which is programmed to erase the trip odometer memory while retaining total accumulated mileage in the odometer.

Figure 13-43 Error message alerts the driver that the odometer function is not operating properly.

If an error occurs in the odometer circuit, the display will change to notify the driver. The form of error message differs between manufacturers. The illustration (Figure 13-43) shows one method that Ford uses.

The **power on/off watchdog circuit** supplies a reset voltage to the microprocessor in the event that pulsating output signals from the microprocessor are interrupted.

Tachometers

The tachometer can be a separate function that is displayed at all times, or a part of a multigauge display. Ford uses a multigauge that sequentially changes the gauge between four different gauges (Figure 13-44). The tachometer receives its voltage signals from the ignition system and displays the readout in a bar graph. The Ford multigauge has a built in power supply that provides a 5-volt reference signal to the other monitored systems for the gauge. Also, the gauge has a “watchdog” circuit incorporated in it.

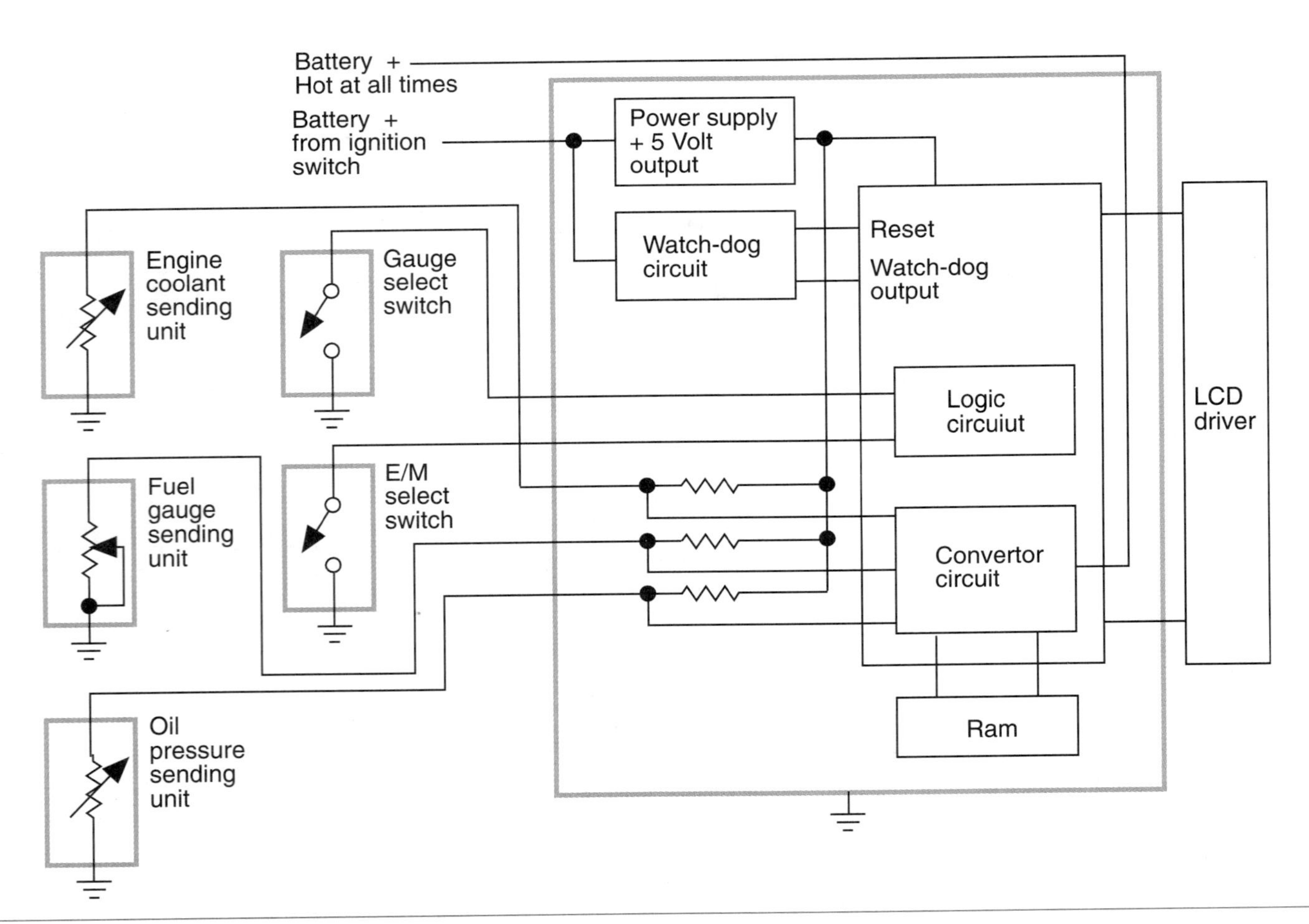

Figure 13-44 Multigauge schematic with tachometer function.

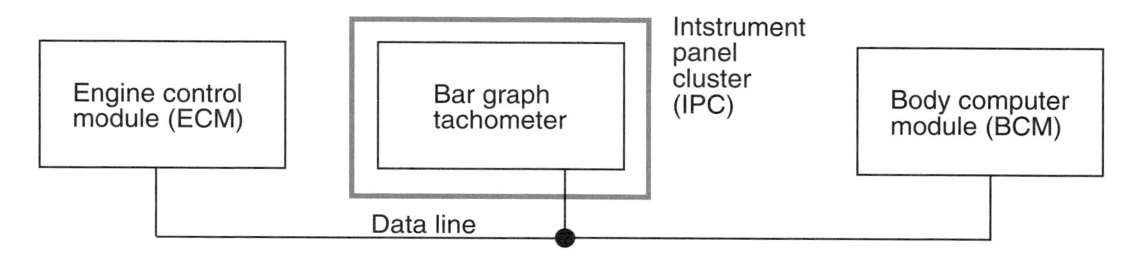

Figure 13-45 The IPC "listens in" on the communication between the ECM and the BCM to gather information on engine speed.

General Motors uses the signal from the direct ignition system (DIS) module to the ECM. The signal is sent from the ECM to the BCM on the serial data link. The IPC does not receive this signal directly, but it eavesdrops on the data link communications (Figure 13-45). The IPC uses this reference signal to calculate and display the tachometer reading.

Electronic Fuel Gauges

Most digital fuel gauges use a fuel level sender that decreases resistance value as the fuel level decreases. This resistance value is converted to voltage values by the microprocessor. A voltage-controlled oscillator changes the signal into a cycles-per-second signal. The microprocessor counts the cycles and sends the appropriate signal to operate the digital display (Figure 13-46).

An **oscillator** creates a rapid back-and-forth movement of voltage.

An *F* is displayed when the tank is full and an *E* is displayed when less than 1 gallon is remaining in the tank. Other warning signals include incandescent lamps, a symbol on the dash, or flashing of the fuel ISO symbol. If the warning is displayed by a bulb, usually a switch is located in the sending unit that closes the circuit. Flashing digital displays are usually controlled by the microprocessor.

WARNING: Not all sending units used for bar graph and digital displays are the same. Some sending units used for bar graph displays increase resistance as the fuel level is decreased.

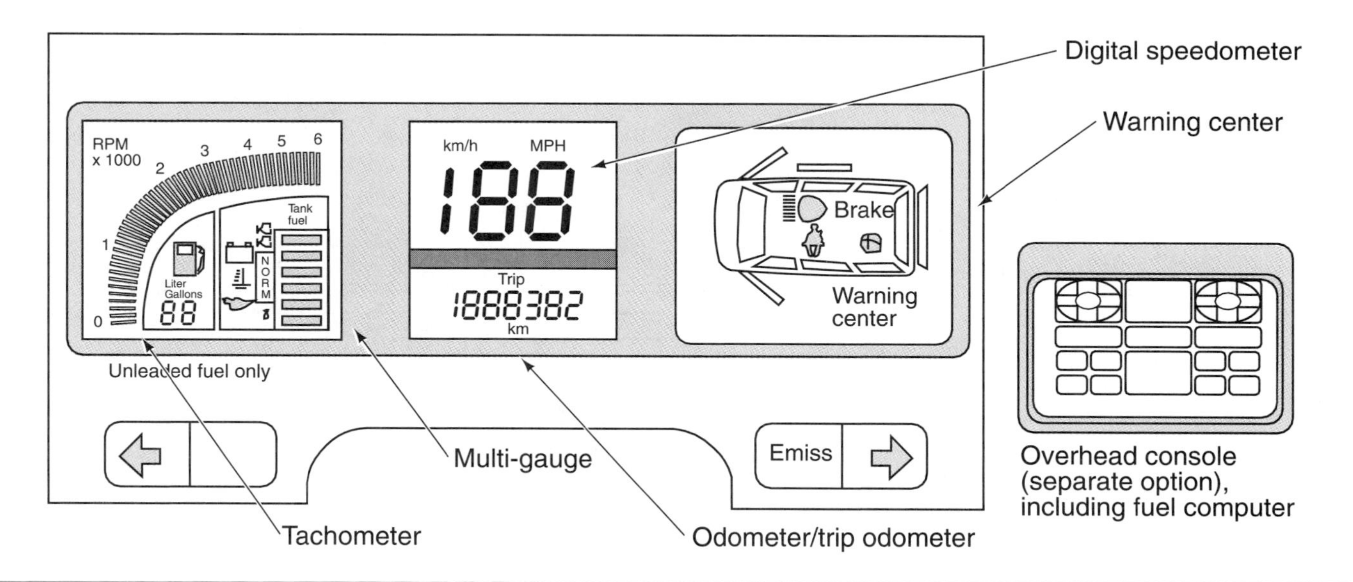

Figure 13-46 Digital fuel gauge displays remaining fuel in gallons or liters. (Reprinted with the permission of Ford Motor Company)

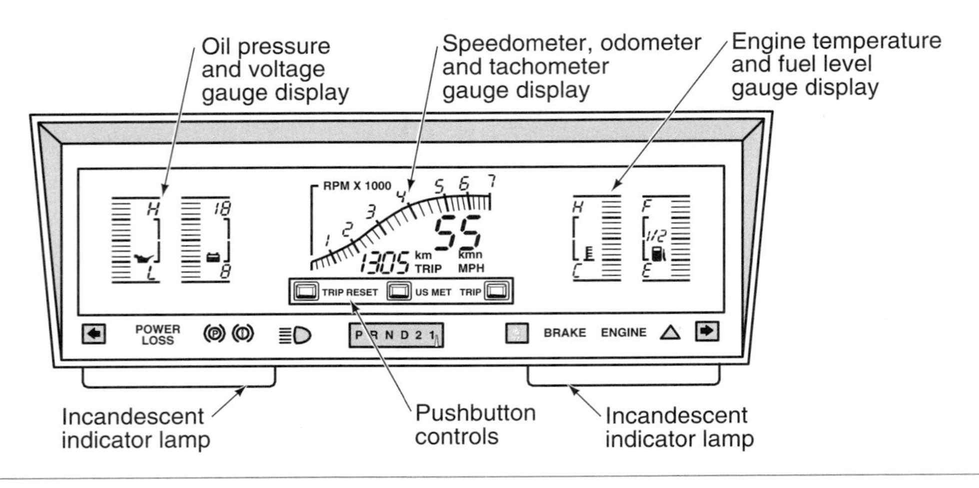

Figure 13-47 Bar graph-style of electronic instrumentation. Each segment represents a different value. (Courtesy of Chrysler Corporation)

The bar graph-style gauge uses segments that represent the amount of fuel remaining in the tank (Figure 13-47). The segments divide the tank into equal levels. The display will also include the F, 1/2, and E symbols along with the ISO fuel symbol. A warning to the driver is displayed when only one bar is lit. The gauge will also alert the driver to problems in the circuit. A common method of indicating an open or short is to flash the F, 1/2, and E symbols while the gauge reads empty.

Shop Manual
Chapter 13,
page 480

Other Digital Gauges

Most of the gauges used to display temperature, oil pressure, and charging voltage are of bar graph design. Another popular method is to use a floating pointer (Figure 13-48).

The temperature gauge will usually receive its input from an NTC thermistor. When the engine is cold, the resistance value of the thermistor is high, resulting in a low voltage input to the microprocessor. This input signal is translated into a low temperature reading on the gauge. As the engine coolant warms, the resistance value drops. At a predetermined resistance level, the microprocessor will activate an alert function to warn the driver of excessive engine temperature.

The voltmeter calculates charging voltage by comparing the voltage supplied to the instrument panel module to a reference voltage signal. The oil pressure gauge uses a piezoresistive sensor that operates like those used for conventional analog gauges.

Digital gauges perform self-tests. If a fault is found, a warning signal will be displayed to the driver. A "CO" indicates the circuit is open, a "CS" indicates the circuit is shorted. The gauge will continue to display these messages until the problem is corrected.

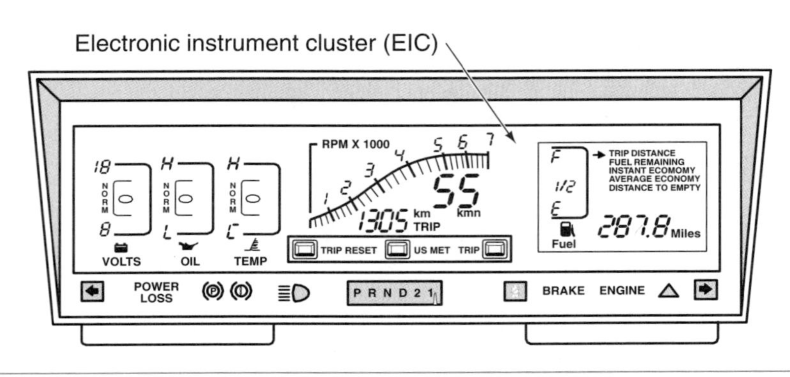

Figure 13-48 Floating pointer indicates the value received from the sensor. (Courtesy of Chrysler Corporation)

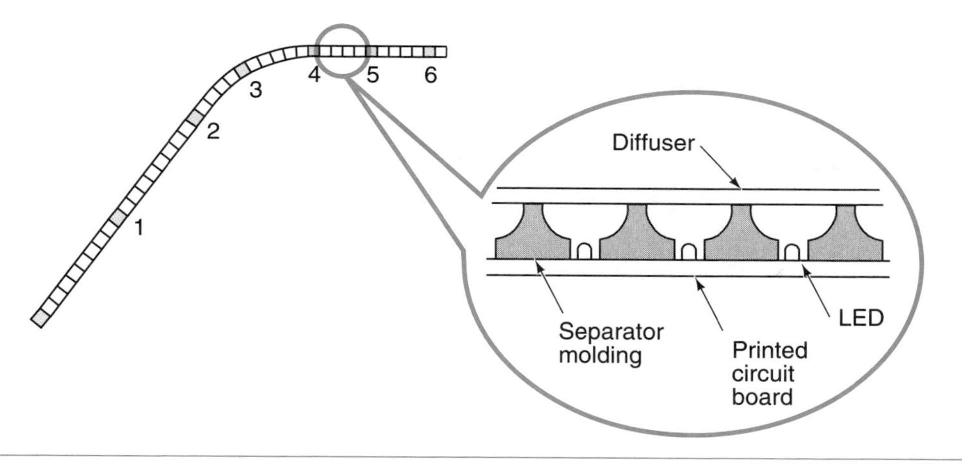

Figure 13-49 LEDs arranged to create a bar graph display.

Digital Displays

The display can be by means of four different methods: Light emitting diodes (LEDs), liquid crystal display (LCDs), vacuum fluorescent display (VFD), and cathode ray tube (CRT).

LED Digital Displays

When first used in the instrument panel, the LED would indicate an on/off status. Chrysler used the LED in conjunction with its gauges to alert the driver of conditions requiring immediate attention. Some manufacturers are using LEDs in bar graph displays (Figure 13-49). Also, the LEDs can be combined to display alphanumeric characters. There are two common methods of using the LED for digital display: (1) seven-segment display (Figure 13-50), and (2) dot matrix display (Figure 13-51).

The **light emitting diode (LED)** is an electroluminescent lamp that converts the energy developed during normal diode operation into light.

A **matrix** is a group of elements arranged in columns and rows.

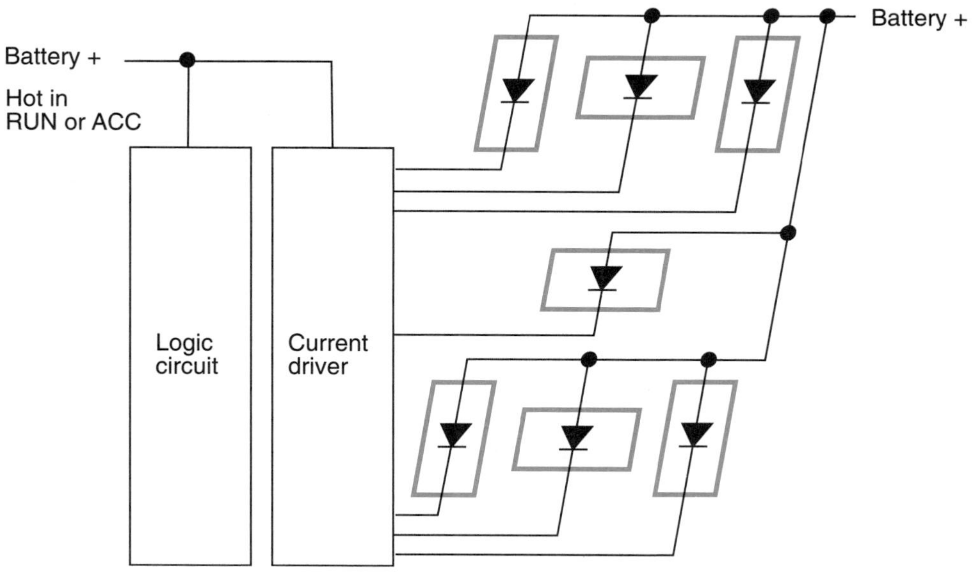

Figure 13-50 Seven-segment display panel.

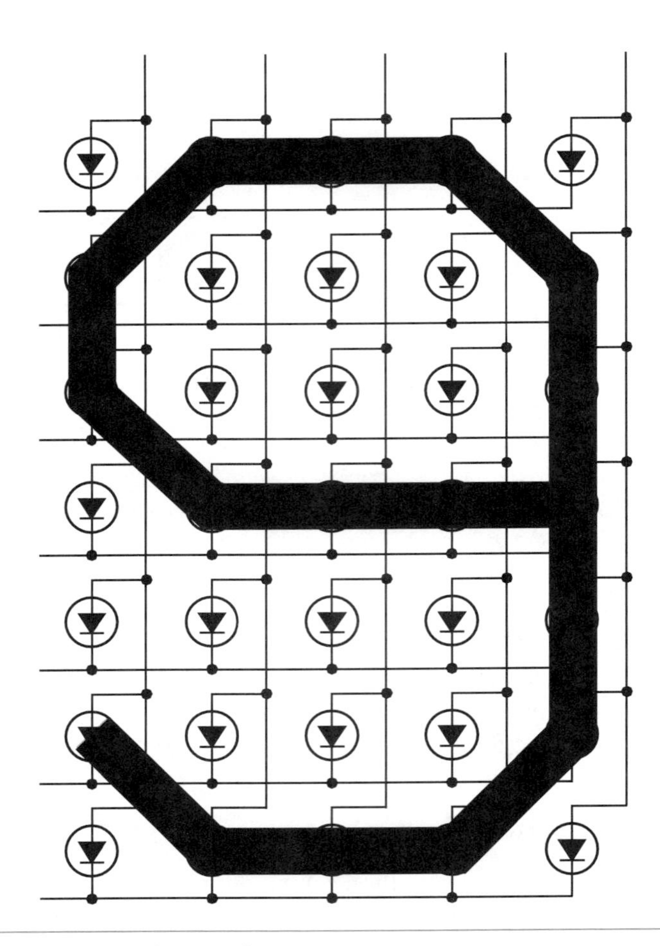

Figure 13-51 Dot matrix display panel.

By activating selected segments or dots, any number or letter can be displayed. LEDs generally are designed to produce a red, green, or yellow light. They work very well in low light conditions. However, they are difficult to see in bright light. Although some manufacturers use LED readout instrument panels, their use is limited due to their comparatively high power requirements.

Liquid Crystal Displays

Liquid crystal displays (LCDs) require an external light source because they do not generate their own (Figure 13-52). The external light source can be supplied by either daylight or by an artificial light. In daylight the segments are activated from the front; the artificial light activates the LCD from the back. The artificial light can be controlled by the headlight switch or turned on whenever the ignition switch is placed in the RUN or ACC position.

The **nematic** fluid is a liquid crystal that has a threadlike form. It has light slots that can be rearranged by applying small amounts of voltage.

The LCD construction consists of a twisted nematic fluid that is sandwiched between two polarized glass sheets. The front polarizer is a vertical polarizer and the rear polarizer is a horizontal polarizer. The display is viewed through the vertical polarizer. The nematic fluid's molecules are arranged in such a manner that they rotate the light from the vertical polarizer 90 degrees (Figure 13-54). The light leaves the fluid in a horizontal waveform. The light continues to pass through the horizontal polarizer to the reflector. The light is then reflected back through the horizontal polarizer to the fluid. The fluid once again rotates the light wave into a vertical position and out the vertical polarizer. When light passes through the LCD in this manner, the display appears light and no pattern is seen.

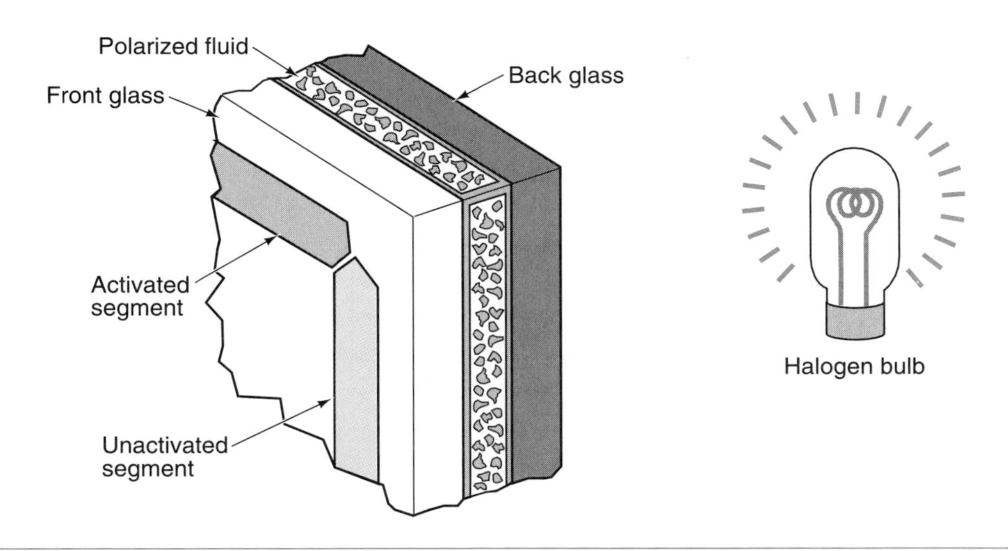

Figure 13-52 LCDs use external light source through a polarized fluid.

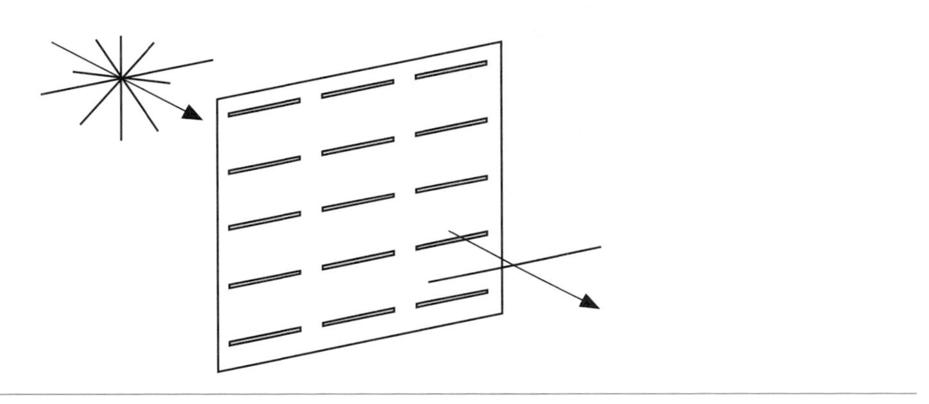

Figure 13-53 The polarizer makes the light waves vibrate in only one direction.

The **polarizers** make light waves vibrate in only one direction. Light is composed of waves that vibrate in several different directions. The polarizer converts the light into polarized light (Figure 13-53).

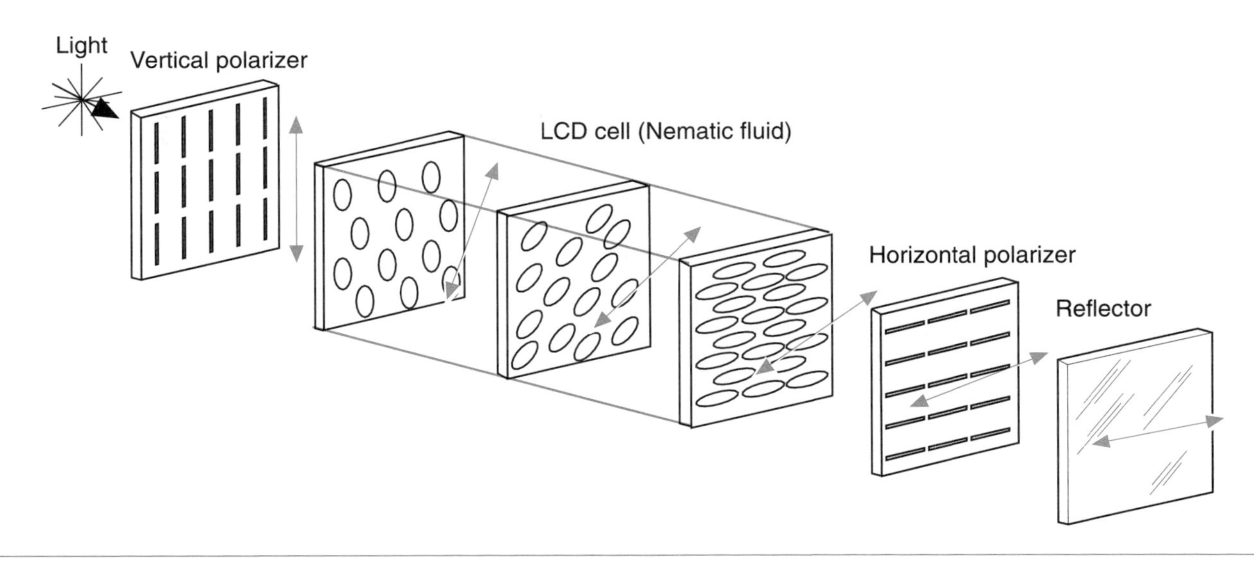

Figure 13-54 The nematic fluid rotates the polarized light wave 90 degrees.

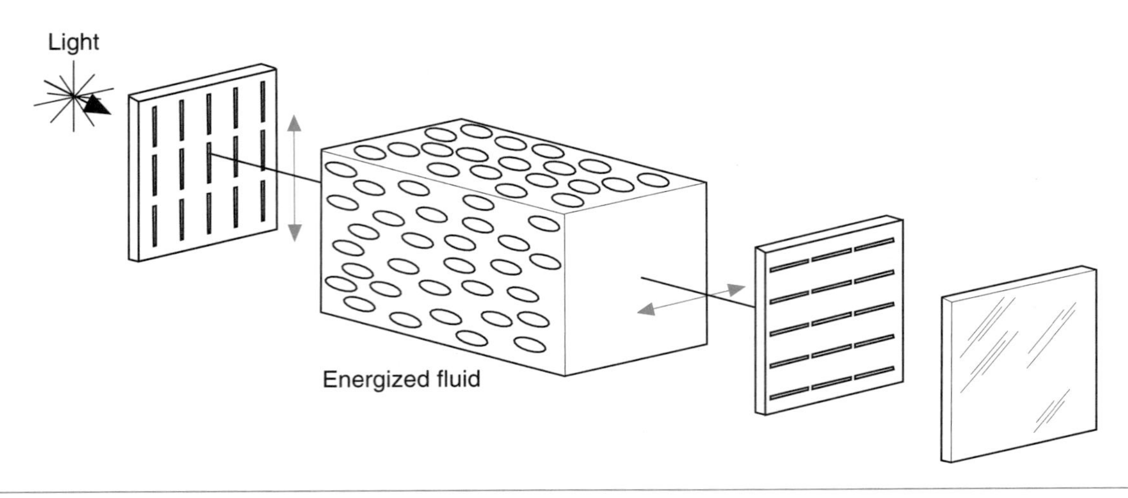

Figure 13-55 When the fluid is energized, the light wave is not rotated.

When a small square wave voltage is applied to the fluid its light slots are rearranged. The fluid will no longer rotate the light waves. The light waves leave the fluid in a vertical plain and cannot pass through the horizontal polarizer to get to the reflector (Figure 13-55). Because the light cannot be reflected back, the display appears dark. Characters are displayed by controlling which segments are dark and which segments remain light (Figure 13-56).

In most instrument panels, the LCD cluster is constantly backlit with halogen lights. But a slightly different principle is used. When the segment is not activated, the light is unable to transmit through the opaque fluid and the segment appears dark (Figure 13-57). When voltage is applied to the fluid, its light slots align and allow the light to pass to the segment. The intensity of the halogen lights is controlled through pulsewidth modulation to provide the correct illumination levels for the LCD under different ambient light conditions. A photocell is used to sense the

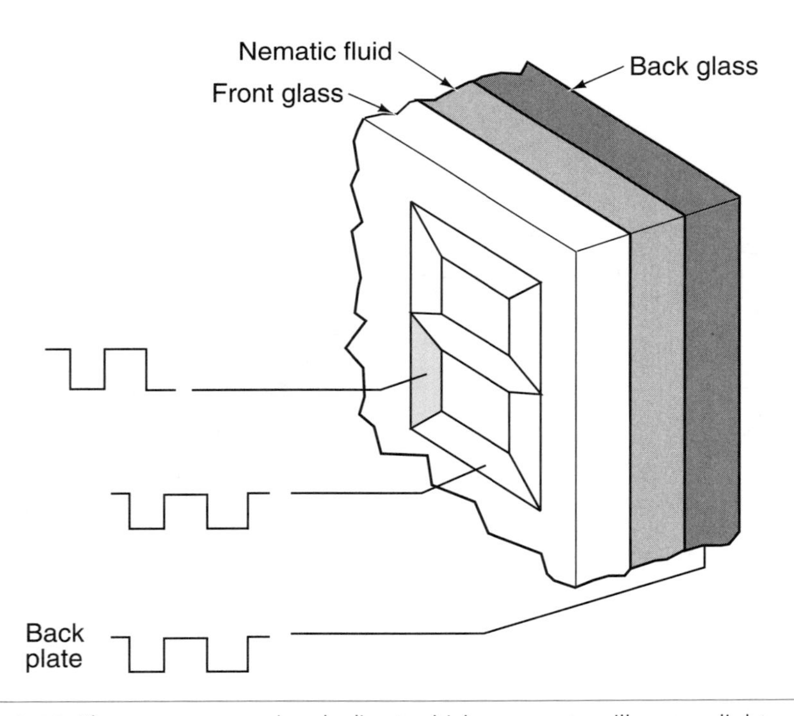

Figure 13-56 The square wave signals direct which segments will appear light and dark.

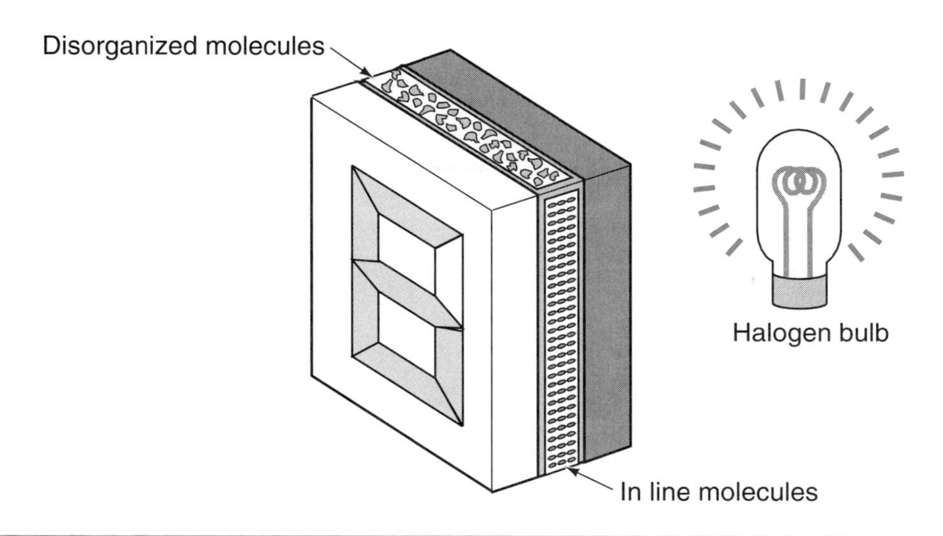

Figure 13-57 Using artificial light to display the LCD.

amount of light intensity inside the vehicle. This information is processed by the microprocessor to determine the correct light intensity for the display. Additional control of light intensity is provided by the driver through the headlight switch rheostat.

The voltage to the fluid is provided through the polarizers that have contacts with each segment in the display. The front glass has metallized shapes where the characters will be displayed. The back glass also is metallized. The color of the display is determined by filters placed in front of the display.

WARNING: LCDs operate off waveform alternating current. Do not apply DC current or the LCD will be destroyed.

Vacuum Fluorescent Displays

The vacuum fluorescent display (VFD) is constructed of a hot cathode of tungsten filaments, a grid, and a phosphorescent screen that is the anode (Figure 13-58). The components are sealed in a flat glass envelope that has been evacuated of oxygen and filled with argon or neon gas.

VFDs are the most commonly used digital display.

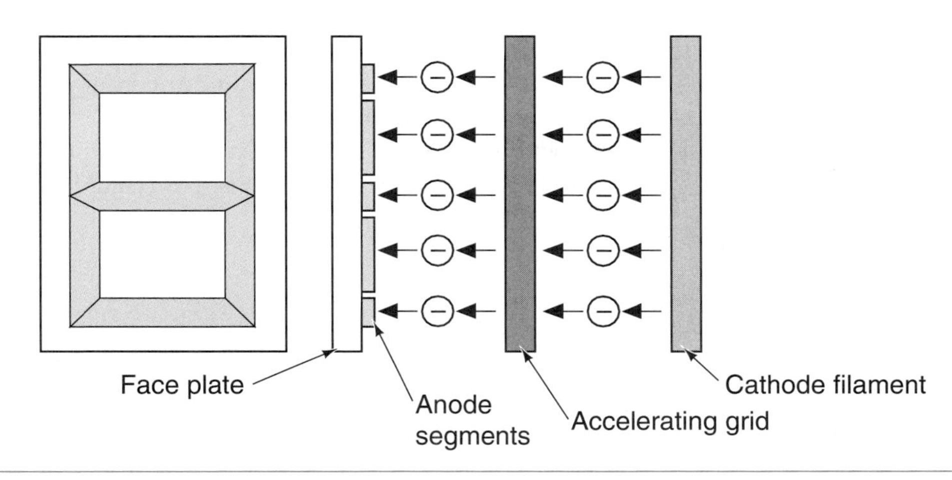

Figure 13-58 VFD construction.

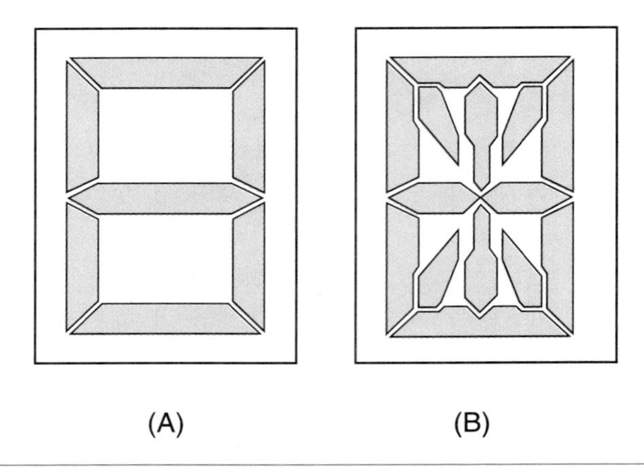

Figure 13-59 (A) Seven-segment display pattern; (B) fourteen-segment display pattern.

A constant voltage is applied to the hot cathode, which results in tungsten electrons being released from the filament wires. The grid is at a higher positive voltage than the cathode. The freed tungsten electrons are accelerated by the positive grid wires and pass through the grid to the anode. The grid ensures that the tungsten electrons will strike the anode uniformly.

The anode is at a higher positive voltage than the grid. The phosphorescent coated anode (screen) will be luminescent when the tungsten electrons strike it. The display is controlled by which segments of the screen are activated by a digital circuit. If the segment is activated, the screen will illuminate. If the segment is not activated, the electrons striking the screen have no effect on the phosphors and the screen remains off.

The segments of a VFD can be arranged in several different patterns. The most common are seven- or fourteen-segment patterns (Figure 13-59). The computer selects the sets of segments that are to emit light for the message.

The VFD display is very bright. Most manufacturers will dim the intensity of the VFD to 75% brightness whenever the headlights are turned on. To provide sufficient brightness in the daylight, with the headlights on, the headlight switch rheostat may have an additional detent to allow bright illumination of the VFD.

CRT Displays

A **cathode ray tube (CRT)** is similar to a television tube. It contains a cathode that emits electrons and an anode that attracts them. The screen will glow at the points that are hit by the electrons. Control plates dictate the direction of the electrons.

A cathode ray tube (CRT) display was first offered as standard equipment on the 1986 Buick Riviera (Figure 13-60). The screen of the CRT is touch sensitive. By touching the button on the screen, the menu can be changed to display different information. The menu-driven instrumentation brings up a screen with a "menu" of items. The driver can select a particular area of vehicle operation. The menu of items includes the radio, climate control, trip computer, and dash instrument information. The technician can access diagnostics through the CRT.

The CRT receives information from the BCM and ECM. It also provides inputs to the BCM and ECM in the form of driver commands to control the various functions (Figure 13-61).

Quartz Analog Instrumentation

Computer-driven quartz swing needle displays are similar in design to the air core electromagnetic gauges used in conventional analog instrument panels (Figure 13-62). A common application for this type of gauge is for the speedometer. However, it may be used for any gauge (Figure 13-63).

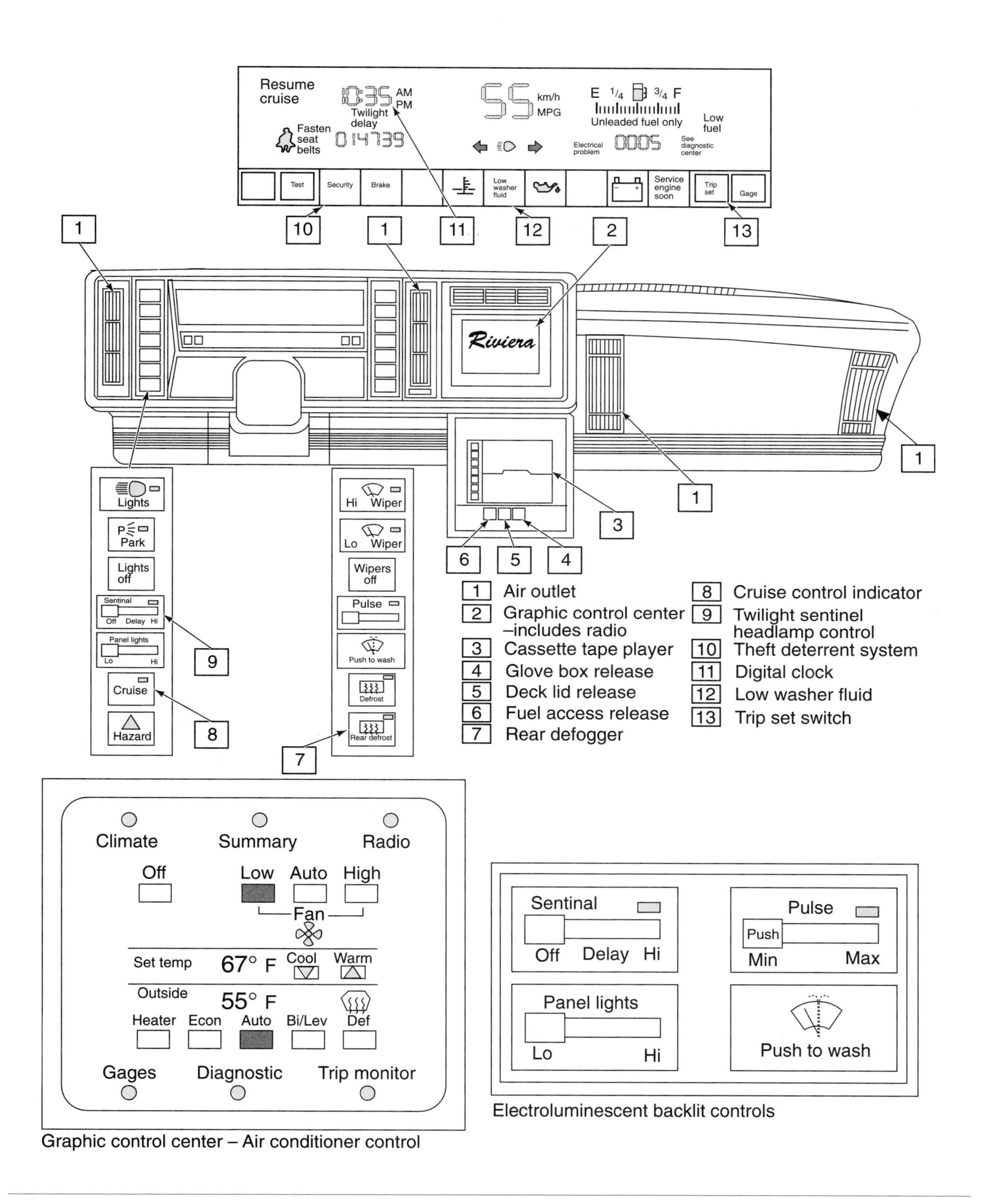

Figure 13-60 CRT display. (Courtesy of General Motors Corporation)

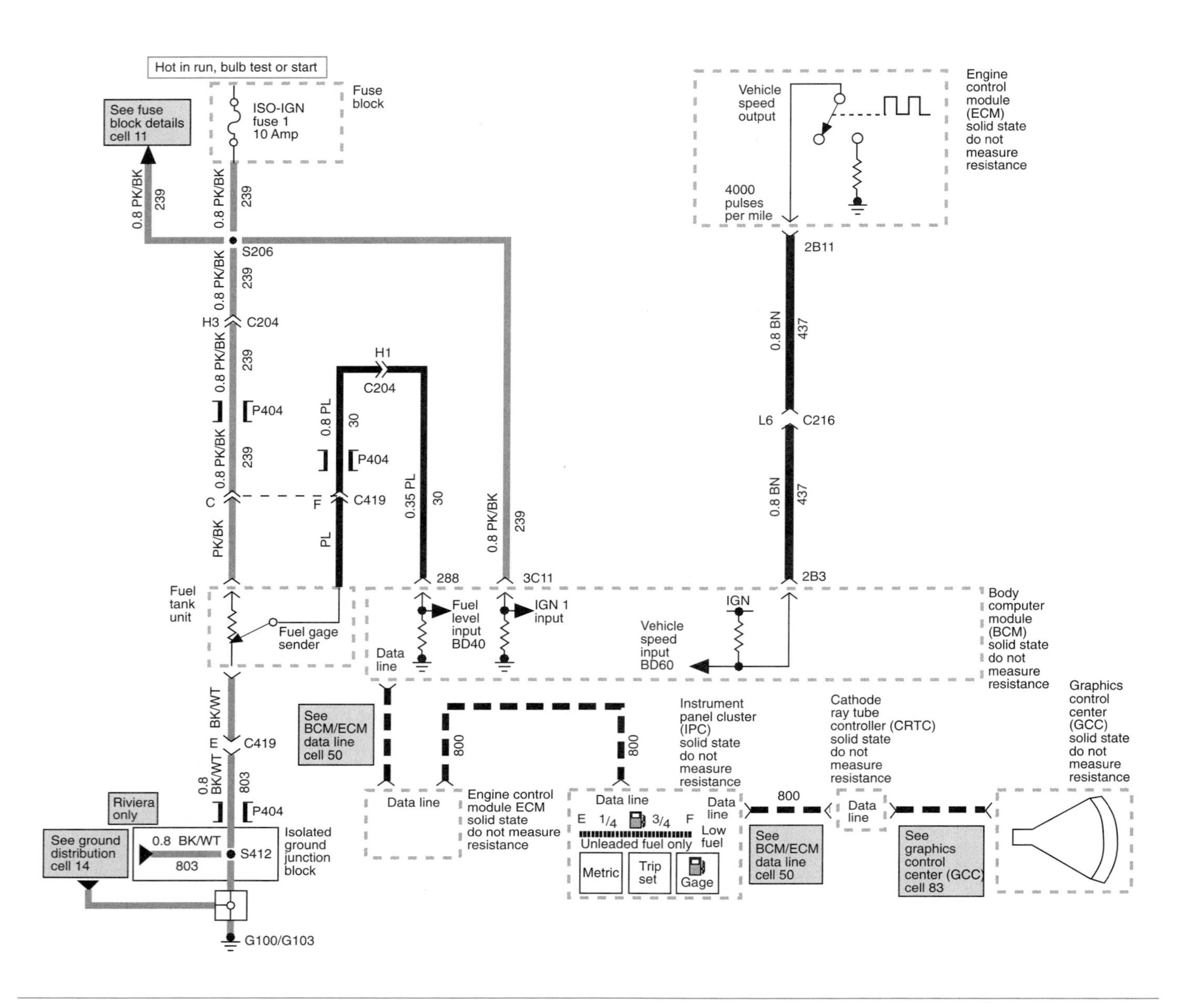

Figure 13-61 The CRT receives input information from the ECM and the BCM. (Courtesy of General Motors Corporation)

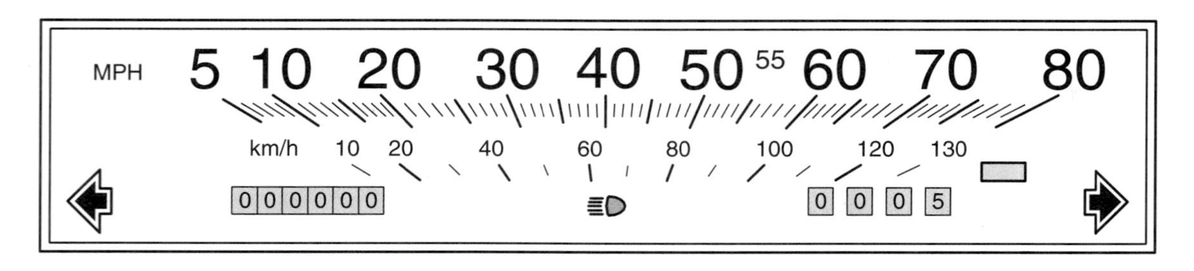

Figure 13-62 Electronic controlled swing needle speedometer. (Courtesy of General Motors Corporation)

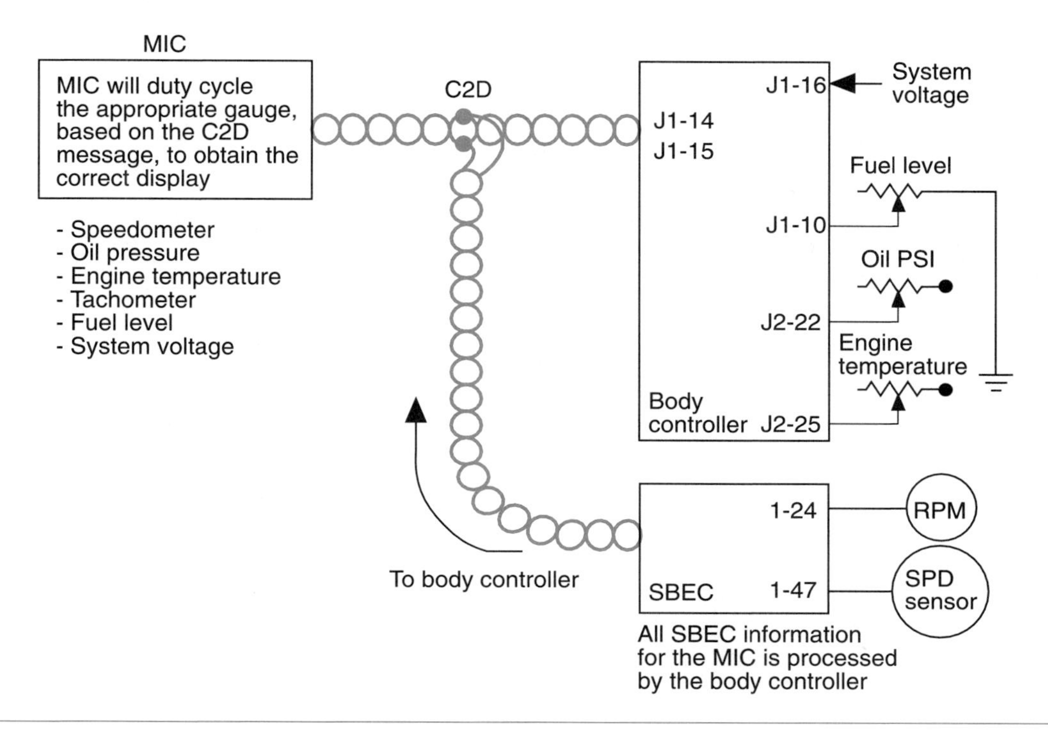

Figure 13-63 The mechanical instrument cluster receives inputs from the body and the engine controllers. (Courtesy of Chrysler Corporation)

In most speedometer gauge systems, a permanent magnet generator sensor is installed in the transaxle or transmission. As the PM generator is rotated, it causes a small AC voltage to be induced in its coil. This AC voltage signal is sent to a buffer, then to the processing unit (Figure 13-65). The signal is passed to a quartz clock circuit, a gain selector circuit, and a driver circuit. The driver circuit sends voltage pulses to the coils of the gauge, the coils operate like conventional air core gauges to move the needle.

The **buffer circuit** changes the AC voltage from the PM generator into a digitalized signal (Figure 13-64).

WARNING: The signal produced by the speed sensor can also be used by the antilock brake and the cruise control systems. If the tire size or gear ratios are changed, the operation of these systems will be affected. The speedometer must be accurately recalibrated if tires or gear ratios are changed from OE.

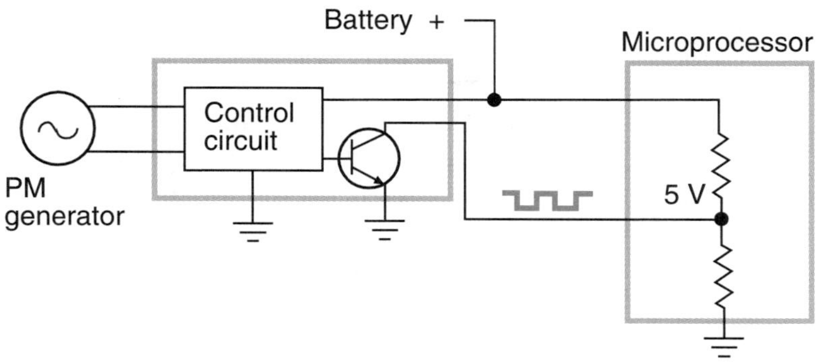

Figure 13-64 A buffer circuit.

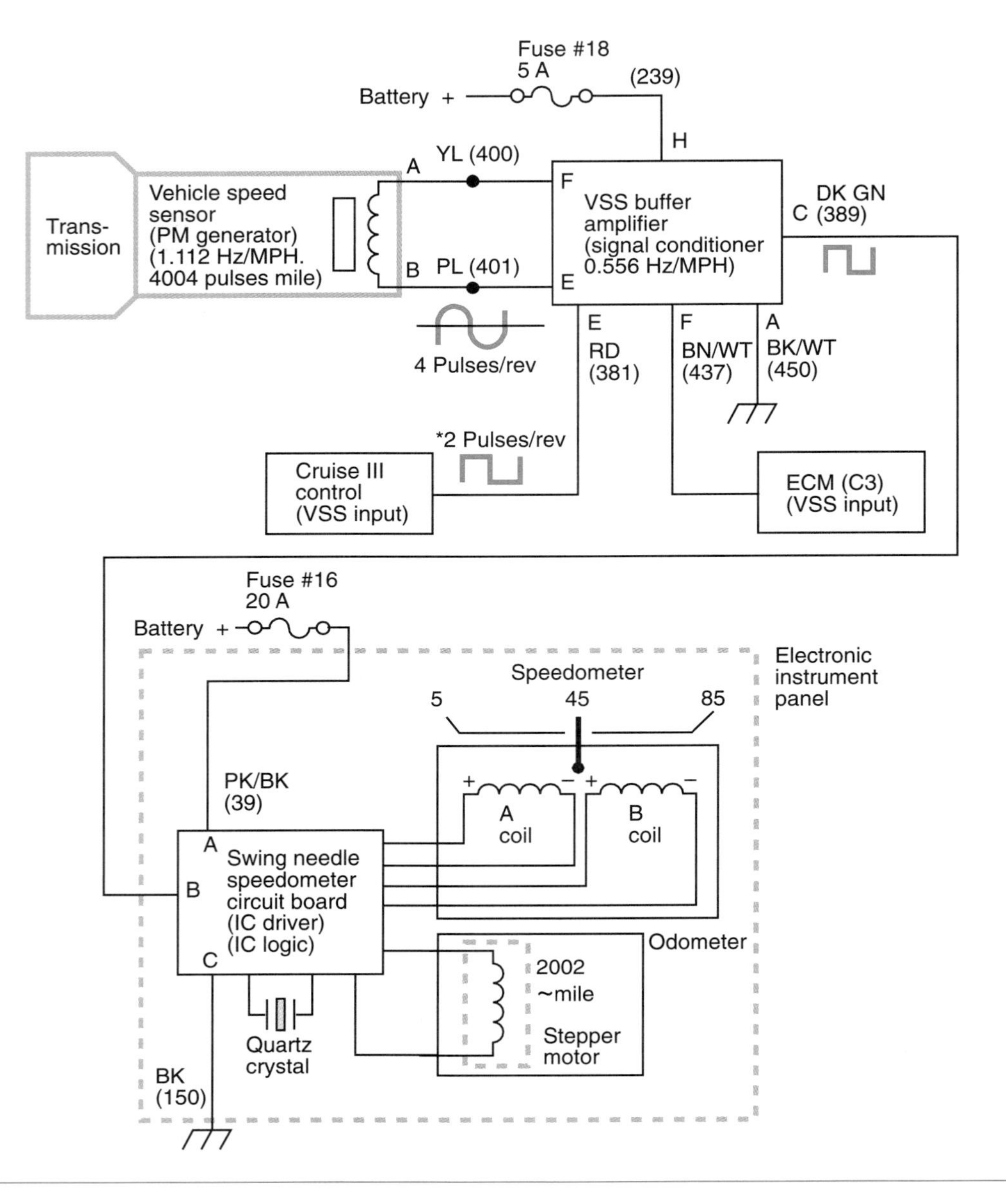

Figure 13-65 Quartz swing needle speedometer schematic. The "A" coil is connected to the system voltage, and the "B" coil receives a voltage that is proportional to input frequency. The magnetic armature reacts to the changing magnetic fields. (Courtesy of Chilton Book Company)

Digital Instrument Panel Dimming

Most conventional and computer-driven analog instrument panels use a direct input from the headlight switch rheostat to control the brightness of the display. Depending on the design and type of digital instrument panel, the display may require constant changes in its intensity so that the driver will be able to read the display under different lighting conditions.

If the vehicle is equipped with a minimum of digital displays, and an analog speedometer, the brightness level of the displays can be controlled by one input from the rheostat (Figure 13-66). In this system the body controller monitors the voltage level supplied by the headlight switch rheostat. Based on the input levels, the pulse width dimming module controls the illumination level.

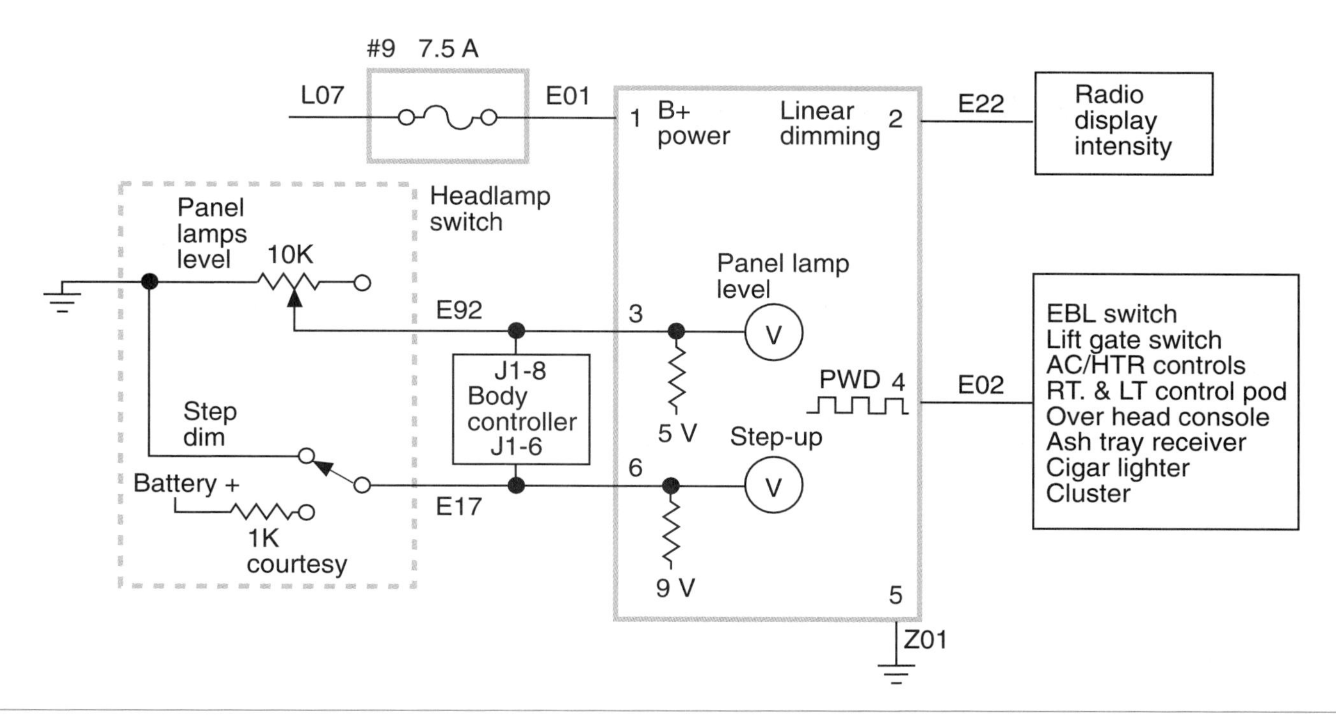

Figure 13-66 Pulse width dimming module. (Courtesy of Chrysler Corporation)

Some digital instrument panel modules also use an ambient light sensor in addition to the rheostat. The ambient sensor will control the display brightness over a 35 to 1 range and the rheostat will control over a 30 to 1 range. When the headlights are turned on, the module compares the values from both inputs and determines the illumination level. When the headlights are off, the module uses only the ambient light sensor for its input.

Head-up Display

Some manufacturers have equipped selected models with a head-up display (HUD) feature. This system displays visual images onto the inside of the windshield in the driver's field of vision (Figure 13-67). With the display located in this area, drivers do not need to remove their eyes from the road to check the instrument panel. The images are projected onto the windshield from a vacuum fluorescent light source, much like a movie projector. The HUD may project speedometer, turn signal indicator, high beam indicator, and low fuel warning displays.

Figure 13-67 The HUD displays various information onto the inside of the windshield. (Courtesy of General Motors Corporation)

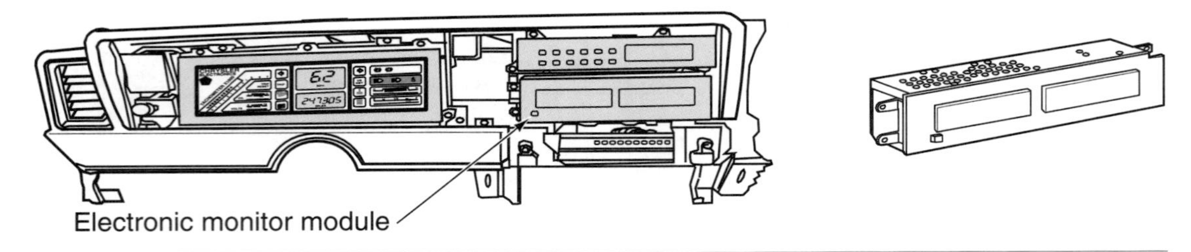

Figure 13-68 Electronic voice monitor. (Courtesy of Chrysler Corporation)

Voice Warning Systems

A **voice synthesizer** uses a computer-controlled phoneme generator capable of reproducing the phonemes used for basic speech. The computer puts the phonemes into the right combination to create words and sentences.

Some warning systems use a voice synthesizer to alert the driver of monitored conditions. The voice warning system can be a basic system that alerts the driver of about six conditions. Or, it may be very complex and monitor several functions.

Chrysler has a 24-function monitor with voice alert (Figure 13-68). This system supplements the warning indicators on the instrument panel and consists of the following components:

1. An alphanumeric readout panel.
2. A car graphic condition/location indicator.
3. An electronic voice alert module.

The alphanumeric readout panel provides a warning message to be displayed (Figure 13-69). The message is displayed until the condition is corrected. The car graphic indicator is a vehicle silhouette that is displayed when the ignition switch is in the RUN position (Figure 13-70). When

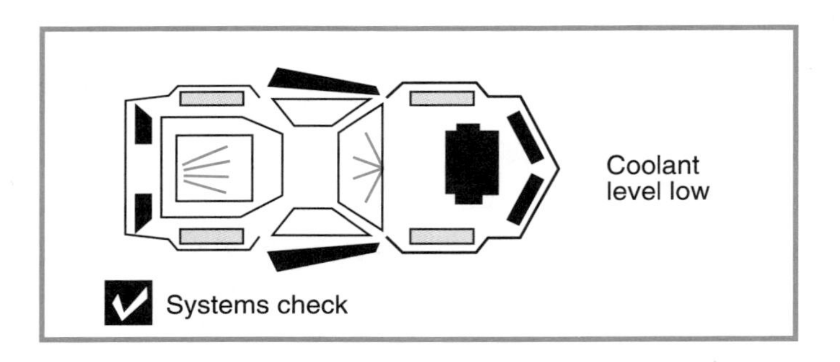

Figure 13-69 The readout panel displays a written message to alert the driver of problems. (Courtesy of Chrysler Corporation)

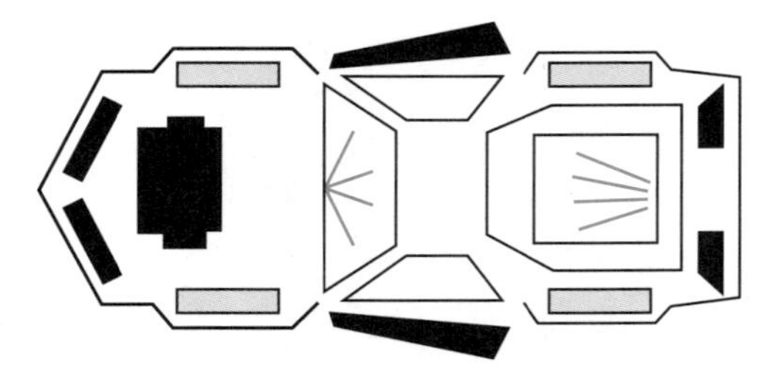

Figure 13-70 The graphic indicator light segments alert the driver to unsafe conditions. (Courtesy of Chrysler Corporation)

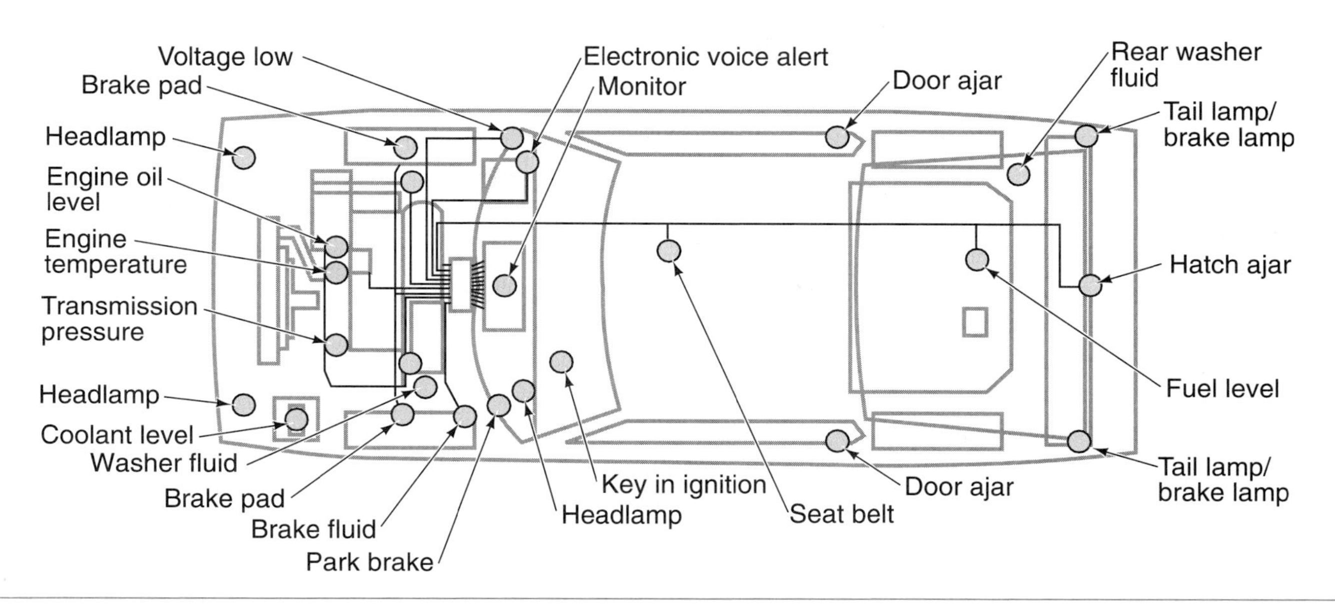

Figure 13-71 Monitored systems of the voice alert module. (Courtesy of Chrysler Corporation)

a condition occurs that requires the driver's attention, a colored indicator will be lighted and remain on until the condition is corrected. The electronic voice alert delivers a verbal message if a new warning condition is detected. The illustration (Figure 13-71) shows the monitored conditions of this system.

Sensors are placed throughout the vehicle to supply information to the microprocessor (Figure 13-72). Four different types of sensors supply information to the microprocessor:

1. Modules that monitor headlights, taillights, and stop lights for proper operation (Figure 13-73).
2. A thermistor that is used to monitor oil levels in the engine. When the temperature increases to a predetermined value due to lack of oil cooling, the microprocessor is sent a warning signal.
3. A voltage sensor that measures charging system output.
4. Normally open switches that provide ground to the microprocessor when there is a component failure or a hazardous condition. These switches are used to monitor door ajar, brake and coolant fluid levels, oil pressure, and brake pad wear.

Additional information is supplied from an internal clock, the coolant temperature sensor, the vehicle speed sensor, and the ignition system for determining engine speed.

Some faults require more than one input to trigger the voice alert. When an unsafe or harmful condition is detected by the sensors, a warning is activated. A warning message is displayed on the readout panel and a tone sounds to alert the driver. The voice alert module delivers a verbal message by interrupting the radio (if on) and delivers the message through the radio speaker closest to the driver. Some systems use a speaker that is built into the module. The illustration (Figure 13-74) shows the verbal messages, their corresponding visual message, and the conditions that initiated the warning. Most systems provide a means of silencing the voice messages if the driver chooses.

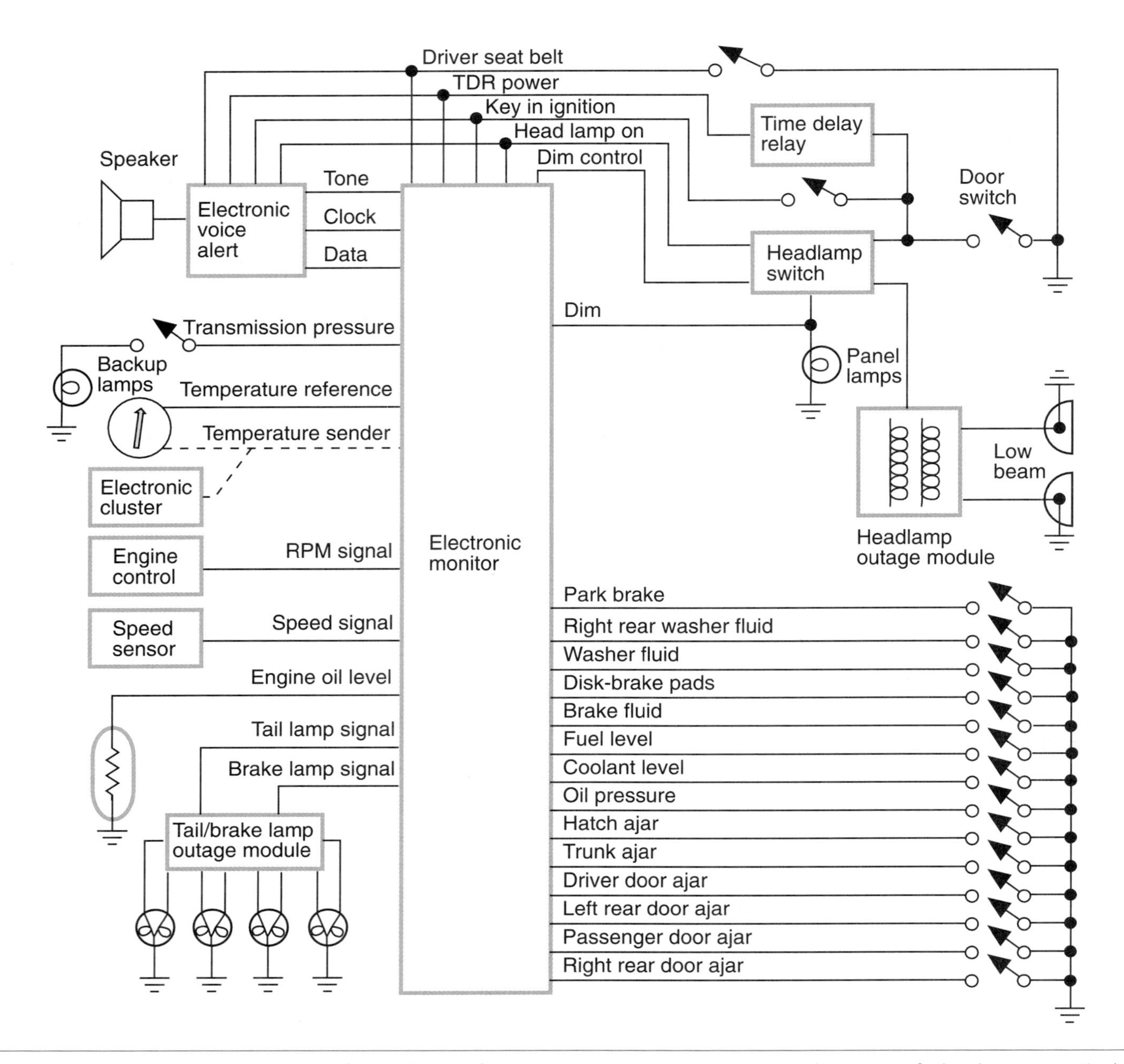

Figure 13-72 Circuit diagram of Chrysler's 24-function voice warning alert system. (Courtesy of Chrysler Corporation)

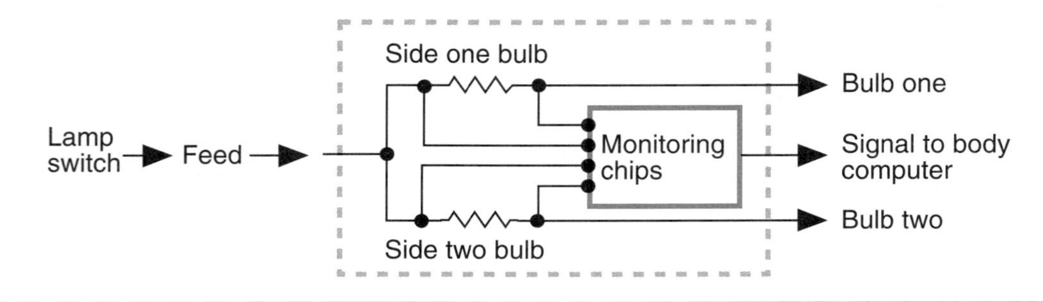

Figure 13-73 Lamp outage module detects voltage drops to determine if lamps are burned out. (Courtesy of Chrysler Corporation)

Condition
When the driver's door is opened while the ignition switch is in the OFF, LOCK, or ACC positions, the system will monitor and, if applicable, report the following:

Verbal Message	Visual Message
"Your keys are in the ignition"	KEYS IN IGNITION
"Your headlamps are on"	EXTERIOR LAMPS ON

An intermittent chime tone will follow these messages and continue until the condition is corrected, or the door is closed.

Condition
When the ignition switch is in the ON position, the system will monitor and, if applicable, report the following:

Verbal Message (1)	Visual Message (1)
"Please close your passenger door"	PASSENGER DOOR AJAR
"Please close your driver door"	DRIVER DOOR AJAR
"Please close your rear hatch"	HATCH AJAR
"Your parking brake is on"	PARK BRAKE ENGAGED

(1) These messages are delivered when a warning condition is detected after the vehicle is in motion. When the condition is corrected, a short tone will sound to acknowledge the action.

Verbal Message (2)	Visual Message (2)
"Your engine oil pressure is critical—Engine damage may occur"	LOW OIL PRESSURE

(2) If this message is delivered when the vehicle is at cruising speeds, immediate attention is required.

If the visual message appears at idle speed, and no verbal message is given, increase the idle speed and the message should go off. If the visual message remains on, and the verbal message is also delivered, immediate attention is required.

Verbal Message (3)	Visual Message (3)
"Your engine temperature is above normal" . . . Followed by . . . "Your engine is overheating —prompt service is required"	ENGINE TEMP HIGH

(3) The first message is delivered when a sensor has determined that the engine coolant is overheating. The second message is delivered approximately 30 seconds after the first if corrective action has not been taken. Immediate attention is required.

Verbal Message (4)	Visual Message (4)
"Please check your engine coolant level"	COOLANT LEVEL LOW
"Please check your fuel level"	FUEL LEVEL LOW
"Please check your brake fluid level"	BRAKE FLUID LEVEL LOW
"Please check your disc brake pads"*	DISC BRAKE PADS WORN
"Your washer fluid is low"	WASHER FLUID LOW
"Your rear washer fluid is low"	RR WASHER FLUID LOW

*Immediate Attention is Required! Dealer inspection of special disc brake pads is recommended.

(4) These messages are delivered when a continuous warning condition exists. Inspection is required. To clear these messages, even after the condition has been corrected, the ignition switch must be turned OFF.

Verbal Message (5)	Visual Message (5)
"Please check your transmission fluid level"*	LOW TRANS PRESSURE*

*Automatic Transaxle only

(5) This message is delivered when a continuous warning exists while the engine is running. Immediate attention is recommended. To clear this message, even after the condition has been corrected, the ignition switch must be turned OFF.

Verbal Message (6)	Visual Message (6)
"Your charging system is malfunctioning—prompt service is required"	VOLTAGE LOW

(6) This message is delivered when a continuous warning condition exists while the vehicle is at cruising speeds. Immediate attention is recommended. To clear this message, even after the condition has been corrected, the ignition switch must be turned OFF.

Verbal Message (7)	Visual Message (7)
"Please fasten your seat belts"	FASTEN SEAT BELTS

(7) This message is delivered if the vehicle has been moved and the driver's seat belt is not fastened.

An intermittent chime tone will also sound for several seconds if the seat belt is not fastened.

Verbal Message (8)	Visual Message (8)
"Please check your headlamp"	HEADLAMP OUT
"Please check your brake lamp"	BRAKE LAMP OUT
"Please check your tail lamp"	TAIL LAMP OUT

(8) These symptoms are constantly monitored with the lamps on or off.

If there is no warning condition to report, a short tone sounds to attract the driver's attention. The message "MONITORED SYSTEMS OK" is displayed.

Figure 13-74 Verbal and visual messages provided by Chrysler's 24-function warning voice system. (Courtesy of Chrysler Corporation)

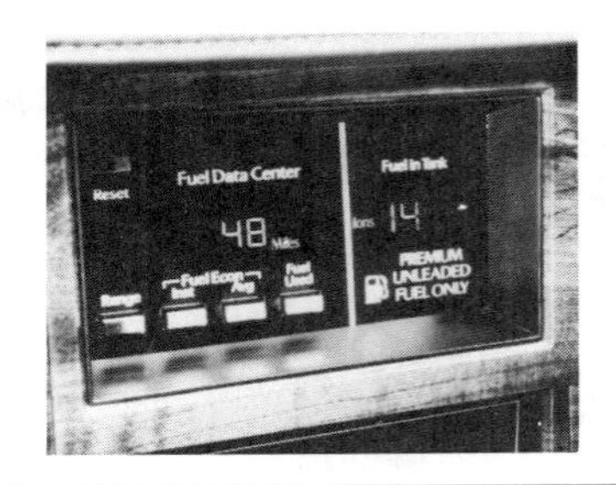

Figure 13-75 General Motors' fuel data display panel. (Courtesy of General Motors Corporation)

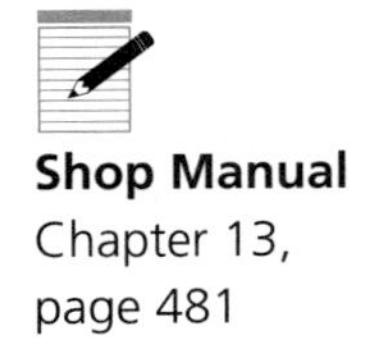

Shop Manual
Chapter 13,
page 481

Travel Information Systems

The travel information system can be a simple calculator that computes fuel economy, distance to empty, and remaining fuel (Figure 13-75). Other systems provide a much larger range of functions (Figure 13-76).

Fuel data centers display the amount of fuel remaining in the tank and provide additional information for the driver (Figure 13-77). By depressing the RANGE button, the body computer calculates the distance until the tank is empty by using the amount of fuel remaining and the average fuel economy. When the INST button is depressed, the fuel data center displays instantaneous fuel economy. The display is updated every 1/2 second and is computed by the body computer.

Depressing the AVG button displays average fuel economy for the total distance traveled since the reset button was last pushed. FUEL USED displays the amount of fuel that has been used since the last time this function was reset. The RESET button resets the average fuel economy and fuel used calculations. The function to be reset must be displayed on the fuel data center.

The system illustrated (Figure 13-78) uses the fuel sender unit and ignition voltage references as inputs for calculating fuel data functions. Other inputs can include speed and fuel flow sensors.

Figure 13-76 The deluxe driver information center provides a wide variety of information to the driver. (Courtesy of General Motors Corporation)

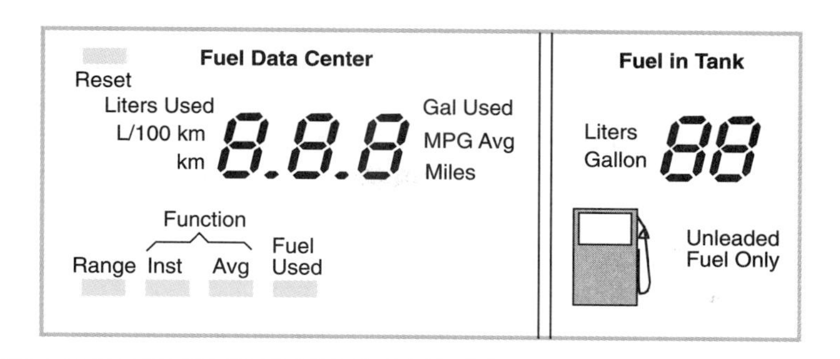

Figure 13-77 Fuel data center. (Courtesy of General Motors Corporation)

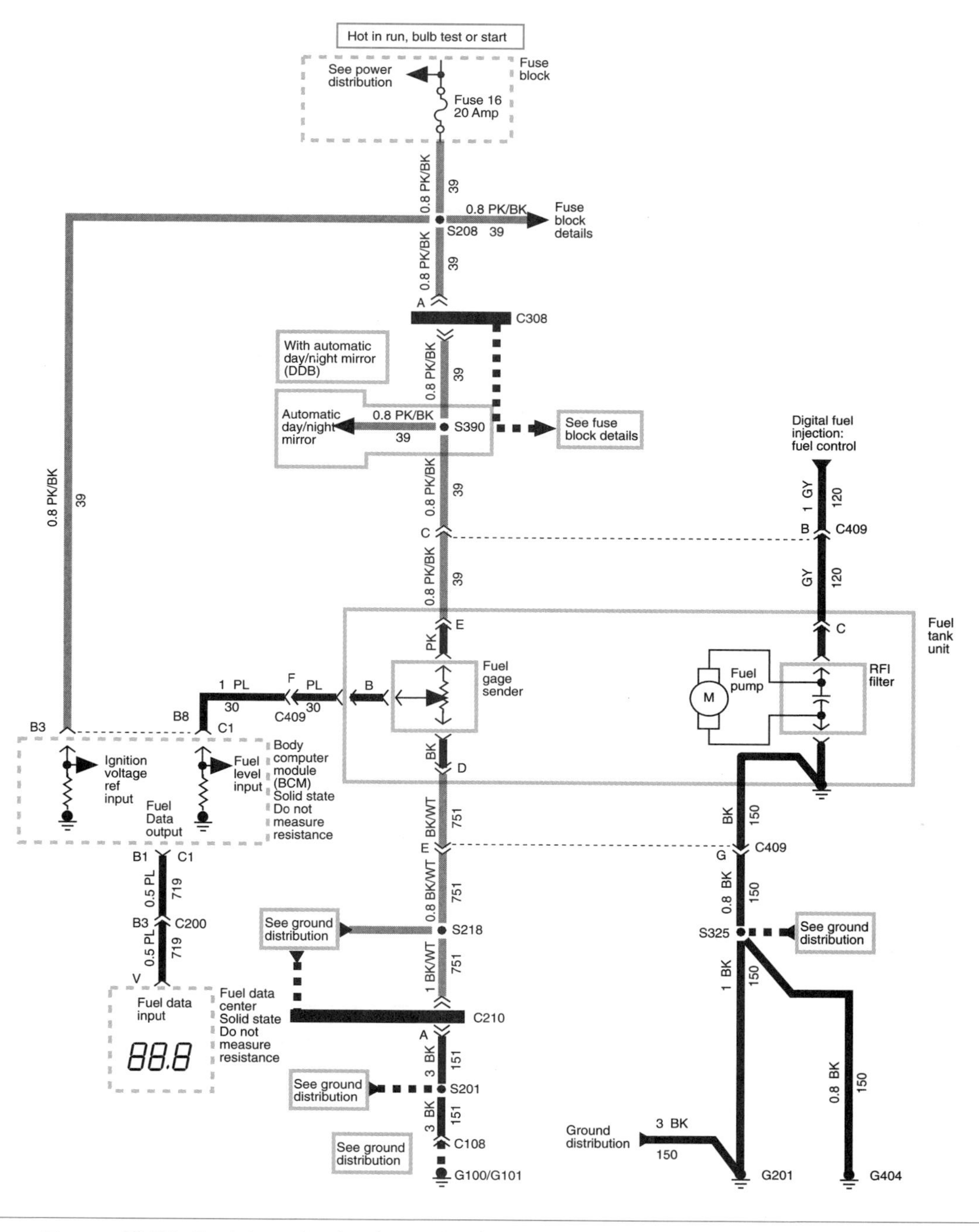

Figure 13-78 Fuel data circuit schematic. (Courtesy of General Motors Corporation)

Deluxe systems may incorporate additional features such as outside temperature, compass, elapsed time, estimated time of arrival, distance to destination, day of the week, time, and average speed. The illustration (Figure 13-79) shows the inputs that are used to determine many of these functions. Fuel system calculations are determined by the sensors shown (Figure 13-80). Injector on time and vehicle speed pulses are used to determine the amount of fuel flow. Some manufacturers use a fuel flow sensor that provides pulse information to the microprocessor concerning fuel consumption (Figure 13-81).

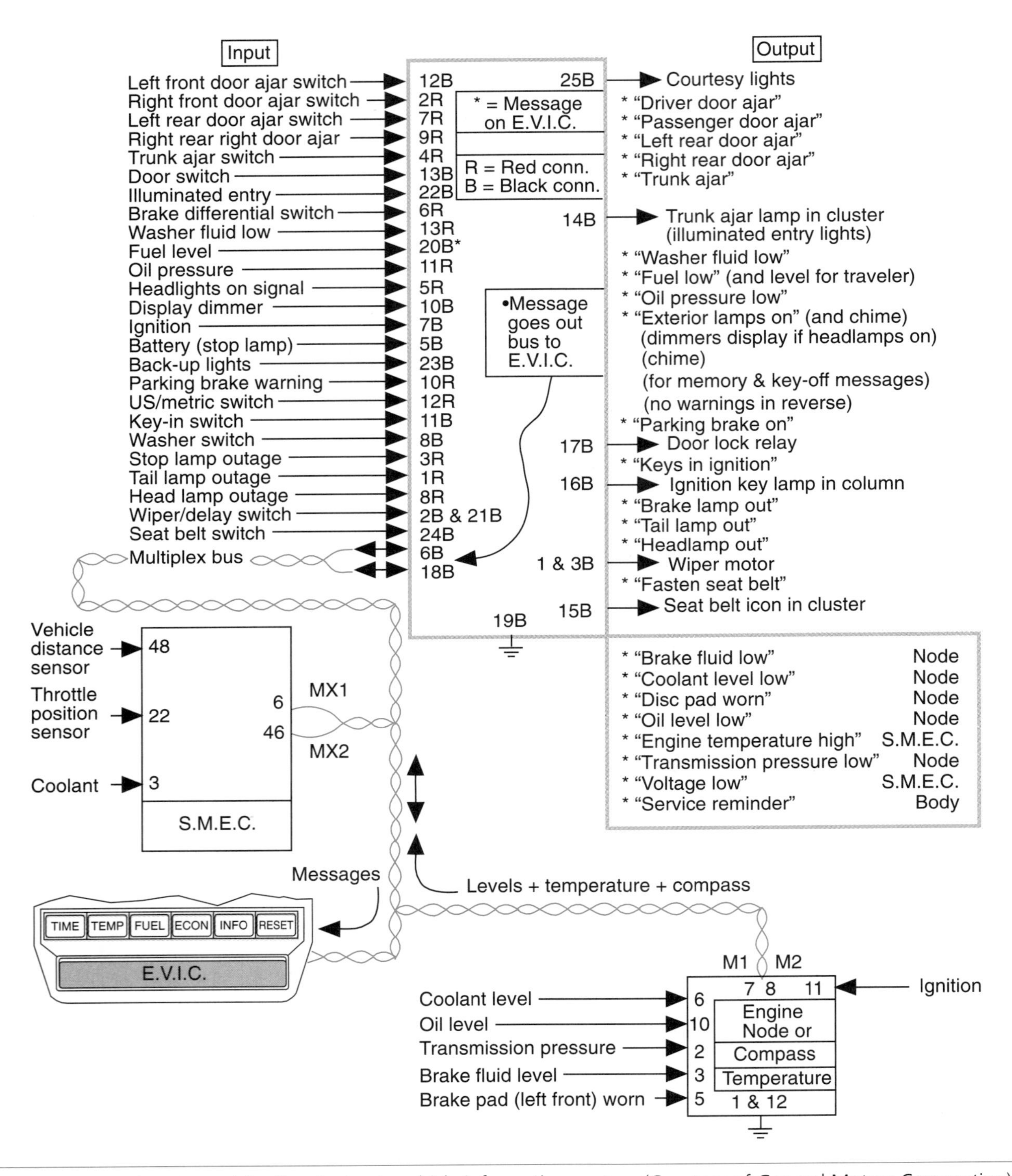

Figure 13-79 Inputs used for the electronic vehicle information center. (Courtesy of General Motors Corporation)

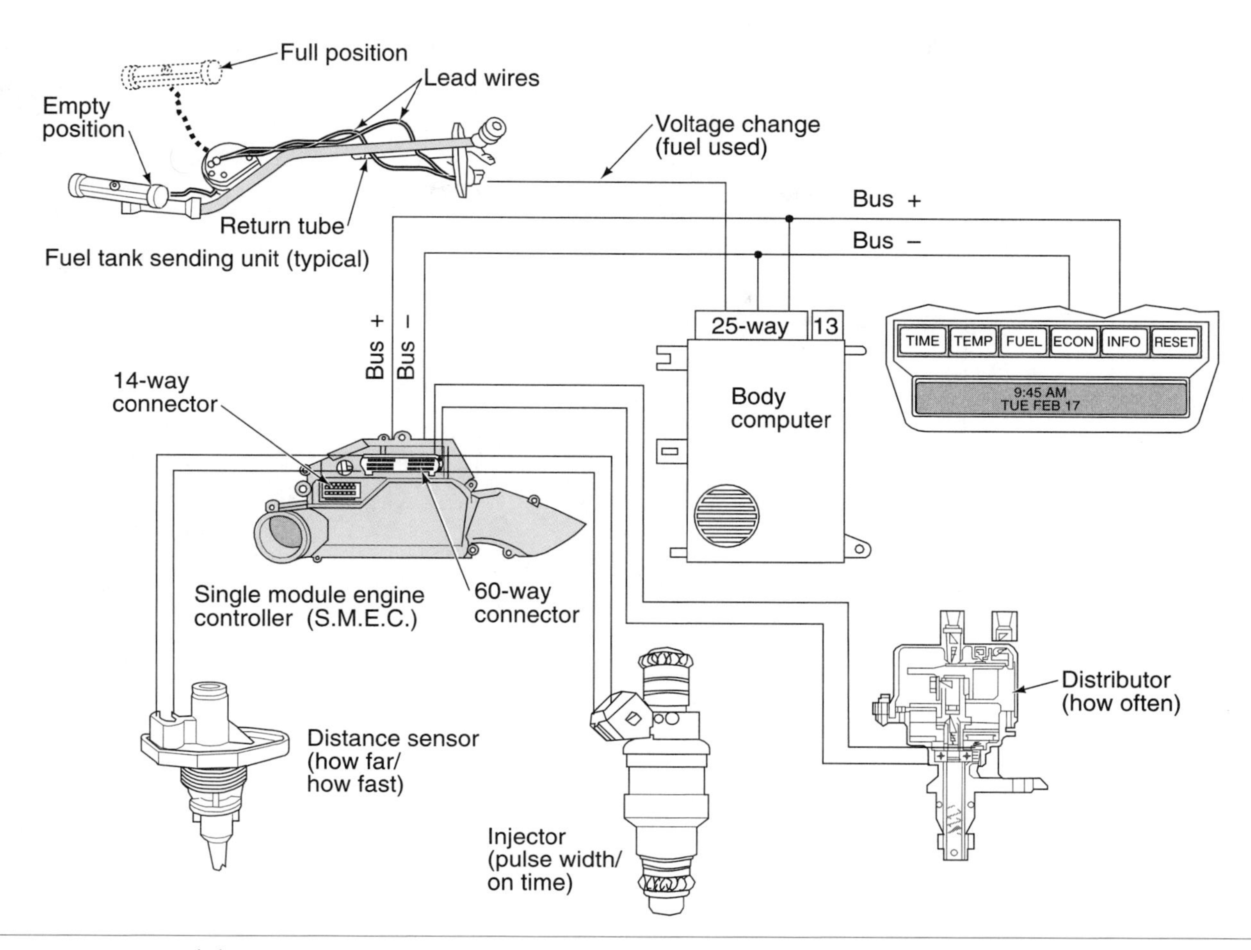

Figure 13-80 Fuel data system inputs. The injector on time is used to calculate the rate of fuel flow. (Courtesy of Chrysler Corporation)

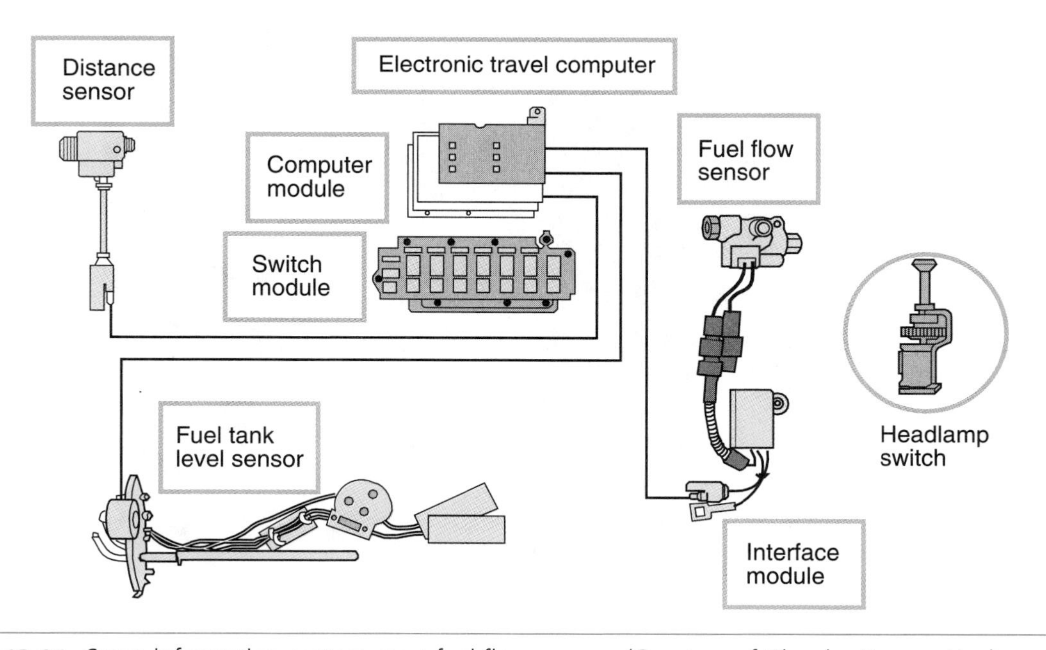

Figure 13-81 Some information centers use a fuel flow sensor. (Courtesy of Chrysler Corporation)

Figure 13-82 Ford's Route Guidance System. (Reprinted with the permission of Ford Motor Company)

In the near future cars will be equipped with navigation systems (Figure 13-82). These computerized route guidance systems will help drivers to their destinations. Using information broadcast by satellites, these systems will display precise road maps and directions to get anywhere on earth. The display unit is mounted to the instrument panel. The information received from the satellites is combined with information stored on a compact disc by the navigational computer. After the driver inputs the intended destination, the system displays the immediate route to get there (Figure 13-83).

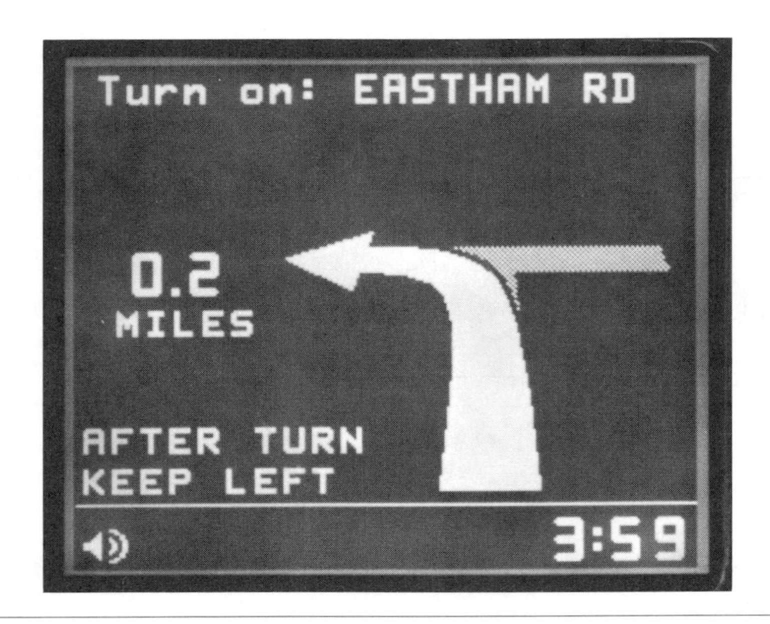

Figure 13-83 Immediate directions are displayed by Ford's Route Guidance System. (Reprinted with the permission of Ford Motor Company)

Summary

- ❑ Most automatic headlight dimming systems consist of a light sensitive photocell and amplifier unit, high-low beam relay, sensitivity control, dimmer switch, flash-to-pass relay, and a wiring harness.
- ❑ The automatic on/off with time delay has two functions: to turn on the headlights automatically when ambient light decreases to a predetermined level and to allow the headlights to remain on for a certain amount of time after the vehicle has been turned off.
- ❑ Fiber optics is the transmission of light through polymethylmethacrylate plastic that keeps the light rays parallel even if there are extreme bends in the plastic.
- ❑ The lamp outage indicator alerts the driver, through an information center on the dash or console, that a light bulb has burned out.
- ❑ Digital instrument clusters use digital and linear displays to notify the driver of monitored system conditions.
- ❑ The most common types of displays used on electronic instrument panels are light emitting diodes (LEDs), liquid crystal displays (LCDs), vacuum fluorescent displays (VFDs), and a Cathode ray tube (CRT)..
- ❑ Computer-driven quartz swing needle displays are similar in design to the air core electromagnetic gauges used in conventional analog instrument panels.
- ❑ A head-up display system displays visual images onto the inside of the windshield in the driver's field of vision.
- ❑ A voice synthesizer uses a computer-controlled phoneme generator that is capable of reproducing the phonemes used for basic speech.

Terms to Know

Automatic headlight dimming
Automatic on/off with time delay
Buffer circuit
Cathode Ray Tube (CRT)
Digital instrument clusters
Fiber optics
H-gate
Illuminated entry
Illuminated entry actuator
Instrument panel dimming
International Standards Organization (ISO)
Lamp outage indicator
Lamp outage module
Light Emitting Diode (LED)
Liquid Crystal Displays (LCDs)
Matrix
Oscillator
Photocell
Polarizers
Quartz analog instrumentation
Sensitivity control
Stepper motor
Timer control
Vacuum Fluorescent Display (VFD)
Wake-up signal

Review Questions

Short Answer Essays

1. Describe the operation of computer-controlled concealed headlight systems.
2. List the common components of the automatic headlight system.
3. Explain the operation of body computer-controlled instrument panel illumination dimming.
4. What is the function of the sensitivity control in the automatic dimmer system?
5. What is the basic operation of the illuminated entry system?
6. Describe the operating principles of the digital speedometer.
7. Explain the operation of IC chip-type odometers.
8. Describe the operation of the electronic fuel gauge.
9. Describe the operation of quartz analog speedometers.
10. What is meant by pulse width dimming?

Fill-in-the-Blanks

1. With body computer-controlled concealed headlights, the computer receives inputs from the ________________ and ________________ switches.

2. The sensitivity control used with automatic dimming sets the sensitivity at which the photocell and amplifier are ________________ .

3. The photocell will have ________________ resistance as the ambient light level increases.

4. In some illuminated entry systems, ________________ ________________ signals the body computer that the courtesy lights are not required.

5. The body computer uses inputs from the ________________ ________________ ________________ and ________________ to determine the illumination level of the instrument panel lights.

6. The body computer dims the illumination lamps by using a ________________ ________________ ________________ signal to the panel lights.

7. Digital instrument clusters use ________________ and ________________ displays to notify the driver of monitored system conditions.

8. Some digital instrument panel modules also use an ________________ ________________ sensor in addition to the rheostat.

9. Most digital fuel gauges use a fuel level sender that ________________ resistance value as the fuel level decreases.

10. Computer-driven quartz swing needle displays are similar in design to the ________________ ________________ electromagnetic gauges used in conventional analog instrument panels.

ASE Style Review Questions

1. The sensitivity control of the automatic headlight dimming system is being discussed.
   Technician A says decreasing the sensitivity means the headlights will switch to the low beams when the approaching vehicle is farther away.
   Technician B says increasing the sensitivity means the headlights will switch to the low beams when the approaching vehicle is closer.
   Who is correct?
   A. A only
   B. B only
   C. Both A and B
   D. Neither A nor B

2. Illuminated entry is being discussed:
   Technician A says when either of the front door handles are lifted a switch in the handle will close the ground path from the actuator.
   Technician B says if the ignition switch is placed in the RUN position before the timer circuit turns off the interior lights, the timer sequence is shut off and the interior lights turn off.
   Who is correct?
   A. A only
   B. B only
   C. Both A and B
   D. Neither A nor B

3. Computer-controlled instrument panel dimming is being discussed:
   Technician A says the body computer dims the illumination lamps by varying resistance through a rheostat that is wired in series to the lights.
   Technician B says the body computer can use inputs from the panel dimming control and photocell to determine the illumination level of the instrument panel lights on certain systems.
   Who is correct?
   A. A only
   B. B only
   C. Both A and B
   D. Neither A nor B

4. Fiber optic applications are being discussed:
   Technician A says fiber optics is the transmission of light through several plastic strands that are sheathed by a polymer.
   Technician B says fiber optics are used only in external lighting applications.
   Who is correct?
   A. A only
   B. B only
   C. Both A and B
   D. Neither A nor B

5. Computer-driven instrumentation is being discussed:
   Technician A says a computer-driven instrument panel uses a microprocessor to process information from various sensors and to control the gauge display.
   Technician B says some manufacturers use the body computer to perform all functions.
   Who is correct?
   A. A only
   B. B only
   C. Both A and B
   D. Neither A nor B

6. The IC chip odometer is being discussed:
   Technician A says if the chip fails, some manufacturers provide for replacement of the chip.
   Technician B says depending on the manufacturer, the new chip may be programmed to display the last odometer reading.
   Who is correct?
   A. A only
   B. B only
   C. Both A and B
   D. Neither A nor B

7. Computer-driven quartz swing needle displays are being discussed:
   Technician A says the "A" coil is connected to system voltage and the "B" coil receives a voltage that is proportional to input frequency.
   Technician B says the quartz swing needle display is similar to air core electromagnetic gauges.
   Who is correct?
   A. A only
   B. B only
   C. Both A and B
   D. Neither A nor B

8. *Technician A* says digital instrumentation displays an average of the readings received from the sensor.
   Technician B says conventional analog instrumentation gives more accurate readings but is not as decorative.
   Who is correct?
   A. A only
   B. B only
   C. Both A and B
   D. Neither A nor B

9. The microprocessor-initiated self check of the electrical instrument cluster is being discussed.
   Technician A says during the first portion of the self test all segments of the speedometer display are lit.
   Technician B says the display should not go blank during any part of the self test.
   Who is correct?
   A. A only
   B. B only
   C. Both A and B
   D. Neither A nor B

10. *Technician A* says bar graph-style gauges do not provide for self tests.
    Technician B says the digital instrument panel will display "CO" to indicate the circuit is shorted.
    Who is correct?
    A. A only
    B. B only
    C. Both A and B
    D. Neither A nor B

Chassis Electronic Control Systems

Upon completion and review of this chapter, you should be able to:

- ❑ Explain the purpose of the automatic temperature control.
- ❑ Explain the purpose and operation of the control assembly in the SATC and EATC systems.
- ❑ List and describe the types of sensors used in SATC and EATC systems.
- ❑ Explain the differences in operation between semiautomatic temperature control and electronic automatic temperature control systems.
- ❑ Define the purpose of the cruise control system.
- ❑ Explain the basic operation of electromechanical cruise control systems.
- ❑ Explain the operating principles of the electronic cruise control system.
- ❑ Explain the purpose and operation of the safety switches used on electromechanical and electronic cruise control systems.
- ❑ Explain the purpose of passive restraint systems.
- ❑ Describe the basic operation of automatic seatbelts.
- ❑ List and explain the function of the components of the air bag module.
- ❑ Explain the functions of the diagnostic module used in air bag systems.
- ❑ Describe the operation of air bag system sensors.
- ❑ Explain the basic operating principles of the emergency tensioning retractor.
- ❑ Explain the operating principles of the memory seat feature.
- ❑ Describe the control concepts of electronically controlled sunroofs.
- ❑ Detail the operation of common antitheft systems.
- ❑ Explain the function of the pass-key security system.
- ❑ Explain the purpose and operation of automatic door lock systems.
- ❑ Detail the operation of the keyless entry system.
- ❑ Explain the operating principles of Ford's and GM's heated windshield systems.
- ❑ Explain the purpose of electronic shift transmissions.
- ❑ Explain the purpose of variable assist steering.
- ❑ Describe the purpose of electronic suspension systems.
- ❑ State the purpose of antilock braking systems.
- ❑ Describe the basic purpose of automatic traction control systems.
- ❑ Describe the purpose and configuration of vehicle sound systems.

Introduction

In this chapter you will learn the operation of the semiautomatic and automatic temperature control systems, cruise control systems, automatic passive restraint systems, and air bag systems. Today's technician will be required to service these systems. The comfort and safety of the driver and/or passengers depend on the technician properly diagnosing and repairing these systems. As with all electrical systems, the technician must have a basic understanding of the operation of these systems before attempting to perform any service.

There are many safety cautions to observe when working on air bag systems. Safe service procedures are accomplished through proper use of the service manual and by understanding the operating principles of these systems.

This chapter also discusses the many electrical accessory systems that have electronic controls added to them to provide additional features and enhancement. These accessories include memory seats, electronic sunroofs, antitheft systems, automatic door locks, keyless entry, and electronic heated windshields.

Introduction to Semiautomatic and Electronic Automatic Temperature Control

Automatic air conditioning systems operate with the same basic components as the conventional systems. The major difference is the automatic temperature control (ATC) system is capable of maintaining a preset level of comfort as selected by the driver. Sensors are used to determine the present temperatures and the system can adjust the level of heating or cooling as required.

The system uses actuators that will open and close air-blend doors to achieve the desired in-vehicle temperature. Some systems will control fan motor speeds to keep the temperature very close to that requested by the driver.

There are two types of automatic temperature control: semiautomatic temperature control (SATC) and electronic automatic temperature control (EATC). The basic difference in the two systems is in self-diagnostic capabilities. Most SATC systems do not provide for storing of trouble codes. EATC systems monitor system operation and set codes in a RAM module for diagnostic use. Other differences include actuator types and the number of sensors used.

Though the systems differ in methods of operation, they are all designed to provide in-car temperatures and humidity conditions at a preset level. The in-car humidity and temperature levels are maintained, regardless of the climate conditions outside the vehicle. The in-car humidity level is maintained at 45 to 55%.

A BIT OF HISTORY

In 1939, Packard introduced a car with air-conditioning. It used refrigeration coils in an air duct behind the rear seat.

Semiautomatic Temperature Control

Shop Manual
Chapter 14, page 488

Basic SATC systems are not much different than manual systems. The primary difference is in the use of a programmer, electric servomotor, and/or control module to operate the actuators. The SATC system maintains a driver-selected comfort level by sensing air temperature and blend-air door positions through the programmer. The driver selects the operating mode and blower speeds manually from the control head.

Common Components

Not all systems will have all of the components described here, but most will have a combination of several of them.

Control Assembly. The control assembly used in SATC systems is similar to that used in manual systems. The main difference is the temperature control has a temperature range imprinted under it or displayed (Figure 14-1). The control assembly is located in the instrument panel and provides the means for driver input to the climate control system. The temperature control lever is a part of a sliding resistor that mechanically converts the lever setting into an electrical resis-

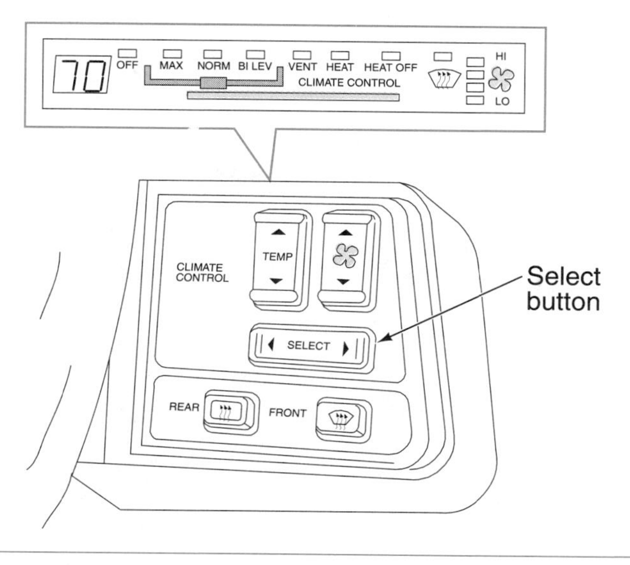

Figure 14-1 Semi-automatic temperature control assembly. (Courtesy of Chrysler Corporation)

The **control assembly** provides for driver input into the automatic temperature control microprocessor. The control assembly is also referred to as the control panel.

tance. The driver selects operating modes (A/C, heat, defogger, and vent) and fan motor speeds through push button selection on the control assembly.

Programmer. The push buttons of the control assembly will input a programmer that directs vacuum to actuators at the air distribution doors (except the blend-air door, which is usually controlled by a servomotor). The programmer also receives electrical input from the in-car and exterior ambient temperature sensors. Based on the inputs from the sensors and the control assembly, the programmer provides output signals to turn the compressor clutch on/off, open/close the heater water valve, and position the mode doors.

The **programmer** controls the blower speed, air mix doors, and vacuum motors of the SATC system. Depending on manufacturer, they are also called servo assemblies.

Sensors. There can be several different sensors used on the SATC system. The most common are the in-car and ambient temperature sensors.

Most **aspirators** are tubular devices that use a venturi effect to draw air from the passenger compartment over the in-car sensor (Figure 14-2). Some manufacturers use a suction motor to draw the air over the sensor.

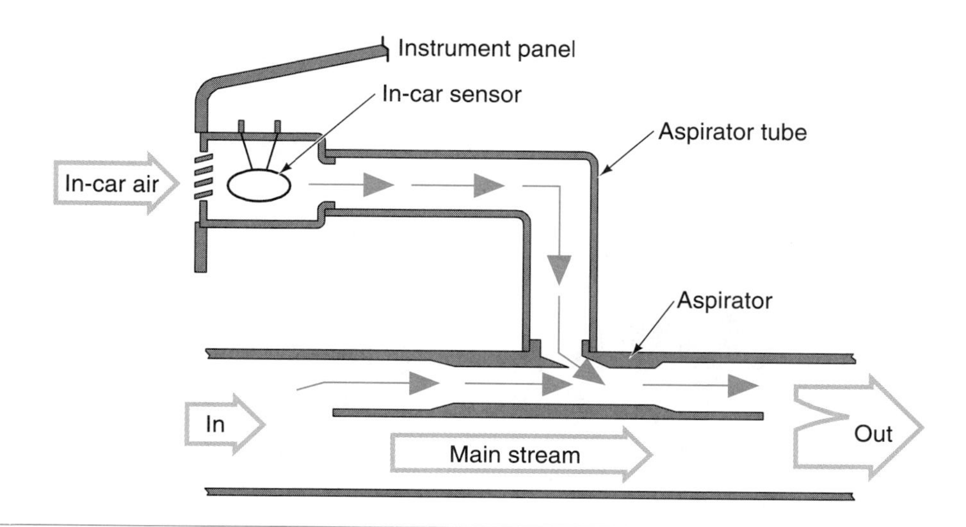

Figure 14-2 A typical aspirator. The main air stream creates a low pressure at the inlet of the aspirator drawing in-car air over the sensor.

Figure 14-3 The in-car sensor is an NTC thermistor located in the aspirator unit.

The in-car sensor contains a temperature sensing NTC thermistor to measure the average temperature inside the vehicle (Figure 14-3). The in-car sensor is located in the aspirator unit.

The ambient sensor is an NTC thermistor used to measure the temperature outside the vehicle (Figure 14-4). The ambient sensor is usually located behind the grill. Due to its location,

Figure 14-4 Ambient temperature sensor.

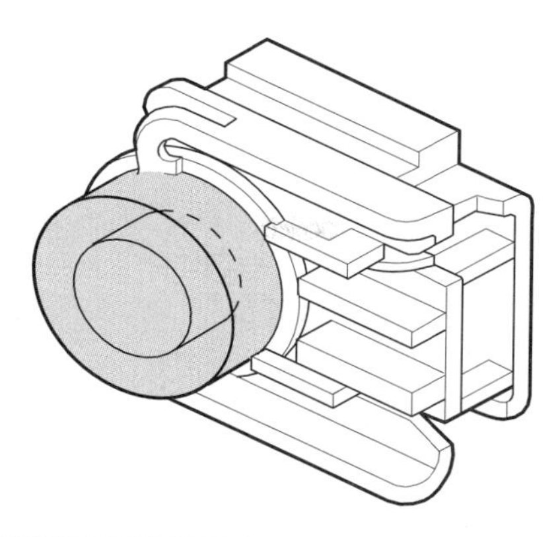

Figure 14-5 The sunload sensor produces a signal proportional to the heat intensity of the sun's heat through the windshield.

and possible influence by engine temperatures, the sensor circuit has several memory features that prevent false input.

Some systems use a sunload sensor (Figure 14-5). The sunload sensor is a photovoltaic diode that sends signals to the programmer concerning the extra generation of heat as the sun beats through the windshield. This sensor is usually located on the dash next to a speaker grill.

Photovoltaic diodes are capable of producing a voltage when exposed to radiant energy.

Electronic Automatic Temperature Control

Shop Manual Chapter 14, page 496

The determining factor separating electronic automatic temperature control (EATC) systems from SATC systems is the ability for self-diagnostics. The body control module (BCM) will set trouble codes that can be accessed by the technician. In addition, the EATC system provides a continuously variable blower speed signal and readjusts interior temperature several times a second.

In addition to the sensors used in the SATC system, the EATC system may also use engine coolant temperature, vehicle speed, and throttle position sensors as inputs. The EATC system may also incorporate a cold engine lock-out switch.

Cold engine lock-out switches signal the BCM or controller to prevent blower motor operation until the air entering the passenger compartment reaches a specified temperature.

There are two categories of EATC systems; BCM-controlled systems and stand-alone systems that use their own computer.

BCM-Controlled Systems

Components

Climate control is one of the primary functions of the BCM on many GM vehicles. The body computer module has central control over the EATC system. The BCM monitors all system sensors and switches, compares the data with programmed instructions, and commands the actuators to provide accurate control of the system.

Shop Manual Chapter 14, page 504

The climate control panel (CCP) contains a circuit board that translates driver inputs into electrical signals (Figure 14-6). The CCP and BCM communicate with each other over a data circuit. The information sent to the CCP can be displayed by the vacuum fluorescent display.

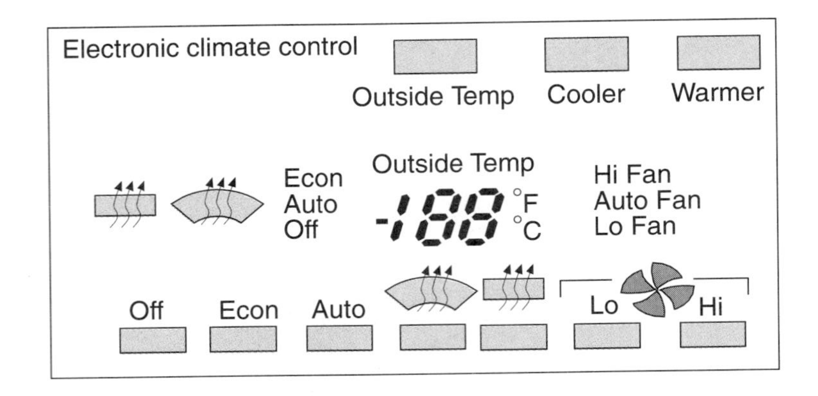

Figure 14-6 Climate control panel. (Courtesy of General Motors Corporation)

The BCM monitors the air conditioning system through several sensors. High side temperature is monitored through a sensor in the pressure line. By monitoring the high side temperature, the BCM is capable of making calculations that translate into pressure. The calculations are based on the pressure-temperature relationship of R-12 or R-134a. The BCM will also monitor low side pressure in the same manner through a low side temperature sensor. The BCM also receives a signal from the low pressure switch in the accumulator. The BCM will shut down the compressor clutch if system operation is not within set parameters.

The programmer uses a bidirectional motor to adjust the blend door. A potentiometer feedback signal is used by the BCM to determine blend door position. The programmer has five vacuum solenoids that control vacuum to the mode doors and heater water valves. All controls are through the BCM.

A power module controls the blower motor (Figure 14-7). The module receives blower drive signals from the BCM and amplifies them to provide variable fan blower speeds.

Operation

The **program number** zero represents maximum cooling, and 100 represents maximum heating. The program number can be observed while in the diagnostic mode.

The BCM calculates a program number, which represents the amount of heating or cooling required to obtain the temperature set by the driver. This number is based on inputs from the control assembly (driver input), ambient temperature, and in-car temperature. Based on this number air delivery mode, fan blower speed, and blend door positioning are determined.

Figure 14-7 Power module amplifies BCM signals to control the blower motor speed.

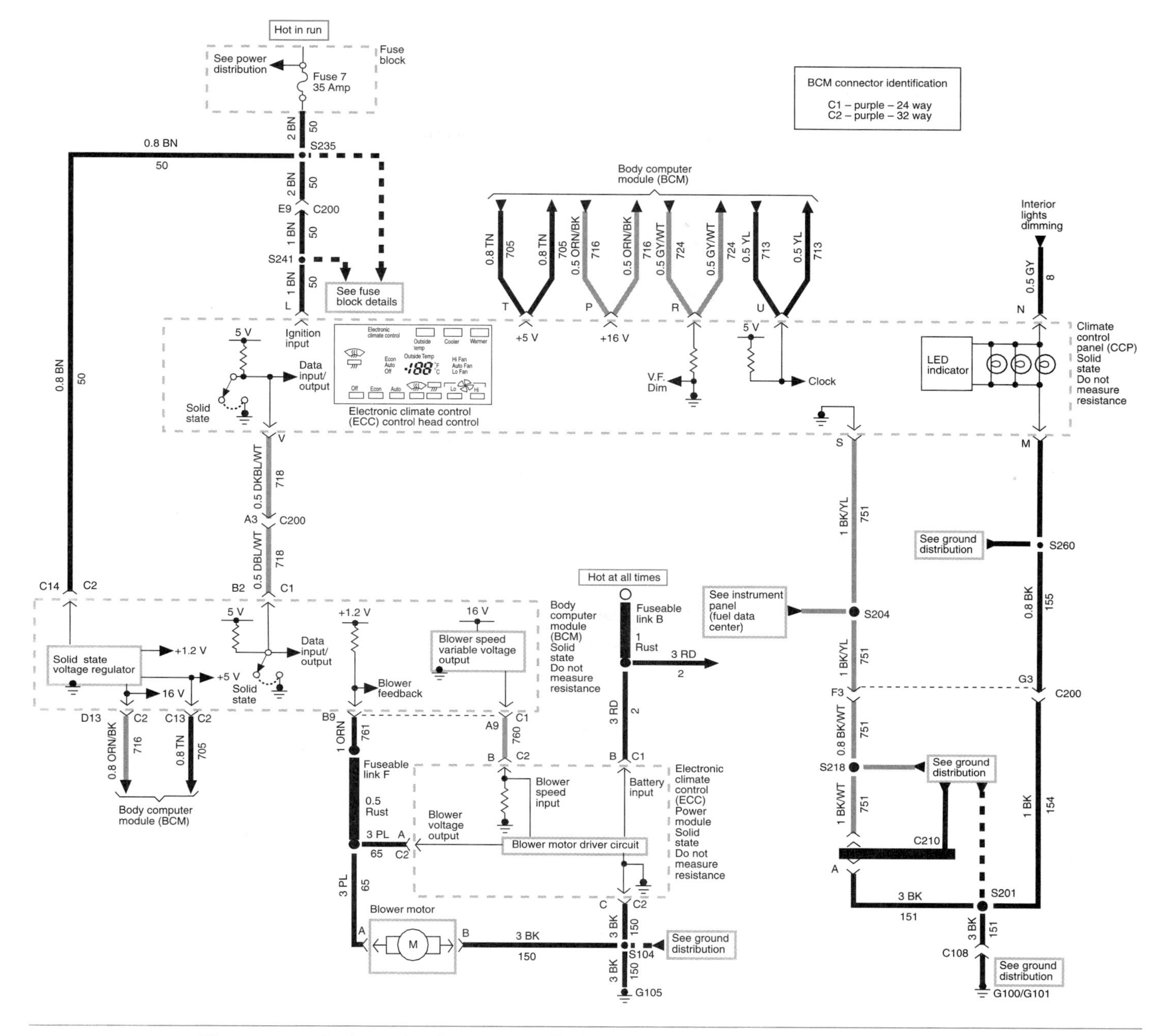

Figure 14-8 Blower control schematic of BCM-controlled EATC. (Courtesy of General Motors Corporation)

To provide the proper mix of inside air temperature entering the passenger compartment, the BCM monitors the ambient and in-car temperatures, the average low side temperature, and the coolant temperature. These inputs are combined with the program number. The BCM commands the programmer to position the blend door for the correct temperature of incoming air. The programmer feedback potentiometer is also monitored by the BCM.

Blower speed is determined by a combination of the program number and driver input temperature. The CCP will signal the BCM over data line 718 (Figure 14-8). The signal is sent

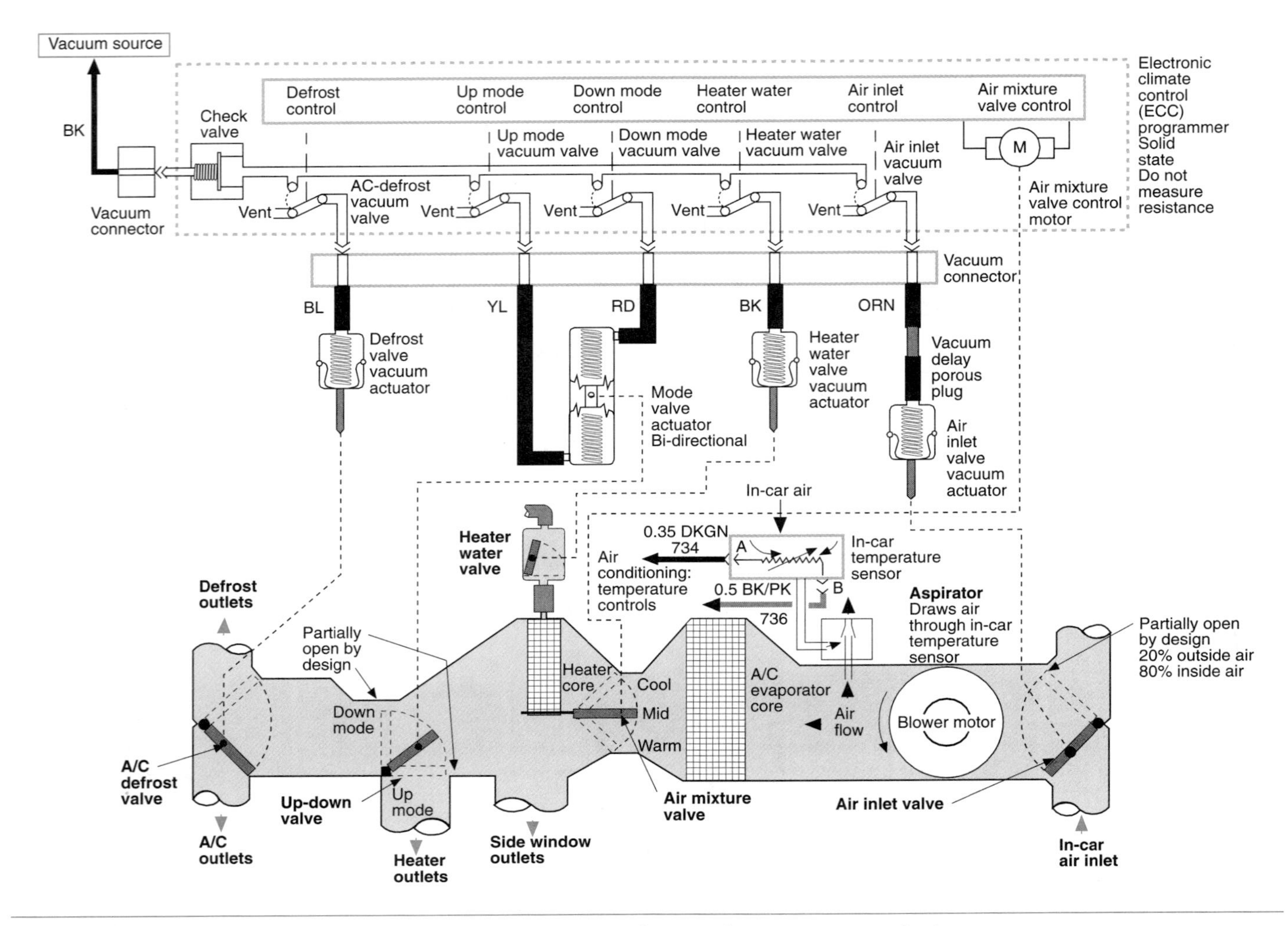

Figure 14-9 Air delivery vacuum schematic. (Courtesy of General Motors Corporation)

from the BCM to the power module from terminal A9. This signal is a constantly variable voltage proportional to blower speed. The power module will amplify the signal then apply it to the blower motor through circuit 65. The BCM monitors the blower speed through a feedback voltage from the motor on circuit 761.

Solenoid valves in the programmer control the air valves that are operated by mechanical and vacuum controls. The air-inlet, up-down, and A/C defrost valves are controlled individually (Figure 14-9). Commands from the BCM to the programmer control the operation of the solenoids.

Air conditioner compressor clutch control is performed by the engine control module (ECM) through inputs from the BCM (Figure 14-10). The ECM cycles the compressor on and off based on input signals from the high side and low side temperature sensors to the BCM. In addition, the ECM will anticipate clutch cycling and will adjust engine idle speed accordingly.

The BCM monitors system inputs and feedback signals. If these voltage signals fall outside programmed parameters, the BCM will turn on the "Service Air Cond" indicator lamp.

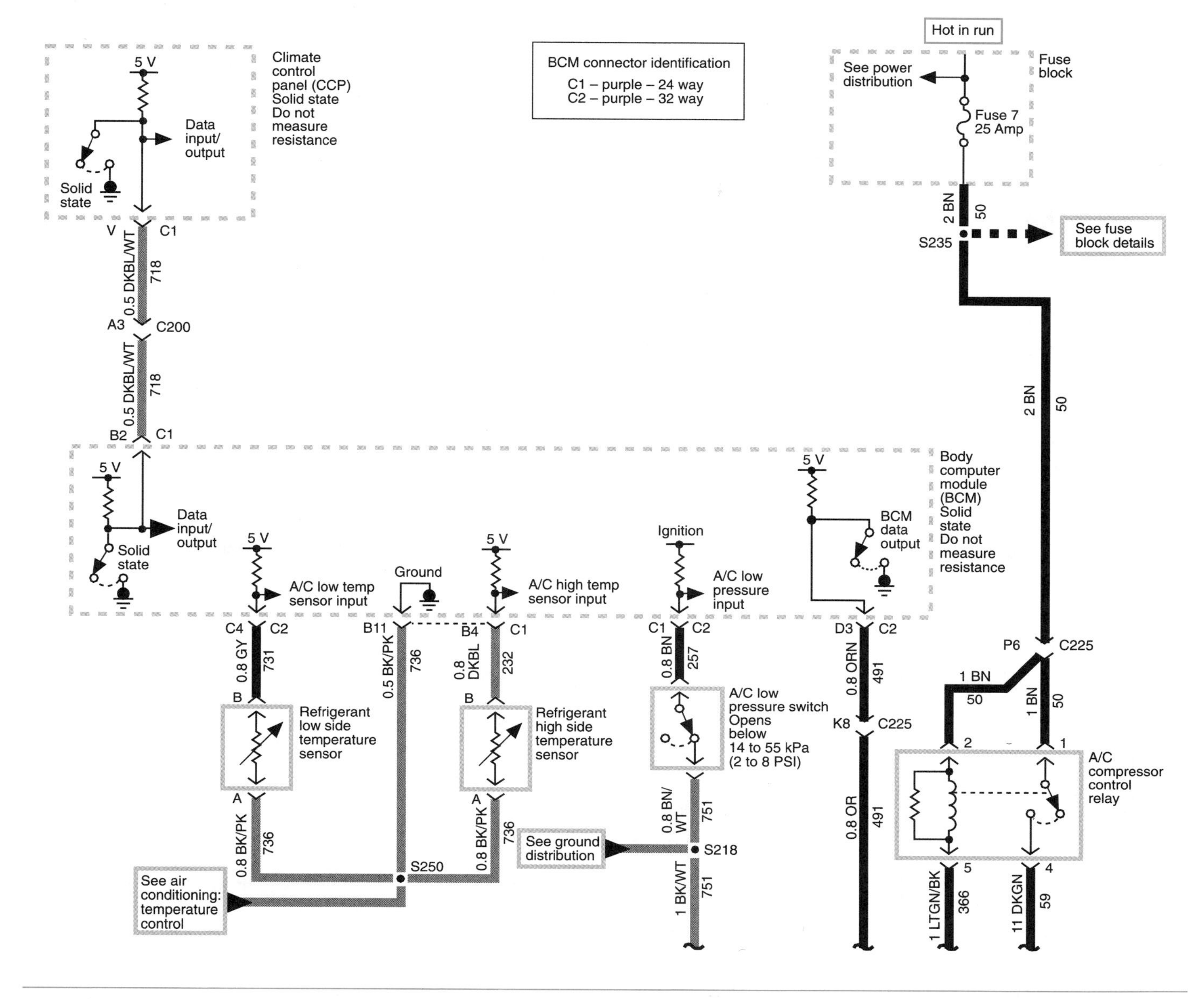

Figure 14-10 Air-conditioning clutch controls. (Courtesy of General Motors Corporation)

Separate Computer-Controlled Systems

Several EATC systems use a separate computer for the sole purpose of climate control. The systems described here are representative examples of these types of systems.

GM Electronic Touch Climate Control

Shop Manual
Chapter 14, page 500

General Motors electronic touch climate control (ETCC) systems are fully automatic and regulate fan speed, air inlet, and outlet positions. The control head contains the microprocessor that will

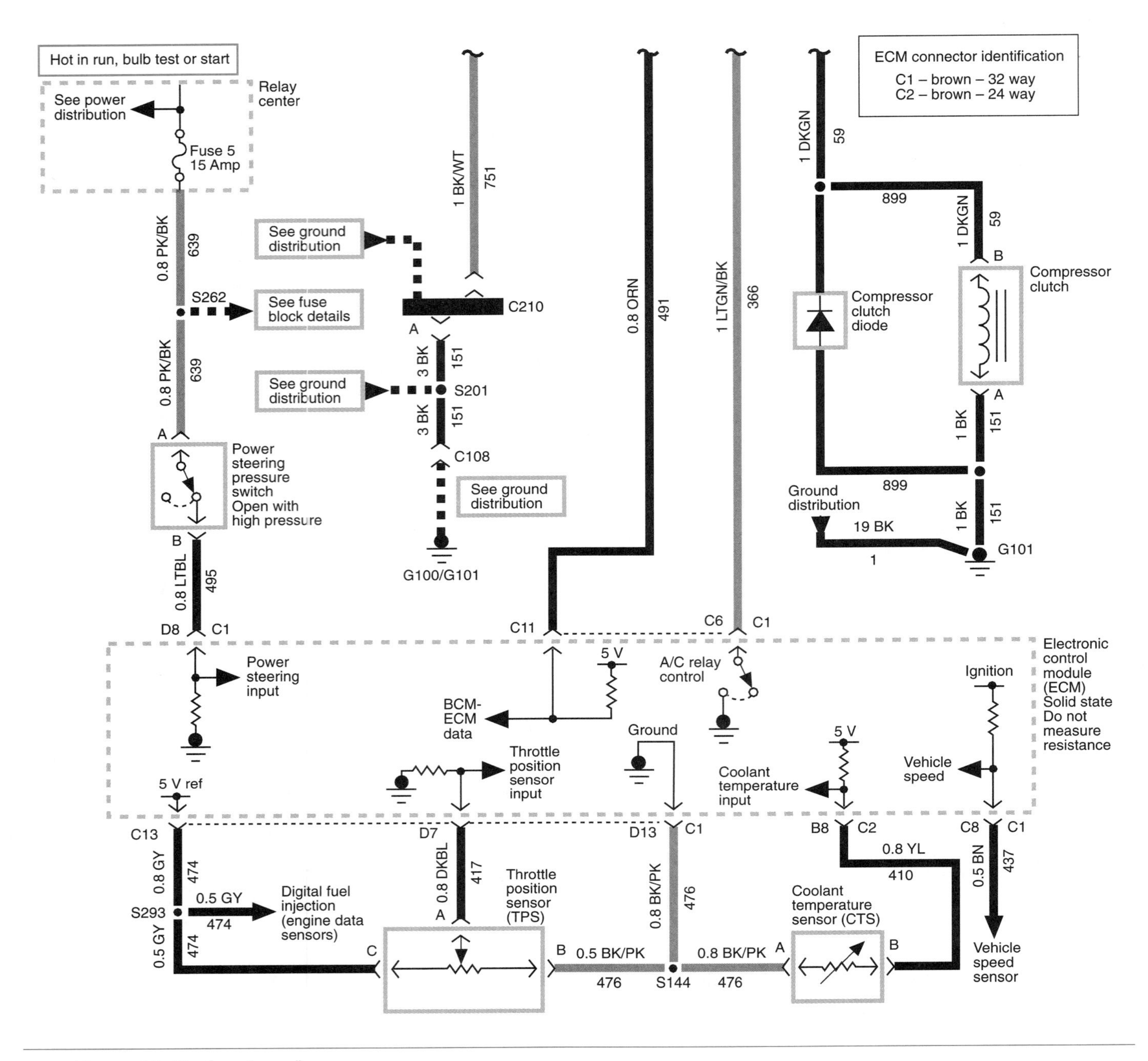

Figure 14-10 (continued)

execute desired selection and remember the last selection (Figure 14-11). The control head receives inputs from ambient temperature, in-car temperature, and setting selection to determine position of the blend door and fan blower speed.

Depending on the model of vehicle the system is installed on, the mode door and water flow actuators can be controlled by either vacuum or electric servomotors (Figure 14-12). All systems use an electrically operated blend door.

Figure 14-11 Electronic touch climate control panel.

Figure 14-12 Electrically operated water valve actuator.

The control head signals the blower and A/C clutch control module to provide for clutch cycling and appropriate blower speeds (Figure 14-13). The blower motor provides up to 256 different speeds.

Ford EATC

Shop Manual
Chapter 14, page 500

The Ford EATC system uses a microprocessor that is built into the control assembly. The control unit controls four DC motor rotary actuators to operate each air distribution door. The sensors and inputs are the same as in previously discussed systems. The control assembly sends continuously variable voltage signals to the blower motor speed controller (BMSC). The BMSC module amplifies the signal to control motor operation. Figure 14-14 is a schematic of one EATC system used by Ford.

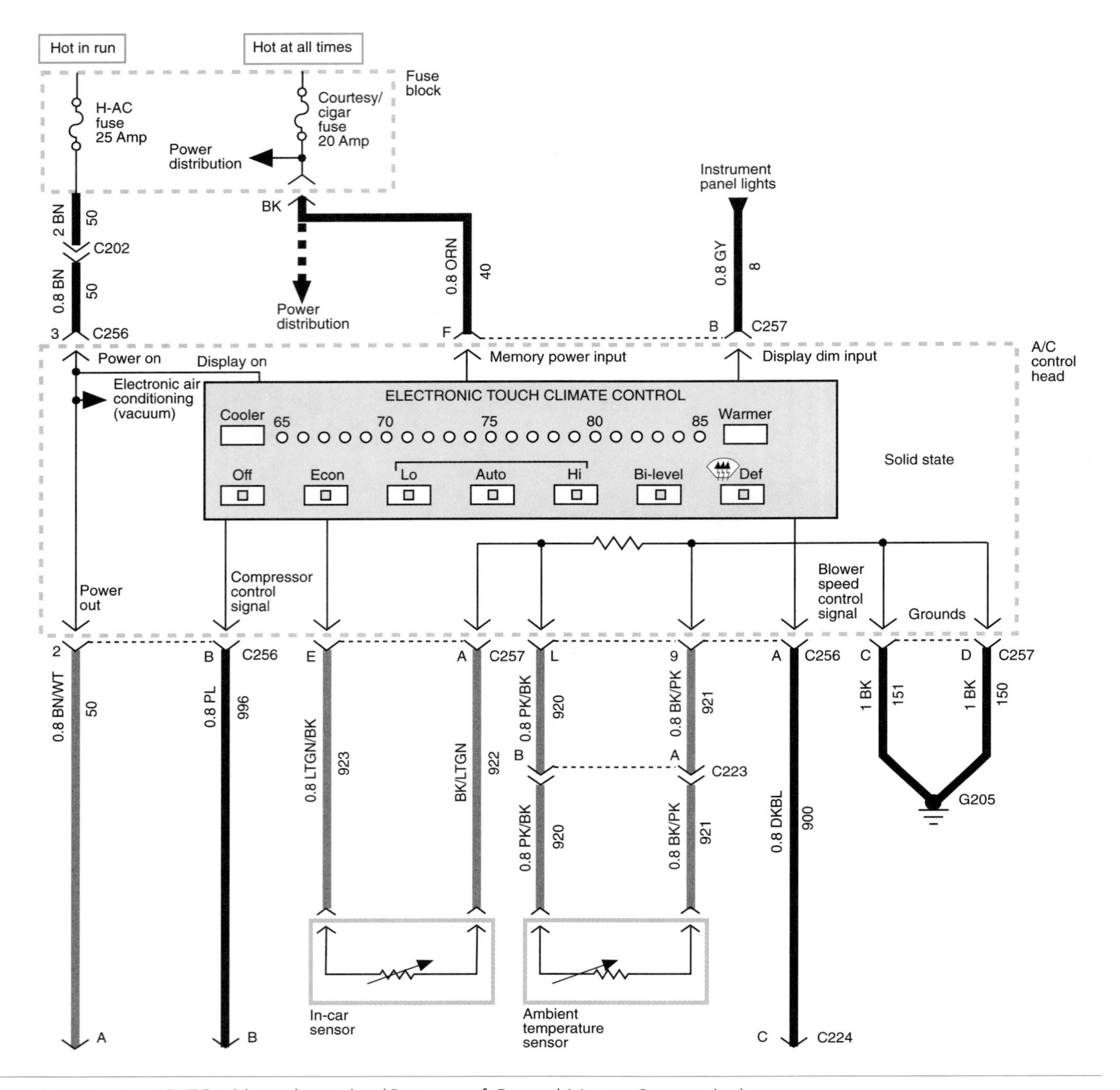

Figure 14-13 EATC wiring schematic. (Courtesy of General Motors Corporation)

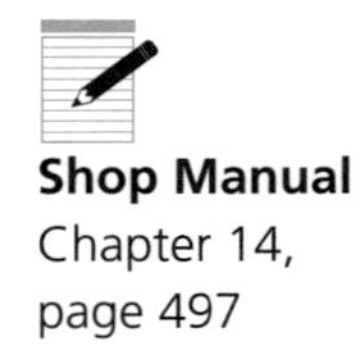

Shop Manual
Chapter 14,
page 497

Chrysler EATC

The Chrysler EATC system uses an individual computer that regulates incoming air temperature and adjusts the system as required every 7 seconds (Figure 14-15). The power-vacuum module (PVM) uses logic signals from the control assembly microprocessor to send a variable voltage signal to the blower motor. In addition, the PVM controls A/C compressor clutch and a voltage signal to the blend door actuator, and vacuum to all other actuators (Figure 14-16)

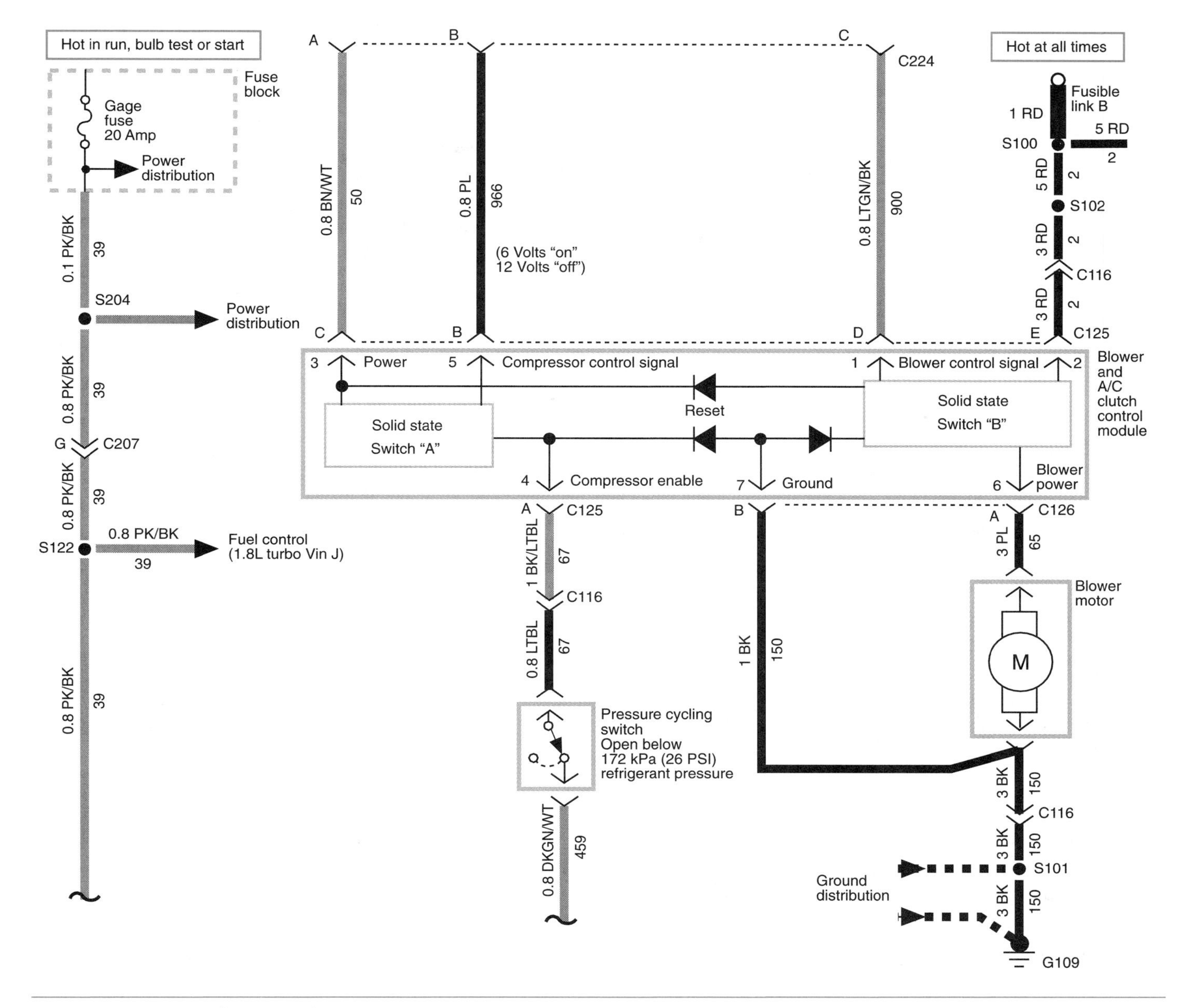

Figure 14-13 (continued)

Introduction to Electronic Cruise Control Systems

Cruise control was first introduced in the 1960s for the purpose of reducing driver fatigue. When engaged, the cruise control system sets the throttle position to maintain the desired vehicle speed.

Most cruise control systems are a combination of electrical and mechanical components. The components used depend on manufacturer and system design. However, the operating principles are similar.

Cruise control is a system that allows the vehicle to maintain a preset speed with the driver's foot off the accelerator.

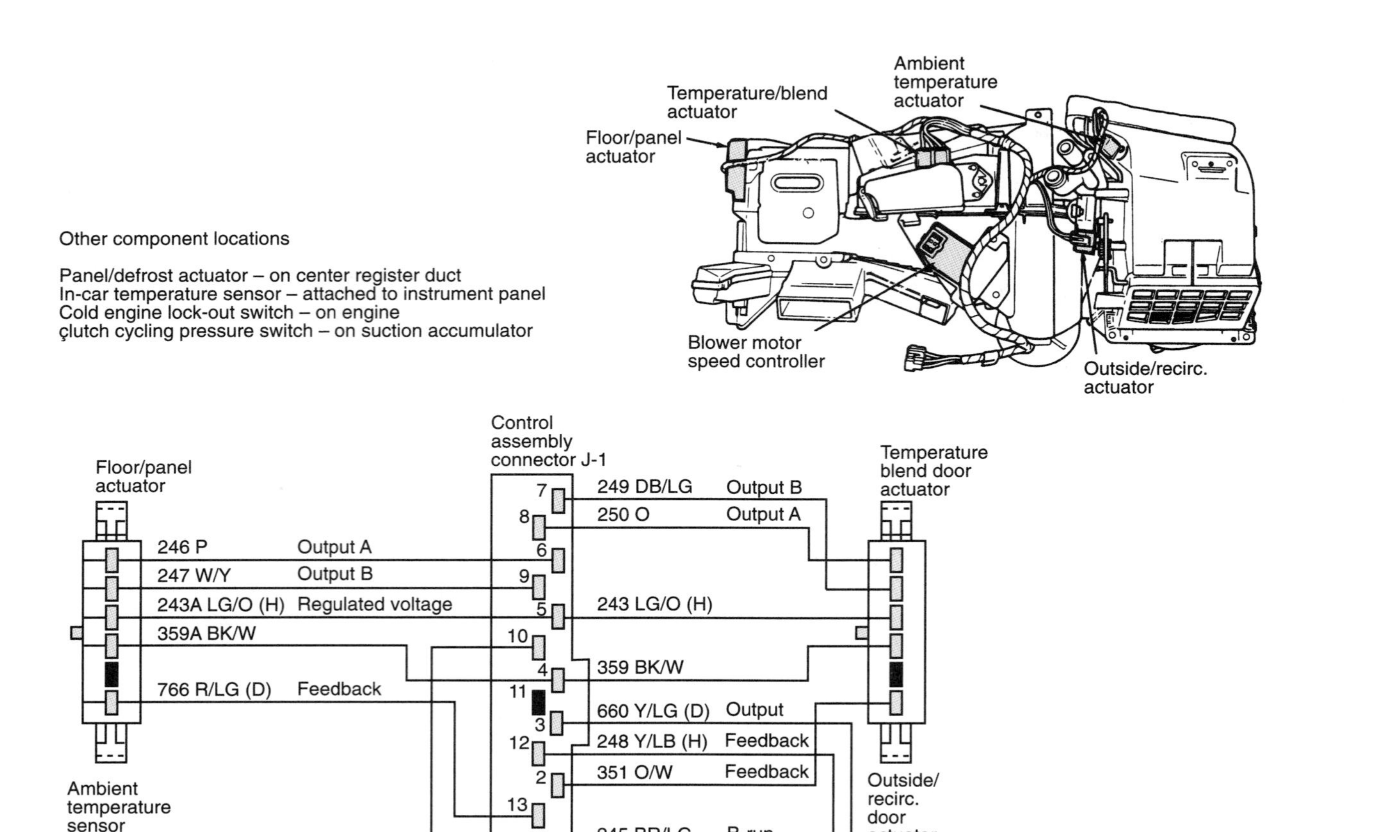

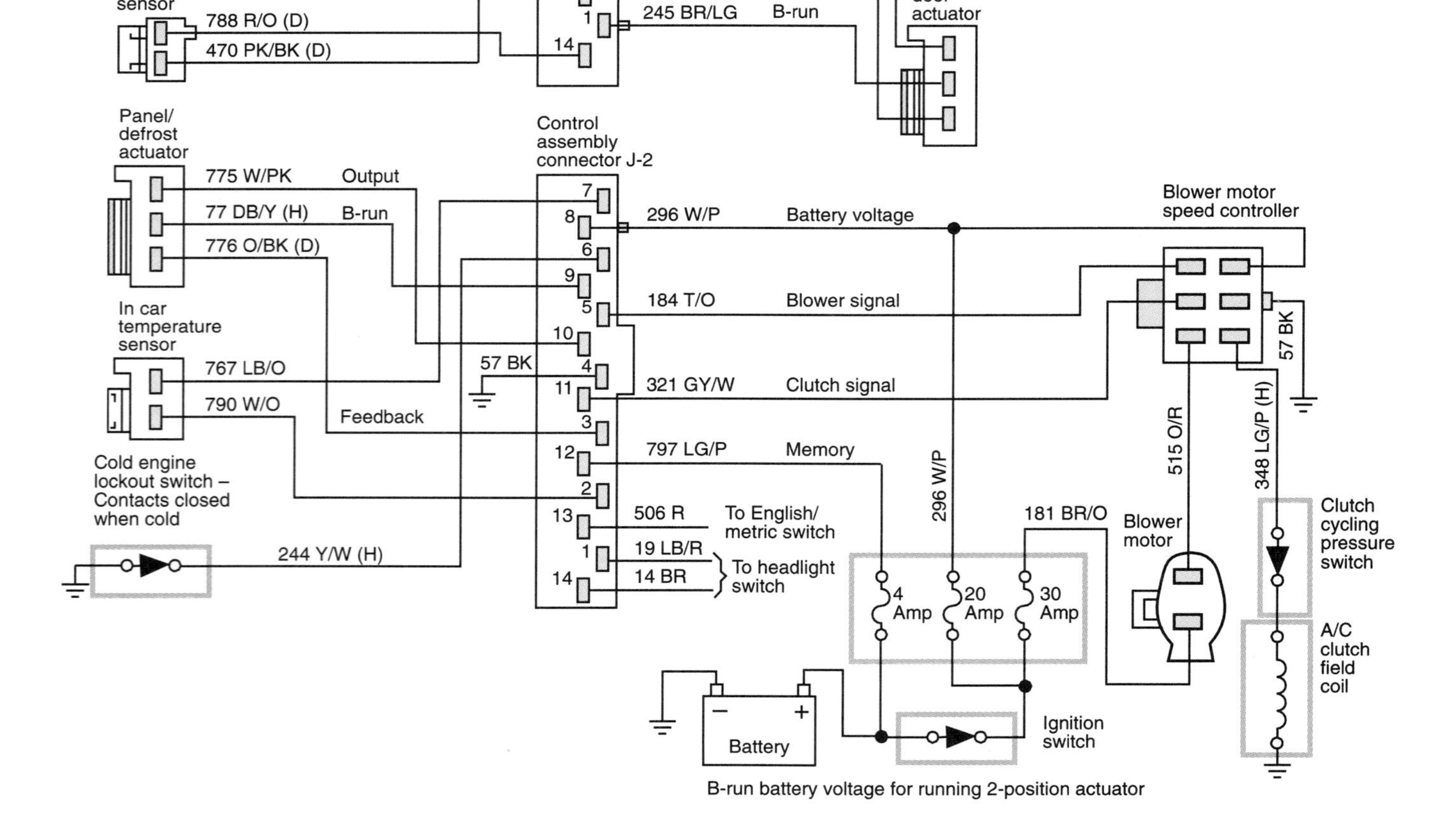

Figure 14-14 Ford's EATC system. (Reprinted with the permission of Ford Motor Company)

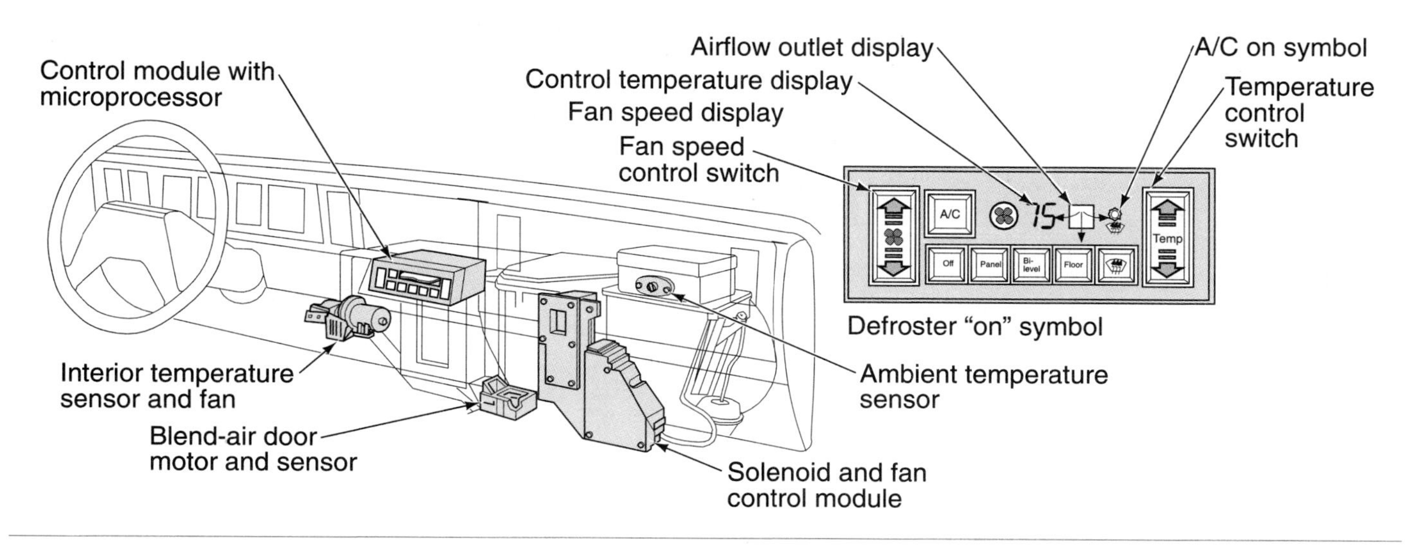

Figure 14-15 Chrysler's EATC system components and control assembly. (Courtesy of Chrysler Corporation)

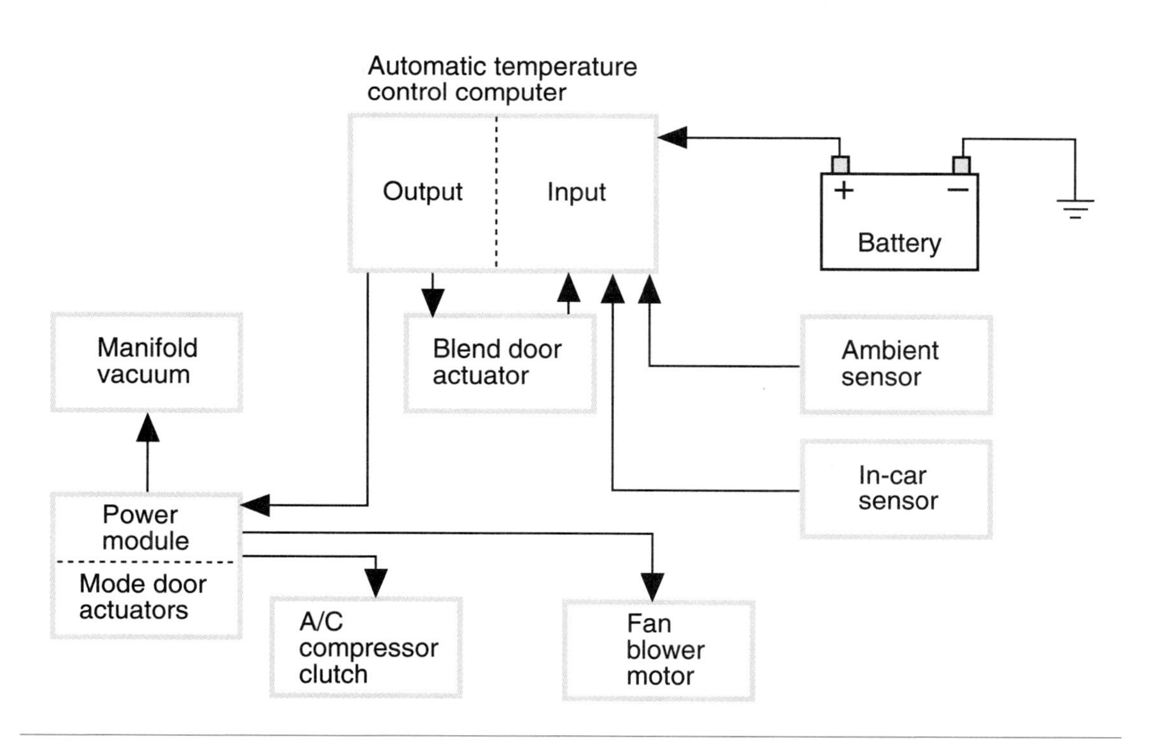

Figure 14-16 Block diagram of Chrysler's EATC system. (Courtesy of Chrysler Corporation)

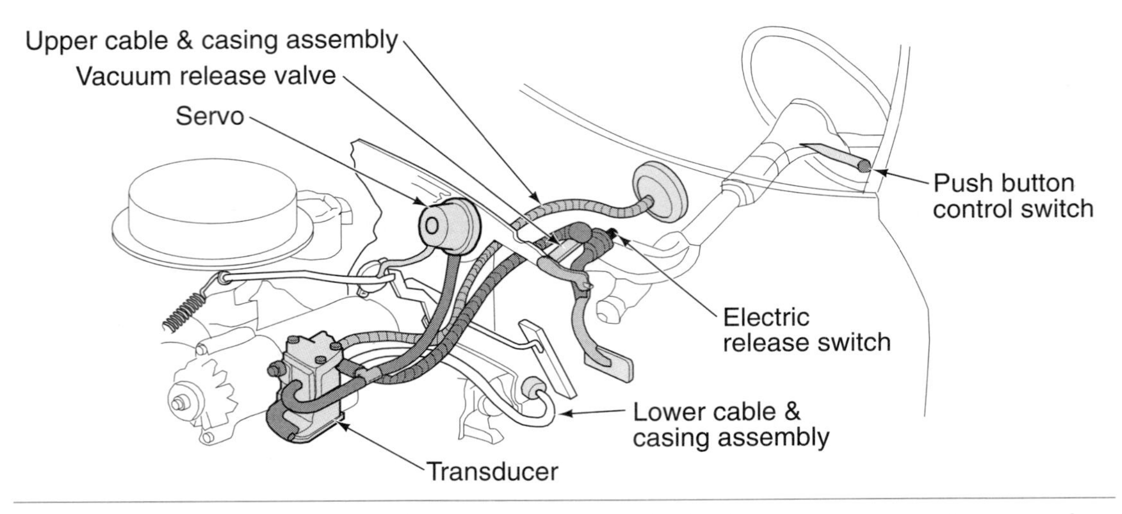

Figure 14-17 Components of typical electromechanical cruise control system. (Courtesy of General Motors Corporation)

Electromechanical Systems

Electromechanical cruise control receives its name because of the two subsystems: the electrical portion and the mechanical portion.

Some manufacturers combine the transducer and servo into one unit. They usually refer to this unit as a **servomotor.**

A review of electromechanical systems is helpful when trying to understand the operation of electronic cruise control. The illustration (Figure 14-17) shows the location of the main components of the electromechanical control system. These components include:

1. Cruise control switch: The control switch is located on the turn signal stock or in the steering wheel (Figure 14-18). The switch assembly is actually a set of driver-operated switches. On most systems, the switches are ON, SET/ACCEL, COAST, and RESUME.
2. Transducer: The transducer receives vehicle speed signals through the speedometer cable. Electrical signals from the control switch, brake switch, or clutch switch are sent to the transducer. In addition, the transducer receives engine manifold vacuum. It regulates the vacuum to the servo through the electrical signals received.

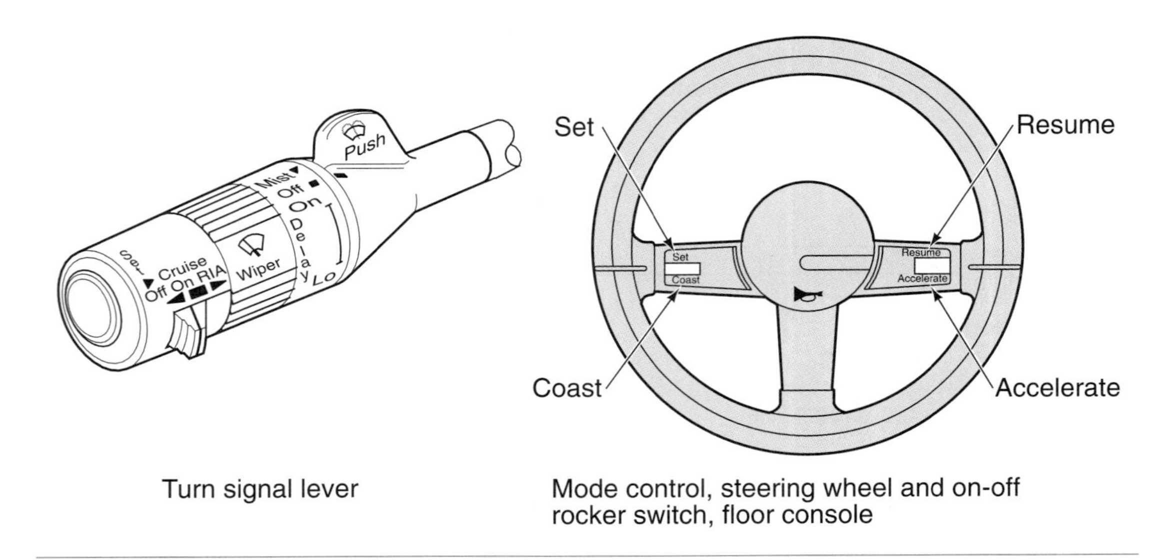

Figure 14-18 The control switch can be mounted on the turn signal stock or into the steering wheel. The switch is used to provide driver inputs for the system. (Courtesy of General Motors Corporation)

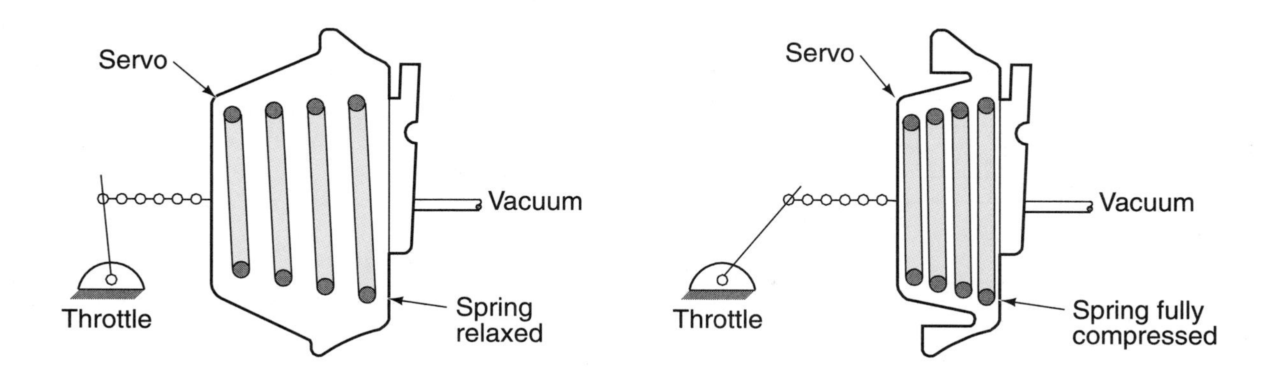

Figure 14-19 Cutaway view of the servo. Vacuum is used to compress the spring and open the throttle. (Courtesy of General Motors Corporation)

3. Servo: The servo controls throttle plate position. It is connected to the throttle plate by a rod, bead chain, or Bowden cable. The servo maintains the set speed by receiving a controlled amount of vacuum from the transducer. When vacuum is applied to the servo, the spring is compressed and the throttle plate is moved to increase speed (Figure 14-19). When the vacuum is released, the spring returns the throttle plate to reduce engine speed.
4. Safety switches: When the brake pedal is depressed, the cruise control system is disengaged through electrical and vacuum switches (Figure 14-20). The switches are usually located on the brake pedal bracket. The two switches provide a fail-safe means of assuring that the cruise control is disengaged when the brakes are applied. If one of the switches fails, the other will still be able to return vehicle speed control over to the driver. Vehicles equipped with manual transmissions may use switches on the clutch pedal to disengage the system whenever the clutch pedal is depressed.

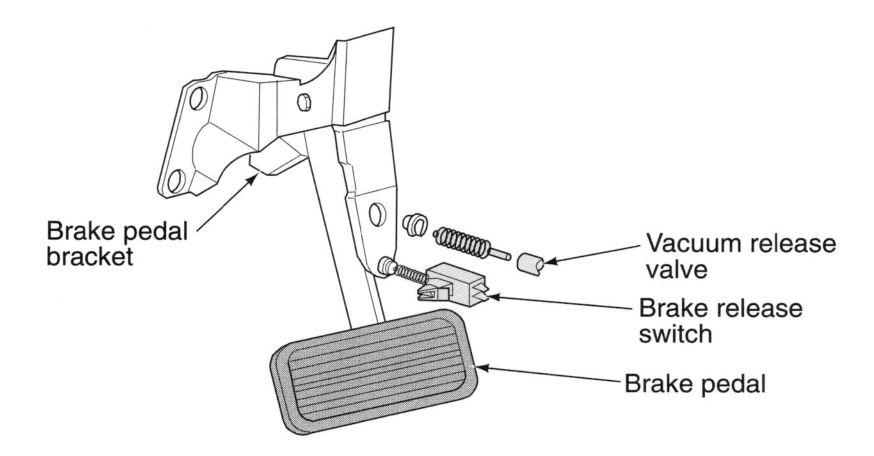

Figure 14-20 The brake release switch and the vacuum release switch work together to disengage the cruise control switch when the brake pedal is depressed. (Courtesy of General Motors Corporation)

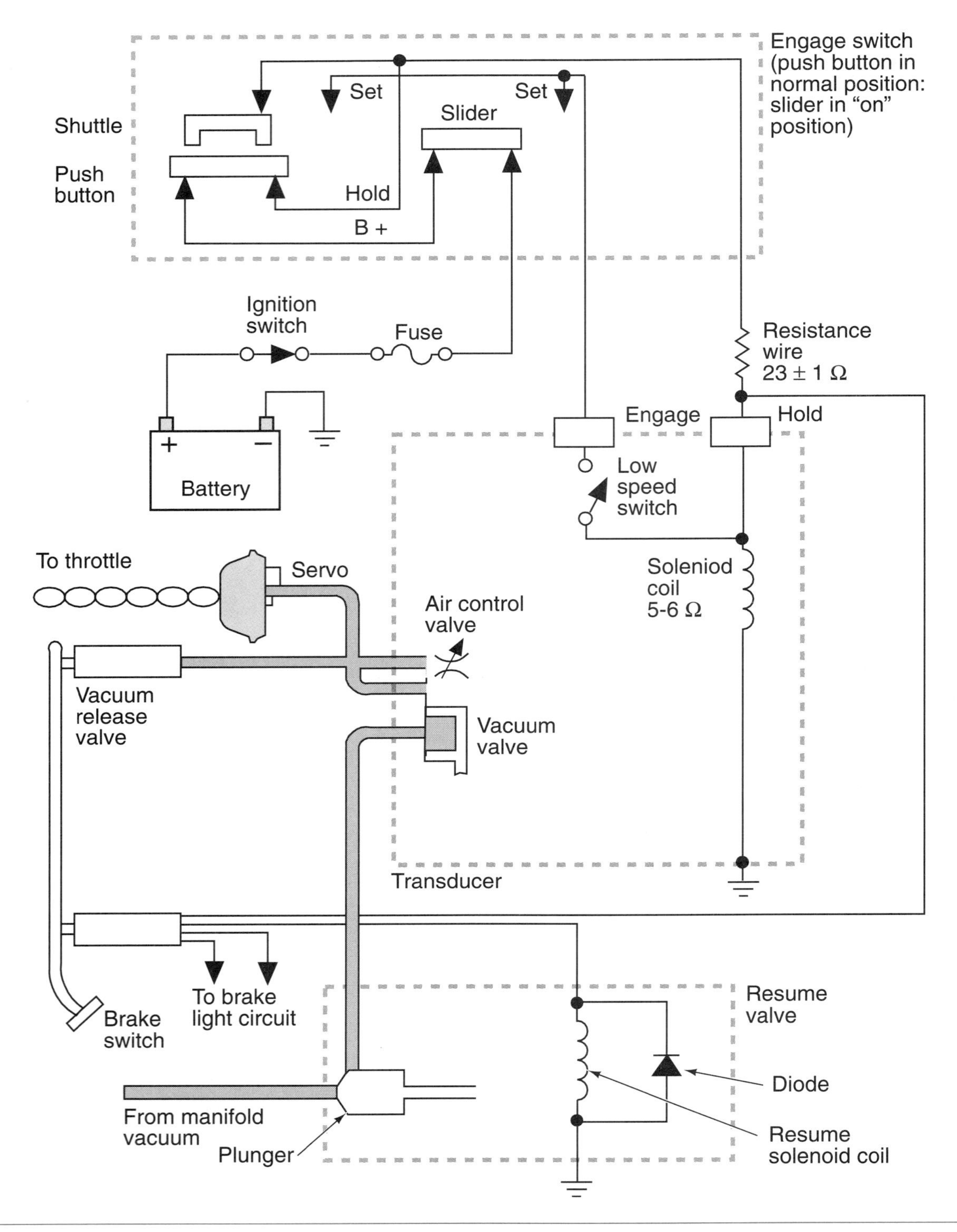

Figure 14-21 Electrical and vacuum schematic of a typical electromechanical cruise control system. (Courtesy of General Motors Corporation)

When the cruise control switch is in the "at rest" position, battery current flows through the switch to the resistance wire to the hold terminal of the transducer (Figure 14-21). Voltage to the hold terminal is too low to activate the solenoid coil because the resistor drops the voltage.

The transducer contains a rubber clutch with an operating arm. At speeds below 30 mph (48 km/h) the rubber clutch arm holds the low speed switch open. When speeds exceed 30 mph

(48 km/h), the clutch arm rotates. The rotation of the arm allows the switch to close and the system can be engaged. By pushing the momentary contact switch in the SET position, the current flow is through the engage terminal of the transducer, through the low speed switch, to the solenoid coil. The resistor is bypassed and sufficient current is applied to the solenoid to activate it. When the momentary switch is released, the current flow is returned through the hold terminal. The current applied to the solenoid coil is sufficient to hold the solenoid in the activated position.

When the solenoid is activated, the vacuum valve opens to allow engine vacuum to the servo and the brake release valve. The air control valve is a variable orifice and is the control mechanism that adjusts vacuum level to the system. At lower speed settings, the air control valve bleeds off vacuum so that less is sent to the servo. At higher set speeds, less vacuum is bled to allow for more throttle plate opening.

If the brake pedal is depressed with the cruise control system engaged, the brake switch provides an alternate path to ground and bypasses the solenoid. With the solenoid deactivated, the vacuum valve closes and returns the throttle control over to the driver. At the same time, the resume solenoid is closed and the vacuum release valve opens to release the vacuum in the system.

Electronic Cruise Control

Shop Manual
Chapter 14, page 505

The electronic cruise control system uses an electronic module to operate the actuators that control throttle position (Figure 14-22). Electronic cruise control offers more precise speed control than the electromechanical system. In addition, other benefits include:

- More frequent throttle adjustments per second.
- More consistent speed increase/decrease when using the tap-up/tap-down feature.

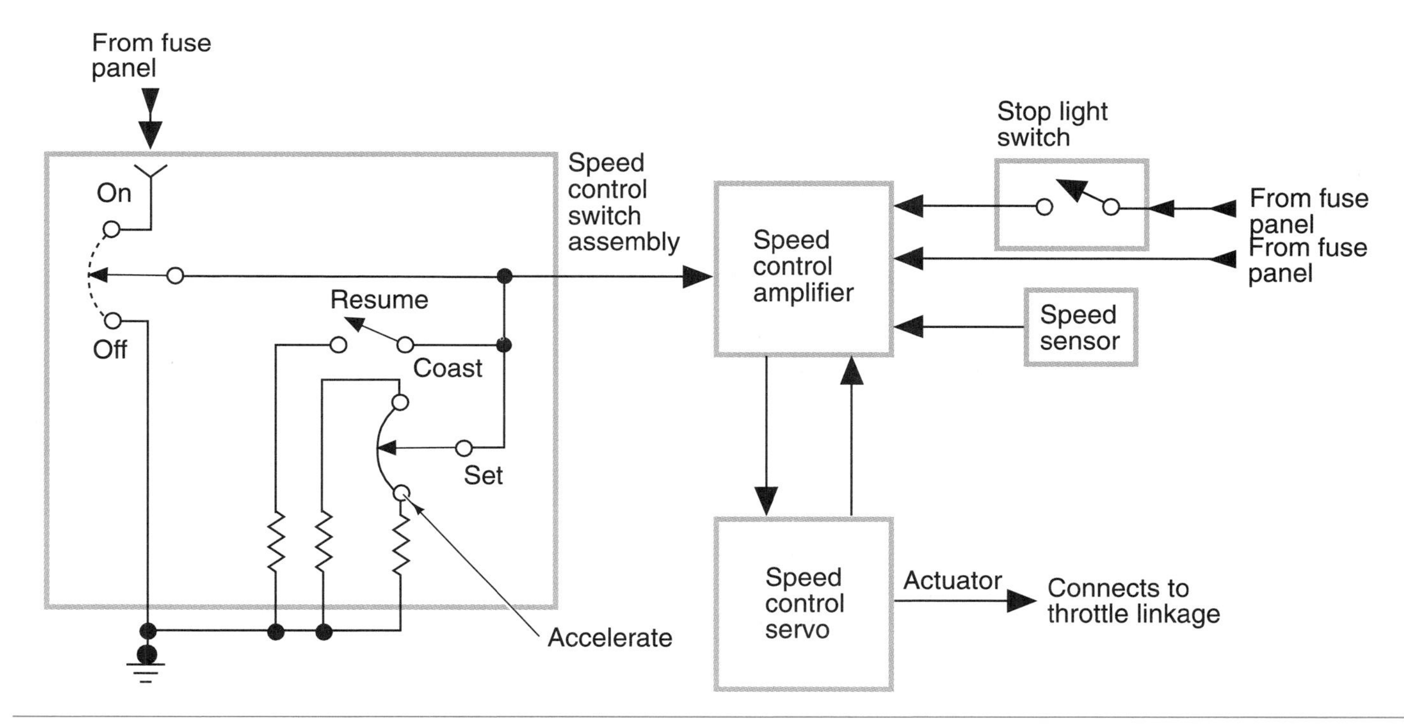

Figure 14-22 Block diagram of electronic cruise control system. (Courtesy of General Motors Corporation)

- ❑ Greater correction of speed variation under loads
- ❑ Rapid deceleration cut-off when deceleration rate exceeds programmed rates.
- ❑ Wheelspin cut-off when acceleration rate exceeds programmed parameters.
- ❑ System malfunction cut-off when the module determines there is a fault in the system.

Common Components

Shop Manual
Chapter 14, page 516

Common components of the electronic cruise control system include:

1. The control module: The module can be a separate cruise control module, the engine control module, or the body control module. The operation of the systems are similar regardless of the module used.
2. The control switch: Depending on system design, the control switch contacts apply the ground circuit through resistors. Because each resistor has a different value, a different voltage is applied to the control module. In some systems the control switch will send a 12-volt signal to different terminals of the control module.
3. The brake or clutch switch.
4. Vacuum release switch.
5. Servo: The servo operates on vacuum that is controlled by supply and vent valves. These operate from controller signals to solenoids.

Depending on system design, the sensors used as inputs to the control module include the vehicle speed sensor, servo position sensor, and throttle position sensor. Other inputs are provided by the brake switch, instrument panel switch, control switch, and the park-neutral switch.

The control module receives signals from the speed sensor and the control switch. When the vehicle speed is fast enough to allow cruise control operation and the driver pushes the SET button on the control switch, an electrical signal is sent to the controller. The voltage level received by the controller is set in memory. This signal is used to create two additional signals. The two signal values are set at 1/4 mph above and below the set speed. The module uses the comparator values to change vacuum levels at the servo to maintain set vehicle speed.

Three safety modes are operated by the control module:

1. Rapid deceleration cutoff: If the module determines that deceleration rate is greater than programmed values, it will disengage the cruise control system and return operation back over to the driver.
2. Wheelspin cutoff: If the control module determines that the acceleration rate is greater than programmed values, it will disengage the system.
3. System malfunction cutoff: The module checks the operation of the switches and circuits. If it determines there is a fault, it will disable the system.

The vacuum-modulated servo is the primary actuator. Vacuum to the servo is controlled by two solenoid valves: supply and vent. The vent valve is a normally open valve and the supply valve is normally closed (Figure 14-23). The servo receives signals from the controller to operate the solenoid valves to maintain a preset throttle position.

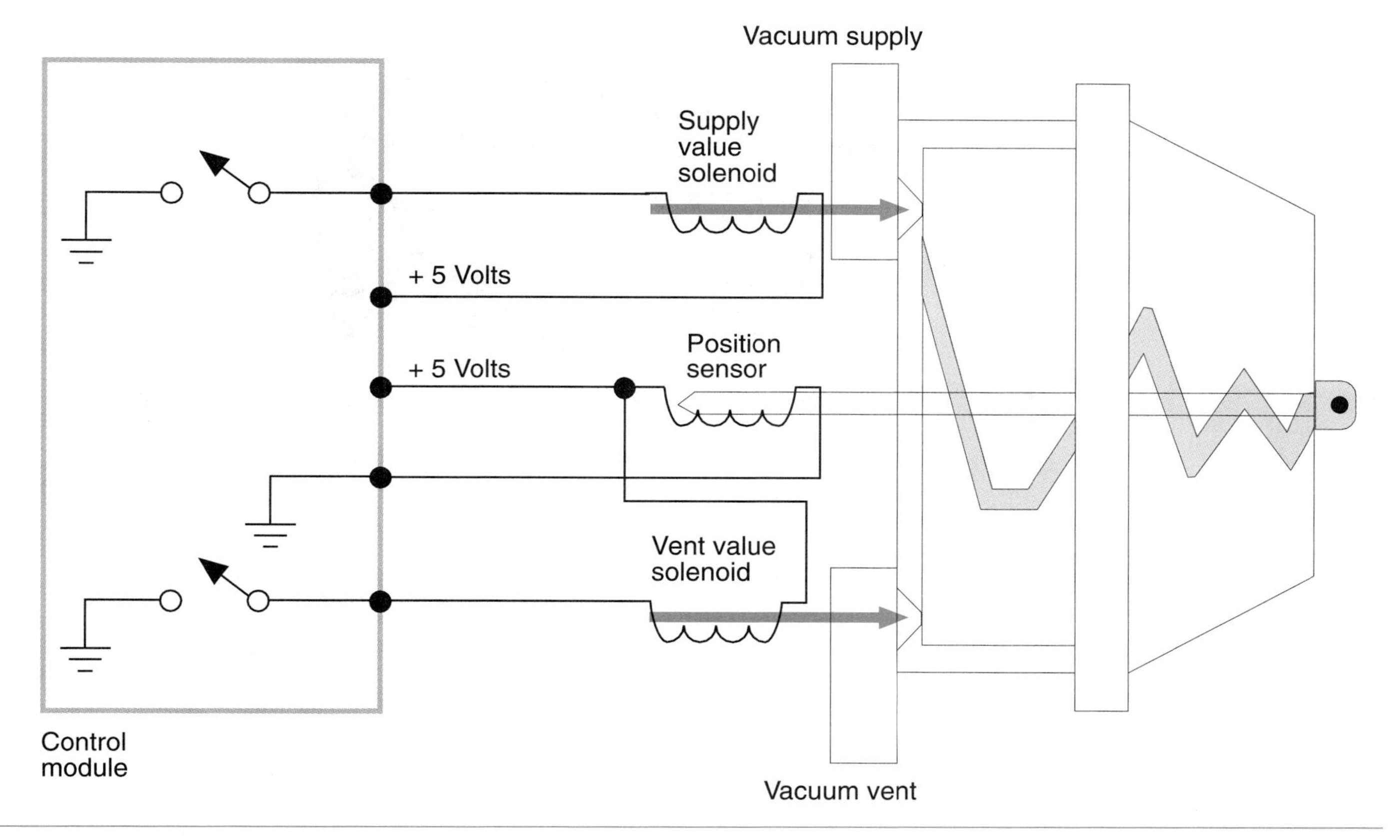

Figure 14-23 Servo valve operation in electronic control system. The servo position sensor informs the controller of servo operation and position.

Principles of Operation

When the driver sends a SET signal to the controller it sets the voltage signals received from the vehicle speed sensor (VSS) into memory. It then determines the high and low comparators.

The controller energizes the supply and vent valves to allow manifold vacuum or atmospheric pressure to enter the servo. The servo uses the vacuum and pressure to move the throttle and maintain the set speed. The vehicle speed is maintained by balancing the vacuum in the servo.

If the voltage signal from the VSS drops below the low comparator value, the control module energizes the supply valve solenoid to allow more vacuum into the servo and increases the throttle opening. When the VSS signal returns to a value within the comparator levels, the supply valve solenoid is de-energized.

If the VSS signal is greater than the high comparator value, the control module de-energizes the vent solenoid valve to release vacuum in the servo. The vehicle speed is reduced until the VSS signals are between the comparator values, at which time the control module will energize the vent valve solenoid again. This constant modulation of the supply and vent valves maintains vehicle speed.

During steady cruise conditions, both valves are closed and a constant vacuum is maintained in the servo.

Classroom Manual
Chapter 14,
page 520

Passive restraints operate automatically with no action required on the part of the driver or occupant (Figure 14-24).

In a two-point system the occupant must manually lock the lap belt.

Inertia lock retractors use a pendulum mechanism to lock the belt tightly during sudden movement (Figure 14-26).

Figure 14-24 Passive automatic seatbelt system operation. (Reprinted with the permission of Ford Motor Company)

Passive Restraints

Federal regulations have mandated the use of automatic passive restraint systems in all vehicles sold in the United States after 1990. Two- or three-point automatic seatbelt and airbag systems are currently offered as a means of meeting this requirement.

The passive seat belt system automatically puts the shoulder and/or lap belt around the driver or occupant. The automatic seat belt system operates by means of DC motors that move the belts by means of carriers on tracks (Figure 14-25).

One end of the seat belt is attached to the carrier: the other end is connected to the inertia lock retractors. When the door is opened, the outer end of the shoulder harness moves forward (to the A-pillar) to allow for easy entry or exit (Figure 14-27). When the door is closed and the ignition switch is placed in the RUN position, the motor moves the outer end of the harness to the locked position in the B-pillar (Figure 14-28).

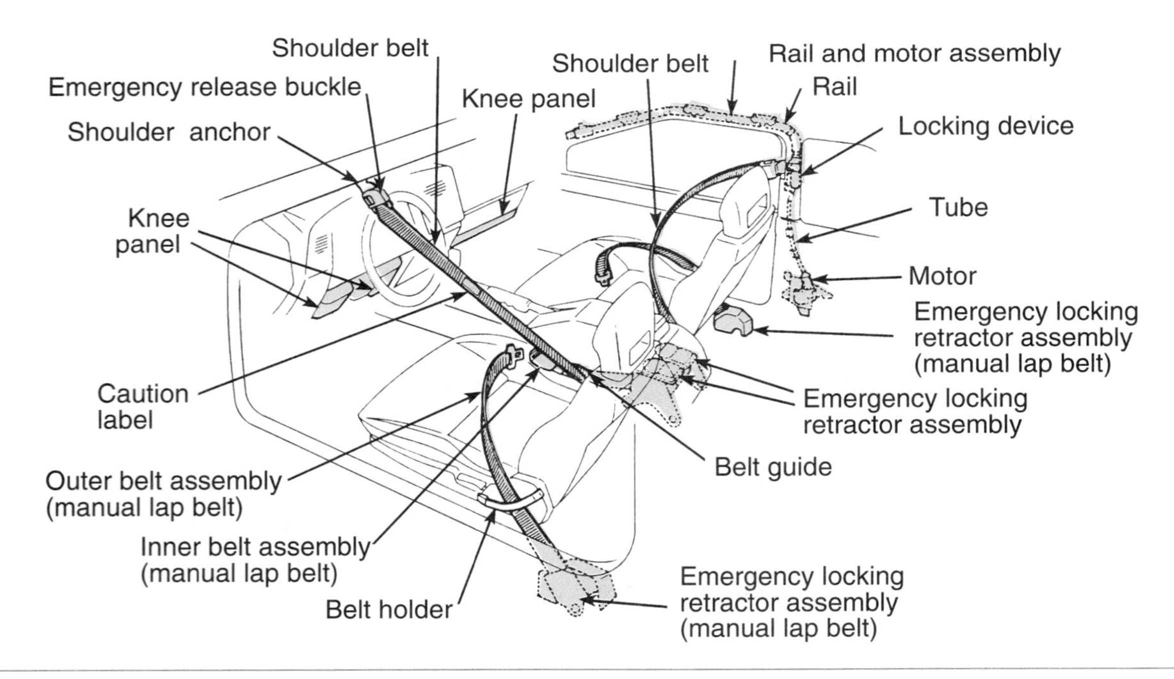

Figure 14-25 Passive seat belt restraint system uses a motor to put the shoulder harness around the occupant. (Reprinted with the permission of Ford Motor Company)

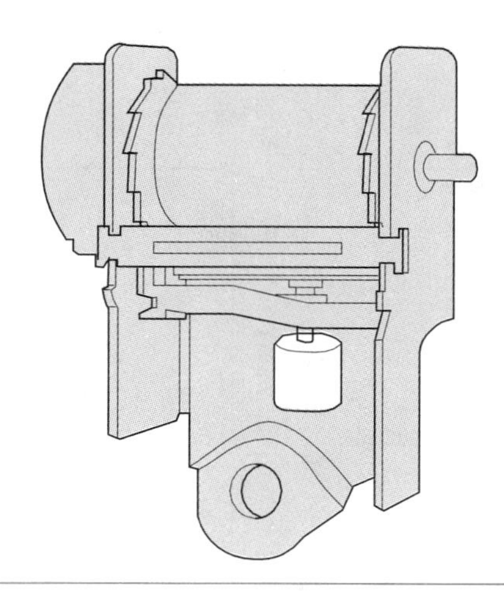

Figure 14-26 Inertia lock seat belt retractor. (Reprinted with the permission of Ford Motor Company)

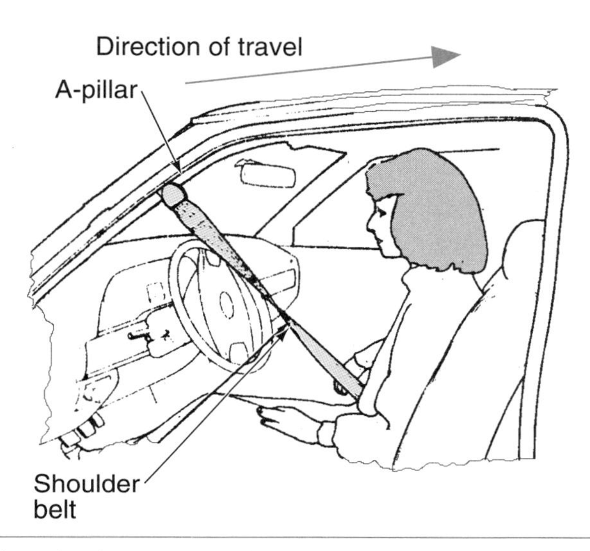

Figure 14-27 When the door is opened the motor pulls the harness to the A-pillar. (Reprinted with the permission of Ford Motor Company)

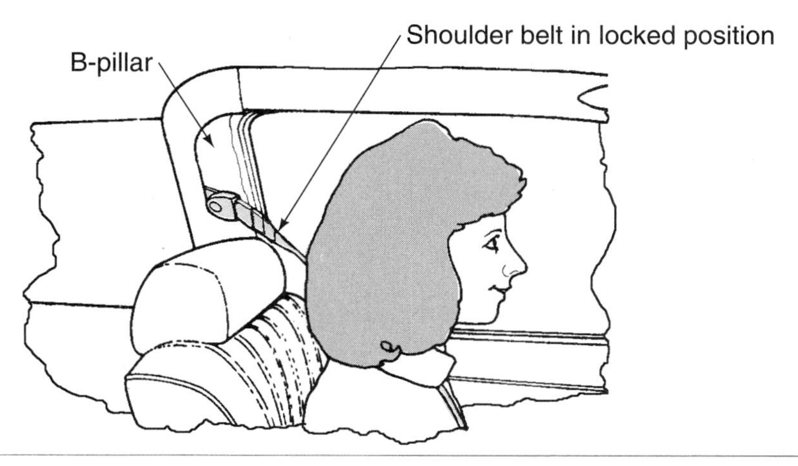

Figure 14-28 When the door is closed and the ignition switch is in the RUN position, the motor draws the harness to its lock position. (Reprinted with the permission of Ford Motor Company)

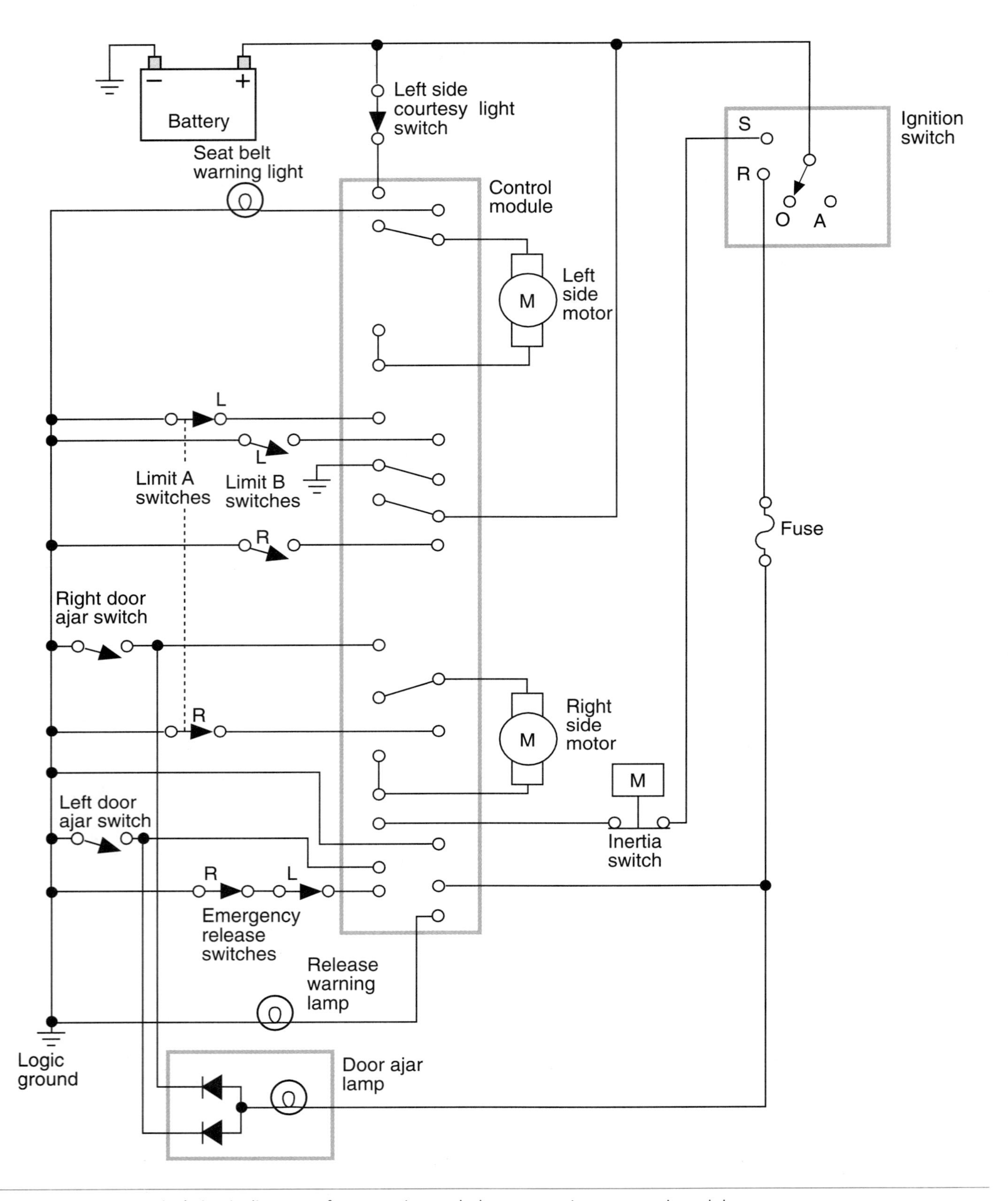

Figure 14-29 Typical circuit diagram of automatic seat belt system using a control module.

The automatic seatbelt system uses a control module to monitor operation (Figure 14-29). The monitor receives inputs from door ajar switches, limit switches, and the emergency release switch.

The door ajar switches signal the position of the door to the module. The switch is open when the door is closed. This signal is used by the control module to activate the motor and move the harness to the lock point behind the occupant's shoulders. If the module receives a signal that the door is open, regardless of ignition switch position, it will activate the motor to move the harness to the forward position.

The limit switches inform the module of the position of the harness. When the harness is moved from the FORWARD position, the front limit switch (limit A) closes. When the harness is located in the LOCK position, the rear limit switch (limit B) opens and the module turns off the power to the motor. When the door is opened, the module reverses the power feed to the motor until the A switch is opened.

An emergency release mechanism is provided in the event that the system fails to operate. The NC emergency release switch is opened whenever the release lever is pulled. The module will turn on the warning lamp in the instrument panel and sound a chime to alert the driver. The opened switch also prevents the harness retractors from locking.

Ford incorporates the fuel pump inertia switch into the automatic seatbelt system. If the module receives a signal that the switch is open, it prevents the harness from moving to the forward position if the door opens.

The **fuel pump inertia switch** is a NC switch that will open if the vehicle is involved in an impact at speeds over 5 mph or rolls over. When the switch opens, it turns off power to the fuel pump. This is a safety feature to prevent fuel from being pumped onto the ground or hot engine components if the engine dies. The switch has to be manually reset if it is triggered (Figure 14-30).

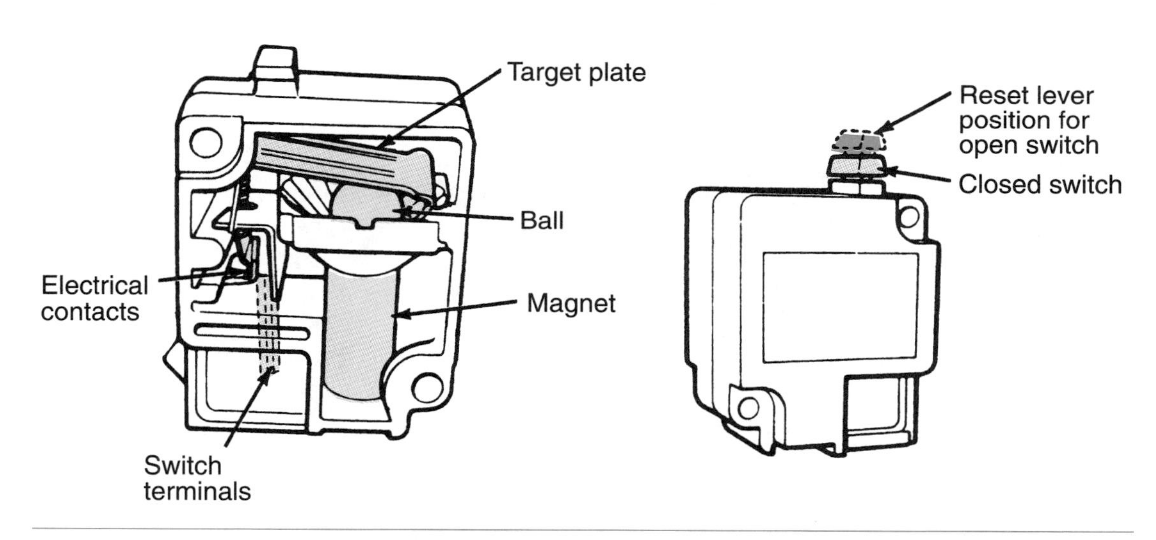

Figure 14-30 Fuel pump inertia switch. (Reprinted with the permission of Ford Motor Company)

Air Bag Systems

The need to supplement the existing restraint system during frontal collisions has led to the development of the supplemental inflatable restraint (SIR) or air bag systems (Figure 14-31). Because most air bag systems are not designed to deploy during side or rear collisions, seatbelts should always be worn in conjunction with the air bag. The air bag is a supplement and the seatbelt is the primary restraint system.

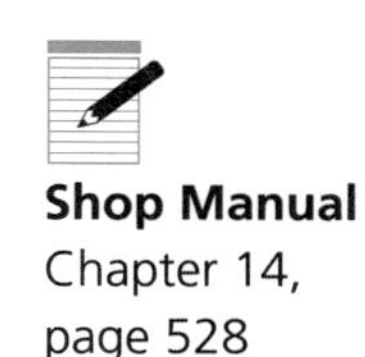

Shop Manual
Chapter 14, page 528

WARNING: Always refer to the specific manufacturer's recommendations. Each system has different safety requirements.

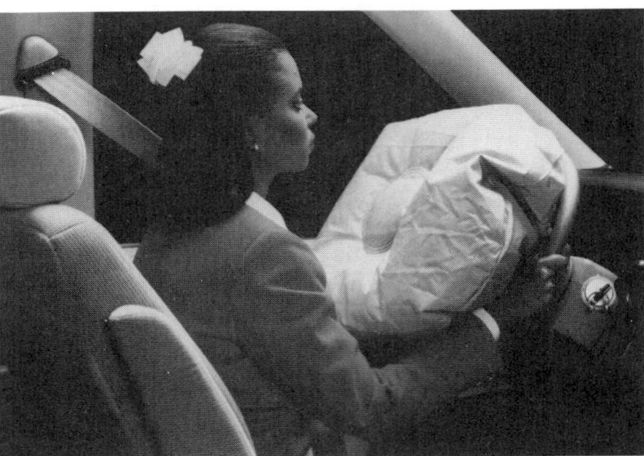

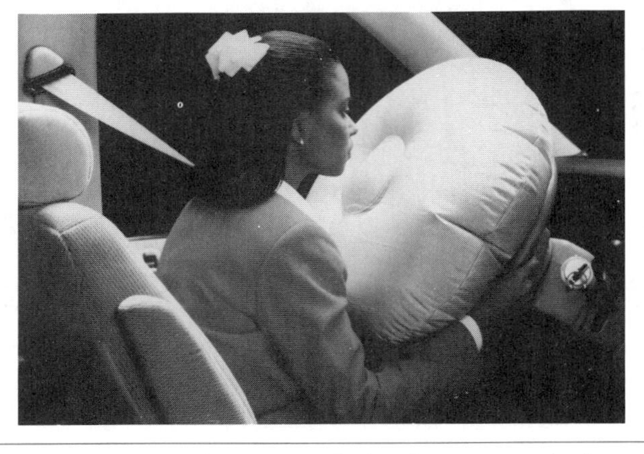

Figure 14-31 Air bag sequence. (Courtesy of Chrysler Corporation)

Shop Manual
Chapter 14,
page 533

Many of the components used for driver side air bags are similar to those used in passenger side air bags. The basic operation of the two systems is the same.

The **air bag** is made of neoprene-coated nylon.

Common Components

A typical air bag system consists of sensors, a diagnostic module, a clock spring, and an air bag module. See the typical location (Figure 14-32) of the common components of the SIR system.

Air Bag Module. The air bag module is the air bag and inflator assembly packaged into a single module. This module is mounted in the center of the steering wheel (Figure 14-33).

The inflation of the air bag is through an explosive release of nitrogen gas. The igniter is an integral component of the inflator assembly. It starts a chemical reaction to inflate the air bag

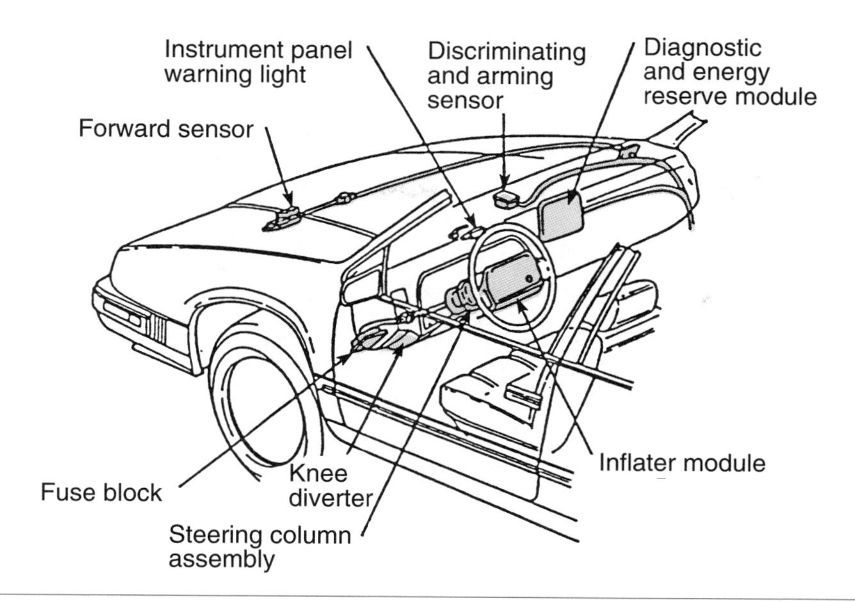

Figure 14-32 Typical location of components of the air bag system. (Courtesy of Chrysler Corporation)

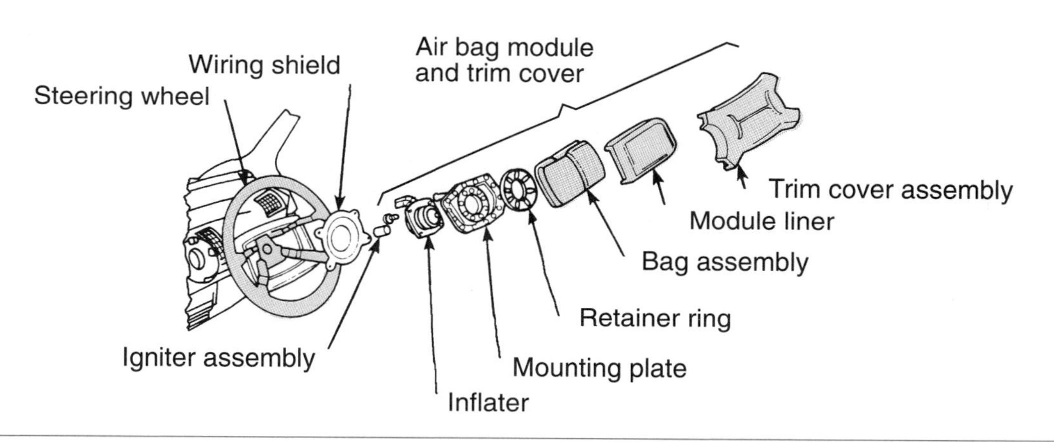

Figure 14-33 Air bag module components. (Courtesy of Chrysler Corporation)

(Figure 14-34). When current is applied to the igniter, it arcs between two pins. The spark ignites a canister of gas generating zerconic potassium perchlorate (ZPP).

The **igniter** is a combustible device that converts electric energy into thermal energy to ignite the inflator propellant.

The inflation assembly is composed of a solid chemical gas generator (generant) containing sodium azide and copper oxide propellent. The ZPP ignites the propellent charge. As the propellent burns, it produces nitrogen gas. The gas passes through a diffuser, where it is filtered and cooled before inflating the air bag.

CAUTION: Wear gloves and eye protection when handling a deployed air bag module. Sodium hydroxide residue may remain on the bag. If this comes in contact with the skin, it can cause irritation.

The air bag module cannot be serviced. If it is has been deployed, or is defective, it must be replaced.

Not all air bag systems use nitrogen gas to inflate the bag; some (such as Chrysler) use compressed argon gas to inflate the passenger side bag.

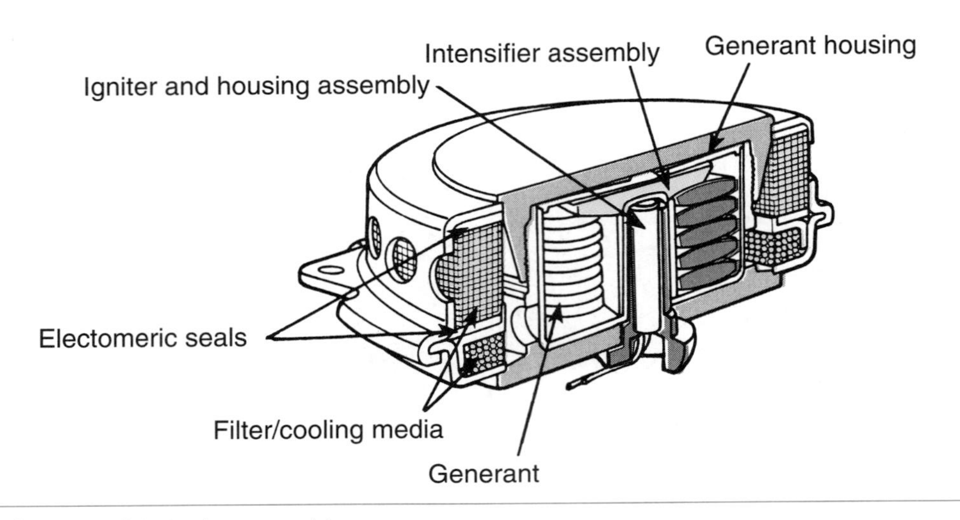

Figure 14-34 Igniter assembly.

The clock spring is also known as coil assembly, cable reel assembly, coil spring unit, and contact reel.

Clock Spring. The clock spring conducts electrical signals to the module while allowing steering wheel rotation (Figure 14-35). The clock spring is located between the column and the steering wheel.

CAUTION: Whenever the air bag is deployed, the heat generated may damage the clock spring. The clock spring should be replaced whenever the air bag is deployed.

Diagnostic Module. The diagnostic module constantly monitors the readiness of the air SIR electrical system. If the module determines that there is a fault in the system, it will illuminate the indicator light. Depending on the fault, the SIR system may be disarmed until the fault is repaired.

The diagnostic module also supplies backup power to the air bag module in the event that the battery or cables are damaged during an accident. The stored charge can last for up to 30 minutes after the battery is disconnected.

CAUTION: Before servicing the air bag system, the backup power supply energy must be depleted. Disconnect the battery and isolate the cable terminal. Wait 5 to 30 minutes before servicing. Refer to the manufacturer's specifications for the accurate amount of time to wait.

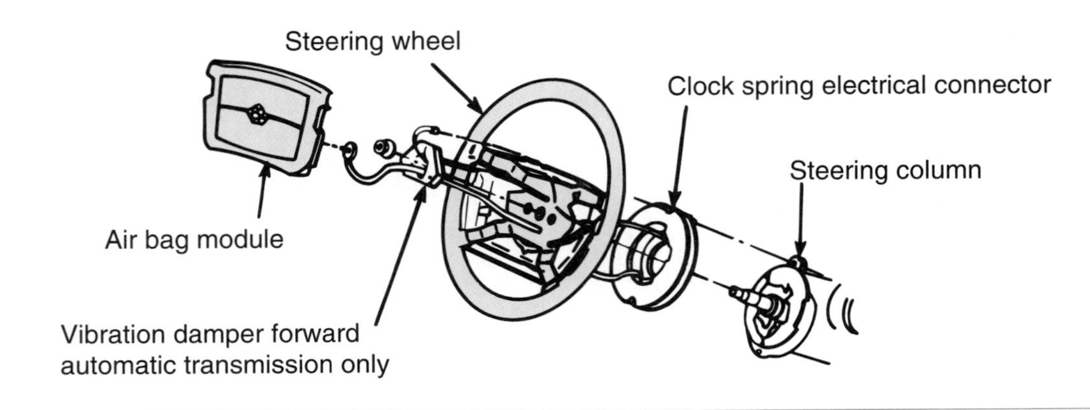

Figure 14-35 The clock spring provides for electrical continuity in all steering wheel positions. (Courtesy of Chrysler Corporation)

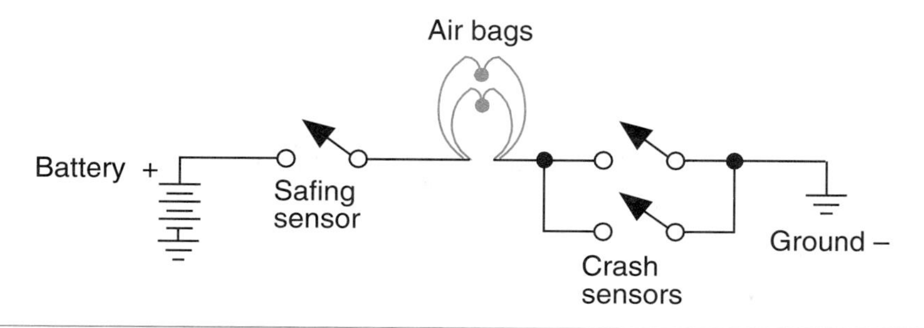

Figure 14-36 Typical sensor wiring circuit diagram.

Crash sensors are normally open electrical switches designed to close when subjected to a predetermined impact.

After impact, the ball will return to its position next to the magnet.

Sensors. To prevent accidental deployment of the air bag, most systems require that at least two sensor switches be closed to deploy the air bag (Figure 14-36). A typical sensor is composed of a gold-plated ball (sensing mass) that is held in place by a magnet (Figure 14-37). At the point of sufficient force, the ball will break loose of the magnet and make contact with the electrical contacts to complete the circuit.

WARNING: SIR system sensors must be installed and tightened with the proper amount of torque and with the arrow pointing to the front of the vehicle for proper operation.

The number of sensors used depends on system design. Some systems use only a single sensor and others use up to five. The name used to identify the different sensors also varies between manufacturers. Usually sensors are located in the engine and passenger compartments. The sensor in the passenger compartment determines if the collision is severe enough to inflate the air bag.

Once the two sensors are closed, the electrical circuit to the igniter is complete. The igniter starts the chemical chain reaction that produces heat. The heat causes the generant to produce nitrogen gas, which fills the air bag.

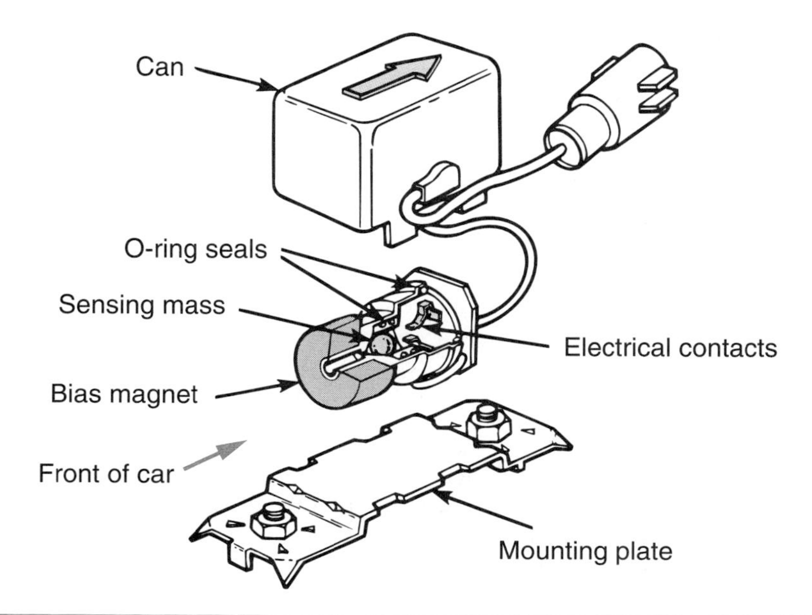

Figure 14-37 Some crash sensors hold the sensing mass by magnetic force. If the impact is severe enough to break the ball free, it will travel forward and close the electrical contacts. (Courtesy of Chrysler Corporation)

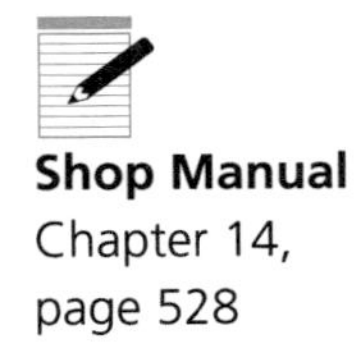

Shop Manual
Chapter 14,
page 528

General Motors' SIR

General Motors has used several different versions of SIR systems. The version discussed here is representative of the system. Changes between models and years require that technicians follow the service manual procedures for the vehicle they are working on.

Refer to the illustration (Figure 14-38) for the location of the components used on the SIR system. The major portions of the SIR system are the deployment loop and the diagnostic energy reserve module (DERM).

The deployment loop supplies current to the inflator module in the steering wheel (Figure 14-39). The components of the deployment loop include the:

arming sensor	inflator module
coil assembly	discriminating sensors

The **arming sensor** is calibrated to close at low level velocity changes. The arming sensor is also referred to as the safing sensor.

The arming sensor switches power to the inflator module on the insulated side of the loop circuit. Either of the discriminating sensors can supply the ground. For the inflator module to ignite, the arming sensor and at least one discriminating sensor must close simultaneously. The complete wiring schematic of this system is illustrated (Figure 14-40).

The **discriminating sensors** are calibrated to close with velocity changes that are severe enough to warrant air bag deployment (velocity changes higher than that of the arming sensor). Discriminating sensors are also known as crash sensors.

There are two discriminating sensors used. One is located in front of the radiator, the other is part of the dual sensor located behind the instrument panel.

The DERM is designed to provide an energy reserve of 36 volts to assure deployment for a few seconds when vehicle voltage is low or lost. The DERM also maintains constant diagnostic monitoring of the electrical system. It will store a code if a fault is found and provide driver notification by illuminating the warning light. The resistor module allows the DERM to monitor the deployment loop for faults and to detect when deployment has occurred. In addition, the DERM records the SIR system status during an accident.

The sequence of events that occur during an impact of a vehicle traveling at 30 mph is as follows:

1. When an accident occurs, the arming sensor is the first to close. It will close due to sudden deceleration caused by braking, or immediately upon impact. One of the discriminating sensors will then close. The amount of time required to close the switches is within 15 milliseconds.
2. Within 40 milliseconds, the igniter module burns the propellant and generates the gas to completely fill the air bag.
3. Within 100 milliseconds, the driver's body has stopped forward movement and the air bag starts to deflate. The air bag deflates by venting the nitrogen gas through holes in the back of the bag.
4. Within 2 seconds the air bag is completely deflated.

The **dual sensor** is a combination of the arming and passenger compartment discriminating sensor.

Mercedes-Benz

The supplemental restraint system (SRS) used on some Mercedes-Benz vehicles combines the air bag with a three-point seatbelt that uses an emergency tensioning retractor (ETR). The air bag operates similar to that discussed. However, it uses only one sensor. The SRS sensor incorporates two ICs and an acceleration pickup to determine the degree and direction of impact. When the longitudinal deceleration is sufficient, the sensor sends a voltage through a bypass filter to an amplifier. The amplifier provides the voltage required to ignite the inflator module.

The SRS sensor uses a mercury switch to disconnect it from the circuit during normal driving.

A voltage convertor is used to keep a constant 12 volts applied to the sensor and energy accumulator. The convertor is capable of maintaining 12 volts even though battery voltage may drop to as low as 4 volts. The energy accumulator operates like a capacitor to provide backup current in the event that battery voltage is totally lost.

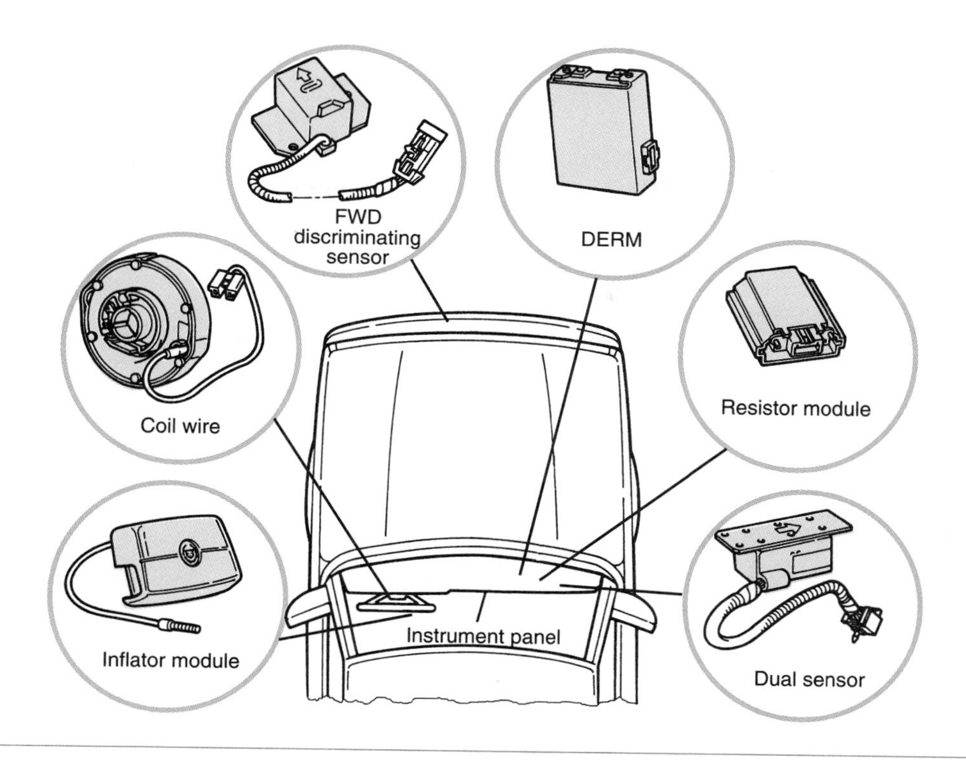

Figure 14-38 Component location of GM's SIR system. (Courtesy of General Motors Corporation)

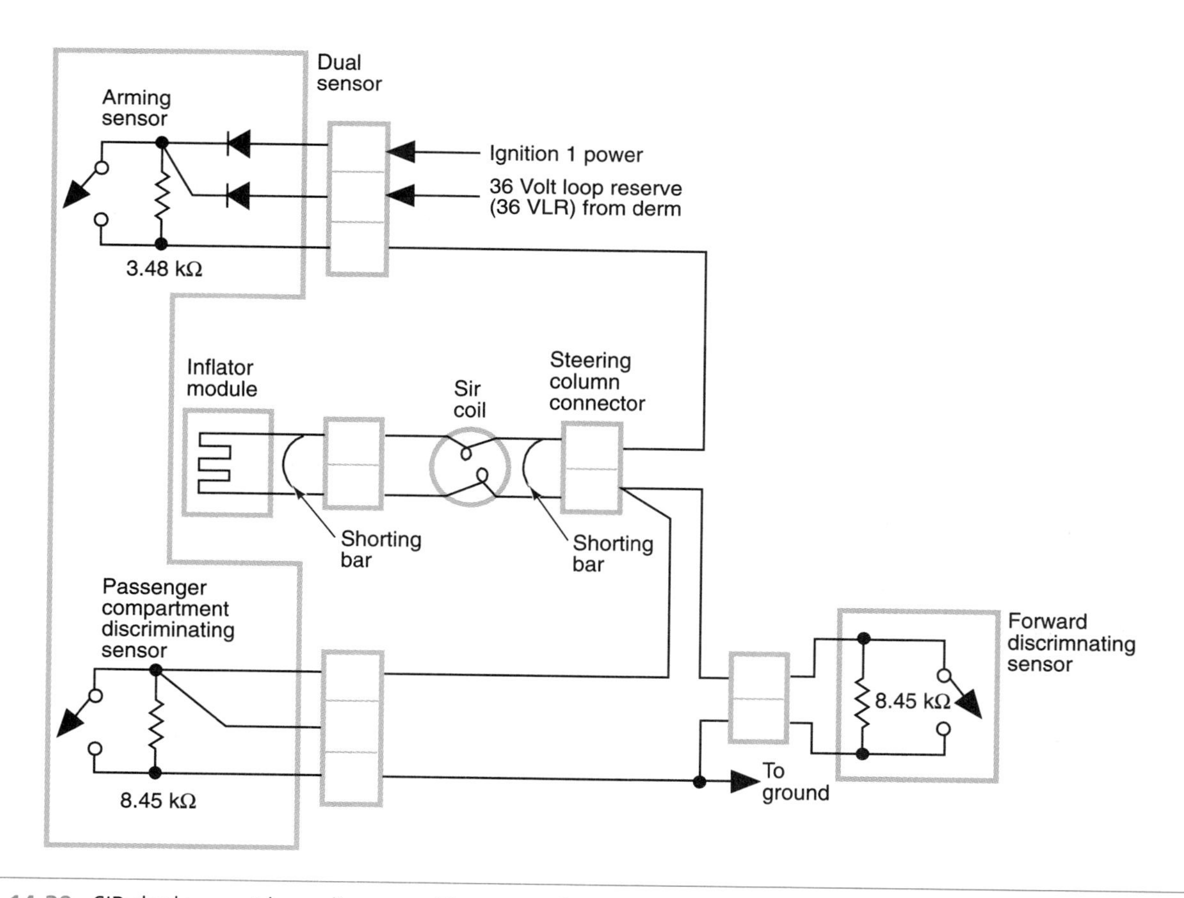

Figure 14-39 SIR deployment loop diagram. (Courtesy of General Motors Corporation)

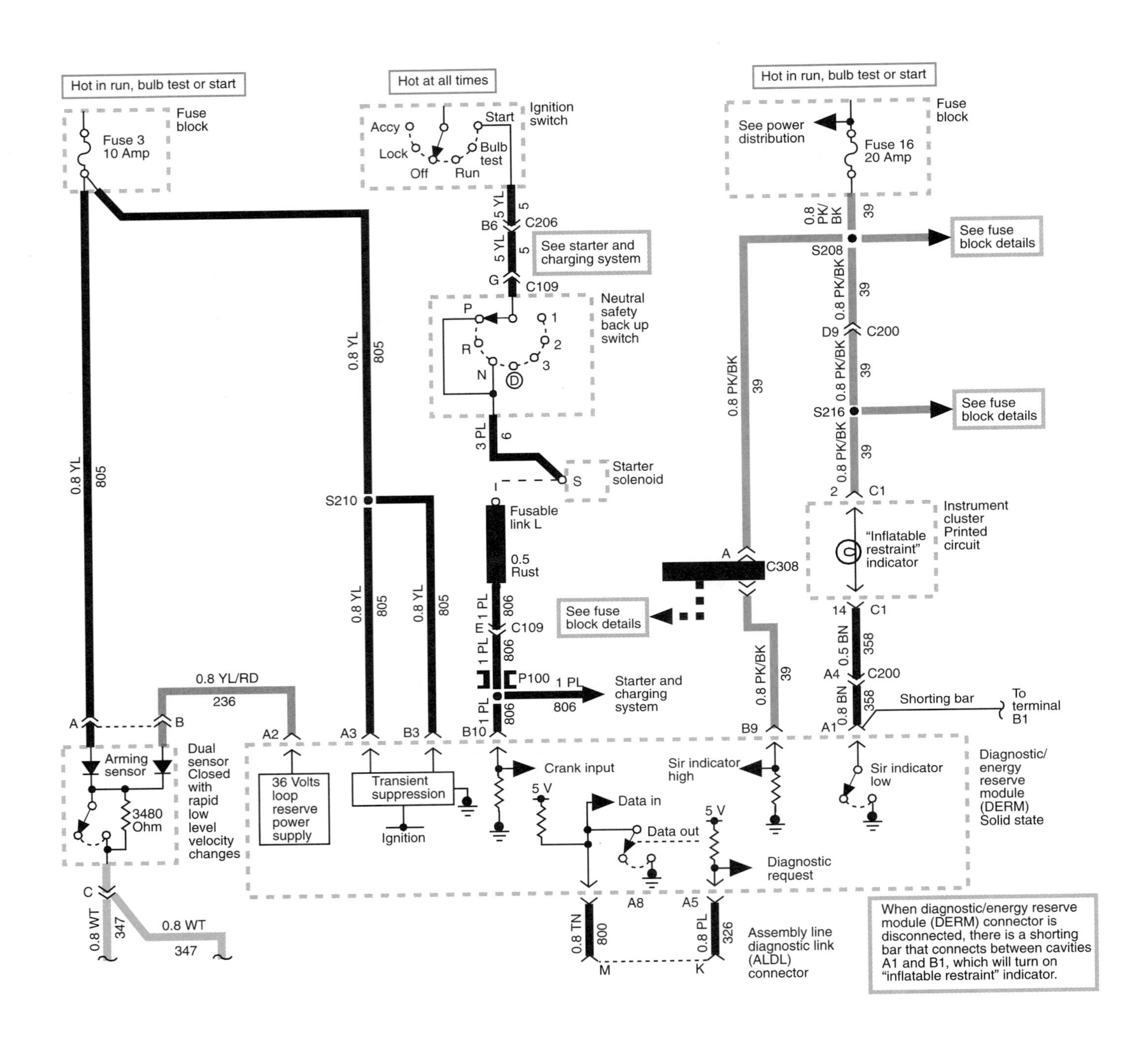

Figure 14-40 General Motors' SIR system wiring diagram. (Courtesy of General Motors Corporation)

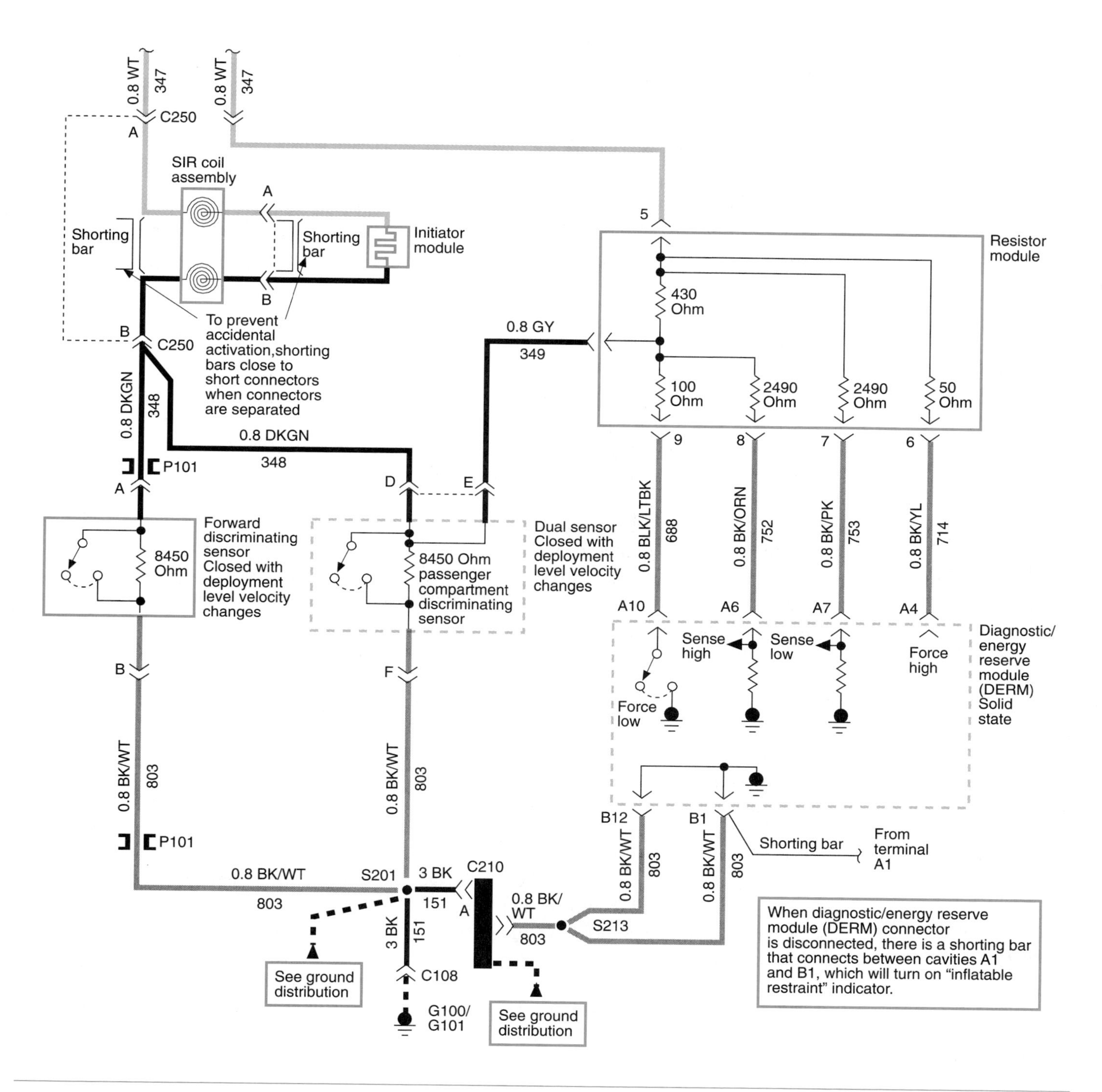

Figure 14-40 (continued)

The **memory seat** feature allows the driver to program different seat positions that can be recalled at the push of a button.

Shop Manual
Chapter 14,
page 535

In Ford's system, variable resistance sensors are used to monitor seat position instead of counting pulses.

The **easy exit** feature is an additional function of the memory seat that provides for easier entrance and exit of the vehicle by moving the seat all the way back and down. Some systems also move the steering wheel up and to full retract.

In some systems, such as the Lincoln Continental, the easy exit feature is automatically activated when the door is opened.

The ETR system is installed on the passenger side. The ETR contains an igniter device that produces a high pressure gas. The gas is used to operate a pulley that pulls the seatbelt harness snugly against the passenger.

CAUTION: Mercedes-Benz, Toyota, and some other manufacturers use mercury in their sensors. If these sensors require replacement, the old sensor must be treated as toxic waste material.

Passenger Side Air Bags

Many manufacturers offer passenger protection with an additional air bag assembly located in the dash (Figure 14-41). These air bags function in the same way as driver side air bags. However, some manufacturers inflate the passenger side air bag with a different gas or by a different set of sensors. In 1996, some model vehicles were available with side impact air bags. These will become more common in the future.

Memory Seats

The memory seat feature is an addition to the basic power seat system. Most memory seat systems share the same basic operating principles. The difference is in programming methods and number of positions that can be programmed.

The power seat system may operate in any gear position. However, the memory seat function will only operate when the transmission is in the PARK position. The purpose of the memory disable feature is to prevent accidental seat movement while the vehicle is being driven. In the PARK position, the seat memory module will receive a 12-volt signal that will enable memory operation. In any other gear selection, the 12-volt signal is removed and the memory function is disabled. This signal can come from the gear selector switch or the neutral safety switch (Figure 14-42).

Most systems provide for two seat positions to be stored in memory. Ford allows for three positions to be stored by pushing both position 1 and 2 buttons together. With the seat in the desired position, depressing the set memory button and moving the memory select switch to either the memory 1 or 2 position will store the seat position into the module's memory (Figure 14-43).

When the seat is moved from its memory position, the seat memory module transmits the voltage applied from the switch to the motors. The module counts the pulses produced by motor operation, and then stores the number of pulses and direction of movement in memory. When the memory switch is closed, the module will operate the seat motors until it counts down to the preset number of pulses.

When the easy exit switch is closed, voltage is applied to both memory 1 and 2 inputs of the module. This signal is interrupted by the module to move the seat to its full down and full back position. As the seat moves to the easy exit position it counts the pulses and stores this information in memory.

Memory is not lost when the ignition switch is turned off. However, it is lost if the battery is disconnected. If memory is lost, the position of the seat at the time power is restored becomes set in memory for both positions.

Electronic Sunroof Concepts

Some manufacturers have introduced electronic control of their electric sunroofs. These systems incorporate a pair of relay circuits and a timer function into the control module. Although there are variations between manufacturers, the systems discussed here provide a study of the two basic types of systems.

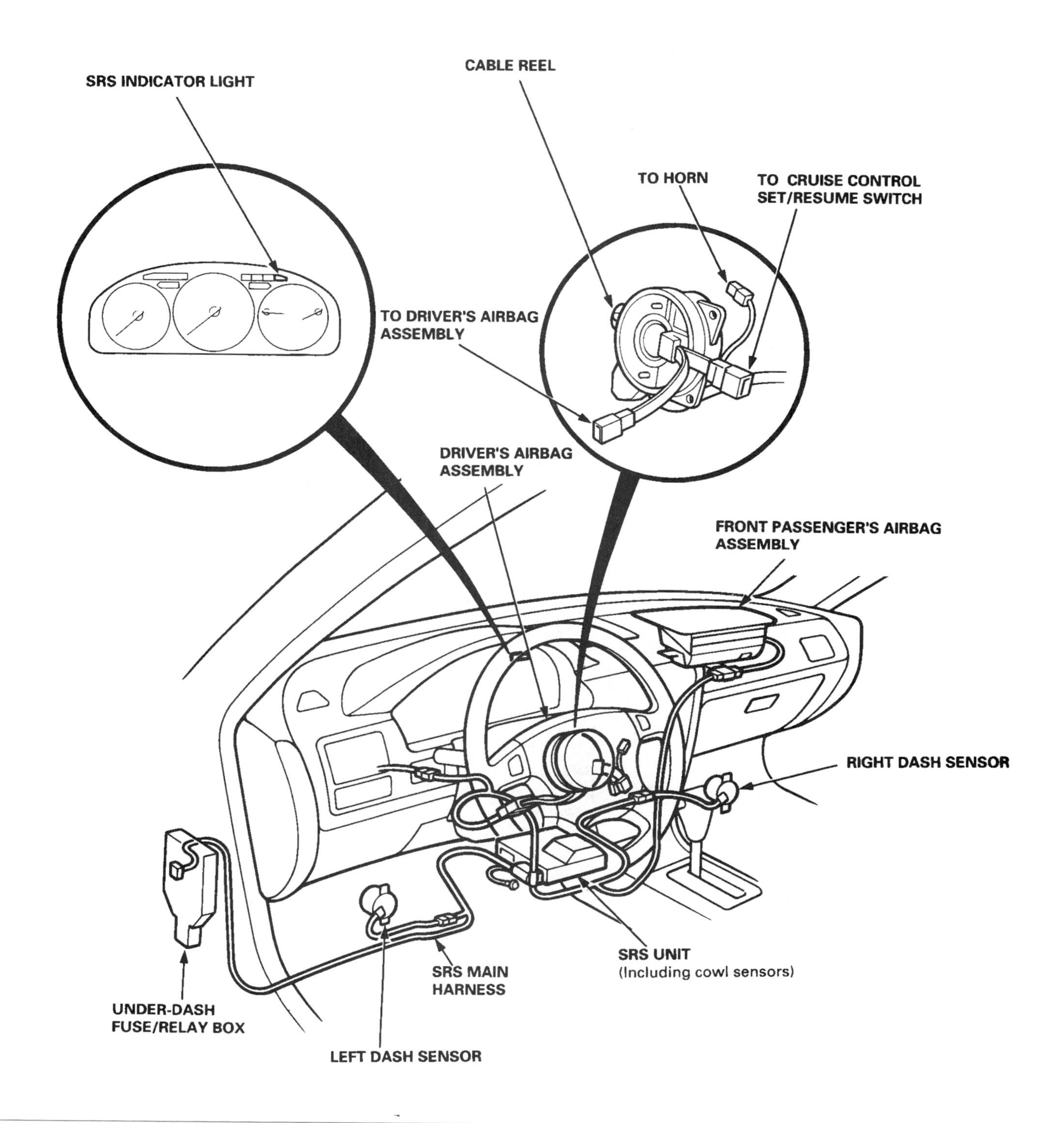

Figure 14-41 Component location for Honda's driver and passenger SRS. (Courtesy of Honda Motor Company)

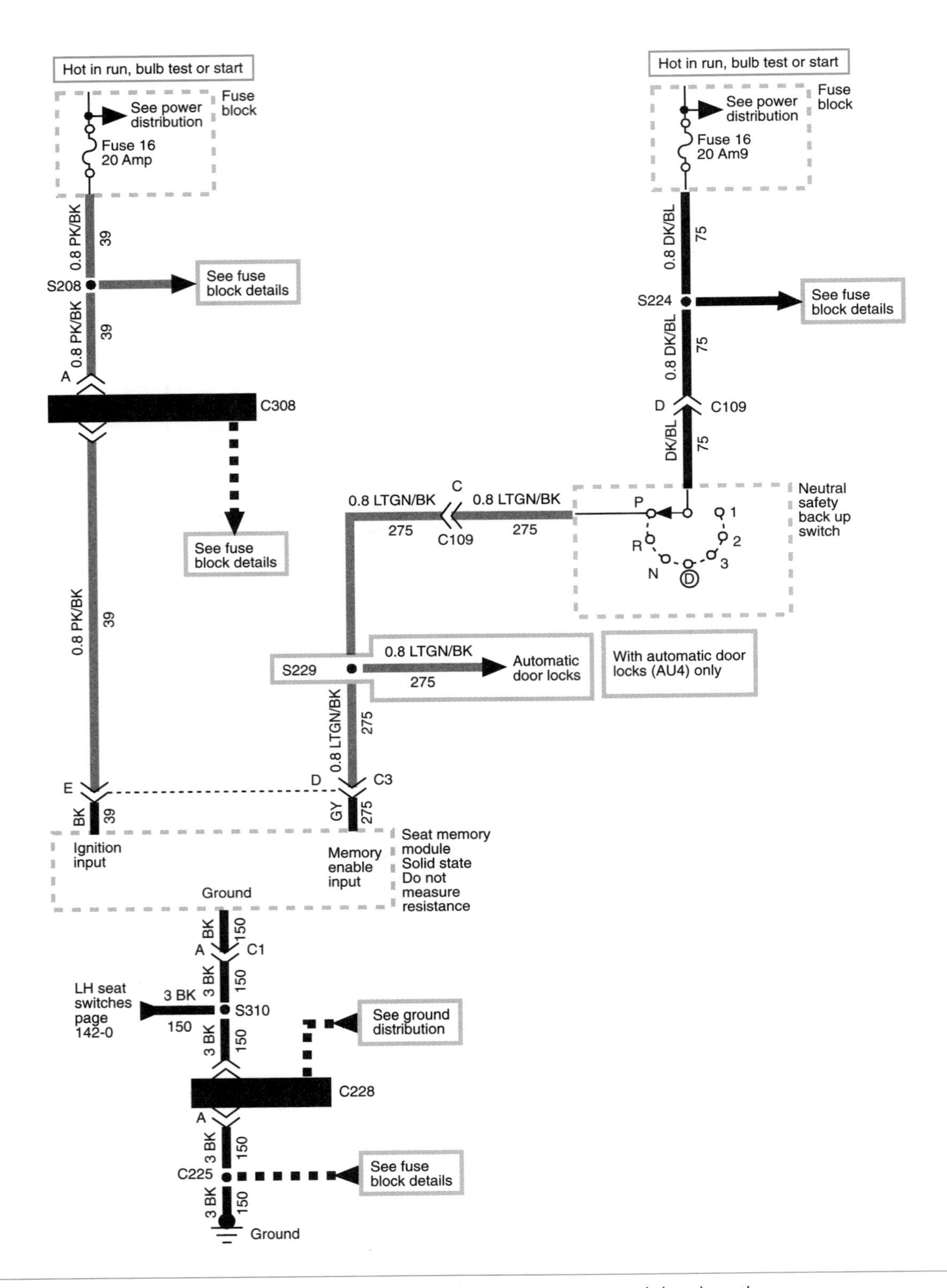

Figure 14-42 The neutral safety/ backup switch signals the seat memory module when the transmission is in PARK. (Courtesy of General Motors Corporation)

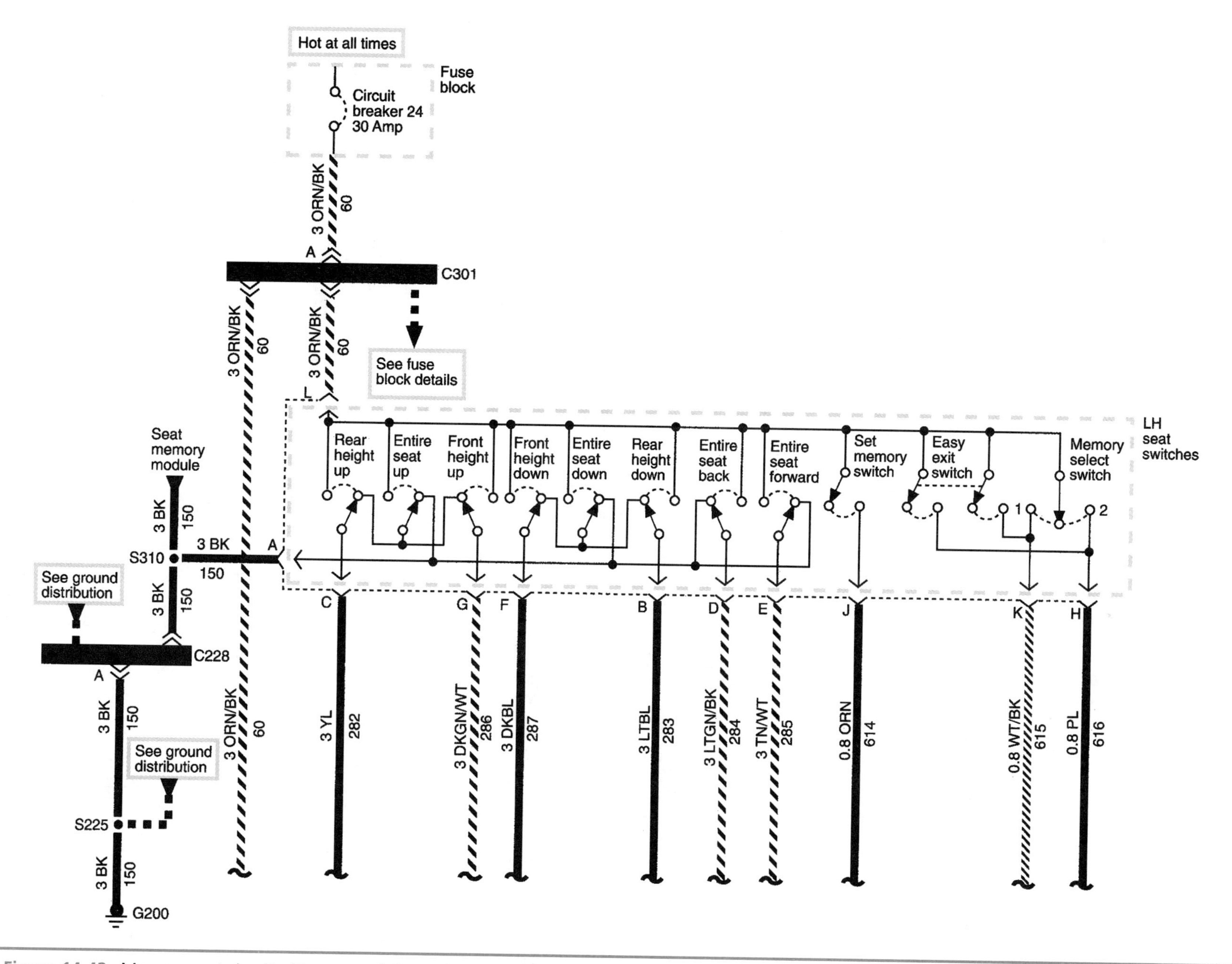

Figure 14-43 Memory seat circuit. (Courtesy of General Motors Corporation)

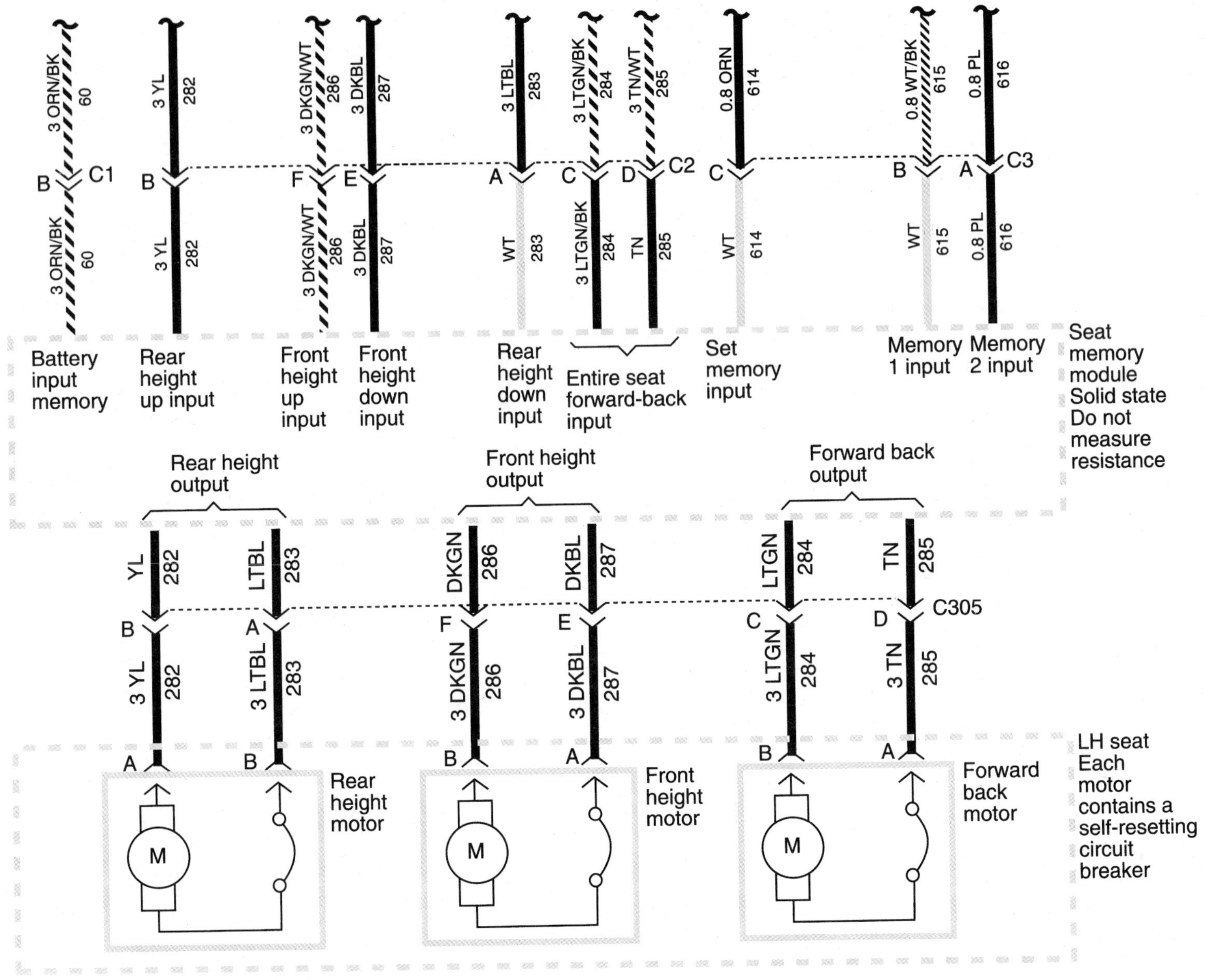

3 ORN/BK 60 — B C1 — 3 ORN/BK 60
3 YL 282 — B — 3 YL 282
3 DKGN/WT 286 — F — 3 DKGN/WT 286
3 DKBL 287 — E — 3 DKBL 287
3 LTBL 283 — A — WT 283
3 LTGN/BK 284 — C — 3 LTGN/BK 284
3 TN/WT 285 — D C2 — TN 285
0.8 ORN 614 — C — WT 614
0.8 WT/BK 615 — B — WT 615
0.8 PL 616 — A C3 — 0.8 PL 616

Battery input memory
Rear height up input
Front height up input
Front height down input
Rear height down input
Entire seat forward-back input
Set memory input
Memory 1 input
Memory 2 input
Seat memory module Solid state Do not measure resistance

Rear height output
Front height output
Forward back output

YL 282 — B — 3 YL 282 — A
LTBL 283 — A — 3 LTBL 283 — B
DKGN 286 — F — 3 DKGN 286 — B
DKBL 287 — E — 3 DKBL 287 — A
LTGN 284 — C — 3 LTGN 284 — B
TN 285 — D C305 — 3 TN 285 — A

M Rear height motor
M Front height motor
M Forward back motor
LH seat Each motor contains a self-resetting circuit breaker

Figure 14-43 (continued)

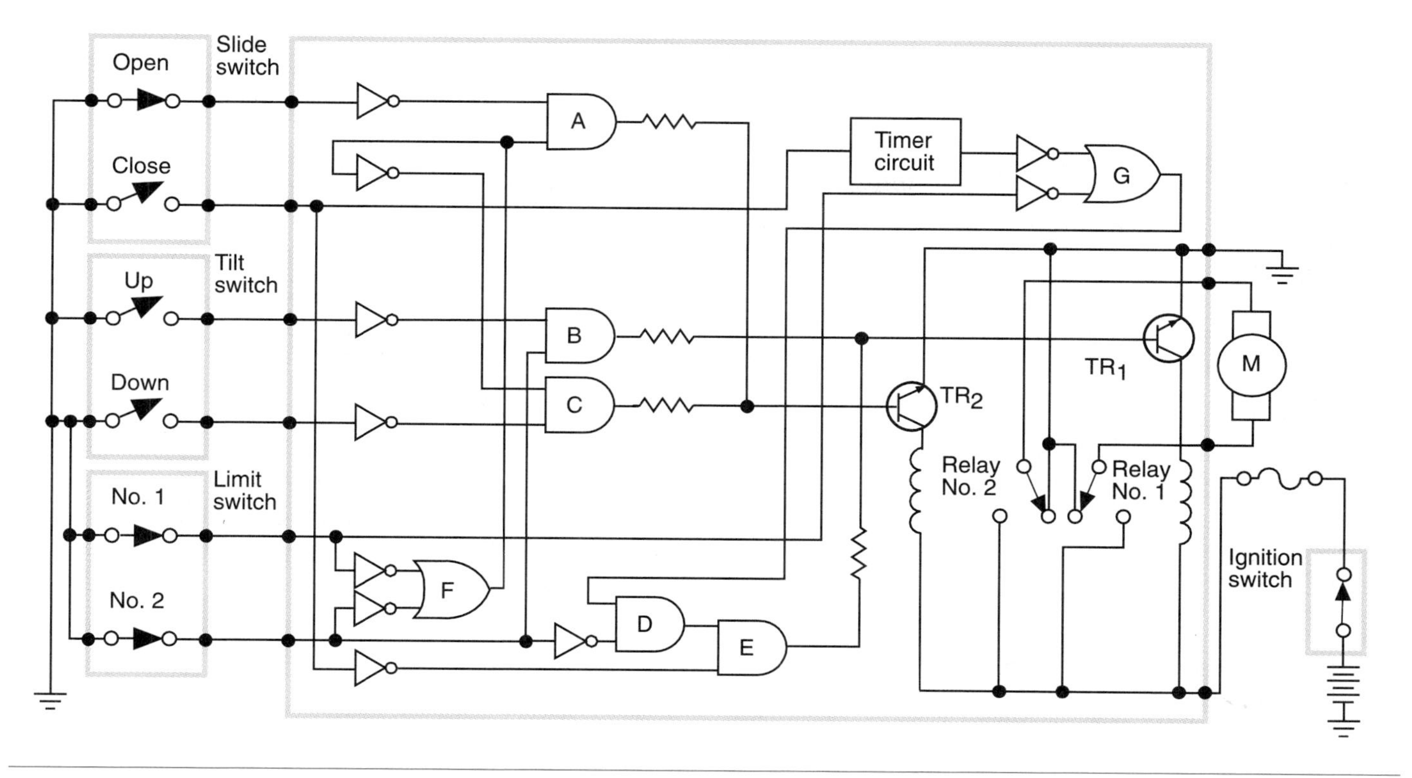

Figure 14-44 Toyota sunroof circuit using electronic controls.

Electronic-Controlled Toyota Sunroof

Refer to schematic (Figure 14-44) of a typical sunroof control circuit used by Toyota. The movement of the sunroof is controlled by the motor that operates a drive gear. The drive gear either pushes or pulls the connecting cable to move the sunroof.

Motor rotation is controlled by relays that are activated according to signals received from the slide, tilt, and limit switches. The limit switches are operated by a cam on the motor (Figure 14-45).

The schematics used to explain the operation of the Toyota sunroof use logic gates. If needed, refer to Chapter 12 of this manual to review the operation of the gates.

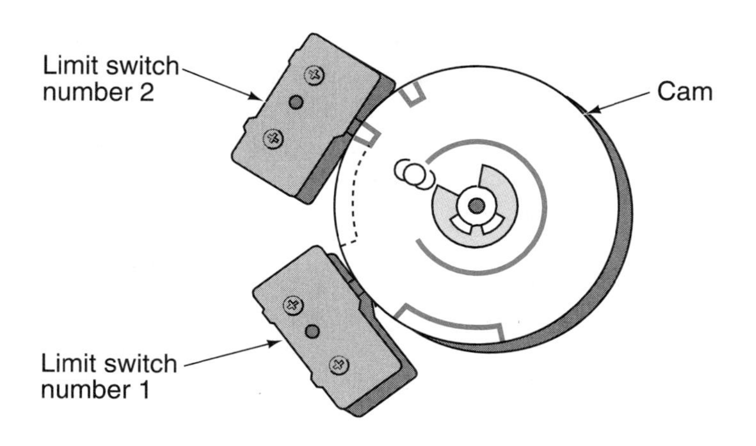

Figure 14-45 The limit switches operate off a cam on the motor.

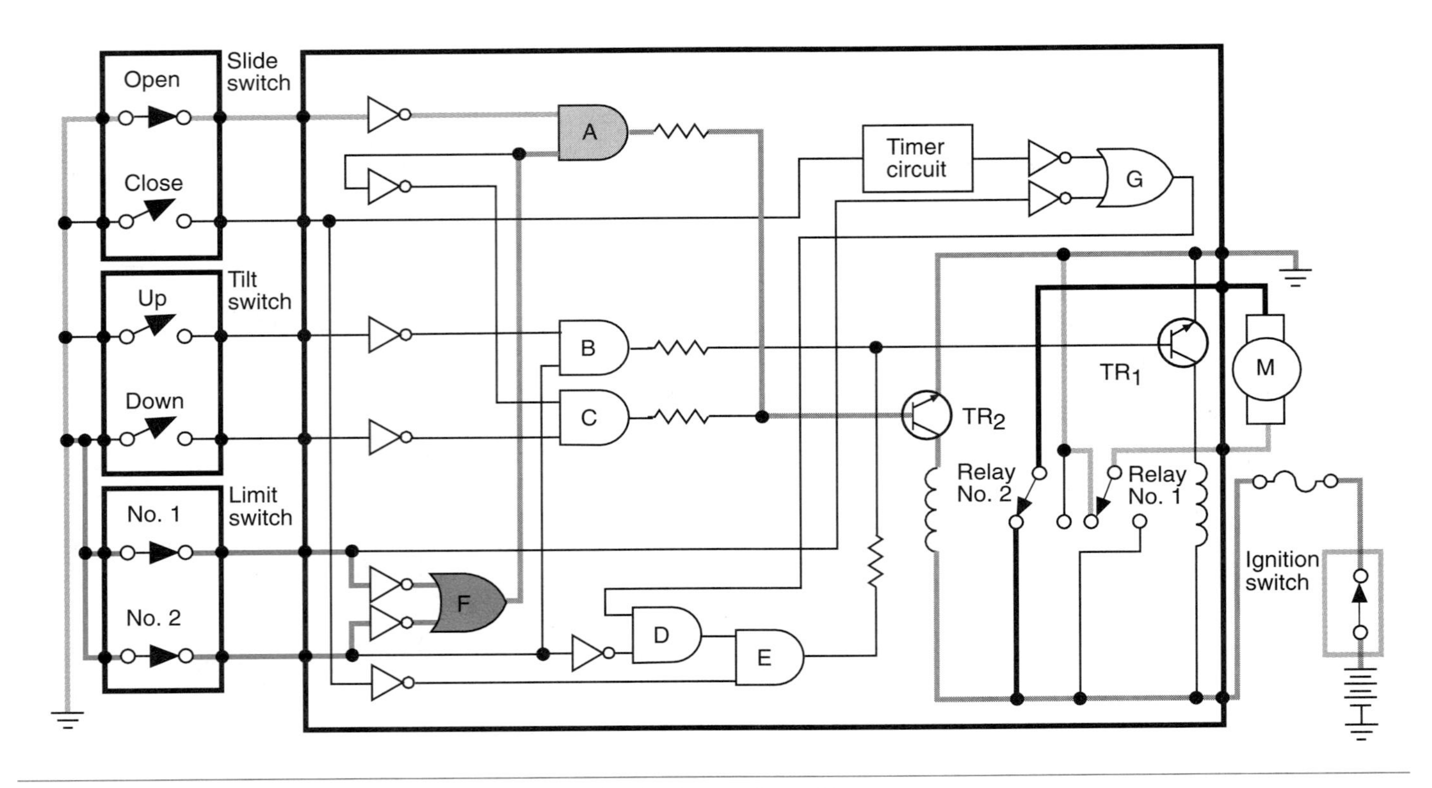

Figure 14-46 Circuit operation when the switch is in the OPEN position.

Negative logic defines the most negative voltage as a logical 1 in the binary code.

The logic gates of this system operate on the principle of negative logic. When the slide switch is moved to the OPEN position, either limit switch 1 or both limit switches are closed (Figure 14-46). Limit switches 1 and 2 provide a negative side signal to the OR gate labeled F. The output from gate F is sent to gate A. Gate A is an AND gate, requiring input from gate F and the open slide switch. The output signal from gate A is used to turn on TR_2. This provides a ground path for the coil in relay 2. Battery voltage is applied to the motor through relay 2; the ground path is provided through the de-energized relay 1. Current is sent to the motor as long as the OPEN switch is depressed. If the OPEN switch is held in this position too long, a clutch in the motor disengages the motor from the drive gear.

Operation of the system during closing depends on how far the sunroof is open. If the sunroof is open more than 7.5 inches and the slide contact is moved to the CLOSE position, an input signal is sent to gate E (Figure 14-47). The other input signal required at gate E is received from the limit switches. The limit switch 1 signal passes through the OR gate G to the AND gate D. Limit switch 2 provides the second signal required by gate D. The output signal from D is the second input signal required by gate E. The output signal from E turns on TR_1. This energizes relay 1 and reverses the current flow through the motor. The motor will operate until the slide switch is opened or limit switch 2 opens.

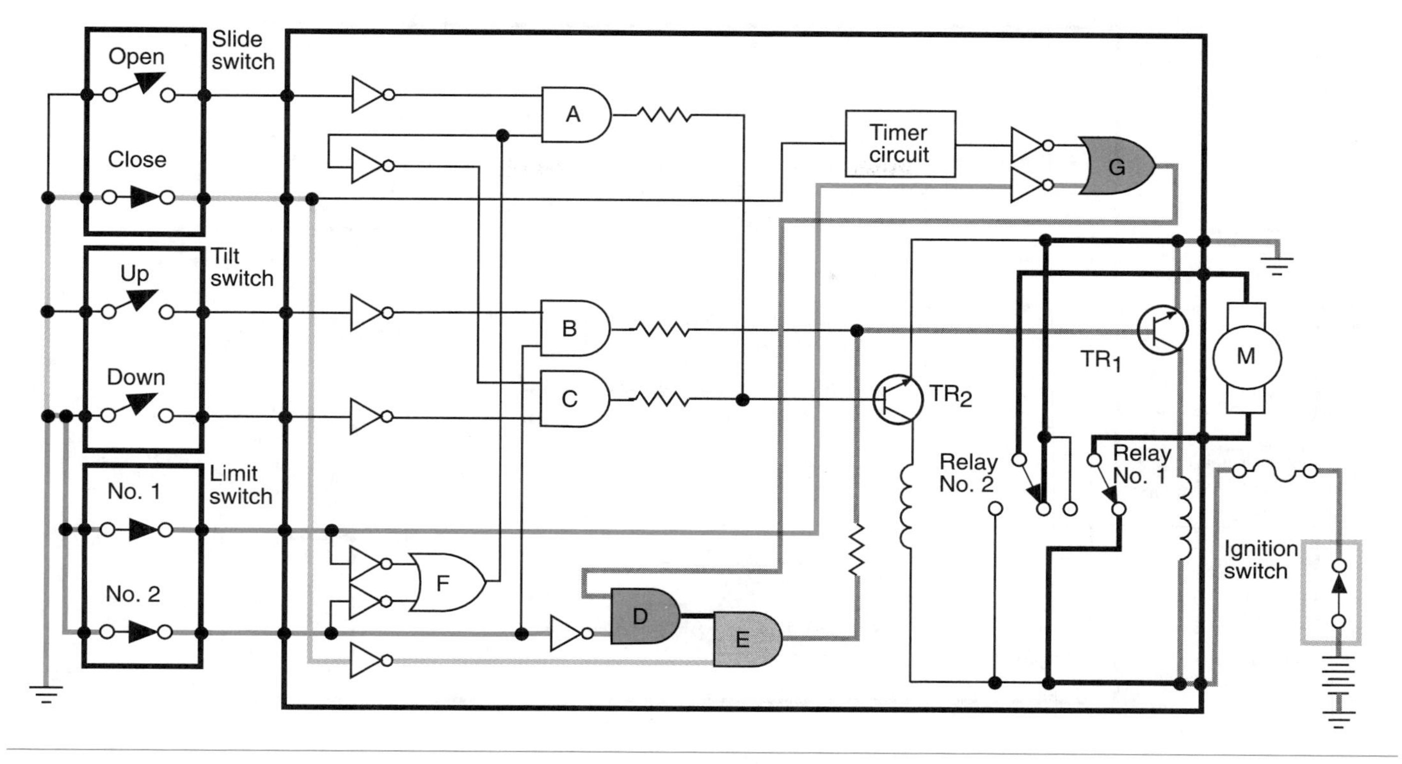

Figure 14-47 Circuit operation when the switch is in the CLOSE position and the sun roof is open more than 7.5 inches.

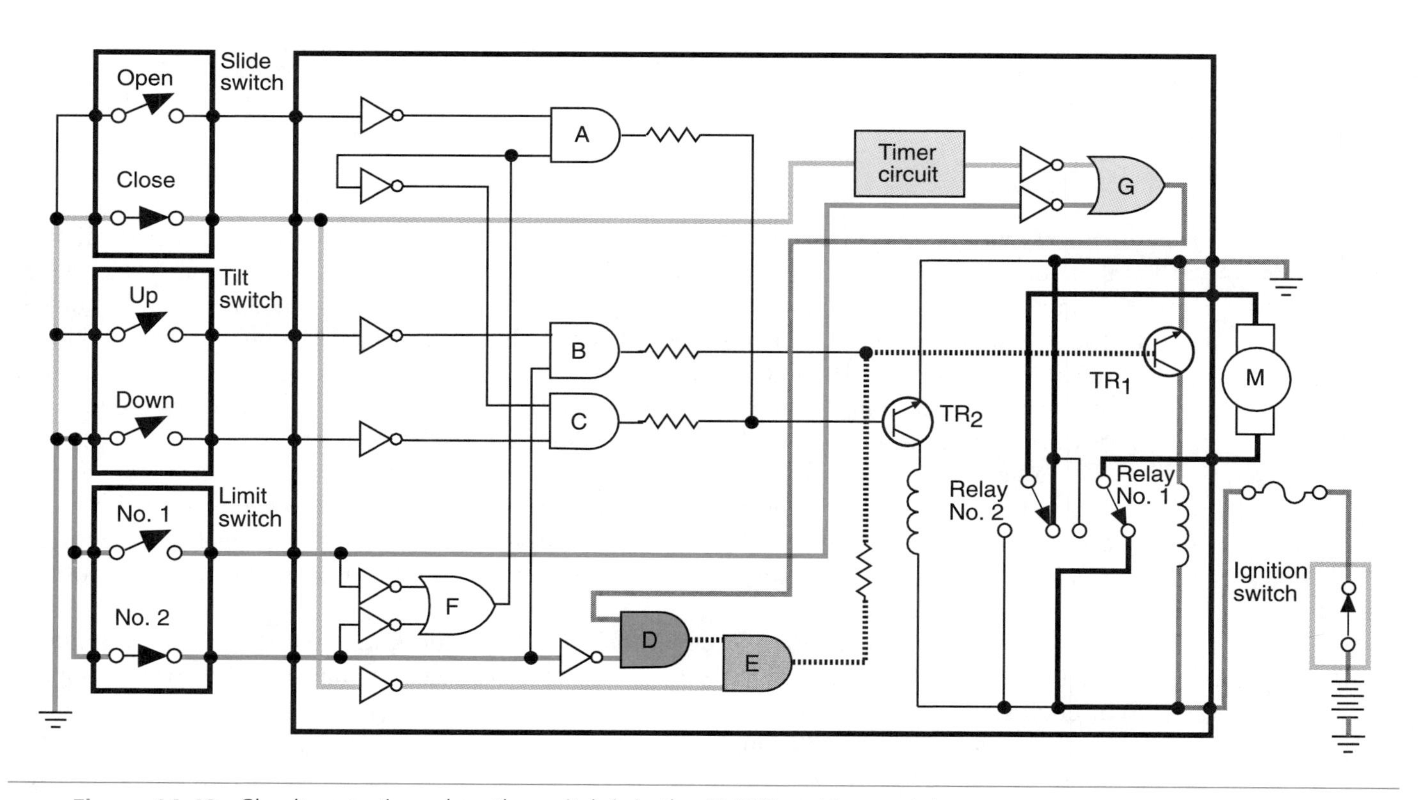

Figure 14-48 Circuit operation when the switch is in the CLOSE position and the sun roof is open less than 7.5 inches.

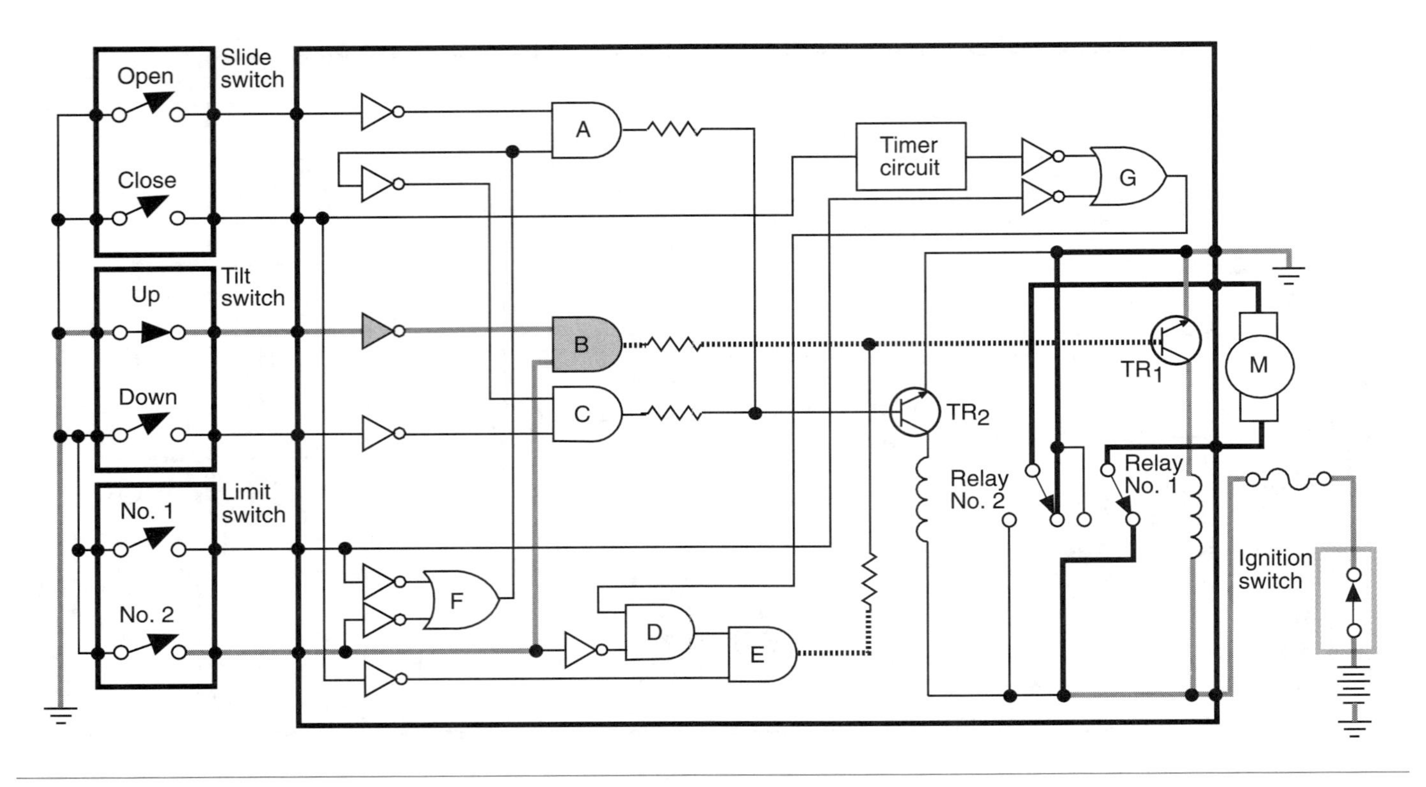

Figure 14-49 Circuit operation when the switch is in the TILT UP position.

If the sunroof is open less than 7.5 inches and the slide switch is placed in the CLOSE position, the timer circuit is activated (Figure 14-48). The CLOSE switch signals the timer and provides an input signal to gate E. Limit switch 1 is open when the sunroof is opened less than 7.5 inches. The second input signal required by gate D is provided by the timer. The timer is activated for .5 second. This turns on TR_1 and operates the motor for .5 second, or long enough for rotation of the motor to close limit switch 1. When limit switch 1 is closed, the operation is the same as described when the sunroof is closed when it is more than 7.5 inches open.

When the tilt switch is located in the UP position, a signal is imposed on gate B (Figure14-49). This signal is inverted by the NOT gate and is equal to the value received from the opened number 2 limit switch. The output signal from gate B turns on TR_1, which energizes relay 1 to turn on the motor. The motor clutch will disengage if the switch is held in the closed position longer than needed.

When the tilt switch is placed in the DOWN position, a signal is imposed on gate C (Figure 14-50). The second signal to gate C is received from the limit switches (both are open) through gate F. The signal from gate F is inverted by the NOT gate and is equal to that from the DOWN switch. The output signal from gate C turns on TR_2 and energizes relay 2 to lower the sunroof. If the DOWN switch is held longer than necessary, limit switch 1 closes. When this switch is closed, the signals received by gate F are not opposite. This results in a mixed input to gate C and turns off the transistor.

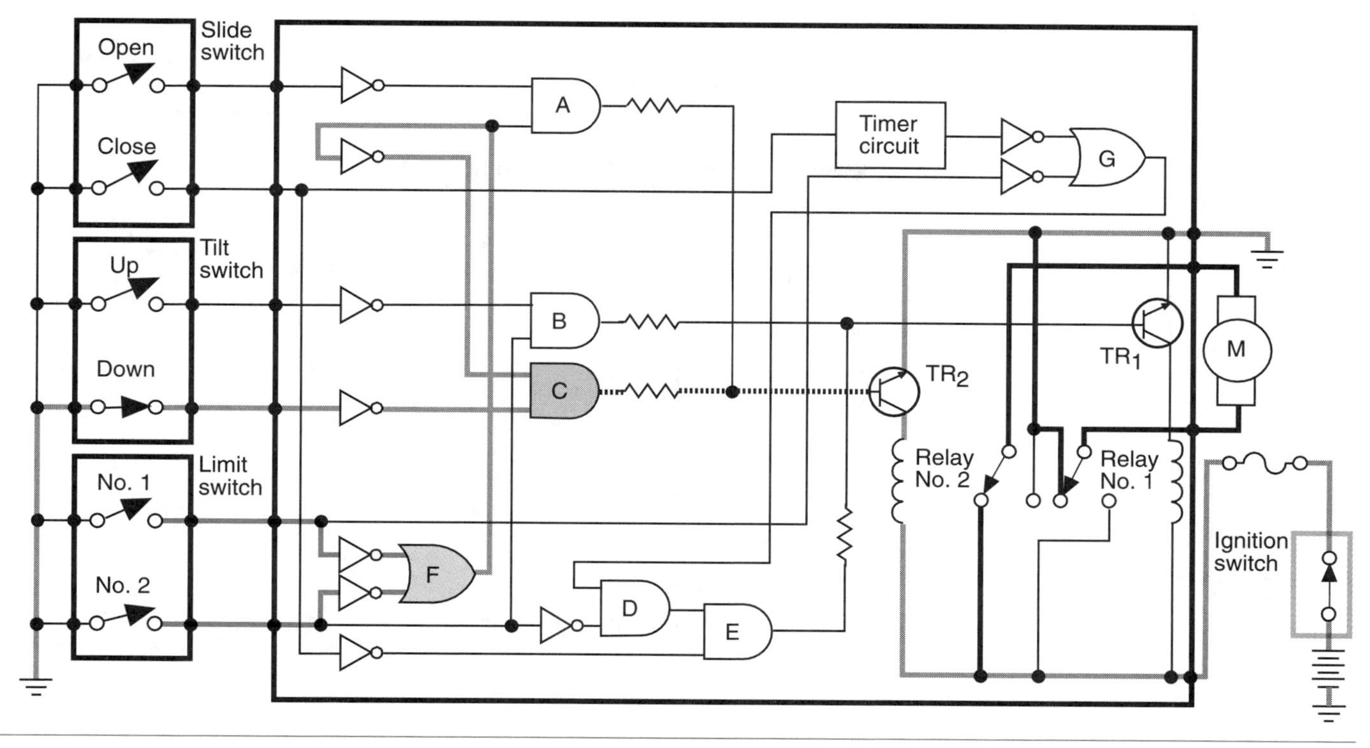

Figure 14-50 Circuit operation when the switch is in the TILT DOWN position.

Electronically Controlled General Motors' Sunroof

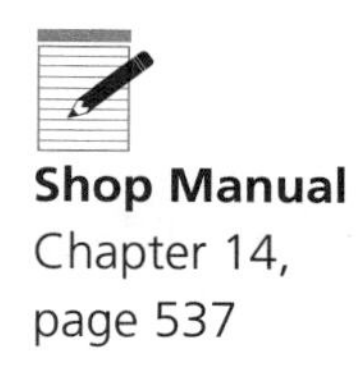

Shop Manual
Chapter 14,
page 537

See the schematic (Figure 14-51) of the sunroof system used on some GM model vehicles. The timing module uses inputs from the control switch and the limit switches to direct current flow to the motor. Depending on the inputs, the relays will be energized to rotate the motor in the proper direction. When the switch is located in the OPEN position, the open relay is

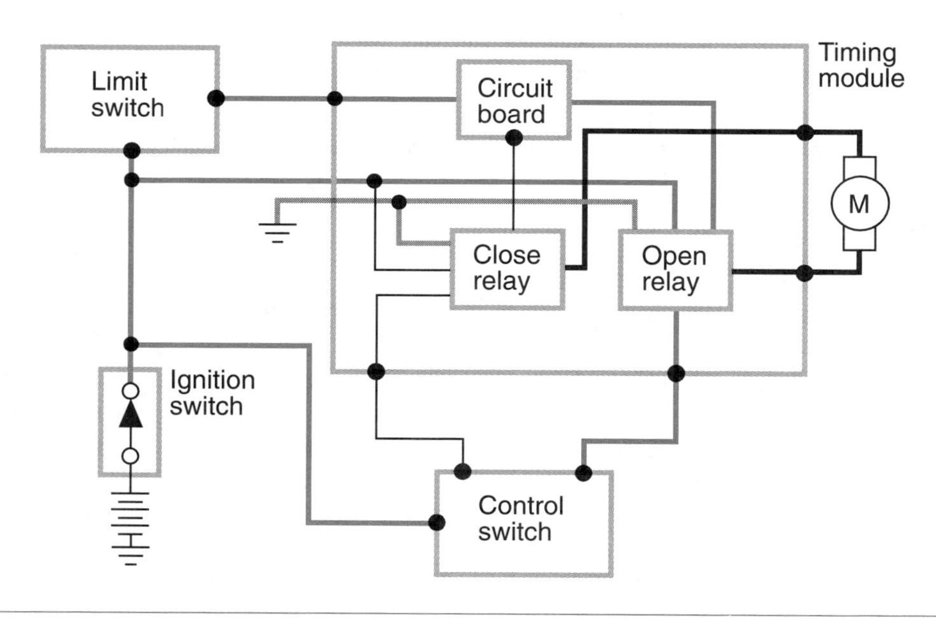

Figure 14-51 Block diagram of the GM sunroof.

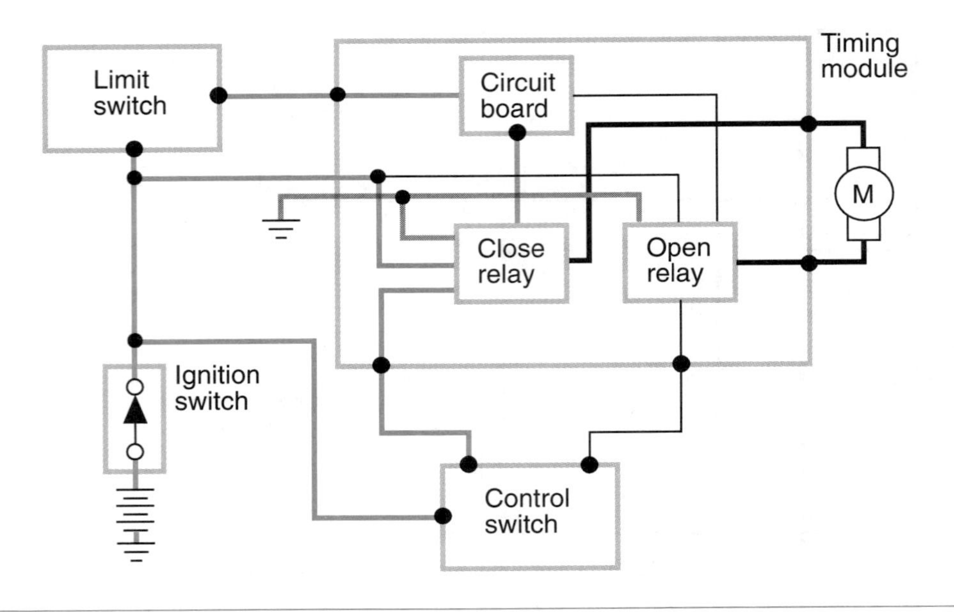

Figure 14-52 Sunroof circuit operation when the control switch is placed in the OPEN position.

energized sending current to the motor (Figure 14-52). The sunroof will continue to retract as long as the switch is held in the OPEN position. When the sunroof reaches its full open position, the limit switch will open and break the circuit to the open relay.

Placing the switch in the CLOSE position will energize the close relay. The current sent to the motor is in the opposite direction to close the sunroof. If the close switch is held until the sunroof reaches the full closed position, the limit switch will open.

Antitheft Systems

Shop Manual
Chapter 14,
page 539

Antitheft systems are deterrent systems designed to scare off would-be thieves by sounding alarms and/or disabling the ignition system.

Arming means placing the alarm system in readiness to detect an illegal entry.

A vehicle is stolen in the United States every 26 seconds. In response to this problem, vehicle manufacturers are offering antitheft systems as optional or standard equipment. The illustration (Figure 14-53) shows many of the common components that are used in an antitheft system. These components include:

1. An electronic control module.
2. Door switches at all doors.
3. Trunk key cylinder switch.
4. Hood switch.
5. Starter inhibitor relay.
6. Horn relay.
7. Alarm.

In addition, many systems incorporate the exterior lights into the system. The lights are flashed if the system is activated.

For the system to operate, it must first be armed. This is done when the ignition switch is turned off and the doors are locked. When the driver's door is shut, a security light will illuminate for approximately 30 seconds to indicate that the system is armed and ready to function. If any other door is open, the system will not arm until it is closed.

The control module monitors the switches. If the doors or trunk are opened, or the key cylinders are rotated, the module will activate the system. The control module will sound the alarm and flash the lights until the timer circuit has counted down. At the end of the timer function, the system will automatically rearm itself.

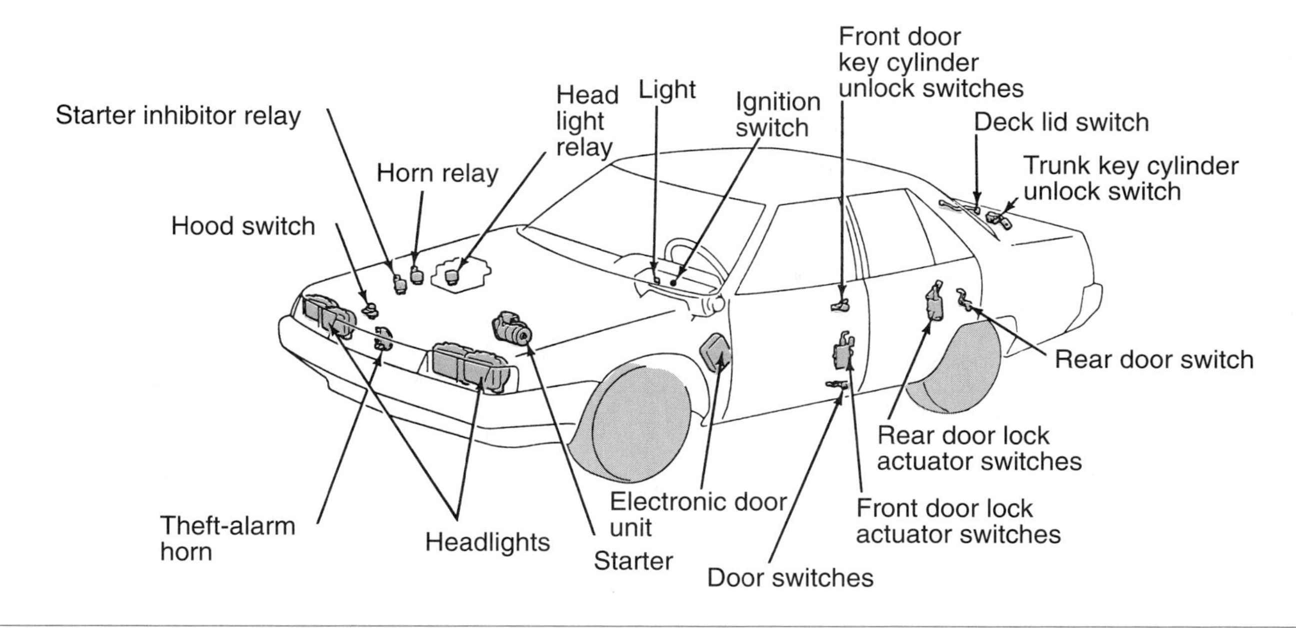

Figure 14-53 Typical components of an antitheft system. (Courtesy of Mitsubishi Motor Sales of America, Inc.)

Some systems use ultrasonic sensors that will signal the control module if someone attempts to enter the vehicle through the door or window (Figure 14-54). The sensors can be placed to sense the parameter of the vehicle and sound the alarm if someone enters within the protected parameter distance.

The system can also use current sensitive sensors that will activate the alarm if there is a change in the vehicle's electrical system. The change can occur if the courtesy lights come on or if an attempt is made to start the engine.

The following systems are provided to give you a sample of the types of antitheft systems used.

Ford Antitheft Protection System

If the system is triggered, it will sound the horn; flash the low-beam headlights, taillights, and parking lamps; and disable the ignition system. See the schematic (Figure 14-55) of the system.

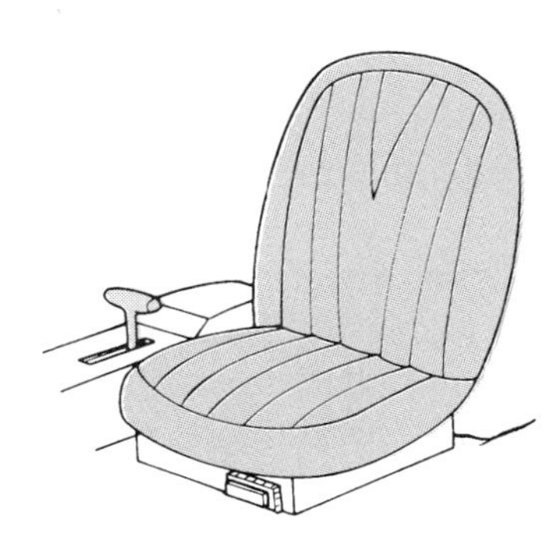

Figure 14-54 The ultrasonic sensor will trigger the alarm if someone enters the vehicle without the alarm being disarmed.

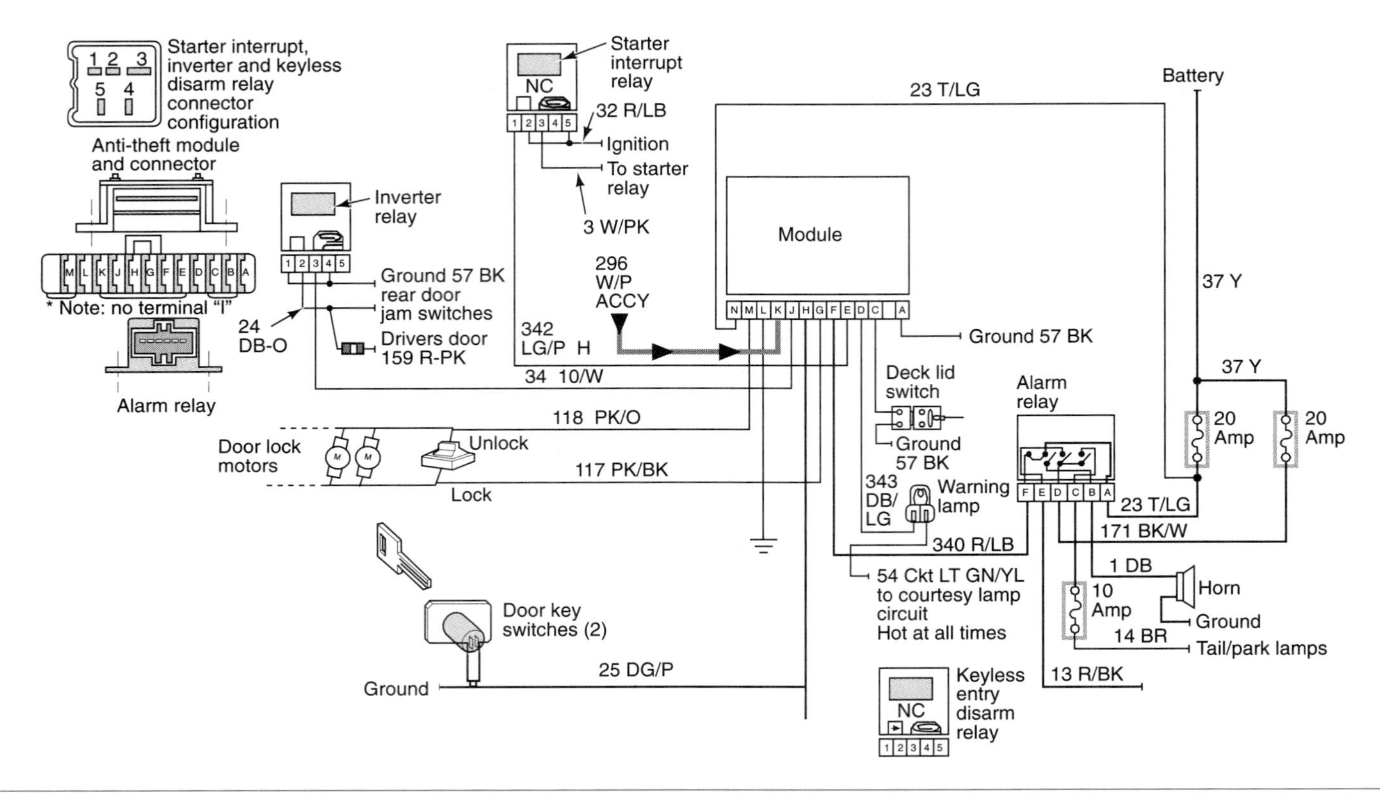

Figure 14-55 Circuit schematic of Ford's antitheft system. (Courtesy of Chilton Book Company)

The arming process is started when the ignition switch is turned off. Voltage provided to the module at terminal K is removed (Figure 14-56). When the door is opened, a voltage is applied to the courtesy lamp circuit 24 through the closed switch. This voltage energizes the inverter relay and provides a ground for module terminal J. This signal is used by the control module to provide an alternating ground at terminal D, causing the indicator lamp to blink. The flashing indicator light alerts the driver that the system is not armed. When the door lock switch is placed in the LOCK position, battery voltage is applied to terminal G of the module. The module uses this signal to apply a steady ground at terminal D, causing the indicator light to stay on continuously. When the door is closed, the door switch is opened. The opened door switch de-energizes the inverter relay coil. Terminal J is no longer grounded and the indicator light goes out after a couple of seconds.

To disarm the system, one of the front doors must be opened with a key or by pressing the correct code into the keyless entry keypad. Unlocking the door closes the lock cylinder switch and grounds terminal H of the module (Figure 14-57). This signal disarms the system.

Once the system is armed, if terminal J and C receive a ground signal the control module will trigger the alarm. Terminal C is grounded if the trunk tamper switch contacts close. Terminal J is grounded when the inverter relay contacts are closed. The inverter relay is controlled by the doorjamb switches. If one of the doors is opened, the switch closes and energizes the relay coil. The contacts close and ground is provided to terminal J (Figure 14-58).

When the alarm is activated, a pulsating ground is provided at module terminal F (Figure 14-59). This pulsating ground energizes and de-energizes the alarm relay. As the relay contacts open and close, a pulsating voltage is sent to the horns and exterior lights.

At the same time, the start interrupt circuit is activated. The start interrupt relay receives battery voltage from the ignition switch when it is in the START position. When the alarm is activated, the module provides a ground through terminal E, causing the relay coil to be energized. The energized relay opens the circuit to the starter system, preventing starter operation.

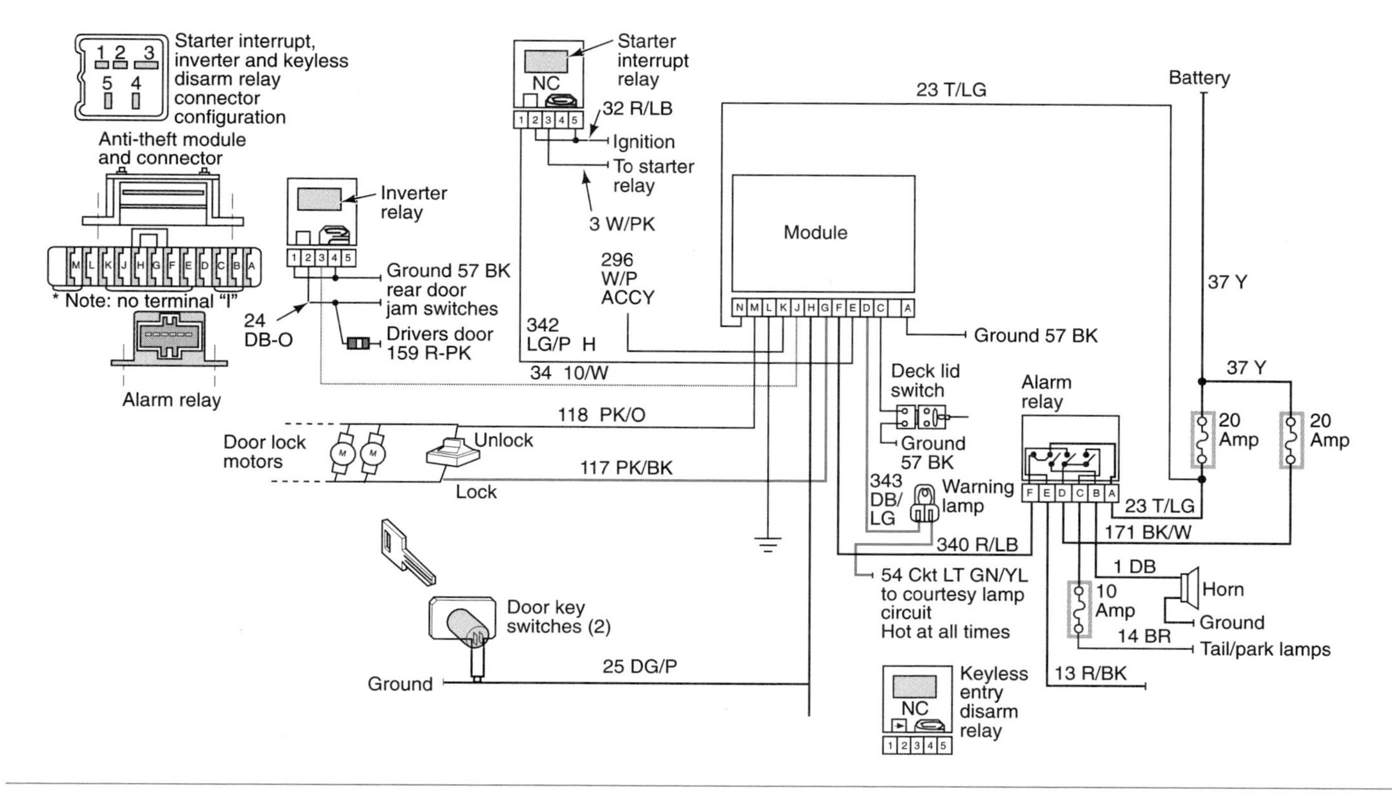

Figure 14-56 Circuit operation during the arming process. (Courtesy of Chilton Book Company)

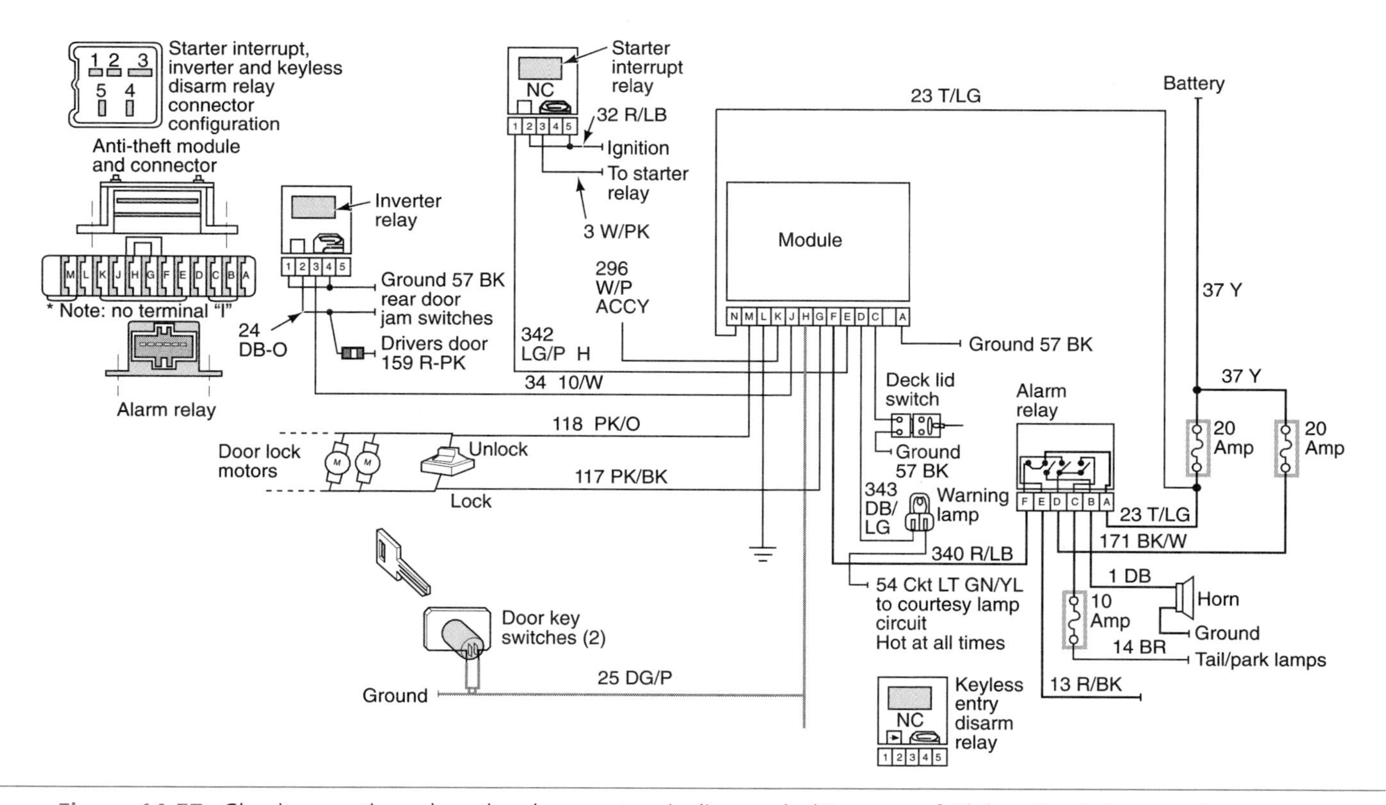

Figure 14-57 Circuit operation when the alarm system is disarmed. (Courtesy of Chilton Book Company)

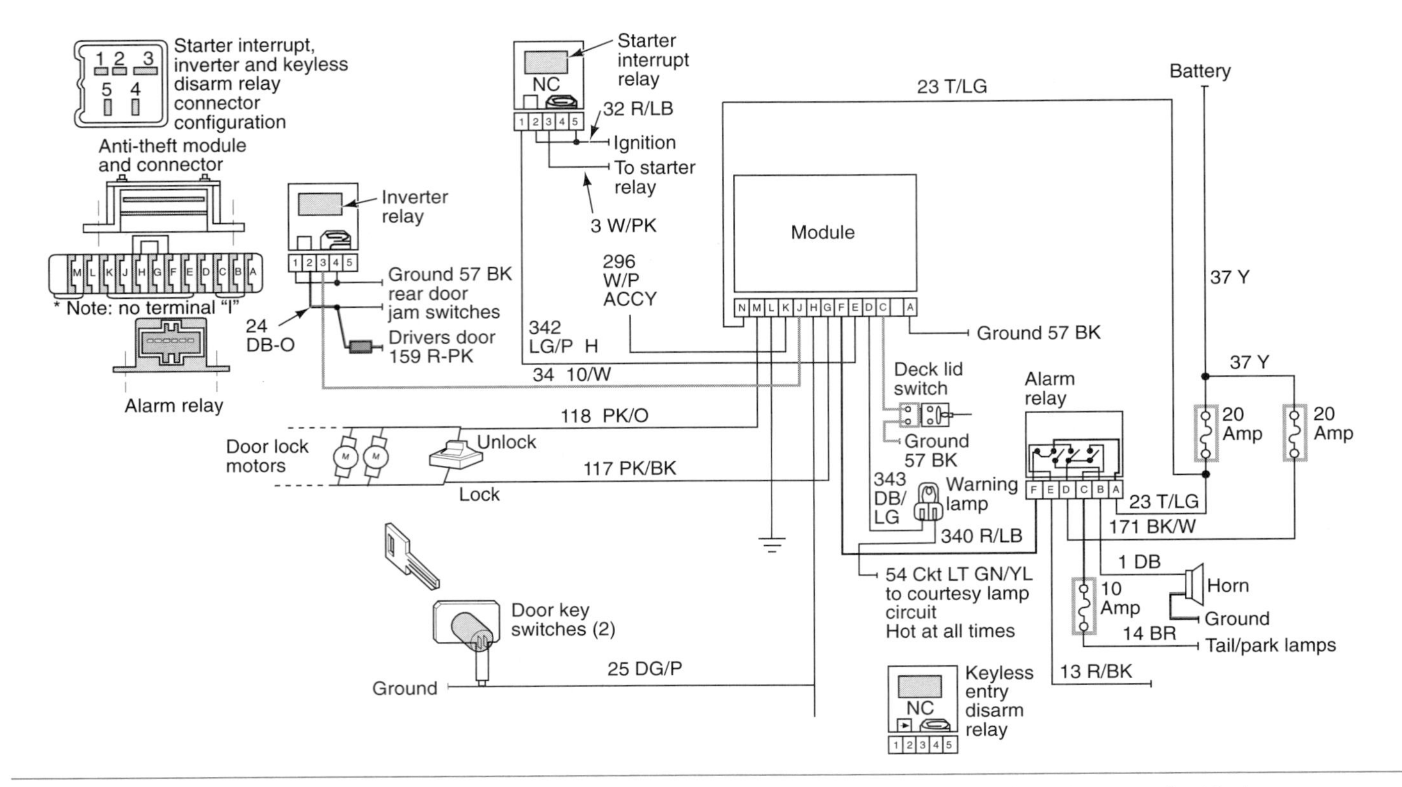

Figure 14-58 Circuit operation indicating the signal to the module when one of the doors is opened with the system armed. (Courtesy of Chilton Book Company)

GM Pass-Key Antitheft System

Pass-key is derived from personal automotive security system.

The basic operation of the GM system is similar to that of the Ford system. An additional feature that GM offers can be used as a stand alone antitheft system or in combination with other systems. This system acts as an engine disable system, by using a pass-key arrangement. The ignition key has an electronic pellet that has a coded resistance value (Figure 14-60). Each of the different pellets used has a specific resistance value that ranges between 380 ohms and 12,300 ohms. The ignition key must be the correct cut to operate the lock and the correct electrical code to close the starter circuit.

When the ignition key is inserted into the cylinder, the pellet makes contact with the resistor sensing contact. When the cylinder is rotated to the START position, battery voltage is sent to a decoder module. In addition, the resistance value of the key pellet is sent to the decoder module. The resistance value is compared to memory. If they match, the starter enable relay is energized. This completes the starter circuit and signals the ECM to start fuel delivery.

If the key pellet resistance does not match, the decoder will prevent starting of the engine for 2 to 4 minutes. Even if the lock cylinder is removed, the engine will not start because the start enable relay will not energize.

Two antitheft devices were introduced in 1996. One is GM's new low-cost pass-lock system, which works like the pass-key system, but without the chip in the key. Pass-lock systems have a Hall-effect sensor in the key cylinder that measures the magnetic properties of the key as it is inserted into the cylinder. The cut pattern of every key has its own magnetic identity. If the wrong key is inserted into the lock cylinder, the car will not start, even if the lock cylinder turns.

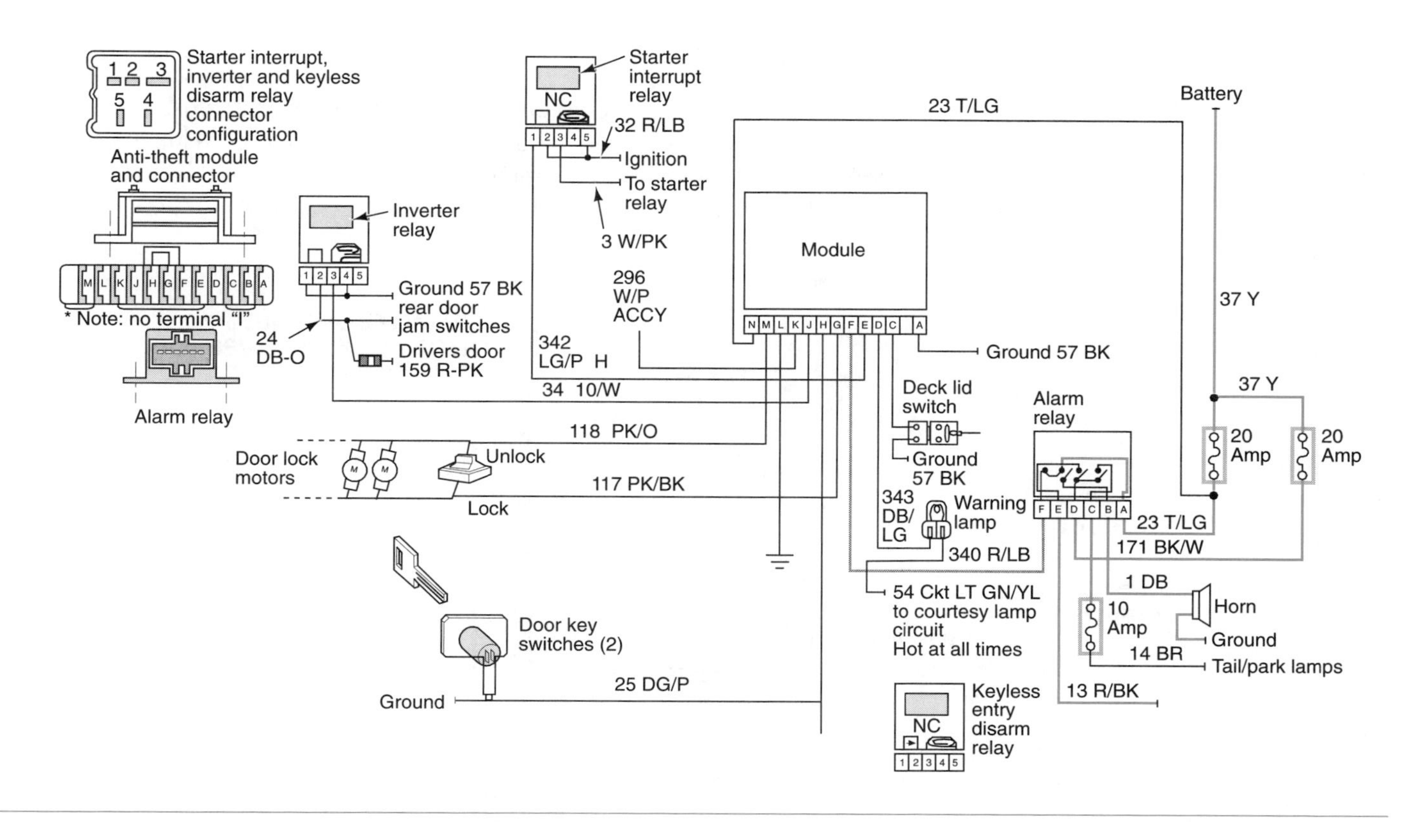

Figure 14-59 The module sends a pulsating signal to the alarm relay to flash the exterior lights and sound the horn. (Courtesy of Chilton Book Company)

Ford's new ignition-disabling antitheft system is based on the use of a key that contains a passive transponder and integrated circuit. When the key is inserted into the lock cylinder and switched ON, a radio transmitter in the lock cylinder sends out a low-power signal. This signal energizers the key circuit, which responds with a code. If the code from the key matches the code from the transmitter, the engine starts. If the code does not match, the engine will quit running after one second.

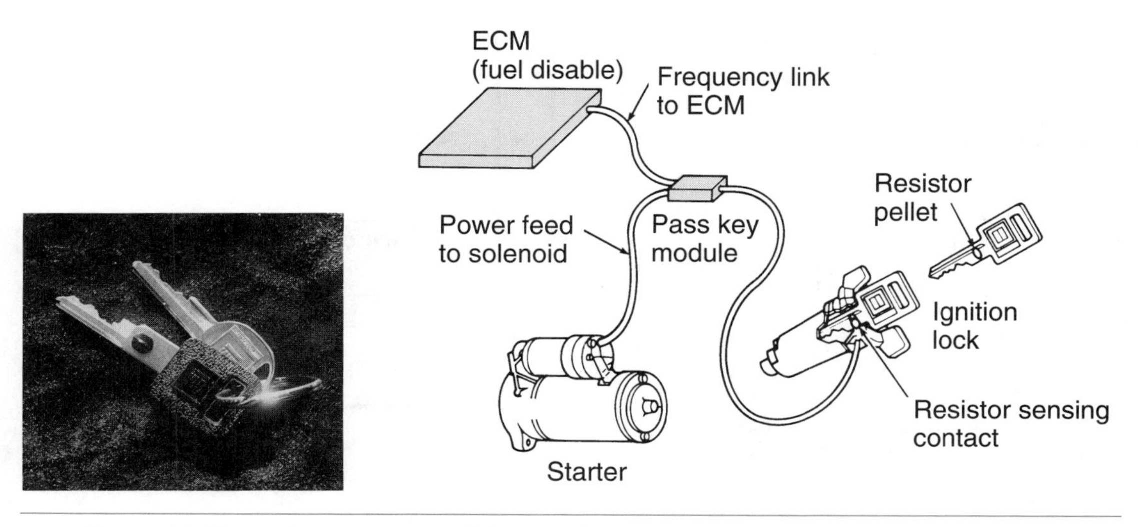

Figure 14-60 Basic components of the pass-key system. (Courtesy of General Motors Corporation)

Remote Keyless Entry

Many new vehicles are equipped with a remote keyless entry system that is used to lock and unlock the doors, turn on the interior lights, and release the trunk latch. A small receiver is installed in the vehicle (Figure 14-61). The transmitter assembly is a handheld item attached to the key ring. It has three buttons that control the functions of the system.

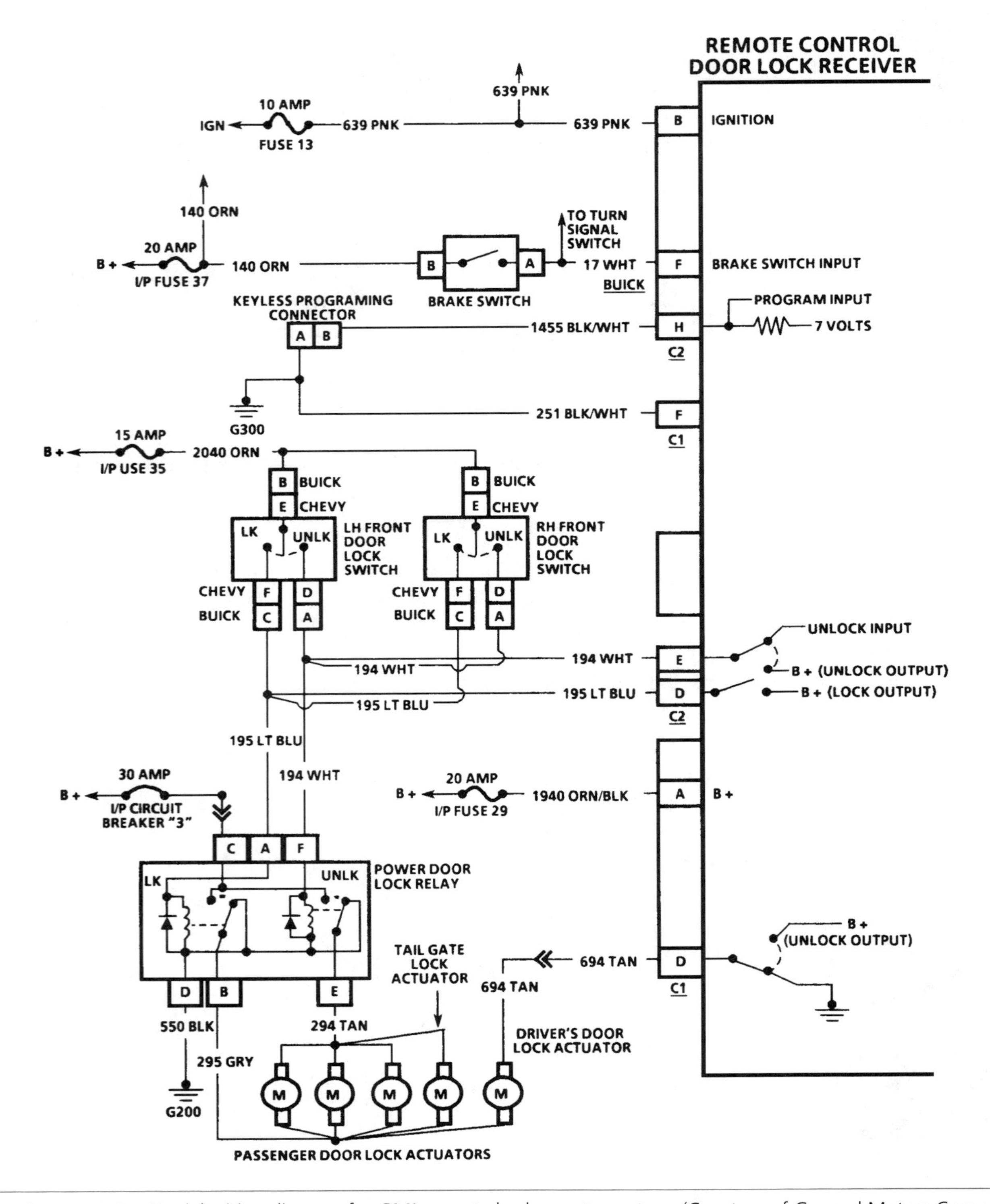

Figure 14-61 Partial wiring diagram for GM's remote keyless entry system. (Courtesy of General Motors Corporation)

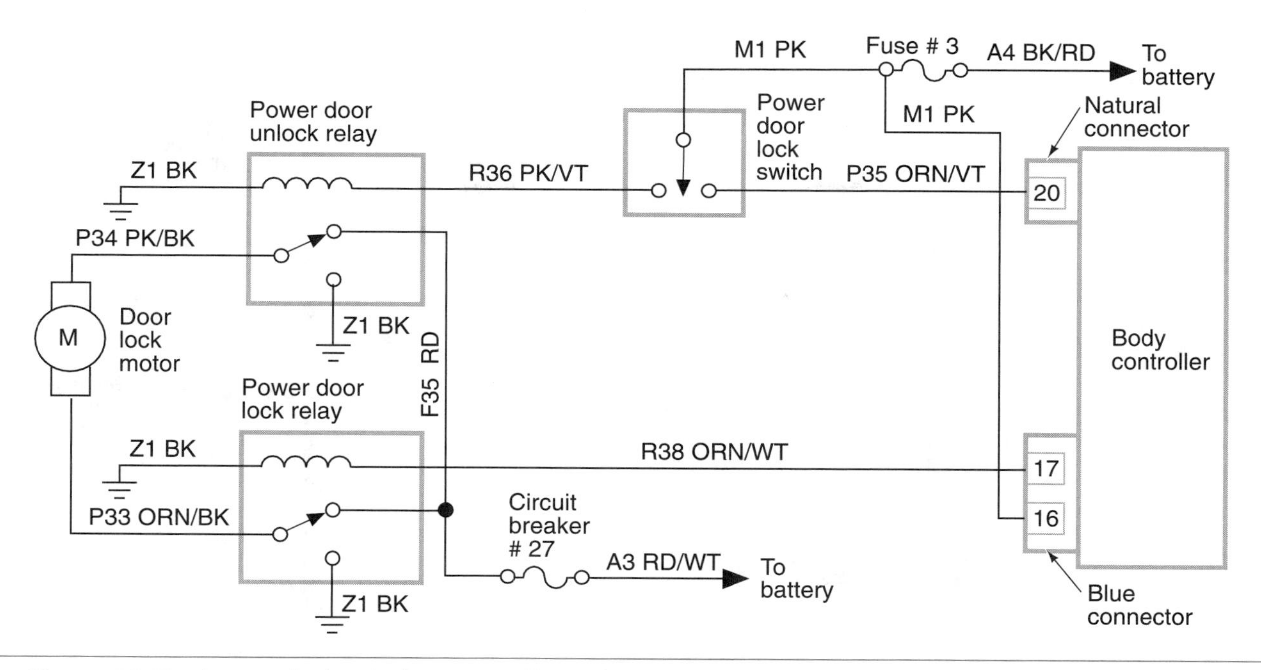

Figure 14-62 Automatic door lock system utilizing the body computer. (Courtsey of Chrysler Corporation)

The system operates at a fixed radio frequency. If the unit does not work from a normal distance, check for two conditions: weak batteries in the remote transmitter or a stronger radio transmitter close by (radio station, airport transmitter, etc.).

Automatic Door Locks

Shop Manual
Chapter 14, page 544

Automatic door locks (ADL) is a passive system used to lock all doors when the required conditions are met.

Many automobile manufacturers are incorporating automatic door locks as an additional safety and convenience system. Most systems lock the doors when the gear selector is placed in drive, the ignition switch is in RUN, and all doors are shut. Some systems will lock the doors when the gear shift selector is passed through the reverse position; others do not lock the doors unless the vehicle is moving 15 mph or faster.

The system may use the body computer to control the door lock relays (Figure 14-62), or a separate controller (Figure 14-63). The controller (or body computer) takes the place of the door lock switches for automatic operation. In order for the door lock controller to lock the doors, the following conditions must be met:

1. Ignition switch is in the RUN position.
2. Seat switch is closed by the driver.
3. All doors are closed (switches are open).
4. Gear selection is not in PARK.
5. Courtesy light switch is off.

When all of the door jam switches are open (doors closed), the ground is removed from the WHT wire to the controller (Figure 14-64). This signals the controller to enable the lock circuit. When the gear selection is moved from the PARK position, the neutral safety switch removes the power signal from the controller. The controller sends voltage through the LH seat switch to the lock relay coil. Current is sent through the motors to lock all doors.

When the gear selector is returned to the PARK position, voltage is applied through the neutral safety switch to the controller (Figure 14-65). The controller then sends power to the unlock relay coil to reverse current flow through the motors.

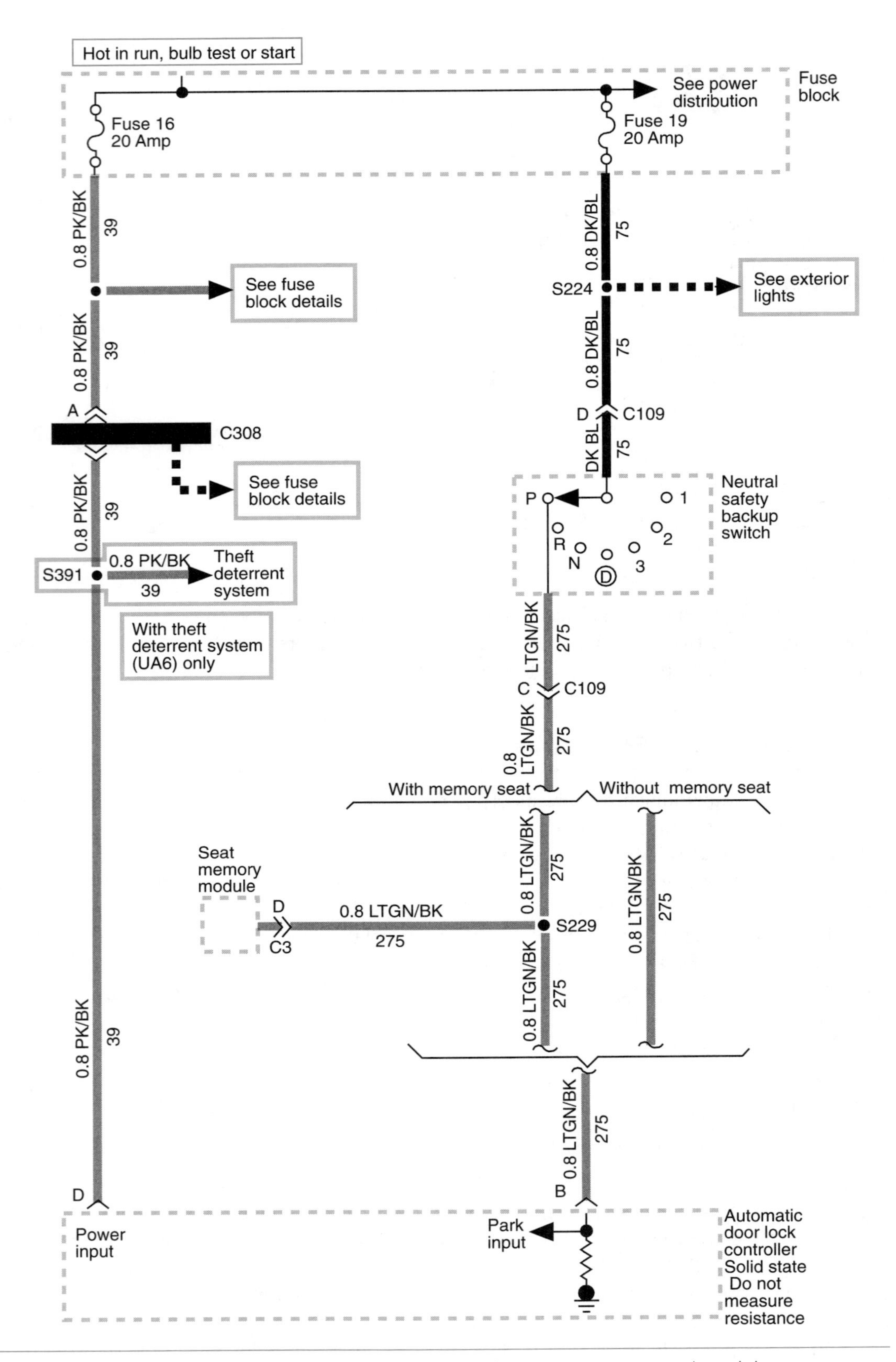

Figure 14-63 Automatic door lock system that utilizes a separate control module. (Courtesy of General Motors Corporation)

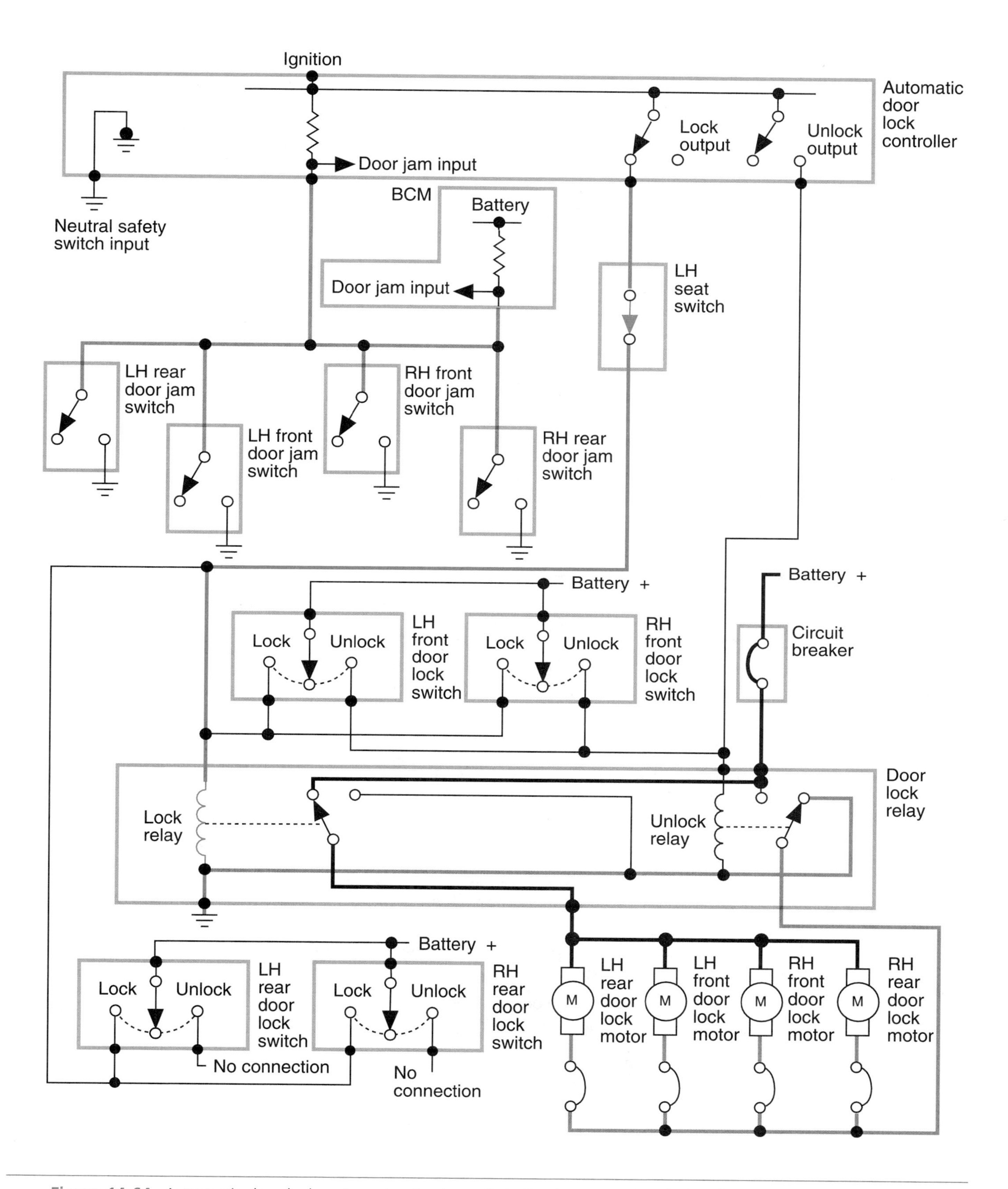

Figure 14-64 Automatic door lock system circuit schematic indicating operation during the lock procedure.

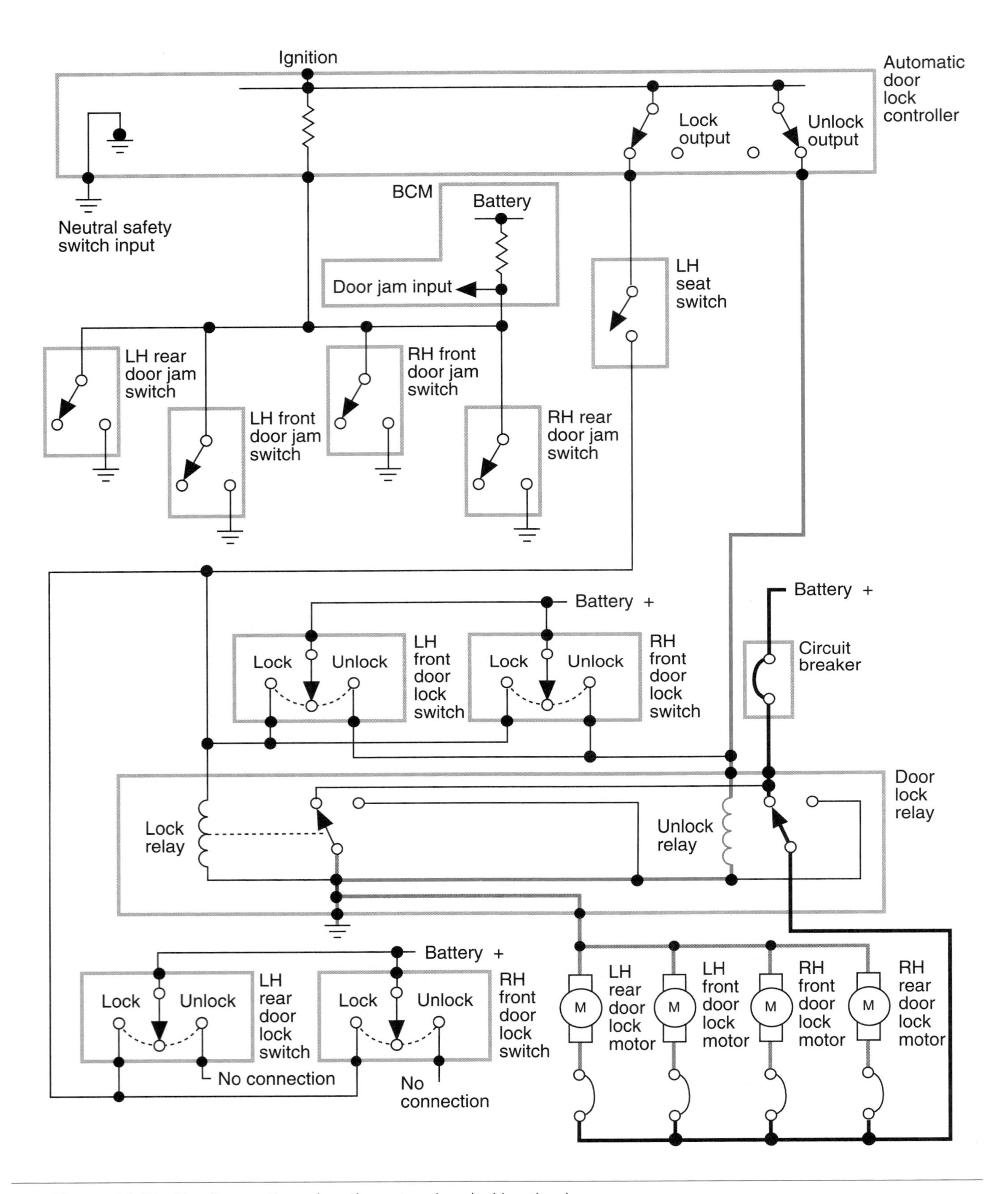

Figure 14-65 Circuit operation when the system is unlocking the doors.

Figure 14-66 Keyless entry system keypad.

Keyless Entry

Shop Manual
Chapter 14,
page 544

The main components of the keyless entry system are the control module, a coded-button keypad located on the driver's door, and the door lock motors.

The keypad consists of five normally open, single-pole, single-throw switches. Each switch represents two numbes: 1-2, 3-4, 5-6, 7-8, and 9-0 (Figure 14-66).

The keypad is wired into the circuit to provide input to the control module (Figure 14-67). The control module is programmed to lock the doors when the 7-8 and 9-0 switches are closed at the same time. The driver's door can be unlocked by entering a five-digit code through the keypad. The unlock code is programmed into the controller at the factory. However, the driver may enter a second code. Either code will operate the system.

The **keyless entry system** allows the driver to unlock the doors or the deck lid (trunk) from outside the vehicle without the use of a key.

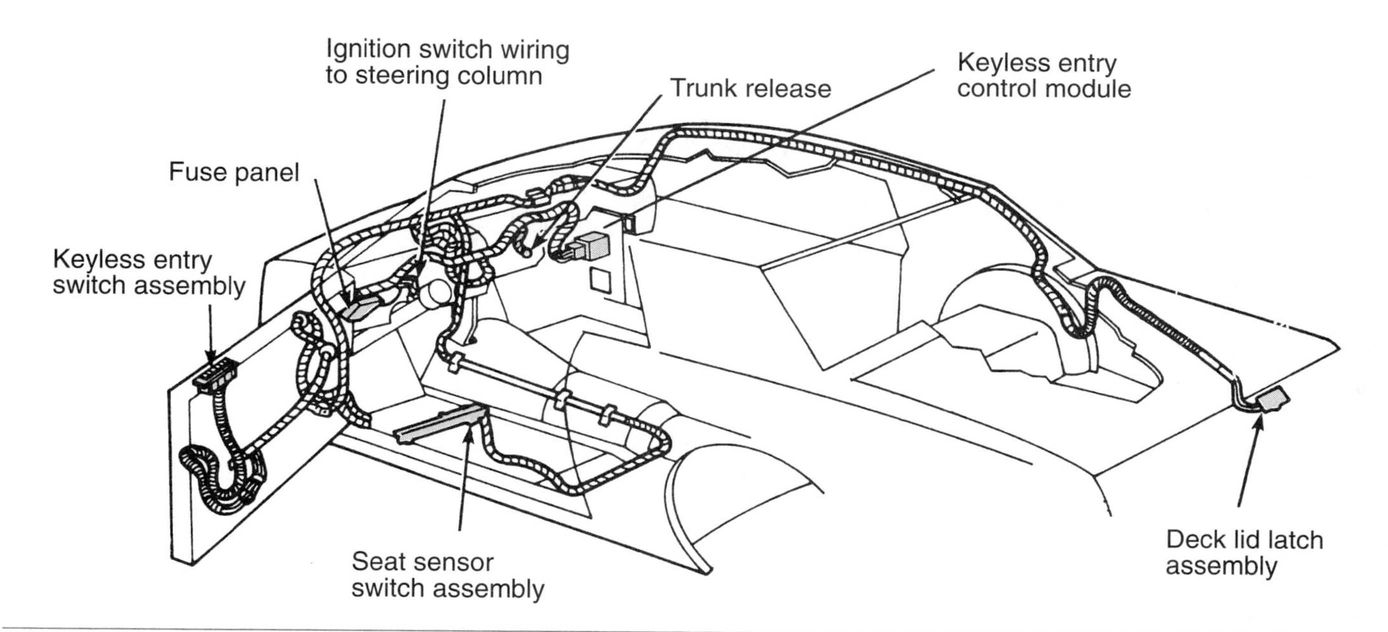

Figure 14-67 Wiring harness and components of a typical keyless entry system. (Reprinted with the permission of Ford Motor Company)

In addition to the aforementioned functions, the keyless entry system also:

1. Unlocks all doors when the 3-4 button is pressed within 5 seconds after the five-digit code has been entered.
2. Releases the deck lid lock if the 5-6 button is pressed within 5 seconds of code entry.
3. Activates the illuminated entry system if one of the buttons is pressed.
4. Operates in conjunction with the automatic door lock system and may share the same control module.

See the schematic (Figure 14-68) of the keyless entry system used by Ford. When the 7-8 and 9-0 buttons on the keypad are pressed, the controller applies battery voltage to all motors through the lock switch (Figure 14-69).

When the five-digit code is entered, the controller closes the driver's switch to apply voltage in the opposite direction to the driver's door motor (Figure 14-70). If the driver presses the 3-4 button, the controller will apply reverse voltage to all motors to unlock the rest of the doors.

Some keyless entry systems can be operated remotely. Pressing a button on a hand-held transmitter will allow operation of the system from distances of 25 to 50 feet (Figure 14-71). When the unlock button is pressed, the driver's door unlocks and the interior lights are illuminated. If a theft deterrent system is installed on the vehicle, it is also disarmed when the unlock button is pressed. A driver exiting the vehicle can activate the door locks and arm the security system by pressing the lock button.

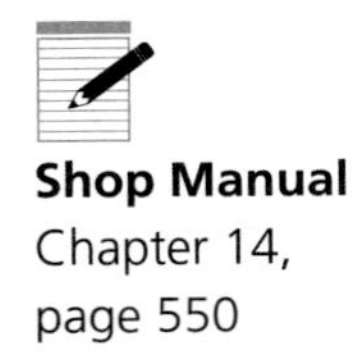

Shop Manual
Chapter 14,
page 550

Electronic Heated Windshield

The heated windshield system is designed to melt ice and frost from the windshield three to five times faster than conventional defroster systems (Figure 14-72). The windshield undergoes a special process during manufacturing to allow for current flow through the glass without interfering with the driver's vision.

The silver and zinc coating gives the windshield a gold tint.

The film used on the heated windshield will block some radio or microwave signals. This may reduce the effective range of garage door openers and radar detectors.

There are two basic methods used to make the heated windshield:

1. Use a layer of plastic laminate that is between two layers of glass. The back of the outer layer is fused with a silver and zinc oxide coating. The coating carries the electrical current. Busbars are attached to the coating at the top and bottom of the windshield (Figure 14-73). A sensor is used to check the condition of the windshield coating. If the windshield has a crack or chip that will affect heating (Figure 14-74), the voltage drop across the resistor will indicate this condition to the control module. If the windshield is damaged, the controller will not allow heated windshield operation.
2. Use a layer of resistive coating sprayed between the inner and outer windshield layers. The coating is transparent and does not provide any tint. A sensor is used to indicate if the coating has been damaged. If a chip or crack is not deep enough to penetrate the coating, it will not affect the system operation.

The two systems discussed here are representative of the methods used to heat the windshield.

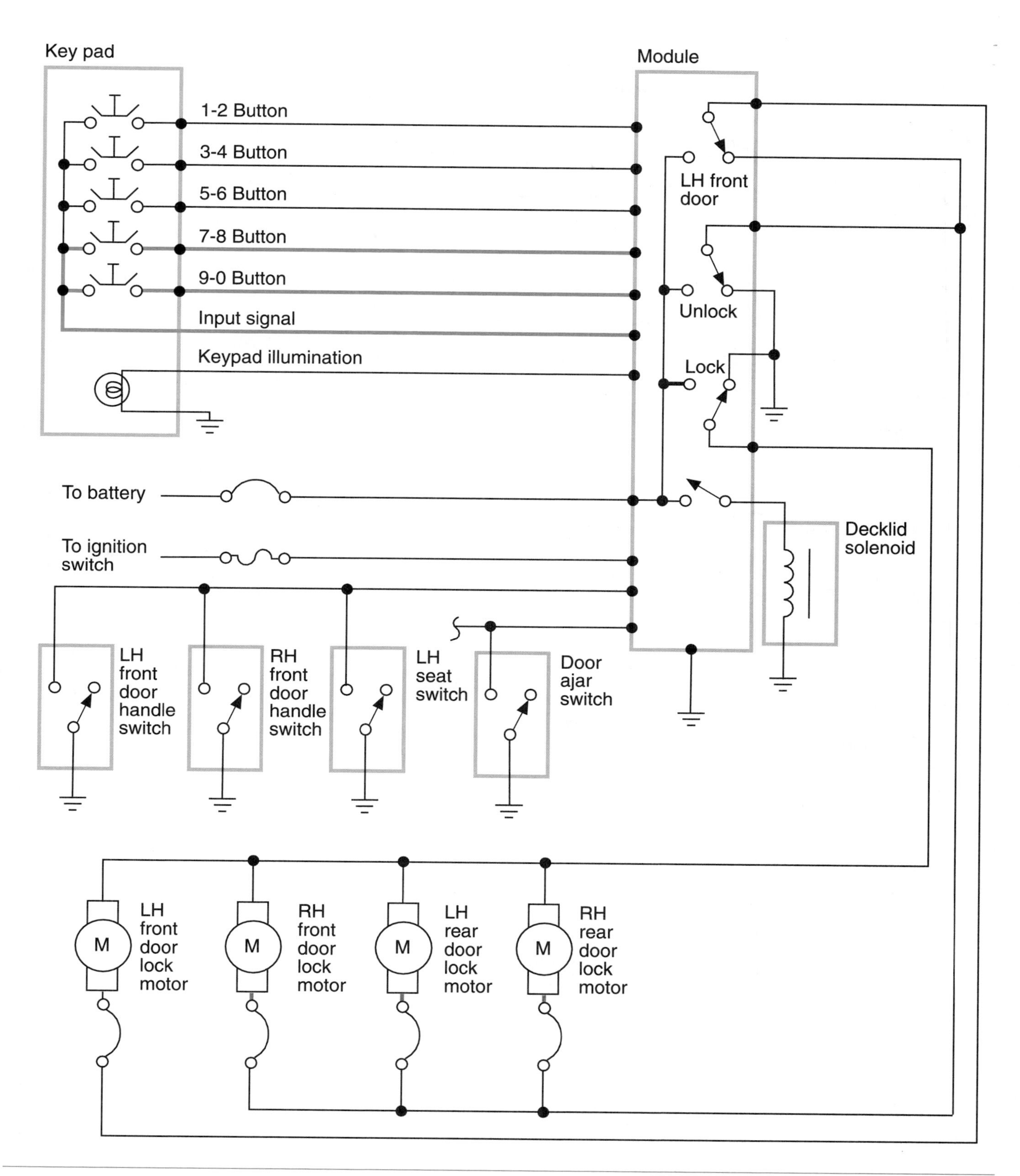

Figure 14-68 Simplified keyless entry system schematic.

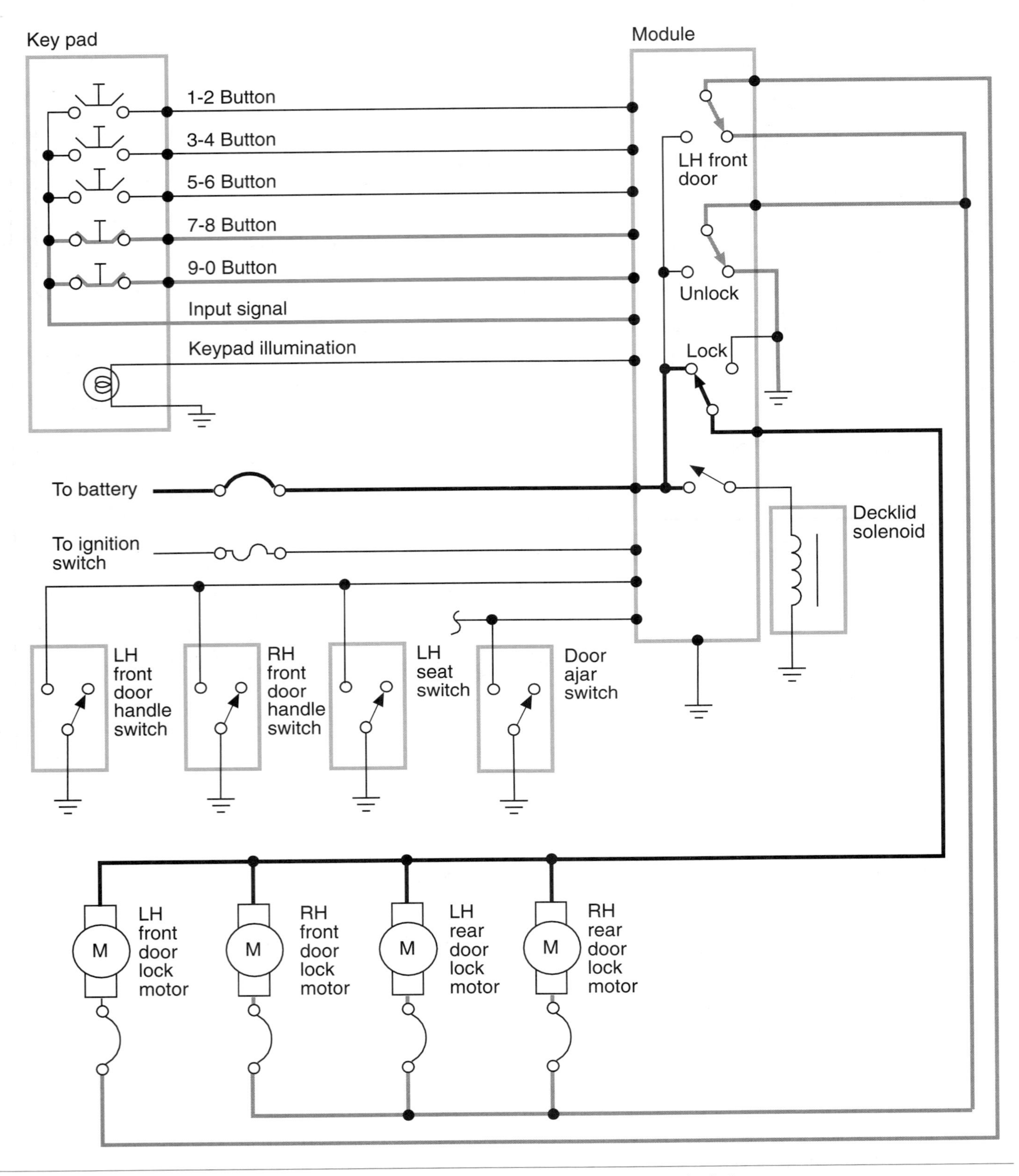

Figure 14-69 Circuit operation when the 7-8 and 9-0 buttons are pressed to lock all doors.

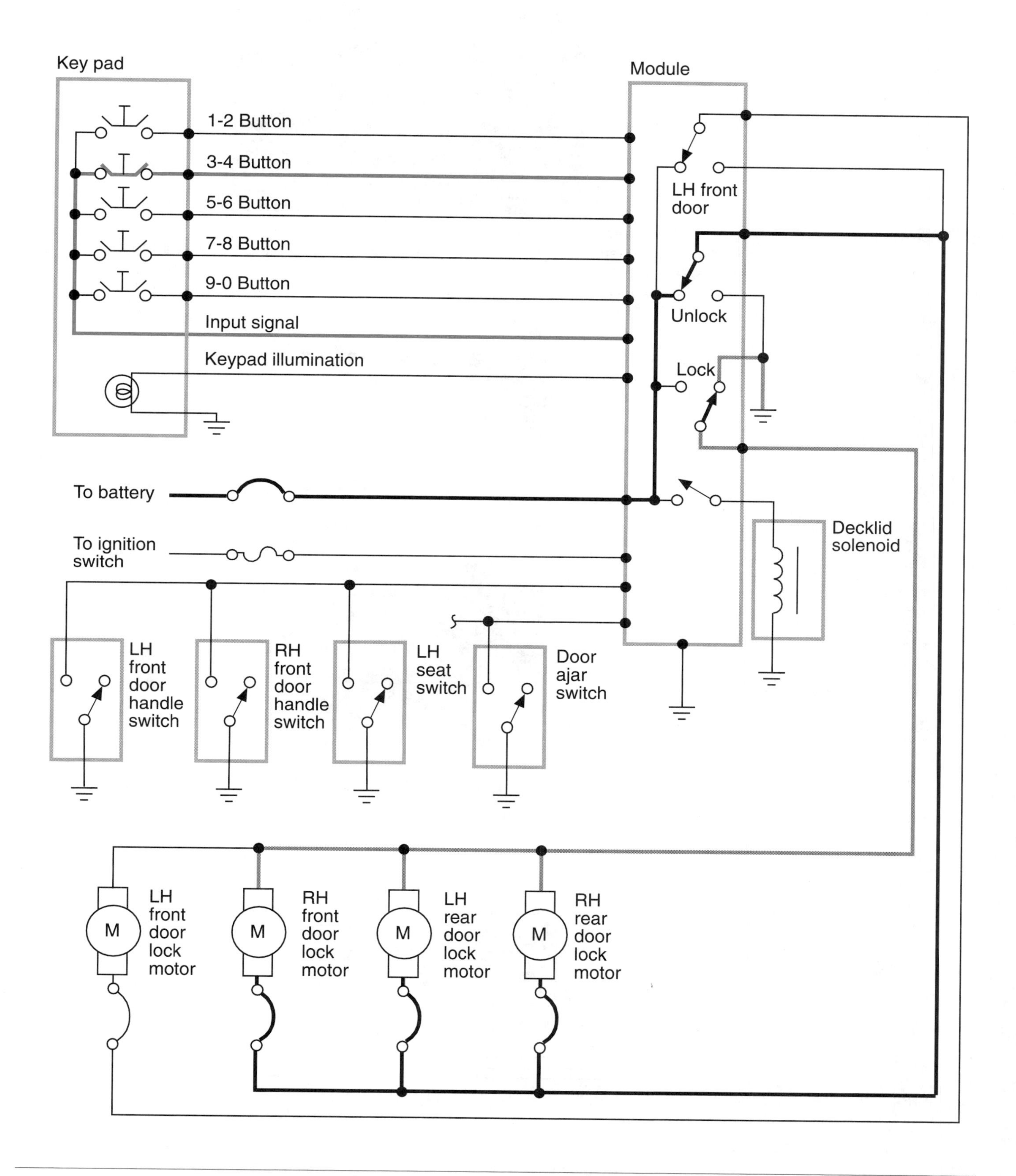

Figure 14-70 Circuit operation when the correct five-digit code is entered to unlock the driver's door.

Figure 14-71 Remote keyless entry system transmitter. (Courtesy of General Motors Corporation)

Figure 14-72 The heated windshield removes ice and frost from the windshield in just a few minutes. (Reprinted with the permission of Ford Motor Company)

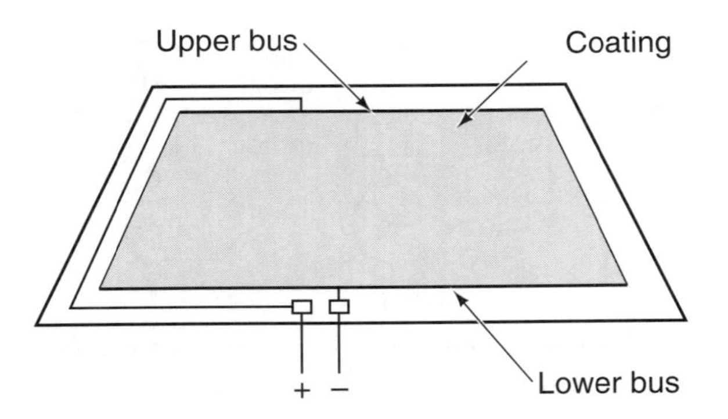

Figure 14-73 The power and ground circuits are connected to the silver and zinc coating through the busbars. (Reprinted with the permission of Ford Motor Company)

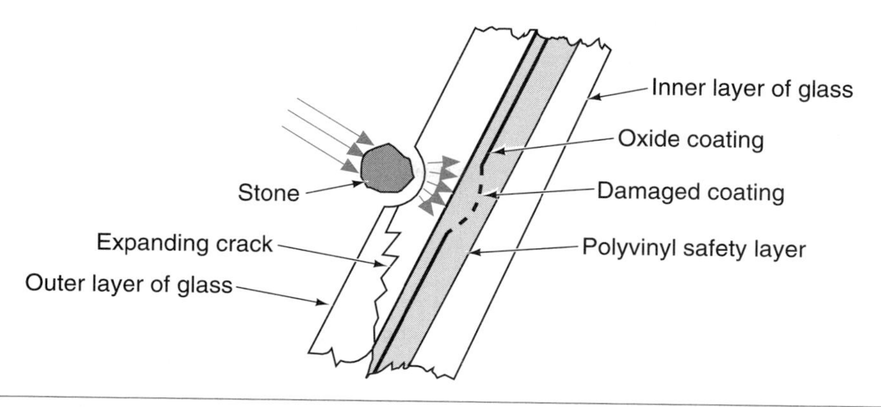

Figure 14-74 An open in the circuit can be caused by a chip or crack in the windshield. A sensor is used to prevent operation if the windshield is damaged. (Reprinted with the permission of Ford Motor Company)

General Motors' Heated Windshield

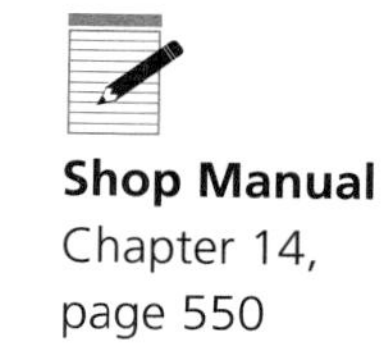

Shop Manual
Chapter 14, page 550

General Motors' heated windshield consists of the following components:

1. The heated windshield: Contains a transparent internal resistive coating that heats when current is applied to it.
2. Special CS 144 generator: There are three special phase terminals to provide AC power to the system's power module (Figure 14-75). The generator can continue to supply its normal DC voltage while AC power is being supplied.
3. The power module: Converts the AC voltage from the generator to a higher DC voltage for use by the windshield.

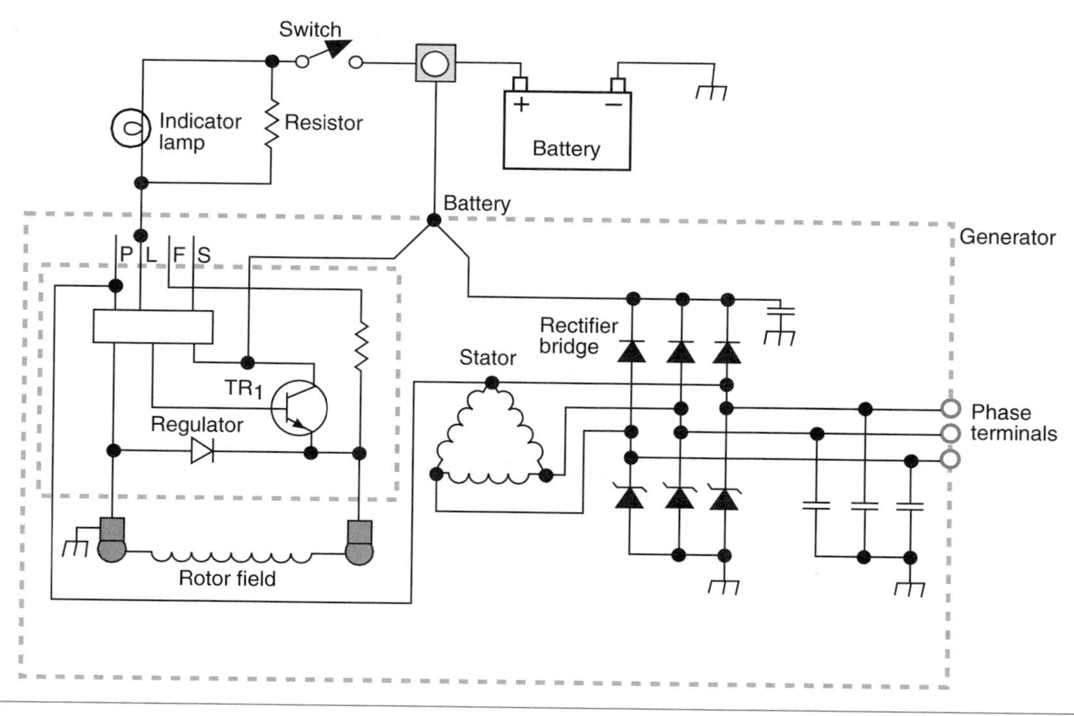

Figure 14-75 Schematic of CS 144 generator used on vehicles equipped with heated windshields. The three terminals provide AC current to the power module.

4. The control module: Controls the heating cycle and provides automatic shut-down at the end of the time cycle, or if a fault is detected in the system.
5. The control switch.

See the schematic (Figure 14-76) of the GM heated windshield. When the system is activated by the driver, the control module starts its turn-on sequence. First it checks that there is more than 11.2 volts present at terminal B6. This assures there will be sufficient voltage to operate other circuits.

The second step for the control module is to check the vehicle's inside temperature. For the system to operate, the temperature must be below 65°F (18°C). Next, the controller checks the windshield sensor to see if there is any damage to the film coating.

The BCM and ECM are used to provide certain functions when the system is activated.

If all of these conditions are met, the control module sends a signal to the BCM to increase the engine speed. The BCM passes the request on to the ECM. If the gear selector is in PARK or NEUTRAL, the ECM will increase the idle speed to approximately 1,400 rpm. The ECM will send a signal back to the BCM to indicate that the speed has been increased. When this feedback signal is received, the control module will turn on the power module relays. The power module will draw AC current from the generator. The current is amplified and rectified by the power module, then sent to the windshield. Voltage at the windshield is between 50 and 90 volts.

Incorporated into the control module is a timer circuit. When the activation switch is turned on for the first time, the control module will operate the system for 3 minutes. If the switch is pressed again, at the end of the first cycle, it will result in a 1-minute cycle. If the switch is pressed while the cycle is still in operation, the system is turned off.

Ford's Heated Windshield System

For the system to be activated, the engine must be running.

The illustration (Figure 14-77) shows the major components of the Ford heated windshield system. For the system to be activated, the engine must be running and inside temperature must be 40°F or less. When the driver activates the system the control module shuts off the voltage regulator and energizes the generator output control relay. This switches the generator output from the electrical system to the windshield circuit (Figure 14-78). After the switch has been completed, the control module turns on the voltage regulator to restore generator output.

With the generator output disconnected from the battery, battery voltage drops below 12 volts. The voltage regulator attempts to charge the battery by full fielding the generator. Because the battery does not receive the generator output, full field voltage reaches 30 to 70 volts. All of the full field power is sent to the windshield.

The control module will monitor the battery voltage and generator output. It will prevent the output from increasing over 70 volts to protect the system. To prevent damage to the battery, if its voltage drops below 11 volts the control module reconnects the electrical system to the generator.

When the system is activated, the control module sends a signal to the EEC controller to increase the idle speed to about 1,400 rpm. If the transmission is placed into a gear selection other than PARK or NEUTRAL, the EEC will return the idle speed to the normal setting.

Intelligent Windshield Wipers

To avoid making the driver select the correct speed of the windshield wipers according to the amount of rain, manufacturers have developed intelligent wiper systems. Two intelligent wiper systems will be discussed here: one senses the amount of rainfall and the other adjusts wiper speed according to vehicle speed.

Cadillac's Rainsense system automatically selects the wiper speed needed to keep the windshield clear by sensing the presence and amount of rain on the windshield. The system relies on a series of eight LEDs that shine at an angle onto the inside of the windshield glass and an equal number of light collectors. The outer surface of a dry windshield will reflect the

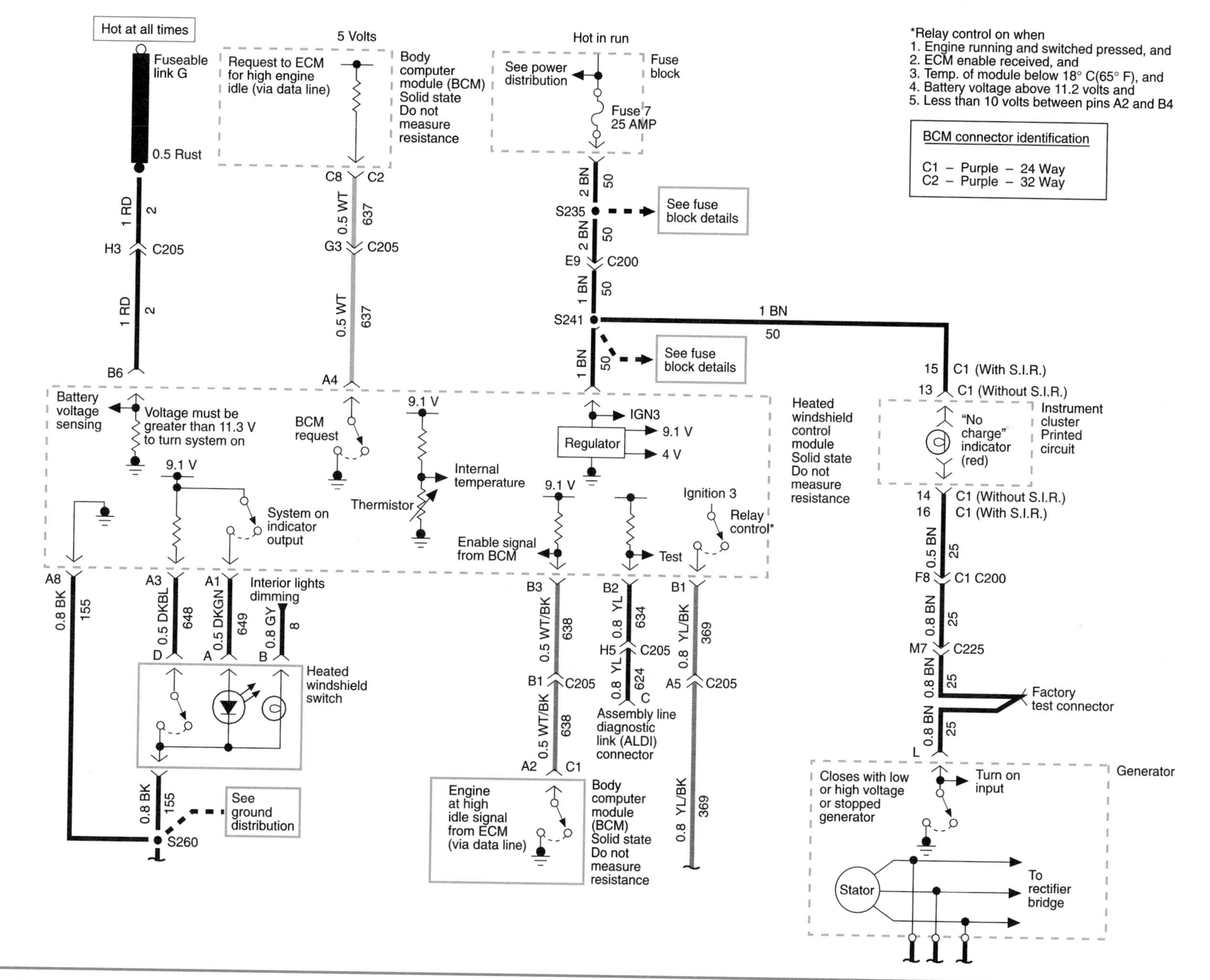

Figure 14-76 General Motors' heated windshield schematic. (Courtesy of General Motors Corporation)

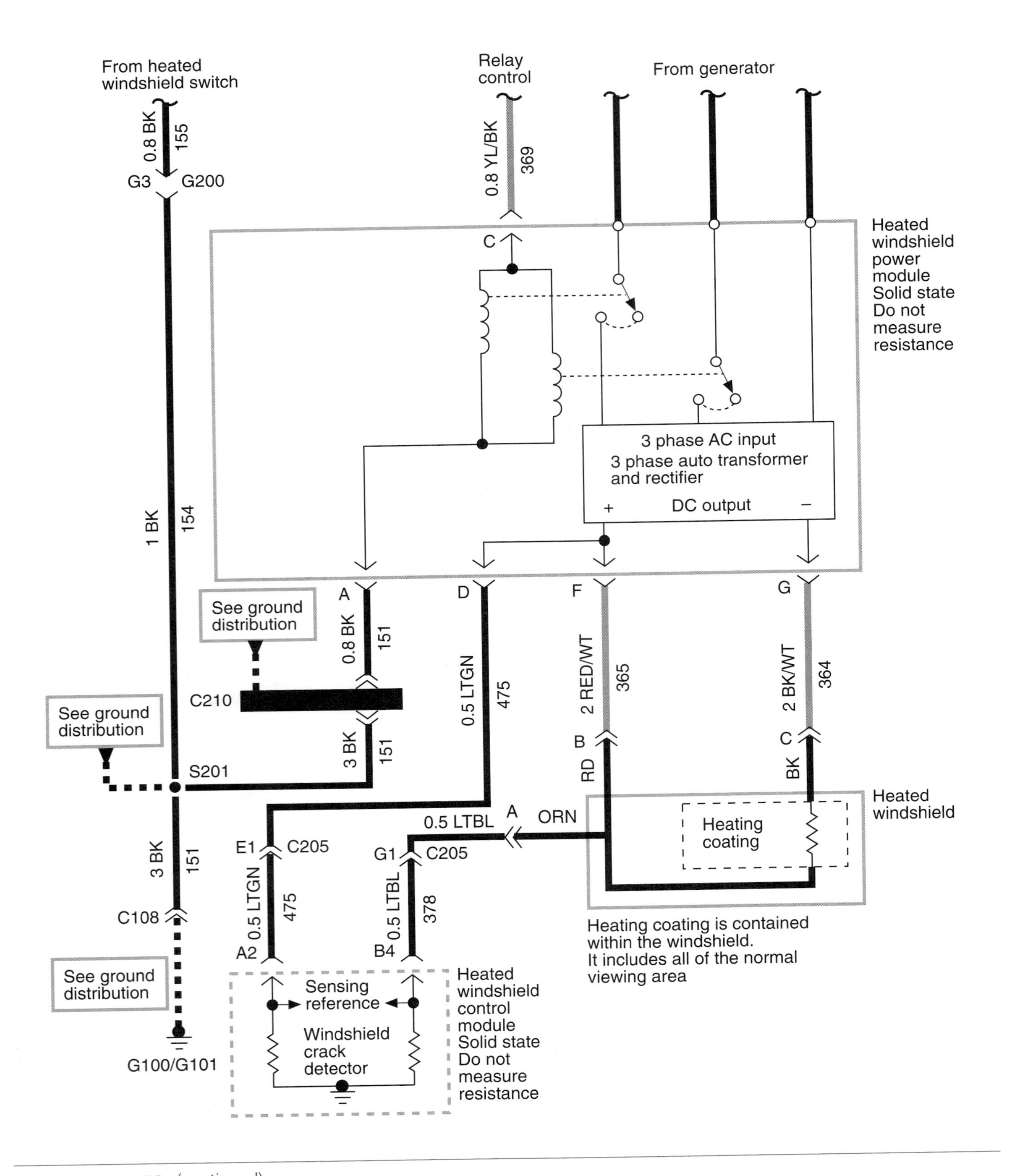

Figure 14-76 (continued)

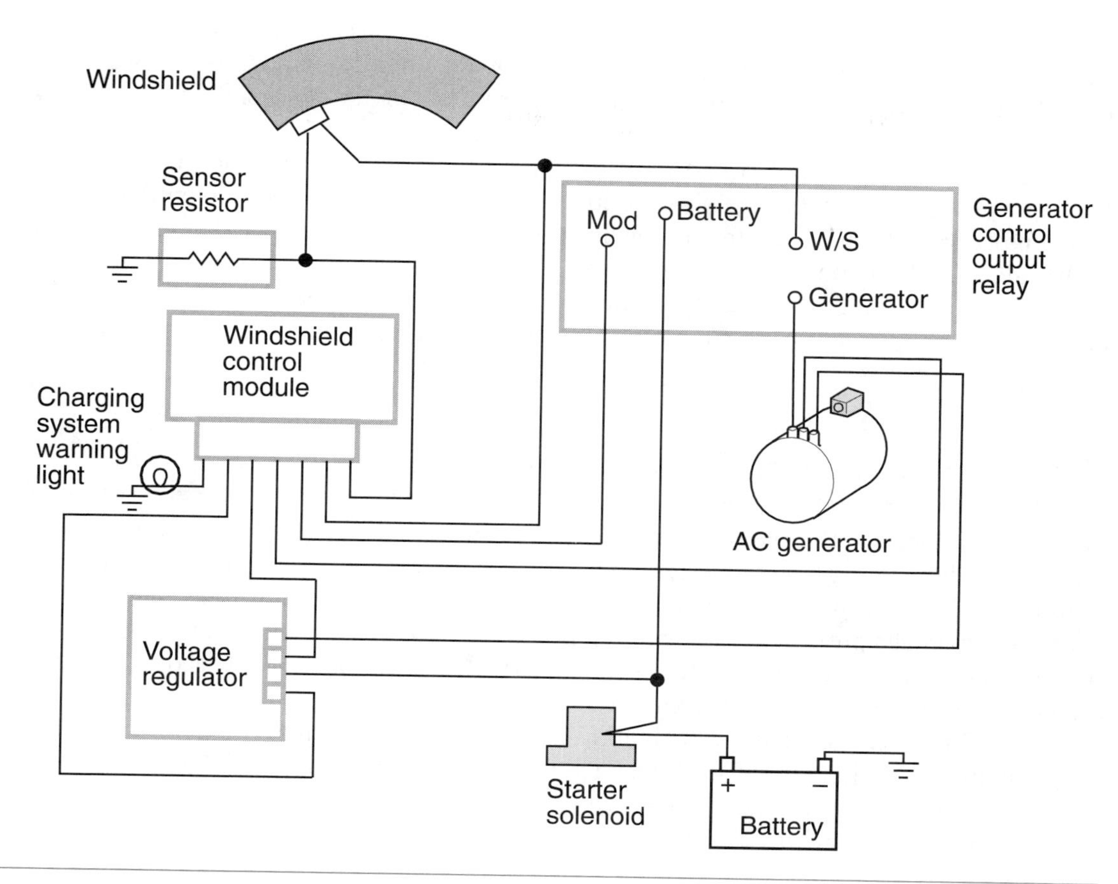

Figure 14-77 Components of Ford's heated windshield system.

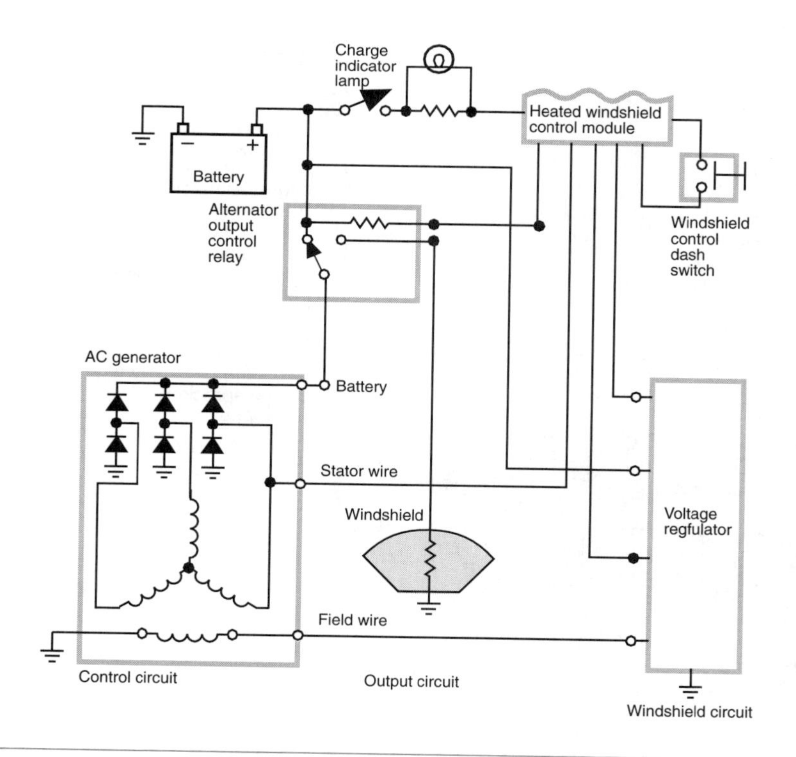

Figure 14-78 Simplified circuit schematic of Ford's heated windshield system.

lights from the LEDs back into a series of collectors. The presence of water on the windshield will refract some of the light away from the collectors. When this happens, the wipers are turned on. If the water is not cleared by one complete travel of the wipers, the wipers operate again. Therefore if the rainfall is heavy, the wipers will operate quickly and often.

AXODE is the model designation given to Ford's electronically shifted automatic transaxle with overdrive.

One of the functions of Ford Motor Company's Generic Electronic Module (GEM) system is front wiper control and a speed-dependent wiper system. Speed-dependent wipers compensate for extra moisture that normally accumulates on the windshield at higher speeds in the rain. At higher speeds, the delay between wipers shortens when the wipers are operating in the interval mode. This delay is automatically adjusted at speeds between 10 and 65 miles per hour. Basically this system functions according to the input the computer receives about vehicle speed.

Electronic Shift Transmissions

Shop Manual
Chapter 14,
page 556

The **valve body** directs fluid oil under pressure to the torque converter, servos, and clutches to control operation.

The use of solenoids and relays in controlling the operation of the engine has been expanded to include the drivetrain. Many of today's vehicles are equipped with electronic shift automatic transmissions. The control module uses several inputs to determine torque converter clutch operation, hydraulic pressure levels, and shift points. The use of electronics within the transmission has improved shift quality and fuel economy. The Ford AXODE transaxle is discussed as an example of the principles used in electronic shift transmissions.

The AXODE transaxle is a fully automatic electronically controlled unit with a lock-up torque converter (Figure 14-79). All major transaxle operations are controlled through the EEC-IV electronic control assembly (ECA). These functions include transaxle shifting, torque converter clutch operation, and line pressure regulation.

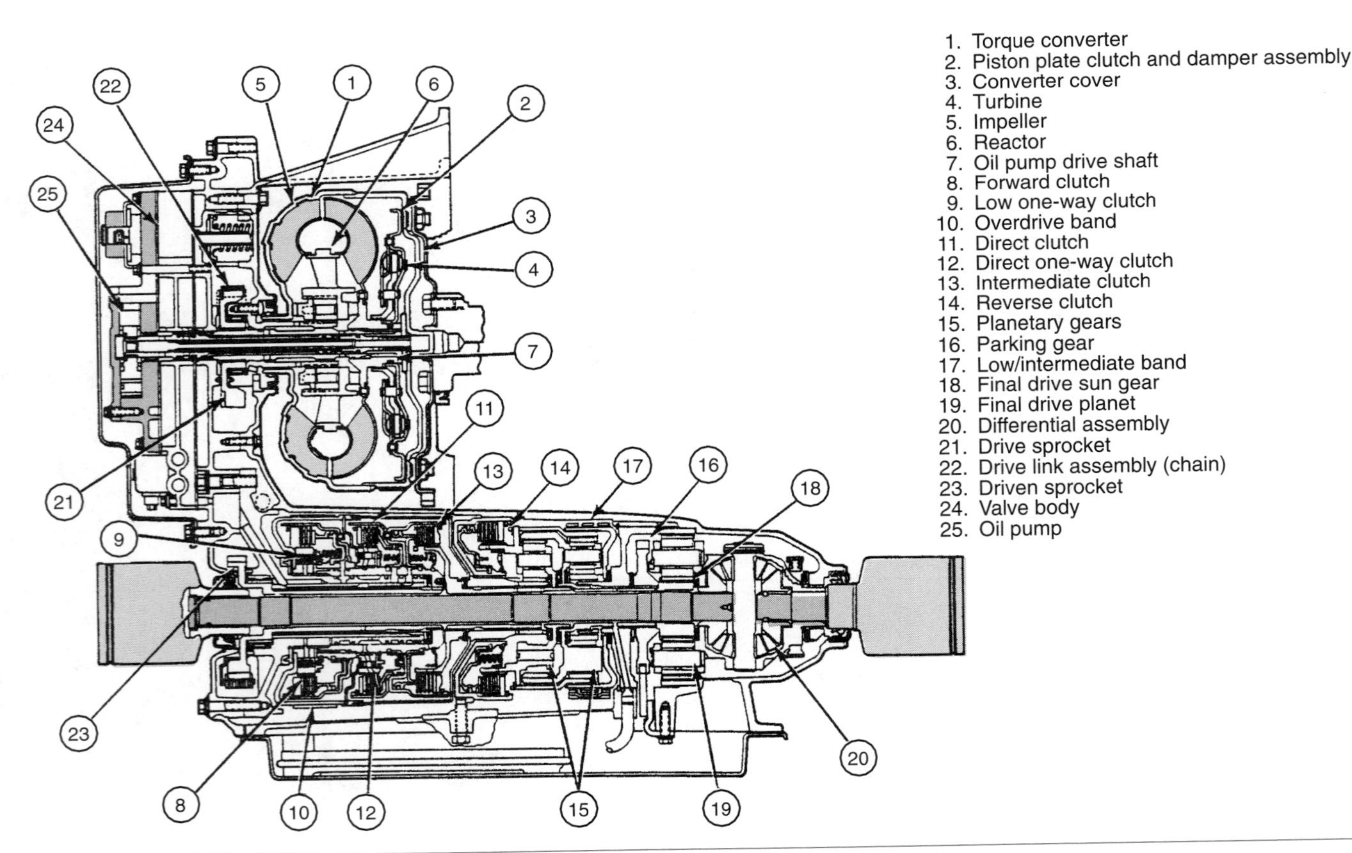

Figure 14-79 AXODE transaxle main components. (Reprinted with the permission of Ford Motor Company)

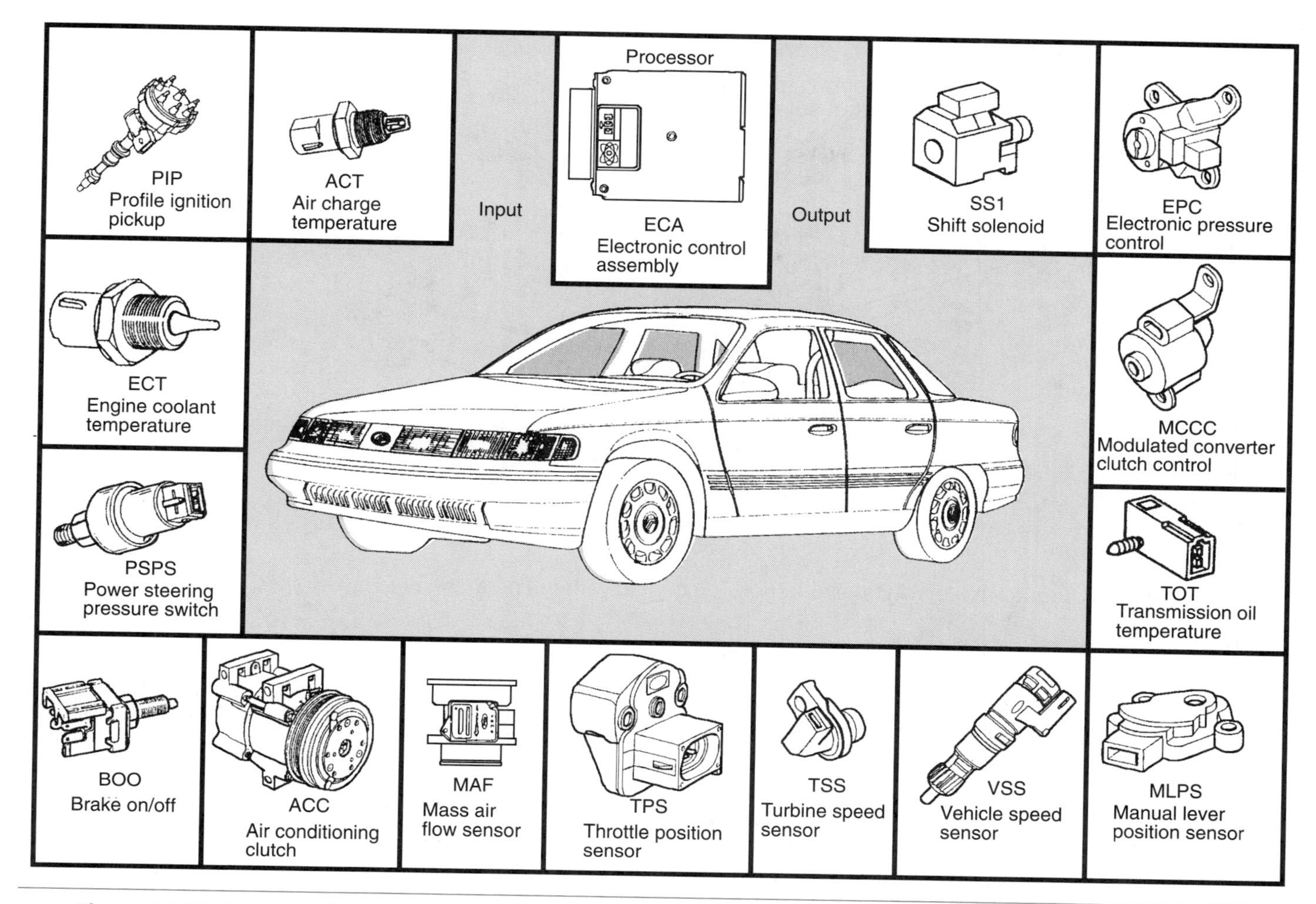

Figure 14-80 Inputs and outputs used to control transmission operation. (Reprinted with the permission of Ford Motor Company)

The ECA receives inputs concerning throttle position, engine speed, torque converter turbine speed, and other drivetrain operations (Figure 14-80). The ECA processes the information then controls transaxle operation through activation of five solenoids located within the valve body (Figure 14-81).

Speed Sensitive Steering

Shop Manual
Chapter 14, page 558

Most vehicles are available with some form of power steering, either as an option or as standard equipment. Conventional power steering systems provide a certain degree of assist to the driver when turning the steering wheel. The disadvantage of conventional power steering is the reduced road feel that the system offers during medium and high speeds. At these speeds it is desirable for a feeling of increased control and performance. Through the use of electronic controls, the advantages of high power assist and excellent road feel can be achieved.

Manufacturers have chosen different methods of accomplishing this task. The most common is to use a means of varying the output of the power steering pump through an electronic

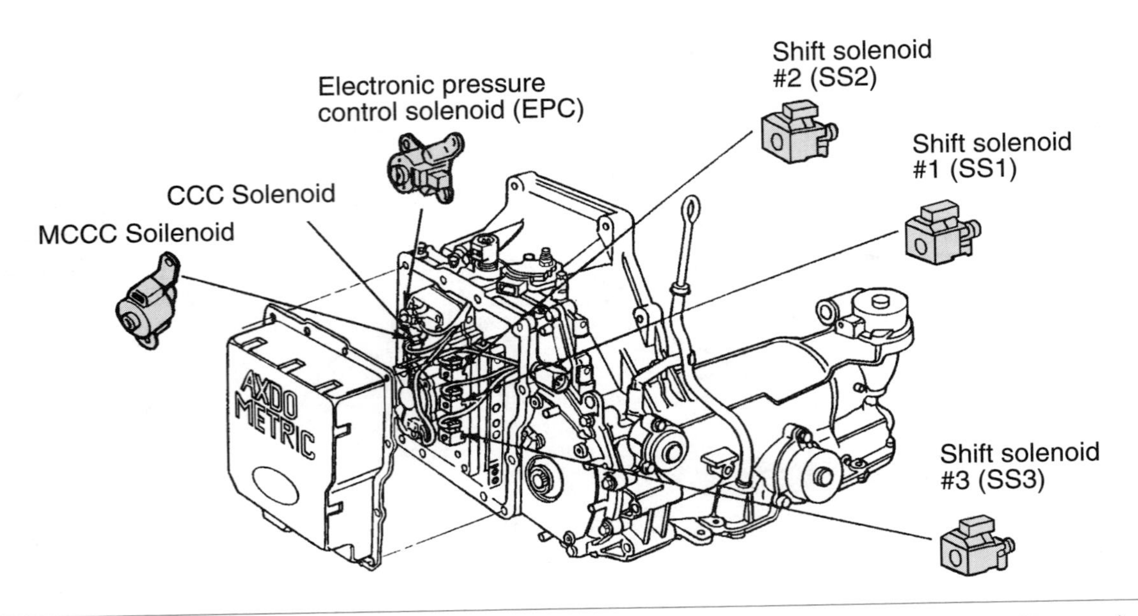

Figure 14-81 Output actuators and their location. (Reprinted with the permission of Ford Motor Company)

variable orifice (Figure 14-82). Other methods include the use of electric motor driven power steering pumps and electric rack and pinion steering (Figures 14-83 and 14-84).

Electronic Suspension Systems

Passive suspension systems have fixed spring rates and shock valving. Adaptive suspension systems are able to change ride characteristics by altering shock damping and ride height continuously. Active suspension systems are controlled by double acting hydraulic cylinders or solenoids (actuators) mounted at each wheel. The actuators support the vehicle's weight, instead of conventional springs or air springs.

Adaptive and active suspensions systems provide additional benefits over conventional passive suspension systems. They are able to change ride height, shock damping, and spring rates in response to changing road and driving conditions. Electronic suspension system types can vary from basic shock damping variations to a complex system of height and ride control that utilizes extensive computer programming.

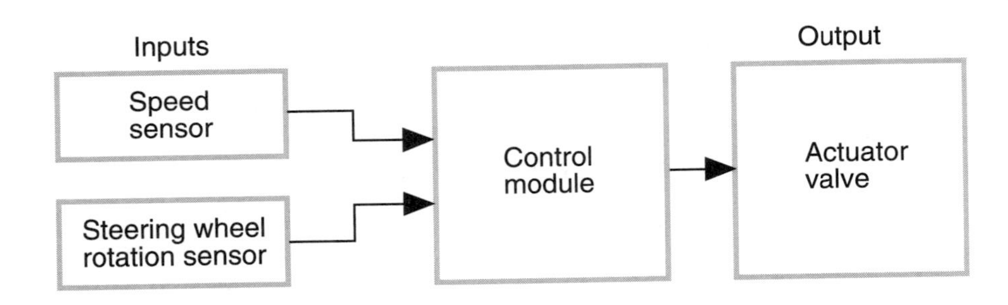

Figure 14-82 The control module receives input signals concerning vehicle speed and steering wheel rotation, then controls the actuator valve. (Reprinted with the permission of Ford Motor Company)

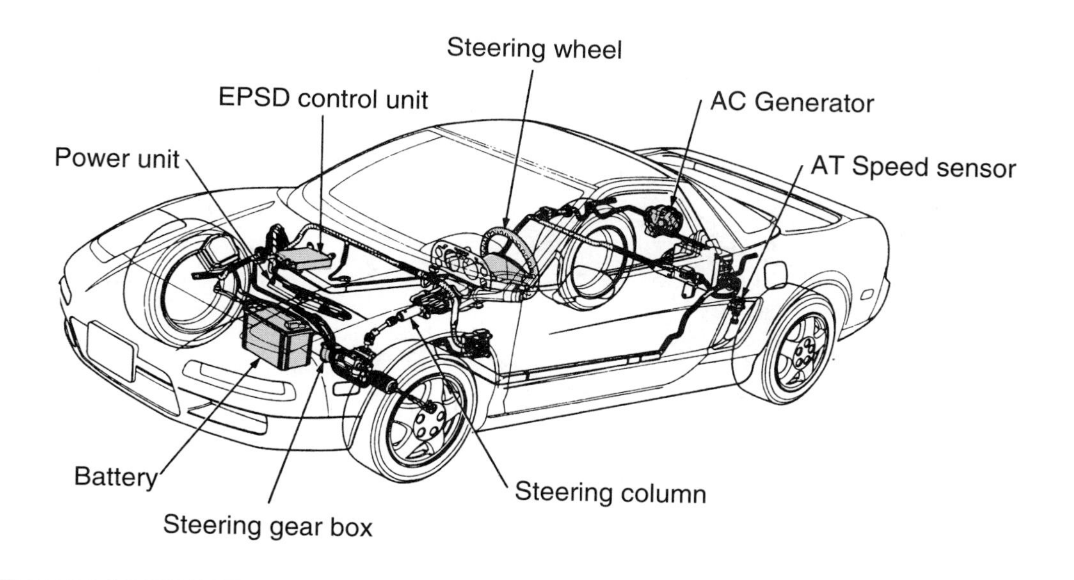

Figure 14-83 The Honda/Acura NSX electric rack and pinion system. (Courtesy of Honda Motor Company)

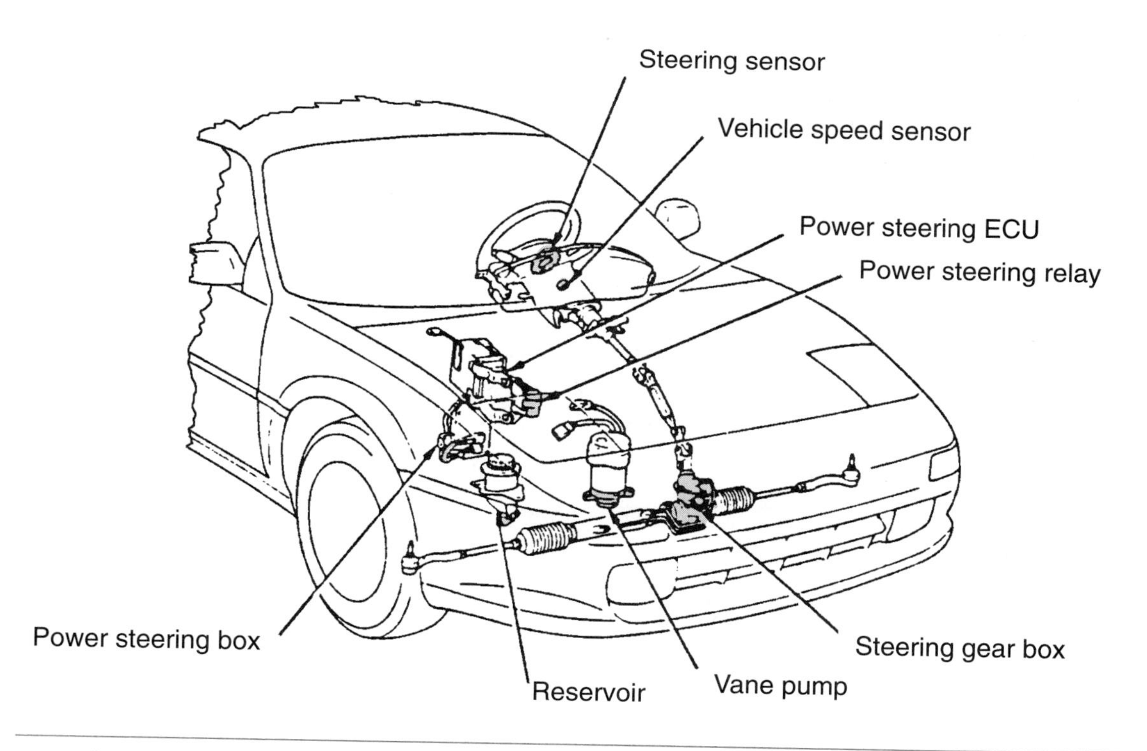

Figure 14-84 Toyota's electrohydraulic steering system has a self-contained motor in the pump. (Courtesy of Toyota Motor Corporation)

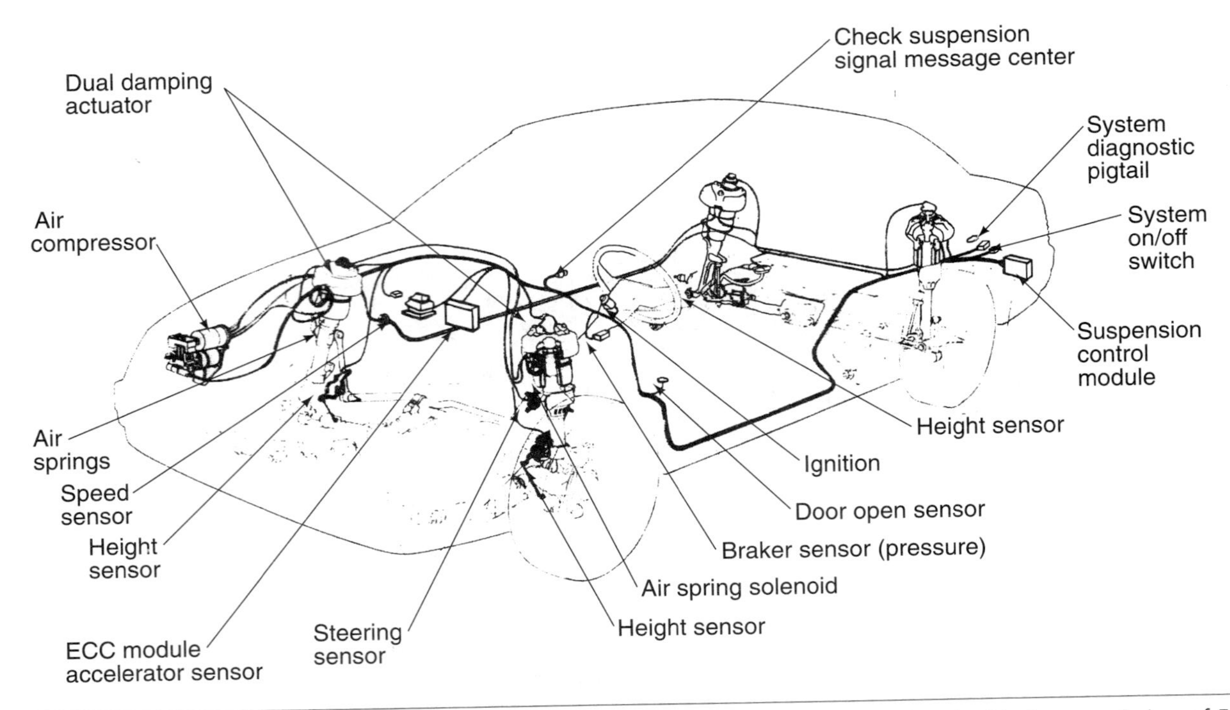

Figure 14-85 Lincoln's air spring automatic ride control (ASARC) components. (Reprinted with the permission of Ford Motor Company)

Adaptive Suspension Systems

Sensors monitor vehicle speed, steering angle, ride height, brake pressure, and vehicle acceleration. They provide this information to the suspension control module. The control module then signals actuators to change shock rates in response to the changing conditions. A small actuator on top of each shock allows a variable range from firm to soft and solenoids on each air spring enable the computer to maintain constant ride height (Figure 14-85).

Depending on system design, other sensors used by the computer include G-sensors, throttle position sensors, speed sensors, roll sensors, lateral acceleration sensors, and brake switches (Figure 14-86).

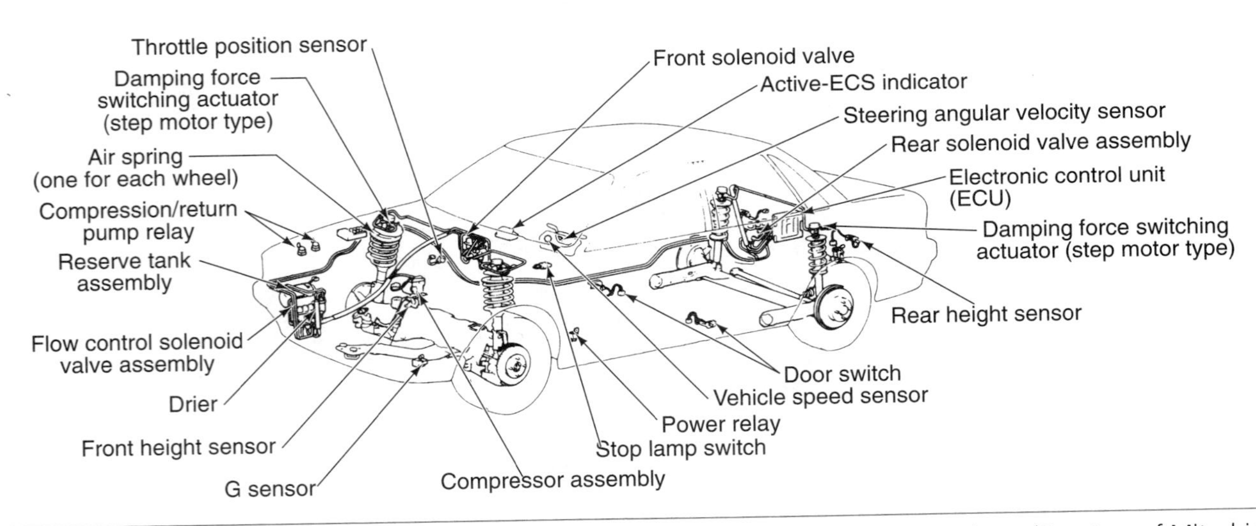

Figure 14-86 Several sensors are used to determine the level of the vehicle at any given time. (Courtesy of Mitsubishi Motor Sales of America, Inc.)

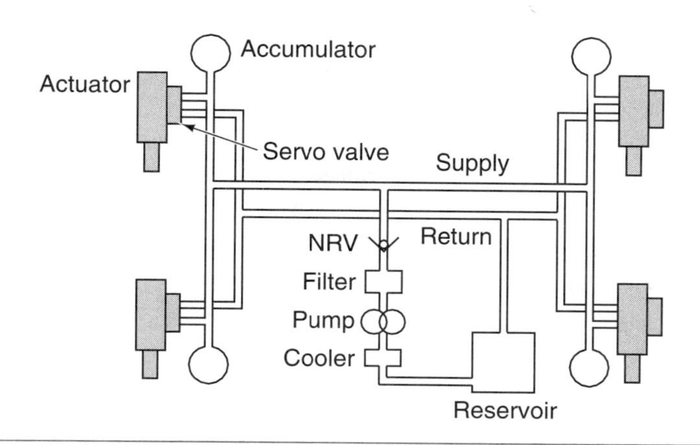

Figure 14-87 Hydraulic schematic of an active suspension system.

Active Suspension Systems

There are several differences between adaptive and active suspension systems. The active system is capable of eliminating body roll and it is faster in its reaction time. The adaptive system is able to reduce body roll, but it cannot eliminate it.

Truly active suspension systems use high pressure hydraulic actuators to carry the vehicle's weight. The actuators combine the function of the shock absorber and the spring into one unit. Through control of the fluid pressure inside the actuators, ride height, body roll, damping, and spring rate can be controlled by the computer. These systems can be programmed to respond almost perfectly to all driving conditions (Figure 14-87).

The system is also able to eliminate body roll on turns by stiffening one side of the vehicle (Figure 14-88). All of these attitude control functions improve vehicle stability and increase tire traction and driver control.

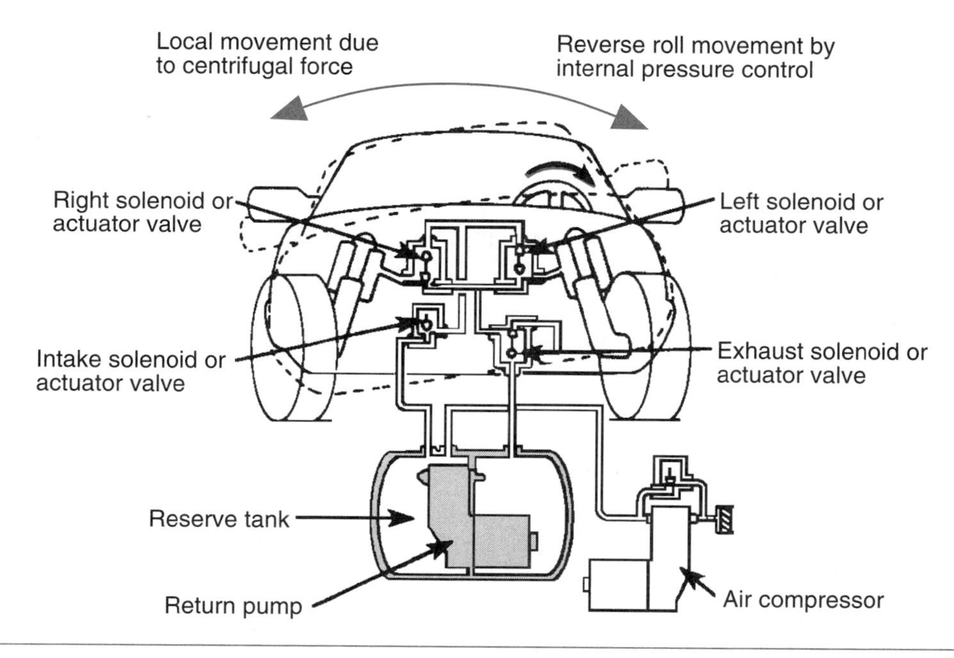

Figure 14-88 System correction to eliminate body roll during a turn. (Courtesy of Mitsubishi Motor Sales of America, Inc.)

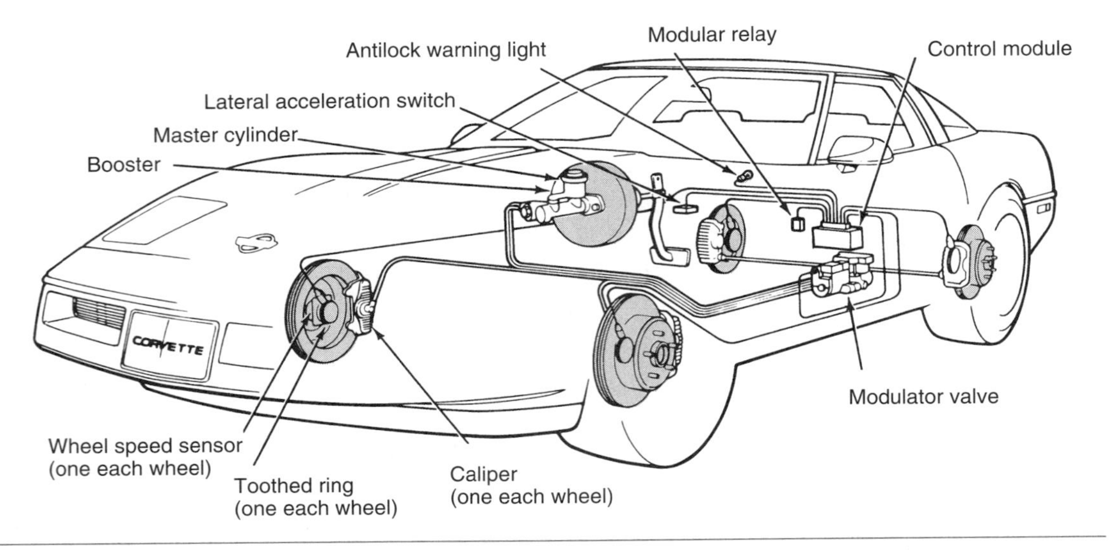

Figure 14-89 Nonintegrated antilock brake system used on Corvettes. (Courtesy of General Motors Corporation)

Introduction to Antilock Brake Systems

The modern brake system is more than adequate to stop the vehicle under normal conditions. However, in approximately 1% of its use it will fail to stop the vehicle safely. This failure is generally the fault of drivers who allow their vehicle to enter an uncontrollable skid. Wheel lockup during braking will increase the stopping distance. A good driver knows that "pumping" the brakes during emergency stops keeps the vehicle from entering into an uncontrollable skid. A tire that is on the verge of slipping produces more friction with respect to the road than one that is locked. The antilock braking system (ABS) is designed to act in a similar manner as a driver pumping the brakes, but with much more control and at a much faster rate.

The ABS system is capable of pumping each brake up to 15 times per second. The control module can pulse the two front brakes separately and the rear brakes separately or as a pair, depending on system configuration. ABS automatically stops the vehicle in the shortest possible distance, without locking a wheel. In addition, ABS maintains directional control on almost any type of road surface or condition. Even though ABS improves vehicle braking, it cannot compensate for worn brake components, worn tires, excessive speed, or driver error.

The exact components of a system depend on manufacturer and system design. Nonintegrated ABS systems use conventional brake master cylinders with a separate hydraulic unit (Figure 14-89). Integrated ABS systems combine the brake master cylinder, hydraulic brake booster and the ABS components in a single hydraulic assembly (Figure 14-90).

Automatic Traction Control

The same technology used for ABS systems is also applied to automatic traction control. An electronic control monitor monitors the wheel speed sensors. If it determines that one of the drive wheels is spinning faster than the other, it will automatically apply the brakes to the spinning wheel. With the brake applied it requires a greater amount of torque to spin the wheel. Because a differential delivers equal torque to each drive wheel, the greater torque requirement is also transferred to the stationary or slower moving wheel. This allows the wheel that has the greatest amount of traction ability to move the vehicle.

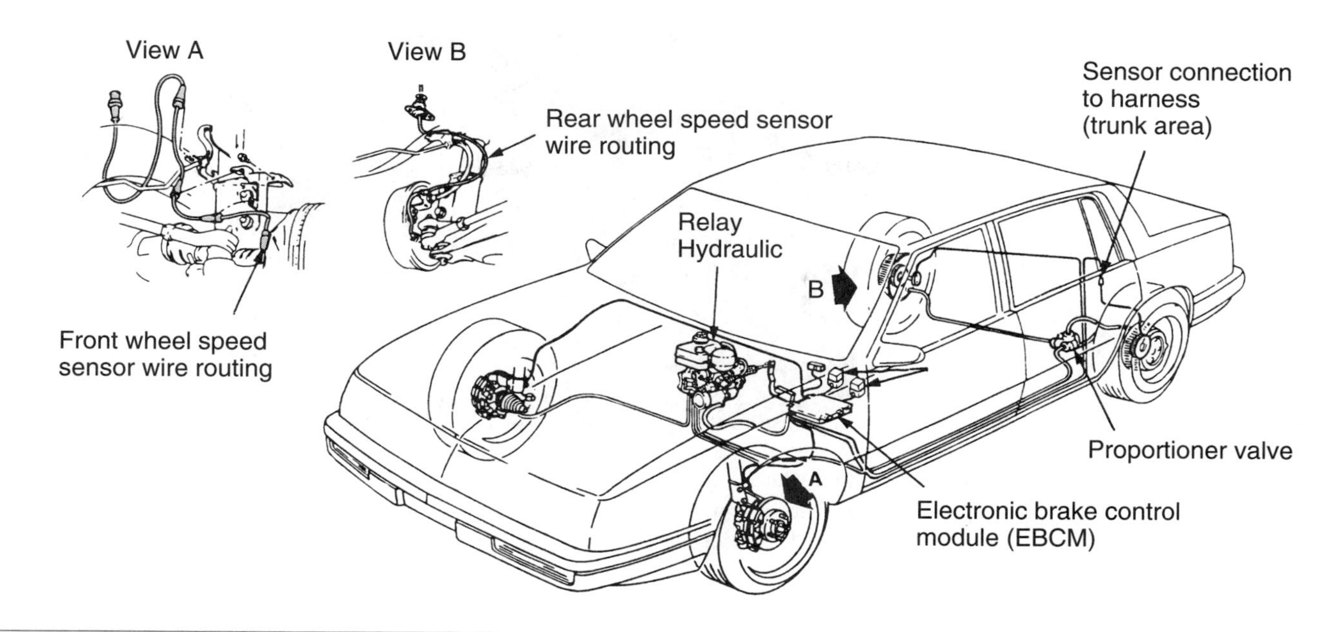

Figure 14-90 Integrated antilock brake system. (Courtesy of General Motors Corporation)

Some manufacturers integrate some engine control functions into the automatic traction control system instead of or in addition to applying the brakes. If the control module senses a loss of traction, it signals the engine control module to retard the engine timing and to decrease the throttle plate position. This action will reduce the engine output in an attempt to reduce the amount of power to the drive wheels. If this action fails to reduce tire slippage, the module will cut fuel delivery to one or more engine cylinders. The reduction of engine power is to prevent overspeeding the engine.

Vehicle Audio Entertainment Systems

Shop Manual
Chapter 14, page 565

When radios were first introduced to the automotive market, they produced little more than "tinny" noise and static. Today's audio sound systems produce music and sound that rivals the best that home sound systems can produce. And with nearly the same or even greater amounts of volume or sound power!

The most common sound system configuration is the all-in-one unit more commonly called a receiver. Housed in this unit is the radio tuner, amplifier, tone controls, and unit controls for all functions. These units may also include internal capabilities such as cassette players, compact disc players, digital audiotape players, and/or graphic equalizers. Most will be electronically tuned with a display that shows all functions being accessed/controlled and digital clock functions (Figure 14-91).

Recent developments by the manufacturers have been made to take individual functions (tape, disc, equalizer, control head, tuner, amplifier, etc.) and put them in individual boxes and call them components. This would allow owners greater flexibility in selecting options to suit their needs and tastes (Figure 14-92). Componentizing has allowed greater dash design flexibility. Some components, such as multiple CD changers, can be remotely mounted in a trunk area for greater security.

Figure 14-91 Radios are actually receivers. They contain the basic elements of a tuner, an amplifier, and a control assembly in one housing and can also contain a tape or CD player.

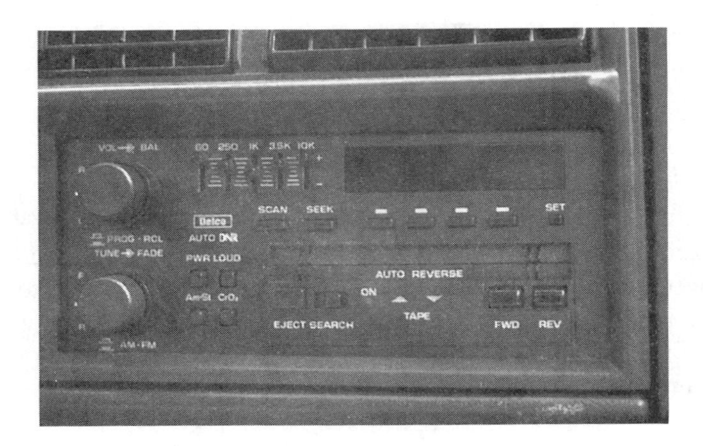

Figure 14-92 Components, like those shown here, allow for a more compact and optimized mounting location within the vehicle.

Wiring diagrams for component systems will be more complex, (Figure 14-93). In addition to power and audio signal wires, note that some systems will have a serial data wire for microprocessor communication between components for the controlling of unit functions. Some functions are shared and integrated with factory-installed cellular phones, such as radio mute. Some systems allow remote control of functions through a control assembly mounted in the steering wheel or alternate passenger compartment location (Figure 14-94).

A BIT OF HISTORY

Radios were introduced in cars by Daimler in 1922. Cars were equipped with Marconi wireless receivers.

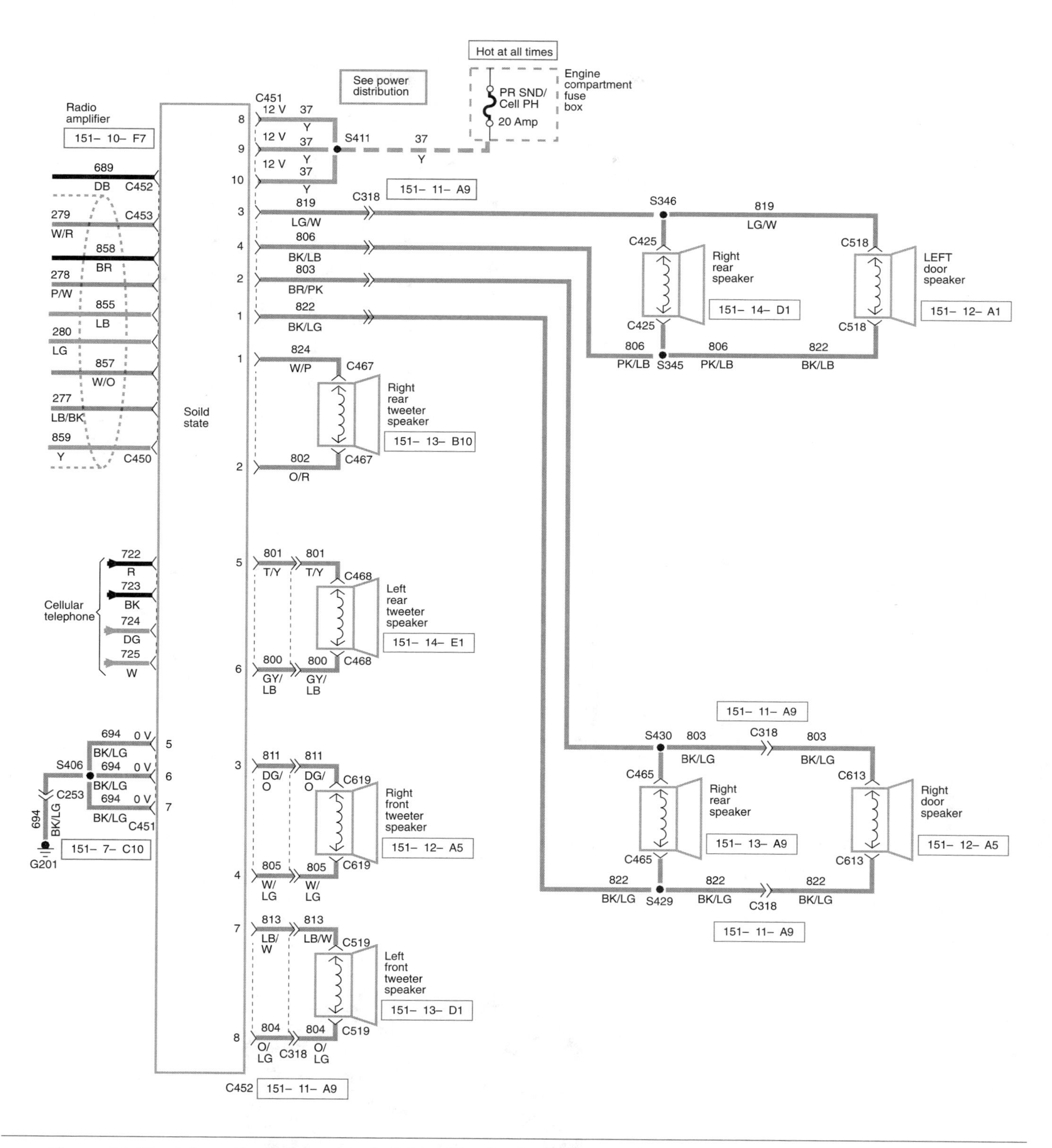

Figure 14-93 Schematic diagram showing the wiring hook-ups for remote-mounted components. (Reprinted with the permission of Ford Motor Company)

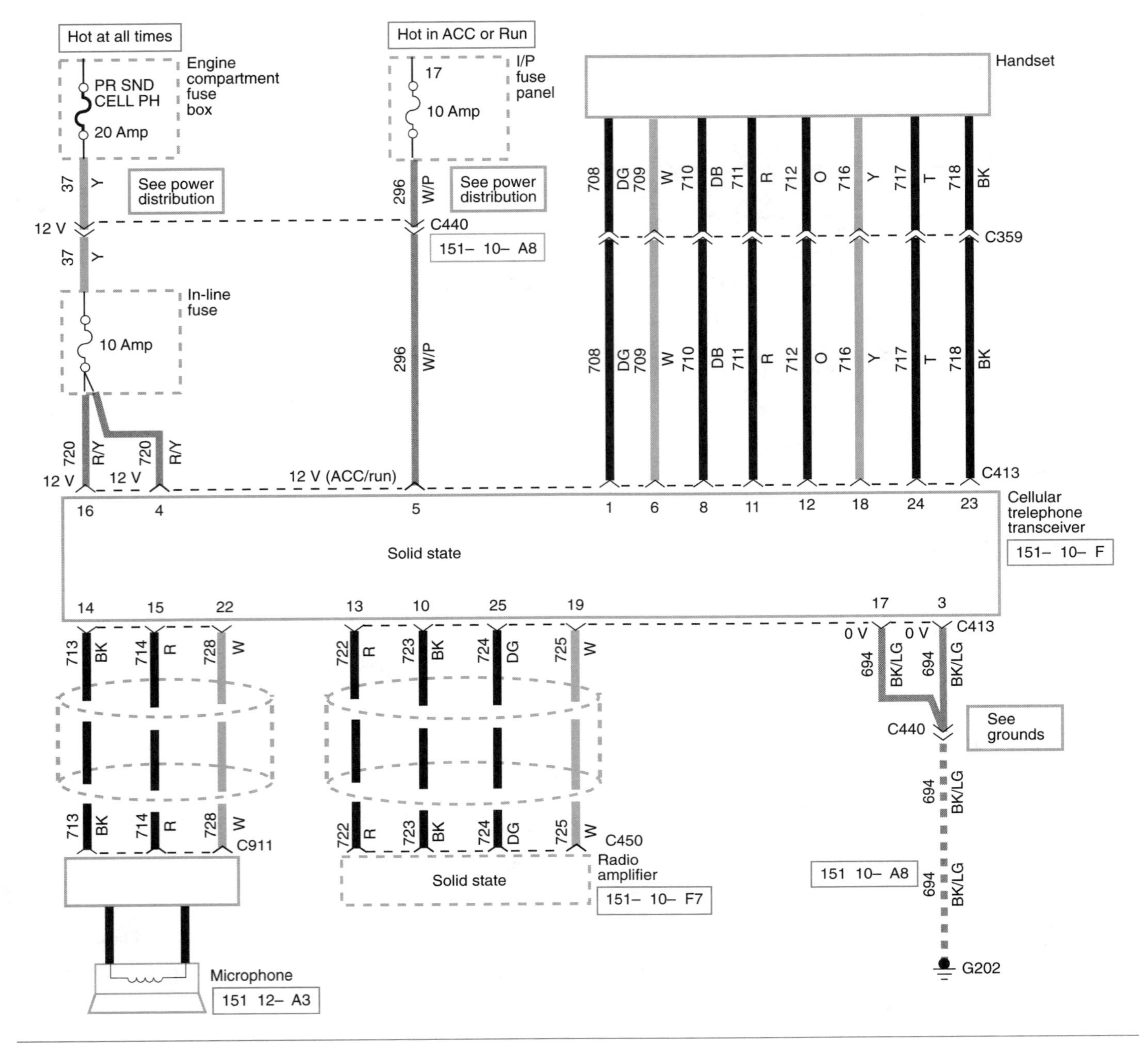

Figure 14-93 (continued)

Figure 14-94 Audio system with steering wheel-mounted remote control.

Summary

- ❑ An automatic temperature control (ATC) system is capable of maintaining a preset level of comfort control as selected by the driver.
- ❑ There are two types of automatic temperature control: semiautomatic temperature control (SATC) and electronic automatic temperature control (EATC).
- ❑ The in-car sensor measures the average temperature inside the vehicle. The ambient sensor measures the temperature outside the vehicle.
- ❑ The sunload sensor produces a signal proportional to the heat intensity of the sun through the windshield.
- ❑ Cruise control is a system that allows the vehicle to maintain a preset speed with the driver's foot off the accelerator.
- ❑ The transducer regulates the vacuum to the servo based on the electrical signals it receives.
- ❑ The servo controls throttle plate position by receiving a controlled amount of vacuum from the transducer.
- ❑ Safety switches return control of vehicle speed to the driver when the brake pedal is depressed.
- ❑ The controller energizes the supply and vent valves to allow manifold vacuum to enter the servo. The servo moves the throttle to maintain the set speed. The vehicle speed is maintained by balancing the vacuum in the servo.
- ❑ Passive restraints operate automatically with no action required on the part of the driver or occupant.
- ❑ An emergency release mechanism is provided in the event that the system fails to operate.
- ❑ The air bag is a supplement. The seatbelt is the primary restraint system.
- ❑ The air bag module is composed of the air bag and inflator assembly. It is packaged in a single module and is mounted in the center of the steering wheel.
- ❑ The diagnostic module constantly monitors the readiness of the SIR electrical system. It supplies backup power to the air bag module in the event that the battery or cables are damaged during an accident.
- ❑ The memory seat feature allows the driver to program different seat positions that can be recalled at the push of a button.
- ❑ The easy exit feature is an additional function of the memory seat that provides for easier entrance and exit of the vehicle by moving the seat all the way back and down.
- ❑ Antitheft systems are deterrent systems designed to scare off would-be thieves by sounding alarms and/or disabling the ignition system.
- ❑ The antitheft control module monitors the switches. If the doors or trunk are opened, or the key cylinders are rotated, the module will activate the system.
- ❑ The pass-key system acts as an engine disable system by using an ignition key that has an electronic pellet containing a coded resistance value.
- ❑ Automatic door locks is a passive system used to lock all doors when the required conditions are met. Many automobile manufacturers are incorporating the system as an additional safety and convenience feature.
- ❑ The keyless entry system allows the driver to unlock the doors or the deck lid from outside the vehicle without the use of a key.
- ❑ The heated windshield system is designed to melt ice and frost from the windshield three to five times faster than conventional defroster systems.

Terms to Know

Active suspension
Adaptive suspension
Air bag
Air bag module
Antitheft
Arming
Arming sensor
Audio system components
Automatic door lock
Automatic traction control
AXODE
Clock spring
Crash sensors
Diagnostic module
Discriminating sensors
Igniter
Keyless entry
Lock-up converter
Memory seat
Negative logic
Passive restraints
Passive suspension
Pass-key
Pressure control solenoid
Radio receiver
Shift solenoid
Torque converter
Valve body

- ❑ With electronically controlled transmissions, shifting, torque converter clutch operation, and line pressure regulation are controlled through computer operation.
- ❑ The shift solenoids provide gear selection, based on engine controller commands, by controlling the pressure to the shift valves.
- ❑ In speed sensitive steering systems, steering effort is controlled based on vehicle speed and rate of steering wheel rotation.
- ❑ Adaptive and active suspension systems provide additional benefits over conventional passive suspension systems by varying ride height, shock damping, and spring rates in response to changing road and driving conditions.
- ❑ Adaptive suspension systems are able to change ride characteristics by altering shock damping and ride height continuously.
- ❑ Active suspension systems are controlled by double acting hydraulic cylinders or solenoids (actuators) mounted at each wheel. Instead of conventional springs or air springs, the actuators support the vehicle's weight.
- ❑ Nonintegrated ABS systems use conventional brake master cylinders with a separate hydraulic unit.
- ❑ Integrated ABS systems combine the brake master cylinder, hydraulic brake booster, and the ABS components in a single hydraulic assembly.
- ❑ Automatic traction control limits the amount of tire spin on slippery road conditions by applying the brakes automatically or reducing engine power automatically.
- ❑ Vehicle audio entertainment systems are generally only a single component (receiver). Recently manufacturers have been developing multiple components (tuner, amplifier, control head, etc.) to allow for greater flexibility.
- ❑ Some audio component systems utilize a serial data line to provide for communication and control of functions between those components.

Review Questions

Short Answer Essay

1. Explain the basic differences between SATC and EATC systems.
2. What is the purpose of highside and lowside temperature sensors in some EATC systems?
3. Explain the basic operating principles of the electronic cruise control system.
4. List and explain the function of the components of the air bag module.
5. List the main components of common antitheft systems.
6. What functions are controlled by the electronic shift transmission?
7. Describe the active suspension system.
8. What are the differences between passive, adaptive, and active suspensions?
9. What is the purpose of antilock braking systems?
10. Explain the difference between a radio receiver unit system and a component radio system.

Fill-in-the-Blanks

1. The ATC system uses ____________________ that will open and close air-blend doors to achieve the desired in-vehicle temperature.

2. The sunload sensor is a ____________________ diode that sends signals to the programmer concerning the extra generation of heat as the sun beats through the windshield.

3. By monitoring the high side temperature, the BCM is capable of making calculations that translate into ____________________ .

4. When the brake pedal is depressed, the cruise control system is disengaged through ____________________ and ____________________ switches.

5. Safe service procedures of air bag systems are accomplished through proper use of the ____________________ ____________________ and by understanding the ____________________ principles of these systems.

6. The ____________________ ____________________ conducts electrical signals to the module while permitting steering wheel rotation.

7. In the EVO system, steering effort is controlled based on vehicle ____________________ and rate of steering wheel ____________________ .

8. ____________________ suspension systems are able to change ride characteristics by altering shock damping and ride height continuously. ____________________ suspension systems are controlled by double acting hydraulic cylinders or solenoids mounted at each wheel.

9. Adaptive and active suspension systems provide additional benefits over conventional passive suspension systems. They are able to change ____________________ , shock ____________________ and ____________________ rates in response to changing road and driving conditions.

10. Radio receiver units usually contain at least the following: an ____________________ , a ____________________ , and function ____________________ .

ASE Style Review Questions

1. EATC systems are being discussed:
   Technician A says the control module or BCM will store trouble codes that can be accessed by the technician.
   Technician B says the EATC system uses fewer sensors than most SATC systems.
   Who is correct?
   A. A only
   B. B only
   C. Both A and B
   D. Neither A nor B

2. A typical electronic engine cruise control system is being discussed:
   Technician A says the system utilizes vent and supply solenoids.
   Technician B says the amount of vacuum applied to the servo is controlled by the power solenoid.
   Who is correct?
   A. A only
   B. B only
   C. Both A and B
   D. Neither A nor B

3. Electronic cruise control systems are being discussed:
Technician A says if the voltage signal from the VSS drops below the low comparator value, the control module energizes the vent valve solenoid.
Technician B says if the VSS signal is greater than the high comparator value, the control module energizes the supply solenoid valve.
Who is correct?
A. A only **C.** Both A and B
B. B only **D.** Neither A nor B

4. Passive restraint systems are being discussed:
Technician A says passive restraints require the manual effort of the occupant.
Technician B says automatic seat belt systems operate by means of DC motors that move the belts by means of carriers on tracks.
Who is correct?
A. A only **C.** Both A and B
B. B only **D.** Neither A nor B

5. The air bag module is being discussed:
Technician A says the module supplies emergency current in the event battery voltage is lost.
Technician B says the module is composed of the air bag and inflator assembly.
Who is correct?
A. A only **C.** Both A and B
B. B only **D.** Neither A nor B

6. *Technician A* says the inflation assembly is composed of a solid chemical gas generator containing sodium azide and copper oxide propellent.
Technician B says the air bag module is serviceable.
Who is correct?
A. A only **C.** Both A and B
B. B only **D.** Neither A nor B

7. The keyless entry system is being discussed:
Technician A says an additional function of the system is that the deck lid lock can be released by pressing the 5-6 button.
Technician B says a second code can be entered into the system.
Who is correct?
A. A only **C.** Both A and B
B. B only **D.** Neither A nor B

8. The heated windshield system is being discussed:
Technician A says if the windshield has a crack or chip that will affect heating, the voltage drop across the resistor will indicate this condition to the control module.
Technician B says if the windshield is damaged, the controller reduces the voltage to the windshield to 20 volts.
Who is correct?
A. A only **C.** Both A and B
B. B only **D.** Neither A nor B

9. Electronic transmission actuators are being discussed:
Technician A says the shift solenoids regulate the amount of line pressure in the transmission.
Technician B says the electronic pressure control solenoid controls provide gear selection by controlling pressure to the shift valves.
Who is correct?
A. A only **C.** Both A and B
B. B only **D.** Neither A nor B

10. While discussing traction control systems:
Technician A says in some systems, if the control unit senses that one drive wheel is spinning faster than the other, it will cause the brakes of that wheel to be applied.
Technician B says the control unit of some systems will order changes to the engine control system to increase its power if the control unit senses that one drive wheel is spinning faster than the other.
Who is correct?
A. A only **C.** Both A and B
B. B only **D.** Neither A nor B

APPENDIX A

Abbreviations

The following abbreviations are some of the more common ones used today in the automotive industry.

TABLE 1—CROSS REFERENCE AND LOOK UP

Existing Usage	Acceptable Usage	Acceptable Acronized Usage
A/C (Air Conditioning)	Air Conditioning	A/C
A/C Cycling Switch	Air Conditioning Cycling Switch	A/C Cycling Switch
A/T (Automatic Transaxle)	Automatic Transaxle[1]	A/T[1]
A/T (Automatic Transmission)	Automatic Transmission[1]	A/T[1]
AAT (Ambient Air Temperature)	**Ambient Air Temperature**	**AAT**
AC (Air Conditioning)	Air Conditioning	A/C
ACC (Air Conditioning Clutch)	Air Conditioning Clutch	A/C Clutch
Accelerator	Accelerator Pedal	AP
Accelerator Pedal Position	**Accelerator Pedal Position[1]**	**APP[1]**
ACCS (Air Conditioning Cyclic Switch)	Air Conditioning Cycling Switch	A/C Cycling Switch
ACH (Air Cleaner Housing)	Air Cleaner Housing[1]	ACL Housing1
ACL (Air Cleaner)	Air Cleaner[1]	ACL[1]
ACL (Air Cleaner) Element	Air Cleaner Element[1]	ACL Element[1]
ACL (Air Cleaner) Housing	Air Cleaner Housing[1]	ACL Housing[1]
ACL (Air Cleaner) Housing Cover	Air Cleaner Housing Cover[1]	ACL Housing Cover[1]
ACS (Air Conditioning System)	Air Conditioning System	A/C System
ACT (Air Charge Temperature)	Intake Air Temperature[1]	IAT[1]
Adaptive Fuel Strategy	Fuel Trim[1]	FT[1]
AFC (Air Flow Control)	Mass Air Flow	MAF
AFC (Air Flow Control(	Volume Air Flow	VAF
AFS (Air Flow Sensor)	Mass Air Flow Sensor	MAF Sensor
AFS (Air Flow Sensor)	Volume Air Flow Sensor	VAF Sensor
After Cooler	Charge Air Cooler[1]	CAC[1]
AI (Air Injection)	Secondary Air Injection[1]	AIR[1]
AIP (Air Injection Pump)	Secondary Air Injection Pump[1]	AIR Pump[1]
AIR (Air Injection Reactor)	Pulsed Secondary Air Injection[1]	PAIR[1]
AIR (Air Injection Reactor)	Secondary Air Injection[1]	AIR[1]
AIRB (Secondary Air Injection Bypass)	Secondary Air Injection Bypass[1]	AIR Bypass[1]
AIRD (Secondary Air Injection Diverter)	Secondary Air Injection Diverter[1]	AIR Diverter[1]
Air Cleaner	Air Cleaner[1]	ACL[1]
Air Cleaner Element	Air Cleaner Element[1]	ACL Element[1]
Air Cleaner Housing	Air Cleaner Housing[1]	ACL Housing[1]
Air Cleaner Housing Cover	Air Cleaner Housing Cover[1]	ACL Housing Cover[1]
Air Conditioning	Air Conditioning	A/C
Air Conditioning Sensor	Air Conditioning Sensor	A/C Sensor
Air Control Valve	Secondary Air Injection Control Valve[1]	AIR Control Valve[1]
Air Flow Meter	Mass Air Flow Sensor[1]	MAF Sensor[1]
Air Flow Meter	Volume Air Flow Sensor[1]	VAF Sensor[1]
Air Intake System	Intake Air System[1]	IA System[1]
Air Flow Sensor	Mass Air Flow Sensor[1]	MAF Sensor[1]
Air Management 1	Secondary Air Injection Bypass[1]	AIR Bypass[1]
Air Management 2	Secondary Air Injection Diverter[1]	AIR Diverter[1]
Air Temperature Sensor	Intake Air Temperature Sensor[1]	IAT Sensor[1]
Air Valve	Idle Air Control Valve[1]	IAC Valve[1]
AIV (Air Injection Valve)	Pulsed Secondary Air Injection[1]	PAIR[1]
ALCL (Assembly Line Communication Link)	Data Link Connector[1]	DLC[1]
Alcohol Concentration Sensor	Flexible Fuel Sensor[1]	FF Sensor[1]
ALDL (Assembly Line Diagnostic Link)	Data Link Connector[1]	DLC[1]

TABLE 1—CROSS REFERENCE AND LOOK UP (CONTINUED)

Existing Usage	Acceptable Usage	Acceptable Acronized Usage
ALT (Alternator)	Generator	GEN
Alternator	Generator	GEN
Ambient Air Temperature	**Ambient Air Temperature**	**AAT**
AM1 (Air Management 1)	Secondary Air Injection Bypass[1]	AIR Bypass[1]
AM2 (Air Management 2)	Secondary Air Injection Diverter[1]	AIR Diverter[1]
APP (Accelerator Pedal Position)	**Accelerator Pedal Position**[1]	**APP**[1]
APS (Absolute Pressure Sensor)	Barometric Pressure Sensor[1]	BARO Sensor[1]
ATS (Air Temperature Sensor)	Intake Air Temperature Sensor[1]	IAT Sensor[1]
Automatic Transaxle	Automatic Transaxle[1]	A/T[1]
Automatic Transmission	Automatic Transmission[1]	A/T[1]
B+ (Battery Positive Voltage)	Battery Positive Voltage	B+
Backpressure Transducer	Exhaust Gas Recirculation Backpressure Transducer[1]	EGR Backpressure Transducer[1]
BARO (Barometric Pressure)	Barometric Pressure[1]	BARO[1]
Barometric Pressure Sensor	Barometric Pressure Sensor[1]	BARO Sensor[1]
Battery Positive Voltage	Battery Positive Voltage	B+
BLM (Block Learn Memory)	Long Term Fuel Trim[1]	Long Term FT[1]
BLM (Block Learn Multiplier)	Long Term Fuel Trim[1]	Long Term FT[1]
BLM (Block Learn Matrix)	Long Term Fuel Trim[1]	Long Term FT[1]
Block Learn Integrator	**Long Term Fuel Trim**[1]	**Long Term FT**[1]
Block Learn Matrix	Long Term Fuel Trim[1]	Long Term FT[1]
Block Learn Memory	Long Term Fuel Trim[1]	Long Term FT[1]
Block Learn Multiplier	Long Term Fuel Trim[1]	Long Term FT
BP (Barometric Pressure) Sensor	Barometric Pressure Sensor[1]	BARO Sensor[1]
BPP (Brake Pedal Position)	**Brake Pedal Position**[1]	**BPP**[1]
Brake Pressure	**Brake Pressure**	**Brake Pressure**
Brake Pedal Position	**Brake Pedal Position**[1]	**BPP**[1]
C3I (Computer Controlled Coil Ignition)	Electronic Ignition[1]	EI[1]
CAC (Charge Air Cooler)	Charge Air Cooler[1]	CAC[1]
Calculated Load Value	**Calculated Load Value**	**LOAD**
Camshaft Position	Camshaft Position[1]	CMP[1]
Camshaft Position Actuator	**Camshaft Position Actuator**[1]	**CMP Actuator**[1]
Camshaft Position Controller	**Camshaft Position Actuator**[1]	**CMP Actuator**[1]
Camshaft Position Sensor	Camshaft Position Sensor[1]	CMP Sensor[1]
Camshaft Sensor	Camshaft Position Sensor[1]	CMP Sensor[1]
Camshaft Timing Actuator	**Camshaft Position Actuator**[1]	**CMP Actuator**[1]
Canister	Canister[1]	Canister[1]
Canister	Evaporative Emission Canister[1]	EVAP Canister[1]
Canister Purge	**Evaporative Emission Canister Purge**[1]	**EVAP Canister Purge**[1]
Canister Purge Vacuum Switching Valve	Evaporative Emission Canister Purge Valve[1]	EVAP Canister Purge Valve[1]
Canister Purge Valve	Evaporative Emission Canister Purge Valve[1]	EVAP Canister Purge Valve[1]
Canister Purge VSV (Vacuum Switching Valve)	Evaporative Emission Canister Purge Valve[1]	EVAP Canister Purge Valve[1]
CANP (Canister Purge)	Evaporative Emission Canister Purge[1]	EVAP Canister Purge[1]
CARB (Carburetor)	Carburetor[1]	CARB[1]
Carburetor	Carburetor[1]	CARB[1]
Catalytic Converter Heater	**Catalytic Converter Heater**	**Catalytic Converter Heater**
CCC (Converter Clutch Control)	Torque Converter Clutch[1]	TCC[1]
CCO (Converter Clutch Override)	Torque Converter Clutch[1]	TCC[1]
CCS (Coast Clutch Solenoid)	**Coast Clutch Solenoid**	**CCS**

TABLE 1—CROSS REFERENCE AND LOOK UP (CONTINUED)

Existing Usage	Acceptable Usage	Acceptable Acronized Usage
CCS (Coast Clutch Solenoid) Valve	**Coast Clutch Solenoid Valve**	**CCS Valve**
CCRM (Constant Control Relay Module)	**Constant Control RM**	**Constant Control RM**
CDI (Capacitive Discharge Ignition)	Distributor Ignition[1]	DI[1]
CDROM (Compact Disc Read Only Memory)	Compact Disc Read Only Memory[1]	CDROM[1]
CES (Clutch Engage Switch)	Clutch Pedal Position Switch	CPP Switch[1]
Central Multiport Fuel Injection	Central Multiport Fuel Injection[1]	Central MFI[1]
Central Sequential Multiport Fuel Injection	**Central Sequential Multiport Fuel Injection**	**Central SFI**
CFI (Continuous Fuel Injection)	Continuous Fuel Injection[1]	CFI[1]
CFI (Central Fuel Injection)	Throttle Body Fuel Injection[1]	TBI[1]
CFV	**Critical Flow Venturi**	**CFV**
Charcoal Canister	Evaporative Emission Canister	EVAP Canister[1]
Charge Air Cooler	Charge Air Cooler	CAC[1]
Check Engine	Service Reminder Indicator[1]	SRI[1]
Check Engine	Malfunction Indicator Lamp[1]	MIL[1]
CID (Cylinder Identification) Sensor	Camshaft Position Sensor	CMP Sensor[1]
CIS (Continuous Injection System)	Continuous Fuel Injection[1]	CFI[1]
CIS-E (Continuous Injection System Electronic)	Continuous Fuel Injection[1]	CFI[1]
CKP (Crankshaft Position)	Crankshaft Position[1]	CKP[1]
CKP (Crankshaft Position) Sensor	Crankshaft Position Sensor[1]	CKP Sensor[1]
CL (Closed Loop)	Closed Loop[1]	CL[1]
Closed Bowl Distributor	Distributor Ignition[1]	DI[1]
Closed Throttle Position	Closed Throttle Position[1]	CTP[1]
Closed Throttle Switch	Closed Throttle Position Switch[1]	CTP Switch[1]
CLS (Closed Loop System)	Closed Loop[1]	CL[1]
CLV	**Calculated Load Value**	**LOAD**
Clutch Engage Switch	Clutch Pedal Position Switch[1]	CPP Switch[1]
Clutch Pedal Position Switch	Clutch Pedal Position Switch[1]	CPP Switch[1]
Clutch Start Switch	Clutch Pedal Position Switch[1]	CPP Switch[1]
Clutch Switch	Clutch Pedal Position Switch[1]	CPP Switch[1]
CMFI (Central Multiport Fuel Injection)	Central Multiport Fuel Injection[1]	Central MFI[1]
CMP (Camshaft Position)	Camshaft Position[1]	CMP[1]
CMP (Camshaft Position) Sensor	Camshaft Position Sensor[1]	CMP Sensor[1]
COC (Continuous Oxidation Catalyst)	Oxidation Catalytic Converter[1]	OC[1]
Coast Clutch Solenoid	**Coast Clutch Solenoid**	**CCS**
Coast Clutch Solenoid Valve	**Coast Clutch Solenoid Valve**	**CCS Valve**
Condenser	Distributor Ignition Capacitor[1]	DI Capacitor[1]
Constant Control Relay Module	**Relay Module**	**RM**
Constant Volume Sampler	**Constant Volume Sampler**	**CVS**
Continuous Fuel Injection	Continuous Fuel Injection[1]	CFI[1]
Continuous Injection System	Continuous Fuel Injection System[1]	CFI System[1]
Continuous Injection System-E	Electronic Continuous Fuel Injection System[1]	Electronic CFI System[1]
Continuous Trap Oxidizer	Continuous Trap Oxidizer[1]	CTOX[1]
Coolant Temperature Sensor	Engine Coolant Temperature Sensor[1]	ECT Sensor[1]
CP (Crankshaft Position)	Crankshaft Position[1]	CKP[1]
CPP (Clutch Pedal Position)	Clutch Pedal Position[1]	CPP[1]
CPP (Clutch Pedal Position) Switch	Clutch Pedal Position Switch	CPP Switch[1]
CPS (Camshaft Position Sensor)	Camshaft Position Sensor[1]	CMP Sensor[1]
CPS (Crankshaft Position Sensor)	Crankshaft Position Sensor[1]	CKP Sensor[1]
Crank Angle Sensor	Crankshaft Position Sensor[1]	CKP Sensor[1]

TABLE 1—CROSS REFERENCE AND LOOK UP (CONTINUED)

Existing Usage	Acceptable Usage	Acceptable Acronized Usage
Crankshaft Position	Crankshaft Position[1]	CKP[1]
Crankshaft Position Sensor	Crankshaft Position Sensor[1]	CKP Sensor[1]
Crankshaft Speed	Engine Speed[1]	RPM[1]
Crankshaft Speed Sensor	Engine Speed Sensor[1]	RPM Sensor[1]
Critical Flow Venturi	**Critical Flow Venturi**	**CFV**
CTO (Continuous Trap Oxidizer)	Continuous Trap Oxidizer[1]	CTOX[1]
CTOX (Continuous Trap Oxidizer)	Continuous Trap Oxidizer[1]	CTOX[1]
CTP (Closed Throttle Position)	Closed Throttle Position[1]	CTP[1]
CTS (Coolant Temperature Sensor)	Engine Coolant Temperature Sensor[1]	ECT Sensor[1]
CTS (Coolant Temperature Switch)	Engine Coolant Temperature Switch[1]	ECT Switch[1]
CVS	**Constant Volume Sampler**	**CVS**
Cylinder ID (Identification) Sensor	Camshaft Position Sensor[1]	CMP Sensor[1]
D-Jetronic	Multiport Fuel Injection[1]	MFI[1]
Data Link Connector	Data Link Connector[1]	DLC[1]
Detonation Sensor	Knock Sensor[1]	KS[1]
DFI (Direct Fuel Injection)	Direct Fuel Injection[1]	DFI[1]
DFI (Digital Fuel Injection)	Multiport Fuel Injection[1]	MFI[1]
DI (Direct Injection)	Direct Fuel Injection[1]	DFI[1]
DI (Distributor Ignition)	Distributor Ignition[1]	DI[1]
DI (Distributor Ignition) Capacitor	Distributor Ignition Capacitor[1]	DI Capacitor[1]
Diagnostic Test Mode	Diagnostic Test Mode[1]	DTM[1]
Diagnostic Trouble Code	Diagnostic Trouble Code[1]	DTC[1]
DID (Direct Injection - Diesel)	Direct Fuel Injection[1]	DFI[1]
Differential Pressure Feedback EGR (Exhaust Gas Recirculation) System	Differential Pressure Feedback Exhaust Gas Recirculation System[1]	Differential Pressure Feedback EGR System[1]
Digital EGR (Exhaust Gas Recirculation)	Exhaust Gas Recirculation[1]	EGR[1]
Direct Fuel Injection	Direct Fuel Injection[1]	DFI[1]
Direct Ignition System	Electronic Ignition System[1]	EI System[1]
DIS (Distributorless Ignition System)	Electronic Ignition System[1]	EI System[1]
DIS (Distributorless Ignition System) Module	Ignition Control Module[1]	ICM[1]
Distance Sensor	Vehicle Speed Sensor[1]	VSS[1]
Distributor Ignition	Distributor Ignition[1]	DI[1]
Distributorless Ignition	Electronic Ignition[1]	EI[1]
DLC (Data Link Connector)	Data Link Connector[1]	DLC[1]
DLI (Distributorless Ignition)	Electronic Ignition[1]	EI[1]
Driver	**Driver**	**Driver**
DS (Detonation Sensor)	Knock Sensor[1]	KS[1]
DTC (Diagnostic Trouble Code)	Diagnostic Trouble Code[1]	DTC[1]
DTM (Diagnostic Test Mode)	Diagnostic Test Mode[1]	DTM[1]
Dual Bed	Three Way + Oxidation Catalytic Converter[1]	TWC+OC[1]
Duty Solenoid for Purge Valve	Evaporative Emission Canister Purge Valve	EVAP Canister Purge Valve[1]
Dynamic Pressure Control	**Dynamic Pressure Control**	**Dynamic PC**
Dynamic Pressure Control Solenoid	**Dynamic Pressure Control Solenoid[1]**	**Dynamic PC Solenoid[1]**
Dynamic Pressure Control Solenoid Valve	**Dynamic Pressure Control Solenoid Valve[1]**	**Dynamic PC Solenoid Valve[1]**
E2PROM (Electrically Erasable Programmable Read Only Memory)	Electrically Erasable Programmable Read Only Memory[1]	EEPROM[1]
Early Fuel Evaporation	Early Fuel Evaporation[1]	EFE[1]
EATX (Electronic Automatic Transmission/ Transaxle)	Automatic Transmission[1]	A/T[1]
EC (Engine Control)	Engine Control[1]	EC[1]

TABLE 1—CROSS REFERENCE AND LOOK UP (CONTINUED)

Existing Usage	Acceptable Usage	Acceptable Acronized Usage
ECA (Electronic Control Assembly)	Powertrain Control Module[1]	PCM[1]
ECL (Engine Coolant Level)	Engine Coolant Level	ECL
ECM (Engine Control Module)	Engine Control Module[1]	ECM[1]
ECT (Engine Coolant Temperature)	Engine Coolant Temperature[1]	ECT[1]
ECT (Engine Coolant Temperature) Sender	Engine Coolant Temperature Sensor[1]	ECT Sensor[1]
ECT (Engine Coolant Temperature) Sensor	Engine Coolant Temperature Sensor[1]	ECT Sensor[1]
ECT (Engine Coolant Temperature) Switch	Engine Coolant Temperature Switch[1]	ECT Switch[1]
ECU4 (Electronic Control Unit 4)	Powertrain Control Module[1]	PCM[1]
EDF (Electro-Drive Fan) Control	Fan Control	FC
EDIS (Electronic Distributor Ignition System)	Distributor Ignition System[1]	DI System[1]
EDIS (Electronic Distributorless Ignition System)	Electronic Ignition System[1]	EI System[1]
EDIS (Electronic Distributor Ignition System) Module	Distributor Ignition Control Module[1]	Distributor ICM[1]
EEC (Electronic Engine Control)	Engine Control[1]	EC[1]
EEC (Electronic Engine Control) Processor	Powertrain Control Module[1]	PCM[1]
EECS (Evaporative Emission Control System)	Evaporative Emission System[1]	EVAP System[1]
EEPROM (Electrically Erasable Programmable Read Only Memory)	Electrically Erasable Programmable Read Only Memory[1]	EEPROM[1]
EFE (Early Fuel Evaporation)	Early Fuel Evaporation[1]	EFE[1]
EFI (Electronic Fuel Injection)	Multiport Fuel Injection[1]	MFI[1]
EFI (Electronic Fuel Injection)	Throttle Body Fuel Injection[1]	TBI[1]
EGO (Exhaust Gas Oxygen) Sensor	Oxygen Sensor[1]	O2S[1]
EGOS (Exhaust Gas Oxygen Sensor)	Oxygen Sensor[1]	O2S[1]
EGR (Exhaust Gas Recirculation)	Exhaust Gas Recirculation[1]	EGR[1]
EGR (Exhaust Gas Recirculation) Diagnostic Valve	Exhaust Gas Recirculation Diagnostic Valve[1]	EGR Diagnostic Valve[1]
EGR (Exhaust Gas Recirculation) System	Exhaust Gas Recirculation System[1]	EGR System[1]
EGR (Exhaust Gas Recirculation) Thermal Vacuum Valve	Exhaust Gas Recirculation Thermal Vacuum Valve[1]	EGR TVV[1]
EGR (Exhaust Gas Recirculation) Valve	Exhaust Gas Recirculation Valve[1]	EGR Valve[1]
EGR TVV (Exhaust Gas Recirculation Thermal Vacuum Valve)	Exhaust Gas Recirculation Thermal Vacuum Valve[1]	EGR TVV[1]
EGRT (Exhaust Gas Recirculation Temperature)	Exhaust Gas Recirculation Temperature	EGRT[1]
EGRT (Exhaust Gas Recirculation Temperature) Sensor	Exhaust Gas Recirculation Temperature Sensor[1]	EGRT Sensor[1]
EGRV (Exhaust Gas Recirculation Valve)	Exhaust Gas Recirculation Valve[1]	EGR Valve[1]
EGRVC (Exhaust Gas Recirculation Valve Control)	Exhaust Gas Recirculation Valve Control[1]	EGR Valve Control[1]
EGS (Exhaust Gas Sensor)	Oxygen Sensor[1]	O2S[1]
EI (Electronic Ignition) (With Distributor)	Distributor Ignition[1]	DI[1]
EI (Electronic Ignition) (Without Distributor)	Electronic Ignition[1]	EI[1]
Electrically Erasable Programmable Read Only Memory	Electrically Erasable Programmable Read Only Memory[1]	EEPROM[1]
Electronic Engine Control	Electronic Engine Control[1]	Electronic EC[1]
Electronic Ignition	Electronic Ignition[1]	EI[1]
Electronic Spark Advance	Ignition Control[1]	IC[1]
Electronic Spark Timing	Ignition Control[1]	IC[1]
EM (Engine Modification)	Engine Modification[1]	EM[1]
EMR (Engine Maintenance Reminder)	Service Reminder Indicator[1]	SRI[1]
Engine Control	Engine Control[1]	EC[1]
Engine Coolant Fan Control	Fan Control	FC
Engine Coolant Level	Engine Coolant Level	ECL
Engine Coolant Level Indicator	Engine Coolant Level Indicator	ECL Indicator

TABLE 1—CROSS REFERENCE AND LOOK UP (CONTINUED)

Existing Usage	Acceptable Usage	Acceptable Acronized Usage
Engine Coolant Temperature	Engine Coolant Temperature[1]	ECT[1]
Engine Coolant Temperature Sender	Engine Coolant Temperature Sensor[1]	ECT Sensor[1]
Engine Coolant Temperature Sensor	Engine Coolant Temperature Sensor[1]	ECT Sensor[1]
Engine Coolant Temperature Switch	Engine Coolant Temperature Switch[1]	ECT Switch[1]
Engine Modification	Engine Modification[1]	EM[1]
Engine Oil Pressure Sender	**Engine Oil Pressure** Sensor	**EOP** Sensor
Engine Oil Pressure Sensor	**Engine Oil Pressure** Sensor	**EOP** Sensor
Engine Oil Pressure Switch	**Engine Oil Pressure** Switch	**EOP** Switch
Engine Oil Temperature	**Engine Oil Temperature**	**EOT**
Engine Speed	Engine Speed[1]	RPM[1]
EOS (Exhaust Oxygen Sensor)	Oxygen Sensor[1]	O2S[1]
EOT (Engine Oil Temperature)	**Engine Oil Temperature**	**EOT**
EP (Exhaust Pressure)	**Exhaust Pressure**	**EP**
EPROM (Erasable Programmable Read Only Memory)	Erasable Programmable Read Only Memory[1]	EPROM[1]
Erasable Programmable Read Only Memory	Erasable Programmable Read Only Memory[1]	EPROM[1]
ESA (Electronic Spark Advance)	Ignition Control[1]	IC[1]
ESAC (Electronic Spark Advance Control)	Distributor Ignition[1]	DI[1]
EST (Electronic Spark Timing)	Ignition Control[1]	IC[1]
EVAP (Evaporate Emission) CANP (Canister Purge)	**Evaporative Emission Canister Purge**[1]	**EVAP Canister Purge**[1]
EVAP (Evaporative Emission)	Evaporative Emission[1]	EVAP[1]
EVAP (Evaporative Emission) Canister	Evaporative Emission Canister[1]	EVAP Canister[1]
EVAP (Evaporative Emission) Purge Valve	Evaporative Emission Canister Purge Valve[1]	EVAP Canister Purge Valve[1]
Evaporative Emission	Evaporative Emission[1]	EVAP[1]
Evaporative Emission Canister	Evaporative Emission Canister[1]	EVAP Canister[1]
EVP (Exhaust Gas Recirculation Valve Position) Sensor	Exhaust Gas Recirculation Valve Position Sensor[1]	EGR Valve Position Sensor[1]
EVR (Exhaust Gas Recirculation Vacuum Regulator) Solenoid	Exhaust Gas Recirculation Vacuum Regulator Solenoid[1]	EGR Vacuum Regulator Solenoid[1]
EVRV (Exhaust Gas Recirculation Vacuum Regulator Valve)	Exhaust Gas Recirculation Vacuum Regulator Valve[1]	EGR Vacuum Regulator Valve[1]
Exhaust Gas Recirculation	Exhaust Gas Recirculation[1]	EGR[1]
Exhaust Gas Recirculation Temperature	Exhaust Gas Recirculation Temperature[1]	EGRT[1]
Exhaust Gas Recirculation Temperature Sensor	Exhaust Gas Recirculation Temperature Sensor[1]	EGRT Sensor[1]
Exhaust Gas Recirculation Vacuum Solenoid Valve Regulator	**Exhaust Gas Recirculation Vacuum Regulator Solenoid Valve**[1]	**EGR Vacuum Regulator Solenoid Valve**[1]
Exhaust Gas Recirculation Vacuum Regulator Valve	**Exhaust Gas Recirculation Vacuum Regulator Valve**[1]	**EGR Vacuum Regulator Valve**[1]
Exhaust Gas Recirculation Valve	Exhaust Gas Recirculation Valve[1]	EGR Valve[1]
Exhaust Pressure	**Exhaust Pressure**	**EP**
4GR (Fourth Gear)	Fourth Gear	4GR
4WD (Four Wheel Drive)	**Full Time Four Wheel Drive**	**F4WD**
4WD (Four Wheel Drive)	**Selectable Four Wheel Drive**	**S4WD**
F4WD	**Full Time Four Wheel Drive**	**F4WD**
Fan Control	Fan Control	FC
Fan Control Module	Fan Control Module	FC Module
Fan Control Relay	Fan Control Relay	FC Relay
Fan Motor Control Relay	Fan Control Relay	FC Relay
Fast Idle Thermo Valve	Idle Air Control Thermal Valve[1]	IAC Thermal Valve[1]
FBC (Feed Back Carburetor)	Carburetor[1]	CARB[1]
FBC (Feed Back Control)	Mixture Control[1]	MC[1]

TABLE 1—CROSS REFERENCE AND LOOK UP (CONTINUED)

Existing Usage	Acceptable Usage	Acceptable Acronized Usage
FC (Fan Control)	Fan Control	FC
FC (Fan Control) Relay	Fan Control Relay	FC Relay
FEEPROM (Flash Electrically Erasable Programmable Read Only Memory)	Flash Electrically Erasable Programmable Read Only Memory[1]	FEEPROM[1]
FEPROM (Flash Erasable Programmable Read Only Memory)	Flash Erasable Programmable Read Only Memory[1]	FEPROM[1]
FF (Flexible Fuel)	Flexible Fuel[1]	FF[1]
FI (Fuel Injection)	Central Multiport Fuel Injection[1]	Central MFI[1]
FI (Fuel Injection)	Continuous Fuel Injection[1]	CFI[1]
FI (Fuel Injection)	Direct Fuel Injection[1]	DFI[1]
FI (Fuel Injection)	Indirect Fuel Injection[1]	IFI[1]
FI (Fuel Injection)	Multiport Fuel Injection[1]	MFI[1]
FI (Fuel Injection)	Sequential Multiport Fuel Injection[1]	SFI[1]
FI (Fuel Injection)	Throttle Body Fuel Injection[1]	TBI[1]
Flame Ionization Detector	**Flame Ionization Detector**	**FID**
Flash EEPROM (Electrically Erasable Programmable Read Only Memory)	Flash Electrically Erasable Programmable Read Only Memory[1]	FEEPROM[1]
Flash EPROM (Erasable Programmable Read Only Memory)	Flash Erasable Programmable Read Only Memory[1]	FEPROM[1]
Flexible Fuel	Flexible Fuel[1]	FF[1]
Flexible Fuel Sensor	Flexible Fuel Sensor[1]	FF Sensor
Fourth Gear	Fourth Gear	4GR
FP (Fuel Pump)	Fuel Pump	FP
FP (Fuel Pump) Module	Fuel Pump Module	FP Module
Freeze Frame	**Freeze Frame**	**See Table 4**
Front Wheel Drive	**Front Wheel Drive**	**FWD**
FRZF (Freeze Frame)	**Freeze Frame**	**See Table 4**
FT (Fuel Trim)	Fuel Trim[1]	FT[1]
Fuel Charging Station	Throttle Body[1]	TB[1]
Fuel Concentration Sensor	Flexible Fuel Sensor[1]	FF Sensor[1]
Fuel Injection	Central Multiport Fuel Injection[1]	Central MFI[1]
Fuel Injection	Continuous Fuel Injection[1]	CFI[1]
Fuel Injection	Direct Fuel Injection[1]	DFI[1]
Fuel Injection	Indirect Fuel Injection[1]	IFI[1]
Fuel Injection	Multiport Fuel Injection[1]	MFI[1]
Fuel Injection	Sequential Multiport Fuel Injection[1]	SFI[1]
Fuel Injection	Throttle Body Fuel Injection[1]	TBI[1]
Fuel Level Sensor	Fuel Level Sensor	Fuel Level Sensor
Fuel Module	Fuel Pump Module	FP Module
Fuel Pressure	Fuel Pressure[1]	Fuel Pressure[1]
Fuel Pressure	**Fuel Pressure**	**See Table 4**
Fuel Pressure Regulator	Fuel Pressure Regulator[1]	Fuel Pressure Regulator[1]
Fuel Pump	Fuel Pump	FP
Fuel Pump Relay	Fuel Pump Relay	FP Relay
Fuel Quality Sensor	Flexible Fuel Sensor[1]	FF Sensor[1]
Fuel Regulator	Fuel Pressure Regulator[1]	Fuel Pressure Regulator[1]
Fuel Sender	Fuel Pump Module	FP Module
Fuel Sensor	Fuel Level Sensor	Fuel Level Sensor
Fuel System Status	**Fuel System Status**	**See Table 4**
FUEL SYS	**Fuel System Status**	**See Table 4**
Fuel Tank Unit	Fuel Pump Module	FP Module

TABLE 1—CROSS REFERENCE AND LOOK UP (CONTINUED)

Existing Usage	Acceptable Usage	Acceptable Acronized Usage
Fuel Trim	Fuel Trim[1]	FT[1]
Full Time Four Wheel Drive	**Full Time Four Wheel Drive**	**F4WD**
Full Throttle	Wide Open Throttle[1]	WOT[1]
FWD	**Front Wheel Drive**	**FWD**
GCM (Governor Control Module)	Governor Control Module	GCM
GEM (Governor Electronic Module)	Governor Control Module	GCM
GEN (Generator)	Generator	GEN
Generator	Generator	GEN
Glow Plug	**Glow Plug[1]**	**Glow Plug[1]**
GND (Ground)	Ground	GND
Governor	Governor	Governor
Governor Control Module	Governor Control Module	GCM
Governor Electronic Module	Governor Control Module	GCM
Gram Per Mile	**Gram Per Mile**	**GPM**
GRD (Ground)	Ground	GND
Ground	Ground	GND
Heated Oxygen Sensor	Heated Oxygen Sensor[1]	HO2S[1]
HEDF (High Electro-Drive Fan) Control	Fan Control	FC
HEGO (Heated Exhaust Gas Oxygen) Sensor	Heated Oxygen Sensor[1]	HO2S[1]
HEI (High Energy Ignition)	Distributor Ignition[1]	DI[1]
High Speed FC (Fan Control) Switch	High Speed Fan Control Switch	High Speed FC Switch
HO2S (Heated Oxygen Sensor)	Heated Oxygen Sensor[1]	HO2S[1]
HOS (Heated Oxygen Sensor)	Heated Oxygen Sensor[1]	HO2S[1]
Hot Wire Anemometer	Mass Air Flow Sensor[1]	MAF Sensor[1]
IA (Intake Air)	Intake Air	IA
IA (Intake Air) Duct	Intake Air Duct	IA Duct
IAC (Idle Air Control)	Idle Air Control[1]	IAC[1]
IAC (Idle Air Control) Thermal Valve	Idle Air Control Thermal Valve[1]	IAC Thermal Valve[1]
IAC (Idle Air Control) Valve	Idle Air Control Valve[1]	IAC Valve[1]
IACV (Idle Air Control Valve)	Idle Air Control Valve[1]	IAC Valve[1]
IAT (Intake Air Temperature)	Intake Air Temperature[1]	IAT[1]
IAT (Intake Air Temperature) Sensor	Intake Air Temperature Sensor[1]	IAT Sensor[1]
IATS (Intake Air Temperature Sensor)	Intake Air Temperature Sensor[1]	IAT Sensor[1]
IC (Ignition Control)	Ignition Control[1]	IC[1]
ICM (Ignition Control Module)	Ignition Control Module[1]	ICM[1]
ICP (Injection Control Pressure)	**Injection Control Pressure[1]**	**ICP[1]**
IDFI (Indirect Fuel Injection)	Indirect Fuel Injection[1]	IFI[1]
IDI (Integrated Direct Ignition)	Electronic Ignition[1]	EI[1]
IDI (Indirect Diesel Injection)	Indirect Fuel Injection[1]	IFI[1]
Idle Air Bypass Control	Idle Air Control[1]	IAC[1]
Idle Air Control	Idle Air Control[1]	IAC[1]
Idle Air Control Valve	Idle Air Control Valve[1]	IAC Valve[1]
Idle Speed Control	Idle Air Control[1]	IAC[1]
Idle Speed Control	Idle Speed Control[1]	ISC[1]
Idle Speed Control Actuator	Idle Speed Control Actuator[1]	ISC Actuator[1]
IFI (Indirect Fuel Injection)	Indirect Fuel Injection[1]	IFI[1]
IFS (Inertia Fuel Shutoff)	Inertia Fuel Shutoff	IFS
Ignition Control	Ignition Control[1]	IC[1]
Ignition Control Module	Ignition Control Module[1]	ICM[1]
I/M (Inspection and Maintenance)	**Inspection and Maintenance**	**I/M**

TABLE 1—CROSS REFERENCE AND LOOK UP (CONTINUED)

Existing Usage	Acceptable Usage	Acceptable Acronized Usage
IMRC (Intake Manifold Runner Control)	**Intake Manifold Runner Control**	**IMRC**
In Tank Module	Fuel Pump Module	FP Module
Indirect Fuel Injection	Indirect Fuel Injection[1]	IFI[1]
Inertia Fuel Shutoff	Inertia Fuel Shutoff	IFS
Inertia Fuel - Shutoff Switch	Inertia Fuel Shutoff Switch	IFS Switch
Inertia Switch	Inertia Fuel Shutoff Switch	IFS Switch
Injection Control Pressure	**Injection Control Pressure**[1]	**ICP**[1]
Input Shaft Speed	**Input Shaft Speed**	**ISS**
INT (Integrator)	Short Term Fuel Trim[1]	Short Term FT[1]
Inspection and Maintenance	**Inspection and Maintenance**	**I/M**
Intake Air	Intake Air	IA
Intake Air Duct	Intake Air Duct	IA Duct
Intake Air Temperature	Intake Air Temperature[1]	IAT[1]
Intake Air Temperature Sensor	Intake Air Temperature Sensor[1]	IAT Sensor[1]
Intake Manifold Absolute Pressure Sensor	Manifold Absolute Pressure Sensor[1]	MAP Sensor[1]
Intake Manifold Runner Control	**Intake Manifold Runner Control**	**IMRC**
Integrated Relay Module	Relay Module	RM
Integrator	Short Term Fuel Trim[1]	Short Term FT[1]
Inter Cooler	Charge Air Cooler[1]	CAC[1]
ISC (Idle Speed Control)	Idle Air Control[1]	IAC[1]
ISC (Idle Speed Control)	Idle Speed Control[1]	ISC[1]
ISC (Idle Speed Control) Actuator	Idle Speed Control Actuator[1]	ISC Actuator[1]
ISC BPA (Idle Speed Control By Pass Air)	Idle Air Control[1]	IAC
ISC (Idle Speed Control) Solenoid Vacuum Valve	Idle Speed Control Solenoid Vacuum Valve[1]	ISC Solenoid Vacuum Valve[1]
ISS (Input Shaft Speed)	**Input Shaft Speed**	**ISS**
K-Jetronic	Continuous Fuel Injection[1]	CFI[1]
KAM (Keep Alive Memory)	Non Volatile Random Access Memory[1]	NVRAM[1]
KAM (Keep Alive Memory)	Keep Alive Random Access Memory[1]	Keep Alive RAM[1]
KE-Jetronic	Continuous Fuel Injection[1]	CFI[1]
KE-Motronic	Continuous Fuel Injection[1]	CFI[1]
Knock Sensor	Knock Sensor[1]	KS[1]
KS (Knock Sensor)	Knock Sensor[1]	KS[1]
L-Jetronic	Multiport Fuel Injection[1]	MFI[1]
Lambda	Oxygen Sensor[1]	O2S[1]
LH-Jetronic	Multiport Fuel Injection[1]	MFI[1]
Light Off Catalyst	Warm Up Three Way Catalytic Converter[1]	WU-TWC[1]
Light Off Catalyst	Warm Up Oxidation Catalytic Converter[1]	WU-OC[1]
Line Pressure Control Solenoid Valve	**Line Pressure Control Solenoid Valve**	**Line PC Solenoid Valve**
LOAD (Calculated Load Value)	**Calculated Load Value**	**LOAD**
Lock Up Relay	Torque converter Clutch Relay[1]	TCC Relay[1]
Long Term FT (Fuel Trim)	Long Term Fuel Trim[1]	Long Term FT[1]
Long Term Fuel Trim	**Long Term FT**	**Long Term FT**
LONG FT	**Long Term Fuel Trim**	**See Table 4**
Low Speed FC (Fan Control) Switch	Low Speed Fan Control Switch	Low Speed FC Switch
LUS (Lock Up Solenoid) Valve	Torque Converter Clutch Solenoid Valve[1]	TCC Solenoid Valve[1]
M/C (Mixture Control)	Mixture Control[1]	MC[1]
MAF (Mass Air Flow)	Mass Air Flow[1]	MAF[1]
MAF (Mass Air Flow) Sensor	Mass Air Flow Sensor[1]	MAF Sensor[1]

TABLE 1—CROSS REFERENCE AND LOOK UP (CONTINUED)

Existing Usage	Acceptable Usage	Acceptable Acronized Usage
Malfunction Indicator Lamp	Malfunction Indicator Lamp[1]	MIL[1]
Manifold Absolute Pressure	Manifold Absolute Pressure[1]	MAP[1]
Manifold Absolute Pressure Sensor	Manifold Absolute Pressure Sensor	MAP Sensor[1]
Manifold Differential Pressure	Manifold Differential Pressure[1]	MDP[1]
Manifold Surface Temperature	Manifold Surface Temperature[1]	MST[1]
Manifold Vacuum Zone	Manifold Vacuum Zone[1]	MVZ[1]
Manual Lever Position Sensor	Transmission Range Sensor[1]	TR Sensor[1]
MAP (Manifold Absolute Pressure)	Manifold Absolute Pressure[1]	MAP[1]
MAP (Manifold Absolute Pressure) Sensor	Manifold Absolute Pressure Sensor[1]	MAP Sensor[1]
MAPS (Manifold Absolute Pressure Sensor)	Manifold Absolute Pressure Sensor[1]	MAP Sensor[1]
Mass Air Flow	Mass Air Flow[1]	MAF[1]
Mass Air Flow Sensor	Mass Air Flow Sensor[1]	MAF Sensor[1]
MAT (Manifold Air Temperature)	Intake Air Temperature[1]	IAT[1]
MATS (Manifold Air Temperature Sensor)	Intake Air Temperature Sensor[1]	IAT Sensor[1]
MC (Mixture Control)	Mixture Control[1]	MC[1]
MCS (Mixture Control Solenoid)	Mixture Control Solenoid[1]	MC Solenoid[1]
MCU (Microprocessor Control Unit)	Powertrain Control Module[1]	PCM[1]
MDP (Manifold Differential Pressure)	Manifold Differential Pressure[1]	MDP[1]
MFI (Multiport Fuel Injection)	Multiport Fuel Injection[1]	MFI[1]
MIL (Malfunction Indicator Lamp)	Malfunction Indicator Lamp[1]	MIL[1]
Mixture Control	Mixture Control[1]	MC[1]
MLPS (Manual Lever Position Sensor)	**Transmission Range Sensor[1]**	**TR Sensor[1]**
Modes	Diagnostic Test Mode[1]	DTM[1]
Mono-Jetronic	**Throttle Body Injection[1]**	**TBI[1]**
Mono-Motronic	**Throttle Body Injection[1]**	**TBI[1]**
Monotronic	Throttle Body Fuel Injection[1]	TBI[1]
Motronic-Pressure	**Multiport Fuel Injection[1]**	**MFI[1]**
Motronic	Multiport Fuel Injection[1]	MFI[1]
MPI (Multipoint Injection)	Multiport Fuel Injection[1]	MFI[1]
MPI (Multiport Injection)	Multiport Fuel Injection[1]	MFI[1]
MRPS (Manual Range Position Switch)	Transmission Range Switch	TR Switch
MST (Manifold Surface Temperature)	Manifold Surface Temperature[1]	MST[1]
Multiport Fuel Injection	Multiport Fuel Injection[1]	MFI[1]
MVZ (Manifold Vacuum Zone)	Manifold Vacuum Zone[1]	MVZ[1]
NDS (Neutral Drive Switch)	Park/Neutral Position Switch[1]	PNP Switch[1]
Neutral Safety Switch	Park/Neutral Position Switch[1]	PNP Switch[1]
NGS (Neutral Gear Switch)	Park/Neutral Position Switch[1]	PNP Switch[1]
Non Dispersive Infrared	**Non Dispersive Infrared**	**NDIR**
Non Volatile Random Access Memory	Non Volatile Random Access Memory[1]	NVRAM[1]
NPS (Neutral Position Switch)	Park/Neutral Position Switch[1]	PNP Switch[1]
NVM (Non Volatile Memory)	Non Volatile Random Access Memory[1]	NVRAM[1]
NVRAM (Non Volatile Random Access Memory)	Non Volatile Random Access Memory[1]	NVRAM[1]
O2 (Oxygen) Sensor	Oxygen Sensor[1]	O2S[1]
O2S (Oxygen Sensor)	Oxygen Sensor[1]	O2S[1]
Oxygen Sensor Location	**Oxygen Sensor Location**	**See Table 4**
OBD (On Board Diagnostic)	On Board Diagnostic[1]	OBD[1]
OBD Status	**OBD Status**	**see Table 4**
OBD STAT	**OBD Status**	**see Table 4**
OC (Oxidation Catalyst)	Oxidation Catalytic Converter[1]	OC[1]

TABLE 1—CROSS REFERENCE AND LOOK UP (CONTINUED)

Existing Usage	Acceptable Usage	Acceptable Acronized Usage
Oil Pressure Sender	Engine Oil Pressure Sensor	EOP Sensor
Oil Pressure Sensor	Engine Oil Pressure Sensor	EOP Sensor
Oil Pressure Switch	Engine Oil Pressure Switch	EOP Switch
OL (Open Loop)	Open Loop[1]	OL[1]
On Board Diagnostic	On Board Diagnostic[1]	OBD[1]
Open Loop	Open Loop[1]	OL[1]
OS (Oxygen Sensor)	Oxygen Sensor[1]	O2S[1]
OSS (Output Shaft Speed) Sensor	**Output Shaft Speed Sensor[1]**	**OSS Sensor[1]**
Output Driver	**Driver**	**Driver**
Output Shaft Speed Sensor	**Output Shaft Speed Sensor[1]**	**OSS Sensor[1]**
Oxidation Catalytic Converter	Oxidation Catalytic Converter[1]	OC[1]
OXS (Oxygen Sensor) Indicator	Service Reminder Indicator[1]	SRI[1]
Oxygen Sensor	Oxygen Sensor[1]	O2S[1]
P/N (Park/Neutral)	Park/Neutral Position[1]	PNP[1]
P/S (Power Steering) Pressure Switch	Power Steering Pressure Switch	PSP Switch
P- (Pressure) Sensor	Manifold Absolute Pressure Sensor[1]	MAP Sensor[1]
PAIR (Pulsed Secondary Air Injection)	Pulsed Secondary Air Injection[1]	PAIR[1]
Parameter Identification	**Parameter Identification**	**PID**
Parameter Identification Supported	**Parameter Identification Supported**	**See Table 4**
Park/Neutral Position	Park/Neutral Position[1]	PNP[1]
PC (Pressure Control) Solenoid Valve	**Pressure Control Solenoid Valve[1]**	**PC Solenoid Valve[1]**
PCM (Powertrain Control Module)	Powertrain Control Module[1]	PCM[1]
PCV (Positive Crankcase Ventilation)	Positive Crankcase Ventilation[1]	PCV[1]
PCV (Positive Crankcase Ventilation) Valve	Positive Crankcase Ventilation Valve[1]	PCV Valve[1]
Percent Alcohol Sensor	Flexible Fuel Sensor[1]	FF Sensor[1]
Periodic Trap Oxidizer	Periodic Trap Oxidizer[1]	PTOX[1]
PFE (Pressure Feedback Exhaust Gas Recirculation Sensor	**Feedback Pressure** Exhaust Gas Recirculation Sensor[1]	**Feedback Pressure** EGR Sensor[1]
PFI (Port Fuel Injection)	Multiport Fuel Injection[1]	MFI[1]
PG (Pulse Generator)	Vehicle Speed Sensor[1]	VSS[1]
PGM-FI (Programmed Fuel Injection)	Multiport Fuel Injection[1]	MFI[1]
PID (Parameter Identification)	**Parameter Identification**	**PID**
PID SUP	**Parameter Identification Supported**	**See Table 4**
PIP (Position Indicator Pulse)	Crankshaft Position[1]	CKP[1]
PNP (Park/Neutral Position)	Park/Neutral Position[1]	PNP[1]
Positive Crankcase Ventilation	Positive Crankcase Ventilation[1]	PCV[1]
Positive Crankcase Ventilation Valve	Positive Crankcase Ventilation Valve[1]	PCV Valve[1]
Power Steering Pressure	Power Steering Pressure	PSP
Power Steering Pressure Switch	Power Steering Pressure Switch	PSP Switch
Powertrain Control Module	Powertrain Control Module[1]	PCM[1]
Pressure Control Solenoid Valve	**Pressure Control Solenoid Valve[1]**	**PC Solenoid Valve[1]**
Pressure Feedback EGR (Exhaust Gas Recirculation)	Feedback Pressure Exhaust Gas Recirculation[1]	Feedback Pressure EGR[1]
Pressure Sensor	Manifold Absolute Pressure Sensor[1]	MAP Sensor[1]
Pressure Feedback EGR (Exhaust Gas **Recirculation) System**	**Feedback Pressure** Exhaust Gas Recirculation System[1]	**Feedback Pressure** EGR System[1]
Pressure Transducer EGR (Exhaust Gas Recirculation) System	Pressure Transducer Exhaust Gas Recirculation System[1]	Pressure Transducer EGR System[1]
PRNDL (Park- Reverse- Neutral- Drive- Low)	Transmission Range	TR
Programmable Read Only Memory	Programmable Read Only Memory[1]	PROM[1]

TABLE 1—CROSS REFERENCE AND LOOK UP (CONTINUED)

Existing Usage	Acceptable Usage	Acceptable Acronized Usage
PROM (Programmable Read Only Memory)	Programmable Read Only Memory[1]	PROM[1]
PSP (Power Steering Pressure)	Power Steering Pressure	PSP
PSP (Power Steering Pressure) Switch	Power Steering Pressure Switch	PSP Switch
PSPS (Power Steering Pressure Switch)	Power Steering Pressure Switch	PSP Switch
PTOX (Periodic Trap Oxidizer)	Periodic Trap Oxidizer[1]	PTOX[1]
Pulsair	Pulsed Secondary Air Injection[1]	PAIR[1]
Pulsed Secondary Air Injection	Pulsed Secondary Air Injection[1]	PAIR[1]
Pulse Width Modulation	**Pulse Width Modulation**	**PWM**
PWM	**Pulse Width Modulation**	**PWM**
QDM (Quad Driver Module)	**Driver**	**Driver**
Quad Driver Module	**Driver**	**Driver**
Radiator Fan Control	Fan Control	FC
Radiator Fan Relay	Fan Control Relay	FC Relay
RAM (Random Access Memory)	Random Access Memory[1]	RAM[1]
Random Access Memory	Random Access Memory[1]	RAM[1]
Read Only Memory	Read Only Memory[1]	ROM[1]
Rear Wheel Drive	**Rear Wheel Drive**	**RWD**
Recirculated Exhaust Gas Temperature Sensor	Exhaust Gas Recirculation Temperature Sensor	EGRT Sensor[1]
Reed Valve	Pulsed Secondary Air Injection Valve[1]	PAIR Valve[1]
REGTS (Recirculated Exhaust Gas Temperature Sensor)	Exhaust Gas Recirculation Temperature Sensor[1]	EGRT Sensor[1]
Relay Module	Relay Module	RM
Remote Mount TFI (Thick Film Ignition)	Distributor Ignition[1]	DI[1]
Revolutions per Minute	Engine Speed[1]	RPM[1]
RM (Relay Module)	Relay Module	RM
ROM (Read Only Memory)	Read Only Memory[1]	ROM[1]
RPM (Revolutions per Minute)	Engine Speed[1]	RPM[1]
RWD	**Rear Wheel Drive**	**RWD**
S4WD	**Selectable Four Wheel Drive**	**S4WD**
SABV (Secondary Air Bypass Valve)	Secondary Air Injection Bypass Valve[1]	AIR Bypass Valve[1]
SACV (Secondary Air Check Valve)	Secondary Air Injection Control Valve[1]	AIR Control Valve[1]
SASV (Secondary Air Switching Valve)	Secondary Air Injection Switching Valve[1]	AIR Switching Valve[1]
SBEC (Single Board Engine Control)	Powertrain Control Module[1]	PCM[1]
SBS (Supercharger Bypass Solenoid)	Supercharger Bypass Solenoid[1]	SCB Solenoid[1]
SC (Supercharger)	Supercharger[1]	SC[1]
Scan Tool	Scan Tool[1]	ST[1]
SCB (Supercharger Bypass)	Supercharger Bypass[1]	SCB[1]
Secondary Air Bypass Valve	Secondary Air Injection Bypass Valve[1]	AIR Bypass Valve[1]
Secondary Air Check Valve	Secondary Air Injection Check Valve[1]	AIR Check Valve[1]
Secondary Air Injection	Secondary Air Injection[1]	AIR[1]
Secondary Air Injection Bypass	Secondary Air Injection Bypass[1]	AIR Bypass[1]
Secondary Air Injection Diverter	Secondary Air Injection Diverter[1]	AIR Diverter[1]
Secondary Air Switching Valve	Secondary Air Injection Switching Valve[1]	AIR Switching Valve[1]
Selectable Four Wheel Drive	**Selectable Four Wheel Drive**	**S4WD**
SEFI (Sequential Electronic Fuel Injection)	Sequential Multiport Fuel Injection[1]	SFI[1]
Self Test	On Board Diagnostic[1]	OBD[1]
Self Test Codes	Diagnostic Trouble Code[1]	DTC[1]
Self Test Connector	Data Link Connector[1]	DLC[1]
Sequential Multiport Fuel Injection	Sequential Multiport Fuel Injection[1]	SFI[1]

TABLE 1—CROSS REFERENCE AND LOOK UP (CONTINUED)

Existing Usage	Acceptable Usage	Acceptable Acronized Usage
Service Engine Soon	Service Reminder Indicator[1]	SRI[1]
Service Engine Soon	Malfunction Indicator Lamp[1]	MIL[1]
Service Reminder Indicator	Service Reminder Indicator[1]	SRI[1]
SFI (Sequential Fuel Injection)	Sequential Multiport Fuel Injection[1]	SFI[1]
Shift Solenoid	**Shift Solenoid**[1]	**SS**[1]
Shift Solenoid Valve	**Shift Solenoid Valve**[1]	**SS Valve**[1]
Short Term FT (Fuel Trim)	Short Term Fuel Trim[1]	Short Term FT[1]
Short Term Fuel Trim	**Short Term Fuel Trim**[1]	**Short Term FT**[1]
SHRT FT	**Short Term Fuel Trim**[1]	**See Table 4**
SLP (Selection Lever Position)	Transmission Range	TR
SMEC (Single Module Engine Control)	Powertrain Control Module[1]	PCM[1]
Smoke Puff Limiter	Smoke Puff Limiter[1]	SPL[1]
SPARK ADV	**Spark Advance**	**See Table 4**
Spark Advance	**Spark Advance**	**See Table 4**
Spark Plug	**Spark Plug**[1]	**Spark Plug**[1]
SPI (Single Point Injection)	Throttle Body Fuel Injection[1]	TBI[1]
SPL (Smoke Puff Limiter)	Smoke Puff Limiter[1]	SPL[1]
SS (Shift Solenoid)	**Shift Solenoid**[1]	**SS**[1]
SRI (Service Reminder Indicator)	Service Reminder Indicator[1]	SRI[1]
SRT (System Readiness Test)	System Readiness Test[1]	SRT[1]
ST (Scan Tool)	Scan Tool[1]	ST[1]
Supercharger	Supercharger[1]	SC[1]
Supercharger Bypass	Supercharger Bypass[1]	SCB[1]
Sync Pickup	Camshaft Position[1]	CMP[1]
System Readiness Test	System Readiness Test[1]	SRT[1]
3-2TS (3-2 Timing Solenoid)	**3-2 Timing Solenoid**	**3-2TS**
3-2TS Valve (3-2 Timing Solenoid)Valve	**3-2 Timing Solenoid Valve**	**3-2TS Valve**
3-2 Timing Solenoid	**3-2 Timing Solenoid**	**3-2TS**
3-2 Timing Solenoid Valve	**3-2 Timing Solenoid Valve**	**3-2TS Valve**
3GR (Third Gear)	Third Gear	3GR
TAB (Thermactor Air Bypass)	Secondary Air Injection Bypass[1]	AIR Bypass[1]
TAC (Throttle Actuator Control)	**Throttle Actuator Control**	**TAC**
TAC (Throttle Actuator Control) Module	**Throttle Actuator Control Module**[1]	**TAC Module**[1]
TAD (Thermactor Air Diverter)	Secondary Air Injection Diverter[1]	AIR Diverter[1]
TB (Throttle Body)	Throttle Body[1]	TB[1]
TBI (Throttle Body Fuel Injection)	Throttle Body Fuel Injection[1]	TBI[1]
TBT (Throttle Body Temperature)	Intake Air Temperature[1]	IAT[1]
TC (Turbocharger)	Turbocharger[1]	TC[1]
TC (Turbocharger) Wastegate	**Turbocharger Wastegate**[1]	**TC Wastegate**[1]
TC (Turbocharger) Wastegate Regulating Valve	**Turbocharger Wastegate Regulating Valve**[1]	**TC Wastegate Regulating Valve**[1]
TCC (Torque Converter Clutch)	Torque Converter Clutch[1]	TCC[1]
TCC (Torque Converter Clutch) Relay	Torque Converter Clutch Relay[1]	TCC Relay[1]
TCC (Torque Converter Clutch) Solenoid	**Torque Converter Clutch Solenoid** [1]	**TCC Solenoid**[1]
TCC (Torque Converter Clutch) Solenoid Valve	**Torque Converter Clutch Solenoid Valve**[1]	**TCC Solenoid Valve**[1]
TCM (Transmission Control Module)	Transmission Control Module	TCM
TCCP (Torque Converter Clutch Pressure)	**Torque Converter Clutch Pressure**	**TCCP**
TFI (Thick Film Ignition)	Distributor Ignition[1]	DI[1]
TFI (Thick Film Ignition) Module	Ignition Control Module[1]	ICM[1]

TABLE 1—CROSS REFERENCE AND LOOK UP (CONTINUED)

Existing Usage	Acceptable Usage	Acceptable Acronized Usage
TFP (Transmission Fluid Pressure)	**Transmission Fluid Pressure**	**TFP**
TFT (Transmission Fluid Temperature) Sensor	**Transmission Fluid Temperature Sensor**	**TFT Sensor**
Thermac	Secondary Air Injection[1]	AIR[1]
Thermac Air Cleaner	Air Cleaner[1]	ACL[1]
Thermactor	Secondary Air Injection[1]	AIR[1]
Thermactor Air Bypass	Secondary Air Injection Bypass[1]	AIR Bypass[1]
Thermactor Air Diverter	Secondary Air Injection Diverter[1]	AIR Diverter[1]
Thermactor II	Pulsed Secondary Air Injection[1]	PAIR[1]
Thermal Vacuum Switch	Thermal Vacuum Valve[1]	TVV[1]
Thermal Vacuum Valve	Thermal Vacuum Valve[1]	TVV[1]
Third Gear	Third Gear	3GR
Three Way + Oxidation Catalytic Converter	Three Way + Oxidation Catalytic Converter[1]	TWC+OC[1]
Three Way Catalytic Converter	Three Way Catalytic Converter[1]	TWC[1]
Throttle Actuator Control	**Throttle Actuator Control**	**TAC**
Throttle Actuator Control Module	**Throttle Actuator Control Module**	**TAC Module**
Throttle Body	Throttle Body[1]	TB[1]
Throttle Body Fuel Injection	Throttle Body Fuel Injection[1]	TBI[1]
Throttle Opener	Idle Speed Control[1]	ISC[1]
Throttle Opener Vacuum Switching Valve	Idle Speed Control Solenoid Vacuum Valve[1]	ISC Solenoid Vacuum Valve[1]
Throttle Opener VSV (Vacuum Switching Valve)	Idle Speed Control Solenoid Vacuum Valve[1]	ISC Solenoid Vacuum Valve[1]
Throttle Position	Throttle Position[1]	TP
Throttle Position Sensor	Throttle Position Sensor[1]	TP Sensor[1]
Throttle Position Switch	Throttle Position Switch[1]	TP Switch[1]
Throttle Potentiometer	Throttle Position Sensor[1]	TP Sensor[1]
TOC (Trap Oxidizer - Continuous)	Continuous Trap Oxidizer[1]	CTOX[1]
TOP (Trap Oxidizer - Periodic)	Periodic Trap Oxidizer[1]	PTOX[1]
Torque Converter Clutch	Torque Converter Clutch[1]	TCC[1]
Torque Converter Clutch Pressure	**Torque Converter Clutch Pressure**	**TCCP**
Torque Converter Clutch Relay	Torque Converter Clutch Relay[1]	TCC Relay[1]
Torque Converter Clutch Solenoid	**Torque Converter Clutch Solenoid[1]**	**TCC Solenoid[1]**
Torque Converter Clutch Solenoid Valve	**Torque Converter Clutch Solenoid Valve[1]**	**TCC Solenoid Valve[1]**
TP (Throttle Position)	Throttle Position[1]	TP[1]
TP (Throttle Position) Sensor	Throttle Position Sensor[1]	TP Sensor[1]
TP (Throttle Position) Switch	Throttle Position Switch[1]	TP Switch[1]
TPI (Tuned Port Injection)	Multiport Fuel Injection[1]	MFI[1]
TPNP (Transmission Park Neutral Position)	**Park/Neutral Position[1]**	**PNP[1]**
TPS (Throttle Position Sensor)	Throttle Position Sensor[1]	TP Sensor[1]
TPS (Throttle Position Switch)	Throttle Position Switch[1]	TP Switch[1]
TR (Transmission Range)	Transmission Range	TR
Track Road Load Horsepower	**Track Road Load Horsepower**	**TRLHP**
Transmission Control Module	Transmission Control Module	TCM
Transmission Fluid Pressure	**Transmission Fluid Pressure**	**TFP**
Transmission Fluid Temperature Sensor	**Transmission Fluid Temperature Sensor**	**TFT Sensor**
Transmission Park Neutral Position	**Park/Neutral Position[1]**	**PNP[1]**
Transmission Position Switch	Transmission Range Switch	TR Switch
Transmission Range Selection	Transmission Range	TR
Transmission Range Sensor	**Transmission Range Sensor**	**TR Sensor**
TRS (Transmission Range Selection)	Transmission Range	TR

TABLE 1—CROSS REFERENCE AND LOOK UP (CONTINUED)

Existing Usage	Acceptable Usage	Acceptable Acronized Usage
TRSS (Transmission Range Selection Switch)	Transmission Range Switch	TR Switch
TSS (Turbine Shaft Speed) Sensor	**Turbine Shaft Speed Sensor**[1]	**TSS Sensor**[1]
Tuned Port Injection	Multiport Fuel Injection[1]	MFI[1]
Turbine Shaft Speed Sensor	**Turbine Shaft Speed Sensor**[1]	**TSS Sensor**[1]
Turbo (Turbocharger)	Turbocharger[1]	TC[1]
Turbocharger	Turbocharger[1]	TC[1]
Turbocharger Wastegate	**Turbocharger Wastegate**[1]	**TC Wastegate**[1]
Turbocharger Wastegate Regulating Valve	**Turbocharger Wastegate Regulating Valve**[1]	**TC Wastegate Regulating Valve**[1]
TVS (Thermal Vacuum Switch)	Thermal Vacuum Valve[1]	TVV[1]
TVV (Thermal Vacuum Valve)	Thermal Vacuum Valve[1]	TVV[1]
TWC (Three Way Catalytic Converter)	Three Way Catalytic Converter[1]	TWC[1]
TWC + OC (Three Way + Oxidation Catalytic Converter)	Three Way + Oxidation Catalytic Converter[1]	TWC+OC[1]
VAC (Vacuum) Sensor	Manifold Differential Pressure Sensor[1]	MDP Sensor[1]
Vacuum Switches	Manifold Vacuum Zone Switch	MVZ Switch[1]
VAF (Volume Air Flow)	Volume Air Flow[1]	VAF[1]
Valve Position EGR (Exhaust Gas Recirculation) System	**Valve Position Exhaust Gas Recirculation System**[1]	**Valve Position EGR System**[1]
Vane Air Flow	Volume Air Flow[1]	VAF[1]
Variable Control Relay Module	**Variable Control Relay Module**	**VCRM**
Variable Fuel Sensor	Flexible Fuel Sensor	FF Sensor[1]
VAT (Vane Air Temperature)	Intake Air Temperature[1]	IAT[1]
VCC (Viscous Converter Clutch)	Torque Converter Clutch[1]	TCC[1]
VCM	**Vehicle Control Module**	**VCM**
VCRM	**Variable Control Relay Module**	**VCRM**
Vehicle Control Module	**Vehicle Control Module**	**VCM**
Vehicle Identification Number	**Vehicle Identification Number**	**VIN**
Vehicle Speed Sensor	Vehicle Speed Sensor[1]	VSS[1]
VIN (Vehicle Identification Number)	**Vehicle Identification Number**	**VIN**
VIP (Vehicle In Process) Connector	Data Link Connector[1]	DLC[1]
Viscous Converter Clutch	Torque Converter Clutch[1]	TCC[1]
Voltage Regulator	Voltage Regulator	VR
Volume Air Flow	Volume Air Flow[1]	VAF[1]
VR (Voltage Regulator)	Voltage Regulator	VR
VSS (Vehicle Speed Sensor)	Vehicle Speed Sensor[1]	VSS[1]
VSV (Vacuum Solenoid Valve) (Canister)	Evaporative Emission Canister Purge Valve[1]	EVAP Canister Purge Valve[1]
VSV (Vacuum Solenoid Valve) (EVAP)	Evaporative Emission Canister Purge Valve[1]	EVAP Canister Purge Valve[1]
VSV (Vacuum Solenoid Valve) (Throttle)	Idle Speed Control Solenoid Vacuum Valve[1]	ISC Solenoid Vacuum Valve[1]
Warm Up Oxidation Catalytic Converter	Warm Up Oxidation Catalytic Converter[1]	WU-OC[1]
Warm Up Three Way Catalytic Converter	Warm Up Three Way Catalytic Converter[1]	WU-OC[1]
Wide Open Throttle	Wide Open Throttle[1]	WOT[1]
WOT (Wide Open Throttle)	Wide Open Throttle[1]	WOT[1]
WOTS (Wide Open Throttle Switch)	Wide Open Throttle Switch[1]	WOT Switch[1]
WU-OC (Warm Up Oxidation Catalytic Converter)	Warm Up Oxidation Catalytic Converter[1]	WU-OC[1]
WU-TWC (Warm Up Three Way Catalytic Converter)	Warm Up Three Way Catalytic Converter[1]	WU-TWC[1]

Recommended Terms and Recommended Acronyms See Table 2

[1] Emission-Related Term

Bold indicates new/revised entry

GLOSSARY

Note: Terms are highlighted in color, followed by Spanish translation in bold.

A circuit A generator regulator circuit that uses an external grounded field circuit. In the A circuit, the regulator is on the ground side of the field coil.
Circuito A Circuito regulador del generador que utiliza un circuito inductor externo puesto a tierra. En el circuito A, el regulador se encuentra en el lado a tierra de la bobina inductora.

Active suspension systems Suspension systems that are controlled by double-acting hydraulic cylinders or solenoids (actuators) mounted at each wheel. The actuators support the vehicle's weight, instead of conventional springs or air springs.
Sistemas activos de suspensión Sistemas de suspensión controlados por cilindros hidráulicos de doble acción o por solenoides (accionadores) montados en cada rueda. Los accionadores apoyan el peso del vehículo, en vez de muelles convencionales o muelles de aire.

Actuators Devices that perform the actual work commanded by the computer. They can be in the form of a motor, relay, switch, or solenoid.
Accionadores Dispositivos que realizan el trabajo efectivo que ordena la computadora. Dichos dispositivos pueden ser un motor, un relé, un conmutador o un solenoide.

Adaptive suspension systems Suspension systems that are able to change ride characteristics by continuously altering shock damping and ride height.
Sistemas adaptadores de suspensión Sistemas de suspensión que pueden cambiar las características del viaje al alterar continuamente el amortiguamiento y la altura del viaje.

Air bag module Composed of the air bag and inflator assembly, which is packaged into a single module.
Unidad del Airbag Formada por el conjunto del Airbag y el inflador. Este conjunto se empaqueta en una sola unidad.

Air core gauge Gauge design that uses the interaction of two electromagnets and the total field effect upon a permanent magnet to cause needle movement.
Calibrador de núcleo de aire Calibrador diseñado para utilizar la interacción de dos electroimanes y el efecto inductor total sobre un imán permanente para generar el movimiento de la aguja.

Alternating current Electrical current that changes direction between positive and negative.
Corriente alterna Corriente eléctrica que recorre un circuito ya sea en dirección positiva o negativa.

Ambient temperature The temperature of the outside air.
Temperatura ambiente Temperatura del aire ambiente.

American wire gauge (AWG) System used to determine wire sizes based on the cross-sectional area of the conductor.
Calibrador americano de alambres Sistema utilizado para determinar el tamaño de los alambres, basado en el área transversal del conductor.

Ammeter A test meter used to measure current draw.
Amperímetro Instrumento de prueba utilizado para medir la intensidad de una corriente.

Amperes *See* current.
Amperios *Véase* corriente.

Analog A voltage signal that is infinitely variable or that can be changed within a given range.
Señal analógica Señal continua y variable que debe traducirse a valores numéricos discontinuos para poder ser trataba por una computadora.

Anode The positive charge electrode in a voltage cell.
Ánodo Electrodo de carga positiva de un generador de electricidad.

Antilock brakes (ABS) A brake system that automatically pulsates the brakes to prevent wheel lockup under panic stop and poor traction conditions.
Frenos antibloqueo Sistema de frenos que pulsa los frenos automáticamente para impedir el bloqueo de las ruedas en casos de emergencia y de tracción pobre.

Antitheft device A device or system that prevents illegal entry or driving of a vehicle. Most are designed to deter entry.
Dispositivo a prueba de hurto Un dispositivo o sistema que previene la entrada o conducción ilícita de un vehículo. La mayoría se diseñan para detener la entrada.

Armature The movable component of an electric motor, which consists of a conductor wound around a laminated iron core and is used to create a magnetic field.
Armadura Pieza móvil de un motor eléctrico, compuesta de un conductor devanado sobre un núcleo de hierro laminado y que se utiliza para producir un campo magnético.

Arming sensor A device that places an alarm system into "ready" to detect an illegal entry.
Sensor de armado Un dispositivo que pone "listo" un sistema de alarma para detectar una entrada ilícita.

Aspirator Tubular device that uses a venturi effect to draw air from the passenger compartment over the in-car sensor. Some manufacturers use a suction motor to draw the air over the sensor.
Aspirador Dispositivo tubular que utiliza un efecto Venturi para extraer aire del compartimiento del pasajero sobre el sensor dentro del vehículo. Algunos fabricantes utilizan un motor de succión para extraer el aire sobre el sensor.

Atom The smallest part of a chemical element that still has all the characteristics of that element.
Átomo Partícula más pequeña de un elemento químico que conserva las cualidades íntegras del mismo.

Audio system The sound system for a vehicle; can include radio, cassette player, CD player, amplifier, and speakers.
Sistema de audio El sistema de sonido de un vehículo; puede incluir el radio, el tocacaset, el toca discos compactos, el amplificador, y las bocinas.

Automatic door locks A system that automatically locks all doors through the activation of one switch.
Cerraduras automáticas de puerta Un sistema que cierra todas las puertas automaticamente al activar un solo conmutador.

Automatic headlight dimming An electronic feature that automatically switches the headlights from high beam to low beam under two different conditions: light from oncoming vehicles strikes the photocell-amplifier; or light from the taillights of a vehicle that is being passed strikes the photocell-amplifier.
Reducción automática de intensidad luminosa de los faros delanteros Característica electrónica que conmuta los faros delanteros automáticamente de luz larga a luz corta dadas las siguientes circunstancias: la luz de los vehículos que se aproximan alcanza el amplificador de fotocélula, o la luz de los faros traseros de un vehículo que se ha rebasado alcanza el amplificador de fotocélula.

Automatic Traction Control A system that prevents slippage of one of the drive wheels. This is done by applying the brake at that wheel and/or decreasing the engine's power output.
Control Automático de Tracción Un sistema que previene el patinaje de una de las ruedas de mando. Esto se efectúa aplicando el freno en esa rueda y/o disminuyendo la salida de potencia del motor.

Balanced atom An atom that has an equal amount of protons and electrons.
Átomo equilibrado Átomo que tiene el mismo número de protones y de electrones.

Ballast resistor A resistance put in series with a power lead to a component. Its purpose is to reduce the voltage applied to the component and to control the amount of current in the circuit.

Resistencia autorreguladora Una regulación de serie con un conectador de alimentación a un componente. Su propósito es de reducir la tensión que se aplica al componente y controlar la cantidad del corriente en el circuito.

Base The center layer of a bipolar transistor.

Base Capa central de un transistor bipolar.

Battery cell The active unit of a battery.

Acumulador de batería Componente activo de una batería.

Battery holddowns Brackets that secure the battery to the chassis of the vehicle.

Portabatería Los sostenes que fijan la batería al chasis del vehículo.

Battery terminals Terminals at the battery to which the positive and the negative battery cables are connected. The terminals may be posts or threaded inserts.

Bornes de la batería Los bornes en la batería a los cuales se conectan los cables positivos y negativos. Los terminales pueden ser postes o piezas roscadas.

Baud rate The measure of computer data transmission speed in bits per second.

Razón de baúd Medida de la velocidad de la transmisión de datos de una computadora en bits por segundo.

B circuit A generator regulator circuit that is internally grounded. In the B circuit, the voltage regulator controls the power side of the field circuit.

Circuito B Circuito regulador del generador puesto internamente a tierra. En el circuito B, el regulador de tensión controla el lado de potencia del circuito inductor.

Bendix drive A type of starter drive that uses the inertia of the spinning starter motor armature to engage the drive gear to the gears of the flywheel. This type starter drive was used on early models of vehicles and is rarely seen today.

Acoplamiento Bendix Un tipo del acoplamiento del motor de arranque que usa la inercia de la armadura del motor de arranque giratorio para endentar el engrenaje de mando con los engrenajes del volante. Este tipo de acoplamiento del motor de arranque se usaba en los modelos vehículos antiguos y se ven raramente.

Bias voltage Voltage applied across a diode.

Tensión polarizadora Tensión aplicada a través de un diodo.

Bimetallic strip A metal contact wiper consisting of two different types of metals. One strip will react quicker to heat than the other, causing the strip to flex in proportion to the amount of current flow.

Banda bimetálica Contacto deslizante de metal compuesto de dos tipos de metales distintos. Una banda reaccionará más rápido al calor que la otra, haciendo que la banda se doble en proporción con la cantidad de flujo de corriente.

Binary code A series of numbers represented by 1's and 0's. Any number and word can be translated into a combination of binary 1's and 0's.

Código binario Serie de números representados por unos y ceros. Cualquier número y palabra puede traducirse en una combinación de unos y ceros binarios.

Bipolar The name used for transistors because current flows through the materials of both polarities.

Bipolar Nombre aplicado a los transistores porque la corriente fluye por conducto de materiales de ambas polaridades.

Bit A binary digit.

Bit Dígito binario.

Brushes Electrically conductive sliding contacts, usually made of copper and carbon.

Escobillas Contactos deslizantes de conduccion eléctrica, por lo general hechos de cobre y de carbono.

Bucking coil One of the coils in a three—coil gauge. It produces a magnetic field that bucks or opposes the low reading coil.

Bobina compensadora Una de las bobinas de un calibre de tres bobinas. Produce un campo magnético que es contrario o en oposición a la bobina de baja lectura.

Buffer A buffer cleans up a voltage signal. These are used with PM generator sensors to change the AC voltage to a digitalized signal.

Separador Un separador aguza una señal del tensión. Estos se usan con los sensores generadores PM para cambiar la tensión de corriente alterna a una señal digitalizado.

Bulkhead connector A large connector that is used when many wires pass through the bulkhead or firewall.

Conectador del tabique Un conectador que se usa al pasar muchos alambres por el tabique o mamparo de encendios.

Bus Bar A common electrical connectión to which all of the fuses in the fuse box are attached. The bus bar is connected to battery voltage.

Barra colectora Conexión eléctrica común a la que se conectan todos los fusibles de la caja de fusibles. La barra colectora se conecta a la tensión de la batería.

Buzzer An audible warning device that is used to warn the driver of possible safety hazards.

Zumbador Dispositivo audible de advertencia utilizado para prevenir al conductor de posibles riesgos a la seguridad.

Capacitance The ability of two conducting surfaces to store voltage.

Capacitancia Propiedad que permite el almacenamiento de electricidad entre dos conductores aislados entre sí.

Carbon monoxide An odorless, colorless, and toxic gas that is produced as a result of combustion.

Monóxido de carbono Gas inodoro, incoloro y tóxico producido como resultado de la combustión.

Cartridge fuses *See* maxi-fuse.

Fusibles cartucho *Véase* maxifusible.

Cathode Negatively charged electrode of a voltage cell.

Cátodo Electrodo de carga negativa de un generador de electricidad.

Cathode ray tube Similar to a television picture tube. It contains a cathode that emits electrons and an anode that attracts them. The screen of the tube will glow at the points that are hit by the electrons.

Tubo de rayos catódicos Parecidos a un tubo de pantalla de televisor. Contiene un cátodo que emite los electrones y un ánodo que los atrae. La pantalla del tubo iluminará en los puntos en donde pegan los electrones.

Cell element The assembly of a positive and negative plate in a battery.

Elemento de pila La asamblea de una placa positiva y negativa en una bateria.

CHMSL The abbreviation for center high mounted stop light, often referred to as the third brake light.

CHMSL La abreviación para el faro de parada montada alto en el centro que suele referirse como el faro de freno tercero.

Choke coil Fine wire wound into a coil used to absorb oscillations in a switched circuit.

Bobina de inducción Alambre fino devanado en una bobina, utilizado para absorber oscilaciones en un circuito conmutado.

Circuit The path of electron flow consisting of the voltage source, conductors, load component, and return path to the voltage source.

Circuito Trayectoria del flujo de electrones, compuesto de la fuente de tensión, los conductores, el componente de carga y la trayectoria de regreso a la fuente de tensión.

Clamping diode A diode that is connected in parallel with a coil to prevent voltage spikes from the coil from reaching other components in the circuit.

Diodo de bloqueo Un diodo que se conecta en paralelo con una bobina para prevenir que los impulsos de tensión lleguen a otros componentes en el circuito.

Clock circuit A crystal that electrically vibrates when subjected to current at certain voltage levels. As a result, the chip produces very regular series of voltage pulses.
Circuito de reloj Cristal que vibra electrónicamente cuando está sujeto a una corriente a ciertos niveles de tensión. Como resultado, el fragmento produce una serie sumamente regular de impulsos de tensión.

Closed circuit A circuit that has no breaks in the path and allows current to flow.
Circuito cerrado Circuito de trayectoria ininterrumpida que permite un flujo continuo de corriente.

Cold cranking amps (CCA) Rating indicates the battery's ability to deliver a specified amount of current to start an engine at low ambient temperatures.
Amperios de arranque en frío Tasa indicativa de la capacidad de la batería para producir una cantidad específica de corriente para arrancar un motor a bajas temperaturas ambiente.

Collector The portion of a bipolar transistor that receives the majority of current carriers.
Dispositivo de toma de corriente Parte del transistor bipolar que recibe la mayoría de los portadores de corriente.

Common connector A connector that is shared by more than one circuit and/or component.
Conector común Un conector que se comparte entre más de un circuito y/o componente.

Commutator A series of conducting segments located around one end of the armature.
Conmutador Serie de segmentos conductores ubicados alrededor de un extremo de la armadura.

Component locator A service manual that lists and describes the exact location of components on a vehicle.
Localizador de componentes Un manual de servicio que cataloga y describe la posición exacta de los componentes en un vehículo.

Composite bulb A headlight assembly that has a replaceable bulb in its housing.
Bombilla compuesta Una asamblea de faros cuyo cárter tiene una bombilla reemplazable.

Compound motor A motor that has the characteristics of a series—wound and a shunt—wound motor.
Motor compuesta Un motor que tiene las características de un motor exitado en serie y uno en derivación.

Computer An electronic device that stores and processes data and is capable of operating other devices.
Computadora Dispositivo electrónico que almacena y procesa datos y que es capaz de ordenar a otros dispositivos.

Condenser A capacitor made from two sheets of metal foil separated by an insulator.
Condensador Capacitador hecho de dos láminas de metal separadas por un medio aislante.

Conduction Bias voltage difference between the base and the emitter has increased to the point that the transistor is switched on. In this condition, the transistor is conducting. Output current is proportional to that of the current through the base.
Conducción La diferencia de la tensión polarizadora entre la base y el emisor ha aumentado hasta el punto que el transistor es conectado. En estas circunstancias, el transistor está conduciendo. La corriente de salida está en proporción con la de la corriente conducida en la base.

Conductor A material in which electrons flow or move easily.
Conductor Una material en la cual los electrones circulen o se mueven fácilmente.

Continuity Refers to the circuit being continuous with no opens.
Continuidad Se refiere al circuito ininterrumpido, sin aberturas.

Conventional theory Electrical theory that states current flows from a positive point to a more negative point.
Teoría convencional Teoría de electricidad la cual enuncia que el corriente fluye desde un punto positivo a un punto más negativo.

Corner lights Lamps that illuminate when the turn signals are activated. They burn steadily when the turn signal switch is in a turn position to provide additional illumination of the road in the direction of the turn.
Faros laterales Lámparas que se encienden cuando se activan las luces indicadoras para virajes. Se encienden de manera continua cuando el conmutador de las luces indicadoras para virajes está conectado para proveer iluminación adicional de la carretera hacia la dirección del viraje.

Counterelectromotive force (CEMF) An induced voltage that opposes the source voltage.
Fuerza cóntraelectromotriz Tensión inducida en oposición a la tensión fuente.

Courtesy lights Lamps that illuminate the vehicle's interior when the doors are open.
Luces interiores Lámparas que iluminan el interior del vehículo cuando las puertas están abiertas.

Covalent bonding When atoms share valence electrons with other atoms.
Enlace covalente Cuando los átomos comparten electrones de valencia con otros átomos.

Crash sensor Normally open electrical switch designed to close when subjected to a predetermined amount of jolting or impact.
Sensor de impacto Un conmutador normalmente abierto diseñado a cerrarse al someterse a un sacudo de una fuerza predeterminada o un impacto.

Cross—fire The undesired firing of a spark plug that results from the firing of another spark plug. This is caused by electromagnetic induction.
Encendido transversal El encendido no deseable de una bujía que resulta del encendido de otra bujía. Esto se causa por la inducción electromagnética.

Crystal A term used to describe a material that has a definite atom structure.
Cristal Término utilizado para describir un material que tiene una estructura atómica definida.

Current The aggregate flow of electrons through a wire. One ampere represents the movement of 6.25 billion billion electrons (or one coulomb) past one point in a conductor in one second.
Corriente Flujo combinado de electrones a través de un alambre. Un amperio representa el movimiento de 6,25 mil millones de mil millones de electrones (o un colombio) que sobrepasa un punto en un conductor en un segundo.

Cutoff When reverse-bias voltage is applied to the base leg of the transistor. In this condition, the transistor is not conducting and no current will flow.
Corte Cuando se aplica tensión polarizadora inversa a la base del transistor. En estas circunstancias, el transistor no está conduciendo y no fluirá ninguna corriente.

Darlington pair An arrangement of transistors that amplifies current by one transistor acting as a preamplifier which creates a larger base current to the second transistor.
Par Darlington Conjunto de transistores que amplifica la corriente. Un transistor actúa como preamplificador y produce una corriente base más ámplia para el segundo transistor.

d'Arsonval gauge A gauge design that uses the interaction of a permanent magnet and an electromagnet, and the total field effect to cause needle movement.
Calibrador d'Arsonval Calibrador diseñado para utilizar la interacción de un imán permanente y de un electroimán, y el efecto inductor total para generar el movimiento de la aguja.

Deep cycling Discharging the battery completely before recharging it.
Operación cíclica completa La descarga completa de la batería previo al recargo.

Delta stator A three-winding ac generator stator with the ends of each winding connected to each other.
Estátor Delta Estátor generador de corriente alterna de devanado triple, con los extremos de cada devanado conectados entre sí.

Diagnostic module Part of an electronic control system that provides self—diagnostics and/or a testing interface.
Módulo de diagnóstico Parte de un sistema controlado electronicamente que provee autodiagnóstico y/o una interfase de pruebas.

Dielectric An insulator material.
Dieléctrico Material aislante.

Digital A voltage signal is either on-off, yes-no, or high-low.
Digital Una señal de tensión está Encendida-Apagada, es Sí-No o Alta-Baja.

Dimmer switch A switch in the headlight circuit that provides the means for the driver to select either high beam or low beam operation, and to switch between the two. The dimmer switch is connected in series within the headlight circuit and controls the current path for high and low beam.
Conmutador reductor Conmutador en el circuito para faros delanteros que le permite al conductor que elegir la luz larga o la luz corta, y conmutar entre las dos. El conmutador reductor se conecta en serie dentro del circuito para faros delanteros y controla la trayectoria de la corriente para la luz larga y la luz corta.

Diode An electrical one-way check valve that will allow current to flow in one direction only.
Diodo Válvula eléctrica de retención, de una vía, que permite que la corriente fluya en una sola dirección.

Diode rectifier bridge A series of diodes that are used to provide a reasonably constant dc voltage to the vehicle's electrical system and battery.
Puente rectificador de diodo Serie de diodos utilizados para proveerles una tensión de corriente continua bastante constante al sistema eléctrico y a la batería del vehículo.

Diode trio Used by some manufacturers to rectify the stator of an ac generator current so that it can be used to create the magnetic field in the field coil of the rotor.
Trío de diodos Utilizado por algunos fabricantes para rectificar el estátor de la corriente de un generador de corriente alterna y poder así utilizarlo para crear el campo magnético en la bobina inductora del rotor.

Direct current (dc) Electric current that flows in one directión.
Corriente continua Corriente eléctrica que fluye en una dirección.

Direct drive A situation where the drive power is the same as the power exerted by the device that is driven.
Transmisión directa Una situación en la cual el poder de mando es lo mismo que la potencia empleada por el dispositivo arrastrado.

Discrete devices Electrical components that are made separately and have wire leads for connections to an integrated circuit.
Dispositivos discretos Componentes eléctricos hechos uno a uno; tienen conductores de alambre para hacer conexiones a un circuito integrado.

Discriminating sensors Part of the air bag circuitry; these sensors are calibrated to close with speed changes that are great enough to warrant air bag deployment. These sensors are also referred to as crash sensors.
Sensores discriminadores Una parte del conjunto de circuitos de Airbag; estos sensores se calibran para cerrar con los cambios de la velocidad que son bastante severas para justificar el despliegue del Airbag. Estos sensores también se llaman los sensores de impacto.

Doping The addition of another element with three or five valence electrons to a pure semiconductor.
Impurificación La adición de otro elemento con tres o cinco electrones de valencia a un semiconductor puro.

Double filament lamp A lamp designed to execute more than one function. It can be used in the stoplight circuit, taillight circuit, and the turn signal circuit combined.
Lámpara con filamento doble Lámpara diseñada para llevar a cabo más de una función. Puede utilizarse en una combinación de los circuitos de faros de freno, de faros traseros y de luces indicadoras para virajes.

Drain The portion of a field-effect transistor that receives the holes or electrons.
Drenador Parte de un transistor de efecto de campo que recibe los agujeros o electrones.

Drive coil A hollowed field coil used in a positive—engagement starter to attract the movable pole shoe of the starter.
Bobina de excitación Una bobina inductora hueca empleada en un encendedor de acoplamiento directo para atraer la pieza polar móvil del encendedor.

Duty cycle The percentage of on time to total cycle time.
Ciclo de trabajo Porcentaje del trabajo efectivo a tiempo total del ciclo.

Eddy currents Small induced currents.
Corriente de Foucault Pequeñas corrientes inducidas.

Electrical load The working device of the circuit.
Carga eléctrica Dispositivo de trabajo del circuito.

Electrically Erasable PROM (EEPROM) Memory chip that allows for electrically changing the information one bit at a time.
Capacidad de borrado electrónico PROM Fragmento de memoria que permite el cambio eléctrico de la información un bit a la vez.

Electrochemical The chemical action of two dissimilar materials in a chemical solution.
Electroquímico Acción química de dos materiales distintos en una solución química.

Electrolysis The producing of chemical changes by passing electrical current through an electrolyte.
Electrólisis La producción de los cambios químicos al pasar un corriente eléctrico por un electrolito.

Electrolyte A solution of 64% water and 36% sulfuric acid.
Electrolito Solución de un 64% de agua y un 36% de ácido sulfúrico.

Electromagnetic gauge Gauge that produces needle movement by magnetic forces.
Calibrador electromagnético Calibrador que genera el movimiento de la aguja mediante fuerzas magnéticas.

Electromagnetic induction The production of voltage and current within a conductor as a result of relative motion within a magnetic field.
Inducción electromagnética Producción de tensión y de corriente dentro de un conductor como resultado del movimiento relativo dentro de un campo magnético.

Electromagnetic interference (EMI) An undesirable creation of electromagnetism whenever current is switched on and off.
Interferencia electromagnética Fenómeno de electromagnetismo no deseable que resulta cuando se conecta y se desconecta la corriente.

Electromagnetism A form of magnetism that occurs when current flows through a conductor.
Electromagnetismo Forma de magnetismo que ocurre cuando la corriente fluye a través de un conductor.

Electromechanical A device that uses electricity and magnetism to cause a mechanical action.
Electromecánico Un dispositivo que causa una acción mecánica por medio de la electricidad y el magnetismo.

Electromotive force (EMF) *See* voltage.
Fuerza electromotriz *Véase* tensión.

Electron Negative-charged particles of an atom.
Electrón Partículas de carga negativa de un átomo.

Electron theory Defines electrical movement as from negative to positive.
Teoría del electrón Define el movimiento eléctrico como el movimiento de lo negativo a lo positivo.

Electrostatic field The field that is between the two oppositely charged plates.
Campo electrostático Campo que se encuentra entre las placas de carga opuesta.

Emitter The outer layer of the transistor, which supplies the majority of current carriers.
Emisor Capa exterior del transistor que suministra la mayor parte de los portadores de corriente.

Equivalent series load (equivalent resistance) The total resistance of a parallel circuit, which is equivalent to the resistance of a single load in series with the voltage source.
Carga en serie equivalente (resistencia equivalente) Resistencia total de un circuito en paralelo, equivalente a la resistencia de una sola carga en serie con la fuente de tensión.

Erasable PROM (EPROM) Similar to PROM except that its contents can be erased to allow for new data to be installed. A piece of Mylar tape covers a window. If the tape is removed, the microcircuit is exposed to ultraviolet light and erases its memory.
Capacidad de borrado PROM Parecido al PROM, pero su contenido puede borrarse para permitir la instalación de nuevos datos. Un trozo de cinta Mylar cubre una ventana; si se remueve la cinta, el microcircuito queda expuesto a la luz ultravioleta y borra la memoria.

Excitation current Current that magnetically excites the field circuit of the ac generator.
Corriente de excitación Corriente que excita magnéticamente al circuito inductor del generador de corriente alterna.

Feedback 1. Data concerning the effects of the computer's commands are fed back to the computer as an input signal. Used to determine if the desired result has been achieved. 2. A condition that can occur when electricity seeks a path of lower resistance, but the alternate path operates a component other than that intended. Feedback can be classified as a short.
Realimentación 1. Datos referentes a los efectos de las órdenes de la computadora se suministran a la misma como señal de entrada. La realimentación se utiliza para determinar si se ha logrado el resultado deseado. 2. Condición que puede ocurrir cuando la electricidad busca una trayectoria de menos resistencia, pero la trayectoria alterna opera otro componente que aquel deseado. La realimentación puede clasificarse como un cortocircuito.

Fiber optics A medium of transmitting for the transmission of light through polymethylmethacrylate plastic that keeps the light rays parallel even if there are extreme bends in the plastic.
Transmisión por fibra óptica Técnica de transmisión de luz por medio de un plástico de polimetacrilato de metilo que mantiene los rayos de luz paralelos aunque el plástico esté sumamente torcido.

Field current The current going to the field windings of a motor or generator.
Corriente inductora El corriente que va a los devanados inductores de un motor o generador.

Field-effect transistor (FET) A unipolar transistor in which current flow is controlled by voltage in a capacitance field.
Transistor de efecto de campo Transistor unipolar en el cual la tensión en un campo de capacitancia controla el flujo de corriente.

Field relay The relay that controls the amount of current going to the field windings of a generator. This is the main output control unit for a charging system.
Relé inductor El relé que controla la cantidad del corriente a los devanados inductores de un generador. Es la unedad principal de potencia de salida de un sistema de carga.

Floor jack A portable hydraulic tool used to raise and lower a vehicle.
Gato de pie Herramienta hidráulica portátil utilizada para levantar y bajar un vehículo.

Flux density The number of flux lines per square centimeter.
Densidad de flujo Número de líneas de flujo por centímetro cuadrado.

Flux lines Magnetic lines of force.
Líneas de flujo Líneas de fuerza magnética.

Forward bias A positive voltage that is applied to the P-type material and negative voltage to the N-type material of a semiconductor.
Polarización directa Tensión positiva aplicada al material P y tensión negativa aplicada al material N de un semiconductor.

Full field Maximum ac generator output.
Campo completo Salida máxima de un generador de corriente alterna.

Full-wave rectification The conversion of a complete AC voltage signal to a DC voltage signal.
Rectificación de onda plena La conversión de una señal completa de tensión de corriente alterna a una señal de tensión de corriente continua.

Fuse A replaceable circuit protection device that will melt should the current passing through it exceed its rating.
Fusible Dispositivo reemplazable de protección del circuito que se fundirá si la corriente que fluye por el mismo excede su valor determinado.

Fuse box A term used to indicate the central location of the fuses contained in a single holding fixture.
Caja de fusibles Término utilizado para indicar la ubicación central de los fusibles contenidos en un solo elemento permanente.

Fusible link A wire made of meltable material with a special heat-resistant insulation. When there is an overload in the circuit, the link melts and opens the circuit.
Cartucho de fusible Alambre hecho de material fusible con aislamiento especial resistente al calor. Cuando ocurre una sobrecarga en el circuito, el cartucho se funde y abre el circuito.

Gain The ratio of amplification in an electronic device.
Ganancia Razón de amplificación en un dispositivo electrónico.

Ganged Refers to a type of switch in which all wipers of the switch move together.
Acoplado en tándem Se refiere a un tipo de conmutador en el cual todos los contactos deslizantes del mismo se mueven juntos.

Gassing The conversion of a battery's electrolyte into hydrogen and oxygen gas.
Burbujeo La conversión del electrolito de una bateria al gas de hidrógeno y oxígeno.

Gate The portion of a field-effect transistor that controls the capacitive field and current flow.
Compuerta Parte de un transistor de efecto de campo que controla el campo capacitivo y el flujo de corriente.

Gauge 1. A device that displays the measurement of a monitored system by the use of a needle or pointer that moves along a calibrated scale. 2. The number that is assigned to a wire to indicate its size. The larger the number, the smaller the diameter of the conductor.
Calibrador 1. Dispositivo que muestra la medida de un sistema regulado por medio de una aguja o indicador que se mueve a través de una escala calibrada. 2. El número asignado a un alambre indica su tamaño. Mientras mayor sea el número, más pequeño será el diámetro del conductor.

Gear reduction Occurs when two different sized gears are in mesh and the driven gear rotates at a lower speed than the drive gear but with greater torque.
Desmultiplicación Ocurre cuando dos engranajes de distinctos tamaños se endentan y el engrenaje arrastrado gira con una velocidad más baja que el engrenaje de mando pero con más par.

Grid growth A condition where the grid grows little metallic fingers that extend through the separators and short out the plates.
Expansión de la rejilla Una condición en la cual la rejilla produce protrusiones metálicas que se extienden por los separadores y causan cortocircuitos en las placas.

Grids The frame structure of a battery that normally has connector tabs at the top. It is generally made of lead alloys.
Rejillas La estructura encuadrador de una batería que normalmente tiene orejas de conexión en la parte superior. Generalmente se fabrica de aleaciones de plomo.

Ground The common negative connection of the electrical system. It is the point of lowest voltage.
Tierra Conexión negativa común del sistema eléctrico. Es el punto de tensión más baja.

Grounded circuit An electrical defect that allows current to return to ground before it has reached the intended load component.
Circuito puesto a tierra Falla eléctrica que permite el regreso de corriente a tierra antes de alcanzar el componente de carga deseado.

Ground side The portion of the circuit that is from the load component to the negative side of the source.
Lado a tierra Parte del circuito que va del componente de carga al lado negativo de la fuente.

Half-field current The current going to the field windings of a motor or generator after it has passed through a resistor in series with the circuit.
Corriente de medio campo El corriente que va a los devanados inductores de un motor o a un generador después de que haya pasado por un resistor conectado en serie con el circuito.

Half-wave rectification Rectification of one-half of an ac voltage.
Rectificación de media onda Rectificación en la que la corriente fluye únicamente durante semiciclos alternados.

Hall-effect switch A sensor that operates on the principle that if a current is allowed to flow through thin conducting material being exposed to a magnetic field, another voltage is produced.
Conmutador de efecto Hall Sensor que funciona basado en el principio de que si se permite el flujo de corriente a través de un material conductor delgado que ha sido expuesto a un campo magnético, se produce otra tensión.

Halogen The term used to identify a group of chemically related nonmetallic elements. These elements include chlorine, fluorine, and iodine.
Halógeno Término utilizado para identificar un grupo de elementos no metálicos relacionados químicamente. Dichos elementos incluyen el cloro, el flúor y el yodo.

Hand tools Tools that use only the force generated from the body to operate. They multiply the force received through leverage to accomplish the work.
Herramientas manuales Herramientas que para funcionar sólo necesitan la fuerza generada por el cuerpo. Para llevar a cabo el trabajo, las herramientas multiplican la fuerza que reciben por medio de la palancada.

Heat sink An object that absorbs and dissipates heat from another object.
Dispersador térmico Objeto que absorbe y disipa el calor de otro objeto.

HID High Intensity Discharge; a lighting system that uses an arc across electrodes instead of a filament.
HID Descarga de Alta Intensidad; un sistema de iluminación que utiliza un arco por dos electrodos en vez de un filamento.

H-gate A set of four transistors that can reverse current.
Compuerta H Juego de cuatro transistores que pueden invertir la corriente.

Hoist A lift that is used to raise the entire vehicle.
Elevador Montacargas utilizado para elevar el vehículo en su totalidad.

Hold—in winding A winding that holds the plunger of a solenoid in place after it moves to engage the starter drive.
Devanado de retención Un devanado que posiciona el núcleo móvil de un solenoide después de que mueva para accionar el acoplamiento del motor de arranque.

Hole The absence of an electron in an element's atom. These holes are said to be positively charged since they have a tendency to attract free electrons into the hole.
Agujero Ausencia de un electrón en el átomo de un elemento. Se dice que dichos agujeros tienen una carga positiva puesto que tienden a atraer electrones libres hacia el agujero.

Hybrid battery A battery that combines the advantages of low maintenance and maintenance—free batteries.
Batería híbrida Una batería que combina las ventajas de las baterías de bajo mantenimiento y de no mantenimiento.

Hydrometer A test instrument used to check the specific gravity of the electrolyte to determine the battery's state of charge.
Hidrómetro Instrumento de prueba utilizado para verificar la gravedad específica del electrolito y así determinar el estado de la carga de la batería.

Igniter A combustible device that converts electric energy into thermal energy to ignite the inflator propellant in an air bag system.
Ignitor Un dispositivo combustible que convierte la energía eléctrica a la energía termal para encender el propelente inflador en un sistema Airbag.

Incandescence The process of changing energy forms to produce light.
Incandescencia Proceso a través del cual se cambian las formas de energía para producir luz.

Induced voltage Voltage that is produced in a conductor as a result of relative motion within magnetic flux lines.
Tensión inducida Tensión producida en un conductor como resultado del movimiento relativo dentro de líneas de flujo magnético.

Induction The magnetic process of producing a current flow in a wire without any actual contact to the wire. To induce 1 volt, 100 million magnetic lines of force must be cut per second.
Inducción Proceso magnético a través del cual se produce un flujo de corriente en un alambre sin contacto real alguno con el alambre. Para inducir 1 voltio, deben producirse 100 millones de líneas de fuerza magnética por segundo.

Inductive reactance The result of current flowing through a conductor and the resultant magnetic field around the conductor that opposes the normal flow of current.
Reactancia inductiva El resultado de un corriente que circule por un conductor y que resulta en un campo magnético alrededor del conductor que opone el flujo normal del corriente.

Inertia engagement A type of starter motor that uses rotating inertia to engage the drive pinion with the engine flywheel.
Conexión por inercia Tipo de motor de arranque que utiliza inercia giratoria para engranar el piñon de mando con el volante de la máquina.

Instrument voltage regulator (IVR) Provides a constant voltage to the gauge, regardless of the voltage output of the charging system.
Instrumento regulador de tensión Le provee tensión constante al calibrador, sin importar cual sea la salida de tensión del sistema de carga.

Insulated side The portion of the circuit from the positive side of the source to the load component.
Lado aislado Parte del circuito que va del lado positivo de la fuente al componente de carga.

Insulator A material that does not allow electrons to flow easily through it.
Aislador Una material que no permite circular fácilmente los electrones.

Integrated circuit (IC chip) A complex circuit of thousands of transistors, diodes, resistors, capacitors, and other electronic devices that are formed onto a small silicon chip. As many as 30,000 transistors can be placed on a chip that is 1/4 inch (6.35 mm) square.
Circuito integrado (Fragmento CI) Circuito complejo de miles de transistores, diodos, resistores, condensadores, y otros dispositivos electrónicos formados en un fragmento pequeño de silicio. En un fragmento de 1/4 de pulgada (6,35 mm) cuadrada, pueden colocarse hasta 30.000 transistores.

Interface Used to protect the computer from excessive voltage levels and to translate input and output signals.
Interfase Utilizada para proteger la computadora de niveles excesivos de tensión y traducir señales de entrada y salida.

Ion An atom or group of atoms that has an electrical charge.
Ion Átomo o grupo de átomos que poseen una carga eléctrica.

ISO An abbreviation for International Standards Organizations.
ISO Una abreviación de las Organizaciones de Normas Internacionales.

Jack stands Support devices used to hold the vehicle off the floor after it has been raised by the floor jack.
Soportes de gato Dispositivos de soporte utilizados para sostener el vehículo sobre el suelo después de haber sido levantado con el gato de pie.

Keyless entry A lock system that allows for locking and unlocking of a vehicle with a touch keypad instead of a key.
Entrada sin llave Un sistema de cerradura que permite cerrar y abrir un vehículo por medio de un teclado en vez de utilizar una llave.

Lamination The process of constructing something with layers of materials that are firmly connected.
Laminación El proceso de construir algo de capas de materiales unidas con mucha fuerza.

Lamp A device that produces light as a result of current flow through a filament. The filament is enclosed within a glass envelope and is a type of resistance wire that is generally made from tungsten.
Lámpara Dispositivo que produce luz como resultado del flujo de corriente a través de un filamento. El filamento es un tipo de alambre de resistencia hecho por lo general de tungsteno, que es encerrado dentro de una bombilla.

Lamp outage module A current-measuring sensor that contains a set of resistors, wired in series with the power supply to the headlights, taillights, and stop lights. If the sensor indicates that a lamp is burned out, the module will alert the driver.
Unidad de avería de la lámpara Sensor para medir corriente que incluye un juego de resistores, alambrado en serie con la fuente de alimentación a los faros delanteros, traseros y a las luces de freno. Si el sensor indica que se ha apagado una lámpara, la unidad le avisará al conductor.

Light-emitting diode (LED) A gallium-arsenide diode that converts the energy developed when holes and electrons collide during normal diode operation into light.
Diodo emisor de luz Diodo semiconductor de galio y arseniuro que convierte en luz la energía producida por la colisión de agujeros y electrones durante el funcionamiento normal del diodo.

Limit switch A switch used to open a circuit when a predetermined value is reached. Limit switches are normally responsive to a mechanical movement or temperature changes.
Disyuntor de seguridad Un conmutador que se emplea para abrir un circuito al alcanzar un valor predeterminado. Los disyuntores de seguridad suelen ser responsivos a un movimiento mecánico o a los cambios de temperatura.

Linearity Refers to the sensor signal being as constantly proportional to the measured value as possible. It is an expression of the sensor's accuracy.
Linealidad Significa que la variación del valor de una magnitud es lo más proporcional posible a la variación del valor de otra magnitud. Expresa la precisión del sensor.

Liquid crystal display (LCD) A display that sandwiches electrodes and polarized fluid between layers of glass. When voltage is applied to the electrodes, the light slots of the fluid are rearranged to allow light to pass through.
Visualizador de cristal líquido Visualizador digital que consta de dos láminas de vidrio selladas, entre las cuales se encuentran los electrodos y el fluido polarizado. Cuando se aplica tensión a los electrodos, se rompe la disposición de las moléculas para permitir la formación de caracteres visibles.

Logic gates Electronic circuits that act as gates to output voltage signals depending on different combinations of input signals.
Compuertas lógicas Circuitos electrónicos que gobiernan señales de tensión de salida, dependiendo de las diferentes combinaciones de señales de entrada.

Magnetic field The area surrounding a magnet where energy is exerted due to the atoms aligning in the material.
Campo magnético Espacio que rodea un imán donde se emplea la energía debido a la alineación de los átomos en el material.

Magnetic flux density The concentration of the magnetic lines of force.
Densidad de flujo magnético Número de líneas de fuerza magnética.

Magnetic pulse generator Sensor that uses the principle of magnetic induction to produce a voltage signal. Magnetic pulse generators are commonly used to send data to the computer concerning the speed of the monitored component.
Generador de impulsos magnéticos Sensor que funciona según el principio de inducción magnética para producir una señal de tensión. Los generadores de impulsos magnéticos se utilizan comúnmente para transmitir datos a la computadora relacionados a la velocidad del componente regulado.

Magnetism An energy form resulting from atoms aligning within certain materials, giving the materials the ability to attract other metals.
Magnetismo Forma de energía que resulta de la alineación de átomos dentro de ciertos materiales y que le da a éstos la capacidad de atraer otros metales.

Material expanders Fillers that can be used in place of the active materials in a battery. They are used to keep the cost of manufacturing low.
Expansores de materias Los rellenos que se pueden usar en vez de las materiales activas de una bateria. Se emplean para mantener bajos los costos de la fabricación.

Matrix A rectangular array of grids.
Matriz Red lógica en una rejilla de forma rectangular.

Maxi-fuse A circuit protection device that looks similar to a blade-type fuse except that it is larger and has a higher amperage capacity. Maxi-fuses are used because they are less likely to cause an underhood fire when there is an overload in the circuit. If the fusible link burned in two, it is possible that the "hot" side of the fuse could come into contact with the vehicle frame and the wire could catch on fire.
Maxifusible Dispositivo de protección del circuito parecido a un fusible de tipo de cuchilla, pero más grande y con mayor capacidad de amperaje. Se utilizan maxifusibles porque existen menos probabilidades de que ocasionen un incendio debajo de la capota cuando ocurra una sobrecarga en el circuito. Si el cartucho de fusible se quemase en dos partes, es posible que el lado "cargado" del fusible entre en contacto con el armazón del vehículo y que el alambre se encienda.

Memory seats Power seats that can be programmed to return or adjust to a point designated by the driver.
Asientos con memoria Los asientos automáticos que se pueden programar a regresar o ajustarse a un punto indicado por el conductor.

Metri—pack connector Special wire connectors used in some computer circuits. They seal the wire terminals from the atmosphere, thereby preventing corrosion and other damage.
Conector metri-pack Los conectores de alambres especiales que se emplean en algunos circuitos de computadoras. Impermealizan los bornes de los alambres, así previniendo la corrosión y otros daños.

Momentary contact A switch type that operates only when held in position.
Contacto momentáneo Tipo de conmutador que funciona solamente cuando se mantiene en su posición.

Multiplexing A means of transmitting information between computers. It is a system in which electrical signals are transmitted by a peripheral serial bus instead of conventional wires, allowing several devices to share signals on a common conductor.
Multiplexaje Medio de transmitir información entre computadoras. Es un sistema en el cual las señales eléctricas son transmitidas por una colectora periférica en serie en vez de por líneas convencionales. Esto permite que varios dispositivos compartan señales en un conductor común.

Mutual induction An induction of voltage in an adjacent coil by changing current in a primary coil.

Inducción mutua Una inducción de la tensión en una bobina adyacente que se efectúa al cambiar la tensión en una bobina primaria.

Negative logic Defines the most negative voltage as a logical 1 in the binary code.

Lógica negativa Define la tensión más negativa como un 1 lógico en el código binario.

Negative temperature coefficient (NTC) thermistors Thermistors that reduce their resistance as the temperature increases.

Termistores con coeficiente negativo de temperatura Termistores que disminuyen su resistencia según aumenta la temperatura.

Neon lights A light that contains a colorless, odorless inert gas called neon. These lamps are discharge lamps.

Luces de neón Una luz que contiene un gas inerto sin color, inodoro llamado neón. Estas lámparas son lámparas de descarga.

Neutral atom *See* balanced atom.

Átomo neutro *Véase* átomo equilibrado.

Neutral junction The center connection to which the common ends of a Y-type stator winding are connected.

Empalme neutro Conexión central a la cual se conectan los extremos comunes de un devanado del estátor de tipo Y.

Neutral safety switch A switch used to prevent the starting of an engine unless the transmission is in PARK or Neutral.

Disyuntor de seguridad en neutral Un conmutador que se emplea para prevenir que arranque un motor al menos de que la transmisión esté en posición PARK o Neutral.

Neutrons Particles of an atom that have no charge.

Neutrones Partículas de un átomo desprovistas de carga.

Nonvolatile RAM RAM memory that will retain its memory if battery voltage is disconnected. NVRAM is a combination of RAM and EEPROM into the same chip. During normal operation, data is written to and read from the RAM portion of the chip. If the power is removed from the chip, or at programmed timed intervals, the data is transferred from RAM to the EEPROM portion of the chip. When the power is restored to the chip, the EEPROM will write the data back to the RAM.

Memoria de acceso aleatorio no volátil [NV RAM] Memoria de acceso aleatorio (RAM) que retiene su memoria si se desconecta la carga de la batería. La NV RAM es una combinación de RAM y EEPROM en el mismo fragmento. Durante el funcionamiento normal, los datos se escriben en y se leen de la parte RAM del fragmento. Si se remueve la alimentación del fragmento, o si se remueve ésta a intervalos programados, se transfieren los datos de la RAM a la parte del EEPROM del fragmento. Cuando se restaura la alimentación en el fragmento, el EEPROM volverá a escribir los datos en la RAM.

Normally closed (NC) switch A switch designation denoting that the contacts are closed until acted upon by an outside force.

Conmutador normalmente cerrado Nombre aplicado a un conmutador cuyos contactos permanecerán cerrados hasta que sean accionados por una fuerza exterior.

Normally open (NO) switch A switch designation denoting that the contacts are open until acted upon by an outside force.

Conmutador normalmente abierto Nombre aplicado a un conmutador cuyos contactos permanecerán abiertos hasta que sean accionados por una fuerza exterior.

N-type material When there are free electrons, the material is called an N-type material. The N means negative and indicates that it is the negative side of the circuit that pushes electrons through the semiconductor and the positive side that attracts the free electrons.

Material tipo N Al material se le llama material tipo N cuando hay electrones libres. La N significa negativo e indica que el lado negativo del circuito empuja los electrones a través del semiconductor y el lado positivo atrae los electrones libres.

Nucleus The core of an atom that contains the protons and neutrons.

Núcleo Parte central de un átomo que contiene los protones y los neutrones.

Occupational safety glasses Eye protection that is designed with special high-impact lens and frames, and provides for side protection.

Gafas de protección para el trabajo Gafas diseñadas con cristales y monturas especiales resistentes y provistas de protección lateral.

Odometer A mechanical counter in the speedometer unit indicating total miles accumulated on the vehicle.

Odómetro Aparato mecánico en la unidad del velocímetro con el que se cuentan las millas totales recorridas por el vehículo.

Ohm Unit of measure for resistance. One ohm is the resistance of a conductor such that a constant current of 1 ampere in it produces a voltage of 1 volt between its ends.

Ohmio Unidad de resistencia eléctrica. Un ohmio es la resistencia de un conductor si una corriente constante de 1 amperio en el conductor produce una tensión de 1 voltio entre los dos extremos.

Ohmmeter A test meter used to measure resistance and continuity in a circuit.

Ohmiómetro Instrumento de prueba utilizado para medir la resistencia y la continuidad en un circuito.

Ohm's law Defines the relationship between current, voltage, and resistance.

Ley de Ohm Define la relación entre la corriente, la tensión y la resistencia.

Open circuit A term used to indicate that current flow is stopped. By opening the circuit, the path for electron flow is broken.

Circuito abierto Término utilizado para indicar que el flujo de corriente ha sido detenido. Al abrirse el circuito, se interrumpe la trayectoria para el flujo de electrones.

Optical horn A name Chrysler uses to describe their "flash—to—pass" headlamp system.

Claxón óptico Un nombre que usa Chrysler para describir su sistema de faros "relampaguea para rebasar."

Oscillate Fast back and forth movement.

Oscilar Moverse rápidamente de atrás para adelante.

Overload Excess current flow in a circuit.

Sobrecarga Flujo de corriente superior a la que tiene asignada un circuito.

Overrunning clutch A clutch assembly on a starter drive used to prevent the engine's flywheel from turning the armature of the starter motor.

Embrague de sobremarcha Una asamblea de embrague en un acoplamiento del motor de arranque que se emplea para prevenir que el volante del motor dé vueltas al armazón del motor de arranque.

Oxygen sensor A voltage generating sensor that measures the amount of oxygen present in an engine's exhaust.

Sensor de oxígeno Un sensor generador de tensión que mide la cantidad del oxígeno presente en el gas de escape de un motor.

Parallel circuit A circuit that provides two or more paths for electricity to flow.

Circuito en paralelo Circuito que provee dos o más trayectorias para que circule la electricidad.

Parasitic loads Electrical loads that are still present when the ignition switch is in the OFF position.

Cargas parásitas Cargas eléctricas que todavía se encuentran presente cuando el botón conmutador de encendido está en la posición OFF.

Park switch Contact points located inside the wiper motor assembly that supply current to the motor after the wiper control switch has been turned to the PARK position. This allows the motor to continue operating until the wipers have reached their PARK position.

Conmutador PARK Puntos de contacto ubicados dentro del conjunto del motor del frotador que le suministran corriente al motor después de que el conmutador para el control de los frotadores haya sido colocado en la posición PARK. Esto permite que el motor continue su funcionamiento hasta que los frotadores hayan alcanzado la posición original.

Pass key A specially designed vehicle key with a coded resistance value. The term pass is derived from Personal Automotive Security System.

Llave maestra Una llave vehícular de diseño especial que tiene un valor de resistencia codificado. El termino pass se derive de las palabras Personal Automotive Security System (sistema personal de seguridad automotriz).

Passive restraints A passenger restraint system that automatically operates to confine the movement of a vehicle's passengers.

Correas passivas Un sistema de resguardo del pasajero que opera automaticamente para limitar el movimiento de los pasajeros en el vehículo.

Passive suspension systems Use fixed spring rates and shock valving.

Sistemas pasivos de suspensión Utilizan elasticidad de muelle constante y dotación con válvulas amortigadoras.

Permeability Term used to indicate the magnetic conductivity of a substance compared with the conductivity of air. The greater the permeability, the greater the magnetic conductivity and the easier a substance can be magnetized.

Permeabilidad Término utilizado para indicar la aptitud de una sustancia en relación con la del aire, de dar paso a las líneas de fuerza magnética. Mientras mayor sea la permeabilidad, mayor será la conductividad magnética y más fácilmente se comunicará a un cuerpo propiedades magnéticas.

Photocell A variable resistor that uses light to change resistance.

Fotocélula Resistor variable que utiliza luz para cambiar la resistencia.

Phototransistor A transistor that is sensitive to light.

Fototransistor Transistor sensible a la luz.

Photovoltaic diodes Diodes capable of producing a voltage when exposed to radiant energy.

Diodos fotovoltaicos Diodos capaces de generar una tensión cuando se encuentran expuestos a la energía de radiación.

Pickup coil The stationary component of the magnetic pulse generator consisting of a weak permanent magnet that has fine wire wound around it. As the timing disc rotates in front of it, the changes of magnetic lines of force generate a small voltage signal in the coil.

Bobina captadora Componente fijo del generador de impulsos magnéticos compuesta de un imán permanente débil devanado con alambre fino. Mientras gira el disco sincronizador enfrente de él, los cambios de las líneas de fuerza magnética generan una pequeña señal de tensión en la bobina.

Piezoelectricity Voltage produced by the application of pressure to certain crystals.

Piezoelectricidad Generación de polarización eléctrica en ciertos cristales a consecuencia de la aplicación de tensiones mecánicas.

Piezoresistive sensor A sensor that is sensitive to pressure changes.

Sensor piezoresistivo Sensor susceptible a los cambios de presión.

Pinion gear A small gear; typically refers to the drive gear of a starter drive assembly or the small drive gear in a differential assembly.

Engranaje de piñon Un engranaje pequeño; tipicamente se refiere al engranaje de arranque de una asamblea de motor de arranque o al engranaje de mando pequeño de la asamblea del diferencial.

Plate straps Metal connectors used to connect the positive or negative plates in a battery.

Abrazaderas de la placa Los conectores metálicos que sirven para conectar las placas positivas o negativas de una batería.

Plates The basic structure of a battery cell; each cell has at least one positive plate and one negative plate.

Placas La estructura básica de una celula de batería; cada celula tiene al menos una placa positiva y una placa negativa.

P-material Silicon or germanium that is doped with boron or gallium to create a shortage of electrons.

Material-P Boro o galio añadidos al silicio o al germanio para crear una insuficiencia de electrones.

PMGR An abbreviation for permanent magnet gear reduction.

PMGR Una abreviación de desmultiplicación del engrenaje del imán permanente.

Pneumatic tools Power tools that are powered by compressed air.

Herrimientas neumáticas Herramientas mecánicas accionadas por aire comprimido.

PN junction The point at which two opposite kinds of semiconductor materials are joined together.

Unión pn Zona de unión en la que se conectan dos tipos opuestos de materiales semiconductores.

Polarizers Glass sheets that make light waves vibrate in only one direction. This converts light into polarized light.

Polarizadores Las láminas de vidrio que hacen vibrar las ondas de luz en un sólo sentido. Esto convierte la luz en luz polarizada.

Polarizing The process of light polarization or of setting one end of a field as a positive or negative point.

Polarizadora El proceso de polarización de la luz o de establecer un lado de un campo como un punto positivo o negativo.

Pole shoes The components of an electric motor that are made of high-magnetic permeability material to help concentrate and direct the lines of force in the field assembly.

Expansión polar Componentes de un motor eléctrico hechos de material magnético de gran permeabilidad para ayudar a concentrar y dirigir las líneas de fuerza en el conjunto inductor.

Positive engagement starter A type of starter that uses the magnetic field strength of a field winding to engage the starter drive into the flywheel.

Acoplamiento de arranque positivo Un tipo de arrancador que utilisa la fuerza del campo magnético del devanado inductor para accionar el acoplamiento del arrancador en el volante.

Positive temperature coefficient (PTC) thermistors Thermistors that increase their resistance as the temperature increases.

Termistores con coeficiente positivo de temperatura Termistores que aumentan su resistencia según aumenta la temperatura.

Potential The ability to do something; typically voltage is referred to as the potential. If you have voltage, you have the potential for electricity.

Potencial La capacidad de efectuar el trabajo; típicamente se refiere a la tensión como el potencial. Si tiene tensión, tiene la potencial para la electricidad.

Potentiometer A variable resistor that acts as a circuit divider to provide accurate voltage drop readings proportional to movement.

Potenciómetro Resistor variable que actúa como un divisor de circuito para obtener lecturas de perdidas de tensión precisas en proporcion con el movimiento.

Power formula A formula used to calculate the amount of electrical power a component uses. The formula is P = I x E, whereas P stands for power (measured in watts), I stands for current, and E stands for voltage.

Formula de potencia Una formula que se emplea para calcular la cantidad de potencia eléctrica utilizada por un componente. La formula es P = I x E, en el que el P quiere decir potencia (medida en wats), I representa el corriente y el E representa la tensión.

Power tools Tools that use forces other than those generated from the body. They can use compressed air, electricity, or hydraulic pressure to generate and multiply force.

Herramientas mecánicas Herramientas que utilizan fuerzas distintas a las generadas por el cuerpo. Dichas fuerzas pueden ser el aire comprimido, la electricidad, o la presión hidráulica para generar y multiplicar la fuerza.

Pressure control solenoid A solenoid used to control the pressure of a fluid, commonly found in electronically controlled transmissions.

Solenoide de control de la presión Un solenoide que controla la presión de un fluido, suele encontrarse en las transmisiones controladas electronicamente.

Primary wiring Conductors that carry low voltage and current. The insulation of primary wires is usually thin.

Hilos primarios Hilos conductores de tensión y corriente bajas. El aislamiento de hilos primarios es normalmente delgado.

Printed circuit Made of thin phenolic or fiberglass board with copper deposited on it to create current paths. These are used to simplify the wiring of circuits.

Circuito impreso Un circuito hecho de un tablero de fenólico delgado o de fibra de vidrio el cual tiene depósitos del cobre para crear los trayectorios para el corriente. Estos se emplean para simplificar el cableado de los circuitos.

Prism lens A light lens designed with crystal—like patterns, which distort, slant, direct, or color the light that passes through it.

Lente prismático Un lente de luz con diseños cristalinos que distorcionan, inclinan, dirigen o coloran la luz que lo atraviesa.

Program A set of instructions that the computer must follow to achieve desired results.

Programa Conjunto de instrucciones que la computadora debe seguir para lograr los resultados deseados.

PROM (programmable read only memory) Memory chip that contains specific data that pertains to the exact vehicle in which the computer is installed. This information may be used to inform the CPU of the accessories that are equipped on the vehicle.

PROM (memoria de sólo lectura programable) Fragmento de memoria que contiene datos específicos referentes al vehículo particular en el que se instala la computadora. Esta información puede utilizarse para informar a la UCP sobre los accesorios de los cuales el vehículo está dotado.

Protection device Circuit protector that is designed to "turn off" the system that it protects. This is done by creating an open to prevent a complete circuit.

Dispositivo de protección Protector de circuito diseñado para "desconectar" el sistema al que provee protección. Esto se hace abriendo el circuito para impedir un circuito completo.

Proton Positively charged particles contained in the nucleus of an atom.

Protón Partículas con carga positiva que se encuentran en el núcleo de todo átomo.

Prove-out circuit A function of the ignition switch that completes the warning light circuit to ground through the ignition switch when it is in the START position. The warning light will be on during engine cranking to indicate to the driver that the bulb is working properly.

Circuito de prueba Función del botón conmutador de encendido que completa el circuito de la luz de aviso para que se ponga a tierra a través del botón conmutador de encendido cuando éste se encuentra en la posición START. La luz de aviso se encenderá durante el arranque del motor para avisarle al conductor que la bombilla funciona correctamente.

Pulse width The length of time in milliseconds that an actuator is energized.

Duración de impulsos Espacio de tiempo en milisegundos en el que se excita un accionador.

Pulse width modulation On/off cycling of a component. The period of time for each cycle does not change, only the amount of on time in each cycle changes.

Modulación de duración de impulsos Modulación de impulsos de un componente. El espacio de tiempo de cada ciclo no varía; lo que varía es la cantidad de trabajo efectivo de cada ciclo.

Radial grid A type of battery grid that has its patterns branching out from a common center.

Rejilla radial Un tipo de rejilla de bateria cuyos diseños extienden de un centro común.

Radio choke Absorbs voltage spikes and prevents static in the vehicle's radio.

Impedancia del radio Absorba los impulsos de la tensión y previene la presencia del estático en el radio del vehículo.

Radiofrequency interference (RFI) Radio and television interference caused by electromagnetic energy.

Interferencia de frecuencia radioeléctrica Interferencia en la radio y en la televisión producida por energía electromagnética.

RAM (random access memory) Stores temporary information that can be read from or written to by the CPU. RAM can be designed as volatile or nonvolatile.

RAM (memoria de acceso aleatorio) Almacena datos temporales que la UCP puede leer o escribir. La RAM puede ser volátil o no volátil.

Ratio A mathematical relationship between two or more things.

Razón Una relación matemática entre dos cosas o más.

Recombination battery A type of battery that is sometimes called a dry-cell battery because it does not use a liquid electrolyte solution.

Batería de recombinación Un tipo de batería que a veces se llama una pila seca porque no requiere una solución líquida de electrolita.

Rectification The converting of ac current to dc current.

Rectificación Proceso a través del cual la corriente alterna es transformada en una corriente continua.

Reflectors A device whose surface reflects or radiates light.

Reflectores Un dispositivo cuyo superficie refleja o irradia la luz.

Relay A device that uses low current to control a high-current circuit. Low current is used to energize the electromagnetic coil, while high current is able to pass over the relay contacts.

Relé Dispositivo que utiliza corriente baja para controlar un circuito de corriente alta. La corriente baja se utiliza para excitar la bobina electromagnética, mientras que la corriente alta puede transmitirse a través de los contactos del relé.

Reluctance A term used to indicate a material's resistance to the passage of flux lines.

Reluctancia Término utilizado para señalar la resistencia ofrecida por un circuito al paso del flujo magnético.

Reserve-capacity rating An indicator, in minutes, of how long the vehicle can be driven, with the headlights on, if the charging system should fail. The reserve-capacity rating is determined by the length of time, in minutes, that a fully charged battery can be discharged at 25 amperes before battery cell voltage drops below 1.75 volts per cell.

Clasificación de capacidad en reserva Indicación, en minutos, de cuánto tiempo un vehículo puede continuar siendo conducido, con los faros delanteros encendidos, en caso de que ocurriese una falla en el sistema de carga. La clasificación de capacidad en reserva se determina por el espacio de tiempo, en minutos, en el que una batería completamente cargada puede descargarse a 25 amperios antes de que la tensión del acumulador de la batería disminuya a un nivel inferior de 1,75 amperios por acumulador.

Resistance Opposition to current flow.

Resistencia Oposición que presenta un conductor al paso de la corriente eléctrica.

Resistance wire A special type of wire that has some resistance built into it. These typically are rated by ohms per foot.

Alambre de resistencia Un tipo de alambre especial que por diseño tiene algo de resistencia. Estos tipicamente tienen un valor nominal de ohm por pie.

Reversed-bias A positive voltage is applied to the N-type material and negative voltage is applied to the P-type material of a semiconductor.

Polarización inversa Tensión positiva aplicada al material N y tensión negativa aplicada al material P de un semiconductor.

Rheostat A two-terminal variable resistor used to regulate the strength of an electrical current.

Reóstato Resistor variable de dos bornes utilizado para regular la resistencia de una corriente eléctrica.

ROM (read only memory) Memory chip that stores permanent information. This information is used to instruct the computer on what to do in response to input data. The CPU reads the information contained in ROM, but it cannot write to it or change it.

ROM (memoria de sólo lectura) Fragmento de memoria que almacena datos en forma permanente. Dichos datos se utilizan para darle instrucciones a la computadora sobre cómo dirigir la ejecución de una operación de entrada. La UCP lee los datos que contiene la ROM, pero no puede escribir en ella o puede cambiarla.

Rotor The component of the ac generator that is rotated by the drive belt and creates the rotating magnetic field of the ac generator.

Rotor Parte rotativa del generador de corriente alterna accionada por la correa de transmisión y que produce el campo magnético rotativo del generador de corriente alterna.

Safety goggles Eye protection device that fits against the face and forehead to seal off the eyes from outside elements.

Gafas de seguridad Dispositivo protector que se coloca delante de los ojos para preservarlos de elementos extraños.

Safety stands *See* Jack stands.

Soportes de seguridad *Véase* soportes de gato.

Saturation 1. The point at which the magnetic strength eventually levels off, and where an additional increase of the magnetizing force current no longer increases the magnetic field strength. 2. The point where forward-bias voltage to the base leg is at a maximum. With bias voltage at the high limits, output current is also at its maximum.

Saturación 1. Máxima potencia posible de un campo magnético, donde un aumento adicional de la corriente de fuerza magnética no logra aumentar la potencia del campo magnético. 2. La tensión de polarización directa a la base está en su máximo. Ya que polarización directa ha alcanzado su límite máximo, la corriente de salida también alcanza éste.

Schmitt trigger An electronic circuit used to convert analog signals to digital signals or vice versa.

Disparador de Schmitt Un circuito electrónico que se emplea para convertir las señales análogas en señales digitales o vice versa.

Sealed-beam headlight A self-contained glass unit that consists of a filament, an inner reflector, and an outer glass lens.

Faro delantero sellado Unidad de vidrio que contiene un filamento, un reflector interior y una lente exterior de vidrio.

Secondary wiring Conductors, such as battery cables and ignition spark plug wires, that are used to carry high voltage or high current. Secondary wires have extra thick insulation.

Hilos secundarios Conductores, tales como cables de batería e hilos de bujías del encendido, utilizados para transmitir tensión o corriente alta. Los hilos secundarios poseen un aislamiento sumamente grueso.

Semiconductor An element that is neither a conductor nor an insulator. Semiconductors are materials that conduct electric current under certain conditions, yet will not conduct under other conditions.

Semiconductor Elemento que no es ni conductor ni aislante. Los semiconductores son materiales que transmiten corriente eléctrica bajo ciertas circunstancias, pero no la transmiten bajo otras.

Sending unit The sensor for the gauge. It is a variable resistor that changes resistance values with changing monitored conditions.

Unidad emisora Sensor para el calibrador. Es un resistor variable que cambia los valores de resistencia según cambian las condiciones reguladas.

Sensitivity controls A potentiometer that allows the driver to adjust the sensitivity of the automatic dimmer system to surrounding ambient light conditions.

Controles de sensibilidad Un potenciómetro que permite que el conductor ajusta la sensibilidad del sistema de intensidad de iluminación automático a las condiciones de luz ambientales.

Sensor Any device that provides an input to the computer.

Sensor Cualquier dispositivo que le transmite información a la computadora.

Separators Normally constructed of glass with a resin coating. These battery plates offer low resistance to electrical flow but high resistance to chemical contamination.

Separadores Normalmente se construyen del vidrio con una capa de resina. Estas placas de la batería ofrecen baja resistencia al flujo de la electricidad pero alta resistencia a la contaminación química.

Series circuit A circuit that provides a single path for current flow from the electrical source through all the circuit's components, and back to the source.

Circuito en serie Circuito que provee una trayectoria única para el flujo de corriente de la fuente eléctrica a través de todos los componentes del circuito, y de nuevo hacia la fuente.

Series-parallel circuit A circuit that has some loads in series and some in parallel.

Circuito en series paralelas Circuito que tiene unas cargas en serie y otras en paralelo.

Series-wound motor A type of motor that has its field windings connected in series with the armature. This type of motor develops its maximum torque output at the time of initial start. Torque decreases as motor speed increases.

Motor con devanados en serie Un tipo de motor cuyos devanados inductores se conectan en serie con la armadura. Este tipo de motor desarrolla la salida máxima de par de torsión en el momento inicial de ponerse en marcha. El par de torsión disminuye al aumentar la velocidad del motor.

Servomotor An electrical motor that produces rotation of less than a full turn. A feedback mechanism is used to position itself to the exact degree of rotation required.

Servomotor Motor eléctrico que genera rotación de menos de una revolución completa. Utiliza un mecanismo de realimentación para ubicarse al grado exacto de la rotación requerida.

Shell The electron orbit around the nucleus of an atom.

Corteza Órbita de electrones alrededor del núcleo del átomo.

Short An unwanted electrical path; sometimes this path goes directly to ground.

Corto Una trayectoria eléctrica no deseable; a veces este trayectoria viaja directamente a tierra.

Shunt circuits The branches of the parallel circuit.

Circuitos en derivación Las ramas del circuito en paralelo.

Shunt—wound motor A type of motor whose field windings are wired in parallel to the armature. This type of motor does not decrease its torque as speed increases.

Motor con devanados en derivación Un tipo de motor cuyos devanados inductores se cablean paralelos a la armadura. Este tipo de motor no disminuya su par de torsión al aumentar la velocidad.

Shutter wheel A metal wheel consisting of a series of alternating windows and vanes. It creates a magnetic shunt that changes the strength of the magnetic field from the permanent magnet of the Hall-effect switch or magnetic pulse generator.

Rueda obturadora Rueda metálica compuesta de una serie de ventanas y aspas alternas. Genera una derivación magnética que cambia la potencia del campo magnético, del imán permanente del conmutador de efecto Hall o del generador de impulsos magnéticos.

Sine wave A waveform that shows voltage changing polarity.

Onda senoidal Una forma de onda que muestra un cambio de polaridad en la tensión.

Single phase voltage The sine wave voltage induced in one conductor of the stator during one revolution of the rotor.

Tensión monofásica La tensión en forma de onda senoidal inducida en un conductor del estator durante una revolución del rotor.

Solenoid An electromagnetic device that uses movement of a plunger to exert a pulling or holding force.

Solenoide Dispositivo electromagnético que utiliza el movimiento de un pulsador para ejercer una fuerza de arrastre o de retención.

Source The portion of a field-effect transistor that supplies the current-carrying holes or electrons

Fuente Terminal de un transistor de efecto de campo que provee los agujeros o electrones portadores de corriente.

Specific gravity The weight of a given volume of a liquid divided by the weight of an equal volume of water.

Gravedad específica El peso de un volumen dado de líquido dividido por el peso de un volumen igual de agua.

Speedometer An instrument panel gauge that indicates the speed of the vehicle.

Velocímetro Calibrador en el panel de instrumentos que marca la velocidad del vehículo.

Starter drive The part of the starter motor that engages the armature to the engine flywheel ring gear.

Transmisión de arranque Parte del motor de arranque que engrana la armadura a la corona del volante de la máquina.

State of charge The condition of a battery's electrolyte and plate materials at any given time.

Estado de carga Condición del electrolito y de los materiales de la placa de una batería en cualquier momento dado.

Static electricity Electricity that is not in motion.

Electricidad estática Electricidad que no está en movimiento.

Static neutral point The point at which the fields of a motor are in balance.

Punto neutral estático El punto en que los campos de un motor estan equilibrados.

Stator The stationary coil of the ac generator where current is produced.

Estátor Bobina fija del generador de corriente alterna donde se genera corriente.

Stator neutral junction The common junction of wye stator windings.

Unión de estátor neutral La unión común de los devanados de un estátor Y.

Stepped resistor A resistor that has two or more fixed resistor values.

Resistor de secciones escalonadas Resistor que tiene dos o más valores de resistencia fija.

Stepper motor An electrical motor that contains a permanent magnet armature with two or four field coils. Can be used to move the controlled device to whatever location is desired. By applying voltage pulses to selected coils of the motor, the armature will turn a specific number of degrees. When the same voltage pulses are applied to the opposite coils, the armature will rotate the same number of degrees in the opposite direction.

Motor paso a paso Motor eléctrico que contiene una armadura magnética fija con dos o cuatro bobinas inductoras. Puede utilizarse para mover el dispositivo regulado a cualquier lugar deseado. Al aplicárseles impulsos de tensión a ciertas bobinas del motor, la armadura girará un número específico de grados. Cuando estos mismos impulsos de tensión se aplican a las bobinas opuestas, la armadura girará el mismo número de grados en la dirección opuesta.

Stranded wire A conductor comprised of many small solid wires twisted together. This type conductor is used to allow the wire to flex without breaking.

Cable trenzado Un conductor que comprende muchos cables sólidos pequeños trenzados. Este tipo de conductor se emplea para permitir que el cable se tuerza sin quebrar.

Sulfation A condition in a battery that reduces its output. The sulfate in the battery that is not converted tends to harden on the plates, and permanent damage to the battery results.

Sulfatación Una condición en una batería que disminuya su potencia de salida. El sulfato en la batería que no se convierte suele endurecerse en las placas y resulta en daños permanentes en la batería.

Tachometer An instrument that measures the speed of the engine in revolutions per minute (rpm).

Tacómetro Instrumento que mide la velocidad del motor en revoluciones por minuto (rpm).

Thermistor A solid-state variable resistor made from a semiconductor material that changes resistance in relation to temperature changes.

Termistor Resistor variable de estado sólido hecho de un material semiconductor que cambia su resistencia en relación con los cambios de temperatura.

Three-coil gauge A gauge design that uses the interaction of three electromagnets and the total field effect upon a permanent magnet to cause needle movement.

Calibrador de tres bobinas Calibrador diseñado para utilizar la interacción de tres electroimanes y el efecto inductor total sobre un imán permanente para producir el movimiento de la aguja.

Throw Term used in reference to electrical switches or relays referring to the number of output circuits from the switch.

Posición activa Término utilizado para conmutadores o relés eléctricos en relación con el número de circuitos de salida del conmutador.

Thyristor A semiconductor switching device composed of alternating N and P layers. It can also be used to rectify current from ac to dc.

Tiristor Dispositivo de conmutación del semiconductor compuesto de capas alternas de N y P. Puede utilizarse también para rectificar la corriente de corriente alterna a corriente continua.

Timer control A potentiometer that is part of the headlight switch in some systems. It controls the amount of time the headlights stay on after the ignition switch is turned off.

Control temporizador Un potenciómetro que es parte del conmutador de los faros en algunos sistemas. Controla la cantidad del tiempo que quedan prendidos los faros después de apagarse la llave del encendido.

Torque converter A hydraulic device found on automatic transmissions. It is responsible for controlling the power flow from the engine to the transmission; works like a clutch to engage and disengage the engine's power to the drive line.

Convertidor de par Un dispositivo hidraúlico en las transmisiones automáticas. Se encarga de controlar el flujo de la potencia del motor a la transmisión; funciona como un embrague para embragar y desembragar la potencia del motor con la flecha motríz.

Transducer A device that changes energy from one form into another.

Transductor Dispositivo que cambia la energía de una forma a otra.

Transistor A three-layer semiconductor used as a very fast switching device.

Transistor Semiconductor de tres capas utilizado como dispositivo de conmutación sumamente rápido.

TVRS An abbreviation for television—radio—suppression cable.

TVRS Una abreviación del cable de supresíon del televisión y radio.

Two-coil gauge A gauge design that uses the interaction of two electromagnets and the total field effect upon an armature to cause needle movement.

Calibrador de dos bobinas Calibrador diseñado para utilizar la interacción de dos electroimanes y el efecto inductor total sobre una armadura para generar el movimiento de la aguja.

Vacuum distribution valve A valve used in vacuum-controlled concealed headlight systems. It controls the direction of vacuum to various vacuum motors or to vent.

Válvula de distribución al vacío Válvula utilizada en el sistema de faros delanteros ocultos controlado al vacío. Regula la dirección del vacío a varios motores al vacío o sirve para dar salida del sistema.

Vacuum fluorescent display (VFD) A display type that uses anode segments coated with phosphor and bombarded with tungsten electrons to cause the segments to glow.

Visualización de fluorescencia al vacío Tipo de visualización que utiliza segmentos ánodos cubiertos de fósforo y bombardeados de electrones de tungsteno para producir la luminiscencia de los segmentos.

Valence ring The outermost orbit of the atom.
Anillo de valencia Órbita más exterior del átomo.

Valve body A unit that consists of many valves and hydraulic circuits. This unit is the central control point for gear shifting in an automatic transmission.
Cuerpo de la válvula Una unedad que consiste de muchas válvulas y circuitos hidráulicos. Esta unedad es el punto central de mando para los cambios de velocidad en una transmisión automática.

Variable resistor A resistor that provides for an infinite number of resistance values within a range.
Resistor variable Resistor que provee un número infinito de valores de resistencia dentro de un margen.

Volatile RAM RAM memory that is erased when it is disconnected from its power source. Also known as Keep Alive Memory.
RAM volátil Memoria RAM cuyos datos se perderán cuando se la desconecta de la fuente de alimentación. Conocida también como memoria de entretenimiento.

Volt The unit used to measure the amount of electrical force.
Voltio Unidad práctica de tensión para medir la cantidad de fuerza eléctrica.

Voltage The difference or potential that indicates an excess of electrons at the end of the circuit the farthest from the electromotive force. It is the electrical pressure that causes electrons to move through a circuit. One volt is the amount of pressure required to move one amp of current through one ohm of resistance.
Tensión Diferencia o potencial que indica un exceso de electrones al punto del circuito que se encuentra más alejado de la fuerza electromotriz. La presión eléctrica genera el movimiento de electrones a través de un circuito. Un voltio equivale a la cantidad de presión requerida para mover un amperio de corriente a través de un ohmio de resistencia.

Voltage drop A resistance in the circuit that reduces the electrical pressure available after the resistance. The resistance can be either the load component, the conductors, any connections, or unwanted resistance.
Caída de tensión Resistencia en el circuito que disminuye la presión eléctrica disponible después de la resistencia. La resistencia puede ser el componente de carga, los conductores, cualquier conexión o resistencia no deseada.

Voltage limiter Connected through the resistor network of a voltage regulator. It determines whether the field will receive high, low, or no voltage. It controls the field voltage for the required amount of charging.
Limitador de tensión Conectado por el red de resistores de un regulador de tensión. Determina si el campo recibirá alta, baja o ninguna tensión. Controla la tensión de campo durante el tiempo indicado de carga.

Voltage regulator Used to control the output voltage of the ac generator, based on charging system demands, by controlling field current.
Regulador de tensión Dispositivo cuya función es mantener la tensión de salida del generador de corriente alterna, de acuerdo a las variaciones en la corriente de carga, controlando la corriente inductora.

Voltmeter A test meter used to read the pressure behind the flow of electrons.
Voltímetro Instrumento de prueba utilizado para medir la presión del flujo de electrones.

Wake-up signal An input signal used to notify the body computer that an engine start and operation of accessories is going to be initiated soon. This signal is used to warm up the circuits that will be processing information.
Señal despertadora Señal de entrada para avisarle a la computadora del vehículo que el arranque del motor y el funcionamiento de los accesorios se iniciarán dentro de poco. Dicha señal se utiliza para calentar los circuitos que procesarán los datos.

Warning light A lamp that is illuminated to warn the driver of a possible problem or hazardous condition.
Luz de aviso Lámpara que se enciende para avisarle al conductor sobre posibles problemas o condiciones peligrosas.

Watt The unit of measure of electrical power, which is the equivalent of horsepower. One horsepower is equal to 746 watts.
Watio Unidad de potencia eléctrica, equivalente a un caballo de vapor. 746 watios equivalen a un caballo de vapor (CV).

Wattage A measure of the total electrical work being performed per unit of time.
Vataje Medida del trabajo eléctrico total realizado por unidad de tiempo.

Weather—pack connector A type of connector that seals the terminal's ends. This type connector is used in electronic circuits.
Conectador impermeable Un tipo de conectador que sella las extremidades de los terminales. Este tipo de conectador se emplea en los circuitos electrónicos.

Wheatstone bridge A series-parallel arrangement of resistors between an input terminal and ground.
Puente de Wheatstone Conjunto de resistores en series paralelas entre un borne de entrada y tierra.

Wiring diagram An electrical schematic that shows a representation of actual electrical or electronic components and the wiring of the vehicle's electrical systems.
Esquema de conexiones Esquema en el que se muestran las conexiones internas de los componentes eléctricos o electrónicos reales y las de los sistemas eléctricos del vehículo.

Wiring harness A group of wires enclosed in a conduit and routed to specific areas of the vehicle.
Cableado preformado Conjunto de alambres envueltos en un conducto y dirigidos hacia áreas específicas del vehículo.

Worm gear A type of gear whose teeth wrap around the shaft. The action of the gear is much like that of a threaded bolt or screw.
Engranaje de tornillo sin fin Un tipo de engranaje cuyos dientes se envuelven alrededor del vástago. El movimiento del engranaje es muy parecido a un perno enroscado o una tuerca.

Wye connection A type of stator winding in which one end of the individual windings are connected at a common point. The structure resembles the letter "Y."
Conexión Y Un tipo de devanado estátor en el cual una extremidad de los devanados individuales se conectan en un punto común. La estructura parece la letra "Y."

Y-type stator A three-winding ac generator that has one end of each winding connected at the neutral junction.
Estátor de tipo Y Generador de corriente alterna de devanado triple; un extremo de cada devanado se conecta al empalme neutro.

Zener diode A diode that allows reverse current to flow above a set voltage limit.
Diodo Zener Diodo que permite que el flujo de corriente en dirección inversa sobrepase el límite de tensión determinado.

Zener voltage The voltage that is reached when a diode conducts in reverse direction.
Tensión de Zener Tensión alcanzada cuando un diodo conduce en una dirección inversa.